Student's Solutions Manual

Elementary and Intermediate Algebra

Concepts and Applications:
A Combined Approach

Student's Solutions Manual

Elementary and Intermediate Algebra
Concepts and Applications:
A Combined Approach

Marvin L. Bittinger
Indiana University – Purdue University at Indianapolis

David J. Ellenbogen
St. Michael's College

Barbara Johnson
Indiana University – Purdue University at Indianapolis

Judith A. Penna

ADDISON-WESLEY PUBLISHING COMPANY
Reading, Massachusetts • Menlo Park, California • New York
Don Mills, Ontario • Wokingham, England • Amsterdam • Bonn
Sydney • Singapore • Tokyo • Madrid • San Juan • Milan • Paris

Reproduced by Addison-Wesley from camera-ready copy supplied by the author.

ISBN   0-201-85205-5

6 7 8 9 ML 9998

Table of Contents

The author thanks Patsy Hammond, David Penna,
Mike Penna, and Pam Smith for their help in
the preparation of this manual.

Student's Solutions Manual

Elementary and Intermediate Algebra

Concepts and Applications:
A Combined Approach

Chapter 1

Introduction to Algebra and Algebraic Expressions

Exercise Set 1.1

1. Substitute 7 for x and multiply.
$$6x = 6 \cdot 7 = 42$$

2. 49

3. Substitute 7 for a and add.
$$9 + a = 9 + 7 = 16$$

4. 4

5. $\dfrac{3p}{q} = \dfrac{3 \cdot 2}{6} = \dfrac{6}{6} = 1$

6. 3

7. $\dfrac{x + y}{5} = \dfrac{10 + 20}{5} = \dfrac{30}{5} = 6$

8. 9

9. $\dfrac{x - y}{8} = \dfrac{20 - 4}{8} = \dfrac{16}{8} = 2$

10. 2

11. $\dfrac{x}{y} = \dfrac{3}{6} = \dfrac{1}{2}$

12. $\dfrac{1}{4}$

13. $\dfrac{5z}{y} = \dfrac{5 \cdot 8}{2} = \dfrac{40}{2} = 20$

14. 2

15. $rt = 55(4 \text{ mi}) = 220 \text{ mi}$

16. $\dfrac{13}{25}$, or 0.52

17. $bh = (6.5 \text{ cm})(15.4 \text{ cm})$
$= (6.5)(15.4)(\text{cm})(\text{cm})$
$= 100.1 \text{ cm}^2$, or 100.1 square centimeters

18. 2.7 hr

19. $5t = 5(30 \text{ sec}) = 150 \text{ sec}$;
$5t = 5(90 \text{ sec}) = 450 \text{ sec}$;
$5t = 5(2 \text{ min}) = 10 \text{ min}$

20. 23; 28; 41

21. $b + 6$, or $6 + b$

22. $t + 8$, or $8 + t$

23. $c - 9$

24. $d - 4$

25. $q + 6$, or $6 + q$

26. $z + 11$, or $11 + z$

27. $a + b$, or $b + a$

28. $d + c$, or $c + d$

29. $y - x$

30. $h - c$

31. $x \div w$, or $\dfrac{x}{w}$

32. $s \div t$, or $\dfrac{s}{t}$

33. $n - m$

34. $q - p$

35. $r + s$, or $s + r$

36. $d + f$, or $f + d$

37. $2x$

38. $3p$

39. $\dfrac{1}{3}t$, or $\dfrac{t}{3}$

40. $\dfrac{1}{4}d$, or $\dfrac{d}{4}$

41. Let n represent "some number." Then we have 97%n, or $0.97n$.

42. 43%n, or $0.43n$

43. $\$d - \$29.95$

44. $65t$ mi

45. $\underline{x + 17 = 32}$ Writing the equation
$15 + 17 \; ? \; 32$ Substituting 15 for x
     $32 \mid 32$ $32 = 32$ is TRUE.

Since the left-hand and right-hand sides are the same, 15 is a solution.

46. No

47. $\underline{x - 7 = 12}$ Writing the equation
$21 - 7 \; ? \; 12$ Substituting 21 for x
     $14 \mid 12$ $14 = 12$ is FALSE.

Since the left-hand and right-hand sides are not the same, 21 is not a solution.

48. Yes

49.
$$\overline{6x = 54}$$
$$6 \cdot 7 \ ? \ 54$$
$$42 \ | \ 54 \qquad 42 = 54 \text{ is FALSE.}$$
7 is not a solution.

50. Yes

51.
$$\overline{\frac{x}{6} = 5}$$
$$\frac{30}{6} \ ? \ 5$$
$$5 \ | \ 5 \qquad 5 = 5 \text{ is TRUE.}$$
5 is a solution.

52. No

53.
$$\overline{5x + 7 = 107}$$
$$5 \cdot 19 + 7 \ ? \ 107$$
$$95 + 7 \ |$$
$$102 \ | \ 107 \qquad 102 = 107 \text{ is FALSE.}$$
19 is not a solution.

54. Yes

55. Let x represent the number.

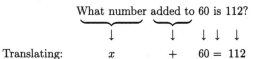

$$x + 60 = 112$$

56. $7w = 2233$

57. Let y represent the number.

Rewording: 42 times $\underbrace{\text{what number}}$ is 2352?

Translating: 42   · y = 2352

$$42y = 2352$$

58. $x + 345 = 987$

59. Let s represent the number of squares your opponent gets.

Rewording: $\underbrace{\text{What number}}$ $\underbrace{\text{added to}}$ 35 is 64?

Translating: s +   35 = 64

$$s + 35 = 64$$

60. $\$80y = \$53,400$

61. Let c = the cost of one box.

Rewording: 4 times $\underbrace{\text{what number}}$ is $7.96?

Translating: 4   · c = $7.96

$$4c = \$7.96$$

62. $d + \$0.2 = \$6.5$, where d is in billions of dollars.

63.

64.

65. $y + 2x$

66. $a + 2 + b$

67. $2x - 3$

68. $a + 5$

69. $b - 2$

70. $x + x$, or $2x$

71. $s + s + s + s$, or $4s$

72. $l + w + l + w$, or $2l + 2w$

73. $y = 8$, $x = 2y = 2 \cdot 8 = 16$;
$$\frac{x + y}{4} = \frac{8 + 16}{4} = \frac{24}{4} = 6$$

74. 5

75. $x = 9$, $y = 3x = 3 \cdot 9 = 27$;
$$\frac{y - x}{3} = \frac{27 - 9}{3} = \frac{18}{3} = 6$$

76. 9

77. The next whole number is one more than $w + 3$:
$$w + 3 + 1 = w + 4$$

78. d

79. If t is the larger number, the other number is 3 less than t, or $t - 3$.

If t is the smaller number, the other number is 3 more than t, or $t + 3$.

80. $n + 10\%n$, or $n + 0.1n$, or $1.1n$

Exercise Set 1.2

1. $5 + y$ Changing the order

2. $6 + x$

3. $ab + 5$

4. $3y + x$

5. $3y + 9x$

6. $7b + 3a$

7. $2(3 + a)$

8. $9(5 + x)$

9. tr Changing the order

10. nm

11. $a5$

12. $b7$

13. $5 + ba$

14. $x + y3$

15. $(a + 3)2$

16. $(x + 5)9$

17. $(x + y) + 2$

18. $a + (3 + b)$

19. $9 + (m + 2)$

20. $(x + 2) + y$

21. $ab + (c + d)$

22. $m + (np + r)$

23. $(5a)b$

24. $7(xy)$

25. $6(mn)$

26. $(9r)p$

27. $(3 \cdot 2)(a + b)$

28. $(5x)(2 + y)$

29. a) $(a + b) + 2 = 2 + (a + b)$ Using the commutative law

        $= 2 + (b + a)$ Using the commutative law again

   b) $(a + b) + 2 = 2 + (a + b)$ Using the commutative law

        $= (2 + a) + b$ Using the associative law

   Answers may vary.

30. $v + (w + 5)$; $(v + 5) + w$; answers may vary.

31. a) $7(ab) = (ab)7$ Using the commutative law

        $= (ba)7$ Using the commutative law again

   b) $7(ab) = (ab)7$ Using the commutative law

        $= a(b7)$ Using the associative law

        $= a(7b)$ Using the commutative law again

   Answers may vary.

32. $(3x)y$; $x(3y)$; answers may vary.

33. $(3a)4 = 4(3a)$ Commutative law

        $= (4 \cdot 3)a$ Associative law

        $= 12a$ Simplifying

34. $(2 + m) + 3 = (m + 2) + 3$ Commutative law

        $= m + (2 + 3)$ Associative law

        $= m + 5$ Simplifying

35. $5 + (2 + x) = (5 + 2) + x$ Associative law

        $= x + (5 + 2)$ Commutative law

        $= x + 7$ Simplifying

36. $(a3)5 = a(3 \cdot 5)$ Associative law

        $= (3 \cdot 5)a$ Commutative law

        $= 15a$ Simplifying

37. $2(b + 5) = 2 \cdot b + 2 \cdot 5 = 2b + 10$

38. $4x + 12$

39. $7(1 + t) = 7 \cdot 1 + 7 \cdot t = 7 + 7t$

40. $6v + 24$

41. $3(x + 1) = 3 \cdot x + 3 \cdot 1 = 3x + 3$

42. $7x + 56$

43. $4(1 + y) = 4 \cdot 1 + 4 \cdot y = 4 + 4y$

44. $9s + 9$

45. $6(5x + 2) = 6 \cdot 5x + 6 \cdot 2 = 30x + 12$

46. $54m + 63$

47. $7(x + 4 + 6y) = 7 \cdot x + 7 \cdot 4 + 7 \cdot 6y = 7x + 28 + 42y$

48. $20x + 32 + 12p$

49. $(a + b)2 = a(2) + b(2) = 2a + 2b$

50. $7x + 14$

51. $(x + y + 2)5 = x(5) + y(5) + 2(5) = 5x + 5y + 10$

52. $12 + 6a + 6b$

53. $2x + 2y = 2(x + y)$ The common factor is 2.

54. $5(y + z)$

55. $5 + 5y = 5 \cdot 1 + 5 \cdot y$ The common factor is 5.

        $= 5(1 + y)$ Using the distributive law

56. $13(1 + x)$

57. $3x + 12y = 3 \cdot x + 3 \cdot 4y = 3(x + 4y)$

58. $5(x + 4y)$

59. $5x + 10 + 15y = 5 \cdot x + 5 \cdot 2 + 5 \cdot 3y = 5(x + 2 + 3y)$

60. $3(1 + 9b + 2c)$

61. $9x + 9 = 9 \cdot x + 9 \cdot 1 = 9(x + 1)$

     Check: $9(x + 1) = 9 \cdot x + 9 \cdot 1 = 9x + 9$

62. $6(x+1)$

63. $9x + 3y = 3 \cdot 3x + 3 \cdot y = 3(3x + y)$

Check: $3(3x + y) = 3 \cdot 3x + 3 \cdot y = 9x + 3y$

64. $5(3x + y)$

65. $2a + 16b + 64 = 2 \cdot a + 2 \cdot 8b + 2 \cdot 32 = 2(a + 8b + 32)$

Check: $2(a + 8b + 32) = 2 \cdot a + 2 \cdot 8b + 2 \cdot 32 = 2a + 16b + 64$

66. $5(1 + 4x + 7y)$

67. $11x + 44y + 121 = 11 \cdot x + 11 \cdot 4y + 11 \cdot 11 = 11(x + 4y + 11)$

Check: $11(x + 4y + 11) = 11 \cdot x + 11 \cdot 4y + 11 \cdot 11 = 11x + 44y + 121$

68. $7(1 + 2b + 8w)$

69. $t - 9$

70. $\dfrac{1}{2}m$, or $\dfrac{m}{2}$

71.

72.

73. Yes; commutative law of addition

74. Yes; associative and distributive laws

75.
$$axy + ax = ax(y + 1) \quad \text{Distributive law}$$
$$= xa(y + 1) \quad \text{Commutative law of multiplication}$$
$$= xa(1 + y) \quad \text{Commutative law of addition}$$

The expressions are equivalent.

76.
$$3a(b + c) = 3[a(b + c)] \quad \text{Associative law of multiplication}$$
$$= 3(ab + ac) \quad \text{Distributive law}$$
$$= (ba + ca)3 \quad \text{Using the commutative law of multiplication 3 times}$$
$$= (ca + ba)3 \quad \text{Commutative law of addition}$$

77.

78. $P(1 + rt)$

79.

Exercise Set 1.3

1. We write several factorizations of 56. There are other correct answers.

$$4 \cdot 14, \; 7 \cdot 8, \; 2 \cdot 4 \cdot 7$$

2. $2 \cdot 51$, $6 \cdot 17$; there are other correct answers.

3. $1 \cdot 93$, $3 \cdot 31$

4. $4 \cdot 36$, $12 \cdot 12$; there are other correct answers.

5. $14 = 2 \cdot 7$

6. $3 \cdot 5$

7. $33 = 3 \cdot 11$

8. $5 \cdot 11$

9. $9 = 3 \cdot 3$

10. $5 \cdot 5$

11. $49 = 7 \cdot 7$

12. $11 \cdot 11$

13. We begin by factoring 18 in any way that we can and continue factoring until each factor is prime.

$$18 = 2 \cdot 9 = 2 \cdot 3 \cdot 3$$

14. $2 \cdot 2 \cdot 2 \cdot 3$

15. We begin by factoring 40 in any way that we can and continue factoring until each factor is prime.

$$40 = 4 \cdot 10 = 2 \cdot 2 \cdot 2 \cdot 5$$

16. $2 \cdot 2 \cdot 2 \cdot 7$

17. We begin by factoring 90 in any way that we can and continue factoring until each factor is prime.

$$90 = 2 \cdot 45 = 2 \cdot 9 \cdot 5 = 2 \cdot 3 \cdot 3 \cdot 5$$

18. $2 \cdot 2 \cdot 2 \cdot 3 \cdot 5$

19. $210 = 2 \cdot 105 = 2 \cdot 3 \cdot 35 = 2 \cdot 3 \cdot 5 \cdot 7$

20. $2 \cdot 3 \cdot 5 \cdot 11$

21. 79 is prime.

22. $11 \cdot 13$

23. $119 = 7 \cdot 17$

24. $13 \cdot 17$

25.
$$\frac{18}{45} = \frac{2 \cdot 9}{5 \cdot 9} \quad \text{Factoring numerator and denominator}$$
$$= \frac{2}{5} \cdot \frac{9}{9} \quad \text{Rewriting as a product of two fractions}$$
$$= \frac{2}{5} \cdot 1 \quad \frac{9}{9} = 1$$
$$= \frac{2}{5} \quad \text{Using the identity property of 1}$$

26. $\dfrac{2}{7}$

27. $\dfrac{49}{14} = \dfrac{7 \cdot 7}{2 \cdot 7} = \dfrac{7}{2} \cdot \dfrac{7}{7} = \dfrac{7}{2} \cdot 1 = \dfrac{7}{2}$

28. $\dfrac{8}{3}$

29. $\dfrac{6}{42} = \dfrac{1 \cdot 6}{7 \cdot 6}$ Factoring and using the identity property of 1 to write 6 as $1 \cdot 6$

$\qquad = \dfrac{1}{7} \cdot \dfrac{6}{6}$

$\qquad = \dfrac{1}{7} \cdot 1 = \dfrac{1}{7}$

30. $\dfrac{1}{8}$

31. $\dfrac{56}{7} = \dfrac{8 \cdot 7}{1 \cdot 7} = \dfrac{8}{1} \cdot \dfrac{7}{7} = \dfrac{8}{1} \cdot 1 = 8$

32. 12

33. $\dfrac{19}{76} = \dfrac{1 \cdot 19}{4 \cdot 19}$ Factoring and using the identity property of 1 to write 19 as $1 \cdot 19$

$\qquad = \dfrac{1 \cdot \cancel{19}}{4 \cdot \cancel{19}}$ Removing a factor of 1: $\dfrac{19}{19} = 1$

$\qquad = \dfrac{1}{4}$

34. $\dfrac{1}{3}$

35. $\dfrac{100}{20} = \dfrac{5 \cdot 20}{1 \cdot 20}$ Factoring and using the identity property of 1 to write 20 as $1 \cdot 20$

$\qquad = \dfrac{5 \cdot \cancel{20}}{1 \cdot \cancel{20}}$ Removing a factor of 1: $\dfrac{20}{20} = 1$

$\qquad = \dfrac{5}{1}$

$\qquad = 5$ Simplifying

36. 6

37. $\dfrac{425}{525} = \dfrac{17 \cdot 25}{21 \cdot 25}$ Factoring the numerator and the denominator

$\qquad = \dfrac{17 \cdot \cancel{25}}{21 \cdot \cancel{25}}$ Removing a factor of 1: $\dfrac{25}{25} = 1$

$\qquad = \dfrac{17}{21}$

38. $\dfrac{25}{13}$

39. $\dfrac{2600}{1400} = \dfrac{2 \cdot 13 \cdot 100}{2 \cdot 7 \cdot 100}$ Factoring

$\qquad = \dfrac{13 \cdot \cancel{2} \cdot \cancel{100}}{7 \cdot \cancel{2} \cdot \cancel{100}}$ Removing a factor of 1: $\dfrac{2 \cdot 100}{2 \cdot 100} = 1$

$\qquad = \dfrac{13}{7}$

40. 3

41. $\dfrac{8 \cdot x}{6 \cdot x} = \dfrac{2 \cdot 4 \cdot x}{2 \cdot 3 \cdot x}$ Factoring

$\qquad = \dfrac{4 \cdot \cancel{2} \cdot \cancel{x}}{3 \cdot \cancel{2} \cdot \cancel{x}}$ Removing a factor of 1: $\dfrac{2 \cdot x}{2 \cdot x} = 1$

$\qquad = \dfrac{4}{3}$

42. $\dfrac{1}{3}$

43. $\dfrac{1}{4} \cdot \dfrac{1}{2} = \dfrac{1 \cdot 1}{4 \cdot 2}$ Multiplying numerators and denominators

$\qquad = \dfrac{1}{8}$

44. $\dfrac{44}{25}$

45. $\dfrac{17}{2} \cdot \dfrac{3}{4} = \dfrac{17 \cdot 3}{2 \cdot 4} = \dfrac{51}{8}$

46. 1

47. $\dfrac{1}{2} + \dfrac{1}{2} = \dfrac{1 + 1}{2}$ Adding numerators; keeping the common denominator

$\qquad = \dfrac{2}{2} = 1$

48. $\dfrac{3}{4}$

49. $\dfrac{4}{9} + \dfrac{13}{18} = \dfrac{4}{9} \cdot \dfrac{2}{2} + \dfrac{13}{18}$ Using 18 as the common denominator

$\qquad = \dfrac{8}{18} + \dfrac{13}{18}$

$\qquad = \dfrac{21}{18}$

$\qquad = \dfrac{7 \cdot \cancel{3}}{6 \cdot \cancel{3}} = \dfrac{7}{6}$ Simplifying

50. $\dfrac{4}{3}$

51. $\dfrac{3}{a} \cdot \dfrac{b}{7} = \dfrac{3b}{7a}$ Multiplying numerators and denominators

52. $\dfrac{xy}{5z}$

53. $\dfrac{3}{x} + \dfrac{2}{x} = \dfrac{5}{x}$ Adding numerators; keeping the common denominator

54. $\dfrac{2}{a}$

55. $\dfrac{3}{10} + \dfrac{8}{15} = \dfrac{3}{10} \cdot \dfrac{3}{3} + \dfrac{8}{15} \cdot \dfrac{2}{2}$ Using 30 as the common denominator

$= \dfrac{9}{30} + \dfrac{16}{30}$

$= \dfrac{25}{30}$

$= \dfrac{5 \cdot \cancel{5}}{6 \cdot \cancel{5}} = \dfrac{5}{6}$ Simplifying

56. $\dfrac{41}{24}$

57. $\dfrac{5}{4} - \dfrac{3}{4} = \dfrac{2}{4}$

$= \dfrac{1 \cdot \cancel{2}}{2 \cdot \cancel{2}} = \dfrac{1}{2}$

58. 2

59. $\dfrac{13}{18} - \dfrac{4}{9} = \dfrac{13}{18} - \dfrac{4}{9} \cdot \dfrac{2}{2}$ Using 18 as the common denominator

$= \dfrac{13}{18} - \dfrac{8}{18}$

$= \dfrac{5}{18}$

60. $\dfrac{31}{45}$

61. $\dfrac{11}{12} - \dfrac{2}{5} = \dfrac{11}{12} \cdot \dfrac{5}{5} - \dfrac{2}{5} \cdot \dfrac{12}{12}$ Using 60 as the common denominator

$= \dfrac{55}{60} - \dfrac{24}{60}$

$= \dfrac{31}{60}$

62. $\dfrac{13}{48}$

63. $\dfrac{7}{6} \div \dfrac{3}{5} = \dfrac{7}{6} \cdot \dfrac{5}{3}$ Multiplying by the reciprocal of the divisor

$= \dfrac{35}{18}$

64. $\dfrac{28}{15}$

65. $\dfrac{8}{9} \div \dfrac{4}{15} = \dfrac{8}{9} \cdot \dfrac{15}{4} = \dfrac{2 \cdot \cancel{4} \cdot \cancel{3} \cdot 5}{\cancel{3} \cdot 3 \cdot \cancel{4}} = \dfrac{10}{3}$

66. $\dfrac{7}{4}$

67. $\dfrac{1}{4} \div \dfrac{1}{2} = \dfrac{1}{4} \cdot \dfrac{2}{1} = \dfrac{1 \cdot \cancel{2}}{\cancel{2} \cdot 2 \cdot 1} = \dfrac{1}{2}$

68. $\dfrac{1}{2}$

69. $\dfrac{\frac{13}{12}}{\frac{39}{5}} = \dfrac{13}{12} \div \dfrac{39}{5} = \dfrac{13}{12} \cdot \dfrac{5}{39} = \dfrac{\cancel{13} \cdot 5}{12 \cdot 3 \cdot \cancel{13}} = \dfrac{5}{36}$

70. $\dfrac{68}{9}$

71. $100 \div \dfrac{1}{5} = \dfrac{100}{1} \cdot \dfrac{5}{1} = \dfrac{500}{1} = 500$

72. 468

73. $\dfrac{3}{4} \div 10 = \dfrac{3}{4} \cdot \dfrac{1}{10} = \dfrac{3}{40}$

74. $\dfrac{1}{18}$

75. $\dfrac{5}{3} \div \dfrac{a}{b} = \dfrac{5}{3} \cdot \dfrac{b}{a} = \dfrac{5b}{3a}$

76. $\dfrac{xy}{28}$

77. $\dfrac{x}{6} - \dfrac{1}{3} = \dfrac{x}{6} - \dfrac{1}{3} \cdot \dfrac{2}{2}$ Using 6 as the common denominator

$= \dfrac{x}{6} - \dfrac{2}{6}$

$= \dfrac{x - 2}{6}$

78. $\dfrac{9 + 5x}{10}$

79. $5(x + 3) = 5(3 + x)$ Commutative law of addition
Answers may vary.

80. $(a + b) + 7$; answers may vary.

81.

82.

83. $\dfrac{128}{192} = \dfrac{2 \cdot \cancel{64}}{3 \cdot \cancel{64}} = \dfrac{2}{3}$

84. $\dfrac{p}{t}$

85. $\dfrac{33sba}{2(11a)} = \dfrac{3 \cdot \cancel{11} \cdot s \cdot b \cdot \cancel{a}}{2 \cdot \cancel{11} \cdot \cancel{a}} = \dfrac{3sb}{2}$

86. $\dfrac{12}{5}$

87. $\dfrac{36 \cdot (2rh)}{8 \cdot (9hg)} = \dfrac{\cancel{4} \cdot \cancel{9} \cdot \cancel{2} \cdot r \cdot \cancel{h}}{\cancel{2} \cdot \cancel{4} \cdot \cancel{9} \cdot \cancel{h} \cdot g} = \dfrac{r}{g}$

88. $\dfrac{5}{2}$

89. We need to find the smallest number that has both 6 and 8 as factors. Starting with 6 we list some numbers with a factor of 6, and starting with 8 we also list some numbers with a factor of 8. Then we find the first number that is on both lists.

6, 12, 18, 24, 30, 36, ...

8, 16, 24, 32, 40, 48, ...

Since 24 is the smallest number that is on both lists, the carton should be 24 in. long.

90.

Product	56	63	36	72	140	96	48
Factor	7	7	2	36	14	8	6
Factor	8	9	18	2	10	12	8
Sum	15	16	20	38	24	20	14

Product	168	110	90	432	63
Factor	21	11	9	24	3
Factor	8	10	10	18	21
Sum	29	21	19	42	24

91. $A = lw = \left(\frac{4}{5}\text{ m}\right)\left(\frac{7}{9}\text{ m}\right)$

$\qquad = \left(\frac{4}{5}\right)\left(\frac{7}{9}\right)(\text{m})(\text{m})$

$\qquad = \frac{28}{45}\text{ m}^2$, or $\frac{28}{45}$ square meters

92. $\frac{25}{28}\text{ m}^2$, or $\frac{25}{28}$ square meters

93. $P = 4s = 4\left(\frac{5}{9}\text{m}\right) = \frac{20}{9}\text{ m}$

94. $\frac{142}{45}$ m

95. ◈

Exercise Set 1.4

1. The integer 5 corresponds to winning 5 points, and the integer -12 corresponds to losing 12 points.

2. 18, -2

3. The integer -170 corresponds to owing \$170, and the integer 950 corresponds to having \$950 in a bank account.

4. 1200, -560

5. The integer -1286 corresponds to 1286 ft below sea level. The integer 29,028 corresponds to 29,028 ft above sea level.

6. Jets: -34, Strikers: 34

7. The integer 750 corresponds to a \$750 deposit, and the integer -125 corresponds to a \$125 withdrawal.

8. $-3,000,000$

9. The integers 20, -150, and 300 correspond to the interception of the missile, the loss of the starship, and the capture of the base, respectively.

10. -10, 235

11. Since $\frac{10}{3} = 3\frac{1}{3}$, its graph is $\frac{1}{3}$ of a unit to the right of 3.

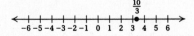

12.

$-\frac{17}{5}$

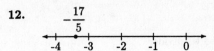

13. The graph of -4.3 is $\frac{3}{10}$ of a unit to the left of -4.

14.

15.

16.

17. We first find decimal notation for $\frac{3}{8}$. Since $\frac{3}{8}$ means $3 \div 8$, we divide.

$$\begin{array}{r} 0.3\,7\,5 \\ 8\,\overline{)3.0\,0\,0} \\ \underline{2\,4} \\ 6\,0 \\ \underline{5\,6} \\ 4\,0 \\ \underline{4\,0} \\ 0 \end{array}$$

Thus $\frac{3}{8} = 0.375$, so $-\frac{3}{8} = -0.375$.

18. -0.125

19. $\frac{5}{3}$ means $5 \div 3$, so we divide.

$$\begin{array}{r} 1.6\,6\,\ldots \\ 3\,\overline{)5.0\,0} \\ \underline{3} \\ 2\,0 \\ \underline{1\,8} \\ 2\,0 \\ \underline{1\,8} \\ 2 \end{array}$$

We have $\frac{5}{3} = 1.\overline{6}$.

20. $0.8\overline{3}$

21. $\frac{7}{6}$ means $7 \div 6$, so we divide.

$$\begin{array}{r} 1.1\,6\,6 \\ 6\,\overline{)7.0\,0\,0} \\ \underline{6} \\ 1\,0 \\ \underline{6} \\ 4\,0 \\ \underline{3\,6} \\ 4\,0 \\ \underline{3\,6} \\ 4 \end{array}$$

We have $\frac{7}{6} = 1.1\overline{6}$.

22. $0.41\overline{6}$

23. $\dfrac{2}{3}$ means $2 \div 3$, so we divide.

$$
\begin{array}{r}
0.6\ 6\ 6 \\
3\overline{)\ 2.0\ 0\ 0} \\
\underline{1\ 8} \\
2\ 0 \\
\underline{1\ 8} \\
2\ 0 \\
\underline{1\ 8} \\
2
\end{array}
$$

We have $\dfrac{2}{3} = 0.\overline{6}$.

24. 0.25

25. We first find decimal notation for $\dfrac{1}{2}$. Since $\dfrac{1}{2}$ means $1 \div 2$, we divide.

$$
\begin{array}{r}
0.5 \\
2\overline{)\ 1.0} \\
\underline{1\ 0} \\
0
\end{array}
$$

Thus, $\dfrac{1}{2} = 0.5$, so $-\dfrac{1}{2} = -0.5$.

26. 0.625

27. $\dfrac{1}{10}$ means $1 \div 10$, so we divide.

$$
\begin{array}{r}
0.1 \\
10\overline{)\ 1.0} \\
\underline{1\ 0} \\
0
\end{array}
$$

We have $\dfrac{1}{10} = 0.1$.

28. -0.35

29. Since 5 is to the right of 0, we have $5 > 0$.

30. $9 > 0$

31. Since -9 is to the left of 5, we have $-9 < 5$.

32. $8 > -8$

33. Since -6 is to the left of 6, we have $-6 < 6$.

34. $0 > -7$

35. Since -8 is to the left of -5, we have $-8 < -5$.

36. $-4 < -3$

37. Since -5 is to the right of -11, we have $-5 > -11$.

38. $-3 > -4$

39. Since -12.5 is to the left of -9.4, we have $-12.5 < -9.4$.

40. $-10.3 > -14.5$

41. Since 2.14 is to the right of 1.24, we have $2.14 > 1.24$.

42. $-3.3 < -2.2$

43. We convert to decimal notation.
$\dfrac{5}{12} = 0.41\overline{6}$ and $\dfrac{11}{25} = 0.44$. Thus, $\dfrac{5}{12} < \dfrac{11}{25}$.

44. $-\dfrac{14}{17} < -\dfrac{27}{35}$

45. $x < -6$ has the same meaning as $-6 > x$.

46. $8 > x$

47. $y \geq -10$ has the same meaning as $-10 \leq y$.

48. $t \leq 12$

49. $-3 \geq -11$ is true, since $-3 > -11$ is true.

50. False

51. $0 \geq 8$ is false, since neither $0 > 8$ nor $0 = 8$ is true.

52. True

53. $-8 \leq -8$ is true because $-8 = -8$ is true.

54. True

55. English: $\underbrace{-5}$ $\underbrace{\text{is greater than}}$ $\underbrace{\text{some number.}}$

   Translation: -5 $\qquad > \qquad$ x

56. $x < -1$

57. English: $\underbrace{\begin{array}{c}\text{A score} \\ \text{of } 120\end{array}}$ $\underbrace{\begin{array}{c}\text{is better} \\ \text{than}\end{array}}$ $\underbrace{\begin{array}{c}\text{a score} \\ \text{of } -20.\end{array}}$

   Translation: 120 $\qquad > \qquad$ -20

58. $20 > -25$

59. English: $\underbrace{\begin{array}{c}\text{A deficit} \\ \text{of} \\ \$500,000\end{array}}$ $\underbrace{\begin{array}{c}\text{is} \\ \text{worse} \\ \text{than}\end{array}}$ $\underbrace{\begin{array}{c}\text{an excess} \\ \text{of} \\ \$1,000,000.\end{array}}$

   Translation: $-500,000$ $\quad < \quad$ $1,000,000$

60. $60 > 20$

61. English: $\underbrace{\begin{array}{c}\text{Alicia's test} \\ \text{score}\end{array}}$ $\underbrace{\text{was at most}}$ $\underbrace{95.}$

   Translation: s $\qquad \leq \qquad$ 95

62. $n \leq -9$

63. English: $\underbrace{\begin{array}{c}\text{Profits} \\ \text{considered} \\ \text{poor}\end{array}}$ $\underbrace{\begin{array}{c}\text{don't} \\ \text{exceed}\end{array}}$ $\underbrace{\$15,000.}$

   Translation: p $\qquad \leq \qquad$ $15,000$

64. $n \leq 0$

65. $|-3| = 3$ since -3 is 3 units from 0.

66. 7

67. $|10| = 10$ since 10 is 10 units from 0.

68. 11

69. $|0| = 0$ since 0 is 0 units from itself.

70. 4

71. $|-24| = 24$ since -24 is 24 units from 0.

72. 325

73. $\left|-\dfrac{2}{3}\right| = \dfrac{2}{3}$ since $-\dfrac{2}{3}$ is $\dfrac{2}{3}$ of a unit from 0.

74. $\dfrac{10}{7}$

75. $|43.9| = 43.9$ since 43.9 is 43.9 units from 0.

76. 14.8

77. When $x = 5$, $|x| = |5| = 5$.

78. $\dfrac{7}{8}$

79. Answers may vary. $-\dfrac{9}{7},\, 0,\, 4\dfrac{1}{2},\, -1.97,\, -491,\, 128,$
$\dfrac{3}{11},\, -\dfrac{1}{7},\, 0.000011,\, -26\dfrac{1}{3}$

80. Answers may vary. $1.26,\, 9\dfrac{1}{5},\, \dfrac{3}{2},\, 0.17,\, \dfrac{6}{11},\, -\dfrac{1}{10,000},$
$-0.1,\, -5.6283,\, -8.3,\, -47\dfrac{1}{2}$

81. Answers may vary. $-\pi,\, \sqrt{42},\, 8.4262262226...$

82. Answers may vary. $-67,892,\, -356,\, -2$

83. $\dfrac{21}{5} \cdot \dfrac{1}{7} = \dfrac{21 \cdot 1}{5 \cdot 7}$ Multiplying numerators and denominators

$= \dfrac{3 \cdot 7 \cdot 1}{5 \cdot 7}$ Factoring the numerator

$= \dfrac{3}{5}$ Removing a factor of 1

84. 42

85. $5 + ab$ is equivalent to $ab + 5$ by the commutative law of addition.

$ba + 5$ is equivalent to $ab + 5$ by the commutative law of multiplication.

$5 + ba$ is equivalent to $ab + 5$ by both commutative laws.

86. $3(x + 3 + 4y)$

87.

88.

89. List the numbers as they occur on the number line, from left to right: $-17, -12, 5, 13$

90. $-23, -17, 0, 4$

91. Converting to decimal notation, we can write
$\dfrac{4}{5}, \dfrac{4}{3}, \dfrac{4}{8}, \dfrac{4}{6}, \dfrac{4}{9}, \dfrac{4}{2}, -\dfrac{4}{3}$ as $0.8, 1.3\overline{3}, 0.5, 0.6\overline{6}, 0.4\overline{4}, 2, -1.3\overline{3}$,
respectively. List the numbers (in fractional form) as they occur on the number line, from left to right:
$-\dfrac{4}{3}, \dfrac{4}{9}, \dfrac{4}{8}, \dfrac{4}{6}, \dfrac{4}{5}, \dfrac{4}{3}, \dfrac{4}{2}$

92. $-\dfrac{5}{6}, -\dfrac{3}{4}, -\dfrac{2}{3}, \dfrac{1}{6}, \dfrac{3}{8}, \dfrac{1}{2}$

93. $|-5| = 5$ and $|-2| = 2$, so $|-5| > |-2|$.

94. $|4| < |-7|$

95. $|-8| = 8$ and $|8| = 8$, so $|-8| = |8|$.

96. $|23| = |-23|$

97. $|-3| = 3$ and $|5| = 5$, so $|-3| < |5|$.

98. $|-19| < |-27|$

99. $|x| = 7$

x represents a number whose distance from 0 is 7. Thus, $x = 7$ or $x = -7$.

100. $-1, 0, 1$

101. a) $0.5555... = \dfrac{5}{6}(0.6666....) = \dfrac{5}{6} \cdot \dfrac{2}{3} = \dfrac{5 \cdot 2}{6 \cdot 3} = \dfrac{5 \cdot 2}{2 \cdot 3 \cdot 3} = \dfrac{5}{9}$

b) $0.1111... = \dfrac{0.3333...}{3} = \dfrac{\frac{1}{3}}{3} = \dfrac{1}{9}$

c) $0.2222... = \dfrac{0.6666...}{3} = \dfrac{\frac{2}{3}}{3} = \dfrac{2}{9}$

d) $0.9999... = 3(0.3333....) = 3 \cdot \dfrac{1}{3} = \dfrac{3}{3} = 1$

Exercise Set 1.5

1. Start at -9. Move 2 units to the right.

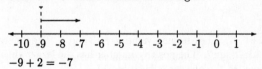

$-9 + 2 = -7$

2. -3

3. Start at -10. Move 6 units to the right.

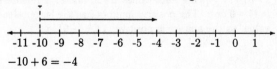

$-10 + 6 = -4$

4. 5

5. Start at -8. Move 8 units to the right.

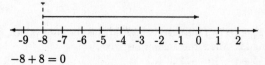

$-8 + 8 = 0$

6. 0

7. Start at -3. Move 5 units to the left.

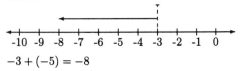

$-3 + (-5) = -8$

8. -10

9. $-7 + 0$ One number is 0. The answer is the other number. $-7 + 0 = -7$

10. -13

11. $0 + (-27)$ One number is 0. The answer is the other number. $0 + (-27) = -27$

12. -35

13. $17 + (-17)$ The numbers have the same absolute value. The sum is 0. $17 + (-17) = 0$

14. 0

15. $-17 + (-25)$ Two negatives. Add the absolute values, getting 42. Make the answer negative. $-17 + (-25) = -42$

16. -41

17. $-18 + 18$ The numbers have the same absolute value. The sum is 0. $-18 + 18 = 0$

18. 0

19. $8 + (-5)$ The absolute values are 8 and 5. The difference is $8 - 5$, or 3. The positive number has the larger absolute value, so the answer is positive. $8 + (-5) = 3$

20. 1

21. $-4 + (-5)$ Two negatives. Add the absolute values, getting 9. Make the answer negative. $-4 + (-5) = -9$

22. -2

23. $13 + (-6)$ The absolute values are 13 and 6. The difference is $13 - 6$, or 7. The positive number has the larger absolute value, so the answer is positive. $13 + (-6) = 7$

24. 11

25. $11 + (-9)$ The absolute values are 11 and 9. The difference is $11 - 9$ or 2. The positive number has the larger absolute value, so the answer is positive. $11 + (-9) = 2$

26. -33

27. $-20 + (-6)$ Two negatives. Add the absolute values, getting 26. Make the answer negative. $-20 + (-6) = -26$

28. 0

29. $-15 + (-7)$ Two negatives. Add the absolute values, getting 22. Make the answer negative. $-15 + (-7) = -22$

30. 18

31. $40 + (-8)$ The absolute values are 40 and 8. The difference is $40 - 8$, or 32. The positive number has the larger absolute value, so the answer is positive. $40 + (-8) = 32$

32. -32

33. $-25 + 25$ A negative and a positive number. The numbers have the same absolute value. The sum is 0. $-25 + 25 = 0$

34. 0

35. $63 + (-18)$ The absolute values are 63 and 18. The difference is $63 - 18$, or 45. The positive number has the larger absolute value, so the answer is positive. $63 + (-18) = 45$

36. 20

37. $-6.5 + 4.7$ The absolute values are 6.5 and 4.7. The difference is $6.5 - 4.7$, or 1.8. The negative number has the larger absolute value, so the answer is negative. $-6.5 + 4.7 = -1.8$

38. -1.7

39. $-2.8 + (-5.3)$ Two negatives. Add the absolute values, getting 8.1. Make the answer negative. $-2.8 + (-5.3) = -8.1$

40. -14.4

41. $-\dfrac{3}{5} + \dfrac{2}{5}$ The absolute values are $\dfrac{3}{5}$ and $\dfrac{2}{5}$. The difference is $\dfrac{3}{5} - \dfrac{2}{5}$, or $\dfrac{1}{5}$. The negative number has the larger absolute value, so the answer is negative.
$$-\frac{3}{5} + \frac{2}{5} = -\frac{1}{5}$$

42. $-\dfrac{2}{3}$

43. $-\dfrac{3}{7} + \left(-\dfrac{5}{7}\right)$ Two negatives. Add the absolute values, getting $\dfrac{8}{7}$. Make the answer negative.
$$-\frac{3}{7} + \left(-\frac{5}{7}\right) = -\frac{8}{7}$$

44. $-\dfrac{10}{9}$

45. $-\dfrac{5}{8} + \dfrac{1}{4}$ The absolute values are $\dfrac{5}{8}$ and $\dfrac{1}{4}$. The difference is $\dfrac{5}{8} - \dfrac{2}{8}$, or $\dfrac{3}{8}$. The negative number has the larger absolute value, so the answer is negative.
$$-\frac{5}{8} + \frac{1}{4} = -\frac{3}{8}$$

46. $-\dfrac{1}{6}$

47. $-\dfrac{3}{7} + \left(-\dfrac{2}{5}\right)$ Two negatives. Add the absolute values,

getting $\dfrac{15}{35} + \dfrac{14}{35}$, or $\dfrac{29}{35}$. Make the answer negative.

$-\dfrac{3}{7} + \left(-\dfrac{2}{5}\right) = -\dfrac{29}{35}$

48. $-\dfrac{23}{24}$

49. $\quad 75 + (-14) + (-17) + (-5)$

$\quad = 75 + [(-14) + (-17) + (-5)]$ Using the
associative law of addition

$\quad = 75 + (-36)$ Adding the negatives

$\quad = 39$ Adding a positive and a negative

50. -62

51. $\quad -44 + \left(-\dfrac{3}{8}\right) + 95 + \left(-\dfrac{5}{8}\right)$

$\quad = \left[-44 + \left(-\dfrac{3}{8}\right) + \left(-\dfrac{5}{8}\right)\right] + 95$

     Using the associative law of addition

$\quad = -45 + 95$ Adding the negatives

$\quad = 50$ Adding a negative and a positive

52. 37.9

53. $\quad 98 + (-54) + 113 + (-998) + 44 + (-612) +$

$\quad\quad (-18) + 334$

$\quad = (98 + 113 + 44 + 334) + [-54 + (-998) +$

$\quad\quad (-612) + (-18)]$

$\quad = 589 + (-1682)$ Adding the positives;
adding the negatives

$\quad = -1093$

54. -1021

55. Rewording:

First try plus second try plus third try plus

$\downarrow \quad\quad \downarrow \quad\quad \downarrow \quad\quad \downarrow \quad\quad \downarrow \quad\quad \downarrow$

Translating: $13 \quad + \quad 0 \quad + \quad (-12) \quad +$

fourth try plus fifth try is the total gain (or loss).

$\downarrow \quad \downarrow \quad \downarrow \quad \downarrow \quad\quad\quad \downarrow$

$21 \quad + \quad (-14) = $ Total gain (or loss).

Since $13 + 0 + (-12) + 21 + (-14)$

$\quad = 13 + (-12) + 21 + (-14)$

$\quad = 1 + 21 + (-14)$

$\quad = 22 + (-14)$

$\quad = 8,$

the total gain was 8 yd.

56. $77,320 profit

57. Rewording:

First change plus second change plus

$\downarrow \quad\quad \downarrow \quad\quad \downarrow \quad\quad \downarrow$

Translating: $-6 \quad + \quad 3 \quad +$

third change plus fourth change is the total change.

$\downarrow \quad \downarrow \quad \downarrow \quad \downarrow \quad\quad \downarrow$

$(-14) \quad + \quad 4 \quad = $ Total change.

Since $-6 + 3 + (-14) + 4$

$\quad = [-6 + (-14)] + (3 + 4)$

$\quad = -20 + 7$

$\quad = -13,$

the total change in pressure is a 13 mb drop.

58. $0

59. Rewording:

Amount owed plus amount paid plus

$\downarrow \quad\quad \downarrow \quad\quad \downarrow \quad\quad \downarrow$

Translating: $-470 \quad + \quad 45 \quad +$

additional charge plus additional payment is amount owed.

$\downarrow \quad \downarrow \quad \downarrow \quad \downarrow \quad\quad \downarrow$

$(-160) \quad + \quad 500 \quad = $ amount owed.

Since $-470 + 45 + (-160) + 500$

$\quad = [-470 + (-160) + (45 + 500)$

$\quad = -630 + 545$

$\quad = -85,$

Kyle owes the company $85.

60. $85 overdrawn

61. $3a + 8a = (3 + 8)a$ Using the distributive law

$\quad\quad = 11a$

62. $12x$

63. $-2x + 15x = (-2 + 15)x$ Using the distributive
law

$\quad\quad\quad = 13x$

64. $-5m$

65. $4x + 7x = (4 + 7)x = 11x$

66. $14a$

67. $7m + (-9m) = [7 + (-9)]m = -2m$

68. $5x$

69. $-6a + 10a = (-6 + 10)a = 4a$

70. $-7n$

71. $-3 + 8x + 4 + (-10x)$

$= -3 + 4 + 8x + (-10x)$ Using the commutative law of addition

$= (-3 + 4) + [8 + (-10)]x$ Using the distributive law

$= 1 - 2x$ Adding

72. $7a + 2$

73. Perimeter $= 7 + 5x + 9 + 6x$

$= 7 + 9 + 5x + 6x$

$= (7 + 9) + (5 + 6)x$

$= 16 + 11x$

74. $10a + 13$

75. Perimeter $= 9 + 3m + 7 + 4m + 2m$

$= 9 + 7 + 3m + 4m + 2m$

$= (9 + 7) + (3 + 4 + 2)m$

$= 16 + 9m$

76. $19n + 11$

77. $7(3z + y + 2) = 7 \cdot 3z + 7 \cdot y + 7 \cdot 2 = 21z + 7y + 14$

78. $\dfrac{28}{3}$

79.

80.

81. Starting with the final value, we "undo" the rise and drop in value by adding their opposites. The result is the original value.

Rewording: $\underbrace{\text{Final value}}$ plus $\underbrace{\text{opposite of rise}}$ plus

$\qquad\qquad\quad \downarrow \qquad \downarrow \qquad\quad \downarrow \qquad\quad \downarrow$

Translating: $\quad 64\frac{3}{8} \qquad + \qquad \left(-2\frac{3}{8}\right) \qquad +$

$\underbrace{\text{opposite of drop}}$ is $\underbrace{\text{original value.}}$

$\quad \downarrow \qquad\quad \downarrow \qquad \downarrow$

$\quad 3\frac{1}{4} \qquad\quad = \text{ original value.}$

Since $64\frac{3}{8} + \left(-2\frac{3}{8}\right) + 3\frac{1}{4} = 62 + 3\frac{1}{4}$

$= 65\frac{1}{4},$

the stock's original value was $\$65\frac{1}{4}$.

82. $\$40.80$

---

83. $7x + \underline{} + (-9x) + (-2y)$

$= 7x + (-9x) + \underline{} + (-2y)$

$= [7 + (-9)]x + \underline{} + (-2y)$

$= -2x + \underline{} + (-2y)$

This expression is equivalent to $-2x - 7y$, so the missing term is the term which yields $-7y$ when added to $-2y$. Since $-5y + (-2y) = -7y$, the missing term is $-5y$.

84. $-11b$

85. $3m + 2n + \underline{} + (-2m)$

$= 2n + \underline{} + (-2m) + 3m$

$= 2n + \underline{} + (-2 + 3)m$

$= 2n + \underline{} + m$

This expression is equivalent to $2n + (-6m)$, so the missing term is the term which yields $-6m$ when added to m. Since $-7m + m = -6m$, the missing term is $-7m$.

86. $-3y$

87. $P = 2l + 2w = 6x + 10$

We know $2l = 2 \cdot 5 = 10$, so $2w$ is $6x$. Then the width is a number which yields $6x$ when added to itself. Since $3x + 3x = 6x$, the width is $3x$.

88. 1 under par

Exercise Set 1.6

1. The opposite of 24 is -24 because $24 + (-24) = 0$.

2. 64

3. The opposite of -9 is 9 because $-9 + 9 = 0$.

4. $-\dfrac{7}{2}$

5. The opposite of -26.9 is 26.9 because $-26.9 + 26.9 = 0$.

6. -48.2

7. If $x = 9$, then $-x = -(9) = -9$. (The opposite of 9 is -9.)

8. 26

9. If $x = -\dfrac{14}{3}$, then $-x = -\left(-\dfrac{14}{3}\right) = \dfrac{14}{3}$.

$\left(\text{The opposite of } -\dfrac{14}{3} \text{ is } \dfrac{14}{3}.\right)$

10. $-\dfrac{1}{328}$

11. If $x = 0.101$, then $-x = -(0.101) = -0.101$.

(The opposite of 0.101 is -0.101.)

12. 0

13. If $x = -65$, then $-(-x) = -[-(-65)] = -65$

(The opposite of the opposite of -65 is -65.)

14. 29

15. If $x = \frac{5}{3}$, then $-(-x) = -\left(-\frac{5}{3}\right) = \frac{5}{3}$.

$\left(\text{The opposite of the opposite of } \frac{5}{3} \text{ is } \frac{5}{3}.\right)$

16. -9.1

17. When we change the sign of -1 we obtain 1.

18. 7

19. When we change the sign of 7 we obtain -7.

20. -10

21. $3 - 7 = 3 + (-7) = -4$

22. -5

23. $0 - 7 = 0 + (-7) = -7$

24. -10

25. $-8 - (-2) = -8 + 2 = -6$

26. 2

27. $-10 - (-10) = -10 + 10 = 0$

28. 0

29. $12 - 16 = 12 + (-16) = -4$

30. -5

31. $20 - 27 = 20 + (-27) = -7$

32. 26

33. $-9 - (-3) = -9 + 3 = -6$

34. 2

35. $-40 - (-40) = -40 + 40 = 0$

36. 0

37. $7 - 7 = 7 + (-7) = 0$

38. 0

39. $7 - (-7) = 7 + 7 = 14$

40. 8

41. $8 - (-3) = 8 + 3 = 11$

42. -11

43. $-6 - 8 = -6 + (-8) = -14$

44. 16

45. $-4 - (-9) = -4 + 9 = 5$

46. -16

47. $-6 - (-5) = -6 + 5 = -1$

48. -1

49. $8 - (-10) = 8 + 10 = 18$

50. 11

51. $0 - 5 = 0 + (-5) = -5$

52. -6

53. $-5 - (-2) = -5 + 2 = -3$

54. -2

55. $-7 - 14 = -7 + (-14) = -21$

56. -25

57. $0 - (-5) = 0 + 5 = 5$

58. 1

59. $-8 - 0 = -8 + 0 = -8$

60. -9

61. $7 - (-5) = 7 + 5 = 12$

62. 35

63. $2 - 25 = 2 + (-25) = -23$

64. -45

65. $-42 - 26 = -42 + (-26) = -68$

66. -81

67. $-71 - 2 = -71 + (-2) = -73$

68. -52

69. $24 - (-92) = 24 + 92 = 116$

70. 121

71. $-50 - (-50) = -50 + 50 = 0$

72. 0

73. $\frac{3}{8} - \frac{5}{8} = \frac{3}{8} + \left(-\frac{5}{8}\right) = -\frac{2}{8} = -\frac{1}{4}$

74. $-\frac{2}{3}$

75. $\frac{3}{4} - \frac{2}{3} = \frac{9}{12} - \frac{8}{12} = \frac{9}{12} + \left(-\frac{8}{12}\right) = \frac{1}{12}$

76. $-\frac{1}{8}$

77. $-\frac{3}{4} - \frac{2}{3} = -\frac{9}{12} - \frac{8}{12} = -\frac{9}{12} + \left(-\frac{8}{12}\right) = -\frac{17}{12}$

78. $-\frac{11}{8}$

79. $-2.8 - 0 = -2.8 + 0 = -2.8$

80. 4.94

81. $0.99 - 1 = 0.99 + (-1) = -0.01$

82. -0.13

83. $\frac{1}{6} - \frac{2}{3} = \frac{1}{6} - \frac{4}{6} = \frac{1}{6} + \left(-\frac{4}{6}\right) = -\frac{3}{6} = -\frac{1}{2}$

84. $\frac{1}{8}$

85. $-\dfrac{4}{7} - \left(-\dfrac{10}{7}\right) = -\dfrac{4}{7} + \dfrac{10}{7} = \dfrac{6}{7}$

86. 0

87. We subtract the smaller number from the larger.
Translate: $1.5 - (-3.5)$
Simplify: $1.5 - (-3.5) = 1.5 + 3.5 = 5$

88. $-2.1 - (-5.9)$; 3.8

89. We subtract the smaller number from the larger.
Translate: $114 - (-79)$
Simplify: $114 - (-79) = 114 + 79 = 193$

90. $23 - (-17)$; 40

91. $-13 - 41 = -13 + (-41) = -54$

92. -26

93. $9 - (-25) = 9 + 25 = 34$

94. 26

95. $-3.2 - 5.8$ is read "negative three point two minus five point eight."
$$-3.2 - 5.8 = -3.2 + (-5.8) = -9$$

96. Negative two point seven minus five point nine; -8.6

97. $-230 - (-500)$ is read "negative two hundred thirty minus negative five hundred."
$$-230 - (-500) = -230 + 500 = 270$$

98. Negative three hundred fifty minus negative one thousand; 650

99. $18 - (-15) - 3 - (-5) + 2 = 18 + 15 + (-3) + 5 + 2 = 37$

100. -22

101. $-31 + (-28) - (-14) - 17 = (-31) + (-28) + 14 + (-17) = -62$

102. 22

103. $-34 - 28 + (-33) - 44 = (-34) + (-28) + (-33) + (-44) = -139$

104. 5

105. $-93 - (-84) - 41 - (-56) = (-93) + 84 + (-41) + 56 = 6$

106. 4

107. $3x - 2y = 3x + (-2y)$, so the terms are $3x$ and $-2y$.

108. $7a$, $-9b$

109. $-5 + 3m - 6mn = -5 + 3m + (-6mn)$, so the terms are $-5, 3m$, and $-6mn$.

110. -9, $-4t$, $10rt$

111. $5 - a - 6b + 2 = 5 + (-a) + (-6b) + 2$, so the terms are $5, -a, -6b$, and 2.

112. -2, $3x$, $-y$, -8

113. $7a - 12a$
$= 7a + (-12a)$ Adding the opposite
$= (7 + (-12))a$ Using the distributive law
$= -5a$

114. $-12x$

115. $-3m - 5 + m$
$= -3m + (-5) + m$ Rewriting as addition
$= -3m + m + (-5)$ Using the commutative law of addition
$= -2m + (-5)$ Adding like terms mentally
$= -2m - 5$ Rewriting as subtraction

116. $9n - 15$

117. $3x + 5 - 9x$
$= 3x + 5 + (-9x)$
$= 3x + (-9x) + 5$
$= -6x + 5$

118. $3a - 5$

119. $2 - 6t - 9 - 2t$
$= 2 + (-6t) + (-9) + (-2t)$
$= 2 + (-9) + (-6t) + (-2t)$
$= -7 - 8t$

120. $-2b - 12$

121. $-5 - (-3x) + 3x + 4x - (-12)$
$= -5 + 3x + 3x + 4x + 12$
$= 3x + 3x + 4x + (-5) + 12$
$= 10x + 7$

122. $7x + 46$

123. $13x - (-2x) + 45 - (-21) = 13x + 2x + 45 + 21 = 15x + 66$

124. $15x + 39$

125. We subtract the amount borrowed from the total assets:
$$\$619.46 - \$950 = \$619.46 + (-\$950) = -\$330.54$$
Your total assets are now $-\$330.54$.

126. \$264

127. We subtract the lower average temperature from the higher average temperature:
$$19 - (-31) = 19 + 31 = 50$$
The difference in the average daily temperature is $50°$ C.

128. $17°$ C

129. We draw a picture of the situation.

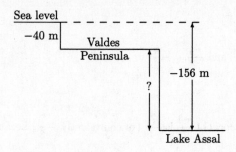

We subtract the lower altitude from the higher altitude:

$$-40 - (-156) = -40 + 156 = 116$$

Lake Assal is 116 m lower than the Valdes Peninsula.

130. 1767 m

131. Area $= lw = (36 \text{ ft})(12 \text{ ft}) = 432 \text{ ft}^2$

132. $2 \cdot 2 \cdot 2 \cdot 2 \cdot 2 \cdot 3 \cdot 3 \cdot 3$

133.

134.

135. See the answer section in the text.

136. False. For example, let $m = -3$ and $n = -5$. Then $-3 > -5$, but $-3 + (-5) = -8 \not> 0$.

137. See the answer section in the text.

138. True. For example, for $m = 4$ and $n = -4$, $4 = -(-4)$ and $4 + (-4) = 0$; for $m = -3$ and $n = 3$, $-3 = -3$ and $-3 + 3 = 0$.

139.

140.

Exercise Set 1.7

1. -16

2. -10

3. -42

4. -18

5. -24

6. -45

7. -72

8. -30

9. 16   Multiplying absolute values

10. 10

11. 42   Multiplying absolute values

12. 18

13. -120

14. 120

15. -238

16. 195

17. 1200

18. -1677

19. 98

20. -203.7

21. -72

22. -63

23. 21.7

24. 12.8

25. $\dfrac{2}{3} \cdot \left(-\dfrac{3}{5}\right) = -\left(\dfrac{2 \cdot 3}{3 \cdot 5}\right) = -\left(\dfrac{2}{5} \cdot \dfrac{3}{3}\right) = -\dfrac{2}{5}$

26. $-\dfrac{10}{21}$

27. $-\dfrac{3}{8} \cdot \left(-\dfrac{2}{9}\right) = \dfrac{\not{3} \cdot \not{2} \cdot 1}{4 \cdot \not{2} \cdot \not{3} \cdot 3} = \dfrac{1}{12}$

28. $\dfrac{1}{4}$

29. -17.01

30. -38.95

31. $-\dfrac{5}{9} \cdot \dfrac{3}{4} = -\dfrac{5 \cdot \not{3}}{\not{3} \cdot 3 \cdot 4} = -\dfrac{5}{12}$

32. -6

33. $7 \cdot (-4) \cdot (-3) \cdot 5 = 7 \cdot 12 \cdot 5 = 7 \cdot 60 = 420$

34. 756

35. $-\dfrac{2}{3} \cdot \dfrac{1}{2} \cdot \left(-\dfrac{6}{7}\right) = -\dfrac{2}{6} \cdot \left(-\dfrac{6}{7}\right) = \dfrac{2 \cdot \not{6}}{7 \cdot \not{6}} = \dfrac{2}{7}$

36. $-\dfrac{3}{160}$

37. $-3 \cdot (-4) \cdot (-5) = 12 \cdot (-5) = -60$

38. -70

39. $-2 \cdot (-5) \cdot (-3) \cdot (-5) = 10 \cdot 15 = 150$

40. 30

41. 0, The product of 0 and any real number is 0.

42. 0

43. $(-8)(-9)(-10) = 72(-10) = -720$

44. 5040

45. $(-6)(-7)(-8)(-9)(-10) = 42 \cdot 72 \cdot (-10) = 3024 \cdot (-10) = -30,240$

46. 151,200

47. $36 \div (-6) = -6$ Check: $-6 \cdot (-6) = 36$

48. -4

49. $\dfrac{26}{-2} = -13$ Check: $-13 \cdot (-2) = 26$

50. -2

51. $\dfrac{-16}{8} = -2$ Check: $-2 \cdot 8 = -16$

52. 11

53. $\dfrac{-48}{-12} = 4$ Check: $4(-12) = -48$

54. 7

55. $\dfrac{-72}{9} = -8$ Check: $-8 \cdot 9 = -72$

56. -2

57. $-100 \div (-50) = 2$ Check: $2(-50) = -100$

58. -25

59. $-108 \div 9 = -12$ Check: $-12 \cdot 9 = -108$

60. $\dfrac{64}{7}$

61. $\dfrac{200}{-25} = -8$ Check: $-8(-25) = 200$

62. $\dfrac{300}{13}$

63. Undefined

64. 0

65. $\dfrac{88}{-9} = -\dfrac{88}{9}$ Check: $-\dfrac{88}{9} \cdot (-9) = 88$

66. Indeterminate

67. $\dfrac{0}{-9} = 0$

68. Undefined

69. $0 \div 0$ is indeterminate.

70. 0

71. $\dfrac{9}{-5} = \dfrac{-9}{5}$ and $\dfrac{9}{-5} = -\dfrac{9}{5}$

72. $\dfrac{12}{-7}, \ -\dfrac{12}{7}$

73. $\dfrac{-36}{11} = \dfrac{36}{-11}$ and $\dfrac{-36}{11} = -\dfrac{36}{11}$

74. $\dfrac{-9}{14}, \ -\dfrac{9}{14}$

75. $-\dfrac{7}{3} = \dfrac{-7}{3}$ and $-\dfrac{7}{3} = \dfrac{7}{-3}$

76. $\dfrac{-4}{15}, \ \dfrac{4}{-15}$

77. $\dfrac{-x}{2} = \dfrac{x}{-2}$ and $\dfrac{-x}{2} = -\dfrac{x}{2}$

78. $\dfrac{-9}{a}, \ -\dfrac{9}{a}$

79. The reciprocal of $\dfrac{-3}{7}$ is $\dfrac{7}{-3}$ $\left($or equivalently, $-\dfrac{7}{3}\right)$ because $\dfrac{-3}{7} \cdot \dfrac{7}{-3} = 1.$

80. $\dfrac{-9}{2}$, or $-\dfrac{9}{2}$

81. The reciprocal of $-\dfrac{47}{13}$ is $-\dfrac{13}{47}$ because $-\dfrac{47}{13} \cdot \left(-\dfrac{13}{47}\right) = 1.$

82. $-\dfrac{12}{31}$

83. The reciprocal of -10 is $\dfrac{1}{-10}$ $\left($or equivalently, $-\dfrac{1}{10}\right)$ because $-10\left(\dfrac{1}{-10}\right) = 1.$

84. $\dfrac{1}{13}$

85. The reciprocal of 4.3 is $\dfrac{1}{4.3}$ because $4.3\left(\dfrac{1}{4.3}\right) = 1.$

86. $\dfrac{1}{-8.5}$, or $-\dfrac{1}{8.5}$

87. The reciprocal of $\dfrac{5}{-3}$ is $\dfrac{-3}{5}$ $\left($or equivalently, $-\dfrac{3}{5}\right)$ because $\dfrac{5}{-3}\left(\dfrac{-3}{5}\right) = 1.$

88. $\dfrac{11}{-6}$, or $-\dfrac{11}{6}$

89. The reciprocal of -1 is $\dfrac{1}{-1}$, or -1 because $(-1)(-1) = 1.$

90. $\dfrac{1}{2}$

91. $\left(-\dfrac{3}{7}\right)\left(\dfrac{2}{-5}\right)$
$= \left(\dfrac{3}{-7}\right)\left(\dfrac{2}{-5}\right)$ Rewriting $-\dfrac{3}{7}$ as $\dfrac{3}{-7}$
$= \dfrac{6}{35}$

92. $\dfrac{8}{27}$

93. $\left(\dfrac{7}{-2}\right)\left(\dfrac{-5}{6}\right) = \left(\dfrac{-7}{2}\right)\left(\dfrac{-5}{6}\right) = \dfrac{35}{12}$

94. $\dfrac{12}{55}$

95. $\dfrac{-4}{5} + \dfrac{7}{-5}$

$= \dfrac{-4}{5} + \dfrac{-7}{5}$ Rewriting $\dfrac{7}{-5}$ with a denominator of 5

$= \dfrac{-11}{5}$, or $-\dfrac{11}{5}$

96. -1

97. $\left(-\dfrac{2}{7}\right)\left(\dfrac{5}{-8}\right) = \left(\dfrac{2}{-7}\right)\left(\dfrac{5}{-8}\right) = \dfrac{10}{56} = \dfrac{\not{2}\cdot 5}{\not{2}\cdot 28} = \dfrac{5}{28}$

98. $\dfrac{18}{7}$

99. $\dfrac{-9}{7} + \left(-\dfrac{4}{7}\right) = \dfrac{-9}{7} + \dfrac{-4}{7} = \dfrac{-13}{7}$, or $-\dfrac{13}{7}$

100. $-\dfrac{8}{11}$

101. $\dfrac{3}{4} \div \left(-\dfrac{2}{3}\right) = \dfrac{3}{4} \cdot \left(-\dfrac{3}{2}\right) = -\dfrac{9}{8}$

102. $-\dfrac{7}{4}$

103. $\dfrac{-5}{12} \cdot \dfrac{7}{15} = -\dfrac{5}{12} \cdot \dfrac{7}{15} = -\dfrac{5 \cdot 7}{12 \cdot 15} = -\dfrac{\not{5}\cdot 7}{12 \cdot \not{5}\cdot 3} = -\dfrac{7}{36}$

104. -12

105. $\left(-\dfrac{12}{5}\right) + \left(-\dfrac{3}{5}\right) = -\dfrac{15}{5} = -3$

106. -3

107. $-\dfrac{5}{4} \div \left(-\dfrac{3}{4}\right) = -\dfrac{5}{4} \cdot \left(-\dfrac{4}{3}\right) = \dfrac{5 \cdot \not{4}}{\not{4}\cdot 3} = \dfrac{5}{3}$

108. $\dfrac{2}{3}$

109. $-6.6 \div 3.3 = -2$ Do the long division. Make the answer negative.

110. 7

111. $\dfrac{-3}{7} - \dfrac{2}{7} = -\dfrac{3}{7} - \dfrac{2}{7} = -\dfrac{3}{7} + \left(-\dfrac{2}{7}\right) = -\dfrac{5}{7}$

112. $-\dfrac{7}{9}$

113. $\dfrac{-5}{9} + \dfrac{2}{-3}$

$= \dfrac{-5}{9} + \dfrac{-2}{3}$

$= \dfrac{-5}{9} + \dfrac{-2}{3} \cdot \dfrac{3}{3}$ Using a common denominator of 6

$= \dfrac{-5}{9} + \dfrac{-6}{9}$

$= \dfrac{-11}{9}$, or $-\dfrac{11}{9}$

114. $-\dfrac{7}{10}$

115. $\left(\dfrac{-3}{5}\right) \div \dfrac{6}{15}$

$= \left(-\dfrac{3}{5}\right) \cdot \dfrac{15}{6}$ Rewriting $\dfrac{-3}{5}$ as $-\dfrac{3}{5}$; multiplying by the reciprocal of the divisor

$= -\dfrac{3 \cdot 15}{5 \cdot 6}$

$= -\dfrac{\not{3}\cdot 3 \cdot \not{5}}{\not{5}\cdot \not{3}\cdot 2}$

$= -\dfrac{3}{2}$

116. $-\dfrac{7}{6}$

117. $\dfrac{4}{9} - \dfrac{1}{-9} = \dfrac{4}{9} - \left(-\dfrac{1}{9}\right) = \dfrac{4}{9} + \dfrac{1}{9} = \dfrac{5}{9}$

118. $\dfrac{6}{7}$

119. $\dfrac{3}{-10} + \dfrac{-1}{5} = \dfrac{-3}{10} + \dfrac{-1}{5} = \dfrac{-3}{10} + \dfrac{-1}{5} \cdot \dfrac{2}{2} =$

$\dfrac{-3}{10} + \dfrac{-2}{10} = \dfrac{-5}{10} = \dfrac{-1 \cdot \not{5}}{2 \cdot \not{5}} = \dfrac{-1}{2}$, or $-\dfrac{1}{2}$

120. $-\dfrac{14}{15}$

121. $\dfrac{-2}{3} - \dfrac{1}{-6} = \dfrac{-2}{3} - \left(-\dfrac{1}{6}\right) = \dfrac{-2}{3} + \dfrac{1}{6} = \dfrac{-2}{3} \cdot \dfrac{2}{2} + \dfrac{1}{6} =$

$\dfrac{-4}{6} + \dfrac{1}{6} = \dfrac{-3}{6} = \dfrac{-1 \cdot \not{3}}{2 \cdot \not{3}} = \dfrac{-1}{2}$, or $-\dfrac{1}{2}$

122. $-\dfrac{1}{2}$

123. $\dfrac{264}{468} = \dfrac{\not{2}\cdot \not{2}\cdot 2 \cdot \not{3}\cdot 11}{\not{2}\cdot \not{2}\cdot \not{3}\cdot 3 \cdot 13} = \dfrac{22}{39}$

124. $12x - 2y - 9$

125. ◈

126. ◈

127. There are none. A reciprocal has the same sign as the number. Zero has no reciprocal.

128. -1 and 1

129. When n is negative, $-n$ is positive, so $\dfrac{-n}{m}$ is the quotient of a positive and a negative number and, thus, is negative.

130. Positive

131. $\dfrac{-n}{m}$ is negative (see Exercise 129), so $-\left(\dfrac{-n}{m}\right)$ is the opposite, or additive inverse, of a negative number and, thus, is positive.

132. Positive

133. When n and m are negative, $-n$ and $-m$ are positive, so $\dfrac{-n}{-m}$ is the quotient of two positive numbers and, thus, is positive. Then, $-\left(\dfrac{-n}{-m}\right)$ is the opposite, or additive inverse, of a positive number and, thus, is negative.

134. a) m and n have different signs;

b) either m or n is zero;

c) m and n have the same sign

135. $a(-b) + ab = a[-b + b]$ Distributive law

$\qquad\qquad\quad = a(0)$ Law of opposites

$\qquad\qquad\quad = 0$ Multiplicative property of 0

Therefore, $a(-b) = -(ab)$.   Law of opposites

136.

Exercise Set 1.8

1. $\underbrace{10 \times 10 \times 10}_{3 \text{ factors}} = 10^3$

2. 6^4

3. $\underbrace{x \cdot x \cdot x \cdot x \cdot x \cdot x \cdot x}_{7 \text{ factors}} = x^7$

4. y^6

5. $3y \cdot 3y \cdot 3y \cdot 3y = (3y)^4$

6. $(5m)^5$

7. $2^4 = 2 \cdot 2 \cdot 2 \cdot 2 = 4 \cdot 4 = 16$

8. 125

9. $(-3)^2 = (-3)(-3) = 9$

10. 49

11. $1^5 = 1 \cdot 1 \cdot 1 \cdot 1 \cdot 1 = 1 \cdot 1 \cdot 1 = 1 \cdot 1 = 1$

12. -1

13. $4^3 = 4 \cdot 4 \cdot 4 = 16 \cdot 4 = 64$

14. 9

15. $(-4)^3 = (-4)(-4)(-4) = 16(-4) = -64$

16. 625

17. $7^1 = 7$ (1 factor)

18. 1

19. $(4a)^2 = (4a)(4a) = 4 \cdot 4 \cdot a \cdot a = 16a^2$

20. $9x^2$

21. $(-7x)^3 = (-7x)(-7x)(-7x) =$
$(-7)(-7)(-7)(x)(x)(x) = -343x^3$

22. $625x^4$

23. $7 + 2 \times 6 = 7 + 12$ Multiplying
$\qquad\qquad\quad = 19$ Adding

24. 27

25. $8 \times 7 + 6 \times 5 = 56 + 30$ Multiplying
$\qquad\qquad\qquad\quad = 86$ Adding

26. 51

27. $19 - 5 \times 3 + 3 = 19 - 15 + 3$ Multiplying
$\qquad\qquad\qquad\quad = 4 + 3$ Subtracting and add-
$\qquad\qquad\qquad\quad = 7$ ing from left to right

28. 9

29. $9 \div 3 + 16 \div 8 = 3 + 2$ Dividing
$\qquad\qquad\qquad\quad = 5$ Adding

30. 28

31. $7 + 10 - 10 \div 2 = 7 + 10 - 5$ Dividing
$\qquad\qquad\qquad\quad = 17 - 5$ Adding and subtract-
$\qquad\qquad\qquad\quad = 12$ ing from left to right

32. 9

33. $(3 - 5)^3 = (-2)^3$ Working within parentheses first
$\qquad\qquad = -8$ Simplifying the exponential expression

34. 24

35. $8 - 2 \cdot 3 - 9 = 8 - 6 - 9$ Multiplying
$\qquad\qquad\qquad = 2 - 9$ Adding and subtracting
$\qquad\qquad\qquad = -7$ from left to right

36. 11

37. $(8 - 2 \cdot 3) - 9 = (8 - 6) - 9$ Multiplying inside the parentheses
$\qquad\qquad\qquad\quad = 2 - 9$ Subtracting inside the parentheses
$\qquad\qquad\qquad\quad = -7$

38. -36

39. $(-24) \div (-3) \cdot \left(-\dfrac{1}{2}\right) = 8 \cdot \left(-\dfrac{1}{2}\right) = -\dfrac{8}{2} = -4$

40. 32

41. $16 \cdot (-24) + 50 = -384 + 50 = -334$

42. -160

43. $2^4 + 2^3 - 10 = 16 + 8 - 10 = 24 - 10 = 14$

44. 23

45. $5^3 + 26 \cdot 71 - (16 + 25 \cdot 3) = 5^3 + 26 \cdot 71 - (16 + 75) =$
$5^3 + 26 \cdot 71 - 91 = 125 + 26 \cdot 71 - 91 = 125 + 1846 - 91 =$
$1971 - 91 = 1880$

46. 305

47. $[2 \cdot (5-3)]^2 = [2 \cdot 2]^2 = 4^2 = 16$

48. 76

49. $\dfrac{7+2}{5^2-4^2} = \dfrac{9}{25-16} = \dfrac{9}{9} = 1$

50. 2

51. $8(-7) + |6(-5)| = -56 + |-30| = -56 + 30 = -26$

52. 49

53. $19 - 5(-3) + 3 = 19 + 15 + 3 = 34 + 3 = 37$

54. 33

55. $9 \div (-3) \cdot 16 \div 8 = -3 \cdot 16 \div 8 = -48 \div 8 = -6$

56. -28

57. $20 + 4^3 \div (-8) \cdot 2 = 20 + 64 \div (-8) \cdot 2 = 20 + (-8) \cdot 2 = 20 + (-16) = 4$

58. -3000

59. $8|(6-13) - 11| = 8|-7-11| = 8|-18| = 8 \cdot 18 = 144$

60. 60

61. $256 \div (-32) \div (-4) = -8 \div (-4)$ Doing the divisions in order
       $= 2$ from left to right

62. 1

63. $\dfrac{5^2 - 4^3 - 3}{9^2 - 2^2 - 1^5} = \dfrac{25 - 64 - 3}{81 - 4 - 1} = \dfrac{-39 - 3}{77 - 1} = \dfrac{-42}{76} =$

$-\dfrac{\cancel{2} \cdot 21}{\cancel{2} \cdot 38} = -\dfrac{21}{38}$

64. $-\dfrac{23}{18}$

65. $\dfrac{20(8-3) - 4(10-3)}{10(2-6) - 2(5+2)} = \dfrac{20 \cdot 5 - 4 \cdot 7}{10(-4) - 2 \cdot 7} =$

$\dfrac{100 - 28}{-40 - 14} = \dfrac{72}{-54} = -\dfrac{\cancel{18} \cdot 4}{\cancel{18} \cdot 3} = -\dfrac{4}{3}$

66. -118

67. $7 - 3x = 7 - 3 \cdot 5 = 7 - 15 = -8$

68. -7

69. $a \div 6 \cdot 2 = 12 \div 6 \cdot 2 = 2 \cdot 2 = 4$

70. 25

71. $-20 \div t^2 - 3(t-1) = -20 \div (-4)^2 - 3((-4) - 1) = -20 \div (-4)^2 - 3(-5) = -20 \div 16 - 3(-5) =$

$\dfrac{-20}{16} + 15 = \dfrac{-5}{4} + 15 = \dfrac{-5}{4} + \dfrac{60}{4} = \dfrac{55}{4}$

72. 20

73. $-x^2 - 5x = -(-3)^2 - 5(-3) = -9 - 5(-3) = -9 + 15 = 6$

74. 24

75. $-(2x + 7) = -2x - 7$ Removing parentheses and changing the sign of each term

76. $-3x - 5$

77. $-(5x - 8) = -5x + 8$ Removing parentheses and changing the sign of each term

78. $-6x + 7$

79. $-(4a - 3b + 7c) = -4a + 3b - 7c$

80. $-5x + 2y + 3z$

81. $-(3x^2 + 5x - 1) = -3x^2 - 5x + 1$

82. $-8x^3 + 6x - 5$

83. $9x - (4x + 3)$
$= 9x - 4x - 3$ Removing parentheses and changing the sign of each term
$= 5x - 3$ Collecting like terms

84. $5y - 9$

85. $2a - (5a - 9) = 2a - 5a + 9 = -3a + 9$

86. $8n + 7$

87. $2x + 7x - (4x + 6) = 2x + 7x - 4x - 6 = 5x - 6$

88. $a - 7$

89. $2x - 4y - 3(7x - 2y) = 2x - 4y - 21x + 6y = -19x + 2y$

90. $-a - 4b$

91. $15x - y - 5(3x - 2y + 5z)$
$= 15x - y - 15x + 10y - 25z$ Multiplying each term in parentheses by -5
$= 9y - 25z$

92. $-16a + 27b - 32c$

93. $3x^2 + 7 - (2x^2 + 5) = 3x^2 + 7 - 2x^2 - 5$
                $= x^2 + 2$

94. $2x^4 + 6x$

95. $9x^3 + x - 2(x^3 + 3x) = 9x^3 + x - 2x^3 - 6x$
                   $= 7x^3 - 5x$

96. $-10x^2 + 17x$

97. $12a^2 - 3ab + 5b^2 - 5(-5a^2 + 4ab - 6b^2)$
$= 12a^2 - 3ab + 5b^2 + 25a^2 - 20ab + 30b^2$
$= 37a^2 - 23ab + 35b^2$

98. $-20a^2 + 29ab + 48b^2$

99. $-7t^3 - t^2 - 3(5t^3 - 3t)$
$= -7t^3 - t^2 - 15t^3 + 9t$
$= -22t^3 - t^2 + 9t$

100. $9t^4 - 45t^3 + 17t$

101.
$$[10(x+3) - 4] + [2(x-1) + 6]$$
$$= [10x + 30 - 4] + [2x - 2 + 6]$$
$$= [10x + 26] + [2x + 4]$$
$$= 10x + 26 + 2x + 4$$
$$= 12x + 30$$

102. $13x - 1$

103.
$$[7(x^2 + 5) - 19] - [4(x^2 - 6) + 10]$$
$$= [7x^2 + 35 - 19] - [4x^2 - 24 + 10]$$
$$= [7x^2 + 16] - [4x^2 - 14]$$
$$= 7x^2 + 16 - 4x^2 + 14$$
$$= 3x^3 + 30$$

104. $x^3 + 41$

105.
$$3\{[7(x-2) + 4] - [2(2x-5) + 6]\}$$
$$= 3\{[7x - 14 + 4] - [4x - 10 + 6]\}$$
$$= 3\{[7x - 10] - [4x - 4]\}$$
$$= 3\{7x - 10 - 4x + 4\}$$
$$= 3\{3x - 6\}$$
$$= 9x - 18$$

106. $-16x + 44$

107.
$$4\{[5(x^3 - 3) + 2] - 3[2(x^3 + 5) - 9]\}$$
$$= 4\{[5x^3 - 15 + 2] - 3[2x^3 + 10 - 9]\}$$
$$= 4\{[5x^3 - 13] - 3[2x^3 + 1]\}$$
$$= 4\{5x^3 - 13 - 6x^3 - 3\}$$
$$= 4\{-x^3 - 16\}$$
$$= -4x^3 - 64$$

108. $-12x^2 - 237$

109. $2x + 9$

110. $\dfrac{1}{2}(x + y)$

111.

112.

113.
$$z - \{2z - [3z - (4z - 5z) - 6z] - 7z\} - 8z$$
$$= z - \{2z - [3z - (-z) - 6z] - 7z\} - 8z$$
$$= z - \{2z - [3z + z - 6z] - 7z\} - 8z$$
$$= z - \{2z - [-2z] - 7z\} - 8z$$
$$= z - \{2z + 2z - 7z\} - 8z$$
$$= z - \{-3z\} - 8z$$
$$= z + 3z - 8z$$
$$= -4z$$

114. $-2x - f$

115.
$$x - \{x - 1 - [x - 2 - (x - 3 - \{x - 4 -$$
$$[x - 5 - (x - 6)]\})]\}$$
$$= x - \{x - 1 - [x - 2 - (x - 3 - \{x - 4 -$$
$$[x - 5 - x + 6]\})]\}$$
$$= x - \{x - 1 - [x - 2 - (x - 3 - \{x - 4 - 1\})]\}$$
$$= x - \{x - 1 - [x - 2 - (x - 3 - \{x - 5\})]\}$$
$$= x - \{x - 1 - [x - 2 - (x - 3 - x + 5)]\}$$
$$= x - \{x - 1 - [x - 2 - 2]\}$$
$$= x - \{x - 1 - [x - 4]\}$$
$$= x - \{x - 1 - x + 4\}$$
$$= x - 3$$

116.

117.

118. False

119. False; $-n + m = -(n - m) \neq -(n + m)$ for $m > 0$

120. True

121. False; $-n - m = -(n + m) \neq -(n - m)$ for $m > 0$

122. False

123. False; $-m(n - m) = -mn + m^2 = -(mn - m^2) \neq -(mn + m^2)$ for $m > 0$

124. True

125. True; $-n(-n - m) = n^2 + nm = n(n + m)$

Chapter 2

Equations, Inequalities, and Problem Solving

Exercise Set 2.1

1. $x + 2 = 6$

$x + 2 - 2 = 6 - 2$ Subtracting 2 on both sides

$x = 4$ Simplifying

Check: $\dfrac{x + 2 = 6}{4 + 2 \ ? \ 6}$

$\qquad\qquad 6 \mid 6$ TRUE

The solution is 4.

2. 3

3. $x + 15 = -5$

$x + 15 - 15 = -5 - 15$ Subtracting 15 on both sides

$x = -20$

Check: $\dfrac{x + 15 = -5}{-20 + 15 \ ? \ -5}$

$\qquad\qquad -5 \mid -5$ TRUE

The solution is -20.

4. 34

5. $x + 6 = -8$

$x + 6 - 6 = -8 - 6$

$x = -14$

Check: $\dfrac{x + 6 = -8}{-14 + 6 \ ? \ -8}$

$\qquad\qquad -8 \mid -8$ TRUE

The solution is -14.

6. -21

7. $-2 = x + 16$

$-2 - 16 = x + 16 - 16$

$-18 = x$

Check: $\dfrac{-2 = x + 16}{-2 \ ? \ -18 + 16}$

$\qquad\quad -2 \mid -2$ TRUE

The solution is -18.

8. -31

9. $x - 9 = 6$

$x - 9 + 9 = 6 + 9$

$x = 15$

Check: $\dfrac{x - 9 = 6}{15 - 9 \ ? \ 6}$

$\qquad\qquad 6 \mid 6$ TRUE

The solution is 15.

10. 13

11. $x - 7 = -21$

$x - 7 + 7 = -21 + 7$

$x = -14$

Check: $\dfrac{x - 7 = -21}{-14 - 7 \ ? \ -21}$

$\qquad\qquad -21 \mid -21$ TRUE

The solution is -14.

12. -11

13. $5 + t = 7$

$-5 + 5 + t = -5 + 7$

$t = 2$

Check: $\dfrac{5 + t = 7}{5 + 2 \ ? \ 7}$

$\qquad\qquad 7 \mid 7$ TRUE

The solution is 2.

14. 4

15. $13 = -7 + y$

$7 + 13 = 7 + (-7) + y$

$20 = y$

Check: $\dfrac{13 = -7 + y}{13 \ ? \ -7 + 20}$

$\qquad 13 \mid 13$ TRUE

The solution is 20.

16. 24

17. $-3 + t = -9$

$3 + (-3) + t = 3 + (-9)$

$t = -6$

Check: $\dfrac{-3 + t = -9}{-3 + (-6) \ ? \ -9}$

$\qquad\qquad -9 \mid -9$ TRUE

The solution is -6.

18. -15

19. $r + \dfrac{1}{3} = \dfrac{8}{3}$

$r + \dfrac{1}{3} - \dfrac{1}{3} = \dfrac{8}{3} - \dfrac{1}{3}$

$r = \dfrac{7}{3}$

Check:

$$r + \frac{1}{3} = \frac{8}{3}$$

$$\frac{7}{3} + \frac{1}{3} \; ? \; \frac{8}{3}$$

$$\frac{8}{3} \; \Big| \; \frac{8}{3} \qquad \text{TRUE}$$

The solution is $\frac{7}{3}$.

20. $\frac{1}{4}$

21.

$$m + \frac{5}{6} = -\frac{11}{12}$$

$$m + \frac{5}{6} - \frac{5}{6} = -\frac{11}{12} - \frac{5}{6}$$

$$m = -\frac{11}{12} - \frac{5}{6} \cdot \frac{2}{2}$$

$$m = -\frac{11}{12} - \frac{10}{12}$$

$$m = -\frac{21}{12} = -\frac{\cancel{3} \cdot 7}{\cancel{3} \cdot 4}$$

$$m = -\frac{7}{4}$$

Check:

$$m + \frac{5}{6} = -\frac{11}{12}$$

$$-\frac{7}{4} + \frac{5}{6} \; ? \; -\frac{11}{12}$$

$$-\frac{21}{12} + \frac{10}{12} \; \Big|$$

$$-\frac{11}{12} \; \Big| \; -\frac{11}{12} \qquad \text{TRUE}$$

The solution is $-\frac{7}{4}$.

22. $-\frac{3}{2}$

23.

$$x - \frac{5}{6} = \frac{7}{8}$$

$$x - \frac{5}{6} + \frac{5}{6} = \frac{7}{8} + \frac{5}{6}$$

$$x = \frac{7}{8} \cdot \frac{3}{3} + \frac{5}{6} \cdot \frac{4}{4}$$

$$x = \frac{21}{24} + \frac{20}{24}$$

$$x = \frac{41}{24}$$

Check:

$$x - \frac{5}{6} = \frac{7}{8}$$

$$\frac{41}{24} - \frac{5}{6} \; ? \; \frac{7}{8}$$

$$\frac{41}{24} - \frac{20}{24} \; \Big| \; \frac{21}{24}$$

$$\frac{21}{24} \; \Big| \; \frac{21}{24} \qquad \text{TRUE}$$

The solution is $\frac{41}{24}$.

24. $\frac{19}{12}$

25.

$$-\frac{1}{5} + z = -\frac{1}{4}$$

$$\frac{1}{5} - \frac{1}{5} + z = \frac{1}{5} - \frac{1}{4}$$

$$z = \frac{1}{5} \cdot \frac{4}{4} - \frac{1}{4} \cdot \frac{5}{5}$$

$$z = \frac{4}{20} - \frac{5}{20}$$

$$z = -\frac{1}{20}$$

Check:

$$-\frac{1}{5} + z = -\frac{1}{4}$$

$$-\frac{1}{5} + \left(-\frac{1}{20}\right) \; ? \; -\frac{1}{4}$$

$$-\frac{4}{20} + \left(-\frac{1}{20}\right) \; \Big| \; -\frac{5}{20}$$

$$-\frac{5}{20} \; \Big| \; -\frac{5}{20} \qquad \text{TRUE}$$

The solution is $-\frac{1}{20}$.

26. $-\frac{5}{8}$

27.

$$x + 2.3 = 7.4$$

$$x + 2.3 + (-2.3) = 7.4 + (-2.3)$$

$$x = 5.1$$

Check:

$$x + 2.3 = 7.4$$

$$5.1 + 2.3 \; ? \; 7.4$$

$$7.4 \; \Big| \; 7.4 \qquad \text{TRUE}$$

The solution is 5.1.

28. 4.7

29.

$$-9.7 = -4.7 + y$$

$$4.7 + (-9.7) = 4.7 + (-4.7) + y$$

$$-5 = y$$

Check:

$$-9.7 = -4.7 + y$$

$$-9.7 \; ? \; -4.7 + (-5)$$

$$-9.7 \; \Big| \; -9.7 \qquad \text{TRUE}$$

The solution is -5.

30. -10.6

31.

$$6x = 36$$

$$\frac{6x}{6} = \frac{36}{6} \qquad \text{Dividing by 6 on both sides}$$

$$1 \cdot x = 6 \qquad \text{Simplifying}$$

$$x = 6 \qquad \text{Identity property of 1}$$

Check:

$$6x = 36$$

$$6 \cdot 6 \; ? \; 36$$

$$36 \; \Big| \; 36 \qquad \text{TRUE}$$

The solution is 6.

32. 13

33. $5x = 45$

$\dfrac{5x}{5} = \dfrac{45}{5}$ Dividing by 5 on both sides

$1 \cdot x = 9$ Simplifying

$x = 9$ Identity property of 1

Check: $\dfrac{5x = 45}{}$

$5 \cdot 9 \ ? \ 45$

$45 \ \big| \ 45$ TRUE

The solution is 9.

34. 8

35. $84 = 7x$

$\dfrac{84}{7} = \dfrac{7x}{7}$ Dividing by 7 on both sides

$12 = 1 \cdot x$

$12 = x$

Check: $\dfrac{84 = 7x}{}$

$84 \ ? \ 7 \cdot 12$

$84 \ \big| \ 84$ TRUE

The solution is 12.

36. 7

37.

$-x = 40$

$-1 \cdot x = 40$

$-1 \cdot (-1 \cdot x) = -1 \cdot 40$

$1 \cdot x = -40$

$x = -40$

Check: $\dfrac{-x = 40}{}$

$-(-40) \ ? \ 40$

$40 \ \big| \ 40$ TRUE

The solution is -40.

38. -100

39.

$-x = -1$

$-1 \cdot x = -1$

$-1 \cdot (-1 \cdot x) = -1 \cdot (-1)$

$1 \cdot x = 1$

$x = 1$

Check: $\dfrac{-x = -1}{}$

$-(1) \ ? \ -1$

$-1 \ \big| \ -1$ TRUE

The solution is 1.

40. 68

41. $7x = -49$

$\dfrac{7x}{7} = \dfrac{-49}{7}$

$1 \cdot x = -7$

$x = -7$

Check: $\dfrac{7x = -49}{}$

$7(-7) \ ? \ -49$

$-49 \ \big| \ -49$ TRUE

The solution is -7.

42. -4

43. $-12x = 72$

$\dfrac{-12x}{-12} = \dfrac{72}{-12}$

$1 \cdot x = -6$

$x = -6$

Check: $\dfrac{-12x = 72}{}$

$-12(-6) \ ? \ 72$

$72 \ \big| \ 72$ TRUE

The solution is -6.

44. -7

45. $-21x = -126$

$\dfrac{-21x}{-21} = \dfrac{-126}{-21}$

$1 \cdot x = 6$

$x = 6$

Check: $\dfrac{-21x = -126}{}$

$-21 \cdot 6 \ ? \ -126$

$-126 \ \big| \ -126$ TRUE

The solution is 6.

46. 8

47. $\dfrac{t}{7} = -9$

$7 \cdot \left(\dfrac{1}{7} t\right) = 7 \cdot (-9)$

$1 \cdot t = -63$

$t = -63$

Check: $\dfrac{\dfrac{t}{7} = -9}{}$

$\dfrac{-63}{7} \ ? \ -9$

$-9 \ \big| \ -9$ TRUE

The solution is -63.

48. -88

49. $\dfrac{3}{4} x = 27$

$\dfrac{4}{3} \cdot \dfrac{3}{4} x = \dfrac{4}{3} \cdot 27$

$1 \cdot x = \dfrac{4 \cdot \not{3} \cdot 3 \cdot 3}{\not{3} \cdot 1}$

$x = 36$

Check: $\dfrac{\dfrac{3}{4} x = 27}{}$

$\dfrac{3}{4} \cdot 36 \ ? \ 27$

$27 \ \big| \ 27$ TRUE

The solution is 36.

50. 20

51.

$$\frac{-t}{3} = 7$$

$$3 \cdot \frac{1}{3} \cdot (-t) = 3 \cdot 7$$

$$-t = 21$$

$$-1 \cdot (-1 \cdot t) = -1 \cdot 21$$

$$1 \cdot t = -21$$

$$t = -21$$

Check: $\dfrac{-t}{3} = 7$

$$\frac{-(-21)}{3} \ ? \ 7$$

$$\frac{21}{3}$$

$$7 \ \Big| \ 7 \qquad \text{TRUE}$$

The solution is -21.

52. -54

53.

$$\frac{1}{5} = -\frac{m}{3}$$

$$\frac{1}{5} = -\frac{1}{3} \cdot m$$

$$-3 \cdot \frac{1}{5} = -3 \cdot \left(-\frac{1}{3} \cdot m \right)$$

$$-\frac{3}{5} = m$$

Check: $\dfrac{1}{5} = -\dfrac{m}{3}$

$$\frac{1}{5} \ ? \ -\frac{-\frac{3}{5}}{3}$$

$$-\left(-\frac{3}{5} \div 3 \right)$$

$$-\left(-\frac{3}{5} \cdot \frac{1}{3} \right)$$

$$-\left(-\frac{1}{5} \right)$$

$$\frac{1}{5} \ \Big| \ \frac{1}{5} \qquad \text{TRUE}$$

The solution is $-\dfrac{3}{5}$.

54. $-\dfrac{7}{9}$

55.

$$-\frac{3}{5}r = -\frac{9}{10}$$

$$-\frac{5}{3} \cdot \left(-\frac{3}{5}r \right) = -\frac{5}{3} \cdot \left(-\frac{9}{10} \right)$$

$$r = \frac{\cancel{5} \cdot \cancel{3} \cdot 3}{\cancel{3} \cdot \cancel{5} \cdot 2}$$

$$r = \frac{3}{2}$$

Check: $-\dfrac{3}{5}r = -\dfrac{9}{10}$

$$-\frac{3}{5} \cdot \frac{3}{2} \ ? \ -\frac{9}{10}$$

$$-\frac{9}{10} \ \Big| \ -\frac{9}{10} \qquad \text{TRUE}$$

The solution is $\dfrac{3}{2}$.

56. $\dfrac{2}{3}$

57.

$$\frac{-3r}{2} = -\frac{27}{4}$$

$$-\frac{3}{2}r = -\frac{27}{4}$$

$$-\frac{2}{3} \cdot \left(-\frac{3}{2}r \right) = -\frac{2}{3} \cdot \left(-\frac{27}{4} \right)$$

$$r = \frac{\cancel{2} \cdot \cancel{3} \cdot 3 \cdot 3}{3 \cdot \cancel{2} \cdot 2}$$

$$r = \frac{9}{2}$$

Check: $\dfrac{-3r}{2} = -\dfrac{27}{4}$

$$-\frac{3}{2} \cdot \frac{9}{2} \ ? \ -\frac{27}{4}$$

$$-\frac{27}{4} \ \Big| \ -\frac{27}{4} \qquad \text{TRUE}$$

The solution is $\dfrac{9}{2}$.

58. -1

59.

$$6.3x = 44.1$$

$$\frac{6.3x}{6.3} = \frac{44.1}{6.3}$$

$$x = 7$$

Check: $6.3x = 44.1$

$$6.3 \cdot 7 \ ? \ 44.1$$

$$44.1 \ \Big| \ 44.1 \qquad \text{TRUE}$$

The solution is 7.

60. 20

61.

$$3.7 + t = 8.2$$

$$3.7 + t - 3.7 = 8.2 - 3.7$$

$$t = 4.5$$

The solution is 4.5.

62. 24

63.

$$18 = -\frac{2}{3}x$$

$$-\frac{3}{2} \cdot 18 = -\frac{3}{2} \left(-\frac{2}{3}x \right)$$

$$-\frac{3 \cdot \cancel{2} \cdot 9}{\cancel{2} \cdot 1} = x$$

$$-27 = x$$

The solution is -27.

64. -5.5

65.
$$17 = y + 29$$
$$17 - 29 = y + 29 - 29$$
$$-12 = y$$

The solution is -12.

66. -128

67.
$$y - \frac{2}{3} = -\frac{1}{6}$$
$$y - \frac{2}{3} + \frac{2}{3} = -\frac{1}{6} + \frac{2}{3}$$
$$y = -\frac{1}{6} + \frac{2}{3} \cdot \frac{2}{2}$$
$$y = -\frac{1}{6} + \frac{4}{6}$$
$$y = \frac{3}{6}$$
$$y = \frac{1}{2}$$

The solution is $\frac{1}{2}$.

68. $-\dfrac{14}{9}$

69.
$$-24 = \frac{8x}{5}$$
$$-24 = \frac{8}{5}x$$
$$\frac{5}{8}(-24) = \frac{5}{8} \cdot \frac{8}{5}x$$
$$-\frac{5 \cdot \cancel{8} \cdot 3}{\cancel{8} \cdot 1} = x$$
$$-15 = x$$

The solution is -15.

70. $-\dfrac{1}{2}$

71.
$$-4.1t = 10.25$$
$$\frac{-4.1t}{-4.1} = \frac{10.25}{-4.1}$$
$$t = -2.5$$

The solution is -2.5.

72. $-\dfrac{19}{23}$

73. $3x + 4x = (3 + 4)x = 7x$

74. $-x + 5$

75. $3x - (4 + 2x) = 3x - 4 - 2x = x - 4$

76. $-5x - 23$

77. ◈

78. ◈

79.
$$-356.788 = -699.034 + t$$
$$699.034 + (-356.788) = 699.034 + (-699.034) + t$$
$$342.246 = t$$

The solution is 342.246.

80. -8655

81. For all x, $0 \cdot x = 0$. There is no solution to $0 \cdot x = 9$.

82. All real numbers

83.
$$4|x| = 48$$
$$|x| = 12$$

x represents a number whose distance from 0 is 12. Thus, $x = -12$ or $x = 12$.

The solution is -12 or 12.

84. No solution

85. For all x, $0 \cdot x = 0$. Thus, the solution is all real numbers.

86. 0

87.
$$x + 4 = 5 + x$$
$$x + 4 - x = 5 + x - x$$
$$4 = 5$$

Since $4 = 5$ is false, the equation has no solution.

88. $-2,\ 2$

89.
$$ax = 5a$$
$$\frac{ax}{a} = \frac{5a}{a}$$
$$x = 5$$

The solution is 5.

90. $a + 4$

91.
$$3x = \frac{b}{a}$$
$$\frac{1}{3} \cdot 3x = \frac{1}{3} \cdot \frac{b}{a}$$
$$x = \frac{b}{3a}$$

The solution is $\frac{b}{3a}$.

92. $\dfrac{a^2 + 1}{c}$

93.
$$1 - c = a + x$$
$$1 - c - a = a + x - a$$
$$1 - c - a = x$$

The solution is $1 - c - a$.

94. $-13,\ 13$

95.
$$x - 4720 = 1634$$
$$x - 4720 + 4720 = 1634 + 4720$$
$$x = 6354$$
$$x + 4720 = 6354 + 4720$$
$$x + 4720 = 11,074$$

96. 250

97.

Exercise Set 2.2

1.
$$5x + 6 = 31$$
$$5x + 6 - 6 = 31 - 6 \qquad \text{Subtracting 6 on both sides}$$
$$5x = 25 \qquad \text{Simplifying}$$
$$\frac{5x}{5} = \frac{25}{5} \qquad \text{Dividing by 5 on both sides}$$
$$x = 5 \qquad \text{Simplifying}$$

Check:
$$\begin{array}{c|c} \multicolumn{2}{c}{5x + 6 = 31} \\ \hline 5 \cdot 5 + 6 \ ? \ 31 & \\ 25 + 6 & \\ 31 & 31 \qquad \text{TRUE} \end{array}$$

The solution is 5.

2. 8

3.
$$8x + 4 = 68$$
$$8x + 4 - 4 = 68 - 4 \qquad \text{Subtracting 4 on both sides}$$
$$8x = 64 \qquad \text{Simplifying}$$
$$\frac{8x}{8} = \frac{64}{8} \qquad \text{Dividing by 8 on both sides}$$
$$x = 8 \qquad \text{Simplifying}$$

Check:
$$\begin{array}{c|c} \multicolumn{2}{c}{8x + 4 = 68} \\ \hline 8 \cdot 8 + 4 \ ? \ 68 & \\ 64 + 4 & \\ 68 & 68 \qquad \text{TRUE} \end{array}$$

The solution is 8.

4. 9

5.
$$4x - 6 = 34$$
$$4x - 6 + 6 = 34 + 6 \qquad \text{Adding 6 on both sides}$$
$$4x = 40$$
$$\frac{4x}{4} = \frac{40}{4} \qquad \text{Dividing by 4 on both sides}$$
$$x = 10$$

Check:
$$\begin{array}{c|c} \multicolumn{2}{c}{4x - 6 = 34} \\ \hline 4 \cdot 10 - 6 \ ? \ 34 & \\ 40 - 6 & \\ 34 & 34 \qquad \text{TRUE} \end{array}$$

The solution is 10.

6. 3

7.
$$3x - 9 = 33$$
$$3x - 9 + 9 = 33 + 9$$
$$3x = 42$$
$$\frac{3x}{3} = \frac{42}{3}$$
$$x = 14$$

Check:
$$\begin{array}{c|c} \multicolumn{2}{c}{3x - 9 = 33} \\ \hline 3 \cdot 14 - 9 \ ? \ 33 & \\ 42 - 9 & \\ 33 & 33 \qquad \text{TRUE} \end{array}$$

The solution is 14.

8. 11

9.
$$7x + 2 = -54$$
$$7x + 2 - 2 = -54 - 2$$
$$7x = -56$$
$$\frac{7x}{7} = \frac{-56}{7}$$
$$x = -8$$

Check:
$$\begin{array}{c|c} \multicolumn{2}{c}{7x + 2 = -54} \\ \hline 7(-8) + 2 \ ? \ -54 & \\ -56 + 2 & \\ -54 & -54 \qquad \text{TRUE} \end{array}$$

The solution is -8.

10. -9

11.
$$-45 = 6y + 3$$
$$-45 - 3 = 6y + 3 - 3$$
$$-48 = 6y$$
$$\frac{-48}{6} = \frac{6y}{6}$$
$$-8 = y$$

Check:
$$\begin{array}{c|c} \multicolumn{2}{c}{-45 = 6y + 3} \\ \hline -45 \ ? \ 6(-8) + 3 & \\ & -48 + 3 \\ -45 & -45 \qquad \text{TRUE} \end{array}$$

The solution is -8.

12. -11

13.
$$-4x + 7 = 35$$
$$-4x + 7 - 7 = 35 - 7$$
$$-4x = 28$$
$$\frac{-4x}{-4} = \frac{28}{-4}$$
$$x = -7$$

Check:
$$\begin{array}{c|c} \multicolumn{2}{c}{-4x + 7 = 35} \\ \hline -4(-7) + 7 \ ? \ 35 & \\ 28 + 7 & \\ 35 & 35 \qquad \text{TRUE} \end{array}$$

The solution is -7.

14. -23

15.
$$-7x - 24 = -129$$
$$-7x - 24 + 24 = -129 + 24$$
$$-7x = -105$$
$$\frac{-7x}{-7} = \frac{-105}{-7}$$
$$x = 15$$

Check: $\dfrac{-7x - 24 = -129}{}$

$-7 \cdot 15 - 24 \ ? \ -129$

$-105 - 24$

$\qquad -129 \ \big| \ -129 \qquad$ TRUE

The solution is 15.

16. 19

17. $5x + 7x = 72$

$\qquad 12x = 72 \qquad$ Collecting like terms

$\qquad \dfrac{12x}{12} = \dfrac{72}{12} \qquad$ Dividing by 12 on both sides

$\qquad x = 6$

Check: $\dfrac{5x + 7x = 72}{}$

$5 \cdot 6 + 7 \cdot 6 \ ? \ 72$

$30 + 42$

$\qquad 72 \ \big| \ 72 \qquad$ TRUE

The solution is 6.

18. 5

19. $8x + 7x = 60$

$\qquad 15x = 60 \qquad$ Collecting like terms

$\qquad \dfrac{15x}{15} = \dfrac{60}{15} \qquad$ Dividing by 15 on both sides

$\qquad x = 4$

Check: $\dfrac{8x + 7x = 60}{}$

$8 \cdot 4 + 7 \cdot 4 \ ? \ 60$

$32 + 28$

$\qquad 60 \ \big| \ 60 \qquad$ TRUE

The solution is 4.

20. 8

21. $4x + 3x = 42$

$\qquad 7x = 42$

$\qquad \dfrac{7x}{7} = \dfrac{42}{7}$

$\qquad x = 6$

Check: $\dfrac{4x + 3x = 42}{}$

$4 \cdot 6 + 3 \cdot 6 \ ? \ 42$

$24 + 18$

$\qquad 42 \ \big| \ 42 \qquad$ TRUE

The solution is 6.

22. 4

23. $-6y - 3y = 27$

$\qquad -9y = 27$

$\qquad \dfrac{-9y}{-9} = \dfrac{27}{-9}$

$\qquad y = -3$

Check: $\dfrac{-6y - 3y = 27}{}$

$-6(-3) - 3(-3) \ ? \ 27$

$18 + 9$

$\qquad 27 \ \big| \ 27 \qquad$ TRUE

The solution is -3.

24. -4

25. $-7y - 8y = -15$

$\qquad -15y = -15$

$\qquad \dfrac{-15y}{-15} = \dfrac{-15}{-15}$

$\qquad y = 1$

Check: $\dfrac{-7y - 8y = -15}{}$

$-7 \cdot 1 - 8 \cdot 1 \ ? \ -15$

$-7 - 8$

$\qquad -15 \ \big| \ -15 \qquad$ TRUE

The solution is 1.

26. 3

27. $10.2y - 7.3y = -58$

$\qquad 2.9y = -58$

$\qquad \dfrac{2.9y}{2.9} = \dfrac{-58}{2.9}$

$\qquad y = -\dfrac{58}{2.9}$

$\qquad y = -20$

Check:

$\dfrac{10.2y - 7.3y = -58}{}$

$10.2(-20) - 7.3(-20) \ ? \ -58$

$-204 + 146$

$\qquad -58 \ \big| \ -58 \qquad$ TRUE

The solution is -20.

28. -20

29. $\qquad x + \dfrac{1}{3}x = 8$

$\qquad \left(1 + \dfrac{1}{3}\right)x = 8$

$\qquad \dfrac{4}{3}x = 8$

$\qquad \dfrac{3}{4} \cdot \dfrac{4}{3}x = \dfrac{3}{4} \cdot 8$

$\qquad x = 6$

Check: $\dfrac{x + \dfrac{1}{3}x = 8}{}$

$6 + \dfrac{1}{3} \cdot 6 \ ? \ 8$

$6 + 2$

$\qquad 8 \ \big| \ 8 \qquad$ TRUE

The solution is 6.

30. 8

31. $8y - 35 = 3y$

$\qquad 8y = 3y + 35 \qquad$ Adding 35 and simplifying

$\qquad 8y - 3y = 35 \qquad$ Subtracting $3y$ and simplifying

$\qquad 5y = 35 \qquad$ Collecting like terms

$\qquad \dfrac{5y}{5} = \dfrac{35}{5} \qquad$ Dividing by 5

$\qquad y = 7$

Check: $\dfrac{8y - 35 = 3y}{}$

$$8 \cdot 7 - 35 \ ? \ 3 \cdot 7$$
$$56 - 35 \ | \ 21$$
$$21 \ | \ 21 \qquad \text{TRUE}$$

The solution is 7.

32. -3

33. $8x - 1 = 23 - 4x$

$8x + 4x = 23 + 1 \qquad$ Adding 1 and $4x$ and simplifying

$12x = 24 \qquad$ Collecting like terms

$\dfrac{12x}{12} = \dfrac{24}{12} \qquad$ Dividing by 12

$x = 2$

Check: $\dfrac{8x - 1 = 23 - 4x}{}$

$$8 \cdot 2 - 1 \ ? \ 23 - 4 \cdot 2$$
$$16 - 1 \ | \ 23 - 8$$
$$15 \ | \ 15 \qquad \text{TRUE}$$

The solution is 2.

34. 5

35. $2x - 1 = 4 + x$

$2x - x = 4 + 1 \qquad$ Adding 1 and $-x$

$x = 5 \qquad$ Collecting like terms

Check: $\dfrac{2x - 1 = 4 + x}{}$

$$2 \cdot 5 - 1 \ ? \ 4 + 5$$
$$10 - 1 \ | \ 9$$
$$9 \ | \ 9 \qquad \text{TRUE}$$

The solution is 5.

36. 2

37. $6x + 3 = 2x + 11$

$6x - 2x = 11 - 3$

$4x = 8$

$\dfrac{4x}{4} = \dfrac{8}{4}$

$x = 2$

Check: $\dfrac{6x + 3 = 2x + 11}{}$

$$6 \cdot 2 + 3 \ ? \ 2 \cdot 2 + 11$$
$$12 + 3 \ | \ 4 + 11$$
$$15 \ | \ 15 \qquad \text{TRUE}$$

The solution is 2.

38. 4

39. $5 - 2x = 3x - 7x + 25$

$5 - 2x = -4x + 25$

$4x - 2x = 25 - 5$

$2x = 20$

$\dfrac{2x}{2} = \dfrac{20}{2}$

$x = 10$

Check:

$$\dfrac{5 - 2x = 3x - 7x + 25}{}$$
$$5 - 2 \cdot 10 \ ? \ 3 \cdot 10 - 7 \cdot 10 + 25$$
$$5 - 20 \ | \ 30 - 70 + 25$$
$$-15 \ | \ -40 + 25$$
$$-15 \ | \ -15 \qquad \text{TRUE}$$

The solution is 10.

40. 10

41. $4 + 3x - 6 = 3x + 2 - x$

$3x - 2 = 2x + 2 \qquad$ Collecting like terms on each side

$3x - 2x = 2 + 2$

$x = 4$

Check: $\dfrac{4 + 3x - 6 = 3x + 2 - x}{}$

$$4 + 3 \cdot 4 - 6 \ ? \ 3 \cdot 4 + 2 - 4$$
$$4 + 12 - 6 \ | \ 12 + 2 - 4$$
$$16 - 6 \ | \ 14 - 4$$
$$10 \ | \ 10 \qquad \text{TRUE}$$

The solution is 4.

42. 0

43. $4y - 4 + y + 24 = 6y + 20 - 4y$

$5y + 20 = 2y + 20$

$5y - 2y = 20 - 20$

$3y = 0$

$y = 0$

Check:

$$\dfrac{4y - 4 + y + 24 = 6y + 20 - 4y}{}$$
$$4 \cdot 0 - 4 + 0 + 24 \ ? \ 6 \cdot 0 + 20 - 4 \cdot 0$$
$$0 - 4 + 0 + 24 \ | \ 0 + 20 - 0$$
$$20 \ | \ 20 \qquad \text{TRUE}$$

The solution is 0.

44. 7

45. $\dfrac{7}{2}x + \dfrac{1}{2}x = 3x + \dfrac{3}{2} + \dfrac{5}{2}x$

The number 2 is the least common denominator, so we multiply by 2 on both sides.

$$2\left(\dfrac{7}{2}x + \dfrac{1}{2}x\right) = 2\left(3x + \dfrac{3}{2} + \dfrac{5}{2}x\right)$$

$$2 \cdot \dfrac{7}{2}x + 2 \cdot \dfrac{1}{2}x = 2 \cdot 3x + 2 \cdot \dfrac{3}{2} + 2 \cdot \dfrac{5}{2}x$$

$$7x + x = 6x + 3 + 5x$$

$$8x = 11x + 3$$

$$8x - 11x = 3$$

$$-3x = 3$$

$$\dfrac{-3x}{-3} = \dfrac{3}{-3}$$

$$x = -1$$

Check:

$$\frac{7}{2}x + \frac{1}{2}x = 3x + \frac{3}{2} + \frac{5}{2}x$$

$\frac{7}{2}(-1) + \frac{1}{2}(-1)$? $3(-1) + \frac{3}{2} + \frac{5}{2}(-1)$	
$-\frac{7}{2} - \frac{1}{2}$	$-3 + \frac{3}{2} - \frac{5}{2}$
$-\frac{8}{2}$	$-\frac{8}{2}$ TRUE

The solution is -1.

46. $\frac{1}{2}$

47. $\frac{2}{3} + \frac{1}{4}t = 6$

The number 12 is the least common denominator, so we multiply by 12 on both sides.

$$12\left(\frac{2}{3} + \frac{1}{4}t\right) = 12 \cdot 6$$
$$12 \cdot \frac{2}{3} + 12 \cdot \frac{1}{4}t = 72$$
$$8 + 3t = 72$$
$$3t = 72 - 8$$
$$3t = 64$$
$$t = \frac{64}{3}$$

Check:

$$\frac{2}{3} + \frac{1}{4}t = 6$$

$\frac{2}{3} + \frac{1}{4}\left(\frac{64}{3}\right)$? 6	
$\frac{2}{3} + \frac{16}{3}$	
$\frac{18}{3}$	
6	6 TRUE

The solution is $\frac{64}{3}$.

48. $-\frac{2}{3}$

49.

$$\frac{2}{3} + 3y = 5y - \frac{2}{15}$$

The number 15 is the least common denominator, so we multiply by 15 on both sides.

$$15\left(\frac{2}{3} + 3y\right) = 15\left(5y - \frac{2}{15}\right)$$
$$15 \cdot \frac{2}{3} + 15 \cdot 3y = 15 \cdot 5y - 15 \cdot \frac{2}{15}$$
$$10 + 45y = 75y - 2$$
$$10 + 2 = 75y - 45y$$
$$12 = 30y$$
$$\frac{12}{30} = y$$
$$\frac{2}{5} = y$$

Check:

$$\frac{2}{3} + 3y = 5y - \frac{2}{15}$$

$\frac{2}{3} + 3 \cdot \frac{2}{5}$? $5 \cdot \frac{2}{5} - \frac{2}{15}$	
$\frac{2}{3} + \frac{6}{5}$	$2 - \frac{2}{15}$
$\frac{10}{15} + \frac{18}{15}$	$\frac{30}{15} - \frac{2}{15}$
$\frac{28}{15}$	$\frac{28}{15}$ TRUE

The solution is $\frac{2}{5}$.

50. -3

51.

$$\frac{5}{3} + \frac{2}{3}x = \frac{25}{12} + \frac{5}{4}x + \frac{3}{4}$$

The number 12 is the least common denominator, so we multiply by 12 on both sides.

$$12\left(\frac{5}{3} + \frac{2}{3}x\right) = 12\left(\frac{25}{12} + \frac{5}{4}x + \frac{3}{4}\right)$$
$$12 \cdot \frac{5}{3} + 12 \cdot \frac{2}{3}x = 12 \cdot \frac{25}{12} + 12 \cdot \frac{5}{4}x + 12 \cdot \frac{3}{4}$$
$$20 + 8x = 25 + 15x + 9$$
$$20 + 8x = 15x + 34$$
$$20 - 34 = 15x - 8x$$
$$-14x = 7x$$
$$\frac{-14}{7} = x$$
$$-2 = x$$

Check:

$$\frac{5}{3} + \frac{2}{3}x = \frac{25}{12} + \frac{5}{4}x + \frac{3}{4}$$

$\frac{5}{3} + \frac{2}{3}(-2)$? $\frac{25}{12} + \frac{5}{4}(-2) + \frac{3}{4}$	
$\frac{5}{3} - \frac{4}{3}$	$\frac{25}{12} - \frac{5}{2} + \frac{3}{4}$
$\frac{1}{3}$	$\frac{25}{12} - \frac{30}{12} + \frac{9}{12}$
	$\frac{4}{12}$
$\frac{1}{3}$	$\frac{1}{3}$ TRUE

The solution is -2.

52. -3

53.
$$2.1x + 45.2 = 3.2 - 8.4x$$
Greatest number of decimal places is 1
$$10(2.1x + 45.2) = 10(3.2 - 8.4x)$$
Multiplying by 10 to clear decimals
$$10(2.1x) + 10(45.2) = 10(3.2) - 10(8.4x)$$
$$21x + 452 = 32 - 84x$$
$$21x + 84x = 32 - 452$$
$$105x = -420$$
$$x = \frac{-420}{105}$$
$$x = -4$$

Check:
$$\frac{2.1x + 45.2 = 3.2 - 8.4x}{}$$
$$\begin{array}{c|c} 2.1(-4) + 45.2 \ ? \ 3.2 - 8.4(-4) \\ -8.4 + 45.2 & 3.2 + 33.6 \\ 36.8 & 36.8 \quad \text{TRUE} \end{array}$$

The solution is -4.

54. $\dfrac{5}{3}$

55.
$$1.03 - 0.6x = 0.71 - 0.2x$$
Greatest number of decimal places is 2
$$100(1.03 - 0.6x) = 100(0.71 - 0.2x)$$
Multiplying by 100 to clear decimals
$$100(1.03) - 100(0.6x) = 100(0.71) - 100(0.2x)$$
$$103 - 60x = 71 - 20x$$
$$32 = 40x$$
$$\frac{32}{40} = x$$
$$\frac{4}{5} = x, \text{ or}$$
$$0.8 = x$$

Check:
$$\frac{1.03 - 0.6x = 0.71 - 0.2x}{}$$
$$\begin{array}{c|c} 1.03 - 0.6(0.8) \ ? \ 0.71 - 0.2(0.8) \\ 1.03 - 0.48 & 0.71 - 0.16 \\ 0.55 & 0.55 \quad \text{TRUE} \end{array}$$

The solution is $\dfrac{4}{5}$, or 0.8.

56. 1

57.
$$\frac{2}{7}x - \frac{1}{2}x = \frac{3}{4}x + 1$$
The least common denominator is 28.
$$28\left(\frac{2}{7}x - \frac{1}{2}x\right) = 28\left(\frac{3}{4}x + 1\right)$$
$$28 \cdot \frac{2}{7}x - 28 \cdot \frac{1}{2}x = 28 \cdot \frac{3}{4}x + 28 \cdot 1$$
$$8x - 14x = 21x + 28$$
$$-6x = 21x + 28$$
$$-6x - 21x = 28$$
$$-27x = 28$$
$$x = -\frac{28}{27}$$

Check:
$$\frac{2}{7}x - \frac{1}{2}x = \frac{3}{4}x + 1$$
$$\begin{array}{c|c} \frac{2}{7}\left(-\frac{28}{27}\right) - \frac{1}{2}\left(-\frac{28}{27}\right) \ ? \ \frac{3}{4}\left(-\frac{28}{27}\right) + 1 \\ -\frac{8}{27} + \frac{14}{27} & -\frac{21}{27} + 1 \\ \frac{6}{27} & \frac{6}{27} \quad \text{TRUE} \end{array}$$

The solution is $-\dfrac{28}{27}$.

58. $\dfrac{32}{7}$

59.
$$3(2y - 3) = 27$$
$$6y - 9 = 27 \qquad \text{Using the distributive law}$$
$$6y = 27 + 9 \qquad \text{Adding 9}$$
$$6y = 36$$
$$y = 6 \qquad \text{Dividing by 6}$$

Check:
$$\frac{3(2y - 3) = 27}{}$$
$$\begin{array}{c|c} 3(2 \cdot 6 - 3) \ ? \ 27 \\ 3(12 - 3) & \\ 3 \cdot 9 & \\ 27 & 27 \quad \text{TRUE} \end{array}$$

The solution is 6.

60. 5

61.
$$40 = 5(3x + 2)$$
$$40 = 15x + 10 \qquad \text{Using the distributive law}$$
$$40 - 10 = 15x$$
$$30 = 15x$$
$$2 = x$$

Check:
$$\frac{40 = 5(3x + 2)}{}$$
$$\begin{array}{c|c} 40 \ ? \ 5(3 \cdot 2 + 2) \\ & 5(6 + 2) \\ & 5 \cdot 8 \\ 40 & 40 \quad \text{TRUE} \end{array}$$

The solution is 2.

62. 1

63.
$$2(3 + 4m) - 9 = 45$$
$$6 + 8m - 9 = 45$$
$$8m - 3 = 45 \qquad \text{Collecting like terms}$$
$$8m = 45 + 3$$
$$8m = 48$$
$$m = 6$$

Check:
$$\frac{2(3 + 4m) - 9 = 45}{}$$
$$\begin{array}{c|c} 2(3 + 4 \cdot 6) - 9 \ ? \ 45 \\ 2(3 + 24) - 9 & \\ 2 \cdot 27 - 9 & \\ 54 - 9 & \\ 45 & 45 \quad \text{TRUE} \end{array}$$

The solution is 6.

64. 9

65. $5r - (2r + 8) = 16$

$\quad\quad 5r - 2r - 8 = 16$

$\quad\quad\quad 3r - 8 = 16 \quad$ Collecting like terms

$\quad\quad\quad\quad\quad 3r = 16 + 8$

$\quad\quad\quad\quad\quad 3r = 24$

$\quad\quad\quad\quad\quad\, r = 8$

Check: $\quad\underline{5r - (2r + 8) = 16}$

$\quad 5 \cdot 8 - (2 \cdot 8 + 8) \; ? \; 16$

$\quad\quad\quad 40 - (16 + 8) \;\bigg|$

$\quad\quad\quad\quad\quad 40 - 24 \;\bigg|$

$\quad\quad\quad\quad\quad\quad\quad 16 \;\bigg|\; 16 \quad$ TRUE

The solution is 8.

66. 8

67. $6 - 2(3x - 1) = 2$

$\quad\quad 6 - 6x + 2 = 2$

$\quad\quad\quad 8 - 6x = 2$

$\quad\quad 8 - 2 = 6x$

$\quad\quad\quad\quad 6 = 6x$

$\quad\quad\quad\quad 1 = x$

Check: $\quad\underline{6 - 2(3x - 1) = 2}$

$\quad\quad 6 - 2(3 \cdot 1 - 1) \; ? \; 2$

$\quad\quad\quad\quad 6 - 2(3 - 1) \;\bigg|$

$\quad\quad\quad\quad\quad\quad 6 - 2 \cdot 2 \;\bigg|$

$\quad\quad\quad\quad\quad\quad\quad 6 - 4 \;\bigg|$

$\quad\quad\quad\quad\quad\quad\quad\quad\quad 2 \;\bigg|\; 2 \quad$ TRUE

The solution is 1.

68. 2

69. $5(d + 4) = 7(d - 2)$

$\quad\quad 5d + 20 = 7d - 14$

$\quad 20 + 14 = 7d - 5d$

$\quad\quad\quad\quad 34 = 2d$

$\quad\quad\quad\quad 17 = d$

Check: $\quad\underline{5(d + 4) = 7(d - 2)}$

$\quad\quad 5(17 + 4) \; ? \; 7(17 - 2)$

$\quad\quad\quad\quad 5 \cdot 21 \;\bigg|\; 7 \cdot 15$

$\quad\quad\quad\quad\quad 105 \;\bigg|\; 105 \quad$ TRUE

The solution is 17.

70. -4

71. $8(2t + 1) = 4(7t + 7)$

$\quad\quad 16t + 8 = 28t + 28$

$\quad 16t - 28t = 28 - 8$

$\quad\quad\quad -12t = 20$

$\quad\quad\quad\quad\quad t = -\dfrac{20}{12}$

$\quad\quad\quad\quad\quad t = -\dfrac{5}{3}$

Check: $\quad\underline{8(2t + 1) = 4(7t + 7)}$

$\quad 8\left(2\left(-\dfrac{5}{3}\right) + 1\right) \; ? \; 4\left(7\left(-\dfrac{5}{3}\right) + 7\right)$

$\quad\quad 8\left(-\dfrac{10}{3} + 1\right) \;\bigg|\; 4\left(-\dfrac{35}{3} + 7\right)$

$\quad\quad\quad\quad 8\left(-\dfrac{7}{3}\right) \;\bigg|\; 4\left(-\dfrac{14}{3}\right)$

$\quad\quad\quad\quad\quad -\dfrac{56}{3} \;\bigg|\; -\dfrac{56}{3} \quad$ TRUE

The solution is $-\dfrac{5}{3}$.

72. -8

73. $3(r - 6) + 2 = 4(r + 2) - 21$

$\quad 3r - 18 + 2 = 4r + 8 - 21$

$\quad\quad\quad 3r - 16 = 4r - 13$

$\quad\quad 13 - 16 = 4r - 3r$

$\quad\quad\quad\quad\quad -3 = r$

Check: $\quad\underline{3(r - 6) + 2 = 4(r + 2) - 21}$

$\quad 3(-3 - 6) + 2 \; ? \; 4(-3 + 2) - 21$

$\quad\quad\quad 3(-9) + 2 \;\bigg|\; 4(-1) - 21$

$\quad\quad\quad\quad -27 + 2 \;\bigg|\; -4 - 21$

$\quad\quad\quad\quad\quad -25 \;\bigg|\; -25 \quad$ TRUE

The solution is -3.

74. -12

75. $19 - (2x + 3) = 2(x + 3) + x$

$\quad 19 - 2x - 3 = 2x + 6 + x$

$\quad\quad 16 - 2x = 3x + 6$

$\quad\quad 16 - 6 = 3x + 2x$

$\quad\quad\quad\quad 10 = 5x$

$\quad\quad\quad\quad\, 2 = x$

Check: $\quad\underline{19 - (2x + 3) = 2(x + 3) + x}$

$\quad 19 - (2 \cdot 2 + 3) \; ? \; 2(2 + 3) + 2$

$\quad\quad\quad 19 - (4 + 3) \;\bigg|\; 2 \cdot 5 + 2$

$\quad\quad\quad\quad\quad 19 - 7 \;\bigg|\; 10 + 2$

$\quad\quad\quad\quad\quad\quad 12 \;\bigg|\; 12 \quad$ TRUE

The solution is 2.

76. 1

77. $\quad\dfrac{1}{3}(6x + 24) - 20 = -\dfrac{1}{4}(12x - 72)$

$\dfrac{1}{3} \cdot 6x + \dfrac{1}{3} \cdot 24 - 20 = -\dfrac{1}{4} \cdot 12x - \dfrac{1}{4}(-72)$

$\quad\quad\quad 2x + 8 - 20 = -3x + 18$

$\quad\quad\quad\quad\quad 2x - 12 = -3x + 18$

$\quad\quad\quad\quad\quad\quad\quad 5x = 30$

$\quad\quad\quad\quad\quad\quad\quad\, x = 6$

The check is left to the student. The solution is 6.

78. 5

79. $2[4 - 2(3 - x)] - 1 = 4[2(4x - 3) + 7] - 25$
$2[4 - 6 + 2x] - 1 = 4[8x - 6 + 7] - 25$
$2[-2 + 2x] - 1 = 4[8x + 1] - 25$
$-4 + 4x - 1 = 32x + 4 - 25$
$4x - 5 = 32x - 21$
$-5 + 21 = 32x - 4x$
$16 = 28x$
$\dfrac{16}{28} = x$
$\dfrac{4}{7} = x$

The check is left to the student. The solution is $\dfrac{4}{7}$.

The solution is $\dfrac{4}{7}$.

80. $-\dfrac{27}{19}$

81. $\dfrac{2}{3}(2x - 1) = 10$

$3 \cdot \dfrac{2}{3}(2x - 1) = 3 \cdot 10$ Multiplying by 3 to
clear the fraction

$2(2x - 1) = 30$
$4x - 2 = 30$
$4x = 30 + 2$
$4x = 32$
$x = 8$

Check: $\dfrac{2}{3}(2x - 1) = 10$

$\dfrac{2}{3}(2 \cdot 8 - 1) \ ? \ 10$

$\dfrac{2}{3}(16 - 1) \ \Big|$

$\dfrac{2}{3} \cdot 15 \ \Big|$

$10 \ \Big| \ 10$ TRUE

The solution is 8.

82. 7

83. $\dfrac{3}{4}\left(3x - \dfrac{1}{2}\right) - \dfrac{2}{3} = \dfrac{1}{3}$

$\dfrac{9}{4}x - \dfrac{3}{8} - \dfrac{2}{3} = \dfrac{1}{3}$

The number 24 is the least common denominator, so we multiply by 24 on both sides.

$24\left(\dfrac{9}{4}x - \dfrac{3}{8} - \dfrac{2}{3}\right) = 24 \cdot \dfrac{1}{3}$

$24 \cdot \dfrac{9}{4}x - 24 \cdot \dfrac{3}{8} - 24 \cdot \dfrac{2}{3} = 8$

$54x - 9 - 16 = 8$
$54x - 25 = 8$
$54x = 8 + 25$
$54x = 33$
$x = \dfrac{33}{54}$
$x = \dfrac{11}{18}$

The check is left to the student. The solution is $\dfrac{11}{18}$.

84. $-\dfrac{5}{32}$

85. $0.7(3x + 6) = 1.1 - (x + 2)$
$2.1x + 4.2 = 1.1 - x - 2$
$10(2.1x + 4.2) = 10(1.1 - x - 2)$ Clearing
decimals
$21x + 42 = 11 - 10x - 20$
$21x + 42 = -10x - 9$
$21x + 10x = -9 - 42$
$31x = -51$
$x = -\dfrac{51}{31}$

The check is left to the student. The solution is $-\dfrac{51}{31}$.

86. $\dfrac{39}{14}$

87. $a + (a - 3) = (a + 2) - (a + 1)$
$a + a - 3 = a + 2 - a - 1$
$2a - 3 = 1$
$2a = 1 + 3$
$2a = 4$
$a = 2$

Check: $\dfrac{a + (a - 3) = (a + 2) - (a + 1)}{}$

$2 + (2 - 3) \ ? \ (2 + 2) - (2 + 1)$
$2 - 1 \ \Big| \ 4 - 3$
$1 \ \Big| \ 1$ TRUE

The solution is 2.

88. -7.4

89. Do the long division. The answer is negative.

$$
\begin{array}{r}
6.5 \\
3.4_\wedge\overline{)2\,2.\,1_\wedge 0} \\
\underline{2\,0\,4} \\
1\,7\,0 \\
\underline{1\,7\,0} \\
0
\end{array}
$$

$-22.1 \div 3.4 = -6.5$

90. $7(x - 3 - 2y)$

91. Since -15 is to the left of -13 on the number line, -15 is less than -13, so $-15 < -13$.

92. -14

93.

94.

95. Since we are using a calculator we will not clear the decimals.

$$0.008 + 9.62x - 42.8 = 0.944x + 0.0083 - x$$
$$9.62x - 42.792 = -0.056x + 0.0083$$
$$9.62x + 0.056x = 0.0083 + 42.792$$
$$9.676x = 42.8003$$
$$x = \frac{42.8003}{9.676}$$
$$x \approx 4.4233464$$

The solution is approximately 4.4233464.

96. -4

97.
$$0 = y - (-14) - (-3y)$$
$$0 = y + 14 + 3y$$
$$0 = 4y + 14$$
$$-14 = 4y$$
$$\frac{-14}{4} = y$$
$$-\frac{7}{2} = y$$

The solution is $-\dfrac{7}{2}$.

98. All real numbers

99.
$$475(54x + 7856) + 9762 = 402(83x + 975)$$
$$25,650x + 3,731,600 + 9762 = 33,366x + 391,950$$
$$25,650x + 3,741,362 = 33,366x + 391,950$$
$$3,741,362 - 391,950 = 33,366x - 25,650x$$
$$3,349,412 = 7716x$$
$$\frac{3,349,412}{7716} = x$$
$$\frac{837,353}{1929} = x$$

The solution is $\dfrac{837,353}{1929}$.

100. -0.000036364

101.
$$x(x - 4) = 3x(x + 1) - 2(x^2 + x - 5)$$
$$x^2 - 4x = 3x^2 + 3x - 2x^2 - 2x + 10$$
$$x^2 - 4x = x^2 + x + 10$$
$$-x^2 + x^2 - 4x = -x^2 + x^2 + x + 10$$
$$-4x = x + 10$$
$$-4x - x = 10$$
$$-5x = 10$$
$$x = -2$$

The solution is -2.

102. -2

103.
$$-2y + 5y = 6y$$
$$3y = 6y$$
$$0 = 3y$$
$$0 = y$$

The solution is 0.

104. 0

105.
$$\frac{5 + 2y}{3} = \frac{25}{12} + \frac{5y + 3}{4}$$

The least common denominator is 12.

$$12\left(\frac{5 + 2y}{3}\right) = 12\left(\frac{25}{12} + \frac{5y + 3}{4}\right)$$
$$4(5 + 2y) = 25 + 3(5y + 3)$$
$$20 + 8y = 25 + 15y + 9$$
$$-7y = 14$$
$$y = -2$$

The solution is -2.

106. $\dfrac{52}{45}$

Exercise Set 2.3

1.
$$A = bh$$
$$\frac{A}{h} = \frac{bh}{h} \qquad \text{Dividing by } h$$
$$\frac{A}{h} = b$$

2. $h = \dfrac{A}{b}$

3.
$$d = rt$$
$$\frac{d}{t} = \frac{rt}{t} \qquad \text{Dividing by } t$$
$$\frac{d}{t} = r$$

4. $t = \dfrac{d}{r}$

5.
$$I = Prt$$
$$\frac{I}{rt} = \frac{Prt}{rt} \qquad \text{Dividing by } rt$$
$$\frac{I}{rt} = P$$

6. $t = \dfrac{I}{Pr}$

7.
$$F = ma$$
$$\frac{F}{m} = \frac{ma}{m} \qquad \text{Dividing by } m$$
$$\frac{F}{m} = a$$

8. $m = \dfrac{F}{a}$

9.
$$P = 2l + 2w$$
$$P - 2l = 2l + 2w - 2l \qquad \text{Subtracting } 2l$$
$$P - 2l = 2w$$
$$\frac{P - 2l}{2} = \frac{2w}{2} \qquad \text{Dividing by } 2$$
$$\frac{P - 2l}{2} = w$$

10. $l = \dfrac{P - 2w}{2}$

11. $A = \pi r^2$

$$\frac{A}{\pi} = \frac{\pi r^2}{\pi}$$

$$\frac{A}{\pi} = r^2$$

12. $\pi = \dfrac{A}{r^2}$

13. $A = \dfrac{1}{2}bh$

$2A = 2 \cdot \dfrac{1}{2}bh$ Multiplying by 2

$2A = bh$

$\dfrac{2A}{h} = \dfrac{bh}{h}$ Dividing by h

$\dfrac{2A}{h} = b$

14. $h = \dfrac{2A}{b}$

15. $E = mc^2$

$\dfrac{E}{c^2} = \dfrac{mc^2}{c^2}$ Dividing by c^2

$\dfrac{E}{c^2} = m$

16. $c^2 = \dfrac{E}{m}$

17. $Q = \dfrac{c + d}{2}$

$2Q = 2 \cdot \dfrac{c + d}{2}$ Multiplying by 2

$2Q = c + d$

$2Q - c = c + d - c$ Subtracting c

$2Q - c = d$

18. $p = 2Q + q$

19. $A = \dfrac{a + b + c}{3}$

$3A = 3 \cdot \dfrac{a + b + c}{3}$ Multiplying by 3

$3A = a + b + c$

$3A - a - c = a + b + c - a - c$ Subtracting a and c

$3A - a - c = b$

20. $c = 3A - a - b$

21. $v = \dfrac{3k}{t}$

$tv = t \cdot \dfrac{3k}{t}$ Multiplying by t

$tv = 3k$

$\dfrac{tv}{v} = \dfrac{3k}{v}$ Dividing by v

$t = \dfrac{3k}{v}$

22. $c = \dfrac{ab}{P}$

23. $Ax + By = C$

$Ax + By - Ax = C - Ax$ Subtracting Ax

$By = C - Ax$

$\dfrac{By}{B} = \dfrac{C - Ax}{B}$ Dividing by B

$y = \dfrac{C - Ax}{B}$

24. $x = \dfrac{C - By}{A}$

25. $A = \dfrac{1}{2}ah + \dfrac{1}{2}bh$

$2A = 2\left(\dfrac{1}{2}ah + \dfrac{1}{2}bh\right)$ Clearing the fractions

$2A = ah + bh$

$2A - ah = bh$ Subtracting ah

$\dfrac{2A - ah}{h} = b$ Dividing by h

26. $a = \dfrac{2A + bh}{h};\ \ h = \dfrac{2A}{a - b}$

27. $Q = 3a + 5ca$

$Q = a(3 + 5c)$ Factoring

$\dfrac{Q}{3 + 5c} = a$ Dividing by $3 + 5c$

28. $m = \dfrac{P}{4 + 7n}$

29. $A = P + Prt$

$A = P(1 + rt)$ Factoring

$\dfrac{A}{1 + rt} = P$ Dividing by $1 + rt$

30. $P = \dfrac{S}{1 - 0.01r}$

31. $A = \dfrac{\pi r^2 S}{360}$

$\dfrac{360}{\pi r^2} \cdot A = \dfrac{360}{\pi r^2} \cdot \dfrac{\pi r^2 S}{360}$

$\dfrac{360}{\pi r^2} = S$

32. $r^2 = \dfrac{360A}{\pi S}$

33. $R = -0.0075t + 3.85$

$R - 3.85 = -0.0075t + 3.85 - 3.85$

$R - 3.85 = -0.0075t$

$\dfrac{R - 3.85}{-0.0075} = t$

34. $C = \dfrac{5F - 160}{9}$, or $\dfrac{5}{9}(F - 32)$

35. We multiply from left to right:

$7(-3)2 = (-21)2 = -42$

36. $-\dfrac{3}{5}$

37. $10 \div (-2) \cdot 5 - 4 = -5 \cdot 5 - 4$ Dividing

$= -25 - 4$ Multiplying

$= -29$ Subtracting

38. 30

39.

40.

41. $\dfrac{\left(\dfrac{y}{z}\right)}{\left(\dfrac{z}{t}\right)} = 1$

$\dfrac{y}{z} = \dfrac{z}{t}$ Multiplying by $\dfrac{z}{t}$

$y = \dfrac{z^2}{t}$ Multiplying by z

42. $F = \dfrac{1 - GE}{G}$, or $\dfrac{1}{G} - E$

43.
$$q = r(s + t)$$
$$q = rs + rt$$
$$q - rs = rt$$
$$\frac{q - rs}{r} = t$$

We could also solve for t as follows.
$$q = r(s + t)$$
$$\frac{q}{r} = s + t$$
$$\frac{q}{r} - s = t$$

44. $c = \dfrac{d}{a - b}$

45.
$$a = c(x + y) + bx$$
$$a = cx + cy + bx$$
$$a - cy = cx + bx$$
$$a - cy = x(c + b)$$
$$\frac{a - cy}{c + b} = x$$

46. $a = \dfrac{c}{3 + b + d}$

47.

48.

Exercise Set 2.4

1. $76\% = 76 \times 0.01$ Replacing % by $\times\, 0.01$
   $= 0.76$

2. 0.54

3. $54.7\% = 54.7 \times 0.01$ Replacing % by $\times\, 0.01$
   $= 0.547$

4. 0.962

5. $100\% = 100 \times 0.01 = 1$

6. 0.01

7. $0.61\% = 0.61 \times 0.01 = 0.0061$

8. 1.25

9. $240\% = 240 \times 0.01 = 2.4$

10. 0.0073

11. $3.25\% = 3.25 \times 0.01 = 0.0325$

12. 0.023

13. 4.54

| First move the decimal point two places to the right; | 4.54. |
| then write a % symbol: | 454% |

14. 100%

15. 0.998

| First move the decimal point two places to the right; | 0.99.8 |
| then write a % symbol: | 99.8% |

16. 73%

17. 2   (Note: 2 = 2.00)

| First move the decimal point two places to the right; | 2.00. |
| then write a % symbol: | 200% |

18. 0.57%

19. 0.072

| First move the decimal point two places to the right; | 0.07.2 |
| then write a % symbol: | 7.2% |

20. 134%

21. 9.2   (Note: 9.2 = 9.20)

| First move the decimal point two places to the right; | 9.20. |
| then write a % symbol: | 920% |

22. 1.3%

23. 0.0068

| First move the decimal point two places to the right; | 0.00.68 |
| then write a % symbol: | 0.68% |

24. 67.5%

25. $\dfrac{1}{8}$ $\left(\text{Note: } \dfrac{1}{8} = 0.125\right)$

| First move the decimal point two places to the right; | 0.12.5 |
| then write a % symbol: | 12.5% |

26. $33.\overline{3}\%$, or $33\dfrac{1}{3}\%$

27. $\frac{17}{25}$ $\left(\text{Note: } \frac{17}{25} = 0.68\right)$

First move the decimal point   0.68.
two places to the right;
then write a % symbol: 68%

28. 55%

29. $\frac{3}{4}$ $\left(\text{Note: } \frac{3}{4} = 0.75\right)$

First move the decimal point   0.75.
two places to the right;
then write a % symbol: 75%

30. 40%

31. $\frac{7}{10}$ $\left(\text{Note: } \frac{7}{10} = 0.7, \text{or } 0.70\right)$

First move the decimal point   0.70.
two places to the right;
then write a % symbol: 70%

32. 80%

33. $\frac{3}{5}$ $\left(\text{Note: } \frac{3}{5} = 0.6, \text{or } 0.60\right)$

First move the decimal point   0.60.
two places to the right;
then write a % symbol: 60%

34. 34%

35. $\frac{2}{3}$ $\left(\text{Note: } \frac{2}{3} = 0.66\overline{6}\right)$

First move the decimal point   0.66.$\overline{6}$
two places to the right;
then write a % symbol: 66.$\overline{6}$%

Since $0.\overline{6} = \frac{2}{3}$, this can also be expressed as $66\frac{2}{3}\%$.

36. 37.5%

37. *Translate.*

What percent of 68 is 17?
$$y \quad \cdot \quad 68 \ = \ 17$$

We solve the equation and then convert to percent notation.
$$y \cdot 68 = 17$$
$$y = \frac{17}{68}$$
$$y = 0.25 = 25\%$$

The answer is 25%.

38. 48%

39. *Translate.*

What percent of 125 is 30?
$$y \quad \cdot \quad 125 \ = \ 30$$

We solve the equation and then convert to percent notation.
$$y \cdot 125 = 30$$
$$y = \frac{30}{125}$$
$$y = 0.24 = 24\%$$

The answer is 24%.

40. 19%

41. *Translate.*

45 is 30% of what number?
$$45 \ = \ 30\% \quad \cdot \qquad y$$

We solve the equation.
$$45 = 0.3y \qquad (30\% = 0.3)$$
$$\frac{45}{0.3} = y$$
$$150 = y$$

The answer is 150.

42. 85

43. *Translate.*

0.3 is 12% of what number?
$$0.3 \ = \ 12\% \quad \cdot \qquad y$$

We solve the equation.
$$0.3 = 0.12y \qquad (12\% = 0.12)$$
$$\frac{0.3}{0.12} = y$$
$$2.5 = y$$

The answer is 2.5.

44. 4

45. *Translate.*

What number is 65% of 840?
$$y \qquad = \ 65\% \quad \cdot \quad 840$$

We solve the equation.
$$y = 0.65 \cdot 840 \qquad (65\% = 0.65)$$
$$y = 546 \qquad\qquad \text{Multiplying}$$

The answer is 546.

46. 10,000

47. *Translate.*

What percent of 80 is 100?
$$y \quad \cdot \quad 80 \ = \ 100$$

We solve the equation and then convert to percent notation.
$$y \cdot 80 = 100$$
$$y = \frac{100}{80}$$
$$y = 1.25 = 125\%$$

The answer is 125%.

48. 2050%

49. *Translate.*

What is 2% of 40?
$$\downarrow \quad \downarrow \quad \downarrow \quad \downarrow \quad \downarrow$$
$$x \quad = \quad 2\% \quad \cdot \quad 40$$

We solve the equation.

$$x = 0.02 \cdot 40 \qquad (2\% = 0.02)$$
$$x = 0.8 \qquad \text{Multiplying}$$

The answer is 0.8.

50. 0.8

51. *Translate.*

2 is what percent of 40?
$$\downarrow \downarrow \qquad \downarrow \qquad \downarrow \quad \downarrow$$
$$2 = \qquad y \qquad \cdot \quad 40$$

We solve the equation and convert to percent notation.

$$2 = y \cdot 40$$
$$\frac{2}{40} = y$$
$$0.05 = y, \text{ or } 5\% = y$$

The answer is 5%.

52. 2000

53. We reword the problem. If $y =$ the percent accepted, we have:

What percent of 16,000 is 600?
$$\downarrow \qquad\quad \downarrow \quad\quad \downarrow \quad \downarrow$$
$$y \qquad \cdot \quad 16,000 = 600$$
$$y = \frac{600}{16,000}$$
$$y = 0.0375 = 3.75\%$$

The FBI accepts 3.75% of the applicants.

54. 7410

55. We reword the problem. If $y =$ the number of bowlers who are left-handed, we have:

What is 17% of 160?
$$\downarrow \quad \downarrow \quad \downarrow \quad \downarrow \quad \downarrow$$
$$y \quad = \quad 17\% \quad \cdot \quad 160$$
$$y = 0.17 \times 160$$
$$y = 27.2$$

You would expect 27 of the bowlers to be left-handed. (We round to the nearest one.)

56. 7%

57. We reword the problem. If $y =$ the percent that were correct, we have:

What percent of 88 is 76?
$$\downarrow \qquad\quad \downarrow \quad \downarrow \quad \downarrow$$
$$y \qquad \cdot \quad 88 = 76$$
$$y = \frac{76}{88}$$
$$y = 0.864 = 86.4\%$$

86.4% were correct.

58. 52%

59. When sales tax is 5%, the total paid is 105% of the price of the merchandise. If $c =$ the cost of the merchandise, we have:

$37.80 is 105% of c.
$$\downarrow \quad \downarrow \quad \downarrow \quad \downarrow \downarrow$$
$$37.80 = 1.05 \cdot c$$
$$\frac{37.80}{1.05} = c$$
$$36 = c$$

The merchandise cost $36 before tax.

60. $940

61. We divide:

$$\begin{array}{r} 0.9\,2 \\ 2\,5\,\overline{)2\,3.0\,0} \\ \underline{2\,2\,5} \\ 5\,0 \\ \underline{5\,0} \\ 0 \end{array}$$

Decimal notation for $\frac{23}{25}$ is 0.92.

62. −90

63. $-45.8 - (-32.6) = -45.8 + 32.6 = -13.2$

64. $-21a + 12b$

65.

66.

67. At Rollie's Music, the total cost is 107% of the price of the disk. If $R =$ the cost at Rollie's, we have:

$$R = 107\%(\$11.99)$$
$$R = 1.07(\$11.99)$$
$$R = \$12.83 \qquad \text{Rounding}$$

At Sound Warp, the total cost is $13.99 − $2.00 plus 7% of the original price of $13.99. If $S =$ the cost at Sound Warp, we have:

$$S = (\$13.99 - \$2.00) + 7\%(\$13.99)$$
$$S = \$11.99 + 0.07(\$13.99)$$
$$S = \$11.99 + \$0.98 \qquad \text{Rounding}$$
$$S = \$12.97$$

68. 40%; 70%; 95%

69. $x = 1.6y$, so $y = \frac{x}{1.6}$, or $\frac{1}{1.6}x$, or $0.625x$.

Since $0.625 = 62.5\%$, y is 62.5% of x.

70. 20%

71. Since the tax rate r is in decimal notation, the total cost is $1 + r$ times the cost of the merchandise. Then we have:

$$T = (1 + r)c$$
$$\frac{T}{1+r} = c$$

Exercise Set 2.5

1. Let x = the number. Then "twice the number" translates to $2x$, and "three less than $2x$" translates to $2x - 3$.

2. $\dfrac{x}{8} - 5$

3. Let x = the number. Then "the product of a number and 7" translates to $x \cdot 7$, or $7x$, and "one half of $7x$" translates to $\dfrac{1}{2} \cdot 7x$.

4. $10n - 2$

5. Let a = the number. Then "the sum of 3 and some number" translates to $a + 3$, and "5 times $a + 3$" translates to $5(a + 3)$.

6. $6(x + y)$

7. The longer piece is 2 ft longer than L, or $L + 2$.

8. $x \div 3$, or $\dfrac{x}{3}$

9. The amount of the reduction is $30\%b$. Then the sale price is $b - 30\%b$, or $b - 0.3b$, or $0.7b$.

10. $p - 20\%p$, or $p - 0.2p$, or $0.8p$

11. Each even integer is 2 greater than the one preceding it. If we let x = the first of the even integers, then $x + 2$ = the second and $x + 4$ = the third. We can express their sum as $x + (x + 2) + (x + 4)$.

12. $x + (x + 1) + (x + 2)$

13. The cost is (Initial charge) plus (mileage charge), or

$34.95 plus	Cost per mile	time	Number of miles driven
↓ ↓	↓	↓	↓
$34.95 +	$0.27	·	m

which is $\$34.95 + \$0.27m$.

14. $\dfrac{t}{2} + 2$

15. a) Twice the width is $2w$.

    b) The width is one-half the length, or $\dfrac{1}{2}l$.

16. $3(h + 5)$

17. Second angle: Three times the first is $3x$

    Third angle: 30° more than the first is $x + 30$

18. $4x$; $5x - 45$

19. **Familiarize.** Let x = the number. Using the result of Exercise 1, we know that "three less than twice a number" translates to $2x - 3$.

    Translate.

    $$\underbrace{\text{Three less than twice a number}}_{2x - 3} \underbrace{\text{is}}_{=} \underbrace{25.}_{25}$$

Carry out. We solve the equation.
$$2x - 3 = 25$$
$$2x = 28 \quad \text{Adding 3}$$
$$x = 14 \quad \text{Dividing by 2}$$

Check. Twice, or two times, 14 is 28. Three less than 28 is 25. The answer checks.

State. The number is 14.

20. 12

21. **Familiarize.** Let a = the number. Using the result of Exercise 5, we know that "five times the sum of 3 and some number" translates to $5(a + 3)$.

    Translate.

    $$\underbrace{\text{Five times the sum of 3 and some number}}_{5(a + 3)} \underbrace{\text{is}}_{=} \underbrace{70.}_{70}$$

Carry out. We solve the equation.
$$5(a + 3) = 70$$
$$5a + 15 = 70 \quad \text{Using the distributive law}$$
$$5a = 55 \quad \text{Subtracting 15}$$
$$a = 11 \quad \text{Dividing by 5}$$

Check. The sum of 3 and 11 is 14, and $5 \cdot 14 = 70$. The answer checks.

State. The number is 11.

22. 13

23. **Familiarize.** Let x = the number. Then six times the number is $6x$.

    Translate. We reword the problem.

    $$\underbrace{\text{6 times a number}}_{6x} \underbrace{\text{less}}_{-} \underbrace{18}_{18} \underbrace{\text{is}}_{=} \underbrace{96.}_{96}$$

Carry out. We solve the equation.
$$6x - 18 = 96$$
$$6x = 114 \quad \text{Adding 18}$$
$$x = 19 \quad \text{Dividing by 6}$$

Check. Six times 19 is 114. When 18 is subtracted from 114 we get 96. The answer checks.

State. The number is 19.

24. 52

25. **Familiarize.** Let y = the number.

    Translate. We reword the problem.

    $$\underbrace{\text{Two}}_{2} \underbrace{\text{times}}_{\cdot} \underbrace{\text{a number}}_{y} \underbrace{\text{plus}}_{+} \underbrace{16}_{16} \underbrace{\text{is}}_{=} \underbrace{\frac{2}{5}}_{\frac{2}{5}} \underbrace{\text{of}}_{\cdot} \underbrace{\text{the number.}}_{y}$$

Carry out. We solve the equation.

$$2y + 16 = \frac{2}{5}y$$

$$5(2y + 16) = 5 \cdot \frac{2}{5}y \qquad \text{Clearing the fraction}$$

$$10y + 80 = 2y$$

$$80 = -8y \qquad \text{Adding } -10y$$

$$-10 = y \qquad \text{Dividing by } -8$$

Check. We double -10 and get -20. Adding 16, we get -4, and $\frac{2}{5}(-10) = -4$. The answer checks.

State. The number is -10.

26. -68

27. *Familiarize.* Let $x =$ the number.

Translate. We reword the problem.

A number	plus	two-fifths	of	the number	is	56.
↓	↓	↓	↓	↓	↓	↓
x	$+$	$\frac{2}{5}$	$\cdot$	x	$=$	56

Carry out. We solve the equation.

$$x + \frac{2}{5}x = 56$$

$$\frac{7}{5}x = 56 \qquad \text{Collecting like terms}$$

$$x = \frac{5}{7} \cdot 56 \qquad \text{Multiplying by } \frac{5}{7}$$

$$x = 40$$

Check. $\frac{2}{5} \cdot 40 = 16$, and $40 + 16 = 56$. The answer checks.

State. The number is 40.

28. 36

29. *Familiarize.* First draw a picture.

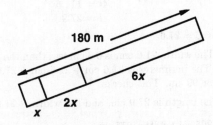

180 m

6x

2x

x

We use x for the first length, $2x$ for the second length, and $3 \cdot 2x$, or $6x$, for the third length.

Translate. The lengths of the three pieces add up to 180 m. This gives us the equation.

Length of 1st piece	plus	Length of 2nd piece	plus	Length of 3rd piece	is	180.
↓	↓	↓	↓	↓	↓	↓
x	$+$	$2x$	$+$	$6x$	$=$	180

Carry out. We solve the equation.

$$x + 2x + 6x = 180$$

$$9x = 180$$

$$x = 20$$

Check. If the first piece is 20 m long, then the second is $2 \cdot 20$ m, or 40 m and the third is $6 \cdot 20$ m, or 120 m. The lengths of these pieces add up to 180 m ($20 + 40 + 120 = 180$). This checks.

State. The first piece measures 20 m. The second measures 40 m, and the third measures 120 m.

30. 30 m, 90 m, 360 m

31. *Familiarize.* The page numbers are consecutive integers. (See Example 3.) If we let $p =$ the smaller number, then $p + 1 =$ the larger number.

Translate. We reword the problem.

First integer	+	Second integer	= 273
↓	↓	↓	↓ ↓
x	$+$	$(x + 1)$	$= 273$

Carry out. We solve the equation.

$$x + (x + 1) = 273$$

$$2x + 1 = 273 \qquad \text{Collecting like terms}$$

$$2x = 272 \qquad \text{Adding } -1$$

$$x = 136 \qquad \text{Dividing by 2}$$

Check. If $x = 136$, then $x + 1 = 137$. These are consecutive integers, and $136 + 137 = 273$. The answer checks.

State. The page numbers are 136 and 137.

32. 140 and 141

33. *Familiarize.* Let $x =$ the smaller even integer. Then $x + 2 =$ the next even integer.

Translate. We reword the problem.

Smaller even integer	+	next even integer	is 114.
↓	↓	↓	↓ ↓
x	$+$	$(x + 2)$	$= 114$

Carry out. We solve the equation.

$$x + (x + 2) = 114$$

$$2x + 2 = 114 \qquad \text{Collecting like terms}$$

$$2x = 112 \qquad \text{Subtracting 2}$$

$$x = 56 \qquad \text{Dividing by 2}$$

If x is 56, then $x + 2$ is 58.

Check. 56 and 58 are consecutive even integers, and their sum is 114. The answer checks.

State. The integers are 56 and 58.

34. 52 and 54

35. *Familiarize.* Let $x =$ the first integer. Then $x + 1 =$ the second integer, and $x + 2 =$ the third.

Translate.

First integer	plus	Second integer	plus	Third integer	is 108.
↓	↓	↓	↓	↓	↓ ↓
x	$+$	$(x + 1)$	$+$	$(x + 2)$	$= 108$

Carry out. We solve the equation.

$$x + (x + 1) + (x + 2) = 108$$
$$3x + 3 = 108$$
$$3x = 105$$
$$x = 35$$

Check. If the first integer is 35, then the second is $35 + 1$, or 36, and the third is $35 + 2$, or 37. They are consecutive integers. Their sum, $35 + 36 + 37$, is 108. This checks.

State. The integers are 35, 36, and 37.

36. 41, 42, and 43

37. *Familiarize.* Let $x =$ the first odd integer. Then $x + 2 =$ the second odd integer and $(x + 2) + 2$, or $x + 4 =$ the third odd integer.

Translate. We reword the problem.

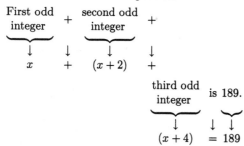

Carry out. We solve the equation.

$$x + (x + 2) + (x + 4) = 189$$
$$3x + 6 = 189 \qquad \text{Collecting like}$$
$$\qquad\qquad\qquad\qquad \text{terms}$$
$$3x = 183 \qquad \text{Subtracting 6}$$
$$x = 61 \qquad \text{Dividing by 3}$$

If x is 61, then $x + 2$ is 63 and $x + 4$ is 65.

Check. 61, 63, and 65 are consecutive odd integers, and their sum is 189. The answer checks.

State. The integers are 61, 63, and 65.

38. 130, 132, and 134

39. *Familiarize.* We draw a picture. Let $w =$ the width of the rectangle. Then $w + 60 =$ the length.

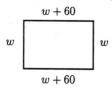

The perimeter of a rectangle is the sum of the lengths of the sides. The area is the product of the length and the width.

Translate. We use the definition of perimeter to write an equation that will allow us to find the width and length.

$$\underbrace{\text{Width}}_{w} + \underbrace{\text{Width}}_{w} + \underbrace{\text{Length}}_{(w+60)} + \underbrace{\text{Length}}_{(w+60)} = \underbrace{\text{Perimeter.}}_{520}$$

Carry out. We solve the equation.

$$w + w + (w + 60) + (w + 60) = 520$$
$$4w + 120 = 520$$
$$4w = 400$$
$$w = 100$$

If $w = 100$, then $w + 60 = 100 + 60 = 160$, and the area is $160(100) = 16{,}000$.

Check. The length is 60 ft more than the width. The perimeter is $100 + 100 + 160 + 160 = 520$ ft. This checks. To check the area we recheck the computation. This also checks.

State. The width of the rectangle is 100 ft; the length is 160 ft; and the area is 16,000 ft^2.

40. Width: 165 ft, length: 265 ft; area: 43,725 ft^2

41. *Familiarize.* We draw a picture. Let $l =$ the length of the paper. Then $l - 6.3 =$ the width.

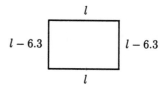

The perimeter is the sum of the lengths of the sides.

Translate. We use the definition of perimeter to write an equation.

$$\underbrace{\text{Width}}_{(l-6.3)} + \underbrace{\text{Width}}_{(l-6.3)} + \underbrace{\text{Length}}_{l} + \underbrace{\text{Length}}_{l} \text{ is } \underbrace{99.}_{= 99}$$

Carry out. We solve the equation.

$$(l - 6.3) + (l - 6.3) + l + l = 99$$
$$4l - 12.6 = 99$$
$$4l = 111.6$$
$$l = 27.9$$

Then $l - 6.3 = 21.6$.

Check. The width, 21.6 cm, is 6.3 cm less than the length, 27.9 cm. The perimeter is 21.6 cm + 21.6 cm + 27.9 cm + 27.9 cm, or 99 cm. This checks.

State. The length is 27.9 cm, and the width is 21.6 cm.

42. Length: 365 mi, width: 275 mi

43. *Familiarize.* We draw a picture. We let $x =$ the measure of the first angle. Then $4x =$ the measure of the second angle, and $(x + 4x) - 45$, or $5x - 45 =$ the measure of the third angle.

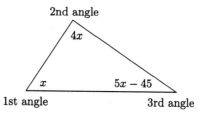

Recall that the measures of the angles of any triangle add up to 180°.

Translate.

$$\underbrace{\text{Measure of first angle}}_{\downarrow} \quad + \quad \underbrace{\text{measure of second angle}}_{\downarrow} \quad +$$

$$x \quad + \quad 4x \quad +$$

$$\underbrace{\text{measure of third angle}}_{\downarrow} \quad \underbrace{\text{is } 180.}_{\downarrow\ \downarrow}$$

$$(5x - 45) \quad = 180$$

Carry out. We solve the equation.

$$x + 4x + (5x - 45) = 180$$
$$10x - 45 = 180$$
$$10x = 225$$
$$x = 22.5$$

Possible answers for the angle measures are as follows:

First angle: $x = 22.5°$
Second angle: $4x = 4(22.5) = 90°$
Third angle: $5x - 45 = 5(22.5) - 45$
$= 112.5 - 45 = 67.5°$

Check. Consider 22.5°, 90°, and 67.5°. The second is four times the first, and the third is 45° less than five times the first. The sum is 180°. These numbers check.

State. The measure of the first angle is 22.5°.

44. 25.625°

45. Familiarize. Let $x =$ the original price. Then, $40\%x =$ the reduction. The sale price is found by subtracting the amount of reduction from the original price.

Translate.

$$\underbrace{\text{Original price}}_{\downarrow} \text{ minus } \underbrace{\text{reduction}}_{\downarrow} \text{ is } \$9.60.$$

$$x \quad - \quad 40\%x \quad = \quad 9.60$$

Carry out. We solve the equation.

$$x - 40\%x = 9.60$$
$$1 \cdot x - 0.40x = 9.60$$
$$(1 - 0.40)x = 9.60$$
$$0.6x = 9.60$$
$$x = \frac{9.60}{0.6}$$
$$x = 16$$

Check. 40% of $16 is $6.40. Subtracting this from $16 we get $9.60. This checks.

State. The original price was $16.

46. $14

47. Familiarize. Let $x =$ the original investment. Interest earned in 1 year is found by taking 6% of the original investment. Then $6\%x =$ the interest. The amount in the account at the end of the year is the sum of the original investment and the interest earned.

Translate.

$$\underbrace{\text{Original investment}}_{\downarrow} \text{ plus } \underbrace{\text{interest earned}}_{\downarrow} \text{ is } \$4664.$$

$$x \quad + \quad 6\%x \quad = \quad 4664$$

Carry out. We solve the equation.

$$x + 6\%x = 4664$$
$$1 \cdot x + 0.06x = 4664$$
$$1.06x = 4664$$
$$x = \frac{4664}{1.06}$$
$$x = 4400$$

Check. 6% of $4400 is $264. Adding this to $4400 we get $4664. This checks.

State. The original investment was $4400.

48. $6540

49. Familiarize. The total cost is the daily charge plus the mileage charge. The mileage charge is the cost per mile times the number of miles driven. Let $m =$ the number of miles that can be driven for $80.

Translate. We reword the problem.

$$\underbrace{\text{Daily rate}}_{\downarrow} \text{ plus } \underbrace{\text{Cost per mile}}_{\downarrow} \text{ times } \underbrace{\text{Number of miles driven}}_{\downarrow} \text{ is}$$

$$34.95 \quad + \quad 0.10 \quad \cdot \quad m \quad =$$

$$\underbrace{\text{Amount}}_{\downarrow}$$

$$80$$

Carry out. We solve the equation.

$$34.95 + 0.10m = 80$$
$$100(34.95 + 0.10m) = 100(80) \quad \text{Clearing decimals}$$
$$3495 + 10m = 8000$$
$$10m = 4505$$
$$m = 450.5$$

Check. The mileage cost is found by multiplying 450.5 by $0.10 obtaining $45.05. Then we add $45.05 to $34.95, the daily rate, and get $80.

State. The businessperson can drive 450.5 mi on the car-rental allotment.

50. 460.5 mi

51. Familiarize. Using the labels on the drawing in the text, we let $x =$ the measure of the first angle. Then $3x =$ the measure of the second angle, and $x + 40 =$ the measure of the third angle. Recall that the sum of measures of the angles of a triangle is 180°.

Translate.

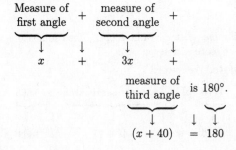

$$\underbrace{\text{Measure of first angle}}_{\downarrow} \quad + \quad \underbrace{\text{measure of second angle}}_{\downarrow} \quad +$$

$$x \quad + \quad 3x \quad +$$

$$\underbrace{\text{measure of third angle}}_{\downarrow} \quad \underbrace{\text{is } 180°.}_{\downarrow\ \downarrow}$$

$$(x + 40) \quad = 180$$

Carry out. We solve the equation.

$$x + 3x + (x + 40) = 180$$
$$5x + 40 = 180$$
$$5x = 140$$
$$x = 28$$

Possible answers for the angle measures are as follows:

First angle: $x = 28°$
Second angle: $3x = 3(28) = 84°$
Third angle: $x + 40 = 28 + 40 = 68°$

Check. Consider 28°, 84°, and 68°. The second angle is three times the first, and the third is 40° more than the first. The sum, 28° + 84° + 68°, is 180°. These numbers check.

State. The measures of the angles are 28°, 84°, and 68°.

52. 5°, 160°, 15°

53. *Familiarize.* We will use the equation $R = -0.028t + 20.8$ where R is in seconds and t is the number of years since 1920. We want to find t when $R = 18.0$ sec.

Translate.

$$\underbrace{\text{Record}}_{\downarrow} \quad \underset{\downarrow}{\text{is}} \quad \underbrace{\text{18.0 sec.}}_{\downarrow}$$
$$-0.028t + 20.8 \quad = \quad 18.0$$

Carry out.

$$-0.028t + 20.8 = 18.0$$
$$1000(-0.028t + 20.8) = 1000(18.0) \quad \text{Clearing the decimals}$$
$$-28t + 20,800 = 18,000$$
$$-28t = -2800$$
$$t = 100$$

Check. Substitute 100 for t in the given equation:

$$R = -0.028(100) + 20.8 = -2.8 + 20.8 = 18.0$$

This checks.

State. The record will be 18.0 sec 100 years after 1920, or in 2020.

54. 160

55. $3x - 12y + 60 = 3 \cdot x - 3 \cdot 4y + 3 \cdot 20$
$$= 3(x - 4y + 20)$$

56. $11x - 13$

57.

59.

59. *Familiarize.* Let s = one score. Then four score = $4s$ and four score and seven = $4s + 7$.

Translate. We reword .

$$\underbrace{1776}_{\downarrow} \quad \underset{\downarrow}{\text{plus}} \quad \underbrace{\text{four score and seven}}_{\downarrow} \quad \underset{\downarrow}{\text{is}} \quad \underbrace{1863}_{\downarrow}$$
$$1776 \quad + \quad (4s + 7) \quad = \quad 1863$$

Carry out. We solve the equation.

$$1776 + (4s + 7) = 1863$$
$$4s + 1783 = 1863$$
$$4s = 80$$
$$s = 20$$

Check. If a score is 20 years, then four score and seven represents 87 years. Adding 87 to 1776 we get 1863. This checks.

State. A score is 20.

60. 16 and 4

61. *Familiarize.* The cost of rental is the daily charge plus the mileage charge. Let c = the cost per mile that will make the total cost equal to the budget amount. Then c is the highest price per mile the person can afford.

Translate.

$$\underbrace{\begin{array}{c}\text{Daily} \\ \text{rate}\end{array}}_{\downarrow} \quad \underset{\downarrow}{\text{plus}} \quad \underbrace{\begin{array}{c}\text{Cost} \\ \text{per} \\ \text{mile}\end{array}}_{\downarrow} \quad \underset{\downarrow}{\text{times}}$$
$$18.90 \quad + \quad c \quad \cdot$$

$$\underbrace{\begin{array}{c}\text{Number of} \\ \text{miles} \\ \text{driven}\end{array}}_{\downarrow} \quad \underset{\downarrow}{\text{is}} \quad \underbrace{\begin{array}{c}\text{Budget} \\ \text{amount.}\end{array}}_{\downarrow}$$
$$190 \quad = \quad 55$$

Carry out.

$$18.90 + 190c = 55$$
$$10(18.90 + 190c) = 10(55) \quad \text{Clearing decimals}$$
$$189 + 1900c = 550$$
$$1900c = 361$$
$$c = 0.19$$

Check. The mileage cost is found by multiplying 190 by \$0.19, obtaining \$36.10. Adding \$36.10 to \$18.90, the daily rate, we get \$55.

State. The cost per mile cannot exceed \$0.19 to stay within a \$55 budget.

62. 19

63. *Familiarize.* We let x = the length of the original rectangle. Then $\frac{3}{4}x$ = the width. We draw a picture of the enlarged rectangle. Each dimension is increased by 2 cm, so $x + 2$ = the length of the enlarged rectangle and $\frac{3}{4}x + 2$ = the width.

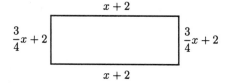

Translate. We use the perimeter of the enlarged rectangle to write an equation.

$$\underbrace{\text{Width}}_{\downarrow} + \underbrace{\text{Width}}_{\downarrow} + \underbrace{\text{Length}}_{\downarrow} +$$

$$\left(\frac{3}{4}x + 2\right) + \left(\frac{3}{4}x + 2\right) + (x + 2) +$$

$$\underbrace{\text{Length}}_{\downarrow} \quad \text{is} \quad \underbrace{\text{Perimeter.}}_{\downarrow}$$

$$(x + 2) \quad = \quad 50$$

Carry out.

$$\left(\frac{3}{4}x + 2\right) + \left(\frac{3}{4}x + 2\right) + (x + 2) + (x + 2) = 50$$

$$\frac{7}{2}x + 8 = 50$$

$$2\left(\frac{7}{2}x + 8\right) = 2 \cdot 50$$

$$7x + 16 = 100$$

$$7x = 84$$

$$x = 12$$

Then $\frac{3}{4}x = \frac{3}{4}(12) = 9$.

Check. If the dimensions of the original rectangle are 12 cm and 9 cm, then the dimensions of the enlarged rectangle are 14 cm and 11 cm. The perimeter of the enlarged rectangle is $11 + 11 + 14 + 14 = 50$ cm. Also, 9 is $\frac{3}{4}$ of 12. These values check.

State. The length is 12 cm, and the width is 9 cm.

64. 120

65. *Familiarize.* Let $x =$ the number of additional games the Falcons will have to play. Then $\frac{x}{2} =$ the number of those games they will win, $15 + \frac{x}{2} =$ the total number of games won, and $20 + x =$ the total number of games played.

Translate.

$$\underbrace{\begin{array}{c}\text{The number of}\\ \text{games won}\end{array}}_{\downarrow} \quad \underbrace{\text{is}}_{\downarrow} \underbrace{60\%}_{\downarrow} \underbrace{\text{of}}_{\downarrow} \quad \underbrace{\begin{array}{c}\text{the number of}\\ \text{games played.}\end{array}}_{\downarrow}$$

$$\left(15 + \frac{x}{2}\right) \quad = \quad 60\% \quad \cdot \quad (20 + x)$$

Carry out.

$$15 + \frac{x}{2} = 60\%(20 + x)$$

$$15 + \frac{x}{2} = 0.6(20 + x)$$

$$15 + 0.5x = 12 + 0.6x \quad \text{Expressing } \frac{1}{2} \text{ as } 0.5$$

$$10(15 + 0.5x) = 10(12 + 0.6x) \quad \text{Clearing decimals}$$

$$150 + 5x = 120 + 6x$$

$$30 = x$$

Check. If 30 more games are played, of which $15\left(\frac{30}{2} = 15\right)$ are won, then the total games played will be $20 + 30$, or 50, and the total games won will be $15 + 15$, or 30. 30 is 60% of 50. The numbers check.

State. The Falcons must play 30 more games.

66. \$600

67. *Familiarize.* Let $h =$ the height of the triangle. We know that the base is 8 in. Recall that the area of a triangle is given by the formula $A = \frac{1}{2}bh$.

Translate.

$$\underbrace{\text{Area}}_{\downarrow} \quad \underbrace{\text{is}}_{\downarrow} \quad \underbrace{2.9047 \text{ in}^2.}_{\downarrow}$$

$$\frac{1}{2} \cdot 8 \cdot h \quad = \quad 2.9047$$

Carry out. We solve the equation.

$$\frac{1}{2} \cdot 8 \cdot h = 2.9047$$

$$4h = 2.9047$$

$$h = 0.726175$$

Check. The area of a triangle whose base is 8 in. and whose height is 0.726175 in. is $\frac{1}{2}(8)(0.726175)$, or 2.9047. The answer checks.

State. The height of the triangle is 0.726175 in.

68. 76

69.

Exercise Set 2.6

1. $x > -4$

a) Since $4 > -4$ is true, 4 is a solution.

b) Since $0 > -4$ is true, 0 is a solution.

c) Since $-4.1 > -4$ is false, -4.1 is not a solution.

d) Since $-3.9 > -4$ is true, -3.9 is a solution.

e) Since $5.6 > -4$ is true, 5.6 is a solution.

2. a) Yes, b) No, c) Yes, d) Yes, e) No

3. $x \geq 6$

a) Since $-6 \geq 6$ is false, -6 is not a solution.

b) Since $0 \geq 6$ is false, 0 is not a solution.

c) Since $6 \geq 6$ is true, 6 is a solution.

d) Since $6.01 \geq 6$ is true, 6.01 is a solution.

e) Since $-3\frac{1}{2} \geq 6$ is false, $-3\frac{1}{2}$ is not a solution.

4. a) Yes, b) Yes, c) Yes, d) No, e) Yes

5. The solutions of $x > 4$ are those numbers greater than 4. They are shown on the graph by shading all points to the right of 4. The open circle at 4 indicates that 4 is not part of the graph.

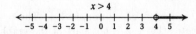

6.

7. The solutions of $t < -3$ are those numbers less than -3. They are shown on the graph by shading all points to the left of -3. The open circle at -3 indicates that -3 is not part of the graph.

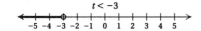

8.

9. The solutions of $m \geq -1$ are are shown by shading the point for -1 and all points to the right of -1. The closed circle at -1 indicates that -1 is part of the graph.

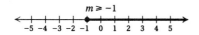

10.

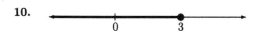

11. In order to be a solution of the inequality $-3 < x \leq 4$, a number must be a solution of both $-3 < x$ and $x \leq 4$. The solution set is graphed as follows:

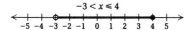

The open circle at -3 means that -3 is not part of the graph. The closed circle at 4 means that 4 is part of the graph.

12.

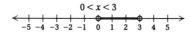

13. In order to be a solution of the inequality $0 < x < 3$, a number must be a solution of both $0 < x$ and $x < 3$. The solution set is graphed as follows:

The open circles at 0 and at 3 mean that 0 and 3 are not part of the graph.

14.

15. All points to the right of -1 are shaded. The open circle at -1 indicates that -1 is not part of the graph. Using set-builder notation we have $\{x|x > -1\}$.

16. $\{x|x < 3\}$

17. The point 2 and all points to the left of 2 are shaded. Using set-builder notation we have $\{x|x \leq 2\}$.

18. $\{x|x \geq -2\}$

19. All points to the left of -2 are shaded. The open circle at -2 indicates that -2 is not part of the graph. Using set-builder notation we have $\{x|x < -2\}$.

20. $\{x|x > 1\}$

21. The point 0 and all points to the right of 0 are shaded. Using set-builder notation we have $\{x|x \geq 0\}$.

22. $\{x|x \leq 0\}$

23.
$$y + 5 > 8$$
$$y + 5 - 5 > 8 - 5 \quad \text{Adding } -5$$
$$y > 3$$

The solution set is $\{y|y > 3\}$.

The graph is as follows:

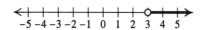

24. $\{y|y > 2\}$

25.
$$x + 8 \leq -10$$
$$x + 8 - 8 \leq -10 - 8 \quad \text{Subtracting } 8$$
$$x \leq -18 \quad \text{Simplifying}$$

The solution set is $\{x|x \leq -18\}$.

The graph is as follows:

26. $\{x|x \leq -21\}$

27.
$$x - 7 < 9$$
$$x - 7 + 7 < 9 + 7$$
$$x < 16$$

The solution set is $\{x|x < 16\}$.

The graph is as follows:

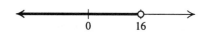

28. $\{x|x < 17\}$

29.
$$x - 6 \geq 2$$
$$x - 6 + 6 \geq 2 + 6$$
$$x \geq 8$$

The solution set is $\{x|x \geq 8\}$.

The graph is as follows:

30. $\{x|x \geq 13\}$

31.
$$y - 7 > -12$$
$$y - 7 + 7 > -12 + 7$$
$$y > -5$$

The solution set is $\{y|y > -5\}$.

The graph is as follows:

32. $\{y|y > -6\}$

33.
$$2x + 3 \leq x + 5$$
$$2x + 3 - 3 \leq x + 5 - 3 \quad \text{Adding } -3$$
$$2x \leq x + 2 \quad \text{Simplifying}$$
$$2x - x \leq x + 2 - x \quad \text{Adding } -x$$
$$x \leq 2 \quad \text{Simplifying}$$

The solution set is $\{x|x \leq 2\}$.

The graph is as follows:

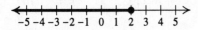

34. $\{x|x \leq 3\}$

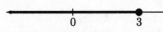

35.
$$3x - 6 \geq 2x + 7$$
$$3x - 6 + 6 \geq 2x + 7 + 6 \quad \text{Adding } 6$$
$$3x \geq 2x + 13$$
$$3x - 2x \geq 2x + 13 - 2x \quad \text{Adding } -2x$$
$$x \geq 13$$

The solution set is $\{x|x \geq 13\}$.

36. $\{x|x \geq 20\}$

37.
$$5x - 6 < 4x - 2$$
$$5x - 6 + 6 < 4x - 2 + 6$$
$$5x < 4x + 4$$
$$5x - 4x < 4x + 4 - 4x$$
$$x < 4$$

The solution set is $\{x|x < 4\}$.

38. $\{x|x < -1\}$

39.
$$7 + c > 7$$
$$-7 + 7 + c > -7 + 7$$
$$c > 0$$

The solution set is $\{c|c > 0\}$.

40. $\{c|c > 18\}$

41.
$$y + \frac{1}{4} \leq \frac{1}{2}$$
$$y + \frac{1}{4} - \frac{1}{4} \leq \frac{1}{2} - \frac{1}{4}$$
$$y \leq \frac{2}{4} - \frac{1}{4} \quad \text{Obtaining a common denominator}$$
$$y \leq \frac{1}{4}$$

The solution set is $\left\{y|y \leq \frac{1}{4}\right\}$.

42. $\left\{y|y \leq \frac{1}{2}\right\}$

43.
$$x - \frac{1}{3} > \frac{1}{4}$$
$$x - \frac{1}{3} + \frac{1}{3} > \frac{1}{4} + \frac{1}{3}$$
$$x > \frac{3}{12} + \frac{4}{12} \quad \text{Obtaining a common denominator}$$
$$x > \frac{7}{12}$$

The solution set is $\left\{x|x > \frac{7}{12}\right\}$.

44. $\left\{x|x > \frac{5}{8}\right\}$

45.
$$-14x + 21 > 21 - 15x$$
$$-14x + 21 + 15x > 21 - 15x + 15x$$
$$x + 21 > 21$$
$$x + 21 - 21 > 21 - 21$$
$$x > 0$$

The solution set is $\{x|x > 0\}$.

46. $\{x|x > 3\}$

47.
$$5x < 35$$
$$\frac{1}{5} \cdot 5x < \frac{1}{5} \cdot 35 \quad \text{Multiplying by } \frac{1}{5}$$
$$x < 7$$

The solution set is $\{x|x < 7\}$. The graph is as follows:

48. $\{x|x \geq 4\}$

49.
$$9y \leq 81$$
$$\frac{1}{9} \cdot 9y \leq \frac{1}{9} \cdot 81 \quad \text{Multiplying by } \frac{1}{9}$$
$$y \leq 9$$

The solution set is $\{y|y \leq 9\}$. The graph is as follows:

50. $\{x|x > 24\}$

51.
$$7x < 13$$
$$\frac{1}{7} \cdot 7x < \frac{1}{7} \cdot 13$$
$$x < \frac{13}{7}$$

The solution set is $\left\{x|x < \frac{13}{7}\right\}$. The graph is as follows:

52. $\left\{y \middle| y < \dfrac{17}{8}\right\}$

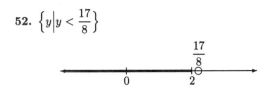

53. $12x > -36$

$$\frac{1}{12} \cdot 12x > \frac{1}{12} \cdot (-36)$$

$$x > -3$$

The solution set is $\{x|x > -3\}$. The graph is as follows:

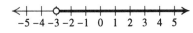

54. $\{x|x < -4\}$

55. $5y \geq -2$

$$\frac{1}{5} \cdot 5y \geq \frac{1}{5} \cdot (-2)$$

$$y \geq -\frac{2}{5}$$

The solution set is $\left\{y \middle| y \geq -\dfrac{2}{5}\right\}$.

56. $\left\{x \middle| x > -\dfrac{4}{7}\right\}$

57. $-2x \leq 12$

$$-\frac{1}{2} \cdot (-2x) \geq -\frac{1}{2} \cdot 12 \quad \text{Multiplyinging by } -\frac{1}{2}$$

$$\text{The symbol has to be reversed.}$$

$$x \geq -6 \qquad \text{Simplifying}$$

The solution set is $\{x|x \geq -6\}$.

58. $\{y|y \geq -5\}$

59. $-4y \geq -16$

$$-\frac{1}{4} \cdot (-4y) \leq -\frac{1}{4} \cdot (-16) \quad \text{Multiplying by } -\frac{1}{4}$$

$$\text{The symbol has to be reversed.}$$

$$y \leq 4 \qquad\qquad \text{Simplifying}$$

The solution set is $\{y|y \leq 4\}$.

60. $\{x|x > 3\}$

61. $-3x < -17$

$$-\frac{1}{3} \cdot (-3x) > -\frac{1}{3} \cdot (-17) \quad \text{Multiplying by } -\frac{1}{3}$$

$$\text{The symbol has to be reversed.}$$

$$x > \frac{17}{3} \qquad\qquad \text{Simplifying}$$

The solution set is $\left\{x \middle| x > \dfrac{17}{3}\right\}$.

62. $\left\{y \middle| y < \dfrac{23}{5}\right\}$

63. $-2y > \dfrac{1}{7}$

$$-\frac{1}{2} \cdot (-2y) < -\frac{1}{2} \cdot \frac{1}{7}$$

$$\text{The symbol has to be reversed.}$$

$$y < -\frac{1}{14}$$

The solution set is $\left\{y \middle| y < -\dfrac{1}{14}\right\}$.

64. $\left\{x \middle| x \geq -\dfrac{1}{36}\right\}$

65. $-\dfrac{6}{5} \leq -4x$

$$-\frac{1}{4} \cdot \left(-\frac{6}{5}\right) \geq -\frac{1}{4} \cdot (-4x)$$

$$\frac{6}{20} \geq x$$

$$\frac{3}{10} \geq x, \text{ or } x \leq \frac{3}{10}$$

The solution set is $\left\{x \middle| \dfrac{3}{10} \geq x\right\}$, or $\left\{x \middle| x \leq \dfrac{3}{10}\right\}$.

66. $\left\{t \middle| t > \dfrac{1}{64}\right\}$

67. $4 + 3x < 28$

$$-4 + 4 + 3x < -4 + 28 \quad \text{Adding } -4$$

$$3x < 24 \qquad\qquad \text{Simplifying}$$

$$\frac{1}{3} \cdot 3x < \frac{1}{3} \cdot 24 \qquad \text{Multiplying by } \frac{1}{3}$$

$$x < 8$$

The solution set is $\{x|x < 8\}$.

68. $\{y|y < 8\}$

69. $6 + 5y \geq 36$

$$-6 + 6 + 5y \geq -6 + 36 \quad \text{Adding } -6$$

$$5y \geq 30$$

$$\frac{1}{5} \cdot 5y \geq \frac{1}{5} \cdot 30 \qquad \text{Multiplying by } \frac{1}{5}$$

$$y \geq 6$$

The solution set is $\{y|y \geq 6\}$.

70. $\{x|x \geq 8\}$

71. $3x - 5 \leq 13$

$$3x - 5 + 5 \leq 13 + 5 \quad \text{Adding } 5$$

$$3x \leq 18$$

$$\frac{1}{3} \cdot 3x \leq \frac{1}{3} \cdot 18 \qquad \text{Multiplying by } \frac{1}{3}$$

$$x \leq 6$$

The solution set is $\{x|x \leq 6\}$.

72. $\{y|y \leq 6\}$

73. $13x - 7 < -46$

$$13x - 7 + 7 < -46 + 7$$

$$13x < -39$$

$$\frac{1}{13} \cdot 13x < \frac{1}{13} \cdot (-39)$$

$$x < -3$$

The solution set is $\{x|x < -3\}$.

74. $\{y|y < -6\}$

75.
$$5x + 3 \geq -7$$
$$5x + 3 - 3 \geq -7 - 3$$
$$5x \geq -10$$
$$\frac{1}{5} \cdot 5x \geq \frac{1}{5} \cdot (-10)$$
$$x \geq -2$$

The solution set is $\{x|x \geq -2\}$.

76. $\{y|y \geq -2\}$

77.
$$13 < 4 - 3y$$
$$13 - 4 < 4 - 3y - 4 \quad \text{Adding } -4$$
$$9 < -3y$$
$$-\frac{1}{3} \cdot 9 > -\frac{1}{3} \cdot (-3y) \quad \text{Multiplying by } -\frac{1}{3}$$
$$\underline{\quad} \text{ The symbol has to be reversed.}$$
$$-3 > y$$

The solution set is $\{y|-3 > y\}$, or $\{y|y < -3\}$.

78. $\{x|x < -2\}$

79.
$$30 > 3 - 9x$$
$$30 - 3 > 3 - 9x - 3 \quad \text{Adding } -3$$
$$27 > -9x$$
$$-\frac{1}{9} \cdot 27 < -\frac{1}{9} \cdot (-9x) \quad \text{Multiplying by } -\frac{1}{9}$$
$$\underline{\quad} \text{ The symbol has to be reversed.}$$
$$-3 < x$$

The solution set is $\{x|-3 < x\}$, or $\{x|x > -3\}$.

80. $\{y|y > -5\}$

81.
$$3 - 6y > 23$$
$$-3 + 3 - 6y > -3 + 23$$
$$-6y > 20$$
$$-\frac{1}{6} \cdot (-6y) < -\frac{1}{6} \cdot 20$$
$$\underline{\quad} \text{ The symbol has to be reversed.}$$
$$y < -\frac{20}{6}$$
$$y < -\frac{10}{3}$$

The solution set is $\left\{y \middle| y < -\frac{10}{3}\right\}$.

82. $\{y|y < -3\}$

83.
$$-3 < 8x + 7 - 7x$$
$$-3 < x + 7 \quad \text{Collecting like terms}$$
$$-3 - 7 < x + 7 - 7$$
$$-10 < x$$

The solution set is $\{x|-10 < x\}$, or $\{x|x > -10\}$.

84. $\{x|x > -13\}$

85.
$$6 - 4y > 4 - 3y$$
$$6 - 4y + 4y > 4 - 3y + 4y \quad \text{Adding } 4y$$
$$6 > 4 + y$$
$$-4 + 6 > -4 + 4 + y \quad \text{Adding } -4$$
$$2 > y, \text{ or } y < 2$$

The solution set is $\{y|2 > y\}$, or $\{y|y < 2\}$.

86. $\{y|y < 2\}$

87.
$$5 - 9y \leq 2 - 8y$$
$$5 - 9y + 9y \leq 2 - 8y + 9y$$
$$5 \leq 2 + y$$
$$-2 + 5 \leq -2 + 2 + y$$
$$3 \leq y, \text{ or } y \geq 3$$

The solution set is $\{y|3 \leq y\}$, or $\{y|y \geq 3\}$.

88. $\{y|y \geq 2\}$

89.
$$21 - 8y < 6y + 49$$
$$21 - 8y + 8y < 6y + 49 + 8y$$
$$21 < 14y + 49$$
$$21 - 49 < 14y + 49 - 49$$
$$-28 < 14y$$
$$\frac{1}{14} \cdot (-28) < \frac{1}{14} \cdot 14y$$
$$-2 < y, \text{ or } y > -2$$

The solution set is $\{y|-2 < y\}$, or $\{y|y > -2\}$.

90. $\{x|x > -4\}$

91.
$$27 - 11x > 14x - 18$$
$$27 - 11x + 11x > 14x - 18 + 11x$$
$$27 > 25x - 18$$
$$27 + 18 > 25x - 18 + 18$$
$$45 > 25x$$
$$\frac{1}{25} \cdot 45 > \frac{1}{25} \cdot 25x$$
$$\frac{45}{25} > x$$
$$\frac{9}{5} > x, \text{ or } x < \frac{9}{5}$$

The solution set is $\left\{x \middle| \frac{9}{5} > x\right\}$, or $\left\{x \middle| x < \frac{9}{5}\right\}$.

92. $\left\{y \middle| y < \frac{61}{28}\right\}$

93.
$$2.1x + 45.2 > 3.2 - 8.4x$$
$$10(2.1x + 45.2) > 10(3.2 - 8.4x) \quad \text{Multiplying by}$$
$$\text{10 to clear decimals}$$
$$21x + 452 > 32 - 84x$$
$$21x + 84x > 32 - 452 \quad \text{Adding } 84x \text{ and}$$
$$-452$$
$$105x > -420$$
$$x > -4 \quad \text{Multiplying by } \frac{1}{105}$$

The solution set is $\{x|x > -4\}$.

94. $\left\{y \middle| y \leq \frac{5}{3}\right\}$

95.
$$0.7n - 15 + n \geq 2n - 8 - 0.4n$$
$$1.7n - 15 \geq 1.6n - 8 \quad \text{Collecting like terms}$$
$$10(1.7n - 15) \geq 10(1.6n - 8) \quad \text{Multiplying by 10}$$
$$17n - 150 \geq 16n - 80$$
$$17n - 16n \geq -80 + 150 \quad \text{Adding } -16n \text{ and}$$
$$150$$
$$n \geq 70$$

The solution set is $\{n|n \geq 70\}$

96. $\{t|t > 1\}$

97.
$$\frac{x}{3} - 2 \le 1$$

$3\left(\dfrac{x}{3} - 2\right) \le 3 \cdot 1$ Multiplying by 3 to to clear the fraction

$x - 6 \le 3$ Simplifying

$x \le 9$ Adding 6

The solution set is $\{x|x \le 9\}$.

98. $\{x|x > 2\}$

99.
$$\frac{y}{5} + 1 \le \frac{2}{5}$$

$5\left(\dfrac{y}{5} + 1\right) \le 5 \cdot \dfrac{2}{5}$ Clearing fractions

$y + 5 \le 2$

$y \le -3$ Adding -5

The solution set is $\{y|y \le -3\}$.

100. $\{x|x \ge -25\}$

101. $3(2y - 3) < 27$

$6y - 9 < 27$ Removing parentheses

$6y < 36$ Adding 9

$y < 6$ Multiplying by $\dfrac{1}{6}$

The solution set is $\{y|y < 6\}$.

102. $\{y|y > 5\}$

103. $5(d + 4) \le 7(d - 2)$

$5d + 20 \le 7d - 14$ Removing parentheses

$5d - 7d \le -14 - 20$ Adding $-7d$ and -20

$-2d \le -34$

$d \ge 17$ Multiplying by $-\dfrac{1}{2}$

⬑___ The symbol has to be reversed.

The solution set is $\{d|d \ge 17\}$.

104. $\{t|t \le -4\}$

105. $8(2t + 1) > 4(7t + 7)$

$16t + 8 > 28t + 28$

$16t - 28t > 28 - 8$

$-12t > 20$

$t < -\dfrac{20}{12}$ Multiplying by $-\dfrac{1}{12}$ and reversing the symbol

$t < -\dfrac{5}{3}$

The solution set is $\{t|t < -\dfrac{5}{3}\}$.

106. $\{x|x > -8\}$

107. $3(r - 6) + 2 < 4(r + 2) - 21$

$3r - 18 + 2 < 4r + 8 - 21$

$3r - 16 < 4r - 13$

$-16 + 13 < 4r - 3r$

$-3 < r$, or $r > -3$

The solution set is $\{r|r > -3\}$.

108. $\{t|t > -12\}$

109.
$$\frac{2}{3}(2x - 1) \ge 10$$

$\dfrac{3}{2} \cdot \dfrac{2}{3}(2x - 1) \ge \dfrac{3}{2} \cdot 10$ Multiplying by $\dfrac{3}{2}$

$2x - 1 \ge 15$

$2x \ge 16$

$x \ge 8$

The solution set is $\{x|x \ge 8\}$.

110. $\{x|x \le 7\}$

111.
$$\frac{3}{4}\left(3x - \frac{1}{2}\right) - \frac{2}{3} < \frac{1}{3}$$

$\dfrac{3}{4}\left(3x - \dfrac{1}{2}\right) < 1$ Adding $\dfrac{2}{3}$

$\dfrac{9}{4}x - \dfrac{3}{8} < 1$ Removing parentheses

$8 \cdot \left(\dfrac{9}{4}x - \dfrac{3}{8}\right) < 8 \cdot 1$ Clearing fractions

$18x - 3 < 8$

$18x < 11$

$x < \dfrac{11}{18}$

The solution set is $\left\{x\middle|x < \dfrac{11}{18}\right\}$.

112. $\left\{x\middle|x > -\dfrac{5}{32}\right\}$

113.
$$10 \div 2 \cdot 5 - 3^2 + (-4)^2$$

$= 10 \div 2 \cdot 5 - 9 + 16$ Evaluating the exponential notation

$= 5 \cdot 5 - 9 + 16$ Dividing

$= 25 - 9 + 16$ Multiplying

$= 32$ Subtracting and adding

114. 98

115. ◈

116. ◈

117. $2[4 - 2(3 - x)] - 1 \ge 4[2(4x - 3) + 7] - 25$

$2[4 - 6 + 2x] - 1 \ge 4[8x - 6 + 7] - 25$

$2[-2 + 2x] - 1 \ge 4[8x + 1] - 25$

$-4 + 4x - 1 \ge 32x + 4 - 25$

$-5 + 4x \ge 32x - 21$

$4x - 32x \ge -21 + 5$

$-28x \ge -16$

$x \le \dfrac{-16}{-28}$

$x \le \dfrac{4}{7}$

The solution set is $\left\{x\middle|x \le \dfrac{4}{7}\right\}$.

118. $\left\{t\middle|t > -\dfrac{27}{19}\right\}$

119.
$$-(x+5) \geq 4a - 5$$
$$-x - 5 \geq 4a - 5$$
$$-x \geq 4a - 5 + 5$$
$$-x \geq 4a$$
$$-1(-x) \leq -1 \cdot 4a$$
$$x \leq -4a$$

The solution set is $\{x | x \leq -4a\}$.

120. $\{x | x > 7\}$

121.
$$y < ax + b \quad \text{Assume } a > 0.$$
$$y - b < ax$$
$$\frac{y-b}{a} < x \quad \begin{array}{l}\text{Since } a > 0, \text{ the inequality} \\ \text{symbol stays the same.}\end{array}$$

The solution set is $\left\{x \big| x > \dfrac{y-b}{a}\right\}$.

122. $\left\{x \big| x < \dfrac{y-b}{a}\right\}$

123. $|x| < 3$

a) Since $|0| = 0$, and $0 < 3$ is true, 0 is a solution.

b) Since $|-2| = 2$ and $2 < 3$ is true, -2 is a solution.

c) Since $|-3| = 3$ and $3 < 3$ is false, -3 is not a solution.

d) Since $|4| = 4$ and $4 < 3$ is false, 4 is not a solution.

e) Since $|3| = 3$ and $3 < 3$ is false, 3 is not a solution.

f) Since $|1.7| = 1.7$ and $1.7 < 3$ is true, 1.7 is a solution.

g) Since $|-2.8| = 2.8$ and $2.8 < 3$ is true, -2.8 is a solution.

124.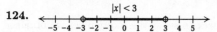

Exercise Set 2.7

1. $x > 4$

2. $x < 7$

3. $x \leq -6$

4. $y \geq 13$

5. $t \leq 80$

6. $w \geq 2$

7. $75 < a < 100$

8. $90 < s < 110$

9. $p \geq 1200$

10. $c \leq \$3457.95$

11. $y \leq 500$

12. $c \geq \$0.99$

13. $3x + 2 < 13$

14. $\dfrac{1}{2}n - 5 > 17$

15. *Familiarize.* The average of the five scores is their sum divided by the number of quizzes, 5. We let s represent the student's score on the last quiz.

Translate. The average of the five scores is given by
$$\frac{73 + 75 + 89 + 91 + s}{5}.$$
Since this average must be at least 85, this means that it must be greater than or equal to 85. Thus, we can translate the problem to the inequality
$$\frac{73 + 75 + 89 + 91 + s}{5} \geq 85.$$

Carry out. We first multiply by 5 to clear the fraction.
$$5\left(\frac{73 + 75 + 89 + 91 + s}{5}\right) \geq 5 \cdot 85$$
$$73 + 75 + 89 + 91 + s \geq 425$$
$$328 + s \geq 425$$
$$s \geq 425 - 328$$
$$s \geq 97$$

Check. Suppose s is a score greater than or equal to 97. Then by successively adding 73, 75, 89, and 91 on both sides of the inequality we get
$$73 + 75 + 89 + 91 + s \geq 425$$
so
$$\frac{73 + 75 + 89 + 91 + s}{5} \geq \frac{425}{5}, \text{ or } 85.$$

State. Any score which is at least 97 will give an average quiz grade of 85. The solution set is $\{s | s \geq 97\}$.

16. $\{s | s \geq 84\}$

17. *Familiarize.* Let m represent the number of miles per day. Then the cost per day for those miles is $\$0.46m$. The total cost is the daily rate plus the daily mileage cost. The total cost cannot exceed $\$200$. In other words the total cost must be less or equal to $\$200$, the daily budget.

Translate.

$$\underbrace{\text{Daily rate}}_{\downarrow} + \underbrace{\text{Mileage cost}}_{\downarrow} \leq \underbrace{\text{Budget}}_{\downarrow}$$
$$42.95 \quad + \quad 0.46m \quad \leq \quad 200$$

Carry out.
$$42.95 + 0.46m \leq 200$$
$$4295 + 46m \leq 20,000 \quad \text{Clearing decimals}$$
$$46m \leq 15,705$$
$$m \leq \frac{15,705}{46}$$
$$m \leq 341.4 \quad \begin{array}{l}\text{Rounding to the near-} \\ \text{est tenth}\end{array}$$

Check. We can check to see if the solution set seems reasonable.

When $m = 342$, the total cost is
$$42.95 + 0.46(342), \text{ or } \$200.27.$$
When $m = 341.4$, the total cost is
$$42.95 + 0.46(341.4), \text{ or } \$199.99.$$

When $m = 341$, the total cost is

$$42.95 + 0.46(341), \text{ or } \$199.81.$$

From these calculations it would appear that the solution is correct considering that rounding occurred.

State. To stay within the budget, the number of miles the family drives must not exceed 341.4. The solution set is $\{m | m \leq 341.4 \text{ mi}\}$.

18. $\{m | m \leq 525.8 \text{ mi}\}$

19. *Familiarize*. We first make a drawing. We let l represent the length.

The area is the length times the width, or $4l$.

Translate.

$$\underbrace{\text{Area}}_{\downarrow} \quad \underbrace{\text{is less than}}_{\downarrow} \quad \underbrace{86 \text{ cm}^2}_{\downarrow}.$$
$$4l \qquad\quad < \qquad\qquad 86$$

Carry out.

$$4l < 86$$
$$l < 21.5$$

Check. We check to see if the solution seems reasonable.

When $l = 22$, the area is $22 \cdot 4$, or 88 cm^2.

When $l = 21.5$, the area is $21.5(4)$, or 86 cm^2.

When $l = 21$, the area is $21 \cdot 4$, or 84 cm^2.

From these calculations, it would appear that the solution is correct.

State. The area will be less than 86 cm^2 for lengths less than 21.5 cm. The solution set is $\{l | l < 21.5 \text{ cm}\}$.

20. $\{l | l \geq 16.5 \text{ yd}\}$

21. *Familiarize*. We let $n =$ the number of half hours that Laura's car is parked. Then the total parking time t, in hours, will be $t = n/2$. We will express all costs in dollars.

Translate.

$$\underbrace{\$0.45 \atop \text{charge}}_{\downarrow} \quad \underbrace{\text{plus}}_{\downarrow} \quad \underbrace{\text{charge for} \atop \text{parking time}}_{\downarrow} \quad \underbrace{\text{is at least}}_{\downarrow} \quad \underbrace{\$2.20}_{\downarrow}$$
$$0.45 \quad + \qquad 0.25n \qquad\qquad \geq \qquad 2.20$$

Carry out. We solve the inequality.

$$0.45 + 0.25n \geq 2.20$$
$$45 + 25n \geq 220 \quad \text{Clearing decimals}$$
$$25n \geq 175$$
$$n \geq 7$$

Note that when $n \geq 7$, $n/2 \geq 7/2$, or 3.5.

Check. We check to see if the solution seems reasonable.

When $n = 6$, the charge is $\$0.45 + \$0.25(6)$, or $\$1.95$.

When $n = 7$, the charge is $\$0.45 + \$0.25(7)$, or $\$2.20$.

When $n = 8$, the charge is $\$0.45 + \$0.25(8)$, or $\$2.45$.

From these calculations it would appear that the solution is correct.

State. The charge is at least $\$2.20$ when the car is parked for at least 3.5 hr. The solution set is $\{t | t \geq 3.5 \text{ hr}\}$.

22. $\{m | m \geq 5 \text{ min}\}$

23. *Familiarize*. $R = -0.075t + 3.85$

In the formula R represents the world record and t represents the years since 1930. When $t = 0$ (1930), the record was $-0.075 \cdot 0 + 3.85$, or 3.85 minutes. When $t = 2$ (1932), the record was $-0.075(2) + 3.85$, or 3.7 minutes. For what values of t will $-0.075t + 3.85$ be less than 3.5?

Translate. The record is to be less than 3.5. We have the inequality

$$R < 3.5.$$

To find the t values which satisfy this condition we substitute $-0.075t + 3.85$ for R.

$$-0.075t + 3.85 < 3.5$$

Carry out.

$$-0.075t + 3.85 < 3.5$$
$$-0.075t < 3.5 - 3.85$$
$$-0.075t < -0.35$$
$$t > \frac{-0.35}{-0.075}$$
$$t > 4\frac{2}{3}$$

Check. We check to see if the solution set we obtained seems reasonable.

When $t = 4\frac{1}{2}$, $R = -0.075(4.5) + 3.85$, or 3.5125.

When $t = 4\frac{2}{3}$, $R = -0.075\left(\frac{14}{3}\right) + 3.85$, or 3.5.

When $t = 4\frac{3}{4}$, $R = -0.075(4.75) + 3.85$, or 3.49375.

Since $r = 3.5$ when $t = 4\frac{2}{3}$ and R decreases as t increases, R will be less than 3.5 when t is greater than $4\frac{2}{3}$.

State. The world record will be less than 3.5 minutes when t is greater than $4\frac{2}{3}$ years $\left(\text{more than } 4\frac{2}{3} \text{ years}\right.$ after $\left.1930\right)$. The solution set is $\{t | t > 1934\}$.

24. $\{t | t > 1984\}$

25. *Familiarize*. Let $w =$ the number of weeks it takes for the puppy's weight to exceed $22\frac{1}{2}$ lb.

Translate.

$$\underbrace{\text{Initial weight}}_{\downarrow} \quad \underbrace{\text{plus}}_{\downarrow} \quad \underbrace{\text{amount gained in } w \text{ weeks}}_{\downarrow} \quad \underbrace{\text{exceeds}}_{\downarrow} \quad \underbrace{22\frac{1}{2} \text{ lb.}}_{\downarrow}$$

$$9 \quad + \quad \frac{3}{4}w \quad > \quad 22\frac{1}{2}$$

Carry out. We solve the inequality.

$$9 + \frac{3}{4}w > 22\frac{1}{2}$$
$$\frac{3}{4}w > 13\frac{1}{2}$$
$$\frac{3}{4}w > \frac{27}{2} \quad \left(13\frac{1}{2} = \frac{27}{2}\right)$$
$$w > 18 \quad \text{Multiplying by } \frac{4}{3}$$

Check. We check to see if the solution seems reasonable.

When $w = 17$, $9 + \frac{3}{4} \cdot 17 = 21\frac{3}{4}$.

When $w = 18$, $9 + \frac{3}{4} \cdot 18 = 22\frac{1}{2}$.

When $w = 19$, $9 + \frac{3}{4} \cdot 19 = 23\frac{1}{4}$.

It would appear that the solution is correct.

State. The puppy's weight will exceed $22\frac{1}{2}$ lb after 18 weeks. The solution set is $\{w | w > 18 \text{ wk}\}$.

26. $\{w | w \geq 6 \text{ wk after July 1}\}$

27. *Familiarize.* We will use the formula $F = \frac{9}{5}C + 32$.

Translate.

$$\underbrace{\text{Fahrenheit temperature}}_{\downarrow} \quad \underbrace{\text{is below}}_{\downarrow} \quad \underbrace{88°.}_{\downarrow}$$

$$F \quad < \quad 88$$

Substituting $\frac{9}{5}C + 32$ for F, we have

$$\frac{9}{5}C + 32 < 88.$$

Carry out. We solve the inequality.

$$\frac{9}{5}C + 32 < 88$$
$$\frac{9}{5}C < 56$$
$$C < \frac{280}{9}$$
$$C < 31.1 \quad \text{Rounding}$$

Check. We check to see if the solution seems reasonable.

When $C = 31$, $\frac{9}{5} \cdot 31 + 32 = 87.8$.

When $C = 31.1$, $\frac{9}{5}(31.1) + 32 = 87.98$.

When $C = 31.2$, $\frac{9}{5}(31.2) + 32 = 88.16$.

It would appear that the solution is correct, considering that rounding occurred.

State. Butter stays solid at Celsius temperatures below about 31.1°. The solution set is $\{C | C < 31.1°\}$.

28. $\{C | C > 37°\}$

29. *Familiarize.* Let n represent the number.

Translate.

$$\underbrace{\text{The number}}_{\downarrow} \quad \underbrace{\text{plus}}_{\downarrow} \, \underbrace{15}_{\downarrow} \quad \underbrace{\text{is less than}}_{\downarrow} \quad \underbrace{4}_{\downarrow} \, \underbrace{\text{times}}_{\downarrow} \quad \underbrace{\text{the number.}}_{\downarrow}$$

$$n \quad + \quad 15 \quad < \quad 4 \quad \cdot \quad n$$

Carry out.

$$n + 15 < 4n$$
$$15 < 3n$$
$$5 < n, \text{ or } n > 5$$

Check. We check to see if the solution seems reasonable.

When $n = 4$, we have $4 + 15 < 4 \cdot 4$, or $19 < 16$. This is false.

When $n = 5$, we have $5 + 15 < 4 \cdot 5$, or $20 < 20$. This is false.

When $n = 6$, we have $6 + 15 < 4 \cdot 6$, or $21 < 24$. This is true.

Since the inequality is false for the numbers less than or equal to 5 that we tried and true for the number greater than 5, it would appear that $n > 5$ is correct.

State. All numbers greater than 5 are solutions. The solution set is $\{n | n > 5\}$.

30. $\{n | n \leq 0\}$

31. *Familiarize.* We first make a drawing. We let w represent the width.

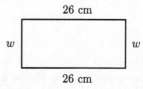

The perimeter is $P = 2l + 2w$, or $2 \cdot 26 + 2w$, or $52 + 2w$.

Translate.

$$\underbrace{\text{The perimeter}}_{\downarrow} \quad \underbrace{\text{is greater than}}_{\downarrow} \quad \underbrace{80 \text{ cm.}}_{\downarrow}$$

$$52 + 2w \quad > \quad 80$$

Carry out.

$$52 + 2w > 80$$
$$2w > 28$$
$$w > 14$$

Check. We check to see if the solution seems reasonable.

When $w = 13$, $P = 52 + 2 \cdot 13$, or 78 cm.

When $w = 14$, $P = 52 + 2 \cdot 14$, or 80 cm.

When $w = 15$, $P = 52 + 2 \cdot 15$, or 82 cm.

From these calculations, it appears that the solution is correct.

State. Widths greater than 14 cm will make the perimeter greater than 80 cm. The solution set is $\{w | w > 14 \text{ cm}\}$.

32. $\{l | l \geq 92 \text{ ft}\}$; $\{l | l \leq 92 \text{ ft}\}$

33. *Familiarize*. We first make a drawing. We let b represent the length of the base. Then the lengths of the other sides are $b - 2$ and $b + 3$.

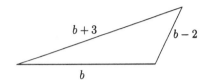

The perimeter is the sum of the lengths of the sides or $b + b - 2 + b + 3$, or $3b + 1$.

Translate.

The perimeter is greater than 19 cm.

$$\downarrow \qquad\qquad \downarrow \qquad\qquad \downarrow$$
$$3b + 1 \qquad\qquad > \qquad\qquad 19$$

Carry out.
$$3b + 1 > 19$$
$$3b > 18$$
$$b > 6$$

Check. We check to see if the solution seems reasonable.

When $b = 5$, the perimeter is $3 \cdot 5 + 1$, or 16 cm.

When $b = 6$, the perimeter is $3 \cdot 6 + 1$, or 19 cm.

When $b = 7$, the perimeter is $3 \cdot 7 + 1$, or 22 cm.

From these calculations, it would appear that the solution is correct.

State. For lengths of the base greater than 6 cm the perimeter will be greater than 19 cm. The solution set is $\{b | b > 6 \text{ cm}\}$.

34. $\left\{w | w \leq \dfrac{35}{3} \text{ ft}\right\}$.

35. *Familiarize*. The average number of calls per week is the sum of the calls for the three weeks divided by the number of weeks, 3. We let c represent the number of calls made during the third week.

Translate. The average of the three weeks is given by
$$\frac{17 + 22 + c}{3}.$$

Since the average must be at least 20, this means that it must be greater than or equal to 20. Thus, we can translate the problem to the inequality
$$\frac{17 + 22 + c}{3} \geq 20.$$

Carry out. We first multiply by 3 to clear the fraction.
$$3\left(\frac{17 + 22 + c}{3}\right) \geq 3 \cdot 20$$
$$17 + 22 + c \geq 60$$
$$39 + c \geq 60$$
$$c \geq 21$$

Check. Suppose c is a number greater than or equal to 21. Then by adding 17 and 22 on both sides of the inequality we get
$$17 + 22 + c \geq 17 + 22 + 21$$
$$17 + 22 + c \geq 60$$

so
$$\frac{17 + 22 + c}{3} \geq \frac{60}{3}, \text{ or } 20.$$

State. Any number of calls which is at least 21 will maintain an average of at least 20 for the three-week period. The solution set is $\{c | c \geq 21\}$.

36. George: more than 12 hours, Joan: more than 15 hours

37. *Familiarize*. Let s represent the amount Angelo can spend on each sweater. Then the total amount he can spend is represented by $\$21.95 + 2s$.

Translate.

Total spent	is less than or equal to	$120.00
$\downarrow$	$\downarrow$	$\downarrow$
$21.95 + 2s$	$\leq$	120

Carry out.
$$21.95 + 2s \leq 120$$
$$2195 + 200s \leq 12{,}000 \quad \text{Clearing decimals}$$
$$200s \leq 9805$$
$$s \leq 49.02 \quad \text{Rounding}$$

Check. We check to see if the solution seems reasonable.

When $s = \$49.01$, Angelo spends
$$21.95 + 2(49.01), \text{ or } \$119.97.$$

When $s = \$49.02$, Angelo spends
$$21.95 + 2(49.02), \text{ or } \$119.99.$$

When $s = \$49.03$, Angelo spends
$$21.95 + 2(49.03), \text{ or } \$120.01.$$

From these calculations, it would appear that the solution is correct.

State. Angelo can spend at most $49.02 for each sweater. The solution set is $\{s | s \leq \$49.02\}$.

38. $\{s | 0 \text{ lb} < s \leq 9 \text{ lb}\}$

39. *Familiarize*. We first make a drawing. We let l represent the length.

l

32 km | Area ≥ 2048 km^2 | 32 km

l

The area is the length times the width, or $32l$.

Translate.

The area is at least 2048 km^2.

$$\downarrow \qquad \downarrow \qquad \downarrow$$
$$32l \qquad \geq \qquad 2048$$

Carry out.
$$32l \geq 2048$$
$$l \geq 64$$

Check. We check to see if the solution seems reasonable.

When $l = 63$, the area is $32 \cdot 63$, or 2016 km^2.

When $l = 64$, the area is $32 \cdot 64$, or 2048 km^2.

When $l = 65$, the area is $32 \cdot 65$, or 2080 km^2.

From these calculations, it would appear that the solution is correct.

State. Lengths of 64 km or more will make the area at least 2048 km^2. The solution set is $\{l | l \geq 64$ km$\}$.

40. $\{b | b \leq 4$ cm$\}$

41. *Familiarize*. We will use the formula
$P = 0.1522Y - 298.592$.

Translate. We have the inequality $P \geq 6$. To find the years that satisfy this condition we substitute $0.1522Y - 298.592$ for P:

$$0.1522Y - 298.592 \geq 6$$

Carry out.

$0.1522Y - 298.592 \geq 6$
$\qquad 0.1522Y \geq 304.592$
$\qquad\qquad Y \geq 2001 \qquad$ Rounding

Check. We check to see if the solution seems reasonable.

When $Y = 2000$, $P = 0.1522(2000) - 298.592$, or about \$5.81.

When $Y = 2001$, $P = 0.1522(2001) - 298.592$, or about \$5.96.

When $Y = 2002$, $P = 0.1522(2002) - 298.592$, or about \$6.11.

From these calculations, it would appear that the solution is correct considering that rounding occurred.

State. From about 2001 on, the average price of a movie ticket will be at least \$6. The solution set is $\{Y | Y \geq 2001\}$.

42. $\left\{ x \,\middle|\, x \leq 215\dfrac{5}{27}\ \text{mi} \right\}$

43.
$\qquad -3 + 2(-5)^2(-3) - 7$
$= -3 + 2(25)(-3) - 7$ Evaluating the exponential expression
$= -3 - 150 - 7$ Multiplying
$= -160$ Subtracting

44. $4a^2 - 2$

45.
$\qquad 9x - 5 + 4x^2 - 2 - 13x$
$= 4x^2 + 9x - 13x - 5 - 2$
$= 4x^2 + (9 - 13)x - 5 - 2$
$= 4x^2 - 4x - 7$

46. $-17x + 18$

47.

48.

49. *Familiarize*. We make a drawing. Let s represent the length of a side of the square.

$$A \leq 64 \text{ cm}^2 \quad s$$

The area s is the square of the length of a side, or s^2.

Translate.

$\underbrace{\text{The area}}$ $\underbrace{\text{is no more than}}$ $\underbrace{64 \text{ cm}^2}$.
$\quad\downarrow \qquad\qquad \downarrow \qquad\qquad \downarrow$
$\quad s^2 \qquad\qquad \leq \qquad\qquad 64$

Carry out.

$$s^2 \leq 64$$
$$s^2 - 64 \leq 0$$
$$(s+8)(s-8) \leq 0$$

We know that $(s+8)(s-8) = 0$ for $s = -8$ or $s = 8$. Now $(s+8)(s-8) < 0$ when the two factors have opposite signs. That is:

$s+8 > 0 \quad and \quad s-8 < 0 \qquad$ or
$\qquad\qquad\qquad\qquad s+8 < 0 \ and \ s-8 > 0$
$\quad s > -8 \ and \qquad s < 8 \qquad$ or
$\qquad\qquad\qquad\qquad s < -8 \ and \qquad s > 8$

This can be expressed $\qquad$ This is not possible.
as $-8 < s < 8$.

Then $(s+8)(s-8) \leq 0$ for $-8 \leq s \leq 8$.

Check. Since the length of a side cannot be negative we only consider positive values of s, or $0 < s \leq 8$. We check to see if this solution seems reasonable.

When $s = 7$, the area is 7^2, or 49 cm^2.

When $s = 8$, the area is 8^2, or 64 cm^2.

When $s = 9$, the area is 9^2, or 81 cm^2.

From these calculations, it appears that the solution is correct.

State. Sides of length 8 or less will allow an area of no more than 64 cm^2. The solution set is $\{s | s \leq 8$ cm and s is positive$\}$, or $\{s | 0$ cm $< s \leq 8$ cm$\}$.

50. 47 and 49

51. *Familiarize*. Let $h = $ the number of hours the car has been parked. Then $h - 1 = $ the number of hours after the first hour.

Translate.

$\underbrace{\text{Charge for first hour}}$	plus	$\underbrace{\begin{array}{c}\text{charge for}\\ \text{additional}\\ \text{hours}\end{array}}$	exceeds	\$16.50.
$\downarrow$	$\downarrow$	$\downarrow$	$\downarrow$	$\downarrow$
4.00	+	$2.50(h-1)$	>	16.50

Carry out. We solve the inequality.

$$4.00 + 2.50(h - 1) > 16.50$$
$$40 + 25(h - 1) > 165 \quad \text{Multiplying by 10 to clear decimals}$$
$$40 + 25h - 25 > 165$$
$$25h + 15 > 165$$
$$25h > 150$$
$$h > 6$$

Check. We check to see if this solution seems reasonable.

When $h = 5$, $4.00 + 2.50(5 - 1) = 14.00$.

When $h = 6$, $4.00 + 2.50(6 - 1) = 16.50$.

When $h = 7$, $4.00 + 2.50(7 - 1) = 19.00$.

It appears that the solution is correct.

State. The charge exceeds $16.50 when the car has been parked for more than 6 hr.

52. $\{s | s < \$20,000\}$

53. ***Familiarize***. We define h as in Exercise 51. Note that the parking charge must be more than $14 *and* also less than $24. We will solve two inequalities and find the solutions that are in both solution sets.

Translate.

Charge for first hour	plus	charge for additional hours	is more than	$14.
↓	↓	↓	↓	↓
4.00	+	2.50(h − 1)	>	14

and

Charge for first hour	plus	charge for additional hours	is at most	$24.
↓	↓	↓	↓	↓
4.00	+	2.50(h − 1)	<	24

Carry out. We solve both inequalities. We get $h > 5$ *and* $h < 9$. The solutions that are in both solution sets are $\{h | 5 < h < 9\}$.

Check. The check is left to the student.

State. The car has been parked between 5 hr and 9 hr.

54. Between $-15°$ and $-9\dfrac{4°}{9}$

55. ◈

56. ◈

Chapter 3

Introduction to Graphing

Exercise Set 3.1

1. We go to the top of the bar that is above the body weight 100 lb. Then we move horizontally from the top of the bar to the vertical scale listing numbers of drinks. It appears approximately 3 drinks will give a 100 lb person a blood-alcohol level of 0.10%.

2. Approximately 5 drinks

3. From $3\frac{1}{2}$ on the vertical scale we move horizontally until we reach a bar whose top is above the horizontal line on which we are moving. The first such bar corresponds to a body weight of 120 lb. Thus, an individual weighs at least 120 lb if $3\frac{1}{2}$ drinks are consumed without reaching a blood-alcohol level of 0.10%.

4. 160 lb

5. The longest bar represents boredom. Thus this was the reason given most often.

6. Work/Military service

7. We go to the right end of the bar representing grades and then go down to the percent scale. We find that approximately 5% dropped out because of grades.

8. Approximately 40%

9. We locate 85 on the vertical scale and then move right until the line is reached. At that point we move down to the horizontal scale and read the information we are seeking. We see that the pulse rate was 85 beats per minute after 1 month of regular exercise.

10. 6 months

11. By observation or by computing the decrease or increase in beats per minute between successive pairs of points, we find that the greatest drop is about 15 beats per minute. This occurs between the second and third points from the left. We move down the horizontal scale and see that these points correspond to 1 month and 2 months. Thus, the greatest drop in pulse rate occurred during the second month.

12. The fifth month

13. We find the portion of the graph labeled medical care and read that 12% of income is spent on medical expenses.

14. $220

15. **Familiarize**. The graph tells us that 28% of the income is spent on food. We let $y =$ the amount spent on food.

    Translate. We reword and translate.

    What is 28% of income?

    $$y = 28\% \cdot \$2400$$

    Carry out. We do the computation.

    $$y = 0.28 \cdot \$2400 = \$672$$

    Check. We go over the computation.

    State. The family would spend $672 on food.

16. 55%

17. We locate 1965 on the horizontal scale and then move up to the line representing public education expenditures. At that point we move left to the vertical scale and read the information we are seeking. Approximately 4% of the GNP was spent on public education in 1965.

18. Approximately 8%

19. We locate 10% on the vertical scale and then move right until the line representing health care expenditures is reached. At that point we can move down to the horizontal scale and read the information we are seeking. Health care costs represented about 10% of the GNP in 1982.

20. 1990

21. We locate 8% on the vertical scale and then move right, noting the three points at which we hit the line representing defense expenditures. From each of the points we move down to the horizontal scale and read the information we are seeking. Defense expenditures were approximately 8% of the GNP in 1953, 1963, and 1970.

22. 1989

23. The highest point on the line representing defense expenditures occurs above 1955 on the horizontal scale. Defense expenditures peaked in 1955.

24. 1966, 1980

25. **Familiarize**. From the graph we read that health care expenditures were about 7% of GNP in 1970 and about 12% of GNP in 1990. We let $y =$ the growth over the years 1970-1990.

    Translate. We reword and translate.

    Growth is 1990 percentage less 1970 percentage.

    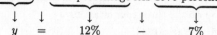

    Carry out. We do the computation.

    $$y = 12\% - 7\% = 5\%$$

Check. We go over the computation.

State. Health care expenditures (as a percentage of GNP) grew approximately 5% over the years 1970-1990.

26. Approximately 1%

27. We find the portion of the graph labeled Jazz and read that 3.7% of all recordings sold are jazz.

28. 9.0%

29. *Familiarize*. Let p = the percent of all recordings sold that are either soul or pop/rock. We will use the graph to find the percent for each type of music and then find their sum.

Translate. We reword the problem.

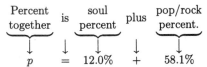

Carry out. We do the computation.

$$p = 12.0\% + 58.1\% = 70.1\%$$

Check. We go over the computation. The answer checks.

State. Together, 70.1% of all recordings sold are either soul or pop/rock.

30. 10.5%

31. *Familiarize*. The graph tells us that 58.1% of all recordings sold are pop/rock, 12.0% are soul, and 9.0% are country. Let p, s, and c represent the number of pop/rock, soul, and country recordings sold, respectively.

Translate. We reword the problem and write an equation for each type of music.

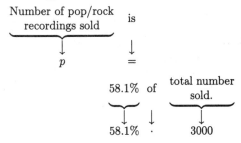

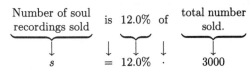

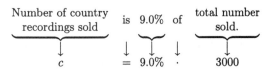

Solve. We do the three computations.

$$p = 0.581 \cdot 3000 = 1743$$

$$s = 0.12 \cdot 3000 = 360$$

$$c = 0.09 \cdot 3000 = 270$$

Check. We go over the computations. The answers check.

State. The store sells 1743 pop/rock, 360 soul, and 270 country recordings.

32. Pop/rock: 1452; classical: 170; gospel: 30

33. $(2, 5)$ is 2 units right and 5 units up.

$(-1, 3)$ is 1 unit left and 3 units up.

$(3, -2)$ is 3 units right and 2 units down.

$(-2, -4)$ is 2 units left and 4 units down.

$(0, 4)$ is 0 units left or right and 4 units up.

$(0, -5)$ is 0 units left or right and 5 units down.

$(5, 0)$ is 5 units right and 0 units up or down.

$(-5, 0)$ is 5 units left and 0 units up or down.

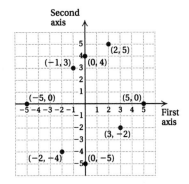

34.

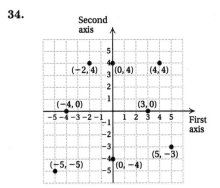

35. Since the first coordinate is negative and the second coordinate positive, the point $(-5, 3)$ is located in quadrant II.

36. II

37. Since the first coordinate is positive and the second coordinate negative, the point $(100, -1)$ is in quadrant IV.

38. IV

39. Since both coordinates are negative, the point $(-6, -29)$ is in quadrant III.

40. III

41. Since both coordinates are positive, the point $(3.8, 9.2)$ is in quadrant I.

42. I

43. In quadrant III, first coordinates are always <u>negative</u> and second coordinates are always <u>negative</u>.

44. Second, first

45.

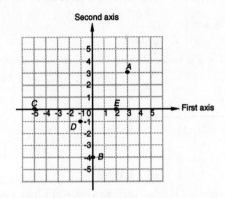

Point A is 3 units right and 3 units up. The coordinates of A are $(3, 3)$.

Point B is 0 units left or right and 4 units down. The coordinates of B are $(0, -4)$.

Point C is 5 units left and 0 units up or down. The coordinates of C are $(-5, 0)$.

Point D is 1 unit left and 1 unit down. The coordinates of D are $(-1, -1)$.

Point E is 2 units right and 0 units up or down. The coordinates of E are $(2, 0)$.

46. A: $(4,1)$, B: $(0, -5)$, C: $(-4, 0)$, D: $(-3, -2)$, E: $(3,0)$

47.
$$\frac{3}{5} \cdot \frac{10}{9} = \frac{3 \cdot 10}{5 \cdot 9}$$
$$= \frac{3 \cdot 2 \cdot 5}{5 \cdot 3 \cdot 3}$$
$$= \frac{\cancel{3} \cdot 2 \cdot \cancel{5}}{\cancel{5} \cdot \cancel{3} \cdot 3}$$
$$= \frac{2}{3}$$

48. $\dfrac{13}{15}$

49.
$$\frac{3}{7} - \frac{4}{5}$$
$$= \frac{3}{7} \cdot \frac{5}{5} - \frac{4}{5} \cdot \frac{7}{7} \quad \text{Using 35 as the common denominator}$$
$$= \frac{15}{35} - \frac{28}{35}$$
$$= -\frac{13}{35}$$

50. $-\dfrac{2}{15}$

51.

52.

53.

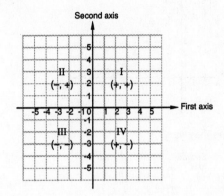

If the first coordinate is positive, then the point must be in either quadrant I or quadrant IV.

54. III or IV

55. If the first and second coordinates are equal, they must either be both positive or both negative. The point must be in either quadrant I (both positive) or quadrant III (both negative).

56. II or IV

57.

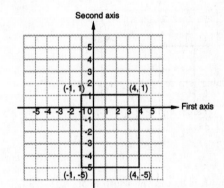

The coordinates of the fourth vertex are $(-1, -5)$.

58. $(5, 2)$, $(-7, 2)$, or $(3, -8)$.

59. Answers may vary.

We select eight points such that the sum of the coordinates for each point is 6.

$(-1, 7)$	$-1 + 7 = 6$
$(0, 6)$	$0 + 6 = 6$
$(1, 5)$	$1 + 5 = 6$
$(2, 4)$	$2 + 4 = 6$
$(3, 3)$	$3 + 3 = 6$
$(4, 2)$	$4 + 2 = 6$
$(5, 1)$	$5 + 1 = 6$
$(6, 0)$	$6 + 0 = 6$

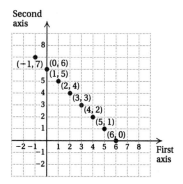

60. Answers may vary.

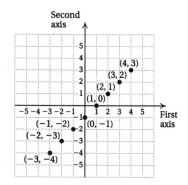

61.

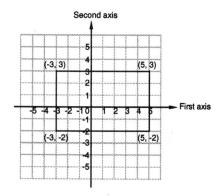

The length is 8, and the width is 5.

$P = 2l + 2w$

$P = 2 \cdot 8 + 2 \cdot 5 = 16 + 10 = 26$

62. $32\frac{1}{2}$

63. Latitude 32.5° North,
Longitude 64.5° West

64. Latitude 27° North,
Longitude 81° West

Exercise Set 3.2

1. $\underline{y = 3x - 1}$

$\quad 5 \ ? \ 3 \cdot 2 - 1$ Substituting 2 for x and 5 for y
$\quad\ \ \big|\ \ 6 - 1$ (alphabetical order of variables)
$\quad 5 \ \big|\ 5$ TRUE

Since $5 = 5$ is true, the pair $(2, 5)$ is a solution.

2. Yes

3. $\underline{3x - y = 4}$

$\quad 3 \cdot 2 - (-3) \ ? \ 4$ Substituting 2 for x and
$\qquad\qquad\qquad\qquad$ -3 for y
$\quad\ \ 6 + 3 \ \big|$
$\qquad 9 \ \big|\ 4$ FALSE

Since $9 = 4$ is false, the pair $(2, -3)$ is not a solution.

4. No

5. $\underline{2c + 2d = -7}$

$\quad 2(-2) + 2(-1) \ ? \ -7$ Substituting -2 for c and
$\qquad\qquad\qquad\qquad$ -1 for d
$\quad\ \ -4 - 2 \ \big|$
$\qquad -6 \ \big|\ -7$ FALSE

Since $-6 = -7$ is false, the pair $(-2, -1)$ is not a solution.

6. No

7. $y = x$

We first make a table of values. We choose *any* number for x and then determine y by substitution.

When $x = 0$, $y = 0$.
When $x = -2$, $y = -2$.
When $x = 3$, $y = 3$.

x	y
0	0
-2	-2
3	3

Since two points determine a line, that is all we really need to graph a line, but you may plot a third point as a check.

Plot these points, draw the line they determine, and label the graph $y = x$.

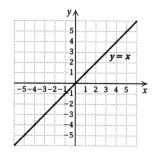

8.

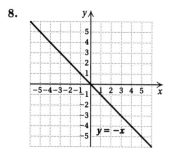

9. $y = -2x$

We first make a table of values.

When $x = 0$, $y = -2 \cdot 0 = 0$.
When $x = 2$, $y = -2 \cdot 2 = -4$.
When $x = -1$, $y = -2(-1) = 2$.

x	y
0	0
2	-4
-1	2

Plot these points, draw the line they determine, and label the graph $y = -2x$.

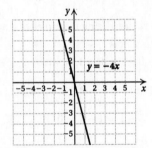

10.

11. $y = \dfrac{1}{3}x$

We first make a table of values. Using multiples of 3 for x avoids fractions.

When $x = 0$, $y = \dfrac{1}{3} \cdot 0 = 0$.

When $x = 6$, $y = \dfrac{1}{3} \cdot 6 = 2$.

When $x = -3$, $y = \dfrac{1}{3}(-3) = -1$.

x	y
0	0
6	2
-3	-1

Plot these points, draw the line they determine, and label the graph $y = \dfrac{1}{3}x$.

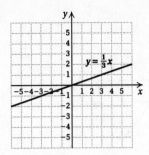

12.

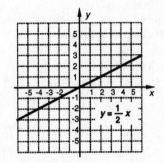

13. $y = -\dfrac{3}{2}x$

We first make a table of values. Using multiples of 2 for x avoids fractions.

When $x = 0$, $y = -\dfrac{3}{2} \cdot 0 = 0$.

When $x = 2$, $y = -\dfrac{3}{2} \cdot 2 = -3$.

When $x = -2$, $y = -\dfrac{3}{2}(-2) = 3$.

x	y
0	0
2	-3
-2	3

Plot these points, draw the line they determine, and label the graph $y = -\dfrac{3}{2}x$.

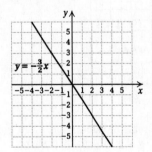

14.

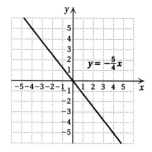

15. $y = x + 1$

We first make a table of values. We choose *any* number for x and then determine y by substitution.

When $x = 0$, $y = 0 + 1 = 1$.
When $x = 3$, $y = 3 + 1 = 4$.
When $x = -5$, $y = -5 + 1 = -4$.

x	y
0	1
3	4
-5	-4

Plot these points, draw the line they determine, and label the graph $y = x + 1$.

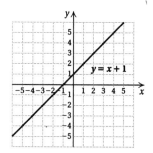

16.

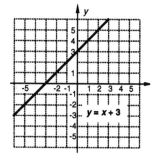

17. $y = 2x + 2$

We first make a table of values.
When $x = 0$, $y = 2 \cdot 0 + 2 = 0 + 2 = 2$.
When $x = -3$, $y = 2(-3) + 2 = -6 + 2 = -4$.
When $x = 1$, $y = 2 \cdot 1 + 2 = 2 + 2 = 4$.

x	y
0	2
-3	-4
1	4

Plot these points, draw the line they determine, and label the graph $y = 2x + 2$.

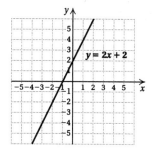

18.

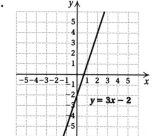

19. $y = \dfrac{1}{3}x - 1$

We first make a table of values. Using multiples of 3 avoids fractions.

When $x = 0$, $y = \dfrac{1}{3} \cdot 0 - 1 = 0 - 1 = -1$.

When $x = -6$, $y = \dfrac{1}{3}(-6) - 1 = -2 - 1 = -3$.

When $x = 3$, $y = \dfrac{1}{3} \cdot 3 - 1 = 1 - 1 = 0$.

x	y
0	-1
-6	-3
3	0

Plot these points, draw the line they determine, and label the graph $y = \dfrac{1}{3}x - 1$.

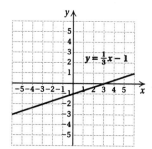

20.

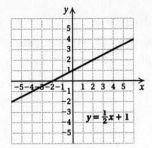

$$y = \frac{1}{2}x + 1$$

21. $y + x = -3$

   $\quad\quad y = -x - 3$ Solving for y

We first make a table of values.

When $x = 0$, $\quad y = -0 - 3 = -3$.
When $x = 1$, $\quad y = -1 - 3 = -4$.
When $x = -5$, $\quad y = -(-5) - 3 = 5 - 3 = 2$.

x	y
0	-3
1	-4
-5	2

Plot these points, draw the line they determine, and label the graph.

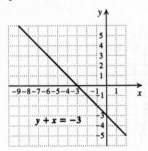

$$y + x = -3$$

22.

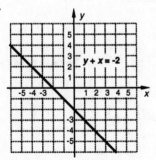

$$y + x = -2$$

23. $y = \dfrac{5}{2}x + 3$

We first make a table of values. Using multiples of 2 for x avoids fractions.

When $x = 0$, $\quad y = \dfrac{5}{2} \cdot 0 + 3 = 0 + 3 = 3$.

When $x = -2$, $\quad y = \dfrac{5}{2}(-2) + 3 = -5 + 3 = -2$.

When $x = -4$, $\quad y = \dfrac{5}{2}(-4) + 3 = -10 + 3 = -7$.

x	y
0	3
-2	-2
-4	-7

Plot these points, draw the line they determine, and label the graph.

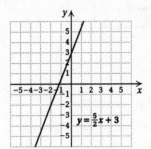

$$y = \frac{5}{2}x + 3$$

24.

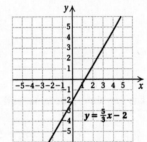

$$y = \frac{5}{3}x - 2$$

25. $y = -\dfrac{5}{2}x - 2$

We first make a table of values. Using multiples of 2 for x avoids fractions.

When $x = 0$, $\quad y = -\dfrac{5}{2} \cdot 0 - 2 = 0 - 2 = -2$.

When $x = -2$, $\quad y = -\dfrac{5}{2}(-2) - 2 = 5 - 2 = 3$.

When $x = 2$, $\quad y = -\dfrac{5}{2}(2) - 2 = -5 - 2 = -7$.

x	y
0	-2
-2	3
2	-7

Plot these points, draw the line they determine, and label the graph.

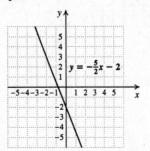

$$y = -\frac{5}{2}x - 2$$

26.

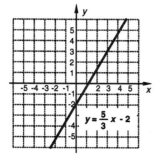

27. $y = \frac{1}{2}x - 5$

The $y-$intercept is $(0, -5)$. We find two other pairs using multiples of 2 for x to avoid fractions.

When $x = 2$, $y = \frac{1}{2} \cdot 2 - 5 = 1 - 5 = -4$.

When $x = 4$, $y = \frac{1}{2} \cdot 4 - 5 = 2 - 5 = -3$.

x	y
0	-5
2	-4
4	-3

Plot these points, draw the line they determine, and label the graph.

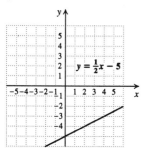

28.

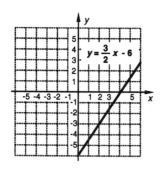

29. $2x + y = 3$

$y = -2x + 3$ Solving for y

The $y-$intercept is $(0, 3)$. We find two other pairs.

When $x = -1$, $y = -2(-1) + 3 = 2 + 3 = 5$.

When $x = 3$, $y = -2 \cdot 3 + 3 = -6 + 3 = -3$.

x	y
0	3
-1	5
3	-3

Plot these points, draw the line they determine, and label the graph.

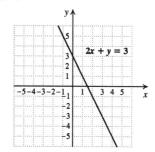

30.

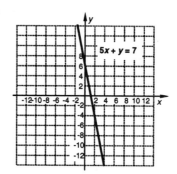

31. $y = x - \frac{1}{2}$

The $y-$intercept is $\left(0, -\frac{1}{2}\right)$. We find two other pairs.

When $x = -3$, $y = -3 - \frac{1}{2} = -3\frac{1}{2}$.

When $x = 4$, $y = 4 - \frac{1}{2} = 3\frac{1}{2}$.

x	y
0	$-\frac{1}{2}$
-3	$-3\frac{1}{2}$
4	$3\frac{1}{2}$

Plot these points, draw the line they determine, and label the graph.

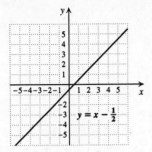

$y = x - \frac{1}{2}$

32.

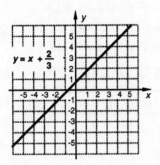

$y = x + \frac{2}{3}$

33. We solve for y.

$$x + 2y = -4$$
$$2y = -x - 4$$
$$y = \frac{1}{2}(-x - 4)$$
$$y = -\frac{1}{2}x - 2$$

The y–intercept is $(0, -2)$. We find two other pairs using multiples of 2 for x to avoid fractions.

When $x = -4$, $\quad y = -\frac{1}{2}(-4) - 2 = 2 - 2 = 0$.

When $x = 4$, $\quad y = -\frac{1}{2} \cdot 4 - 2 = -2 - 2 = -4$.

x	y
0	-2
-4	0
4	-4

Plot these points, draw the line they determine, and label the graph.

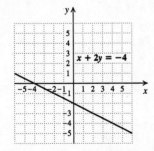

$x + 2y = -4$

34.

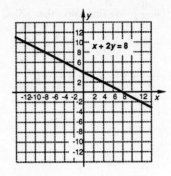

$x + 2y = 8$

35. We solve for y.

$$6x - 3y = 9$$
$$-3y = -6x + 9$$
$$y = -\frac{1}{3}(-6x + 9)$$
$$y = 2x - 3$$

The y–intercept is $(0, -3)$. We find two other pairs.

When $x = -1$, $\quad y = 2(-1) - 3 = -2 - 3 = -5$.

When $x = 3$, $\quad y = 2 \cdot 3 - 3 = 6 - 3 = 3$.

x	y
0	-3
-1	-5
3	3

Plot these points, draw the line they determine, and label the graph.

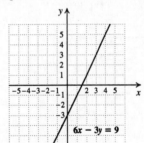

$6x - 3y = 9$

36.

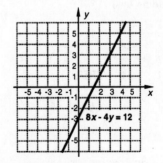

$8x - 4y = 12$

37. We solve for y.

$$6y + 2x = 8$$
$$6y = -2x + 8$$
$$y = \frac{1}{6}(-2x + 8)$$
$$y = -\frac{1}{3}x + \frac{4}{3}$$

The y-intercept is $\left(0, \dfrac{4}{3}\right)$. We find two other pairs.

When $x = -2$, $y = -\dfrac{1}{3}(-2) + \dfrac{4}{3} = \dfrac{2}{3} + \dfrac{4}{3} = \dfrac{6}{3} = 2$.

When $x = 4$, $y = -\dfrac{1}{3} \cdot 4 + \dfrac{4}{3} = -\dfrac{4}{3} + \dfrac{4}{3} = 0$.

x	y
0	$\dfrac{4}{3}$
-2	2
4	0

Plot these points, draw the line they determine, and label the graph.

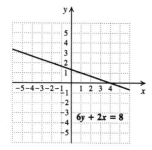

38.

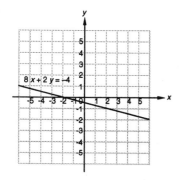

39. $3x - 7 = -34$

$\qquad 3x = -27 \quad$ Adding 7

$\qquad\ x = -9 \quad$ Dividing by 3

Check:
$$\begin{array}{c|c} \multicolumn{2}{c}{3x - 7 = -34} \\ \hline 3(-9) - 7 \ ? \ -34 & \\ -27 - 7 & \\ -34 & -34 \quad \text{TRUE} \end{array}$$

The solution is -9.

40. $\dfrac{16}{5}$

41. $Ax + By = C$

$\qquad By = C - Ax \quad$ Subtracting Ax

$\qquad y = \dfrac{C - Ax}{B} \quad$ Dividing by B

42. $Q = 2A - T$

43.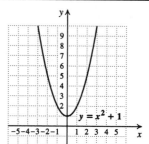

44.

45. $y = x^2 + 1$

When $x = 0$, $y = 0^2 + 1 = 0 + 1 = 1$.

When $x = -1$, $y = (-1)^2 + 1 = 1 + 1 = 2$.

When $x = 1$, $y = 1^2 + 1 = 1 + 1 = 2$.

When $x = -2$, $y = (-2)^2 + 1 = 4 + 1 = 5$.

When $x = 2$, $y = 2^2 + 1 = 4 + 1 = 5$.

When $x = -3$, $y = (-3)^2 + 1 = 9 + 1 = 10$.

When $x = 3$, $y = 3^2 + 1 = 9 + 1 = 10$.

x	0	-1	1	-2	2	-3	3
y	1	2	2	5	5	10	10

46. $(0,6)$, $(1,5)$, $(2,4)$, $(3,3)$, $(4,2)$, $(5,1)$, $(6,0)$

47. $x + 3y = 15$

Since $3y$ and 15 are both multiples of 3, x must also be a multiple of 3.

When $x = 0$, $0 + 3y = 15$
$\qquad\qquad\qquad 3y = 15$
$\qquad\qquad\qquad\ y = 5$

When $x = 3$, $3 + 3y = 15$
$\qquad\qquad\qquad 3y = 12$
$\qquad\qquad\qquad\ y = 4$

When $x = 6$, $6 + 3y = 15$
$\qquad\qquad\qquad 3y = 9$
$\qquad\qquad\qquad\ y = 3$

When $x = 9$, $9 + 3y = 15$
$\qquad\qquad\qquad 3y = 6$
$\qquad\qquad\qquad\ y = 2$

When $x = 12$, $12 + 3y = 15$
$\qquad\qquad\qquad 3y = 3$
$\qquad\qquad\qquad\ y = 1$

When $x = 15$, $15 + 3y = 15$
$\qquad\qquad\qquad 3y = 0$
$\qquad\qquad\qquad\ y = 0$

The whole number solutions are $(0,5)$, $(3,4)$, $(6,3)$, $(9,2)$, $(12,1)$, $(15,0)$.

48. $5n + 10d = 195$, or $0.05n + 0.1d = 1.95$; $(10,19)$, $(0,39)$, and $(15,9)$; answers may vary.

49. The value of n nickels is $0.05n$.

The value of q quarters is $0.25q$.

Thus, $0.05n + 0.25q = 2.35$

or $\quad 5n + 25q = 235 \qquad$ Clearing decimals

$$\begin{aligned} \text{When } n = 42, \quad 5 \cdot 42 + 25q &= 235 \\ 210 + 25q &= 235 \\ 25q &= 25 \\ q &= 1 \end{aligned}$$

$$\begin{aligned} \text{When } n = 7, \quad 5 \cdot 7 + 25q &= 235 \\ 35 + 25q &= 235 \\ 25q &= 200 \\ q &= 8 \end{aligned}$$

$$\begin{aligned} \text{When } n = 27, \quad 5 \cdot 27 + 25q &= 235 \\ 135 + 25q &= 235 \\ 25q &= 100 \\ q &= 4 \end{aligned}$$

Solutions are ordered pairs of the form (n, q). Three are $(42, 1)$, $(7, 8)$, and $(27, 4)$. Answers may vary.

50. Answers may vary. $(-3, 3)$, $(2, 2)$, $(0, 0)$

51. See the answer section in the text.

52. $y = 4.5x + 2.1$

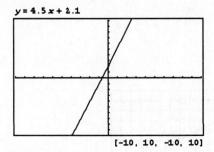

$[-10, 10, -10, 10]$

53. See the answer section in the text.

54. $y = -\dfrac{33}{8}x - \dfrac{45}{7}$

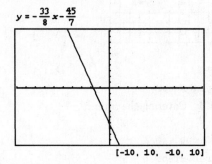

$[-10, 10, -10, 10]$

55. Yes; when we solve $2x + 5y = 15$ for y we get $y = -\dfrac{2}{5}x + 3$.

Exercise Set 3.3

1. (a) The graph crosses the y-axis at $(0, 3)$, so the y-intercept is $(0, 3)$.

(b) The graph crosses the x-axis at $(4, 0)$, so the x-intercept is $(4, 0)$.

2. (a) $(0, 5)$; (b) $(2, 0)$

3. (a) The graph crosses the y-axis at $(0, 5)$, so the y-intercept is $(0, 5)$.

(b) The graph crosses the x-axis at $(-3, 0)$, so the x-intercept is $(-3, 0)$.

4. (a) $(0, -4)$; (b) $(3, 0)$

5. $2x + 5y = 20$

(a) To find the y-intercept, let $x = 0$. This is the same as covering the x-term and then solving.

$$\begin{aligned} 5y &= 20 \\ y &= 4 \end{aligned}$$

The y-intercept is $(0, 4)$.

(b) To find the x-intercept, let $y = 0$. This is the same as covering the y-term and then solving.

$$\begin{aligned} 2x &= 20 \\ x &= 10 \end{aligned}$$

The x-intercept is $(10, 0)$.

6. (a) $(0, 5)$; (b) $(3, 0)$

7. $4x - 3y = 24$

(a) To find the y-intercept, let $x = 0$. This is the same as covering the x-term and then solving.

$$\begin{aligned} -3y &= 24 \\ y &= -8 \end{aligned}$$

The y-intercept is $(0, -8)$.

(b) To find the x-intercept, let $y = 0$. This is the same as covering the y-term and then solving.

$$\begin{aligned} 4x &= 24 \\ x &= 6 \end{aligned}$$

The x-intercept is $(6, 0)$.

8. (a) $(0, -4)$; (b) $(14, 0)$

9. $-6x + y = 8$

(a) To find the y-intercept, let $x = 0$. This is the same as covering the x-term and then solving.

$$y = 8$$

The y-intercept is $(0, 8)$.

(b) To find the x-intercept, let $y = 0$. This is the same as covering the y-term and then solving.

$$\begin{aligned} -6x &= 8 \\ x &= -\dfrac{4}{3} \end{aligned}$$

The x-intercept is $\left(-\dfrac{4}{3}, 0\right)$.

10. (a) $(0, 10)$; (b) $\left(-\dfrac{5}{4}, 0\right)$

11. $\quad 2y - 4 = 6x$

$\quad\quad -6x + 2y = 4 \quad$ Writing the equation in the
$\quad\quad\quad\quad\quad\quad\quad$ form $Ax + By = C$

(a) To find the y-intercept, let $x = 0$. This is the same as covering the x-term and then solving.

$$2y = 4$$
$$y = 2$$

The y-intercept is $(0, 2)$.

(b) To find the x-intercept, let $y = 0$. This is the same as covering the y-term and then solving.

$$-6x = 4$$
$$x = -\frac{2}{3}$$

The x-intercept is $\left(-\frac{2}{3}, 0\right)$.

12. (a) $(0, -2)$; (b) $\left(\frac{2}{3}, 0\right)$

13. $3x + 2y = 12$

Find the y-intercept:

$$2y = 12 \quad \text{Covering the } x\text{-term}$$
$$y = 6$$

The y-intercept is $(0, 6)$.

Find the x-intercept:

$$3x = 12 \quad \text{Covering the } y\text{-term}$$
$$x = 4$$

The x-intercept is $(4, 0)$.

To find a third point we replace x with 2 and solve for y.

$$3 \cdot 2 + 2y = 12$$
$$6 + 2y = 12$$
$$2y = 6$$
$$y = 3$$

The point $(2, 3)$ appears to line up with the intercepts, so we draw the graph.

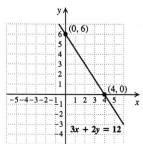

14.

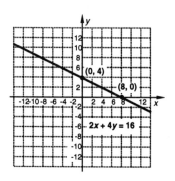

15. $x + 3y = 6$

Find the y-intercept:

$$3y = 6 \quad \text{Covering the } x\text{-term}$$
$$y = 2$$

The y-intercept is $(0, 2)$.

Find the x-intercept:

$$x = 6 \quad \text{Covering the } y\text{-term}$$

The x-intercept is $(6, 0)$.

To find a third point we replace x with 3 and solve for y.

$$3 + 3y = 6$$
$$3y = 3$$
$$y = 1$$

The point $(3, 1)$ appears to line up with the intercepts, so we draw the graph.

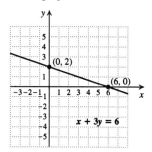

16.

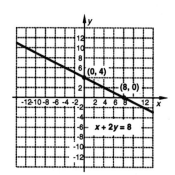

17. $-x + 2y = 4$

Find the y-intercept:

$$2y = 4 \quad \text{Covering the } x\text{-term}$$
$$y = 2$$

The y-intercept is $(0, 2)$.

Find the x-intercept:

$$-x = 4 \quad \text{Covering the } y\text{-term}$$
$$x = -4$$

The x-intercept is $(-4, 0)$.

To find a third point we replace x with 4 and solve for y.

$$-4 + 2y = 4$$
$$2y = 8$$
$$y = 4$$

The point $(4, 4)$ appears to line up with the intercepts, so we draw the graph.

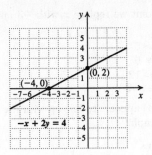

18.

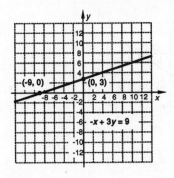

19. $3x + y = 9$

Find the y-intercept:

   $y = 9$ Covering the x-term

The y-intercept is $(0, 9)$.

Find the x-intercept:

   $3x = 9$ Covering the y-term
   $x = 3$

The x-intercept is $(3, 0)$.

To find a third point we replace x with 2 and solve for y.

   $3 \cdot 2 + y = 9$
   $6 + y = 9$
   $y = 3$

The point $(2, 3)$ appears to line up with the intercepts, so we draw the graph.

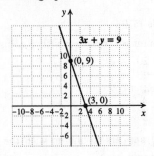

20.

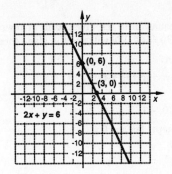

21. $2y - 2 = 6x$

We can leave the equation in the given form or rewrite it in the form $Ax + By = C$. We will use the given form.

To find the y-intercept, let $x = 0$.

   $2y - 2 = 6 \cdot 0$
   $2y - 2 = 0$
   $2y = 2$
   $y = 1$

The y-intercept is $(0, 1)$.

Find the x-intercept:

   $-2 = 6x$ Covering the y-term
   $-\dfrac{1}{3} = x$

The x-intercept is $\left(-\dfrac{1}{3}, 0\right)$.

To find a third point we replace x with 1 and solve for y.

   $2y - 2 = 6 \cdot 1$
   $2y - 2 = 6$
   $2y = 8$
   $y = 4$

The point $(1, 4)$ appears to line up with the intercepts, so we draw the graph.

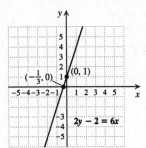

22.

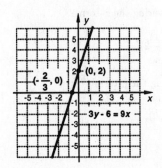

23. $3x - 9 = 3y$

We can leave the equation in the given form or rewrite it in the form $Ax + By = C$. We will use the given form.

Find the y-intercept:

$$-9 = 3y \quad \text{Covering the } x\text{-term}$$
$$-3 = y$$

The y-intercept is $(0, -3)$.

To find the x-intercept, let $y = 0$.

$$3x - 9 = 3 \cdot 0$$
$$3x - 9 = 0$$
$$3x = 9$$
$$x = 3$$

The x-intercept is $(3, 0)$.

To find a third point we replace x with 1 and solve for y.

$$3 \cdot 1 - 9 = 3y$$
$$3 - 9 = 3y$$
$$-6 = 3y$$
$$-2 = y$$

The point $(1, -2)$ appears to line up with the intercepts, so we draw the graph.

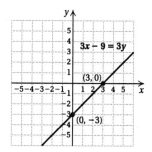

24.

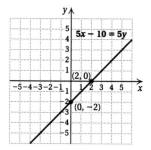

25. $2x - 3y = 6$

Find the y-intercept:

$$-3y = 6 \quad \text{Covering the } x\text{-term}$$
$$y = -2$$

The y-intercept is $(0, -2)$.

Find the x-intercept:

$$2x = 6 \quad \text{Covering the } y\text{-term}$$
$$x = 3$$

The x-intercept is $(3, 0)$.

To find a third point we replace x with -3 and solve for y.

$$2(-3) - 3y = 6$$
$$-6 - 3y = 6$$
$$-3y = 12$$
$$y = -4$$

The point $(-3, -4)$ appears to line up with the intercepts, so we draw the graph.

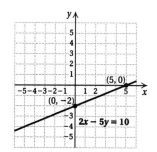

26.

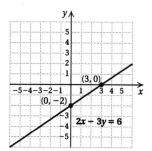

27. $4x + 5y = 20$

Find the y-intercept:

$$5y = 20 \quad \text{Covering the } x\text{-term}$$
$$y = 4$$

The y-intercept is $(0, 4)$.

Find the x-intercept:

$$4x = 20 \quad \text{Covering the } y\text{-term}$$
$$x = 5$$

The x-intercept is $(5, 0)$.

To find a third point we replace x with 4 and solve for y.

$$4 \cdot 4 + 5y = 20$$
$$16 + 5y = 20$$
$$5y = 4$$
$$y = \frac{4}{5}$$

The point $\left(4, \frac{4}{5}\right)$ appears to line up with the intercepts, so we draw the graph.

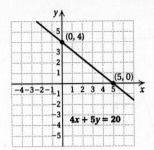

4x + 5y = 20

28.

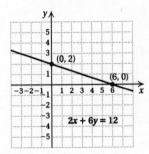

2x + 6y = 12

29. $2x + 3y = 8$

Find the y-intercept:

$$3y = 8 \qquad \text{Covering the } x\text{-term}$$
$$y = \frac{8}{3}$$

The y-intercept is $\left(0, \frac{8}{3}\right)$.

Find the x-intercept:

$$2x = 8 \qquad \text{Covering the } y\text{-term}$$
$$x = 4$$

The x-intercept is $(4, 0)$.

To find a third point we replace x with 1 and solve for y.

$$2 \cdot 1 + 3y = 8$$
$$2 + 3y = 8$$
$$3y = 6$$
$$y = 2$$

The point $(1, 2)$ appears to line up with the intercepts, so we draw the graph.

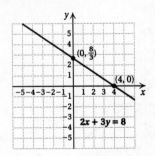

2x + 3y = 8

30.

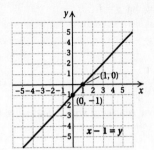

x − 1 = y

31. $x - 3 = y$

We can leave the equation in the given form or rewrite it in the form $Ax + By = C$. We will use the given form.

Find the y-intercept:

$$-3 = y \qquad \text{Covering the } x\text{-term}$$

The y-intercept is $(0, -3)$.

To find the x-intercept, let $y = 0$.

$$x - 3 = 0$$
$$x = 3$$

The x-intercept is $(3, 0)$.

To find a third point we replace x with -2 and solve for y.

$$-2 - 3 = y$$
$$-5 = y$$

The point $(-2, -5)$ appears to line up with the intercepts, so we draw the graph.

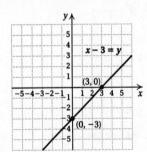

32.

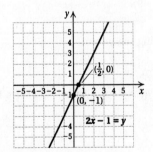

2x − 1 = y

33. $3x - 2 = y$

We can leave the equation in the given form or rewrite it in the form $Ax + By = C$. We will use the given form.

Find the y-intercept:

$$-2 = y \qquad \text{Covering the } x\text{-term}$$

The y-intercept is $(0, -2)$.

To find the x-intercept, let $y = 0$.

$$3x - 2 = 0$$
$$3x = 2$$
$$x = \frac{2}{3}$$

The x-intercept is $\left(\frac{2}{3}, 0\right)$.

To find a third point we replace x with 2 and solve for y.

$$3 \cdot 2 - 2 = y$$
$$6 - 2 = y$$
$$4 = y$$

The point $(2, 4)$ appears to line up with the intercepts, so we draw the graph.

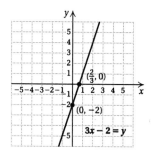

34.

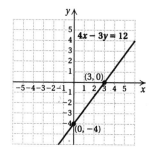

35. $6x - 2y = 18$

Find the y-intercept:

$$-2y = 18 \quad \text{Covering the } x\text{-term}$$
$$y = -9$$

The y-intercept is $(0, -9)$.

Find the x-intercept:

$$6x = 18 \quad \text{Covering the } y\text{-term}$$
$$x = 3$$

The x-intercept is $(3, 0)$.

To find a third point we replace x with 1 and solve for y.

$$6 \cdot 1 - 2y = 18$$
$$6 - 2y = 18$$
$$-2y = 12$$
$$y = -6$$

The point $(1, -6)$ appears to line up with the intercepts, so we draw the graph.

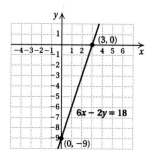

36.

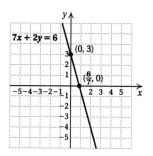

37. $3x + 4y = 5$

Find the y-intercept:

$$4y = 5 \quad \text{Covering the } x\text{-term}$$
$$y = \frac{5}{4}$$

The y-intercept is $\left(0, \frac{5}{4}\right)$.

Find the x-intercept:

$$3x = 5 \quad \text{Covering the } y\text{-term}$$
$$x = \frac{5}{3}$$

The x-intercept is $\left(\frac{5}{3}, 0\right)$.

To find a third point we replace x with 3 and solve for y.

$$3 \cdot 3 + 4y = 5$$
$$9 + 4y = 5$$
$$4y = -4$$
$$y = -1$$

The point $(3, -1)$ appears to line up with the intercepts, so we draw the graph.

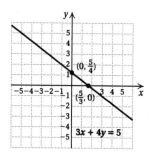

38.

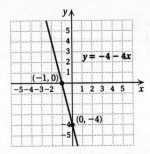

39. $y = -3 - 3x$

We can leave the equation in the given form or rewrite it in the form $Ax + By = C$. We will use the given form.

Find the y-intercept:

$y = -3$ Covering the x-term

The y-intercept is $(0, -3)$.

To find the x-intercept, let $y = 0$.

$0 = -3 - 3x$
$3x = -3$
$x = -1$

The x-intercept is $(-1, 0)$.

To find a third point we replace x with -2 and solve for y.

$y = -3 - 3 \cdot (-2)$
$y = -3 + 6$
$y = 3$

The point $(-2, 3)$ appears to line up with the intercepts, so we draw the graph.

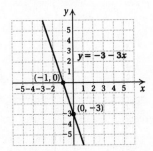

40.

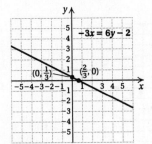

41. $-4x = 8y - 5$

We can leave the equation in the given form or rewrite it in the form $Ax + By = C$. We will use the given form.

To find the y-intercept, let $x = 0$.

$-4 \cdot 0 = 8y - 5$
$0 = 8y - 5$
$5 = 8y$
$\frac{5}{8} = y$

The y-intercept is $\left(0, \frac{5}{8}\right)$.

Find the x-intercept:

$-4x = -5$ Covering the y-term
$x = \frac{5}{4}$

The x-intercept is $\left(\frac{5}{4}, 0\right)$.

To find a third point we replace x with -5 and solve for y.

$-4(-5) = 8y - 5$
$20 = 8y - 5$
$25 = 8y$
$\frac{25}{8} = y$

The point $\left(-5, \frac{25}{8}\right)$ appears to line up with the intercepts, so we draw the graph.

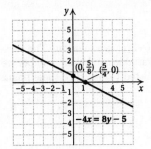

42.

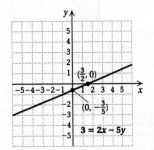

43. $y - 3x = 0$

Find the y-intercept:

$y = 0$ Covering the x-term

The y-intercept is $(0, 0)$. Note that this is also the x-intercept.

In order to graph the line, we will find a second point.

When $x = 1$, $y - 3 \cdot 1 = 0$
$y - 3 = 0$
$y = 3$

To find a third point we replace $x = -1$ and solve for y.

$y - 3(-1) = 0$
$y + 3 = 0$
$y = -3$

The point $(-1, -3)$ appears to line up with the other two points, so we draw the graph.

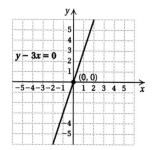

44.

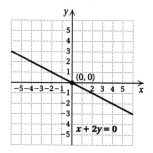

45. $x = -2$

Any ordered pair $(-2, y)$ is a solution. The variable x must be -2, but the y variable can be any number we choose. A few solutions are listed below. Plot these points and draw the line.

x	y
-2	-2
-2	0
-2	4

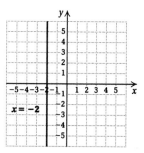

46.

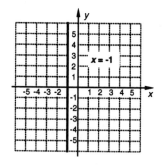

47. $y = 2$

Any ordered pair $(x, 2)$ is a solution. The variable y must be 2, but the x variable can be any number we choose. A few solutions are listed below. Plot these points and draw the line.

x	y
-3	2
0	2
2	2

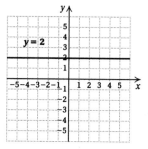

48.

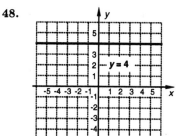

49. $x = 7$

Any ordered pair $(7, y)$ is a solution. The variable x must be 7, but the y variable can be any number we choose. A few solutions are listed below. Plot these points and draw the line.

x	y
7	-1
7	4
7	5

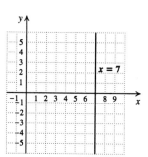

50.

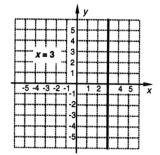

51. $y = 0$

Any ordered pair $(x, 0)$ is a solution. The variable y must be 0, but the x variable can be any number we choose. A few solutions are listed below. Plot these points and draw the line.

x	y
-5	0
-1	0
3	0

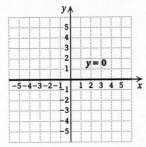

52.

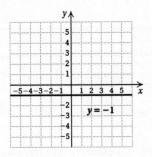

53. $x = \dfrac{3}{2}$

Any ordered pair $\left(\dfrac{3}{2}, y\right)$ is a solution. The variable x must be $\dfrac{3}{2}$, but the y variable can be any number we choose. A few solutions are listed below. Plot these points and draw the line.

x	y
$\dfrac{3}{2}$	-2
$\dfrac{3}{2}$	0
$\dfrac{3}{2}$	4

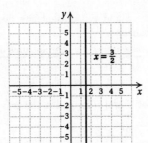

54.

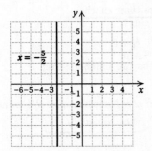

55. $3y = -5$

$\qquad y = -\dfrac{5}{3} \qquad$ Solving for y

Any ordered pair $\left(x, -\dfrac{5}{3}\right)$ is a solution. A few solutions are listed below. Plot these points and draw the line.

x	y
-3	$-\dfrac{5}{3}$
0	$-\dfrac{5}{3}$
2	$-\dfrac{5}{3}$

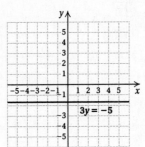

56.

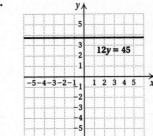

57. $4x + 3 = 0$

$\qquad 4x = -3$

$\qquad x = -\dfrac{3}{4} \qquad$ Solving for x

Any ordered pair $\left(-\dfrac{3}{4}, y\right)$ is a solution. A few solutions are listed below. Plot these points and draw the line.

x	y
$-\dfrac{3}{4}$	-2
$-\dfrac{3}{4}$	0
$-\dfrac{3}{4}$	3

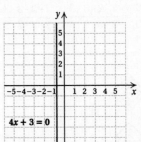

58.

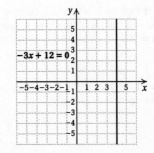

59. $18 - 3y = 0$

$$-3y = -18$$
$$y = 6 \qquad \text{Solving for } y$$

Any ordered pair $(x, 6)$ is a solution. A few solutions are listed below. Plot these points and draw the line.

x	y
-4	6
0	6
2	6

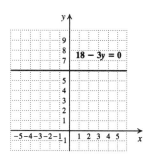

60.

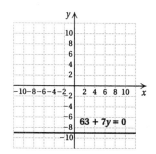

61. We begin by factoring 98 in any way we can:

$$98 = 2 \cdot 49$$

The factor 49 is not prime, so we factor it again:

$$98 = 2 \cdot 49 = 2 \cdot 7 \cdot 7$$

Both 2 and 7 are prime, so the prime factorization of 98 is $2 \cdot 7 \cdot 7$.

62. $2 \cdot 2 \cdot 2 \cdot 2 \cdot 3 \cdot 5$

63. $\dfrac{36}{90} = \dfrac{2 \cdot 2 \cdot 3 \cdot 3}{2 \cdot 3 \cdot 3 \cdot 5} = \dfrac{\cancel{2} \cdot 2 \cdot \cancel{3} \cdot \cancel{3}}{\cancel{2} \cdot \cancel{3} \cdot \cancel{3} \cdot 5} = \dfrac{2}{5}$

64. $\dfrac{1}{7}$

65.

66. ◈

67. The y-axis is a vertical line, so it is of the form $x = a$. All points on the y-axis are of the form $(0, y)$, so a must be 0 and the equation is $x = 0$.

68. $y = 0$

69. The x-coordinate must be -3, and the y-coordinate must be 6. The point of intersection is $(-3, 6)$.

70. $y = -5$

71. A line parallel to the y-axis has an equation of the form $x = a$. Since the line is 13 units to the right of the y-axis, all points on the line are of the form $(13, y)$. Thus, a is 13, and the equation is $x = 13$.

72. $y = 2.8$

73. We substitute 2 for x and 0 for y and solve for m:

$$y = mx + 3$$
$$0 = m(2) + 3$$
$$-3 = 2m$$
$$-\frac{3}{2} = m$$

74. -8

75.

Exercise Set 3.4

1. Let x and y represent the two numbers; then $x + y = 27$.

2. Let x and y represent the two numbers; then $x + y = 53$.

3. Let x and y represent the two numbers. We translate.

$$
\begin{array}{ccccc}
\underline{\text{A number}} & \text{plus} & \overbrace{\substack{\text{twice} \\ \text{another number}}} & \text{is} & 65. \\
\downarrow & \downarrow & \downarrow & \downarrow & \downarrow \\
x & + & 2y & = & 65
\end{array}
$$

4. Let x and y represent the two numbers; then $x + 2y = 93$.

5. Let x and y represent the two numbers. We reword and translate.

$$
\begin{array}{ccc}
\underline{\text{One number}} & \text{is} & \underline{\text{three times another number.}} \\
\downarrow & \downarrow & \downarrow \\
x & = & 3y
\end{array}
$$

6. Let x and y represent the two numbers; then $x = \dfrac{1}{2}y$.

7. Let x and y represent the two numbers. We translate.

$$
\begin{array}{ccc}
\underline{\text{One number}} & \text{is} & \underline{\text{5 more than another.}} \\
\downarrow & \downarrow & \downarrow \\
x & = & y + 5
\end{array}
$$

8. Let x and y represent the two numbers; then $x = y - 7$.

9. Let $h =$ Hank's age and $n =$ Nanette's age. We translate.

$$
\begin{array}{ccccc}
\underline{\text{Hank's age}} & \text{plus} & 7 & \text{is} & \underline{\text{twice Nanette's age.}} \\
\downarrow & \downarrow & \downarrow & & \downarrow \\
h & + & 7 & = & 2n
\end{array}
$$

10. Let $a =$ Lisa's age and $b =$ Lou's age; then $a = 2b - 5$.

11. Let x = Lois' salary and y = Roberta's salary. We reword and translate.

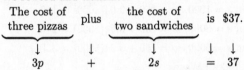

$$x = 3y + 170$$

12. Let x = Evelyn's salary and y = Eric's salary; then $x + 200 = 4y$.

13. Let n = the time of the nonstop flight and d = the time of the direct flight. We reword and translate.

$$n = \frac{1}{2}d + \frac{3}{4}$$

14. Let d = the length of the delay and t = the flight time; then $d = \frac{1}{2}t + 5$.

15. Let p = the cost of a pizza and s = the cost of sandwich. We reword and translate.

$$3p + 2s = 37$$

16. Let m = the cost of an entree and d the cost of a dessert; then $m = 2d$.

17. a) We substitute and calculate.

1 hr: $d = 55 \cdot 1 = 55$ mi
2 hr: $d = 55 \cdot 2 = 110$ mi
5 hr: $d = 55 \cdot 5 = 275$ mi
10 hr: $d = 55 \cdot 10 = 550$ mi

b) We plot the points found in part (a) and any others that we may calculate and draw the line they determine.

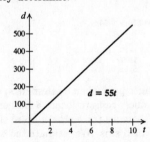

18. a) 1 in.; 2.5 in.; 4 in.; 6 in.

b)

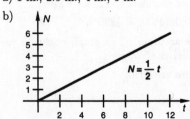

19. a) We substitute and calculate.

Size 4: $y = 4 - 2 = 2$
Size 5: $y = 5 - 2 = 3$
Size 6: $y = 6 - 2 = 4$
Size 7: $y = 7 - 2 = 5$
Size 8: $y = 8 - 2 = 6$

b) We plot the points found in part (a) and any others that we may calculate and draw the line they determine.

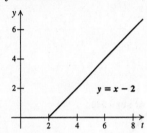

20. a) \$0.46; \$2.89; \$4.24

b)

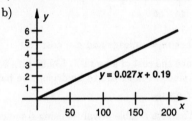

21. a) We substitute and calculate.

$$2\frac{1}{2} + w = 15$$
$$w = 12\frac{1}{2}$$

Sandy is $12\frac{1}{2}$ times more likely to die from lung cancer than Polly.

b) We use the point found in part (a) and others that we calculate to make a table and draw the graph.

When $t = 1$, $1 + w = 15$, or $w = 14$.

When $t = 5$, $5 + w = 15$, or $w = 10$.

t	w
1	14
$2\frac{1}{2}$	$12\frac{1}{2}$
5	10

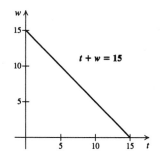

22. a) $33\frac{1}{3}$ lb

b)

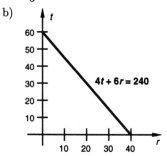

23. Familiarize. Let m = mileage and c = cost.

Translate. Since the cost of a rental is $39.95 plus 55¢ for each mile, and since m miles are to be driven, we have

$$c = 39.95 + 0.55m$$

Carry out. We make a table of values using some convenient choices for m, and then we draw the graph.

When $m = 50$, $c = 39.95 + 0.55(50) = 67.45$.

When $m = 150$, $c = 39.95 + 0.55(150) = 122.45$.

When $m = 300$, $c = 39.95 + 0.55(300) = 204.95$.

Mileage	Cost
50	$67.45
150	$122.45
300	$204.95

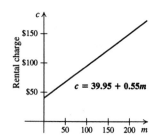

To estimate the cost of renting a 20-ft truck for one day and driving 180 mi, we locate 180 on the horizontal axis. From there we trace a path up to the line and then left to the vertical axis. We estimate the cost of the rental at $140.

Check. We could calculate the exact cost.

$$c = 39.95 + 0.55(180) = \$138.95$$

Our estimate is close enough to serve as a good approximation. The rental firm could use the graph for other quick cost estimates.

State. The cost is about $140.

24.

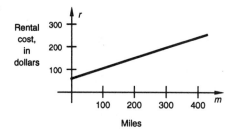

About $210

25. Familiarize. Let s = weekly sales and w = weekly wages.

Translate. Since the wages are $150 plus 4% of sales, and since the amount of sales is s, we have

$$w = 150 + 4\%s, \text{ or } w = 150 + 0.04s$$

Carry out. We make a table of values using some convenient choices for s and then we draw the graph.

When $s = 1000$, $w = 150 + 0.04(1000) = 190$.

When $s = 3000$, $w = 150 + 0.04(3000) = 270$.

When $s = 5000$, $w = 150 + 0.04(5000) = 350$.

Sales	Wages
$1000	$190
$3000	$270
$5000	$350

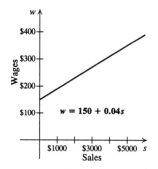

To estimate the wages paid when a salesperson sells $4500 in merchandise in one week, we locate $4500 on the horizontal axis. From there we trace a path up to the line and then left to the vertical axis. We estimate the wages to be $330.

Check. We could calculate the exact wages.

$$w = 150 + 0.04(4500) = \$330$$

Our estimate is accurate. The graph could be used for other quick estimates of wages.

State. The wages are $330.

26.

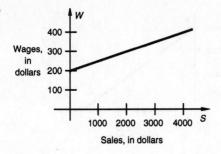

About $375

27. *Familiarize*. Let t = the number 15-min units of time and c = the cost.

***Translate*.** Since the cost is $35 plus $10 for each 15-min unit of time, and since there are t 15-min units of time, we have

$$c = 35 + 10t.$$

***Carry out*.** We make a table of values using some convenient choices for t and then we draw the graph.

When $t = 2$, $c = 35 + 10 \cdot 2 = 55$.

When $t = 5$, $c = 35 + 10 \cdot 5 = 85$.

When $t = 8$, $c = 35 + 10 \cdot 8 = 115$.

15-min Time Units	Cost
2	$55
5	$85
8	$115

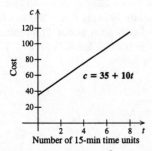

To estimate the cost of a $1\frac{1}{2}$-hr road call, we first determine that there are six 15-min units of time

in $1\frac{1}{2}$ hr $\left(1\frac{1}{2} \text{ hr} \div \frac{1}{4} \text{ hr} = 6\right)$. We locate 6 on the

horizontal axis; next we trace a path up to the line and then left to the vertical axis. We estimate the cost to the $95.

***Check*.** We could calculate the exact cost.

$$c = 35 + 10 \cdot 6 = \$95$$

Our estimate is accurate. The graph could be used for other quick estimates of the cost of a road call.

***State*.** The cost of a $1\frac{1}{2}$-hr road call is $95.

28.

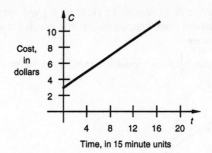

About $10

29. *Familiarize*. Let n = the number of people and p = the number of pounds of cheese needed. Then $n - 10$ = the number of people in excess of 10.

***Translate*.** Since the number of pounds of cheese needed is 3 lb plus $\frac{2}{9}$ lb for each person in excess of 10, and since the number of people in excess of 10 is $n - 10$, we have

$$p = 3 + \frac{2}{9}(n - 10), \text{ or}$$

$$p = 3 + \frac{2}{9}n - \frac{20}{9}, \text{ or}$$

$$p = \frac{2}{9}n + \frac{7}{9}, \ n \geq 10.$$

***Carry out*.** We make a table of values using some convenient choices for n and then we draw the graph.

When $n = 12$, $p = \frac{2}{9} \cdot 12 + \frac{7}{9} = \frac{31}{9}$, or $3\frac{4}{9}$.

When $n = 18$, $p = \frac{2}{9} \cdot 18 + \frac{7}{9} = \frac{43}{9}$, or $4\frac{7}{9}$.

When $n = 24$, $p = \frac{2}{9} \cdot 24 + \frac{7}{9} = \frac{55}{9}$, or $6\frac{1}{9}$.

Number of people	Pounds of Cheese
12	$3\frac{4}{9}$
18	$4\frac{7}{9}$
24	$6\frac{1}{9}$

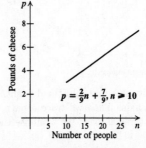

To estimate the amount of cheese needed for a party of 21, we locate 21 on the horizontal axis. Then we trace a path up to the line and left to the vertical axis. We estimate the amount to be $5\frac{1}{2}$ lb.

***Check*.** We could calculate the exact amount.

$$p = \frac{2}{9} \cdot 21 + \frac{7}{9} = \frac{49}{9}, \text{ or } 5\frac{4}{9} \text{ lb.}$$

Our estimate is close enough to serve as a good approximation. The catering firm could use the graph for other quick estimates.

State. About $5\frac{1}{2}$ lb of cheese is needed for a party of 21.

30.

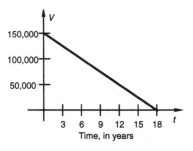

About $80,000

31. ***Familiarize***. Let t = the time of the descent and a = the altitude.

Translate. The altitude is 32,000 ft less 3000 ft for each minute of the descent. We have

$$a = 32,000 - 3000t.$$

Carry out. We make a table of values using some convenient choices for t and then we draw the graph.

When $t = 2$, $a = 32,000 - 3000 \cdot 2 = 26,000$.

When $t = 6$, $a = 32,000 - 3000 \cdot 6 - 14,000$.

When $t = 10$, $a = 32,000 - 3000 \cdot 10 = 2000$.

Time of Descent	Altitude
2	26,000
6	14,000
10	2000

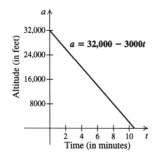

To estimate the altitude 8 min into the descent, we locate 8 on the horizontal axis. Then we trace a path up to the line and left to the vertical axis. We estimate the altitude to be 8000 ft.

Check. We could calculate the exact altitude.

$$a = 32,000 - 3000 \cdot 8 = 8000 \text{ ft}$$

Our estimate is accurate. The graph could be used for other quick estimates of altitude.

State. The altitude 8 min into the descent is 8000 ft.

32.

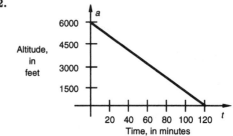

About 5000 ft

33.
$$\begin{aligned} s &= vt + d \\ s - d &= vt \qquad \text{Subtracting } d \\ \frac{s-d}{v} &= t \qquad \text{Dividing by } v \end{aligned}$$

34. 25

35.

36.

37. Let t = flight time and a = altitude. While the plane is climbing at a rate of 6500 ft/min, the equation $a = 6500t$ describes the situation. Solving $34,000 = 6500t$, we find that the cruising altitude of 34,000 ft is reached after about 5.23 min. Thus we graph $a = 6500t$ for $0 \leq t \leq 5.23$.

The plane cruises at 34,000 ft for 3 min, so we graph $a = 34,000$ for $5.23 < t \leq 8.23$. After 8.23 min the plane descends at a rate of 3500 ft/min and lands. The equation $a = 34,000 - 3500(t - 8.23)$, or $a = -3500t + 62,805$, describes this situation. Solving $0 = -3500t + 62,805$, we find that the plane lands after about 17.94 min. Thus we graph $a = -3500t + 62,805$ for $8.23 < t \leq 17.94$. The entire graph is show below.

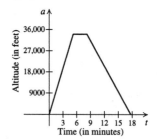

38.

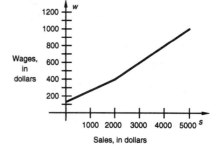

About $550

39. Let s = Paul's weekly salary. We reword and translate.

$$\underbrace{\text{Peggy's salary}}_{\downarrow \atop p} \quad \underset{\downarrow \atop =}{\text{is}} \quad \underbrace{\text{twice Paul's salary}}_{\downarrow \atop 2s} \quad \underset{-}{\text{less \$150.}} \atop 150$$

$$\underbrace{\text{Paul's salary}}_{\downarrow \atop s} \quad \underset{\downarrow \atop =}{\text{is}} \quad \underset{\downarrow \atop 70}{\text{\$70}} \quad \underset{\downarrow \atop +}{\underline{\text{more than}}} \quad \underbrace{\text{half of Jenna's salary.}}_{\downarrow \atop \frac{1}{2}j}$$

We solve the second equation for j.

$$s = 70 + \frac{1}{2}j$$
$$s - 70 = \frac{1}{2}j$$
$$2(s - 70) = j$$
$$2s - 140 = j$$

We make a table of values by choosing values for s and finding the corresponding values of j and p. Then we draw the graph by plotting the points (j, p) and drawing a line through them.

s	j	p
100	60	50
150	160	150
250	360	350

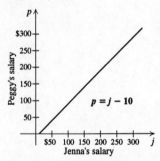

From the graph it appears that Peggy's salary is related to Jenna's by the equation $p = j - 10$. To verify this we can substitute $70 + \frac{1}{2}j$ for s in the first equation and simplify.

$$p = 2s - 150$$
$$p = 2\left(70 + \frac{1}{2}j\right) - 150 \quad \text{Substituting}$$
$$p = 140 + j - 150$$
$$p = j - 10$$

40.

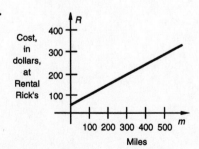

$$c = 38 + 0.475m$$

41. \$280.13

42. \$376.40

Chapter 4

Polynomials

1. $2^4 \cdot 2^3 = 2^{4+3} = 2^7$

2. 3^7

3. $8^5 \cdot 8^9 = 8^{5+9} = 8^{14}$

4. n^{23}

5. $x^4 \cdot x^3 = x^{4+3} = x^7$

6. y^{16}

7. $9^{17} \cdot 9^{21} = 9^{17+21} = 9^{38}$

8. t^{16}

9. $(3y)^4(3y)^8 = (3y)^{4+8} = (3y)^{12}$

10. $(2t)^{25}$

11. $(7y)^1(7y)^{16} = (7y)^{1+16} = (7y)^{17}$

12. $8x$

13. $(a^2b^7)(a^3b^2) = a^2b^7a^3b^2$ Using an associative law
$= a^2a^3b^7b^2$ Using a commutative law
$= a^5b^9$ Adding exponents

14. $m^{11}n^6$

15. $(xy^9)(x^3y^5) = xy^9x^3y^5$
$= xx^3y^9y^5$
$= x^4y^{14}$

16. $a^{12}b^4$

17. $r^3 \cdot r^7 \cdot r^2 = r^{3+7+2} = r^{12}$

18. s^{11}

19. $x^3(xy^4)(xy) = x^3xy^4xy$
$= x^3 \cdot x \cdot x \cdot y^4 \cdot y$
$= x^5y^5$

20. a^8b^2

21. $\dfrac{7^5}{7^2} = 7^{5-2} = 7^3$ Subtracting exponents

22. 4^4

23. $\dfrac{8^{12}}{8^6} = 8^{12-6} = 8^6$ Subtracting exponents

24. 9^{12}

25. $\dfrac{y^9}{y^5} = y^{9-5} = y^4$

26. x

27. $\dfrac{(5a^7)}{(5a^6)} = (5a)^{7-6} = (5a)^1 = 5a$

28. $3m$

29. $\dfrac{6^5x^8}{6^2x^3} = 6^{5-2}x^{8-3} = 6^3x^5$

30. 3^5a^2

31. $\dfrac{18m^5}{6m^2} = \dfrac{18}{6}m^{5-2} = 3m^3$

32. $5n^4$

33. $\dfrac{a^9b^7}{a^2b} = a^{9-2}b^{7-1} = a^7b^6$

34. r^7s^7

35. $\dfrac{m^9n^8}{m^0n^4} = m^{9-0}n^{8-4} = m^9n^4$

36. a^8b^9

37. When $x = -12$, $x^0 = (-12)^0 = 1$. (Any nonzero number raised to the 0 power is 1.)

38. 1

39. When $x = -4$, $5x^0 = 5(-4)^0 = 5 \cdot 1 = 5$.

40. 7

41. When $n \neq 0$, $n^0 = 1$. (Any nonzero number raised to the 0 power is 1.)

42. 1

43. $10^0 = 1$

44. 1

45. $5^1 - 5^0 = 5 - 1 = 4$

46. -7

47. $(x^3)^4 = x^{3 \cdot 4} = x^{12}$ Multiplying exponents

48. a^{24}

49. $(2^3)^8 = 2^{3 \cdot 8} = 2^{24}$ Multiplying exponents

50. 5^{21}

51. $(m^7)^5 = m^{7 \cdot 5} = m^{35}$

52. n^{18}

53. $(a^{25})^3 = a^{25 \cdot 3} = a^{75}$

54. a^{75}

55. $(3x)^2 = 3^2x^2$ Raising each factor to the second power

$= 9x^2$ Simplifying

56. $25a^2$

57. $(-2a)^3 = (-2)^3a^3 = -8a^3$

58. $-27x^3$

59. $(4m^3)^2 = 4^2(m^3)^2 = 16m^6$

60. $25n^8$

61. $(3a^2b)^3 = 3^3(a^2)^3b^3 = 27a^6b^3$

62. $125x^3y^6$

63. $(a^3b^2)^5 = (a^3)^5(b^2)^5 = a^{15}b^{10}$

64. $m^{24}n^{30}$

65. $(-5x^4y^5)^2 = (-5)^2(x^4)^2(y^5)^2 = 25x^8y^{10}$

66. $81a^{20}b^{28}$

67. $\left(\dfrac{a}{4}\right)^3 = \dfrac{a^3}{4^3} = \dfrac{a^3}{64}$ Raising the numerator and the denominator to the third power

68. $\dfrac{81}{x^4}$

69. $\left(\dfrac{7}{5a}\right)^2 = \dfrac{7^2}{(5a)^2} = \dfrac{49}{5^2a^2} = \dfrac{49}{25a^2}$

70. $\dfrac{64x^3}{27}$

71. $\left(\dfrac{a^2}{b^3}\right)^4 = \dfrac{(a^2)^4}{(b^3)^4} = \dfrac{a^8}{b^{12}}$

72. $\dfrac{x^{15}}{y^{20}}$

73. $\left(\dfrac{y^3}{2}\right)^2 = \dfrac{(y^3)^2}{2^2} = \dfrac{y^6}{4}$

74. $\dfrac{a^{15}}{27}$

75. $\left(\dfrac{5x^2}{y^3}\right)^3 = \dfrac{(5x^2)^3}{(y^3)^3} = \dfrac{5^3(x^2)^3}{y^9} = \dfrac{125x^6}{y^9}$

76. $\dfrac{49y^{10}}{x^8}$

77. $\left(\dfrac{a^3}{-2b^5}\right)^4 = \dfrac{(a^3)^4}{(-2b^5)^4} = \dfrac{a^{12}}{(-2)^4(b^5)^4} = \dfrac{a^{12}}{16b^{20}}$

78. $\dfrac{x^{20}}{81y^{12}}$

79. $\left(\dfrac{2a^2}{3b^4}\right)^3 = \dfrac{(2a^2)^3}{(3b^4)^3} = \dfrac{2^3(a^2)^3}{3^3(b^4)^3} = \dfrac{8a^6}{27b^{12}}$

80. $\dfrac{9x^{10}}{16y^6}$

81. $\left(\dfrac{4x^3y^5}{3z^7}\right)^2 = \dfrac{(4x^3y^5)^2}{(3z^7)^2} = \dfrac{4^2(x^3)^2(y^5)^2}{3^2(z^7)^2} = \dfrac{16x^6y^{10}}{9z^{14}}$

82. $\dfrac{125a^{21}}{8b^{15}c^3}$

83. $3s + 3t + 24 = 3s + 3t + 3 \cdot 8 = 3(s + t + 8)$

84. $-7(x + 2)$

85. $9x + 2y - 4x - 2y = (9 - 4)x + (2 - 2)y = 5x$

86. 37.5%

87.

88.

89. $\dfrac{w^{50}}{w^x} = w^x$

$w^{50-x} = w^x$ Using the quotient rule

$50 - x = x$ The exponents are the same.

$50 = 2x$

$25 = x$

90. y^{5x}

91. $a^{5k} \div a^{3k} = a^{5k-3k} = a^{2k}$

92. a^{4t}

93. $\dfrac{\left(\frac{1}{2}\right)^4}{\left(\frac{1}{2}\right)^5} = \dfrac{\left(\frac{1}{2}\right)^4}{\left(\frac{1}{2}\right)^4\left(\frac{1}{2}\right)} = \dfrac{\left(\frac{1}{2}\right)^{4-4}}{\frac{1}{2}} = \dfrac{\left(\frac{1}{2}\right)^0}{\frac{1}{2}} = \dfrac{1}{\frac{1}{2}} =$

$1 \cdot \dfrac{2}{1} = 2$

94. 1

95. Since the bases are the same, the expression with the larger exponent is larger. Thus, $3^5 > 3^4$.

96. $4^2 < 4^3$

97. Since the exponents are the same, the expression with the larger base is larger. Thus, $4^3 < 5^3$.

98. $4^3 < 3^4$

99. Choose any number except 0. For example, let $x = 1$.

$3x^2 = 3 \cdot 1^2 = 3 \cdot 1 = 3$, but

$(3x)^2 = (3 \cdot 1)^2 = 3^2 = 9$.

100. Choose any number except 0.

For example, let $a = 1$. Then $(a+3)^2 = (1+3)^2 = 4^2 = 16$, but $a^2 + 3^2 = 1^2 + 3^2 = 1 + 9 = 10$.

101. Choose any number except 2. For example, let $x = 1$.

Then $\dfrac{x+2}{2} = \dfrac{1+2}{2} = \dfrac{3}{2}$, but $x = 1$.

102. Choose any number except 0 or 1.

For example, let $y = 2$. Then $\dfrac{y^6}{y^3} = \dfrac{2^6}{2^3} = \dfrac{64}{8} = 8$, but $y^2 = 2^2 = 4$.

103. $A = P(1 + r)^t$

We substitute 10,400 for P, 8.5% for r, and 5 for t.

$A = 10,400(1 + 8.5\%)^5$
$A = 10,400(1 + 0.085)^5$
$A = 10,400(1.085)^5$
$A \approx 15,638.03$ Using a calculator to do the computations

There is \$15,638.03 in the account at the end of 5 years.

104. \$27,087.01

105. $A = s^2 = (5x)^2 = 5^2 x^2 = 25x^2$

106. $343a^3$

Exercise Set 4.2

1. Three monomials are added, so $x^2 - 10x + 25$ is a trinomial.

2. Monomial

3. The polynomial $x^3 - 7x^2 + 2x - 4$ is none of these because it is composed of four monomials.

4. Binomial

5. Two monomials are added, so $4x^2 - 25$ is a binomial.

6. None of these

7. The polynomial $40x$ is a monomial because it is the product of a constant and a variable raised to a whole number power.

8. Trinomial

9. $2 - 3x + x^2 = 2 + (-3x) + x^2$

The terms are 2, $-3x$, and x^2.

10. $2x^2$, $3x$, -4

11. $5x^3 + 6x^2 - 3x^2$

Like terms: $6x^2$ and $-3x^2$ Same variable and exponent

12. $3x^2$ and $-2x^2$

13. $2x^4 + 5x - 7x - 3x^4$

Like terms: $2x^4$ and $-3x^4$ Same variable and
Like terms: $5x$ and $-7x$ exponent

14. $-3t$ and $-2t$, t^3 and $-5t^3$

15. $-3x + 6$

The coefficient of $-3x$, the first term, is -3.

The coefficient of 6, the second term, is 6.

16. 2, -4

17. $5x^2 + 3x + 3$

The coefficient of $5x^2$, the first term, is 5.

The coefficient of $3x$, the second term, is 3.

The coefficient of 3, the third term, is 3.

18. 3, -5, 2

19. $-7x^3 + 6x^2 + 3x + 7$

The coefficient of $-7x^3$, the first term, is -7.

The coefficient of $6x^2$, the second term, is 6.

The coefficient of $3x$, the third term, is 3.

The coefficient of 7, the fourth term, is 7.

20. 5, 1, -1, 2

21. $-5x^4 + 6x^3 - 3x^2 + 8x - 2$

The coefficient of $-5x^4$, the first term, is -5.

The coefficient of $6x^3$, the second term, is 6.

The coefficient of $-3x^2$, the third term, is -3.

The coefficient of $8x$, the fourth term, is 8.

The coefficient of -2, the fifth term, is -2.

22. 7, -4, -4, 5

23. $2x - 4 = 2x^1 - 4x^0$

The degree of $2x$ is 1.

The degree of -4 is 0.

The degree of the polynomial is 1, the largest exponent.

24. 0, 1; 1

25. $3x^2 - 5x + 2 = 3x^2 - 5x^1 + 2x^0$

The degree of $3x^2$ is 2.

The degree of $-5x$ is 1.

The degree of 2 is 0.

The degree of the polynomial is 2, the largest exponent.

26. 3, 2, 0; 3

27. $-7x^3 + 6x^2 + 3x + 7 = -7x^3 + 6x^2 + 3x^1 + 7x^0$

The degree of $-7x^3$ is 3.

The degree of $6x^2$ is 2.

The degree of $3x$ is 1.

The degree of 7 is 0.

The degree of the polynomial is 3, the largest exponent.

28. 4, 2, 1, 0; 4

29. $x^2 - 3x + x^6 - 9x^4 = x^2 - 3x^1 + x^6 - 9x^4$

The degree of x^2 is 2.

The degree of $-3x$ is 1.

The degree of x^6 is 6.

The degree of $-9x^4$ is 4.

The degree of the polynomial is 6, the largest exponent.

30. 1, 2, 0, 3; 3

31. See the answer section in the text.

32. $3x^2 + 8x^5 - 46x^3 + 6x - 2.4 - \frac{1}{2}x^4$

Term	Coefficient	Degree of Term	Degree of Polynomial
$8x^5$	8	5	
$-\frac{1}{2}x^4$	$-\frac{1}{2}$	4	
$-46x^3$	-46	3	5
$3x^2$	3	2	
$6x$	6	1	
-2.4	-2.4	0	

33. $2x - 5x = (2 - 5)x = -3x$

34. $10x^2$

35. $x - 9x = 1x - 9x = (1 - 9)x = -8x$

36. $-4x$

37. $5x^3 + 6x^3 + 4 = (5 + 6)x^3 + 4 = 11x^3 + 4$

38. $4x^4 + 5$

39. $5x^3 + 6x - 4x^3 - 7x = (5 - 4)x^3 + (6 - 7)x =$
$1x^3 + (-1)x = x^3 - x$

40. $4a^4$

41. $6b^5 + 3b^2 - 2b^5 - 3b^2 = (6 - 2)b^5 + (3 - 3)b^2 =$
$4b^5 + 0b^2 = 4b^5$

42. $6x^2 - 3x$

43. $\frac{1}{4}x^5 - 5 + \frac{1}{2}x^5 - 2x - 37 =$
$\left(\frac{1}{4} + \frac{1}{2}\right)x^5 - 2x + (-5 - 37) = \frac{3}{4}x^5 - 2x - 42$

44. $\frac{1}{6}x^3 + 2x - 12$

45. $6x^2 + 2x^4 - 2x^2 - x^4 - 4x^2 =$
$6x^2 + 2x^4 - 2x^2 - 1x^4 - 4x^2 =$
$(6 - 2 - 4)x^2 + (2 - 1)x^4 = 0x^2 + 1x^4 =$
$0 + x^4 = x^4$

46. $-x^3$

47. $\frac{1}{4}x^3 - x^2 - \frac{1}{6}x^2 + \frac{3}{8}x^3 + \frac{5}{16}x^3 =$
$\frac{1}{4}x^3 - 1x^2 - \frac{1}{6}x^2 + \frac{3}{8}x^3 + \frac{5}{16}x^3 =$
$\left(\frac{1}{4} + \frac{3}{8} + \frac{5}{16}\right)x^3 + \left(-1 - \frac{1}{6}\right)x^2 =$
$\left(\frac{4}{16} + \frac{6}{16} + \frac{5}{16}\right)x^3 + \left(-\frac{6}{6} - \frac{1}{6}\right)x^2 = \frac{15}{16}x^3 - \frac{7}{6}x^2$

48. 0

49. $3x^4 - 5x^6 - 2x^4 + 6x^6 = x^4 + x^6 = x^6 + x^4$

50. $x^4 - 2x^3 + 1$

51. $-2x + 4x^3 - 7x + 9x^3 + 8 = -9x + 13x^3 + 8 =$
$13x^3 - 9x + 8$

52. $x^2 - 4x + 1$

53. $3x + 3x + 3x - x^2 - 4x^2 = 9x - 5x^2 = -5x^2 + 9x$

54. $-4x^3 - 6x$

55. $-x + \frac{3}{4} + 15x^4 - x - \frac{1}{2} - 3x^4 = -2x + \frac{1}{4} + 12x^4 =$
$12x^4 - 2x + \frac{1}{4}$

56. $4x^3 + x - \frac{1}{2}$

57. $-5x + 2 = -5 \cdot 4 + 2 = -20 + 2 = -18$

58. -11

59. $2x^2 - 5x + 7 = 2 \cdot 4^2 - 5 \cdot 4 + 7 = 2 \cdot 16 - 20 + 7 =$
$32 - 20 + 7 = 19$

60. 59

61. $x^3 - 5x^2 + x = 4^3 - 5 \cdot 4^2 + 4 = 64 - 5 \cdot 16 + 4 =$
$64 - 80 + 4 = -12$

62. 51

63. $3x + 5 = 3(-1) + 5 = -3 + 5 = 2$

64. 8

65. $x^2 - 2x + 1 = (-1)^2 - 2(-1) + 1 = 1 + 2 + 1 = 4$

66. -10

67. $-3x^3 + 7x^2 - 3x - 2 =$
$-3(-1)^3 + 7(-1)^2 - 3(-1) - 2 =$
$-3(-1) + 7 \cdot 1 + 3 - 2 = 3 + 7 + 3 - 2 = 11$

68. -4

69. $0.4r^2 - 40r + 1039 = 0.4(18)^2 - 40(18) + 1039 =$
$0.4(324) - 720 + 1039 = 129.6 - 720 + 1039 =$
448.6

There are approximately 449 accidents daily involving an 18-year-old driver.

70. 399

71. $11.12t^2 = 11.12(10)^2 = 11.12(100) = 1112$

A skydiver has fallen approximately 1112 ft 10 seconds after jumping from a plane.

72. 3091 ft

73. Evaluate the polynomial for $x = 75$:
$280x - 0.4x^2 = 280 \cdot 75 - 0.4(75)^2 =$
$21,000 - 0.4(5625) = 21,000 - 2250 = 18,750$
The total revenue is \$18,750.

74. $24,000

75. Evaluate the polynomial for $x = 500$:

$5000 + 0.6(500)^2 = 5000 + 0.6(250,000) =$

$5000 + 150,000 = 155,000$

The total cost is $155,000.

76. $258,500

77. $2\pi r = 2(3.14)(10)$ Substituting 3.14 for π
and 10 for r

$\qquad = 62.8$

The circumference is 62.8 cm.

78. 31.4 ft

79. $\pi r^2 = 3.14(5)^2$ Substituting 3.14 for π
and 5 for r

$\qquad = 3.14(25)$

$\qquad = 78.5$

The area is 78.5 m^2.

80. 314 in^2

81. When $t = 1$, $-t^2 + 6t - 4 = -1^2 + 6 \cdot 1 - 4 =$
$-1 + 6 - 4 = 1$.
When $t = 2$, $-t^2 + 6t - 4 = -2^2 + 6 \cdot 2 - 4 =$
$-4 + 12 - 4 = 4$.
When $t = 3$, $-t^2 + 6t - 4 = -3^2 + 6 \cdot 3 - 4 =$
$-9 + 18 - 4 = 5$.
When $t = 4$, $-t^2 + 6t - 4 = -4^2 + 6 \cdot 4 - 4 =$
$-16 + 24 - 4 = 4$.
When $t = 5$, $-t^2 + 6t - 4 = -5^2 + 6 \cdot 5 - 4 =$
$-25 + 30 - 4 = 1$.

We complete the table. Then we plot the points and connect them with a smooth curve.

t	$-t^2 + 6t - 4$
1	1
2	4
3	5
4	4
5	1

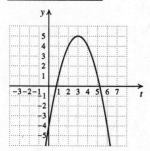

82.

t	$-t^2 + 10t - 18$
3	3
4	6
5	7
6	6
7	3

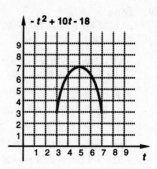

83. *Familiarize*. The page numbers on facing pages are consecutive integers. If $x =$ the smaller number, then $x + 1 =$ the larger number.

Translate. We reword and translate.

First integer plus second integer is 549.

$\qquad x \qquad + \qquad (x + 1) \qquad = \qquad 549$

Carry out. We solve the equation.

$x + (x + 1) = 549$

$2x + 1 = 549$

$2x = 548$

$x = 274$

If x is 274, then $x + 1$ is $274 + 1$, or 275.

Check. 274 and 275 are consecutive integers, and their sum is 549. The numbers check.

State. The page numbers are 274 and 275.

84. $0.125, or 12.5¢

85. ⬩

86. ⬩

87.

$\dfrac{9}{2}x^8 + \dfrac{1}{9}x^2 + \dfrac{1}{2}x^9 + \dfrac{9}{2}x + \dfrac{9}{2}x^9 + \dfrac{8}{9}x^2 +$

$\qquad \dfrac{1}{2}x - \dfrac{1}{2}x^8$

$= \left(\dfrac{1}{2} + \dfrac{9}{2}\right)x^9 + \left(\dfrac{9}{2} - \dfrac{1}{2}\right)x^8 + \left(\dfrac{1}{9} + \dfrac{8}{9}\right)x^2 +$

$\qquad \left(\dfrac{9}{2} + \dfrac{1}{2}\right)x$

$= \dfrac{10}{2}x^9 + \dfrac{8}{2}x^8 + \dfrac{9}{9}x^2 + \dfrac{10}{2}x$

$= 5x^9 + 4x^8 + x^2 + 5x$

88. $3x^6$

89. For $s = 18$:
$$s^2 - 50s + 675 = 18^2 - 50(18) + 675 =$$
$$324 - 900 + 675 = 99$$

$$-s^2 + 50s - 675 = -(18)^2 + 50(18) - 675 =$$
$$-324 + 900 - 675 = -99$$

For $s = 25$:
$$s^2 - 50s + 675 = 25^2 - 50(25) + 675 =$$
$$625 - 1250 + 675 = 50$$

$$-s^2 + 50s - 675 = -(25)^2 + 50(25) - 675 =$$
$$-625 + 1250 - 675 = -50$$

For $s = 32$:
$$s^2 - 50s + 675 = 32^2 - 50(32) + 675 =$$
$$1024 - 1600 + 675 = 99$$

$$-s^2 + 50s - 675 = -(32)^2 + 50(32) - 675 =$$
$$-1024 + 1600 - 675 = -99$$

90. 50

91. Answers may vary. Use an ax^5-term, where a is an integer, and 3 other terms with different degrees each less than degree 5, and integer coefficients. Three answers are $-6x^5 + 14x^4 - x^2 + 11$, $x^5 - 8x^3 + 3x + 1$, and $23x^5 + 2x^4 - x^2 + 5x$.

92. Answers may vary.
$$0.2y^4 - y + \frac{5}{2}, \; -\frac{8}{7}y^4 + 5.5y^3 - 2y^2, \; 2.9y^4 - 4y^2 - \frac{11}{3}$$

93. $(5m^5)^2 = 5^2 m^{5 \cdot 2} = 25m^{10}$

The degree is 10.

94. Answers may vary.
$$9y^4, \; -\frac{3}{2}y^4, \; 4.2y^4$$

95. When $d = 0$, $-0.0064d^2 + 0.8d + 2 =$
$$-0.0064(0)^2 + 0.8(0) + 2 = 0 + 0 + 2 = 2.$$
When $d = 30$, $-0.0064(30)^2 + 0.8(30) + 2 =$
$$-5.76 + 24 + 2 = 20.24.$$
When $d = 60$, $-0.0064(60)^2 + 0.8(60) + 2 =$
$$-23.04 + 48 + 2 = 26.96.$$
When $d = 90$, $-0.0064(90)^2 + 0.8(90) + 2 =$
$$-51.84 + 72 + 2 = 22.16.$$
When $d = 120$, $-0.0064(120)^2 + 0.8(120) + 2 =$
$$-92.16 + 96 + 2 = 5.84.$$
We complete the table. Then we plot the points and connect them with a smooth curve.

d	$-0.0064d^2 + 0.8d + 2$
0	2
30	20.24
60	26.96
90	22.16
120	5.84

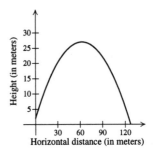

96. $x^3 - 2x^2 - 6x + 3$

Exercise Set 4.3

---

1. $(3x + 2) + (-4x + 3) = (3 - 4)x + (2 + 3) = -x + 5$

2. $-x + 3$

3. $(-6x + 2) + (x^2 + x - 3) =$
$$x^2 + (-6 + 1)x + (2 - 3) = x^2 - 5x - 1$$

4. $x^2 + 3x - 5$

5. $(x^2 - 9) + (x^2 + 9) = (1 + 1)x^2 + (-9 + 9) = 2x^2$

6. $3x^3 - 4x^2$

7. $(3x^2 - 5x + 10) + (2x^2 + 8x - 40) =$
$$(3 + 2)x^2 + (-5 + 8)x + (10 - 40) = 5x^2 + 3x - 30$$

8. $6x^4 + 3x^3 + 4x^2 - 3x + 2$

9. $(1.2x^3 + 4.5x^2 - 3.8x) + (-3.4x^3 - 4.7x^2 + 23) =$
$$(1.2 - 3.4)x^3 + (4.5 - 4.7)x^2 - 3.8x + 23 =$$
$$-2.2x^3 - 0.2x^2 - 3.8x + 23$$

10. $2.8x^4 - 0.6x^2 + 1.8x - 3.2$

11. $(1 + 4x + 6x^2 + 7x^3) + (5 - 4x + 6x^2 - 7x^3) =$
$$(1 + 5) + (4 - 4)x + (6 + 6)x^2 + (7 - 7)x^3 =$$
$$6 + 0x + 12x^2 + 0x^3 = 6 + 12x^2, \text{ or } 12x^2 + 6$$

12. $3x^4 - 4x^3 + x^2 + x + 4$

13. $(9x^8 - 7x^4 + 2x^2 + 5) + (8x^7 + 4x^4 - 2x) =$
$$9x^8 + 8x^7 + (-7 + 4)x^4 + 2x^2 - 2x + 5 =$$
$$9x^8 + 8x^7 - 3x^4 + 2x^2 - 2x + 5$$

14. $4x^5 + 9x^2 + 1$

15. $\left(\frac{1}{4}x^4 + \frac{2}{3}x^3 + \frac{5}{8}x^2 + 7\right) + \left(-\frac{3}{4}x^4 + \frac{3}{8}x^2 - 7\right) =$
$$\left(\frac{1}{4} - \frac{3}{4}\right)x^4 + \frac{2}{3}x^3 + \left(\frac{5}{8} + \frac{3}{8}\right)x^2 + (7 - 7) =$$
$$-\frac{2}{4}x^4 + \frac{2}{3}x^3 + \frac{8}{8}x^2 + 0 =$$
$$-\frac{1}{2}x^4 + \frac{2}{3}x^3 + x^2$$

16. $\frac{2}{15}x^9 - \frac{2}{5}x^5 + \frac{1}{4}x^4 + \frac{1}{4}x^2 + \frac{15}{2}$

17. $(0.02x^5 - 0.2x^3 + x + 0.08) + (-0.01x^5 + x^4 - 0.8x - 0.02) =$
$(0.02 - 0.01)x^5 + x^4 - 0.2x^3 + (1 - 0.8)x + (0.08 - 0.02) =$
$0.01x^5 + x^4 - 0.2x^3 + 0.2x + 0.06$

18. $0.10x^6 + 0.02x^3 + 0.22x + 0.55$

19. $\begin{aligned} -3x^4 + 6x^2 + 2x - 1 \\ -3x^2 + 2x + 1 \\ \hline -3x^4 + 3x^2 + 4x + 0 \\ -3x^4 + 3x^2 + 4x \end{aligned}$

20. $-4x^3 + 4x^2 + 6x$

21. Rewrite the problem so the coefficients of like terms have the same number of decimal places.

$$\begin{aligned}
0.15x^4 + 0.10x^3 - 0.90x^2 \qquad\qquad\quad \\
- 0.01x^3 + 0.01x^2 + x \qquad\quad \\
1.25x^4 \qquad\quad + 0.11x^2 \qquad + 0.01 \\
0.27x^3 \qquad\qquad\qquad + 0.99 \\
-0.35x^4 \qquad\qquad + 15.00x^2 \qquad - 0.03 \\
\hline
1.05x^4 + 0.36x^3 + 14.22x^2 + x + 0.97
\end{aligned}$$

22. $1.3x^4 + 0.35x^3 + 9.53x^2 + 2x + 0.96$

23. Two equivalent expressions for the opposite of $-5x$ are
a) $-(-5x)$ and
b) $5x$. (Changing the sign)

24. $-(x^2 - 3x)$, $-x^2 + 3x$

25. Two equivalent expressions for the opposite of $-x^2 + 10x - 2$ are
a) $-(-x^2 + 10x - 2)$ and
b) $x^2 - 10x + 2$. (Changing the sign of every term)

26. $-(-4x^3 - x^2 - x)$, $4x^3 + x^2 + x$

27. Two equivalent expressions for the opposite of $12x^4 - 3x^3 + 3$ are
a) $-(12x^4 - 3x^3 + 3)$ and
b) $-12x^4 + 3x^3 - 3$. (Changing the sign of every term)

28. $-(4x^3 - 6x^2 - 8x + 1)$, $-4x^3 + 6x^2 + 8x - 1$

29. We change the sign of every term inside parentheses.
$-(3x - 7) = -3x + 7$

30. $2x - 4$

31. We change the sign of every term inside parentheses.
$-(4x^2 - 3x + 2) = -4x^2 + 3x - 2$

32. $6a^3 - 2a^2 + 9a - 1$

33. We change the sign of every term inside parentheses.
$-\left(-4x^4 + 6x^2 + \dfrac{3}{4}x - 8\right) = 4x^4 - 6x^2 - \dfrac{3}{4}x + 8$

34. $5x^4 - 4x^3 + x^2 - 0.9$

35. $\quad (3x + 2) - (-4x + 3)$
$= 3x + 2 + 4x - 3 \quad$ Changing the sign of every term inside parentheses
$= 7x - 1$

36. $13x - 1$

37. $(-6x + 2) - (x^2 + x - 3) = -6x + 2 - x^2 - x + 3$
$\qquad\qquad\qquad\qquad\qquad = -x^2 - 7x + 5$

38. $x^2 - 13x + 13$

39. $(x^2 - 9) - (x^2 + 9) = x^2 - 9 - x^2 - 9 = -18$

40. $-x^3 + 6x^2$

41. $\quad (6x^4 + 3x^3 - 1) - (4x^2 - 3x + 3)$
$= 6x^4 + 3x^3 - 1 - 4x^2 + 3x - 3$
$= 6x^4 + 3x^3 - 4x^2 + 3x - 4$

42. $-3x^3 + x^2 + 2x - 3$

43. $\quad (1.2x^3 + 4.5x^2 - 3.8x) - (-3.4x^3 - 4.7x^2 + 23)$
$= 1.2x^3 + 4.5x^2 - 3.8x + 3.4x^3 + 4.7x^2 - 23$
$= 4.6x^3 + 9.2x^2 - 3.8x - 23$

44. $-1.8x^4 - 0.6x^2 - 1.8x + 4.6$

45. $\quad (5x^2 + 6) - (3x^2 - 8) = 5x^2 + 6 - 3x^2 + 8$
$\qquad\qquad\qquad\qquad\quad = 2x^2 + 14$

46. $7x^3 - 9x^2 - 2x + 10$

47. $\quad (6x^5 - 3x^4 + x + 1) - (8x^5 + 3x^4 - 1)$
$= 6x^5 - 3x^4 + x + 1 - 8x^5 - 3x^4 + 1$
$= -2x^5 - 6x^4 + x + 2$

48. $-x^2 - 2x + 4$

49. $\quad (6x^2 + 2x) - (-3x^2 - 7x + 8)$
$= 6x^2 + 2x + 3x^2 + 7x - 8$
$= 9x^2 + 9x - 8$

50. $7x^3 + 3x^2 + 2x - 1$

51. $\quad \dfrac{5}{8}x^3 - \dfrac{1}{4}x - \dfrac{1}{3} - \left(-\dfrac{1}{8}x^3 + \dfrac{1}{4}x - \dfrac{1}{3}\right)$
$= \dfrac{5}{8}x^3 - \dfrac{1}{4}x - \dfrac{1}{3} + \dfrac{1}{8}x^3 - \dfrac{1}{4}x + \dfrac{1}{3}$
$= \dfrac{6}{8}x^3 - \dfrac{2}{4}x$
$= \dfrac{3}{4}x^3 - \dfrac{1}{2}x$

52. $\dfrac{3}{5}x^3 - 0.11$

53. $\quad (0.08x^3 - 0.02x^2 + 0.01x) - (0.02x^3 + 0.03x^2 - 1)$
$= 0.08x^3 - 0.02x^2 + 0.01x - 0.02x^3 - 0.03x^2 + 1$
$= 0.06x^3 - 0.05x^2 + 0.01x + 1$

54. $0.1x^4 - 0.9$

55. $x^2 + 5x + 6$
           $\underline{-(x^2 + 2x \quad\quad)}$

$\begin{array}{ll} x^2 \;+\; 5x \;+\; 6 & \text{Changing signs and} \\ \underline{-x^2 \;-\; 2x} & \text{removing parentheses} \\ \quad\quad 3x \;+\; 6 & \text{Adding} \end{array}$

56. $-x^2 + 1$

57. $5x^4 + 6x^3 - 9x^2$
           $\underline{-(-6x^4 - 6x^3 \quad\quad + 8x + 9)}$

$\begin{array}{ll} 5x^4 + 6x^3 \;-9x^2 & \text{Changing signs and} \\ \underline{6x^4 + 6x^3 \quad\quad -8x-9} & \text{removing parentheses} \\ 11x^4 + 12x^3 - 9x^2 - 8x - 9 & \text{Adding} \end{array}$

58. $5x^4 - 6x^3 - x^2 + 5x + 15$

59. $3x^4 + 6x^2 + 8x - 1$
           $\underline{-(4x^5 - 6x^4 \quad\quad - 8x - 7)}$

$\begin{array}{ll} \quad\quad 3x^4 + 6x^2 + 8x \;-1 & \text{Changing signs and} \\ \underline{-4x^5 + 6x^4 \quad\quad + 8x \;+7} & \text{removing parentheses} \\ -4x^5 + 9x^4 + 6x^2 + 16x + 6 & \text{Adding} \end{array}$

60. $-4x^5 - 6x^3 + 8x^2 - 5x - 2$

61. $x^5 \quad\quad\quad\quad - 1$
           $\underline{-(x^5 - x^4 + x^3 - x^2 + x - 1)}$

$\begin{array}{ll} x^5 \quad\quad\quad\quad -1 & \text{Changing signs and} \\ \underline{-x^5 + x^4 - x^3 + x^2 - x + 1} & \text{removing parentheses} \\ \quad\quad x^4 - x^3 + x^2 - x & \text{Adding} \end{array}$

62. $2x^4 - 2x^3 + 2x^2$

63. a)

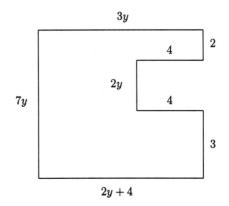

Familiarize. The area of a rectangle is the product of the length and the width.

Translate. The sum of the areas is found as follows:

$\begin{array}{ccccccc} \text{Area} & & \text{Area} & & \text{Area} & & \text{Area} \\ \text{of } A & + & \text{of } B & + & \text{of } C & + & \text{of } D \\ = 3x \cdot x & + & x \cdot x & + & x \cdot x & + & 4 \cdot x \end{array}$

Carry out. We collect like terms.

$3x^2 + x^2 + x^2 + 4x = 5x^2 + 4x$

Check. We can go over our calculations. We can also assign some value to x, say 2, and carry out the computation of the area in two ways.

Sum of areas: $3 \cdot 2 \cdot 2 + 2 \cdot 2 + 2 \cdot 2 + 4 \cdot 2 =$
$12 + 4 + 4 + 8 = 28$

Substituting in the polynomial:
$5(2)^2 + 4 \cdot 2 = 20 + 8 = 28$

Since the results are the same, our solution is probably correct.

State. A polynomial for the sum of the areas is $5x^2 + 4x$.

b)   For $x = 3$: $5x^2 + 4x = 5 \cdot 3^2 + 4 \cdot 3 =$
           $5 \cdot 9 + 4 \cdot 3 = 45 + 12 = 57$

When $x = 3$, the sum of the areas is 57 square units.
       For $x = 8$: $5x^2 + 4x = 5 \cdot 8^2 + 4 \cdot 8 =$
           $5 \cdot 64 + 4 \cdot 8 = 320 + 32 = 352$

When $x = 8$, the sum of the areas is 352 square units.

64. a) $r^2\pi + 13\pi$

  b) 38π; 140.69π

65.

```
                3y
   ┌──────────────────────┐
   │                      │  2
   │              ┌───────┘
   │          4   │
   │  2y  ┌───────┘
   │      │   4
7y │      └───────┐
   │              │  3
   │              │
   └──────────────┘
         2y + 4
```

Familiarize. The perimeter is the sum of the lengths of the sides.

Translate. The sum of the lengths is found as follows:

$3y + 7y + (2y + 4) + 3 + 4 + 2y + 4 + 2$

Carry out. We collect like terms. $(3 + 7 + 2 + 2)y + (4 + 3 + 4 + 4 + 2) = 14y + 17$

Check. We can go over our calculations. We can also assign some value to y, say 3, and carry out the computation of the perimeter in two ways.

Sum of lengths: $3 \cdot 3 + 7 \cdot 3 + (2 \cdot 3 + 4) + 3 + 4 + 2 \cdot 3 + 4 + 2 =$
$9 + 21 + 10 + 3 + 4 + 6 + 4 + 2 = 59$

Substituting in the polynomial:
$14 \cdot 3 + 17 = 42 + 17 = 59$

Since the results are the same, our solution is probably correct.

State. A polynomial for the perimeter of the figure is $14y + 17$.

66. $11\frac{1}{2}a + 10$

67.

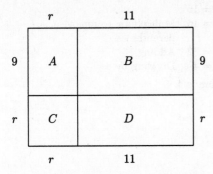

The area of the figure can be found by adding the areas of the four rectangles A, B, C, and D. The area of a rectangle is the product of the length and the width.

$$\begin{array}{cccccccc} & \text{Area} & & \text{Area} & & \text{Area} & & \text{Area} \\ & \text{of } A & + & \text{of } B & + & \text{of } C & + & \text{of } D \\ = & 9 \cdot r & + & 11 \cdot 9 & + & r \cdot r & + & 11 \cdot r \\ = & 9r & + & 99 & + & r^2 & + & 11r \end{array}$$

An algebraic expression for the area of the figure is $9r + 99 + r^2 + 11r$.

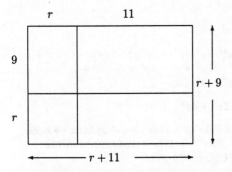

The length and width of the figure can be expressed as $r + 11$ and $r + 9$, respectively. The area of this figure (a rectangle) is the product of the length and width. An algebraic expression for the area is $(r + 11) \cdot (r + 9)$.

The algebraic expressions $9r + 99 + r^2 + 11r$ and $(r + 11) \cdot (r + 9)$ represent the same area.

68. $(x + 3)^2$; $x^2 + 3x + 3x + 9$

69.

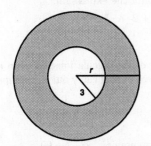

Familiarize. Recall that the area of a circle is the product of π and the square of the radius, r^2: $A = \pi r^2$.

Translate.

$$\begin{array}{ccc} \text{Area of circle} & \text{Area of circle} & \text{Shaded} \\ \text{with radius } r & - \quad \text{with radius 3} & = \quad \text{area} \\ \pi \cdot r^2 & - \quad \pi \cdot 3^2 & = \text{Shaded area} \end{array}$$

Carry out. We simplify the expression.

$$\pi \cdot r^2 - \pi \cdot 3^2 = \pi r^2 - 9\pi$$

Check. We can go over our calculations. We can also assign some value to r, say 5, and carry out the computation in two ways.

Difference of areas: $\pi \cdot 5^2 - \pi \cdot 3^2 = 25\pi - 9\pi = 16\pi$

Substituting in the polynomial: $\pi \cdot 5^2 - 9\pi = 25\pi - 9\pi = 16\pi$

Since the results are the same, our solution is probably correct.

State. A polynomial for the shaded area is $\pi r^2 - 9\pi$.

70. $m^2 - 28$

71.

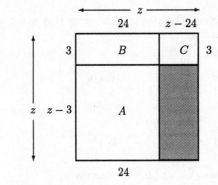

Familiarize. We label the sides of A, B, and C with additional information. The area of the square is $z \cdot z$, or z^2. The area of the shaded section is z^2 minus the areas of sections A, B, and C.

Translate.

$$\begin{array}{cccccccc} \text{Area of} & & \text{Area} & & \text{Area} & & \text{Area} & & \text{Area} \\ \text{shaded} & = & \text{of} & - & \text{of} & - & \text{of} & - & \text{of} \\ \text{section} & & \text{square} & & A & & B & & C \end{array}$$

$$\begin{array}{l} \text{Area of} \\ \text{shaded} \ = z \cdot z - 24(z - 3) - 3 \cdot 24 - 3(z - 24) \\ \text{section} \end{array}$$

Carry out. We simplify the expression.

$$z^2 - 24z + 72 - 72 - 3z + 72 = z^2 - 27z + 72$$

Check. We can go over our calculations. We can also assign some value to z, say 30, and carry out the computation in two ways.

Difference of areas: $30 \cdot 30 - 24 \cdot 27 - 3 \cdot 24 - 3 \cdot 6 = 900 - 648 - 72 - 18 = 162$

Substituting in the polynomial:

$$30^2 - 27 \cdot 30 + 72 = 900 - 810 + 72 = 162$$

Since the results are the same, our solution is probably correct.

State. A polynomial for the shaded area is $z^2 - 27z + 72$.

72. $\pi x^2 - 2x^2$, or $(\pi - 2)x^2$

73.

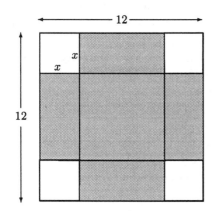

Familiarize. The area of each corner that is not shaded is $x \cdot x$, or x^2.

Translate.

Area of shaded section	=	Area of square	−	Area of four corners
Area of shaded section	=	12^2	−	$4 \cdot x^2$

Carry out. We simplify the expression.

$$12 \cdot 12 - 4 \cdot x^2 = 144 - 4x^2$$

Check. We can go over the calculations. We can also assign some value to x, say 2, and carry out the computation in two ways.

Difference of areas: $12^2 - 4 \cdot 2^2 = 144 - 16 = 128$

Substituting in the polynomial: $144 - 4 \cdot 2^2 = 144 - 16 = 128$

Since the results are the same, our solution is probably correct.

State. A polynomial for the shaded area is $144 - 4x^2$.

74. $y^2 - 4y + 4$

75.
$$1.5x - 2.7x = 23 - 5.6x$$
$$10(1.5x - 2.7x) = 10(23 - 5.6x) \quad \text{Clearing decimals}$$
$$15x - 27x = 230 - 56x$$
$$-12x = 230 - 56x \quad \text{Collecting like terms}$$
$$44x = 230 \quad \text{Adding } 56x$$
$$x = \frac{230}{44} \quad \text{Dividing by 44}$$
$$x = \frac{115}{22} \quad \text{Simplifying}$$

The solution is $\frac{115}{22}$.

76. 1

77.
$$8(x - 2) = 16$$
$$8x - 16 = 16 \quad \text{Multiplying to remove parentheses}$$
$$8x = 32 \quad \text{Adding 16}$$
$$x = 4 \quad \text{Dividing by 8}$$

The solution is 4.

78. $-\dfrac{76}{3}$

79.

80.

81.
$$(4a^2 - 3a) + (7a^2 - 9a - 13) - (6a - 9)$$
$$= 4a^2 - 3a + 7a^2 - 9a - 13 - 6a + 9$$
$$= 11a^2 - 18a - 4$$

82. $5x^2 - 9x - 1$

83.
$$(-8y^2 - 4) - (3y + 6) - (2y^2 - y)$$
$$= -8y^2 - 4 - 3y - 6 - 2y^2 + y$$
$$= -10y^2 - 2y - 10$$

84. $4x^3 - 5x^2 + 6$

85.
$$(-y^4 - 7y^3 + y^2) + (-2y^4 + 5y - 2) - (-6y^3 + y^2)$$
$$= -y^4 - 7y^3 + y^2 - 2y^4 + 5y - 2 + 6y^3 - y^2$$
$$= -3y^4 - y^3 + 5y - 2$$

86. $2 + x + 2x^2 + 4x^3$

87.
$$(345.099x^3 - 6.178x) - (-224.508x^3 + 8.99x)$$
$$= 345.099x^3 - 6.178x + 224.508x^3 - 8.99x$$
$$= 569.607x^3 - 15.168x$$

88. $28x + 2x^2$

89. *Familiarize.* The surface area is $2lw + 2lh + 2wh$, where $l = $ length, $w = $ width, and $h = $ height of the rectangular solid. Here we have $l = 9$, $w = a$, and $h = 5$.

Translate. We substitute in the formula above.

$$2 \cdot 9 \cdot a + 2 \cdot 9 \cdot 5 + 2 \cdot a \cdot 5$$

Carry out. We simplify the expression.

$$2 \cdot 9 \cdot a + 2 \cdot 9 \cdot 5 + 2 \cdot a \cdot 5$$
$$= 18a + 90 + 10a$$
$$= 28a + 90$$

Check. We can go over the calculations. We can also assign some value to a, say 7, and carry out the computation in two ways.

Using the formula: $2 \cdot 9 \cdot 7 + 2 \cdot 9 \cdot 5 + 2 \cdot 7 \cdot 5 = 126 + 90 + 70 = 286$

Substituting in the polynomial: $28 \cdot 7 + 90 = 196 + 90 = 286$

Since the results are the same, our solution is probably correct.

State. A polynomial for the surface area is $28a + 90$.

90.

91. a) $R - C = 280x - 0.4x^2 - (5000 + 0.6x^2)$
$\qquad\qquad = 280x - 0.4x^2 - 5000 - 0.6x^2$
$\qquad\qquad = -x^2 + 280x - 5000$

A polynomial for total profit is $-x^2 + 280x - 5000$.

b) Evaluate the polynomial for $x = 75$:
$\qquad -x^2 + 280x - 5000$
$\qquad = -(75)^2 + 280(75) - 5000$
$\qquad = -5625 + 21,000 - 5000$
$\qquad = 10,375$

The total profit on the production and sale of 75 stereos is \$10,375.

c) Evaluate the polynomial for $x = 100$:
$\qquad -x^2 + 280x - 5000$
$\qquad = -(100)^2 + 280(100) - 5000$
$\qquad = -10,000 + 28,000 - 5000$
$\qquad = 13,000$

The total profit on the production and sale of 100 stereos is \$13,000.

Exercise Set 4.4

1. $(6x^2)(7) = (6 \cdot 7)x^2 = 42x^2$

2. $-10x^2$

3. $(-x^3)(-x) = (-1x^3)(-1x) = (-1)(-1)(x^3 \cdot x) = x^4$

4. $-x^6$

5. $(-x^5)(x^3) = (-1x^5)(1x^3) = (-1)(1)(x^5 \cdot x^3) = -x^8$

6. x^8

7. $(3x^4)(2x^2) = (3 \cdot 2)(x^4 \cdot x^2) = 6x^6$

8. $20x^8$

9. $(7t^5)(4t^3) = (7 \cdot 4)(t^5 \cdot t^3) = 28t^8$

10. $30a^4$

11. $(-0.1x^6)(0.2x^4) = (-0.1)(0.2)(x^6 \cdot x^4) = -0.02x^{10}$

12. $-0.12x^9$

13. $\left(-\frac{1}{5}x^3\right)\left(-\frac{1}{3}x\right) = \left(-\frac{1}{5}\right)\left(-\frac{1}{3}\right)(x^3 \cdot x) = \frac{1}{15}x^4$

14. $-\dfrac{1}{20}x^{12}$

15. $(-4x^2)(0) = 0$ Any number multiplied by 0 is 0.

16. $4m^5$

17. $(3x^2)(-4x^3)(2x^6) = (3)(-4)(2)(x^2 \cdot x^3 \cdot x^6) = -24x^{11}$

18. $60y^{12}$

19. $3x(-x + 5) = 3x(-x) + 3x(5)$
$\qquad\qquad\quad = -3x^2 + 15x$

20. $8x^2 - 12x$

21. $4x(x + 1) = 4x(x) + 4x(1)$
$\qquad\qquad\quad = 4x^2 + 4x$

22. $3x^2 + 6x$

23. $(x + 7)5x = x \cdot 5x + 7 \cdot 5x$
$\qquad\qquad\quad = 5x^2 + 35x$

24. $3x^2 - 18x$

25. $x^2(x^3 + 1) = x^2(x^3) + x^2(1)$
$\qquad\qquad\quad\ = x^5 + x^2$

26. $-2x^5 + 2x^3$

27. $3x(2x^2 - 6x + 1) = 3x(2x^2) + 3x(-6x) + 3x(1)$
$\qquad\qquad\qquad\qquad = 6x^3 - 18x^2 + 3x$

28. $-8x^4 + 24x^3 + 20x^2 - 4x$

29. $4x^2(3x + 6) = 4x^2(3x) + 4x^2(6)$
$\qquad\qquad\qquad = 12x^3 + 24x^2$

30. $-10x^3 + 5x^2$

31. $-6x^2(x^2 + x) = -6x^2(x^2) - 6x^2(x)$
$\qquad\qquad\qquad\ = -6x^4 - 6x^3$

32. $-4x^4 + 4x^3$

33. $3y^2(6y^4 + 8y^3) = 3y^2(6y^4) + 3y^2(8y^3)$
$\qquad\qquad\qquad\ = 18y^6 + 24y^5$

34. $4y^7 - 24y^6$

35. $\qquad 3x^4(14x^{50} + 20x^{11} + 6x^{57} + 60x^{15})$
$\qquad = 3x^4(14x^{50}) + 3x^4(20x^{11}) + 3x^4(6x^{57}) +$
$\qquad\qquad 3x^4(60x^{15})$
$\qquad = 42x^{54} + 60x^{15} + 18x^{61} + 180x^{19}$

36. $20x^{38} - 50x^{25} + 25x^{14}$

37. $(x + 6)(x + 3) = (x + 6)x + (x + 6)3$
$\qquad\qquad\qquad\ = x \cdot x + 6 \cdot x + x \cdot 3 + 6 \cdot 3$
$\qquad\qquad\qquad\ = x^2 + 6x + 3x + 18$
$\qquad\qquad\qquad\ = x^2 + 9x + 18$

38. $x^2 + 7x + 10$

39. $(x + 5)(x - 2) = (x + 5)x + (x + 5)(-2)$
$\qquad\qquad\qquad\ = x \cdot x + 5 \cdot x + x(-2) + 5(-2)$
$\qquad\qquad\qquad\ = x^2 + 5x - 2x - 10$
$\qquad\qquad\qquad\ = x^2 + 3x - 10$

40. $x^2 + 4x - 12$

41. $(x - 4)(x - 3) = (x - 4)x + (x - 4)(-3)$
$\qquad\qquad\qquad\ = x \cdot x - 4 \cdot x + x(-3) - 4(-3)$
$\qquad\qquad\qquad\ = x^2 - 4x - 3x + 12$
$\qquad\qquad\qquad\ = x^2 - 7x + 12$

42. $x^2 - 10x + 21$

43. $(x+3)(x-3) = (x+3)x + (x+3)(-3)$
$$= x \cdot x + 3 \cdot x + x(-3) + 3(-3)$$
$$= x^2 + 3x - 3x - 9$$
$$= x^2 - 9$$

44. $x^2 - 36$

45. $(5-x)(5-2x) = (5-x)5 + (5-x)(-2x)$
$$= 5 \cdot 5 - x \cdot 5 + 5(-2x) - x(-2x)$$
$$= 25 - 5x - 10x + 2x^2$$
$$= 25 - 15x + 2x^2$$

46. $18 + 12x + 2x^2$

47. $(2x+5)(2x+5) = (2x+5)2x + (2x+5)5$
$$= 2x \cdot 2x + 5 \cdot 2x + 2x \cdot 5 + 5 \cdot 5$$
$$= 4x^2 + 10x + 10x + 25$$
$$= 4x^2 + 20x + 25$$

48. $9x^2 - 24x + 16$

49. $(3y-4)(3y+4) = (3y-4)3y + (3y-4)4$
$$= 3y \cdot 3y - 4 \cdot 3y + 3y \cdot 4 - 4 \cdot 4$$
$$= 9y^2 - 12y + 12y - 16$$
$$= 9y^2 - 16$$

50. $4y^2 - 1$

51. $\left(x - \dfrac{5}{2}\right)\left(x + \dfrac{2}{5}\right) = \left(x - \dfrac{5}{2}\right)x + \left(x - \dfrac{5}{2}\right)\dfrac{2}{5}$
$$= x \cdot x - \frac{5}{2} \cdot x + x \cdot \frac{2}{5} - \frac{5}{2} \cdot \frac{2}{5}$$
$$= x^2 - \frac{5}{2}x + \frac{2}{5}x - 1$$
$$= x^2 - \frac{25}{10}x + \frac{4}{10}x - 1$$
$$= x^2 - \frac{21}{10}x - 1$$

52. $x^2 + \dfrac{17}{6}x + 2$

53. $(x^2 + x + 1)(x - 1)$
$$= (x^2 + x + 1)x + (x^2 + x + 1)(-1)$$
$$= x^2 \cdot x + x \cdot x + 1 \cdot x + x^2(-1) + x(-1) + 1(-1)$$
$$= x^3 + x^2 + x - x^2 - x - 1$$
$$= x^3 - 1$$

54. $x^3 + 3x^2 + 4$

55. $(2x+1)(2x^2 + 6x + 1)$
$$= 2x(2x^2 + 6x + 1) + 1(2x^2 + 6x + 1)$$
$$= 2x \cdot 2x^2 + 2x \cdot 6x + 2x \cdot 1 + 1 \cdot 2x^2 + 1 \cdot 6x + 1 \cdot 1$$
$$= 4x^3 + 12x^2 + 2x + 2x^2 + 6x + 1$$
$$= 4x^3 + 14x^2 + 8x + 1$$

56. $12x^3 - 10x^2 - x + 1$

57. $(y^2 - 3)(3y^2 - 6y + 2)$
$$= y^2(3y^2 - 6y + 2) - 3(3y^2 - 6y + 2)$$
$$= y^2 \cdot 3y^2 + y^2(-6y) + y^2 \cdot 2 - 3 \cdot 3y^2 - 3(-6y) - 3 \cdot 2$$
$$= 3y^4 - 6y^3 + 2y^2 - 9y^2 + 18y - 6$$
$$= 3y^4 - 6y^3 - 7y^2 + 18y - 6$$

58. $3y^4 + 18y^3 - 18y - 3$

59. $(x^3 + x^2)(x^3 + x^2 - x)$
$$= x^3(x^3 + x^2 - x) + x^2(x^3 + x^2 - x)$$
$$= x^3 \cdot x^3 + x^3 \cdot x^2 + x^3(-x) + x^2 \cdot x^3 + x^2 \cdot x^2 + x^2(-x)$$
$$= x^6 + x^5 - x^4 + x^5 + x^4 - x^3$$
$$= x^6 + 2x^5 - x^3$$

60. $x^6 - 2x^5 + 2x^4 - x^3$

61. $(-5x^3 - 7x^2 + 1)(2x^2 - x)$
$$= (-5x^3 - 7x^2 + 1)2x^2 + (-5x^3 - 7x^2 + 1)(-x)$$
$$= -5x^3 \cdot 2x^2 - 7x^2 \cdot 2x^2 + 1 \cdot 2x^2 - 5x^3(-x) - 7x^2(-x) + 1(-x)$$
$$= -10x^5 - 14x^4 + 2x^2 + 5x^4 + 7x^3 - x$$
$$= -10x^5 - 9x^4 + 7x^3 + 2x^2 - x$$

62. $-20x^5 + 25x^4 - 4x^3 - 5x^2 - 2$

63.

$$
\begin{array}{l}
1 + x + x^2 \qquad \text{Line up like terms} \\
\underline{-\,1 - x + x^2} \qquad \text{in columns} \\
x^2 + x^3 + x^4 \qquad \text{Multiplying the top row} \\
 \text{by } x^2 \\
-\,x - x^2 - x^3 \text{Multiplying by } -x \\
\underline{-1 - x - x^2} \text{Multiplying by } -1 \\
-1 - 2x - x^2 + x^4 \quad \text{Collecting like terms}
\end{array}
$$

This result can also be written in descending order:
$x^4 - x^2 - 2x - 1$

64. $x^4 - 2x^3 + 3x^2 - 2x + 1$

65.

$$
\begin{array}{l}
2x^2 + 3x - 4 \\
\underline{2x^2 + x - 2} \\
-4x^2 - 6x + 8 \qquad \text{Multiplying by } -2 \\
2x^3 + 3x^2 - 4x \text{Multiplying by } x \\
\underline{4x^4 + 6x^3 - 8x^2} \text{Multiplying by } 2x^2 \\
4x^4 + 8x^3 - 9x^2 - 10x + 8 \quad \text{Collecting like terms}
\end{array}
$$

66. $4x^4 - 12x^3 - 5x^2 + 17x + 6$

67. We will multiply horizontally while still aligning like terms.

$$(x+1)(x^3 + 7x^2 + 5x + 4)$$

$$
\begin{array}{l}
= x^4 + 7x^3 + 5x^2 + 4x \text{Multiplying by } x \\
\underline{+\,x^3 + 7x^2 + 5x + 4} \quad \text{Multiplying by } 1 \\
= x^4 + 8x^3 + 12x^2 + 9x + 4
\end{array}
$$

68. $x^4 + 7x^3 + 19x^2 + 21x + 6$

69.

$$
\begin{array}{l}
2x^2 + x - 2 \\
\underline{-2x^2 + 4x - 5} \\
-10x^2 - 5x + 10 \qquad \text{Multiplying by } -5 \\
8x^3 + 4x^2 - 8x \text{Multiplying by } 4x \\
\underline{-4x^4 - 2x^3 + 4x^2} \text{Multiplying by } -2x^2 \\
-4x^4 + 6x^3 - 2x^2 - 13x + 10
\end{array}
$$

70. $-6x^4 + 4x^3 + 36x^2 - 20x + 2$

71. We will multiply horizontally, while still aligning like terms.

$$(2x + 1)(x^3 - 4x^2 + 3x - 2)$$

$$
\begin{aligned}
&= 2x^4 - 8x^3 + 6x^2 - 4x \\
&\quad\; + \; x^3 - 4x^2 + 3x - 2 \\
\hline
&= 2x^4 - 7x^3 + 2x^2 - x - 2
\end{aligned}
$$

72. $4x^4 - 5x^3 + 14x^2 + 11x - 3$

73.
$$
\begin{array}{r}
x^3 + x^2 + x + 1 \\
x - 1 \\
\hline
-x^3 - x^2 - x - 1 \\
x^4 + x^3 + x^2 + x \\
\hline
x^4 \qquad\qquad\quad - 1
\end{array}
$$

74. $x^4 - 3x^3 + 3x^2 - 4x + 4$

75.
$$
\begin{array}{r}
x^3 + x^2 - x - 3 \\
x - 3 \\
\hline
-3x^3 - 3x^2 + 3x + 9 \\
x^4 + x^3 - x^2 - 3x \\
\hline
x^4 - 2x^3 - 4x^2 \qquad + 9
\end{array}
$$

76. $x^4 + 3x^3 - 5x^2 + 16$

77. $-\dfrac{1}{4} - \dfrac{1}{2} = -\dfrac{1}{4} - \dfrac{1}{2} \cdot \dfrac{2}{2} = -\dfrac{1}{4} - \dfrac{2}{4} = -\dfrac{3}{4}$

78. $4(4x - 6y + 9)$

79.

80.

81.

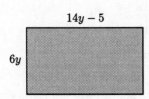

The shaded area is the product of the length and width of the rectangle:

$$
\begin{aligned}
6y(14y - 5) &= 6y \cdot 14y + 6y(-5) \\
&= 84y^2 - 30y
\end{aligned}
$$

82. $78t^2 + 40t$

83.

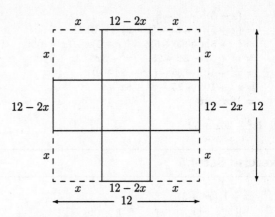

The dimensions of the box are $12 - 2x$ by $12 - 2x$ by x. The volume is the product of the dimensions (volume $=$ length $\times$ width $\times$ height):

$$
\begin{aligned}
\text{Volume} &= (12 - 2x)(12 - 2x)x \\
&= (144 - 48x + 4x^2)x \\
&= 144x - 48x^2 + 4x^3, \text{ or} \\
&\quad\; 4x^3 - 48x^2 + 144x
\end{aligned}
$$

The outside surface area is the sum of the area of the bottom and the areas of the four sides. The dimensions of the bottom are $12 - 2x$ by $12 - 2x$, and the dimensions of each side are x by $12 - 2x$.

$$
\begin{aligned}
\text{Surface area} &= \begin{array}{l} \text{Area of bottom} + \\ \quad 4 \cdot \text{Area of each side} \end{array} \\
&= (12 - 2x)(12 - 2x) + 4 \cdot x(12 - 2x) \\
&= 144 - 24x - 24x + 4x^2 + 48x - 8x^2 \\
&= 144 - 48x + 4x^2 + 48x - 8x^2 \\
&= 144 - 4x^2, \text{ or } -4x^2 + 144
\end{aligned}
$$

84. $x^3 - 5x^2 + 8x - 4$

85. Let $b =$ the length of the base. The $b + 4 =$ the height. Let A represent the area.

$$
\begin{aligned}
\text{Area} &= \frac{1}{2} \times \text{base} \times \text{height} \\
A &= \frac{1}{2} \cdot b \cdot (b + 4) \\
A &= \frac{1}{2}b(b + 4) \\
A &= \frac{1}{2}b^2 + 2b
\end{aligned}
$$

86. 8 ft by 16 ft

87.
$$
\begin{aligned}
&(x + 3)(x + 6) + (x + 3)(x + 6) \\
&= (x + 3)x + (x + 3)6 + (x + 3)x + (x + 3)6 \\
&= x^2 + 3x + 6x + 18 + x^2 + 3x + 6x + 18 \\
&= 2x^2 + 18x + 36
\end{aligned}
$$

88. $2x^2 - 18x + 28$

89.
$$(x+5)^2 - (x-3)^2$$
$$= (x+5)(x+5) - (x-3)(x-3)$$
$$= (x+5)x + (x+5)5 - [(x-3)x - (x-3)3]$$
$$= x^2 + 5x + 5x + 25 - (x^2 - 3x - 3x + 9)$$
$$= x^2 + 10x + 25 - (x^2 - 6x + 9)$$
$$= x^2 + 10x + 25 - x^2 + 6x - 9$$
$$= 16x + 16$$

90. $2x^2 - 20x + 52$

Exercise Set 4.5

1. $(x+1)(x^2+3)$
  $\quad$ F $\quad$ O $\quad$ I $\quad$ L
  $$= x \cdot x^2 + x \cdot 3 + 1 \cdot x^2 + 1 \cdot 3$$
  $$= x^3 + 3x + x^2 + 3, \text{or } x^3 + x^2 + 3x + 3$$

2. $x^3 - x^2 - 3x + 3$

3. $(x^3+2)(x+1)$
  $\quad$ F $\quad$ O $\quad$ I $\quad$ L
  $$= x^3 \cdot x + x^3 \cdot 1 + 2 \cdot x + 2 \cdot 1$$
  $$= x^4 + x^3 + 2x + 2$$

4. $x^5 + 12x^4 + 2x + 24$

5. $(y+2)(y-3)$
  $\quad$ F $\quad$ O $\quad$ I $\quad$ L
  $$= y \cdot y + y \cdot (-3) + 2 \cdot y + 2 \cdot (-3)$$
  $$= y^2 - 3y + 2y - 6$$
  $$= y^2 - y - 6$$

6. $a^2 + 4a + 4$

7. $(3x+2)(3x+3)$
  $\quad$ F $\quad$ O $\quad$ I $\quad$ L
  $$= 3x \cdot 3x + 3x \cdot 3 + 2 \cdot 3x + 2 \cdot 3$$
  $$= 9x^2 + 9x + 6x + 6$$
  $$= 9x^2 + 15x + 6$$

8. $8x^2 + 10x + 2$

9. $(5x-6)(x+2)$
  $\quad$ F $\quad$ O $\quad$ I $\quad$ L
  $$= 5x \cdot x + 5x \cdot 2 + (-6) \cdot x + (-6) \cdot 2$$
  $$= 5x^2 + 10x - 6x - 12$$
  $$= 5x^2 + 4x - 12$$

10. $x^2 - 64$

11. $(3t-1)(3t+1)$
  $\quad$ F $\quad$ O $\quad$ I $\quad$ L
  $$= 3t \cdot 3t + 3t \cdot 1 + (-1) \cdot 3t + (-1) \cdot 1$$
  $$= 9t^2 + 3t - 3t - 1$$
  $$= 9t^2 - 1$$

12. $4m^2 + 12m + 9$

13. $(4x-2)(x-1)$
  $\quad$ F $\quad$ O $\quad$ I $\quad$ L
  $$= 4x \cdot x + 4x \cdot (-1) + (-2) \cdot x + (-2) \cdot (-1)$$
  $$= 4x^2 - 4x - 2x + 2$$
  $$= 4x^2 - 6x + 2$$

14. $6x^2 - x - 1$

15. $\left(p - \frac{1}{4}\right)\left(p + \frac{1}{4}\right)$
  $\quad$ F $\quad$ O $\quad$ I $\quad$ L
  $$= p \cdot p + p \cdot \frac{1}{4} + \left(-\frac{1}{4}\right) \cdot p + \left(-\frac{1}{4}\right) \cdot \frac{1}{4}$$
  $$= p^2 + \frac{1}{4}p - \frac{1}{4}p - \frac{1}{16}$$
  $$= p^2 - \frac{1}{16}$$

16. $q^2 + \frac{3}{2}q + \frac{9}{16}$

17. $(x-0.1)(x+0.1)$
  $\quad$ F $\quad$ O $\quad$ I $\quad$ L
  $$= x \cdot x + x \cdot (0.1) + (-0.1) \cdot x + (-0.1)(0.1)$$
  $$= x^2 + 0.1x - 0.1x - 0.01$$
  $$= x^2 - 0.01$$

18. $x^2 - 0.1x - 0.12$

19. $(2x^2+6)(x+1)$
  $\quad$ F $\quad$ O $\quad$ I $\quad$ L
  $$= 2x^3 + 2x^2 + 6x + 6$$

20. $4x^3 - 2x^2 + 6x - 3$

21. $(-2x+1)(x+6)$
  $\quad$ F $\quad$ O $\quad$ I $\quad$ L
  $$= -2x^2 - 12x + x + 6$$
  $$= -2x^2 - 11x + 6$$

22. $6x^2 - 4x - 16$

23. $(a+7)(a+7)$
  $\quad$ F $\quad$ O $\quad$ I $\quad$ L
  $$= a^2 + 7a + 7a + 49$$
  $$= a^2 + 14a + 49$$

24. $4y^2 + 20y + 25$

25. $(1+2x)(1-3x)$
  $\quad$ F $\quad$ O $\quad$ I $\quad$ L
  $$= 1 - 3x + 2x - 6x^2$$
  $$= 1 - x - 6x^2$$

26. $-3x^2 - 5x - 2$

27. $(x^2+3)(x^3-1)$
  $\quad$ F $\quad$ O $\quad$ I $\quad$ L
  $$= x^5 - x^2 + 3x^3 - 3, \text{ or } x^5 + 3x^3 - x^2 - 3$$

28. $2x^5 + x^4 - 6x - 3$

29. $(3x^2-2)(x^4-2)$
  $\quad$ F $\quad$ O $\quad$ I $\quad$ L
  $$= 3x^6 - 6x^2 - 2x^4 + 4, \text{ or } 3x^6 - 2x^4 - 6x^2 + 4$$

30. $x^{20} - 9$

31. $(3x^5 + 2)(2x^2 + 6)$
         F   O   I   L
     $= 6x^7 + 18x^5 + 4x^2 + 12$

32. $1 + 3x^2 - 2x - 6x^3$, or $-6x^3 + 3x^2 - 2x + 1$

33. $(8x^3 + 1)(x^3 + 8)$
         F   O   I   L
     $= 8x^6 + 64x^3 + x^3 + 8$
     $= 8x^6 + 65x^3 + 8$

34. $20 - 8x^2 - 10x + 4x^3$, or $4x^3 - 8x^2 - 10x + 20$

35. $(4x^2 + 3)(x - 3)$
         F   O   I   L
     $= 4x^3 - 12x^2 + 3x - 9$

36. $14x^2 - 53x + 14$

37. $(4y^4 + y^2)(y^2 + y)$
         F   O   I   L
     $= 4y^6 + 4y^5 + y^4 + y^3$

38. $10y^{12} + 16y^9 + 6y^6$

39. $(x + 4)(x - 4)$ Product of sum and differ-
                          ence of the same two terms
     $= x^2 - 4^2$
     $= x^2 - 16$

40. $x^2 - 1$

41. $(2x + 1)(2x - 1)$ Product of sum and differ-
                           ence of the same two terms
     $= (2x)^2 - 1^2$
     $= 4x^2 - 1$

42. $x^4 - 1$

43. $(5m - 2)(5m + 2)$ Product of sum and diff-
                           erence of the same two terms
     $= (5m)^2 - 2^2$
     $= 25m^2 - 4$

44. $9x^8 - 4$

45. $(2x^2 + 3)(2x^2 - 3)$ Product of sum and diff-
                               erence of the same two terms
     $= (2x^2)^2 - 3^2$
     $= 4x^4 - 9$

46. $36x^{10} - 25$

47. $(3x^4 - 4)(3x^4 + 4)$
     $= (3x^4)^2 - 4^2$
     $= 9x^8 - 16$

48. $t^4 - 0.04$

49. $(x^6 - x^2)(x^6 + x^2)$
     $= (x^6)^2 - (x^2)^2$
     $= x^{12} - x^4$

50. $4x^6 - 0.09$

51. $(x^4 + 3x)(x^4 - 3x)$
     $= (x^4)^2 - (3x)^2$
     $= x^8 - 9x^2$

52. $\dfrac{9}{16} - 4x^6$

53. $(x^{12} - 3)(x^{12} + 3)$
     $= (x^{12})^2 - 3^2$
     $= x^{24} - 9$

54. $144 - 9x^4$

55. $(2y^8 + 3)(2y^8 - 3)$
     $= (2y^8)^2 - 3^2$
     $= 4y^{16} - 9$

56. $m^2 - \dfrac{4}{9}$

57. $(x + 2)^2$
     $= x^2 + 2 \cdot x \cdot 2 + 2^2$ Square of a binomial
     $= x^2 + 4x + 4$

58. $4x^2 - 4x + 1$

59. $(3x^2 + 1)$ Square of a binomial
     $= (3x^2)^2 + 2 \cdot 3x^2 \cdot 1 + 1^2$
     $= 9x^4 + 6x^2 + 1$

60. $9x^2 + \dfrac{9}{2}x + \dfrac{9}{16}$

61. $\left(a - \dfrac{1}{2}\right)^2$ Square of a binomial
     $= a^2 - 2 \cdot a \cdot \dfrac{1}{2} + \left(\dfrac{1}{2}\right)^2$
     $= a^2 - a + \dfrac{1}{4}$

62. $4a^2 - \dfrac{4}{5}a + \dfrac{1}{25}$

63. $(3 + x)^2 = 3^2 + 2 \cdot 3 \cdot x + x^2$
         $= 9 + 6x + x^2$

64. $x^6 - 2x^3 + 1$

65. $(x^2 + 1)^2 = (x^2)^2 + 2 \cdot x^2 \cdot 1 + 1^2$
         $= x^4 + 2x^2 + 1$

66. $64x^2 - 16x^3 + x^4$

67. $(2 - 3x^4)^2 = 2^2 - 2 \cdot 2 \cdot 3x^4 + (3x^4)^2$
         $= 4 - 12x^4 + 9x^8$

68. $36x^6 - 24x^3 + 4$

69. $(5 + 6t^2)^2 = 5^2 + 2 \cdot 5 \cdot 6t^2 + (6t^2)^2$
         $= 25 + 60t^2 + 36t^4$

70. $9p^4 - 6p^3 + p^2$

71. $(7x - 0.3)^2 = (7x)^2 - 2(7x)(0.3) + (0.3)^2$
$$= 49x^2 - 4.2x + 0.09$$

72. $16a^2 - 4.8a + 0.36$

73. $5a^3(2a^2 - 1)$
$= 5a^3 \cdot 2a^2 - 5a^3 \cdot 1$ Multiplying each term of
 the binomial by the monomial
$= 10a^5 - 5a^3$

74. $a^3 - a^2 - 10a + 12$

75. $(x^2 - 5)(x^2 + x - 1)$

$= x^4 + x^3 - x^2$ Multiplying horizontally
$-5x^2 - 5x + 5$ and aligning like terms
$\overline{= x^4 + x^3 - 6x^2 - 5x + 5}$

76. $27x^6 - 9x^5$

77. $(3 - 2x^3)^2$
$= 3^2 - 2 \cdot 3 \cdot 2x^3 + (2x^3)^2$ Squaring a binomial
$= 9 - 12x^3 + 4x^6$

78. $x^2 - 8x^4 + 16x^6$

79. $4x(x^2 + 6x - 3)$
$= 4x \cdot x^2 + 4x \cdot 6x + 4x(-3)$ Multiplying each
 term of the trinomial
 by the monomial
$= 4x^3 + 24x^2 - 12x$

80. $-8x^6 + 48x^3 + 72x$

81. $\left(2x^2 - \dfrac{1}{2}\right)\left(2x^2 - \dfrac{1}{2}\right)$ Squaring a binomial
$= (2x^2)^2 - 2 \cdot 2x^2 \cdot \dfrac{1}{2} + \left(\dfrac{1}{2}\right)^2$
$= 4x^4 - 2x^2 + \dfrac{1}{4}$

82. $x^4 - 2x^2 + 1$

83. $(-1 + 3p)(1 + 3p)$
$= (3p - 1)(3p + 1)$ Product of the sum and
 difference of the same two terms
$= (3p)^2 - 1^2$
$= 9p^2 - 1$

84. $-9q^2 + 4$, or $4 - 9q^2$

85. $3t^2(5t^3 - t^2 + t)$
$= 3t^2 \cdot 5t^3 + 3t^2(-t^2) + 3t^2 \cdot t$ Multiplying each
 term of the trinomial
 by the monomial
$= 15t^5 - 3t^4 + 3t^3$

86. $-6x^5 - 48x^3 + 54x^2$

87. $(6x^4 + 4)^2$ Squaring a binomial
$= (6x^4)^2 + 2 \cdot 6x^4 \cdot 4 + 4^2$
$= 36x^8 + 48x^4 + 16$

88. $64a^2 + 80a + 25$

89. $(3x + 2)(4x^2 + 5)$ Product of two
 binomials; use FOIL
$= 3x \cdot 4x^2 + 3x \cdot 5 + 2 \cdot 4x^2 + 2 \cdot 5$
$= 12x^3 + 15x + 8x^2 + 10$, or
$ 12x^3 + 8x^2 + 15x + 10$

90. $6x^4 - 3x^2 - 63$

91. $(8 - 6x^4)^2$ Squaring a binomial
$= 8^2 - 2 \cdot 8 \cdot 6x^4 + (6x^4)^2$
$= 64 - 96x^4 + 36x^8$

92. $\dfrac{3}{25}x^4 + 4x^2 - 63$

93. $t^2 + t + 1$ Using columns to multiply a
$t - 1$ binomial and a trinomial
$\overline{-t^2 - t - 1}$
$\underline{t^3 + t^2 + t}$
$t^3 - 1$

94. $y^3 + 125$

95. $3^2 + 4^2 = 9 + 16 = 25$
$(3 + 4)^2 = 7^2 = 49$

96. 85; 169

97. $9^2 - 5^2 = 81 - 25 = 56$
$(9 - 5)^2 = 4^2 = 16$

98. 105; 49

99.

We can find the shaded area in two ways.

Method 1: The figure is a square with side $x + 3$, so the area is $(x + 3)^2 = x^2 + 6x + 9$.

Method 2: We add the areas of A, B, C, and D.

$3 \cdot x + 3 \cdot 3 + 3 \cdot x + x \cdot x = 3x + 9 + 3x + x^2 = x^2 + 6x + 9$.

Either way we find that the total shaded area is $x^2 + 6x + 9$.

100. $a^2 + 2a + 1$

101.

3	A	B
t	D	C
	t	4

We can find the shaded area in two ways.

Method 1: The figure is a rectangle with dimensions $t+4$ by $t+3$, so the area is $(t+4)(t+3) =$ $t^2 + 3t + 4t + 12 = t^2 + 7t + 12$.

Method 2: We add the areas of A, B, C, and D. $3 \cdot t + 3 \cdot 4 + 4 \cdot t + t \cdot t = 3t + 12 + 4t + t^2 = t^2 + 7t + 12$.

Either way, we find that the area is $t^2 + 7t + 12$.

102. $x^2 + 7x + 10$

103.

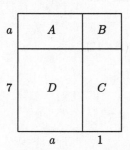

We can find the shaded area in two ways.

Method 1: The figure is a rectangle with dimensions $a+7$ by $a+1$, so the area is $(a+7)(a+1) = a^2 + a + 7a + 7 = a^2 + 8a + 7$.

Method 2: We add the areas of A, B, C, and D. $a \cdot a + a \cdot 1 + 7 \cdot 1 + 7 \cdot a = a^2 + a + 7 + 7a = a^2 + 8a + 7$.

Either way, we find that the area is $a^2 + 8a + 7$.

104. $t^2 + 13t + 36$

105. *Familiarize*. Let $t = $ the number of watts used by the television set. Then $10t = $ the number of watts used by the lamps, and $40t = $ the number of watts used by the air conditioner.

Translate.

$$\underbrace{\text{Lamp watts}}_{\downarrow\ \downarrow} + \underbrace{\text{Air conditioner watts}}_{\downarrow\ \downarrow} + \underbrace{\text{Television watts}}_{\downarrow\ \downarrow} = \underbrace{\text{Total watts}}_{\downarrow\ \downarrow}$$

$$10t \quad + \quad 40t \quad + \quad t \quad = \quad 2550$$

Solve. We solve the equation.

$$10t + 40t + t = 2550$$
$$51t = 2550$$
$$t = 50$$

The possible solution is:

Television, t: 50 watts

Lamps, $10t$: $10 \cdot 50$, or 500 watts

Air conditioner, $40t$: $40 \cdot 50$, or 2000 watts

Check. The number of watts used by the lamps, 500, is 10 times 50, the number used by the television. The number of watts used by the air conditioner, 2000, is 40 times 50, the number used by the television. Also, $50 + 500 + 2000 = 2550$, the total wattage used.

State. The television uses 50 watts, the lamps use 500 watts, and the air conditioner uses 2000 watts.

106. $\dfrac{28}{27}$

107. ◆

108. ◆

109. $4y(y+5)(2y+8)$
$= 4y(2y^2 + 8y + 10y + 40)$
$= 4y(2y^2 + 18y + 40)$
$= 8y^3 + 72y^2 + 160y$

110. $80x^3 + 24x^2 - 216x$

111. $[(3x-2)(3x+2)](9x^2+4)$
$= (9x^2 - 4)(9x^2 + 4)$ Finding the product of the sum and difference of the same two terms
$= 81x^4 - 16$ Finding the product of the sum and difference of the same two terms again

112. $16x^4 - 1$

113. $(5t^3 - 3)^2(5t^3 + 3)^2$
$= [(5t^3 - 3)(5t^3 + 3)][(5t^3 - 3)(5t^3 + 3)]$
$= (25t^6 - 9)(25t^6 - 9)$
$= (25t^6)^2 - 2 \cdot 25t^6 \cdot 9 + 9^2$
$= 625t^{12} - 450t^6 + 81$

114. $5a^2 + 12a - 9$

115. $(67.58x + 3.225)^2$
$= (67.58x)^2 + 2(67.58x)(3.225) + (3.225)^2$
$= 4567.0564x^2 + 435.891x + 10.400625$

116. $x^2 - 8x + 16$

117. $18 \times 22 = (20-2)(20+2) = 400 - 4 = 396$

118. 9951

119. $(x+2)(x-5) = (x+1)(x-3)$
$x^2 - 5x + 2x - 10 = x^2 - 3x + x - 3$
$x^2 - 3x - 10 = x^2 - 2x - 3$
$-3x - 10 = -2x - 3$ Adding $-x^2$
$-3x + 2x = 10 - 3$ Adding $2x$ and 10
$-x = 7$
$x = -7$

The solution is -7.

120. 0

121. If $w =$ the width, then $w + 1 =$ the length, and $(w + 1) + 1$, or $w + 2 =$ the height.

$$
\begin{aligned}
\text{Volume} &= \text{length} \times \text{width} \times \text{height} \\
&= (w + 1) \cdot w \cdot (w + 2) \\
&= (w^2 + w)(w + 2) \\
&= w^3 + 2w^2 + w^2 + 2w \\
&= w^3 + 3w^2 + 2w
\end{aligned}
$$

122. $l^3 - l$

123. If $h =$ the height, then $h - 1 =$ the length, and $(h - 1) - 1$, or $h - 2 =$ the width.

$$
\begin{aligned}
\text{Volume} &= \text{length} \times \text{width} \times \text{height} \\
&= (h - 1) \cdot (h - 2) \cdot h \\
&= (h^2 - 2h - h + 2)h \\
&= (h^2 - 3h + 2)h \\
&= h^3 - 3h^2 + 2h
\end{aligned}
$$

124. $Q(Q - 14) - 5(Q - 14)$, or $(Q - 5)(Q - 14)$

125.

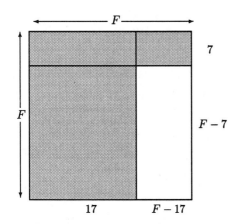

The area of the entire figure is $F \cdot F$, or F^2.

The area of the section not shaded is $(F - 7)(F - 17)$

$$
\underbrace{\begin{array}{c}\text{Area of}\\\text{shaded}\\\text{region}\end{array}}_{} = \underbrace{\begin{array}{c}\text{Area of}\\\text{entire}\\\text{figure}\end{array}}_{} - \underbrace{\begin{array}{c}\text{Area of}\\\text{section}\\\text{not shaded}\end{array}}_{}
$$

$$
\begin{array}{c}\text{Area of}\\\text{shaded}\\\text{region}\end{array} = \quad F^2 \quad - (F - 7)(F - 17)
$$

Now we find another expression for the shaded region.

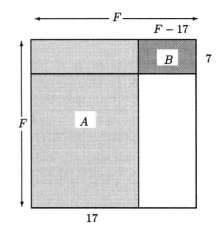

$$
\underbrace{\begin{array}{c}\text{Area of}\\\text{shaded}\\\text{region}\end{array}}_{} = \underbrace{\begin{array}{c}\text{Area of}\\A\end{array}}_{} + \underbrace{\begin{array}{c}\text{Area of}\\B\end{array}}_{}
$$

$$
\begin{aligned}
\begin{array}{c}\text{Area of}\\\text{shaded}\\\text{region}\end{array} &= \quad 17F \quad + 7(F - 17) \\
&= 17F + 7F - 119 \\
&= 24F - 119
\end{aligned}
$$

126. $(y + 1)(y - 1)$, or $y(y + 1) - y - 1$

127. a)

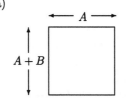

The area of the entire rectangle is $A(A + B)$, or $A^2 + AB$.

b)

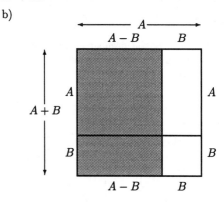

The sum of the areas of the two unshaded rectangles is $A \cdot B + B \cdot B$, or $AB + B^2$.

c)   Area in part (a) - area in part (b)

$$
\begin{aligned}
&= A^2 + AB - (AB + B^2) \\
&= A^2 + AB - AB - B^2 \\
&= A^2 - B^2
\end{aligned}
$$

d) The area of the shaded region is $(A+B)(A-B) = A^2-B^2$. This is the same as the polynomial found in part (c).

128. 10, 11, and 12

129. $(10x+5)^2 = (10x)^2 + 2 \cdot 10x \cdot 5 + 5^2$
$= 100x^2 + 100x + 25$
$= 100(x^2 + x) + 25$

To square any two-digit number ending in 5 mentally, add the first digit to its square, multiply by 100, and add 25.

Exercise Set 4.6

1. We replace x by 3 and y by -2.
$x^2 - y^2 + xy = 3^2 - (-2)^2 + 3(-2) = 9 - 4 - 6 = -1$

2. 19

3. We replace x by 2, y by -3, and z by -1.
$xyz^2 + z = 2(-3)(-1)^2 + (-1) = -6 - 1 = -7$

4. -1

5. Evaluate the polynomial for $P = 10,000$ and $i = 0.08$.
$A = P(1+i)^2 = 10,000(1 + 0.08)^2$
$= 10,000(1.08)^2$
$= 10,000(1.1664)$
$= 11,664$

At 8% interest for 2 years, $10,000 will grow to $11,664.

6. $11,449

7. Evaluate the polynomial for $P = 10,000$ and $i = 0.08$.
$A = P(1+i)^3 = 10,000(1 + 0.08)^3$
$= 10,000(1.08)^3$
$= 10,000(1.259712)$
$= 12,597.12$

At 8% interest for 3 years, $10,000 will grow to $12,597.12.

8. $12,250.43

9. Evaluate the polynomial for $h = 4$, $r = \frac{3}{4}$, and $\pi \approx 3.14$.

$2\pi rh + \pi r^2 \approx 2(3.14)\left(\frac{3}{4}\right)(4) + (3.14)\left(\frac{3}{4}\right)^2$

$\approx 2(3.14)\left(\frac{3}{4}\right)(4) + (3.14)\left(\frac{9}{16}\right)$

$\approx 18.84 + 1.76625$

≈ 20.60625

The surface area is about 20.60625 in^2.

10. 63.78125 in^2

11. $x^3y - 2xy + 3x^2 - 5$

Term	Coefficient	Degree	
x^3y	1	4	(Think: $x^3y = x^3y^1$)
$-2xy$	-2	2	(Think: $-2xy = -2x^1y^1$)
$3x^2$	3	2	
-5	-5	0	(Think: $-5 = -5x^0$)

The degree of the polynomial is the degree of the term of highest degree.

The term of highest degree is x^3y. Its degree is 4. The degree of the polynomial is 4.

12. Coefficients: 5, -1, 15, 1

Degrees: 3, 2, 1, 0; 3

13. $17x^2y^3 - 3x^3yz - 7$

Term	Coefficient	Degree	
$17x^2y^3$	17	5	
$-3x^3yz$	-3	5	(Think: $-3x^3yz = -3x^3y^1z^1$)
-7	-7	0	(Think: $-7 = -7x^0$)

The terms of highest degree are $17x^2y^3$ and $-3x^3yz$. Each has degree 5. The degree of the polynomial is 5.

14. Coefficients: 6, -1, 8, -1

Degrees: 0, 2, 4, 5; 5

15. $a + b - 2a - 3b = (1-2)a + (1-3)b = -a - 2b$

16. $y - 7$

17. $3x^2y - 2xy^2 + x^2$

There are <u>no</u> like terms, so none of the terms can be collected.

18. $m^3 + 2m^2n - 3m^2 + 3mn^2$

19. $2u^2v - 3uv^2 + 6u^2v - 2uv^2$
$= (2+6)u^2v + (-3-2)uv^2$
$= 8u^2v - 5uv^2$

20. $-2x^2 - 4xy - 2y^2$

21. $6au + 3av + 14au + 7av$
$= (6+14)au + (3+7)av$
$= 20au + 10av$

22. $3x^2y + 3z^2y + 3xy^2$

23. $(2x^2 - xy + y^2) + (-x^2 - 3xy + 2y^2)$
$= (2-1)x^2 + (-1-3)xy + (1+2)y^2$
$= x^2 - 4xy + 3y^2$

24. $6 - z$

25. $(r^3 + 3rs - 5s^2) - (5r^3 + rs + 4s^2)$
$= r^3 + 3rs - 5s^2 - 5r^3 - rs - 4s^2$ Adding the opposite
$= (1-5)r^3 + (3-1)rs + (-5-4)s^2$
$= -4r^3 + 2rs - 9s^2$

26. $-2a^4 - 8ab + 7ab^2$

27. $\quad (r - 2s + 3) + (2r + 3s - 7)$
$\quad = (1 + 2)r + (-2 + 3)s + (3 - 7)$
$\quad = 3r + s - 4$

28. $-3b^3a^2 - b^2a^3 + 5ba + 3$

29. $\quad (2x^2 - 3xy + y^2) + (-4x^2 - 6xy - y^2) +$
$\quad\quad\quad\quad\quad\quad (x^2 + xy - y^2)$
$\quad = (2 - 4 + 1)x^2 + (-3 - 6 + 1)xy + (1 - 1 - 1)y^2$
$\quad = -x^2 - 8xy - y^2$

30. $3x^3 - x^2y + xy^2 - 3y^3$

31. $\quad (xy - ab) - (xy - 3ab)$
$\quad = xy - ab - xy + 3ab$
$\quad = (1 - 1)xy + (-1 + 3)ab$
$\quad = 0xy + 2ab$
$\quad = 2ab$

32. $x^4y^2 + y + 2x$

33. $\quad (-2a + 7b - c) + (-3b + 4c - 8d)$
$\quad = -2a + (7 - 3)b + (-1 + 4)c - 8d$
$\quad = -2a + 4b + 3c - 8d$

34. $15a^2b - 4ab$

35. $\quad (4x + 5y) + (-5x + 6y) - (7x + 3y)$
$\quad = 4x + 5y - 5x + 6y - 7x - 3y$
$\quad = (4 - 5 - 7)x + (5 + 6 - 3)y$
$\quad = -8x + 8y$

36. $-5b$

37. $\quad\quad\quad\quad\quad\quad \text{F}\quad\quad \text{O}\quad\quad \text{I}\quad\quad\quad \text{L}$
$(3z - u)(2z + 3u) = 6z^2 + 9zu - 2uz - 3u^2$
$\quad\quad\quad\quad\quad\quad = 6z^2 + 7zu - 3u^2$

38. $a^3 + a^2b - ab^2 - b^3$

39. $\quad\quad\quad\quad\quad\quad \text{F}\quad\quad\quad \text{O}\quad\quad\quad \text{I}\quad\quad \text{L}$
$(a^2b - 2)(a^2b - 5) = a^4b^2 - 5a^2b - 2a^2b + 10$
$\quad\quad\quad\quad\quad\quad = a^4b^2 - 7a^2b + 10$

40. $x^2y^2 + 3xy - 28$

41. $\quad (a^3 + bc)(a^3 - bc) = (a^3)^2 - (bc)^2$
$\quad\quad\quad\quad [(A + B)(A - B) = A^2 - B^2]$
$\quad\quad\quad\quad\quad\quad\quad = a^6 - b^2c^2$

42. $m^4 + m^2n^2 + n^4$

43.
$$\begin{array}{r} y^4x + y^2 + 1 \\ \underline{y^2 + 1} \\ y^4x + y^2 + 1 \\ \underline{y^6x + y^4 \quad\quad + y^2} \\ y^6x + y^4 + y^4x + 2y^2 + 1 \end{array}$$

44. $a^3 - b^3$

45. $\quad (3xy - 1)(4xy + 2)$
$\quad\quad\quad \text{F}\quad\quad \text{O}\quad\quad \text{I}\quad\quad \text{L}$
$\quad = 12x^2y^2 + 6xy - 4xy - 2$
$\quad = 12x^2y^2 + 2xy - 2$

46. $m^6n^2 + 2m^3n - 48$

47. $\quad (3 - c^2d^2)(4 + c^2d^2)$
$\quad\quad\quad \text{F}\quad\quad \text{O}\quad\quad \text{I}\quad\quad\quad \text{L}$
$\quad = 12 + 3c^2d^2 - 4c^2d^2 - c^4d^4$
$\quad = 12 - c^2d^2 - c^4d^4$

48. $30x^2 - 28xy + 6y^2$

49. $\quad (m^2 - n^2)(m + n)$
$\quad\quad\quad \text{F}\quad\quad \text{O}\quad\quad \text{I}\quad\quad\quad \text{L}$
$\quad = m^3 + m^2n - mn^2 - n^3$

50. $0.4p^2q^2 - 0.02pq - 0.02$

51. $\quad (xy + x^5y^5)(x^4y^4 - xy)$
$\quad\quad\quad\quad \text{F}\quad\quad\quad \text{O}\quad\quad\quad \text{I}\quad\quad\quad \text{L}$
$\quad = x^5y^5 - x^2y^2 + x^9y^9 - x^6y^6$
$\quad = x^9y^9 - x^6y^6 + x^5y^5 - x^2y^2$

52. $x^2 + xy^3 - 2y^6$

53. $\quad (x + h)^2$
$\quad = x^2 + 2xh + h^2 \quad [(A + B)^2 = A^2 + 2AB + B^2]$

54. $9a^2 + 12ab + 4b^2$

55. $\quad (r^3t^2 - 4)^2$
$\quad = (r^3t^2)^2 - 2 \cdot r^3t^2 \cdot 4 + 4^2$
$\quad\quad\quad\quad\quad [(A - B)^2 = A^2 - 2AB + B^2]$
$\quad = r^6t^4 - 8r^3t^2 + 16$

56. $9a^4b^2 - 6a^2b^3 + b^4$

57. $\quad (p^4 + m^2n^2)^2$
$\quad = (p^4)^2 + 2 \cdot p^4 \cdot m^2n^2 + (m^2n^2)^2$
$\quad\quad\quad\quad\quad [(A + B)^2 = A^2 + 2AB + B^2]$
$\quad = p^8 + 2p^4m^2n^2 + m^4n^4$

58. $a^2b^2 + 2abcd + c^2d^2$

59. $(2a - b)(2a + b) = (2a)^2 - b^2 = 4a^2 - b^2$

60. $x^2 - y^2$

61. $\quad (c^2 - d)(c^2 + d) = (c^2)^2 - d^2$
$\quad\quad\quad\quad\quad\quad\quad = c^4 - d^2$

62. $p^6 - 25q^2$

63. $\quad (ab + cd^2)(ab - cd^2) = (ab)^2 - (cd^2)^2$
$\quad\quad\quad\quad\quad\quad\quad = a^2b^2 - c^2d^4$

64. $x^2y^2 - p^2q^2$

65. $\quad (x + y - 3)(x + y + 3)$
$\quad = [(x + y) - 3][(x + y) + 3]$
$\quad = (x + y)^2 - 3^2$
$\quad = x^2 + 2xy + y^2 - 9$

66. $p^2 + 2pq + q^2 - 16$

67.
$$[x + y + z][x - (y + z)]$$
$$= [x + (y + z)][x - (y + z)]$$
$$= x^2 - (y + z)^2$$
$$= x^2 - (y^2 + 2yz + z^2)$$
$$= x^2 - y^2 - 2yz - z^2$$

68. $a^2 - b^2 - 2bc - c^2$

69.
$$(a + b + c)(a - b - c)$$
$$= [a + (b + c)][a - (b + c)]$$
$$= a^2 - (b + c)^2$$
$$= a^2 - (b^2 + 2bc + c^2)$$
$$= a^2 - b^2 - 2bc - c^2$$

70. $9x^2 + 12x + 4 - 25y^2$

71. The figure is a square with side $x + y$. Thus area is $(x + y)^2 = x^2 + 2xy + y^2$.

72. $a^2 + ac + ab + bc$

73. The figure is a parallelogram with base $x + z$ and height $x - z$. Thus the area is $(x + z)(x - z) = x^2 - z^2$.

74. $\frac{1}{2}a^2b^2 - 2$

75. Locate December, 1989, on the horizontal scale. Then move up to the line representing white office paper and left to the vertical scale to read the information being sought. In December, 1989, the price being paid for white office paper was $60 per ton.

76. December, 1988

77. Locate the highest point on the line representing newsprint. Then move down to the horizontal scale to read the information being sought. The value of newsprint peaked in December, 1987.

78. December, 1990

79.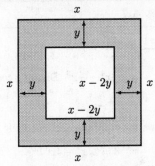

80.

81. It is helpful to add additional labels to the figure.

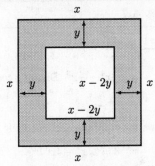

The area of the large square is $x \cdot x$, or x^2. The area of the small square is $(x - 2y)(x - 2y)$, or $(x - 2y)^2$.

| Area of shaded region | = | Area of large square | − | Area of small square |

| Area of shaded region | = | x^2 | − | $(x - 2y)^2$ |

$$= x^2 - (x^2 - 4xy + 4y^2)$$
$$= x^2 - x^2 + 4xy - 4y^2$$
$$= 4xy - 4y^2$$

82. $2\pi ab - \pi b^2$

83. It is helpful to add additional labels to the figure.

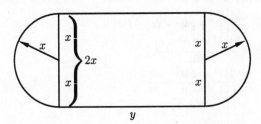

The two semicircles make a circle with radius x. The area of that circle is πx^2. The area of the rectangle is $2x \cdot y$. The sum of the two regions, $\pi x^2 + 2xy$, is the area of the shaded region.

84. $a^2 - 4b^2$

85. Evaluate the polynomial for $h = 165$ and $A = 30$:
$$0.041h - 0.018A - 2.69$$
$$= 0.041(165) - 0.018(30) - 2.69$$
$$= 6.765 - 0.54 - 2.69$$
$$= 3.535$$

The lung capacity is 3.535 liters.

86. 33

87.

Exercise Set 4.7

1. $\dfrac{24x^4 - 4x^3}{8} = \dfrac{24x^4}{8} - \dfrac{4x^3}{8}$

$\qquad\qquad = \dfrac{24}{8}x^4 - \dfrac{4}{8}x^3$ Dividing coefficients

$\qquad\qquad = 3x^4 - \dfrac{1}{2}x^3$

To check, we multiply the quotient by 8:

$$\left(3x^4 - \dfrac{1}{2}x^3\right)8 = 24x^4 - 4x^3$$

The answer checks.

2. $2a^4 - \dfrac{1}{2}a^2$

3. $\qquad \dfrac{u - 2u^2 - u^5}{u}$

$\qquad = \dfrac{u}{u} - \dfrac{2u^2}{u} - \dfrac{u^5}{u}$

$\qquad = 1 - 2u - u^4$

Check: We multiply.

$$\begin{array}{r} 1 - 2u - u^4 \\ u \\ \hline u - 2u^2 - u^5 \end{array}$$

4. $50x^4 - 7x^3 + x$

5. $\qquad (15t^3 + 24t^2 - 6t) \div (3t)$

$\qquad = \dfrac{15t^3 + 24t^2 - 6t}{3t}$

$\qquad = \dfrac{15t^3}{3t} + \dfrac{24t^2}{3t} - \dfrac{6t}{3t}$

$\qquad = 5t^2 + 8t - 2$

Check: We multiply.

$$\begin{array}{r} 5t^2 + 8t - 2 \\ 3t \\ \hline 15t^3 + 24t^2 - 6t \end{array}$$

6. $5t^2 + 3t - 6$

7. $\qquad (20x^6 - 20x^4 - 5x^2) \div (-5x^2)$

$\qquad = \dfrac{20x^6 - 20x^4 - 5x^2}{-5x^2}$

$\qquad = \dfrac{20x^6}{-5x^2} - \dfrac{20x^4}{-5x^2} - \dfrac{5x^2}{-5x^2}$

$\qquad = -4x^4 - (-4x^2) - (-1)$

$\qquad = -4x^4 + 4x^2 + 1$

Check: We multiply.

$$\begin{array}{r} -4x^4 + 4x^2 + 1 \\ -5x^2 \\ \hline 20x^6 - 20x^4 - 5x^2 \end{array}$$

8. $-3x^4 - 4x^3 + 1$

9. $\qquad (24x^5 - 40x^4 + 6x^3) \div (4x^3)$

$\qquad = \dfrac{24x^5 - 40x^4 + 6x^3}{4x^3}$

$\qquad = \dfrac{24x^5}{4x^3} - \dfrac{40x^4}{4x^3} + \dfrac{6x^3}{4x^3}$

$\qquad = 6x^2 - 10x + \dfrac{3}{2}$

Check: We multiply.

$$\begin{array}{r} 6x^2 - 10x + \dfrac{3}{2} \\ 4x^3 \\ \hline 24x^5 - 40x^4 + 6x^3 \end{array}$$

10. $2x^3 - 3x^2 - \dfrac{1}{3}$

11. $\qquad \dfrac{8x^2 - 3x + 1}{2}$

$\qquad = \dfrac{8x^2}{2} - \dfrac{3x}{2} + \dfrac{1}{2}$

$\qquad = 4x^2 - \dfrac{3}{2}x + \dfrac{1}{2}$

Check: We multiply.

$$\begin{array}{r} 4x^2 - \dfrac{3}{2}x + \dfrac{1}{2} \\ 2 \\ \hline 8x^2 - 3x + 1 \end{array}$$

12. $2x^2 + x - \dfrac{2}{3}$

13. $\qquad \dfrac{2x^3 + 6x^2 + 4x}{2x}$

$\qquad = \dfrac{2x^3}{2x} + \dfrac{6x^2}{2x} + \dfrac{4x}{2x}$

$\qquad = x^2 + 3x + 2$

Check: We multiply.

$$\begin{array}{r} x^2 + 3x + 2 \\ 2x \\ \hline 2x^3 + 6x^2 + 4x \end{array}$$

14. $2x^2 - 3x + 5$

15. $\qquad \dfrac{9r^2s^2 + 3r^2s - 6rs^2}{-3rs}$

$\qquad = \dfrac{9r^2s^2}{-3rs} + \dfrac{3r^2s}{-3rs} - \dfrac{6rs^2}{-3rs}$

$\qquad = -3rs - r + 2s$

Check: We multiply.

$$\begin{array}{r} -3rs - r + 2s \\ -3rs \\ \hline 9r^2s^2 + 3r^2s - 6rs^2 \end{array}$$

16. $1 - 2x^2y + 3x^4y^5$

17.
$$x + 2 \overline{\smash{\big)}\, x^2 + 4x + 4} \quad \leftarrow \frac{x+2}{}$$

$$\begin{array}{r} x + 2 \\ x+2 \overline{\smash{\big)}\, x^2+4x+4} \\ \underline{x^2+2x} \\ 2x+4 \leftarrow (x^2+4x)-(x^2+2x)=2x \\ \underline{2x+4} \\ 0 \leftarrow (2x+4)-(2x+4) \end{array}$$

The answer is $x + 2$.

18. $x - 3$

19.
$$\begin{array}{r} x - 5 \\ x-5 \overline{\smash{\big)}\, x^2-10x-25} \\ \underline{x^2-5x} \\ -5x-25 \leftarrow (x^2-10x)-(x^2-5x)= \\ -5x \\ \underline{-5x+25} \\ -50 \leftarrow (-5x-25)-(-5x+25) \end{array}$$

The answer is $x - 5 + \dfrac{-50}{x-5}$, or $x - 5 - \dfrac{50}{x-5}$.

20. $x + 4 - \dfrac{32}{x+4}$

21.
$$\begin{array}{r} x - 2 \\ x+6 \overline{\smash{\big)}\, x^2+4x-14} \\ \underline{x^2+6x} \\ -2x-14 \leftarrow (x^2+4x)-(x^2+6x)= \\ -2x \\ \underline{-2x-12} \\ -2 \leftarrow (-2x-14)-(-2x-12) \end{array}$$

The answer is $x - 2 + \dfrac{-2}{x+6}$, or $x - 2 - \dfrac{2}{x+6}$.

22. $x + 7 + \dfrac{5}{x-2}$

23.
$$\begin{array}{r} x - 3 \\ x+3 \overline{\smash{\big)}\, x^2+0x-9} \leftarrow \text{Filling in the missing term} \\ \underline{x^2+3x} \\ -3x-9 \leftarrow x^2-(x^2+3x)=-3x \\ \underline{-3x-9} \\ 0 \leftarrow (-3x-9)-(-3x-9) \end{array}$$

The answer is $x - 3$.

24. $x - 5$

25.
$$\begin{array}{r} x^4-x^3+x^2-x+1 \\ x+1 \overline{\smash{\big)}\, x^5+0x^4+0x^3+0x^2+0x+1} \leftarrow \text{Filling in} \\ \text{missing terms} \\ \underline{x^5+x^4} \\ -x^4 \leftarrow x^5-(x^5+x^4) \\ \underline{-x^4-x^3} \\ x^3 \leftarrow -x^4-(-x^4-x^3) \\ \underline{x^3+x^2} \\ -x^2 \leftarrow x^3-(x^3+x^2) \\ \underline{-x^2-x} \\ x+1 \leftarrow -x^2- \\ \underline{x+1} \quad (-x^2-x) \\ 0 \leftarrow (x+1)- \\ (x+1) \end{array}$$

The answer is $x^4 - x^3 + x^2 - x + 1$.

26. $x^4 + x^3 + x^2 + x + 1$

27.
$$\begin{array}{r} 2x^2-7x+4 \\ 4x+3 \overline{\smash{\big)}\, 8x^3-22x^2-5x+12} \\ \underline{8x^3+6x^2} \\ -28x^2-5x \leftarrow (8x^3-22x^2)- \\ (8x^3+6x^2)=-28x^2 \\ \underline{-28x^2-21x} \\ 16x+12 \leftarrow (-28x^2-5x)- \\ (-28x^2-21x)=16x \\ \underline{16x+12} \\ 0 \leftarrow (16x+12)-(16x+12) \end{array}$$

The answer is $2x^2 - 7x + 4$.

28. $x^2 - 3x + 1$

29.
$$\begin{array}{r} x^3-6 \\ x^3-7 \overline{\smash{\big)}\, x^6-13x^3+42} \\ \underline{x^6-7x^3} \\ -6x^3+42 \leftarrow (x^6-13x^3)- \\ (x^6-7x^3)=-6x^3 \\ \underline{-6x^3+42} \\ 0 \leftarrow (-6x^3+42)-(-6x^3+42) \end{array}$$

The answer is $x^3 - 6$.

30. $x^3 + 8$

31.
$$\begin{array}{r} x^3+2x^2+4x+8 \\ x-2 \overline{\smash{\big)}\, x^4+0x^3+0x^2+0x-16} \\ \underline{x^4-2x^3} \\ 2x^3 \leftarrow x^4-(x^4-2x^3)=2x^3 \\ \underline{2x^3-4x^2} \\ 4x^2 \leftarrow 2x^3-(2x^3-4x^2)=4x^2 \\ \underline{4x^2-8x} \\ 8x-16 \leftarrow 4x^2-(4x^2-8x)= \\ \underline{8x-16} \quad 8x \\ 0 \leftarrow (8x-16)-(8x-16) \end{array}$$

The answer is $x^3 + 2x^2 + 4x + 8$.

32. $x^3 + 3x^2 + 9x + 27$

33.
$$\begin{array}{r} t^2+1 \\ t-1 \overline{\smash{\big)}\, t^3-t^2+t-1} \\ \underline{t^3-t^2} \\ 0+t-1 \leftarrow (t^3-t^2)-(t^3-t^2)=0 \\ \underline{t-1} \\ 0 \leftarrow (t-1)-(t-1) \end{array}$$

The answer is $t^2 + 1$.

34. $t^2 - 2t + 3 - \dfrac{4}{t+1}$

35. *Familiarize.* Let $w =$ the width. Then $w + 15 =$ the length. We draw a picture.

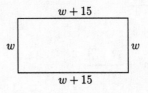

We will use the fact that the perimeter is 640 ft to find w (the width). Then we can find $w + 15$ (the length) and multiply the length and the width to find the area.

Translate.

Width+Width+ Length + Length =Perimeter
$\quad w \quad + \quad w \quad +(w+15)+(w+15)= \quad 640$

Carry out.

$$w + w + (w + 15) + (w + 15) = 640$$
$$4w + 30 = 640$$
$$4w = 610$$
$$w = 152.5$$

If the width is 152.5, then the length is $152.5+15$, or 167.5. The area is $(167.5)(152.5)$, or $25,543.75$ ft^2.

Check. The length, 167.5 ft, is 15 ft greater than the width, 152.5 ft. The perimeter is $152.5 + 152.5 + 167.5 + 167.5$, or 640 ft. We should also recheck the computation we used to find the area. The answer checks.

State. The area is 25,543.75 ft^2.

36. $\left\{ x \middle| x < -\dfrac{12}{5} \right\}$

37. To plot $(4, -1)$ we start at the origin and move 4 units to the right and then down 1 unit. To plot $(0, 5)$ we start at the origin and move 0 units horizontally and up 5 units. To plot $(-2, 3)$ we start at the origin and move 2 units to the left and then up 3 units. To plot $(-3, 0)$ we start at the origin and move 3 units to the left and 0 units vertically.

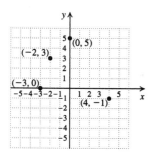

38. III

39.

40.

41.

$$
\begin{array}{r}
x^2 + 5 \\
x^2 + 4\overline{\smash{)}x^4 + 9x^2 + 20} \\
\underline{x^4 + 4x^2} \\
5x^2 + 20 \\
\underline{5x^2 + 20} \\
0
\end{array}
$$

The answer is $x^2 + 5$.

42. $y^3 - ay^2 + a^2y - a^3 + \dfrac{a^2 + a^4}{y + a}$

43.

$$
\begin{array}{r}
a + 3 \\
5a^2 - 7a - 2\overline{\smash{)}5a^3 + 8a^2 - 23a - 1} \\
\underline{5a^3 - 7a^2 - 2a} \\
15a^2 - 21a - 1 \\
\underline{15a^2 - 21a - 6} \\
5
\end{array}
$$

The answer is $a + 3 + \dfrac{5}{5a^2 - 7a - 2}$.

44. $5y + 2 + \dfrac{-10y + 11}{3y^2 - 5y - 2}$

45.
$$(4x^5 - 14x^3 - x^2 + 3)+$$
$$(2x^5 + 3x^4 + x^3 - 3x^2 + 5x)$$
$$= 6x^5 + 3x^4 - 13x^3 - 4x^2 + 5x + 3$$

$$
\begin{array}{r}
2x^2 + x - 3 \\
3x^3 - 2x - 1\overline{\smash{)}6x^5 + 3x^4 - 13x^3 - 4x^2 + 5x + 3} \\
\underline{6x^5 - 4x^3 - 2x^2} \\
3x^4 - 9x^3 - 2x^2 + 5x \\
\underline{3x^4 - 2x^2 - x} \\
-9x^3 + 6x + 3 \\
\underline{-9x^3 + 6x + 3} \\
0
\end{array}
$$

The answer is $2x^2 + x - 3$.

46. $5x^5 + 5x^4 - 8x^2 - 8x + 2$

47.

$$
\begin{array}{r}
3a^{2h} + 2a^h - 5 \\
2a^h + 3\overline{\smash{)}6a^{3h} + 13a^{2h} - 4a^h - 15} \\
\underline{6a^{3h} + 9a^{2h}} \\
4a^{2h} - 4a^h \\
\underline{4a^{2h} + 6a^h} \\
-10a^h - 15 \\
\underline{-10a^h - 15} \\
0
\end{array}
$$

The answer is $3a^{2h} + 2a^h - 5$.

48. -5

49.

$$
\begin{array}{r}
2x + (3c + 2) \\
x - 1\overline{\smash{)}2x^2 + 3cx - 8} \\
\underline{2x^2 - 2x} \\
(3c + 2)x - 8 \\
\underline{(3c + 2)x - (3c + 2)} \\
-8 + (3c + 2)
\end{array}
$$

We set the remainder equal to 0:

$$-8 + 3c + 2 = 0$$
$$3c - 6 = 0$$
$$3c = 6$$
$$c = 2$$

Thus, c must be 2.

50. 1

Exercise Set 4.8

1. $(x^3 - 2x^2 + 2x - 5) \div (x - 1)$

$$
\begin{array}{r|rrrr}
1 & 1 & -2 & 2 & -5 \\
  &   & 1 & -1 & 1 \\
\hline
  & 1 & -1 & 1 & -4
\end{array}
$$

The answer is $x^2 - x + 1$ with R -4, or

$x^2 - x + 1 + \dfrac{-4}{x - 1}$.

2. $x^2 - 3x + 5$, R -10, or $x^2 - 3x + 5 + \dfrac{-10}{x + 1}$

3. $(a^2 + 11a - 19) \div (a + 4) =$

$(a^2 + 11a - 19) \div [a - (-4)]$

$$
\begin{array}{r|rrr}
-4 & 1 & 11 & -19 \\
   &   & -4 & -28 \\
\hline
   & 1 & 7 & -47
\end{array}
$$

The answer is $a + 7$ with R -47, or $a + 7 + \dfrac{-47}{a + 4}$.

4. $a + 15$, R 41, or $a + 15 + \dfrac{41}{a - 4}$

5. $(x^3 - 7x^2 - 13x + 3) \div (x - 2)$

$$
\begin{array}{r|rrrr}
2 & 1 & -7 & -13 & 3 \\
  &   & 2 & -10 & -46 \\
\hline
  & 1 & -5 & -23 & -43
\end{array}
$$

The answer is $x^2 - 5x - 23$ with R -43, or

$x^2 - 5x - 23 + \dfrac{-43}{x - 2}$.

6. $x^2 - 9x + 5$, R -7, or $x^2 - 9x + 5 + \dfrac{-7}{x + 2}$

7. $(3x^3 + 7x^2 - 4x + 3) \div (x + 3) =$

$(3x^3 + 7x^2 - 4x + 3) \div [x - (-3)]$

$$
\begin{array}{r|rrrr}
-3 & 3 & 7 & -4 & 3 \\
   &   & -9 & 6 & -6 \\
\hline
   & 3 & -2 & 2 & -3
\end{array}
$$

The answer is $3x^2 - 2x + 2$ with R -3, or

$3x^2 - 2x + 2 + \dfrac{-3}{x + 3}$.

8. $3x^2 + 16x + 44$, R 135, or $3x^2 + 16x + 44 + \dfrac{135}{x - 3}$

9. $(y^3 - 3y + 10) \div (y - 2) =$

$(y^3 + 0y^2 - 3y + 10) \div (y - 2)$

$$
\begin{array}{r|rrrr}
2 & 1 & 0 & -3 & 10 \\
  &   & 2 & 4 & 2 \\
\hline
  & 1 & 2 & 1 & 12
\end{array}
$$

The answer is $y^2 + 2y + 1$ with R 12, or

$y^2 + 2y + 1 + \dfrac{12}{y - 2}$.

10. $x^2 - 4x + 8$, R -8, or $x^2 - 4x + 8 + \dfrac{-8}{x + 2}$

11. $(3x^4 - 25x^2 - 18) \div (x - 3) =$

$(3x^4 + 0x^3 - 25x^2 + 0x - 18) \div (x - 3)$

$$
\begin{array}{r|rrrrr}
3 & 3 & 0 & -25 & 0 & -18 \\
  &   & 9 & 27 & 6 & 18 \\
\hline
  & 3 & 9 & 2 & 6 & 0
\end{array}
$$

The answer is $3x^3 + 9x^2 + 2x + 6$ with R 0, or $3x^3 + 9x^2 + 2x + 6$.

12. $6y^3 - 3y^2 + 9y + 1$, R 3, or $6y^3 - 3y^2 + 9y + 1 + \dfrac{3}{y + 3}$

13. $(x^3 - 27) \div (x - 3) =$

$(x^3 + 0x^2 + 0x - 27) \div (x - 3)$

$$
\begin{array}{r|rrrr}
3 & 1 & 0 & 0 & -27 \\
  &   & 3 & 9 & 27 \\
\hline
  & 1 & 3 & 9 & 0
\end{array}
$$

The answer is $x^2 + 3x^3 + 9$ with R 0, or $x^2 + 3x + 9$.

14. $y^2 - 3y + 9$

15. $(y^5 - 1) \div (y - 1) =$

$(y^5 + 0y^4 + 0y^3 + 0y^2 + 0y - 1) \div (y - 1)$

$$
\begin{array}{r|rrrrrr}
1 & 1 & 0 & 0 & 0 & 0 & -1 \\
  &   & 1 & 1 & 1 & 1 & 1 \\
\hline
  & 1 & 1 & 1 & 1 & 1 & 0
\end{array}
$$

The answer is $y^4 + y^3 + y^2 + y + 1$ with R 0, or $y^4 + y^3 + y^2 + y + 1$.

16. $x^4 + 2x^3 + 4x^2 + 8x + 16$

17. $(3x^4 + 8x^3 + 2x^2 - 7x - 4) \div (x + 2) =$

$(3x^4 + 8x^3 + 2x^2 - 7x - 4) \div [(x - (-2)]$

$$
\begin{array}{r|rrrrr}
-2 & 3 & 8 & 2 & -7 & -4 \\
   &   & -6 & -4 & 4 & 6 \\
\hline
   & 3 & 2 & -2 & -3 & 2
\end{array}
$$

The answer is $3x^3 + 2x^2 - 2x - 3$ with R 2, or

$3x^3 + 2x^2 - 2x - 3 + \dfrac{2}{x + 2}$.

18. $2x^3 - 3x^2 - 2x + 3$, R 4, or

$2x^3 - 3x^2 - 2x + 3 + \dfrac{4}{x + 1}$.

19. $(3x^3 + 7x^2 - x + 1) \div \left(x + \dfrac{1}{3}\right) =$

$(3x^3 + 7x^2 - x + 1) \div \left[x - \left(-\dfrac{1}{3}\right)\right]$

$$
\begin{array}{r|rrrr}
-\frac{1}{3} & 3 & 7 & -1 & 1 \\
             &   & -1 & -2 & 1 \\
\hline
             & 3 & 6 & -3 & 2
\end{array}
$$

The answer is $3x^2 + 6x - 3$ with R 2, or

$3x^2 + 6x - 3 + \dfrac{2}{x + \frac{1}{3}}$.

20. $8x^2 - 2x + 6$, R 2, or $8x^2 - 2x + 6 + \dfrac{2}{x - \frac{1}{2}}$.

21. $3y + (-10y) = (3 + (-10))y = -7y$

22. x

23. $\begin{aligned} 10m + 5 + m + 1 &= 10m + m + 5 + 1 \\ &= (10 + 1)m + 5 + 1 \\ &= 11m + 6 \end{aligned}$

24. $13x - 6$

25. ◈

26. ◈

27. a) $\begin{array}{r|rrrr} 4 & 1 & -5 & 5 & -4 \\ & & 4 & -4 & 4 \\ \hline & 1 & -1 & 1 & 0 \end{array}$

The remainder is 0. Therefore, we know that
$x^3 - 5x^2 + 5x - 4 = (x - 4)(x^2 - x + 1)$.

b) $f(x) = (x - 4) \cdot p(x)$ for some polynomial $p(x)$,
so $f(4) = (4 - 4)p(x) = 0$.

c) $\begin{aligned} f(4) &= 4^3 - 5 \cdot 4^2 + 5 \cdot 4 - 4 \\ &= 64 - 80 + 20 - 4 \\ &= 0 \end{aligned}$

28. a) $\begin{array}{r|rrrrr} 2 & 2 & 7 & -4 & -27 & -18 \\ & & 4 & 22 & 36 & 18 \\ \hline & 2 & 11 & 18 & 9 & 0 \end{array}$

The remainder is 0. Therefore, we know that
$2x^4 + 7x^3 - 4x^2 - 27x - 18 =$
$(x - 2)(2x^3 + 11x^2 + 18x + 9)$.

b) $P = (x - 2)(2x^3 + 11x^2 + 18x + 9)$

c) $\begin{array}{r|rrrr} -3 & 2 & 11 & 18 & 9 \\ & & -6 & -15 & -9 \\ \hline & 2 & 5 & 3 & 0 \end{array}$

The remainder is 0. Therefore, we know that
$2x^3 + 11x^2 + 18x + 9 = (x + 3)(2x^2 + 5x + 3)$.

d) $P = (x - 2)(x + 3)(2x^2 + 5x + 3)$

e) $\begin{array}{r|rrr} -1 & 2 & 5 & 3 \\ & & -2 & -3 \\ \hline & 2 & 3 & 0 \end{array}$

The remainder is 0. Therefore, we know that
$2x^2 + 5x + 3 = (x + 1)(2x + 3)$.

f) $P = (x - 2)(x + 3)(x + 1)(2x + 3)$

Exercise Set 4.9

1. $3^{-2} = \dfrac{1}{3^2} = \dfrac{1}{9}$

2. $\dfrac{1}{2^3} = \dfrac{1}{8}$

3. $10^{-4} = \dfrac{1}{10^4} = \dfrac{1}{10,000}$

4. $\dfrac{1}{5^6} = \dfrac{1}{15,625}$

5. $7^{-3} = \dfrac{1}{7^3} = \dfrac{1}{343}$

6. $\dfrac{1}{5^2} = \dfrac{1}{25}$

7. $a^{-3} = \dfrac{1}{a^3}$

8. $\dfrac{1}{x^2}$

9. $\dfrac{1}{y^{-4}} = y^4$

10. t^7

11. $\dfrac{1}{z^{-n}} = z^n$

12. h^m

13. $2^{-1} = \dfrac{1}{2^1} = \dfrac{1}{2}$

14. $\dfrac{3}{2}$

15. $\left(\dfrac{1}{4}\right)^{-2} = \dfrac{1}{\left(\dfrac{1}{4}\right)^2} = \dfrac{1}{\dfrac{1}{16}} = 1 \cdot \dfrac{16}{1} = 16$

16. $\dfrac{1}{\left(\dfrac{4}{5}\right)^2} = \dfrac{25}{16}$

17. $\dfrac{1}{4^3} = 4^{-3}$

18. 5^{-2}

19. $\dfrac{1}{x^3} = x^{-3}$

20. y^{-2}

21. $\dfrac{1}{a^4} = a^{-4}$

22. t^{-5}

23. $\dfrac{1}{p^n} = p^{-n}$

24. m^{-n}

25. $\dfrac{1}{5} = \dfrac{1}{5^1} = 5^{-1}$

26. 8^{-1}

27. $\dfrac{1}{t} = \dfrac{1}{t^1} = t^{-1}$

28. m^{-1}

29. $3^{-5} \cdot 3^8 = 3^{-5+8} = 3^3$

30. 5

31. $x^{-2} \cdot x = x^{-2+1} = x^{-1}$, or $\dfrac{1}{x}$

32. 1

33. $x^{-7} \cdot x^{-6} = x^{-13}$, or $\dfrac{1}{x^{13}}$

34. y^{-13}, or $\dfrac{1}{y^{13}}$

35. $\dfrac{m^6}{m^{12}} = m^{6-12} = m^{-6}$, or $\dfrac{1}{m^6}$

36. p^{-1}, or $\dfrac{1}{p}$

37. $\dfrac{(8x)^6}{(8x)^{10}} = (8x)^{6-10} = (8x)^{-4}$, or $\dfrac{1}{(8x)^4}$

38. $(9t)^{-7}$, or $\dfrac{1}{(9t)^7}$

39. $\dfrac{18^9}{18^9} = 18^{9-9} = 18^0 = 1$

40. 1

41. $(a^{-3}b^{-5})(a^{-4}b^{-6}) = a^{-3+(-4)}b^{-5+(-6)} = a^{-7}b^{-11}$, or $\dfrac{1}{a^7b^{11}}$

42. $x^{-5}y^{-9}$, or $\dfrac{1}{x^5y^9}$

43. $\dfrac{x^7}{x^{-2}} = x^{7-(-2)} = x^9$

44. t^{11}

45. $\dfrac{z^{-6}}{z^{-2}} = z^{-6-(-2)} = z^{-4}$, or $\dfrac{1}{z^4}$

46. y^{-4}, or $\dfrac{1}{y^4}$

47. $\dfrac{x^{-5}}{x^{-8}} = x^{-5-(-8)} = x^3$

48. y^5

49. $\dfrac{x}{x^{-1}} = x^{1-(-1)} = x^2$

50. x^5

51. $(a^{-3})^5 = a^{-3\cdot 5} = a^{-15}$, or $\dfrac{1}{a^{15}}$

52. x^{-30}, or $\dfrac{1}{x^{30}}$

53. $(5^2)^{-3} = 5^{2(-3)} = 5^{-6}$, or $\dfrac{1}{5^6}$

54. 9^{-12}, or $\dfrac{1}{9^{12}}$

55. $(x^{-3})^{-4} = x^{(-3)(-4)} = x^{12}$

56. a^{30}

57. $(m^{-3})^7 = m^{-3\cdot 7} = m^{-21}$, or $\dfrac{1}{m^{21}}$

58. n^{-16}, or $\dfrac{1}{n^{16}}$

59. $(ab)^{-3} = a^{-3}b^{-3}$, or $\dfrac{1}{a^3b^3}$

60. $m^{-5}n^{-5}$, or $\dfrac{1}{m^5n^5}$

61. $(5ab)^{-2} = 5^{-2}a^{-2}b^{-2}$, or $\dfrac{1}{5^2a^2b^2}$, or $\dfrac{1}{25a^2b^2}$

62. $4^{-2}x^{-2}y^{-2}$, or $\dfrac{1}{16x^2y^2}$

63. $(6x^{-5})^2 = 6^2x^{-10} = 36x^{-10}$, or $\dfrac{36}{x^{10}}$

64. $81a^{-16}$, or $\dfrac{81}{a^{16}}$

65. $(x^4y^5)^{-3} = (x^4)^{-3}(y^5)^{-3} = x^{4(-3)}y^{5(-3)} = x^{-12}y^{-15} = \dfrac{1}{x^{12}y^{15}}$

66. $t^{-20}x^{-12}$, or $\dfrac{1}{t^{20}x^{12}}$

67. $(x^{-6}y^{-2})^{-4} = (x^{-6})^{-4}(y^{-2})^{-4}$ $= x^{(-6)(-4)}y^{(-2)(-4)} = x^{24}y^8$

68. $x^{10}y^{35}$

69. $(3x^3y^{-8}z^{-3})^2 = 3^2(x^3)^2(y^{-8})^2(z^{-3})^2 = 9x^6y^{-16}z^{-6} = \dfrac{9x^6}{y^{16}z^6}$

70. $8a^6y^{-12}z^{-15}$, or $\dfrac{8a^6}{y^{12}z^{15}}$

71. $(x^3y^{-4}z^{-5})(x^{-4}y^{-2}z^9) = x^{3+(-4)}y^{-4+(-2)}z^{-5+9} = x^{-1}y^{-6}z^4$, or $\dfrac{z^4}{xy^6}$

72. $a^{-8}b^5c^4$, or $\dfrac{b^5c^4}{a^8}$

73. $(m^{-4}n^7p^3)(m^9n^{-2}p^{-10}) = m^{-4+9}n^{7+(-2)}p^{3+(-10)} = m^5n^5p^{-7}$, or $\dfrac{m^5n^5}{p^7}$

74. $t^{-14}p^3m^6$, or $\dfrac{p^3m^6}{t^{14}}$

75. $\left(\dfrac{y^2}{2}\right)^{-3} = \dfrac{(y^2)^{-3}}{2^{-3}} = \dfrac{y^{-6}}{2^{-3}} = \dfrac{2^3}{y^6} = \dfrac{8}{y^6}$

76. $\dfrac{9}{a^8}$

77. $\left(\dfrac{3}{a^2}\right)^3 = \dfrac{3^3}{(a^2)^3} = \dfrac{27}{a^6}$

78. $\dfrac{49}{x^{14}}$

79. $\left(\dfrac{x^2y}{z}\right)^3 = \dfrac{(x^2)^3y^3}{z^3} = \dfrac{x^6y^3}{z^3}$

80. $\dfrac{m^3}{n^{12}p^3}$

81. $\left(\dfrac{a^2b}{cd^3}\right)^{-2} = \dfrac{(a^2)^{-2}b^{-2}}{c^{-2}(d^3)^{-2}} = \dfrac{a^{-4}b^{-2}}{c^{-2}d^{-6}} = \dfrac{c^2d^6}{a^4b^2}$

82. $\dfrac{27b^{12}}{8a^6}$

83. 2.14×10^3

Since the exponent is positive, the decimal point will move to the right.

2.140.   The decimal point moves right 3 places.

$2.14 \times 10^3 = 2140$

84. 892

85. 6.92×10^{-3}

Since the exponent is negative, the decimal point will move to the left.

.006.92   The decimal point moves left 3 places.

$6.92 \times 10^{-3} = 0.00692$

86. 0.000726

87. 7.84×10^8

Since the exponent is positive, the decimal point will move to the right.

7.84000000.

$\qquad$ 8 places

$7.84 \times 10^8 = 784,000,000$

88. $13,500,000$

89. 8.764×10^{-10}

Since the exponent is negative, the decimal point will move to the left.

0.0000000008.764

$\qquad$ 10 places

$8.764 \times 10^{-10} = 0.0000000008764$

90. 0.009043

91. $10^8 = 1 \times 10^8$

Since the exponent is positive, the decimal point will move to the right.

1.00000000.

$\qquad$ 8 places

$10^8 = 100,000,000$

92. $10,000$

93. $10^{-4} = 1 \times 10^{-4}$

Since the exponent is negative, the decimal point will move to the left.

.0001.

$\qquad$ 4 places

$10^{-4} = 0.0001$

94. 0.0000001

95. $25,000 = 2.5 \times 10^n$

To write 2.5 as 25,000 we move the decimal point 4 places to the right. Thus, n is 4 and

$$25,000 = 2.5 \times 10^4.$$

96. 7.15×10^4

97. $0.00371 = 3.71 \times 10^n$

To write 3.71 as 0.00371 we move the decimal point 3 places to the left. Thus, n is -3 and

$$0.00371 = 3.71 \times 10^{-3}.$$

98. 8.14×10^{-2}

99. $78,000,000,000 = 7.8 \times 10^n$

To write 7.8 as 78,000,000,000 we move the decimal point 10 places to the right. Thus, n is 10 and

$$78,000,000,000 = 7.8 \times 10^{10}.$$

100. 3.7×10^{12}

101. $907,000,000,000,000,000 = 9.07 \times 10^n$

To write 9.07 as 907,000,000,000,000,000 we move the decimal point 17 places to the right. Thus, n is 17 and

$$907,000,000,000,000,000 = 9.07 \times 10^{17}.$$

102. 1.68×10^{14}

103. $0.00000374 = 3.74 \times 10^n$

To write 3.74 as 0.00000374 we move the decimal point 6 places to the left. Thus, n is -6 and

$$0.00000374 = 3.74 \times 10^{-6}.$$

104. 2.75×10^{-10}

105. $0.000000018 = 1.8 \times 10^n$

To write 1.8 as 0.000000018 we move the decimal point 8 places to the left. Thus, n is -8 and

$$0.000000018 = 1.8 \times 10^{-8}.$$

106. 2×10^{-11}

107. $10,000,000 = 1 \times 10^n$, or 10^n

To write 1 as 10,000,000 we move the decimal point 7 places to the right. Thus, n is 7 and

$$10,000,000 = 10^7.$$

108. 10^{11}

109. $0.000000001 = 1 \times 10^n$, or 10^n

To write 1 as 0.000000001 we move the decimal point 9 places to the right. Thus, n is -9 and

$$0.000000001 = 10^{-9}.$$

110. 10^{-7}

111. $(3 \times 10^4)(2 \times 10^5) = (3 \cdot 2) \times (10^4 \cdot 10^5)$
$$= 6 \times 10^{4+5} \quad \text{Adding exponents}$$
$$= 6 \times 10^9$$

112. 6.46×10^5

113. $(5.2 \times 10^5)(6.5 \times 10^{-2}) = (5.2 \cdot 6.5) \times (10^5 \cdot 10^{-2})$
$$= 33.8 \times 10^3$$

The answer is not yet in scientific notation since 33.8 is not a number between 1 and 10. We convert to scientific notation.

$$33.8 \times 10^3 = (3.38 \times 10) \times 10^3 = 3.38 \times 10^4$$

114. 6.106×10^{-11}

115. $(9.9 \times 10^{-6})(8.23 \times 10^{-8}) = (9.9 \cdot 8.23) \times (10^{-6} \cdot 10^{-8})$
$$= 81.477 \times 10^{-14}$$

The answer is not yet in scientific notation because 81.477 is not between 1 and 10. We convert to scientific notation.
$$81.477 \times 10^{-14} = (8.1477 \times 10) \times 10^{-14} =$$
$$8.1477 \times 10^{-13}$$

116. 1.123×10^{-5}

117. $\dfrac{8.5 \times 10^8}{3.4 \times 10^{-5}} = \dfrac{8.5}{3.4} \times \dfrac{10^8}{10^{-5}}$
$$= 2.5 \times 10^{8-(-5)}$$
$$= 2.5 \times 10^{13}$$

118. 2.24×10^{-7}

119. $(3.0 \times 10^6) \div (6.0 \times 10^9) = \dfrac{3.0 \times 10^6}{6.0 \times 10^9}$
$$= \dfrac{3.0}{6.0} \times \dfrac{10^6}{10^9}$$
$$= 0.5 \times 10^{6-9}$$
$$= 0.5 \times 10^{-3}$$

The answer is not yet in scientific notation because 0.5 is not between 1 and 10. We convert to scientific notation.
$$0.5 \times 10^{-3} = (5.0 \times 10^{-1}) \times 10^{-3} =$$
$$5.0 \times 10^{-4}$$

120. 9.375×10^2

121. $\dfrac{7.5 \times 10^{-9}}{2.5 \times 10^{12}} = \dfrac{7.5}{2.5} \times \dfrac{10^{-9}}{10^{12}}$
$$= 3.0 \times 10^{-9-12}$$
$$= 3.0 \times 10^{-21}$$

122. 5×10^{-24}

123. *Familiarize*. We express 3064 and 249 million in scientific notation.
$$3064 = 3.064 \times 10^n$$

To write 3.064 as 3064 we move the decimal point 3 places to the right, so n is 3 and $3064 = 3.064 \times 10^3$.
$$249,000,000 = 2.49 \times 10^n$$

To write 2.49 as 249,000,000 we move the decimal point 8 places to the right, so n is 8 and $249,000,000 = 2.49 \times 10^8$.

Let p = the part of the population that are members of the Professional Bowlers Association.

Translate. We reword the problem.

What is	number of members	divided by	population of the U.S.?
↓ ↓	↓	↓	↓
p =	(3.064×10^3)	÷	(2.49×10^8)

Carry out. We do the computation.
$$p = (3.064 \times 10^3) \div (2.49 \times 10^8)$$
$$p = (3.064 \div 2.49) \times (10^3 \div 10^8)$$
$$p \approx 1.231 \times 10^{3-8}$$
$$p \approx 1.231 \times 10^{-5}$$

Check. We review our computation. Also, the answer seems reasonable since it is smaller than either of the original numbers.

State. Approximately 1.231×10^{-5} of the population are members of the Professional Bowlers Association.

124. 3.3×10^{-2}

125. *Familiarize*. There are 365 days in one year. Express 6.5 million and 365 in scientific notation.
$$6.5 \text{ million} = 6,500,000 = 6.5 \times 10^n$$

To write 6.5 as 6,500,000 we move the decimal point 6 places to the right, so n is 6 and $6,500,000 = 6.5 \times 10^6$.
$$365 = 3.65 \times 10^n$$

To write 3.65 as 365 we move the decimal point 2 places to the right, so n is 2 and $365 = 3.65 \times 10^2$.

Let p = the amount of popcorn Americans eat in one year.

Translate. We reword the problem.

What	is	daily consumption	times	number of days in a year?
↓	↓	↓	↓	↓
p	=	(6.5×10^6)	×	(3.65×10^2)

Carry out. We do the computation.
$$p = (6.5 \times 10^6) \times (3.65 \times 10^2)$$
$$p = (6.5 \times 3.65) \times (10^6 \times 10^2)$$
$$p = 23.725 \times 10^8$$
$$p = (2.3725 \times 10) \times 10^8$$
$$p = 2.3725 \times 10^9$$

Check. We review the computation. Also, the answer seems reasonable since it is larger than 6.5 million.

State. Americans eat 2.3725×10^9 gal of popcorn each year.

126. 1.095×10^9 gal

127. *Familiarize*. There are 60 seconds in one minute and 60 minutes in one hour, so there are 60(60), or 3600 seconds in one hour. There are 24 hours in one day and 365 days in one year, so there are 3600(24)(365), or 31,536,000 seconds in one year.

We express 3600, 31,536,000 and 4,200,000 in scientific notation:
$$3600 = 3.6 \times 10^n$$

To write 3.6 as 3600 we move the decimal point 3 places to the right, so n is 3 and $3600 = 3.6 \times 10^3$.
$$31,536,000 = 3.1536 \times 10^n$$

To write 3.1536 as 31,536,000 we move the decimal point 7 places to the right, so n is 7 and $31,536,000 = 3.1536 \times 10^7$.
$$4,200,000 = 4.2 \times 10^n$$

To write 4.2 as 4,200,000 we move the decimal point 6 places to the right, so n is 6 and $4,200,000 = 4.2 \times 10^6$.

Let $h =$ the discharge in one hour and $y =$ the discharge in one year.

Translate. We reword and write two equations.

To find the discharge in one hour:

$$\underset{\underset{h}{\downarrow}}{\underbrace{\text{What}}} \quad \underset{\underset{=}{\downarrow}}{\text{is}} \quad \underset{\underset{(3.6 \times 10^3)}{\downarrow}}{\underbrace{\begin{array}{c}\text{number of} \\ \text{seconds in} \\ \text{one hour}\end{array}}} \quad \underset{\underset{\times}{\downarrow}}{\text{times}} \quad \underset{\underset{(4.2 \times 10^6)}{\downarrow}}{\underbrace{\begin{array}{c}\text{discharge} \\ \text{per second?}\end{array}}}$$

To find the discharge in one year:

$$\underset{\underset{h}{\downarrow}}{\underbrace{\text{What}}} \quad \underset{\underset{=}{\downarrow}}{\text{is}} \quad \underset{\underset{(3.1536 \times 10^7)}{\downarrow}}{\underbrace{\begin{array}{c}\text{number of} \\ \text{seconds in} \\ \text{one year}\end{array}}} \quad \underset{\underset{\times}{\downarrow}}{\text{times}} \quad \underset{\underset{(4.2 \times 10^6)}{\downarrow}}{\underbrace{\begin{array}{c}\text{discharge} \\ \text{per second?}\end{array}}}$$

Carry out. We do the computations.

$$h = (3.6 \times 10^3) \times (4.2 \times 10^6)$$
$$h = (3.6 \times 4.2) \times (10^3 \times 10^6)$$
$$h = 15.12 \times 10^9 = (1.512 \times 10) \times 10^9$$
$$h = 1.512 \times 10^{10}$$

$$y = (3.1536 \times 10^7) \times (4.2 \times 10^6)$$
$$y = (3.1536 \times 4.2) \times (10^7 \times 10^6)$$
$$y = 13.24512 \times 10^{13} = (1.324512 \times 10) \times 10^{13}$$
$$y = 1.324512 \times 10^{14}$$

Check. We can review the computations. Also, the answers seem reasonable since they are both larger than the numbers we started with.

State. In one hour 1.512×10^{10} cu ft of water is discharged. In one year 1.324512×10^{14} cu ft of water is discharged.

128. 6.7×10^{-2}

129. $-9a + 17a = (-9 + 17)a = 8a$

130. $-17x$

131. To plot $(-4, 1)$ we start at the origin and move 4 units to the left and then up 1 unit. To plot $(-3, -2)$ we start at the origin and move 3 units to the left and then down 2 units. To plot $(5, 2)$ we start at the origin and move 5 units to the right and then up 2 units. To plot $(-1, 4)$ we start at the origin and move 1 unit to the left and then up 4 units.

132. I and IV

133.
$$\frac{(5.2 \times 10^6)(6.1 \times 10^{-11})}{1.28 \times 10^{-3}}$$
$$= \frac{(5.2 \cdot 6.1)}{1.28} \times \frac{(10^6 \cdot 10^{-11})}{10^{-3}}$$
$$= 24.78125 \times 10^{6+(-11)-(-3)}$$
$$= 24.78125 \times 10^{-2}$$
$$= (2.478125 \times 10) \times 10^{-2}$$
$$= 2.478125 \times 10^{-1}$$

134. 1.5234375×10^7

135.
$$\{2.1 \times 10^6 [(2.5 \times 10^{-3}) \div (5.0 \times 10^{-5})]\} \div$$
$$(3.0 \times 10^{17})$$
$$= \{2.1 \times 10^6 [0.5 \times 10^2]\} \div (3.0 \times 10^{17})$$
$$\qquad\qquad \text{Dividing inside the brackets first}$$
$$= \{1.05 \times 10^8\} \div (3.0 \times 10^{17}) \quad \text{Multiplying}$$
$$\qquad\qquad\qquad\qquad \text{inside the braces}$$
$$= 0.35 \times 10^{-9} \quad \text{Dividing}$$
$$= (3.5 \times 10^{-1}) \times 10^{-9} \quad \text{Writing 0.35 in scientific}$$
$$\qquad\qquad\qquad\qquad \text{notation}$$
$$= 3.5 \times 10^{-10} \quad \text{Simplifying}$$

136. (a) 1.6×10^2

(b) 2.5×10^{-11}

137. $4^{-3} \cdot 8 \cdot 16 = (2^2)^{-3} \cdot 2^3 \cdot 2^4 = 2^{-6} \cdot 2^3 \cdot 2^4 = 2$

138. 4

139. $(5^{-12})^2 5^{25} = 5^{-24} 5^{25} = 5$

140. 7

141. $\left(\dfrac{1}{a}\right)^{-n} = \dfrac{1^{-n}}{a^{-n}} = \dfrac{1}{a^{-n}} = a^n$

142. 2.5

143. False; let $x = 2$, $y = 3$, $m = 4$, and $n = 2$:
$$2^4 \cdot 3^2 = 16 \cdot 9 = 144, \text{ but}$$
$$(2 \cdot 3)^{4 \cdot 2} = 6^8 = 1,679,616$$

144. False

145. False; let $x = 5$, $y = 3$, and $m = 2$:
$$(5 - 3)^2 = 2^2 = 4, \text{ but}$$
$$5^2 - 3^2 = 25 - 9 = 16$$

Chapter 5

Polynomials and Factoring

Exercise Set 5.1

1. Answers may vary. $6x^3 = (6x)(x^2) = (3x^2)(2x) = (2x^2)(3x)$

2. Answers may vary. $(3x^2)(3x^2)$, $(9x)(x^3)$, $(3x)(3x^3)$

3. Answers may vary. $-9x^5 = (-3x^2)(3x^3) = (-x)(9x^4) = (3x^2)(-3x^3)$

4. Answers may vary. $(-4x)(3x^5)$, $(-6x^2)(2x^4)$, $(12x^3)(-x^3)$

5. Answers may vary. $24x^4 = (6x)(4x^3) = (-3x^2)(-8x^2) = (2x^3)(12x)$

6. Answers may vary. $(3x)(5x^4)$, $(x^3)(15x^2)$, $(3x^2)(5x^3)$

7. $x^2 - 4x = x \cdot x - x \cdot 4$
$$= x(x - 4)$$

8. $x(x + 8)$

9. $2x^2 + 6x = 2x \cdot x + 2x \cdot 3$
$$= 2x(x + 3)$$

10. $3x(x - 1)$

11. $x^3 + 6x^2 = x^2 \cdot x + x^2 \cdot 6$
$$= x^2(x + 6)$$

12. $x^2(4x^2 + 1)$

13. $8x^4 - 24x^2 = 8x^2 \cdot x^2 - 8x^2 \cdot 3$
$$= 8x^2(x^2 - 3)$$

14. $5x^3(x^2 + 2)$

15. $2x^2 + 2x - 8 = 2 \cdot x^2 + 2 \cdot x - 2 \cdot 4$
$$= 2(x^2 + x - 4)$$

16. $3(2x^2 + x - 5)$

17. $17x^5y^3 + 34x^3y^2 + 51xy$
$$= 17xy \cdot x^4y^2 + 17xy \cdot 2x^2y + 17xy \cdot 3$$
$$= 17xy(x^4y^2 + 2x^2y + 3)$$

18. $16xy^2(x^5y^2 - 2x^4y - 3)$

19. $6x^4 - 10x^3 + 3x^2 = x^2 \cdot 6x^2 - x^2 \cdot 10x + x^2 \cdot 3$
$$= x^2(6x^2 - 10x + 3)$$

20. $x(5x^4 + 10x - 8)$

21. $x^5y^5 + x^4y^3 + x^3y^3 - x^2y^2$
$$= x^2y^2 \cdot x^3y^3 + x^2y^2 \cdot x^2y + x^2y^2 \cdot xy + x^2y^2(-1)$$
$$= x^2y^2(x^3y^3 + x^2y + xy - 1)$$

22. $x^3y^3(x^6y^3 - x^4y^2 + xy + 1)$

23. $2x^7 - 2x^6 - 64x^5 + 4x^3$
$$= 2x^3 \cdot x^4 - 2x^3 \cdot x^3 - 2x^3 \cdot 32x^2 + 2x^3 \cdot 2$$
$$= 2x^3(x^4 - x^3 - 32x^2 + 2)$$

24. $5(2x^3 + 5x^2 + 3x - 4)$

25. $1.6x^4 - 2.4x^3 + 3.2x^2 + 6.4x$
$$= 0.8x(2x^3) - 0.8x(3x^2) + 0.8x(4x) + 0.8x(8)$$
$$= 0.8x(2x^3 - 3x^2 + 4x + 8)$$

26. $0.5x^2(5x^4 - x^2 + 10x + 20)$

27. $\frac{5}{3}x^6 + \frac{4}{3}x^5 + \frac{1}{3}x^4 + \frac{1}{3}x^3$
$$= \frac{1}{3}x^3(5x^3) + \frac{1}{3}x^3(4x^2) + \frac{1}{3}x^3(x) + \frac{1}{3}x^3(1)$$
$$= \frac{1}{3}x^3(5x^3 + 4x^2 + x + 1)$$

28. $\frac{1}{7}x(5x^6 + 3x^4 - 6x^2 - 1)$

29. $y(y + 3) + 4(y + 3)$
$$= (y + 3)(y + 4) \qquad \text{Factoring out the common binomial factor } y+3$$

30. $(b - 5)(b - 3)$

31. $x^2(x + 3) + 2(x + 3)$
$$= (x + 3)(x^2 + 2) \quad \text{Factoring out the common binomial factor } x + 3$$

32. $(2z + 1)(3z^2 + 1)$

33. $y^2(y + 8) + (y + 8) = y^2(y + 8) + 1(y + 8)$
$$= (y + 8)(y^2 + 1) \quad \text{Factoring out the common factor}$$

34. $(x - 7)(x^2 - 3)$

35. $x^3 + 3x^2 + 2x + 6$
$$= (x^3 + 3x^2) + (2x + 6)$$
$$= x^2(x + 3) + 2(x + 3) \quad \text{Factoring each binomial}$$
$$= (x + 3)(x^2 + 2) \quad \text{Factoring out the common factor } x + 3$$

36. $(2z + 1)(3z^2 + 1)$

37. $2x^3 + 6x^2 + x + 3$

   $= (2x^3 + 6x^2) + (x + 3)$

   $= 2x^2(x + 3) + 1(x + 3)$ Factoring each
   binomial

   $= (x + 3)(2x^2 + 1)$

38. $(3x + 2)(x^2 + 1)$

39. $8x^3 - 12x^2 + 6x - 9 = 4x^2(2x - 3) + 3(2x - 3)$

   $= (2x - 3)(4x^2 + 3)$

40. $(2x - 5)(5x^2 + 2)$

41. $12x^3 - 16x^2 + 3x - 4$

   $= 4x^2(3x - 4) + 1(3x - 4)$ Factoring 1 out of
   the second binomial

   $= (3x - 4)(4x^2 + 1)$

42. $(6x - 7)(3x^2 + 5)$

43. $x^3 + 8x^2 - 3x - 24 = x^2(x + 8) - 3(x + 8)$

   $= (x + 8)(x^2 - 3)$

44. $(x + 6)(2x^2 - 5)$

45. $w^3 - 7w^2 + 4w - 28 = w^2(w - 7) + 4(w - 7)$

   $= (w - 7)(w^2 + 4)$

46. $(y + 8)(y^2 - 2)$

47. $x^3 - x^2 - 2x + 5 = x^2(x - 1) - 1(2x - 5)$

   This polynomial is not factorable using factoring by grouping.

48. Not factorable by grouping

49. $2x^3 - 8x^2 - 9x + 36 = 2x^2(x - 4) - 9(x - 4)$

   $= (x - 4)(2x^2 - 9)$

50. $(5g - 1)(4g^2 - 5)$

51. Graph: $y = x - 6$

   The equation is in the form $y = mx + b$, so we know the
   y-intercept is $(0, -6)$. We find two other pairs.

   When $x = 5$, $y = 5 - 6 = -1$.

   When $x = 2$, $y = 2 - 6 = -4$.

x	y
0	-6
5	-1
2	-4

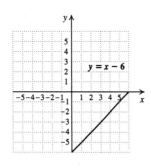

$y = x - 6$

52. $\left\{ x \mid x \le \dfrac{14}{5} \right\}$

53. $-13 - (-25)$

   $= -13 + 25$ Adding an opposite

   $= 12$

54. $p = 2A - q$

55. $(y + 5)(y + 7) = y^2 + 7y + 5y + 35$ Using FOIL

   $= y^2 + 12y + 35$

56. $y^2 + 14y + 49$

57. $(y + 7)(y - 7) = y^2 - 7^2 = y^2 - 49$

   $[(A + B))(A - B) = A^2 - B^2]$

58. $y^2 - 14y + 49$

59. ◇

60. ◇

61. $4x^5 + 6x^3 + 6x^2 + 9 = 2x^3(2x^2 + 3) + 3(2x^2 + 3)$

   $= (2x^2 + 3)(2x^3 + 3)$

62. $(x^2 + 1)(x^4 + 1)$

63. $x^{12} + x^7 + x^5 + 1 = x^7(x^5 + 1) + (x^5 + 1)$

   $= (x^5 + 1)(x^7 + 1)$

64. Not factorable by grouping

65. $p^3 - p^2 + 3p + 3 = p^2(p - 1) + 3(p + 1)$

   This polynomial is not factorable using factoring by grouping.

66. $(x^2 + 2x + 3)(a + 1)$

67. ◇

68. ◇

Exercise Set 5.2

1. $x^2 + 8x + 15$

   Since the constant term and coefficient of the middle term
   are both positive, we look for a factorization of 15 in which
   both factors are positive. Their sum must be 8.

Pairs of factors	Sums of factors
1, 15	16
3, 5	8

   The numbers we want are 3 and 5.

   $x^2 + 8x + 15 = (x + 3)(x + 5)$.

2. $(x + 2)(x + 3)$

3. $x^2 + 7x + 12$

Since the constant term is positive and the coefficient of the middle term is positive, we look for a factorization of 12 in which both factors are positive. Their sum must be 7.

Pairs of factors	Sums of factors
1, 12	13
2, 6	8
3, 4	7

The numbers we want are 3 and 4.

$x^2 + 7x + 12 = (x + 3)(x + 4)$.

4. $(x + 1)(x + 8)$

5. $x^2 - 6x + 9$

Since the constant term is positive and the coefficient of the middle term is negative, we look for a factorization of 9 in which both factors are negative. Their sum must be -6.

Pairs of factors	Sums of factors
$-1, -9$	-10
$-3, -3$	-6

The numbers we want are -3 and -3.

$x^2 - 6x + 9 = (x - 3)(x - 3)$, or $(x - 3)^2$.

6. $(y + 4)(y + 7)$

7. $x^2 + 9x + 14$

Since the constant term is positive and the coefficient of the middle term is positive, we look for a factorization of 14 in which both factors are positive. Their sum must be 9.

Pairs of factors	Sums of factors
1, 14	15
2, 7	9

The numbers we want are 2 and 7.

$x^2 + 9x + 14 = (x + 2)(x + 7)$.

8. $(a + 5)(a + 6)$

9. $b^2 + 5b + 4$

Since the constant term is positive and the coefficient of the middle term is positive, we look for a factorization of 4 in which both factors are positive. Their sum must be 5.

Pairs of factors	Sums of factors
1, 4	5
2, 2	4

The numbers we want are 1 and 4.

$b^2 + 5b + 4 = (b + 1)(b + 4)$.

10. $\left(x - \dfrac{1}{5}\right)^2$

11. $x^2 + \dfrac{2}{3}x + \dfrac{1}{9}$

Since the constant term is positive and the coefficient of the middle term is positive, we look for a factorization of $\dfrac{1}{9}$ in which both factors are positive. Their sum must be $\dfrac{2}{3}$.

Pairs of factors	Sums of factors
$1, \dfrac{1}{9}$	$\dfrac{10}{9}$
$\dfrac{1}{3}, \dfrac{1}{3}$	$\dfrac{2}{3}$

The numbers we want are $\dfrac{1}{3}$ and $\dfrac{1}{3}$.

$x^2 + \dfrac{2}{3}x + \dfrac{1}{9} = \left(x + \dfrac{1}{3}\right)\left(x + \dfrac{1}{3}\right)$, or $\left(x + \dfrac{1}{3}\right)^2$.

12. $(z - 1)(z - 7)$

13. $d^2 - 7d + 10$

Since the constant term is positive and the coefficient of the middle term is negative, we look for a factorization of 10 in which both factors are negative. Their sum must be -7.

Pairs of factors	Sums of factors
$-1, -10$	-11
$-2, -5$	-7

The numbers we want are -2 and -5.

$d^2 - 7d + 10 = (d - 2)(d - 5)$.

14. $(x - 3)(x - 5)$

15. $y^2 - 11y + 10$

Since the constant term is positive and the coefficient of the middle term is negative, we look for a factorization of 10 in which both factors are negative. Their sum must be -11.

Pairs of factors	Sums of factors
$-1, -10$	-11
$-2, -5$	-7

The numbers we want are -1 and -10.

$y^2 - 11y + 10 = (y - 1)(y - 10)$.

16. $(x + 3)(x - 5)$

17. $x^2 + x - 42$

Since the constant term is negative, we look for a factorization of -42 in which one factor is positive and one factor is negative. Their sum must be 1, the coefficient of the middle term.

Pairs of factors	Sums of factors
−1, 42	41
1, −42	−41
−2, 21	19
2, −21	−19
−3, 14	11
3, −14	−11
−6, 7	1
6, −7	−1

The numbers we want are −6 and 7.

$x^2 + x - 42 = (x - 6)(x + 7)$.

18. $(x - 3)(x + 5)$

19. $2x^2 - 14x - 36 = 2(x^2 - 7x - 18)$

After factoring out the common factor, 2, we consider $x^2 - 7x - 18$. Since the constant term is negative, we look for a factorization of −18 in which one factor is positive and one factor is negative. Their sum must be −7, the coefficient of the middle term, so the negative factor must have the larger absolute value. Thus we consider only pairs of factors in which the negative factor has the larger absolute value.

Pairs of factors	Sums of factors
1, −18	−17
2, −9	−7
3, −6	−3

The numbers we want are 2 and −9. The factorization of $x^2 - 7x - 18$ is $(x + 2)(x - 9)$. We must not forget the common factor, 2. The factorization of $2x^2 - 14x - 36$ is $2(x + 2)(x - 9)$.

20. $3(y + 4)(y - 7)$

21. $x^3 - 6x^2 - 16x = x(x^2 - 6x - 16)$

After factoring out the common factor, x, we consider $x^2 - 6x - 16$. Since the constant term is negative, we look for a factorization of −16 in which one factor is positive and one factor is negative. Their sum must be −6, the coefficient of the middle term, so the negative factor must have the larger absolute value. Thus we consider only pairs of factors in which the negative factor has the larger absolute value.

Pairs of factors	Sums of factors
1, −16	−15
2, −8	−6

The numbers we want are 2 and −8.

Then $x^2 - 6x - 16 = (x + 2)(x - 8)$, so $x^3 - 6x^2 - 16x = x(x + 2)(x - 8)$.

22. $x(x + 6)(x - 7)$

23. $y^2 - 4y - 45$

Since the constant term is negative, we look for a factorization of −45 in which one factor is positive and one factor is negative. Their sum must be −4, the coefficient of the

middle term, so the negative factor must have the larger absolute value. Thus we consider only pairs of factors in which the negative factor has the larger absolute value.

Pairs of factors	Sums of factors
1, −45	−44
3, −15	−12
5, −9	−4

The numbers we want are 5 and −9.

$y^2 - 4y - 45 = (y + 5)(y - 9)$.

24. $(x + 5)(x - 12)$

25. $-2x - 99 + x^2 = x^2 - 2x - 99$

Since the constant term is negative, we look for a factorization of −99 in which one factor is positive and one factor is negative. Their sum must be −2, the coefficient of the middle term, so the negative factor must have the larger absolute value. Thus we consider only pairs of factors in which the negative factor has the larger absolute value.

Pairs of factors	Sums of factors
1, −99	−98
3, −33	−30
9, −11	−2

The numbers we want are 9 and −11.

$-2x - 99 + x^2 = (x + 9)(x - 11)$.

26. $(x - 6)(x + 12)$

27. $c^4 + c^3 - 56c^2 = c^2(c^2 + c - 56)$

After factoring out the common factor, c^2, we consider $c^2 + c - 56$. Since the constant term is negative, we look for a factorization of −56 in which one factor is positive and one factor is negative. Their sum must be 1, so the positive factor must have the larger absolute value. Thus we consider only pairs of factors in which the positive factor has the larger absolute value.

Pairs of factors	Sums of factors
−1, 56	55
−2, 28	26
−4, 14	10
−7, 8	1

The numbers we want are −7 and 8. The factorization of $c^2 + c - 56$ is $(c - 7)(c + 8)$, so $c^4 + c^3 - 56c^2 = c^2(c - 7)(c + 8)$.

28. $5(b - 3)(b + 8)$

29. $2a^2 + 4a - 70 = 2(a^2 + 2a - 35)$

After factoring out the common factor, 2, we consider $a^2 + 2a - 35$. Since the constant term is negative, we look for a factorization of −35 in which one factor is positive and one factor is negative. Their sum must be 2, so the positive factor must have the larger absolute value. Thus we consider only pairs of factors in which the positive factor has the larger absolute value.

Pairs of factors	Sums of factors
−1, 35	34
−5, 7	2

The numbers we want are −5 and 7. The factorization of $a^2 + 2a - 35$ is $(a - 5)(a + 7)$, so $2a^2 + 4a - 70 = 2(a - 5)(a + 7)$.

30. $x^3(x + 2)(x - 1)$

31. $x^2 + x + 1$

Since the constant term and the coefficient of the middle term are both positive, we look for a factorization of 1 in which both factors are positive. Their sum must be 1. The only possible pair of factors is 1 and 1, but their sum is not 1. Thus, this polynomial is not factorable into polynomials with integer coefficients.

32. Not factorable

33. $7 - 2p + p^2 = p^2 - 2p + 7$

Since the constant term is positive and the coefficient of the middle term is negative, we look for a factorization of 7 in which both factors are negative. Their sum must be −2. The only possible pair of factors is −1 and −7, but their sum is not −2. Thus, this polynomial is not factorable into polynomials with integer coefficients.

34. Not factorable

35. $x^2 + 20x + 100$

We look for two factors, both positive, whose product is 100 and whose sum is 20.

They are 10 and 10. $10 \cdot 10 = 100$ and $10 + 10 = 20$.

$x^2 + 20x + 100 = (x + 10)(x + 10)$, or $(x + 10)^2$.

36. $(x + 9)(x + 11)$

37. $3x^3 - 63x^2 - 300x = 3x(x^2 - 21x - 100)$

After factoring out the common factor, $3x$, we consider $x^2 - 21x - 100$. We look for two factors, one positive and one negative, whose product is −100 and whose sum is −21.

They are 4 and −25. $4 \cdot (-25) = -100$ and $4 + (-25) = -21$.

$x^2 - 21x - 100 = (x + 4)(x - 25)$, so $3x^3 - 63x^2 - 300x = 3x(x + 4)(x - 25)$.

38. $2x(x - 8)(x - 12)$

39. $x^2 - 21x - 72$

We look for two factors, one positive and one negative, whose product is −72 and whose sum is −21. They are 3 and −24.

$x^2 - 21x - 72 = (x + 3)(x - 24)$.

40. $4(x + 5)^2$

41. $x^2 - 25x + 144$

We look for two factors, both negative, whose product is 144 and whose sum is −25. They are −9 and −16.

$x^2 - 25x + 144 = (x - 9)(x - 16)$.

42. $(y - 9)(y - 12)$

43. $a^4 + a^3 - 132a^2 = a^2(a^2 + a - 132)$

After factoring out the common factor, a^2, we consider $a^2 + a - 132$. We look for two factors, one positive and one negative, whose product is −132 and whose sum is 1. They are −11 and 12.

$a^2 + a - 132 = (a - 11)(a + 12)$, so $a^4 + a^3 - 132a^2 = a^2(a - 11)(a + 12)$.

44. $a^4(a - 6)(a + 15)$

45. $120 - 23x + x^2 = x^2 - 23x + 120$

We look for two factors, both negative, whose product is 120 and whose sum is −23. They are −8 and −15.

$x^2 - 23x + 120 = (x - 8)(x - 15)$.

46. $(d + 6)(d + 16)$

47. First write the polynomial in descending order and factor out −1.

$108 - 3x - x^2 = -x^2 - 3x + 108 = -1(x^2 + 3x - 108)$

Now we factor the polynomial $x^2 + 3x - 108$. We look for two factors, one positive and one negative, whose product is −108 and whose sum is 3. They are −9 and 12.

$x^2 + 3x - 108 = (x - 9)(x + 12)$

The final answer must include −1 which was factored out above.

$-x^2 - 3x + 108 = -1(x - 9)(x + 12)$, or $-(x - 9)(x + 12)$

Using the distributive law to find $-1(x - 9)$, we see that $-1(x - 9)(x + 12)$ can also be expressed as $(-x + 9)(x + 12)$, or $(9 - x)(12 + x)$.

48. $-1(y - 16)(y + 7)$, or $(16 - y)(7 + y)$

49. $y^2 - 0.2y - 0.08$

We look for two factors, one positive and one negative, whose product is −0.08 and whose sum is−0.2. They are −0.4 and 0.2 .

$y^2 - 0.2y - 0.08 = (y - 0.4)(y + 0.2)$

50. $(t - 0.5)(t + 0.2)$

51. $p^2 + 3pq - 10q^2 = p^2 + 3pq - 10q^2$

Think of $3q$ as a "coefficient" of p. Then we look for factors of $-10q^2$ whose sum is $3q$. They are $5q$ and $-2q$.

$p^2 + 3pq - 10q^2 = (p + 5q)(p - 2q)$.

52. $(a - 3b)(a + b)$

53. $m^2 + 5mn + 5n^2 = m^2 + 5nm + 5n^2$

We look for factors of $5n^2$ whose sum is $5n$. The only reasonable possibilities are shown below.

Pairs of factors	Sums of factors
$5n$, n	$6n$
$-5n$, $-n$	$-6n$

There are no factors whose sum is $5n$. Thus, the polynomial is not factorable into polynomials with integer coefficients.

54. $(x - 8y)(x - 3y)$

55. $s^2 - 2st - 15t^2 = s^2 - 2ts - 15t^2$

We look for factors of $-15t^2$ whose sum is $-2t$. They are $-5t$ and $3t$.

$s^2 - 2st - 15t^2 = (s - 5t)(s + 3t)$

56. $(b + 10c)(b - 2c)$

57. $2x^3 - 10x^2 + 12x = 2x(x^2 - 5x + 6)$

After factoring out the common factor, $2x$, we consider $x^2 - 5x + 6$. We look for two factors, both negative, whose product is 6 and whose sum is -5. They are -3 and -2.

$x^2 - 5x + 6 = (x - 3)(x - 2)$, so $2x^3 - 10x^2 + 12x = 2x(x - 3)(x - 2)$.

58. $3a^4(a - 6)(a - 2)$

59. $7a^9 - 28a^8 - 35a^7 = 7a^7(a^2 - 4a - 5)$

After factoring out the common factor, $7a^7$, we consider $a^2 - 4a - 5$. We look for two factors, one positive and one negative, whose product is -5 and whose sum is -4. They are -5 and 1.

$a^2 - 4a - 5 = (a - 5)(a + 1)$, so $7a^9 - 28a^8 - 35a^7 = 7a^7(a - 5)(a + 1)$.

60. $6x^8(x - 7)(x + 2)$

61. $(x + 6)(3x + 4)$

$= 3x^2 + 4x + 18x + 24$ Using FOIL

$= 3x^2 + 22x + 24$

62. $49w^2 + 84w + 36$

63. *Familiarize*. Let $n =$ the number of people arrested the year before.

Translate. We reword the problem.

$$\underbrace{\begin{array}{c}\text{Number} \\ \text{arrested the} \\ \text{year before}\end{array}}\ \text{less } 1.2\% \text{ of }\underbrace{\begin{array}{c}\text{that} \\ \text{number}\end{array}}\ \text{is } 29,090.$$

$$n \quad - \quad 1.2\% \cdot \quad n \quad = 29,090$$

Carry out. We solve the equation.

$n - 1.2\% \cdot n = 29,090$

$1 \cdot n - 0.012n = 29,090$

$0.988n = 29,090$

$n \approx 29,443$ Rounding

Check. 1.2% of 29,443 is $0.012(29,443) \approx 353$ and $29,443 - 353 = 29,090$. The answer checks.

State. Approximately 29,443 people were arrested the year before.

64. $100°, 25°, 55°$

65.

66.

67. $y^2 + my + 50$

We look for pairs of factors whose product is 50. The sum of each pair is represented by m.

Pairs of factors whose product is −50	Sums of factors
1, 50	51
−1, −50	−51
2, 25	27
−2, −25	−27
5, 10	15
−5, −10	−15

The polynomial $y^2 + my + 50$ can be factored if m is 51, -51, 27, -27, 15, or -15.

68. $49, -49, 23, -23, 5, -5$

69. $x^2 - \dfrac{1}{2}x - \dfrac{3}{16}$

We look for two factors, one positive and one negative, whose product is $-\dfrac{3}{16}$ and whose sum is $-\dfrac{1}{2}$.

They are $-\dfrac{3}{4}$ and $\dfrac{1}{4}$.

$-\dfrac{3}{4} \cdot \dfrac{1}{4} = -\dfrac{3}{16}$ and $-\dfrac{3}{4} + \dfrac{1}{4} = -\dfrac{2}{4} = -\dfrac{1}{2}$.

$x^2 - \dfrac{1}{2}x - \dfrac{3}{16} = \left(x - \dfrac{3}{4}\right)\left(x + \dfrac{1}{4}\right)$

70. $\left(x - \dfrac{1}{2}\right)\left(x + \dfrac{1}{4}\right)$

71. $x^2 + \dfrac{30}{7}x - \dfrac{25}{7}$

We look for two factors, one positive and one negative, whose product is $-\dfrac{25}{7}$ and whose sum is $\dfrac{30}{7}$.

They are 5 and $-\dfrac{5}{7}$.

$5 \cdot \left(-\dfrac{5}{7}\right) = -\dfrac{25}{7}$ and $5 + \left(-\dfrac{5}{7}\right) = \dfrac{35}{7} + \left(-\dfrac{5}{7}\right) = \dfrac{30}{7}$.

$x^2 + \dfrac{30}{7}x - \dfrac{25}{7} = (x + 5)\left(x - \dfrac{5}{7}\right)$

72. $\dfrac{1}{3}x(x + 3)(x - 2)$

73. $b^{2n} + 7b^n + 10$

Consider this trinomial as $(b^n)^2 + 7b^n + 10$. We look for numbers p and q such that $b^{2n} + 7b^n + 10 = (b^n + p)(b^n + q)$. We find two factors, both positive, whose product is 10 and whose sum is 7. They are 5 and 2.

$b^{2n} + 7b^n + 10 = (b^n + 5)(b^n + 2)$

74. $(a^m - 7)(a^m - 4)$

75. $\quad (x+1)a^2 + (x+1)3a + (x+1)2$

$\quad = (x+1)(a^2 + 3a + 2)$

After factoring out the common factor $x+1$, we consider $a^2 + 3a + 2$. We look for two factors, whose product is 2 and whose sum is 3. they are 1 and 2.

$a^2 + 3a + 2 = (a+1)(a+2)$, so
$(x+1)a^2 + (x+1)3a + (x+1)2 =$
$(x+1)(a+1)(a+2)$.

76. $(a-5)(x+9)(x-1)$

77. We first label the drawing with additional information.

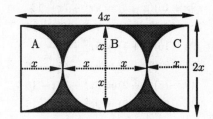

$4x$ represents the length of the rectangle and $2x$ the width. The area of the rectangle is $4x \cdot 2x$, or $8x^2$.

The area of semicircle A is $\frac{1}{2}\pi x^2$.

The area of circle B is πx^2.

The area of semicircle C is $\frac{1}{2}\pi x^2$.

$$\begin{array}{ccccc} \text{Area of} & & \text{Area} & \text{Area} & \text{Area} \\ \text{shaded} = & \text{Area of} & \text{of} - & \text{of} - & \text{of} \\ \text{region} & \text{rectangle} & A & B & C \end{array}$$

$$\begin{array}{ccccc} \text{Area of} \\ \text{shaded} = & 8x^2 & -\frac{1}{2}\pi x^2 & -\pi x^2 & -\frac{1}{2}\pi x^2 \\ \text{region} \end{array}$$

$\quad\quad\quad = 8x^2 - 2\pi x^2$

$\quad\quad\quad = 2x^2(4-\pi)$

The shaded area can be represented by $2x^2(4-\pi)$.

78. $x^2(\pi - 1)$

Exercise Set 5.3

1. $2x^2 - 7x - 4$

(1) Look for a common factor. There is none (other than 1 or -1).

(2) Because $2x^2$ can be factored as $2x \cdot x$, we have this possibility:

$\quad (2x + \quad)(x + \quad)$

(3) There are 3 pairs of factors of -4 and they can be listed two ways:

$\quad\quad -4,1 \quad 4,-1 \quad 2,-2$

$\quad$ and $\quad 1,-4 \quad -1,4 \quad -2,2$

(4) Look for Outer and Inner products resulting from steps (2) and (3) for which the sum is the middle term, $-7x$. We try some possibilities:

$(2x-4)(x+1) = 2x^2 - 2x - 4$

$(2x+4)(x-1) = 2x^2 + 2x - 4$

$(2x+2)(x-2) = 2x^2 - 2x - 4$

$(2x+1)(x-4) = 2x^2 - 7x - 4$

The factorization is $(2x+1)(x-4)$.

2. $(3x-4)(x+1)$

3. $5x^2 + x - 18$

(1) There is no common factor (other than 1 or -1).

(2) Because $5x^2$ can be factored as $5x \cdot x$, we have this possibility:

$\quad (5x + \quad)(x + \quad)$

(3) There are 6 pairs of factors of -18 and they can be listed two ways:

$\quad\quad -18,1 \quad 18,-1 \quad -9,2 \quad 9,-2$

$\quad\quad -6,3 \quad\quad 6,-3$

$\quad$ and $\quad 1,-18 \quad -1,18 \quad 2,-9 \quad -2,9$

$\quad\quad 3,-6 \quad\quad -3,6$

(4) Look for Outer and Inner products resulting from steps (2) and (3) for which the sum is x. We try some possibilities:

$(5x-18)(x+1) = 5x^2 - 13x - 18$

$(5x+18)(x-1) = 5x^2 + 13x - 18$

$(5x-9)(x+2) = 5x^2 + x - 18$

The factorization is $(5x-9)(x+2)$.

4. $(3x+5)(x-3)$

5. $6x^2 + 23x + 7$

(1) There is no common factor (other than 1 or -1).

(2) Because $6x^2$ can be factored as $6x \cdot x$ or $3x \cdot 2x$, we have these possibilities:

$\quad (6x + \quad)(x + \quad)$ and $(3x + \quad)(2x + \quad)$

(3) There are 2 pairs of factors of 7 and they can be listed two ways:

$\quad\quad 7,1 \quad -7,-1$

$\quad$ and $\quad 1,7 \quad -1,-7$

(4) Look for Outer and Inner products resulting from steps (2) and (3) for which the sum is $23x$. Since all signs are positive, we need consider only plus signs. We try some possibilities:

$(6x+7)(x+1) = 6x^2 + 13x + 7$

$(3x+7)(2x+1) = 6x^2 + 17x + 7$

$(6x+1)(x+7) = 6x^2 + 43x + 7$

$(3x+1)(2x+7) = 6x^2 + 23x + 7$

The factorization is $(3x+1)(2x+7)$.

6. $(2x+3)(3x+2)$

7. $3x^2 + 4x + 1$

(1) There is no common factor (other than 1 or -1).

(2) Because $3x^2$ can be factored as $3x \cdot x$, we have this possibility:

$$(3x + \quad)(x + \quad)$$

(3) There are 2 pairs of factors of 1. In this case they can be listed only one way:

$$1, 1 \quad -1, -1$$

(4) Look for Outer and Inner products resulting from steps (2) and (3) for which the sum is the middle term, $4x$. Since all signs are positive, we need consider only plus signs. There is only one such possibility:

$$(3x + 1)(x + 1) = 3x^2 + 4x + 1$$

The factorization is $(3x + 1)(x + 1)$.

8. $(7x + 1)(x + 2)$

9. $4x^2 + 4x - 15$

(1) There is no common factor (other than 1 or -1).

(2) Because $4x^2$ can be factored as $4x \cdot x$ or $2x \cdot 2x$, we have these possibilities:

$$(4x + \quad)(x + \quad) \text{ and } (2x + \quad)(2x + \quad)$$

(3) There are 4 pairs of factors of -15 and they can be listed two ways:

$$15, -1 \quad -15, 1 \quad 5, -3 \quad -5, 3$$
$$\text{and} \quad -1, 15 \quad 1, -15 \quad -3, 5 \quad 3, -5$$

(4) We try some possibilities:

$$(4x + 15)(x - 1) = 4x^2 + 11x - 15$$
$$(2x + 15)(2x - 1) = 4x^2 + 28x - 15$$
$$(4x - 15)(x + 1) = 4x^2 - 11x - 15$$
$$(2x - 15)(2x + 1) = 4x^2 - 28x - 15$$
$$(4x + 5)(x - 3) = 4x^2 - 7x - 15$$
$$(2x + 5)(2x - 3) = 4x^2 + 4x - 15$$

The factorization is $(2x + 5)(2x - 3)$.

10. $(3x - 2)(3x + 4)$

11. $2x^2 - x - 1$

(1) There is no common factor (other than 1 or -1).

(2) Because $2x^2$ can be factored as $2x \cdot x$, we have this possibility:

$$(2x + \quad)(x + \quad)$$

(3) There is 1 pair of factors of -1 and it can be listed two ways:

$$-1, 1 \quad 1, -1$$

(4) We try some possibilities:

$$(2x - 1)(x + 1) = 2x^2 + x - 1$$
$$(2x + 1)(x - 1) = 2x^2 - x - 1$$

The factorization is $(2x + 1)(x - 1)$.

12. $(3x - 5)(5x + 2)$

13. $9x^2 + 18x - 16$

(1) There is no common factor (other than 1 or -1).

(2) Because $9x^2$ can be factored as $9x \cdot x$ or $3x \cdot 3x$, we have these possibilities:

$$(9x + \quad)(x + \quad) \text{ and } (3x + \quad)(3x + \quad)$$

(3) There are 5 pairs of factors of -16 and they can be listed two ways:

$$16, -1 \quad -16, 1 \quad 8, -2 \quad -8, 2 \quad 4, -4$$
$$\text{and} \quad -1, 16 \quad 1, -16 \quad -2, 8 \quad 2, -8 \quad -4, 4$$

(4) We try some possibilities:

$$(9x + 16)(x - 1) = 9x^2 + 7x - 16$$
$$(3x + 16)(3x - 1) = 9x^2 + 45x - 16$$
$$(9x - 16)(x + 1) = 9x^2 - 7x - 16$$
$$(3x - 16)(3x + 1) = 9x^2 - 45x - 16$$
$$(9x + 8)(x - 2) = 9x^2 - 10x - 16$$
$$(3x + 8)(3x - 2) = 9x^2 + 18x - 16$$

The factorization is $(3x + 8)(3x - 2)$.

14. $(2x + 1)(x + 2)$

15. $3x^2 - 5x - 2$

(1) There is no common factor (other than 1 or -1).

(2) Because $3x^2$ can be factored as $3x \cdot x$, we have this possibility:

$$(3x + \quad)(x + \quad)$$

(3) There are 2 pairs of factors of -2 and they can be listed two ways:

$$2, -1 \quad -2, 1$$
$$\text{and} \quad -1, 2 \quad 1, -2$$

(4) We try some possibilities:

$$(3x + 2)(x - 1) = 3x^2 - x - 2$$
$$(3x - 2)(x + 1) = 3x^2 + x - 2$$
$$(3x - 1)(x + 2) = 3x^2 + 5x - 2$$
$$(3x + 1)(x - 2) = 3x^2 - 5x - 2$$

The factorization is $(3x + 1)(x - 2)$.

16. $(6x - 5)(3x + 2)$

17. $12x^2 + 31x + 20$

(1) There is no common factor (other than 1 or -1).

(2) Because $12x^2$ can be factored as $12x \cdot x$, $6x \cdot 2x$ or $4x \cdot 3x$, we have these possibilities:

$$(12x + \quad)(x + \quad) \text{ and } (6x + \quad)(2x + \quad) \text{ and}$$
$$(4x + \quad)(3x + \quad)$$

(3) Since all signs are positive, we need consider only positive pairs of factors of 20. There are 3 such pairs and they can be listed two ways:

$$20, 1 \quad 10, 2 \quad 5, 4$$
$$\text{and} \quad 1, 20 \quad 2, 10 \quad 4, 5$$

(4) We can immediately reject all possibilities in which either factor has a common factor, such as $(12x + 20)$ or $(6x + 4)$, because we determined at the outset that there are no common factors. We try some of the remaining possibilities:

$$(12x + 1)(x + 20) = 12x^2 + 241x + 20$$
$$(12x + 5)(x + 4) = 12x^2 + 53x + 20$$
$$(6x + 1)(2x + 20) = 12x^2 + 122x + 20$$
$$(4x + 5)(3x + 4) = 12x^2 + 31x + 20$$

The factorization is $(4x + 5)(3x + 4)$.

18. $(3x + 5)(5x - 2)$

19. $14x^2 + 19x - 3$

(1) There is no common factor (other than 1 or -1).

(2) Because $14x^2$ can be factored as $14x \cdot x$ or $7x \cdot 2x$, we have these possibilities:

$$(14x + \quad)(x + \quad) \text{ and } (7x + \quad)(2x + \quad)$$

(3) There are 2 pairs of factors of -3 and they can be listed two ways:

$$-1, 3 \quad -3, 1$$
$$\text{and} \quad 3, -1 \quad 1, -3$$

(4) We try some possibilities:

$$(14x - 1)(x + 3) = 14x^2 + 41x - 3$$
$$(7x - 1)(2x + 3) = 14x^2 + 19x - 3$$

The factorization is $(7x - 1)(2x + 3)$.

20. $(7x + 4)(5x + 2)$

21. $9x^2 + 18x + 8$

(1) There is no common factor (other than 1 or -1).

(2) Because $9x^2$ can be factored as $9x \cdot x$ or $3x \cdot 3x$, we have these possibilities:

$$(9x + \quad)(x + \quad) \text{ and } (3x + \quad)(3x + \quad)$$

(3) Since all signs are positive, we need consider only positive pairs of factors of 8. There are 2 such pairs and they can be listed in two ways:

$$8, 1 \quad 4, 2$$
$$\text{and} \quad 1, 8 \quad 2, 4$$

(4) We try some possibilities:

$$(9x + 8)(x + 1) = 9x^2 + 17x + 8$$
$$(3x + 8)(3x + 1) = 9x^2 + 27x + 8$$
$$(9x + 4)(x + 2) = 9x^2 + 22x + 8$$
$$(3x + 4)(3x + 2) = 9x^2 + 18x + 8$$

The factorization is $(3x + 4)(3x + 2)$.

22. Prime

23. $49 - 42x + 9x^2 = 9x^2 - 42x + 49$

(1) There is no common factor (other than 1 or -1).

(2) Because $9x^2$ can be factored as $9x \cdot x$ or $3x \cdot 3x$, we have these possibilities:

$$(9x + \quad)(x + \quad) \text{ and } (3x + \quad)(3x + \quad)$$

(3) Since 49 is positive and the middle term is negative, we need consider only negative pairs of factors of 49. There are 2 such pairs and one pair can be listed two ways:

$$-49, -1 \quad -7, -7$$
$$\text{and} \quad -1, -49$$

(4) We try some possibilities:

$$(9x - 49)(x - 1) = 9x^2 - 58x + 49$$
$$(3x - 49)(3x - 1) = 9x^2 - 150x + 49$$
$$(9x - 7)(x - 7) = 9x^2 - 70x + 49$$
$$(3x - 7)(3x - 7) = 9x^2 - 42x + 49$$

The factorization is $(3x - 7)(3x - 7)$ or $(3x - 7)^2$. This can also be expressed as follows:

$$(3x-7)^2 = (-1)^2(3x-7)^2 = [-1 \cdot (3x-7)]^2 = (-3x+7)^2,$$
or $(7 - 3x)^2$

24. $(5x + 4)^2$

25. $24x^2 + 47x - 2$

(1) There is no common factor (other than 1 or -1).

(2) Because $24x^2$ can be factored as $24x \cdot x$, $12x \cdot 2x$, $6x \cdot 4x$, or $3x \cdot 8x$, we have these possibilities:

$$(24x + \quad)(x + \quad) \text{ and } (12x + \quad)(2x + \quad) \text{ and }$$
$$(6x + \quad)(4x + \quad) \text{ and } (3x + \quad)(8x + \quad)$$

(3) There are 2 pairs of factors of -2 and they can be listed two ways:

$$2, -1 \quad -2, 1$$
$$\text{and} \quad -1, 2 \quad 1, -2$$

(4) We can immediately reject all possibilities in which either a factor has a common factor, such as $(24x + 2)$ or $(12x - 2)$, because we determined at the outset that there are no common factors. We try some of the remaining possibilities:

$$(24x - 1)(x + 2) = 24x^2 + 47x - 2$$

The factorization is $(24x - 1)(x + 2)$.

26. $(8a + 3)(2a + 9)$

27. $35x^2 - 57x - 44$

(1) There is no common factor (other than 1 or -1).

(2) Because $35x^2$ can be factored as $35x \cdot x$ or $7x \cdot 5x$, we have these possibilities:

$$(35x + \quad)(x + \quad) \text{ and } (7x + \quad)(5x + \quad)$$

(3) There are 6 pairs of factors of -44 and they can be listed two ways:

$$1, -44 \quad -1, 44 \quad 2, -22 \quad -2, 22$$
$$4, -11 \quad -4, 11$$
$$\text{and} \quad -44, 1 \quad 44, -1 \quad -22, 2 \quad 22, -2$$
$$-11, 4 \quad 11, -4$$

(4) We try some possibilities:

$$(35x + 1)(x - 44) = 35x^2 - 1539x - 44$$
$$(7x + 1)(5x - 44) = 35x^2 - 303x - 44$$
$$(35x + 2)(x - 22) = 35x^2 - 768x - 44$$
$$(7x + 2)(5x - 22) = 35x^2 - 144x - 44$$
$$(35x + 4)(x - 11) = 35x^2 - 381x - 44$$
$$(7x + 4)(5x - 11) = 35x^2 - 57x - 44$$

The factorization is $(7x + 4)(5x - 11)$.

28. $(3a - 1)(3a + 5)$

29. $2x^2 - 6x - 19$

(1) There is no common factor (other than 1 or -1).

(2) Because $2x^2$ can be factored as $2x \cdot x$ we have this possibility:

$$(2x + \ )(x + \ )$$

(3) There are 2 pairs of factors of -19 and they can be listed two ways:

$$19, -1 \quad -19, 1$$
$$\text{and} \quad -1, 19 \quad 1, -19$$

(4) We try some possibilities:

$$(2x + 19)(x - 1) = 2x^2 + 17x - 19$$
$$(2x - 1)(x + 19) = 2x^2 + 37x - 19$$

The other two possibilities will only change the sign of the middle terms in these trials.

We must conclude that $2x^2 - 6x - 19$ is prime.

30. $(2x + 5)(x - 3)$

31. $12x^2 + 28x - 24$

(1) We factor out the common factor, 4:

$$4(3x^2 + 7x - 6)$$

Then we factor the trinomial $3x^2 + 7x - 6$.

(2) Because $3x^2$ can be factored as $3x \cdot x$ we have this possibility:

$$(3x + \ )(x + \ )$$

(3) There are 4 pairs of factors of -6 and they can be listed two ways:

$$6, -1 \quad -6, 1 \quad 3, -2 \quad -3, 2$$
$$\text{and} \quad -1, 6 \quad 1, -6 \quad -2, 3 \quad 2, -3$$

(4) We can immediately reject all possibilities in which either factor has a common factor, such as $(3x + 6)$ or $(3x - 3)$, because we factored out the largest common factor at the outset. We try some of the remaining possibilities:

$$(3x - 1)(x + 6) = 3x^2 + 17x - 6$$
$$(3x - 2)(x + 3) = 3x^2 + 7x - 6$$

The factorization of $3x^2 + 7x - 6$ is $(3x - 2)(x + 3)$. We must include the common factor in order to get a factorization of the original trinomial.

$$12x^2 + 28x - 24 = 4(3x - 2)(x + 3)$$

32. $3(2x + 1)(x + 5)$

33. $30x^2 - 24x - 54$

(1) Factor out the common factor, 6:

$$6(5x^2 - 4x - 9)$$

Then we factor the trinomial $5x^2 - 4x - 9$.

(2) Because $5x^2$ can be factored as $5x \cdot x$ we have this possibility:

$$(5x + \ )(x + \ )$$

(3) There are 3 pairs of factors of -9 and they can be listed two ways:

$$9, -1 \quad -9, 1 \quad 3, -3$$
$$\text{and} \quad -1, 9 \quad 1, -9 \quad -3, 3$$

(4) We try some possibilities:

$$(5x + 9)(x - 1) = 5x^2 + 4x - 9$$
$$(5x - 9)(x + 1) = 5x^2 - 4x - 9$$

The factorization of $5x^2 - 4x - 9$ is $(5x - 9)(x + 1)$. We must include the common factor in order to get a factorization of the original trinomial.

$$30x^2 - 24x - 54 = 6(5x - 9)(x + 1)$$

34. $5(4x - 1)(x - 1)$

35. $4x + 6x^2 - 10 = 6x^2 + 4x - 10$

(1) Factor out the common factor, 2:

$$2(3x^2 + 2x - 5)$$

Then we factor the trinomial $3x^2 + 2x - 5$.

(2) Because $3x^2$ can be factored as $3x \cdot x$ we have this possibility:

$$(3x + \ )(x + \ )$$

(3) There are 2 pairs of factors of -5 and they can be listed two ways:

$$5, -1 \quad -5, 1$$
$$\text{and} \quad -1, 5 \quad 1, -5$$

(4) We try some possibilities:

$$(3x + 5)(x - 1) = 3x^2 + 2x - 5$$

Then $3x^2 + 2x - 5 = (3x + 5)(x - 1)$, so $6x^2 + 4x - 10 = 2(3x + 5)(x - 1)$.

36. $3(2x - 3)(3x + 1)$

37. $3x^2 - 4x + 1$

(1) There is no common factor (other than 1 or -1).

(2) Because $3x^2$ can be factored as $3x \cdot x$ we have this possibility:

$$(3x + \ )(x + \ )$$

(3) Since 1 is positive and the middle term is negative, we need to consider only negative factor pairs of 1. The only such pair is $-1, -1$.

(4) There is only one possibility:

$$(3x - 1)(x - 1) = 3x^2 - 4x + 1$$

The factorization is $(3x - 1)(x - 1)$.

38. $(2x - 3)(3x - 2)$

39. $12x^2 - 28x - 24$

(1) Factor out the common factor, 4:

$$4(3x^2 - 7x - 6)$$

Then we factor the trinomial $3x^2 - 7x - 6$.

(2) Because $3x^2$ can be factored as $3x \cdot x$ we have this possibility:

$$(3x + \quad)(x + \quad)$$

(3) There are 4 pairs of factors of -6 and they can be listed two ways:

$$6, -1 \quad -6, 1 \quad 3, -2 \quad -3, 2$$
$$\text{and} \quad -1, 6 \quad 1, -6 \quad -2, 3 \quad 2, -3$$

(4) We can immediately reject all possibilities in which either factor has a common factor, such as $(3x - 6)$ or $(3x+3)$, because we factored out the largest common factor at the outset. We try some of the remaining possibilities:

$$(3x - 1)(x + 6) = 3x^2 + 17x - 6$$
$$(3x - 2)(x + 3) = 3x^2 + 7x - 6$$
$$(3x + 2)(x - 3) = 3x^2 - 7x - 6$$

Then $3x^2 - 7x - 6 = (3x + 2)(x - 3)$, so
$12x^2 - 28x - 24 = 4(3x + 2)(x - 3)$.

40. $3(2x - 1)(x - 5)$

41. $-1 + 2x^2 - x = 2x^2 - x - 1$

(1) There is no common factor (other than 1 or -1).

(2) Because $2x^2$ can be factored as $2x \cdot x$ we have this possibility:

$$(2x + \quad)(x + \quad)$$

(3) There is 1 pair of factors of -1 and it can be listed two ways:

$$1, -1 \quad -1, 1$$

(4) We try some possibilities:

$$(2x + 1)(x - 1) = 2x^2 - x - 1$$

The factorization is $(2x + 1)(x - 1)$.

42. $(5x - 3)(3x - 2)$

43. $9x^2 - 18x - 16$

(1) There is no common factor (other than 1 or -1).

(2) Because $9x^2$ can be factored as $9x \cdot x$ or $3x \cdot 3x$, we have these possibilities:

$$(9x + \quad)(x + \quad) \text{ and } (3x + \quad)(3x + \quad)$$

(3) There are 5 pairs of factors of -16 and they can be listed two ways:

$$16, -1 \quad -16, 1 \quad 8, -2 \quad -8, 2 \quad 4, -4$$
$$\text{and} \quad -1, 16 \quad 1, -16 \quad -2, 8 \quad 2, -8 \quad -4, 4$$

(4) We try some possibilities:

$$(9x + 16)(x - 1) = 9x^2 + 7x - 16$$
$$(3x + 16)(3x - 1) = 9x^2 + 45x - 16$$
$$(9x + 8)(x - 2) = 9x^2 - 10x - 16$$
$$(3x + 8)(3x - 2) = 9x^2 + 18x - 16$$
$$(3x - 8)(3x + 2) = 9x^2 - 18x - 16$$

The factorization is $(3x - 8)(3x + 2)$.

44. $7(2x + 1)(x + 2)$

45. $15x^2 - 25x - 10$

(1) Factor out the common factor, 5:

$$5(3x^2 - 5x - 2)$$

Then we factor the trinomial $3x^2 - 5x - 2$. This was done in Exercise 15. We know that $3x^2 - 5x - 2 = (3x+1)(x-2)$, so $15x^2 - 25x - 10 = 5(3x + 1)(x - 2)$.

46. $(6x + 5)(3x - 2)$

47. $12x^3 + 31x^2 + 20x$

(1) Factor out the common factor, x:

$$x(12x^2 + 31x + 20)$$

Then we factor the trinomial $12x^2 + 31x + 20$. This was done in Exercise 17. We know that $12x^2 + 31x + 20 = (3x + 4)(4x + 5)$, so $12x^3 + 31x^2 + 20x = x(3x + 4)(4x + 5)$.

48. $x(5x - 2)(3x + 5)$

49. $14x^4 + 19x^3 - 3x^2$

(1) Factor out the common factor, x^2:

$$x^2(14x^2 + 19x - 3)$$

Then we factor the trinomial $14x^2 + 19x - 3$. This was done in Exercise 19. We know that $14x^2 + 19x - 3 = (7x - 1)(2x + 3)$, so $14x^4 + 19x^3 - 3x^2 = x^2(7x - 1)(2x + 3)$.

50. $2x^2(5x + 2)(7x + 4)$

51. $168x^3 - 45x^2 + 3x$

(1) Factor out the common factor, $3x$:

$$3x(56x^2 - 15x + 1)$$

Then we factor the trinomial $56x^2 - 15x + 1$.

(2) Because $56x^2$ can be factored as $56x \cdot x$, $28x \cdot 2x$, $14x \cdot 4x$, or $7x \cdot 8x$, we have these possibilities:

$$(56x + \quad)(x + \quad) \text{ and } (28x + \quad)(2x + \quad) \text{ and }$$
$$(14x + \quad)(4x + \quad) \text{ and } (7x + \quad)(8x + \quad)$$

(3) Since 1 is positive and the middle term is negative we need consider only the negative factor pair $-1, -1$.

(4) We try some possibilities:

$$(56x - 1)(x - 1) = 56x^2 - 57x + 1$$
$$(28x - 1)(2x - 1) = 56x^2 - 30x + 1$$
$$(14x - 1)(4x - 1) = 56x^2 - 18x + 1$$
$$(7x - 1)(8x - 1) = 56x^2 - 15x + 1$$

Then $56x^2 - 15x + 1 = (7x - 1)(8x - 1)$, so
$168x^3 - 45x^2 + 3x = 3x(7x - 1)(8x - 1)$.

52. $24x^3(3x+2)(2x+1)$

53. $15x^2 - 19x + 6$

(1) There is no common factor (other than 1 or -1).

(2) Because $15x^2$ can be factored as $15x \cdot x$ or $5x \cdot 3x$, we have these possibilities:

$$(15x +\ \ )(x +\ \ ) \text{ and } (5x +\ \ )(3x +\ \ )$$

(3) Since 6 is positive and the middle term is negative, we need consider only negative factor pairs. There are 2 such pairs and they can be listed two ways:

$$-6, -1 \quad -3, -2$$
$$\text{and} \quad -1, -6 \quad -2, -3$$

(4) We can immediately reject all possibilities in which either factor has a common factor, such as $(15x - 6)$ or $(3x - 3)$, because we determined at the outset that there is no common factor. We try some of the remaining possibilities:

$$(15x - 1)(x - 6) = 15x^2 - 91x + 6$$
$$(15x - 2)(x - 3) = 15x^2 - 47x + 6$$
$$(5x - 6)(3x - 1) = 15x^2 - 23x + 6$$
$$(5x - 3)(3x - 2) = 15x^2 - 19x + 6$$

The factorization is $(5x - 3)(3x - 2)$.

54. $(3x+2)(3x+4)$

55. $25t^2 + 80t + 64$

(1) There is no common factor (other than 1 or -1).

(2) Because $25t^2$ can be factored as $25t \cdot t$ or $5t \cdot 5t$, we have these possibilities:

$$(25t +\ \ )(t +\ \ ) \text{ and } (5t +\ \ )(5t +\ \ )$$

(3) Since all signs are positive, we need consider only positive pairs of factors. There are 4 such pairs and 3 of them can be listed two ways:

$$64, 1 \quad 32, 2 \quad 16, 4 \quad 8, 8$$
$$\text{and} \quad 1, 64 \quad 2, 32 \quad 4, 16$$

(4) We try some possibilities:

$$(25t + 64)(t + 1) = 25t^2 + 89t + 64$$
$$(5t + 32)(5t + 2) = 25t^2 + 170t + 64$$
$$(25t + 16)(t + 4) = 25t^2 + 116t + 64$$
$$(5t + 8)(5t + 8) = 25t^2 + 80t + 64$$

The factorization is $(5t + 8)(5t + 8)$ or $(5t + 8)^2$.

56. $(3x - 7)^2$

57. $6x^3 + 4x^2 - 10x$

(1) Factor out the common factor, $2x$:

$$2x(3x^2 + 2x - 5)$$

Then we factor the trinomial $3x^2 + 2x - 5$. We did this in Exercise 35 (after we factored 2 out of the original trinomial). We know that $3x^2 + 2x - 5 = (3x + 5)(x - 1)$, so $6x^3 + 4x^2 - 10x = 2x(3x + 5)(x - 1)$.

58. $3x(3x + 1)(2x - 3)$

59. $25x^2 + 89x + 64$

We follow the same procedure as in Exercise 55. The factorization is the first possibility we tried in step (4): $(25x + 64)(x + 1)$.

60. Prime

61. $x^2 + 3x - 7$

(1) There is no common factor (other than 1 or -1).

(2) Because x^2 can be factored as $x \cdot x$, we have this possibility:

$$(x +\ \ )(x +\ \ )$$

(3) There are 2 pairs of factors of -7. Since the coefficient of x^2 is 1, we only list them one way

$$7, -1 \quad -7, 1$$

(4) We try some possibilities:

$$(x + 7)(x - 1) = x^2 + 6x - 7$$
$$(x - 7)(x + 1) = x^2 - 6x - 7$$

Neither possibility works. Thus, $x^3 + 3x - 7$ is prime. (Note that we could have also used the method of Section 5.2 since the leading coefficient of this trinomial is 1.)

62. Prime

63. $12m^2 + mn - 20n^2$

(1) There is no common factor (other than 1 or -1).

(2) Because $12m^2$ can be factored as $12m \cdot m$, $6m \cdot 2m$, or $3m \cdot 4m$, we have these possibilities:

$$(12m +\ \ )(m +\ \ ) \text{ and } (6m +\ \ )(2m +\ \ )$$
$$\text{and } (3m +\ \ )(4m +\ \ )$$

(3) There are 6 pairs of factors of $-20n^2$ and they can be listed two ways:

$$20n, -n \quad -20n, n \quad 10n, -2n \quad -10n, 2n$$
$$-n, 20n \quad n, -20n \quad -2n, 10n \quad 2n, -10n$$
$$5n, -4n \quad -5n, 4n$$
$$-4n, 5n \quad 4n, -5n$$

(4) We can immediately reject all possibilities in which either factor has a common factor, such as $(12m + 20n)$ or $(4m - 2n)$, because we determined at the outset that there is no common factor. We try some of the remaining possibilities:

$$(12m - n)(m + 20n) = 12m^2 + 239mn - 20n^2$$
$$(12m + 5n)(m - 4n) = 12m^2 - 43mn - 20n^2$$
$$(3m - 20n)(4m + n) = 12m^2 - 77mn - 20n^2$$
$$(3m - 4n)(4m + 5n) = 12m^2 - mn - 20n^2$$
$$(3m + 4n)(4m - 5n) = 12m^2 + mn - 20n^2$$

The factorization is $(3m + 4n)(4m - 5n)$.

64. $(4a + 3b)(3a + 2b)$

65. $6a^2 - ab - 15b^2$

(1) There is no common factor (other than 1 or -1).

(2) Because $6a^2$ can be factored as $6a \cdot a$ or $3a \cdot 2a$, we have these possibilities:

$$(6a + \quad)(a + \quad) \text{ and } (3a + \quad)(2a + \quad)$$

(3) There are 4 pairs of factors of $-15b^2$ and they can be listed two ways:

$$15b, -b \quad -15b, b \quad 5b, -3b \quad -5b, 3b$$
$$\text{and} \quad -b, 15b \quad b, -15b \quad -3b, 5b \quad 3b, -5b$$

(4) We can immediately reject all possibilities in which either factor has a common factor, such as $(6a + 15b)$ or $(3a - 3b)$, because we determined at the outset that there is no common factor. We try some of the remaining possibilities:

$$(6a - b)(a + 15b) = 6a^2 + 89ab - 15b^2$$
$$(3a - b)(2a + 15b) = 6a^2 + 43ab - 15b^2$$
$$(6a + 5b)(a - 3b) = 6a^2 - 13ab - 15b^2$$
$$(3a + 5b)(2a - 3b) = 6a^2 + ab - 15b^2$$
$$(3a - 5b)(2a + 3b) = 6a^2 - ab - 15b^2$$

The factorization is $(3a - 5b)(2a + 3b)$.

66. $(3p + 2q)(p - 6q)$

67. $9a^2 + 18ab + 8b^2$

(1) There is no common factor (other than 1 or -1).

(2) Because $9a^2$ can be factored as $9a \cdot a$ or $3a \cdot 3a$, we have these possibilities:

$$(9a + \quad)(a + \quad) \text{ and } (3a + \quad)(3a + \quad)$$

(3) Since all signs are positive, we need consider only pairs of factors of $8b^2$ with positive coefficients. There are 2 such pairs and they can be listed two ways:

$$8b, b \quad 4b, 2b$$
$$\text{and} \quad b, 8b \quad 2b, 4b$$

(4) We try some possibilities:

$$(9a + 8b)(a + b) = 9a^2 + 17ab + 8b^2$$
$$(3a + 8b)(3a + b) = 9a^2 + 27ab + 8b^2$$
$$(9a + 4b)(a + 2b) = 9a^2 + 22ab + 8b^2$$
$$(3a + 4b)(3a + 2b) = 9a^2 + 18ab + 8b^2$$

The factorization is $(3a + 4b)(3a + 2b)$.

68. $2(5s - 3t)(s + t)$

69. $35p^2 + 34pq + 8q^2$

(1) There is no common factor (other than 1 or -1).

(2) Because $35p^2$ can be factored as $35p \cdot p$ or $7p \cdot 5p$, we have these possibilities:

$$(35p + \quad)(p + \quad) \text{ and } (7p + \quad)(5p + \quad)$$

(3) Since all the signs are positive, we need consider only pairs of factors of $8q^2$ with positive coefficients. There are 2 such pairs and they can be listed two ways:

$$8q, q \quad 4q, 2q$$
$$\text{and} \quad q, 8q \quad 2q, 4q$$

(4) We try some possibilities:

$$(35p + 8q)(p + q) = 35p^2 + 43pq + 8q^2$$
$$(7p + 8q)(5p + q) = 35p^2 + 47pq + 8q^2$$
$$(35p + 4q)(p + 2q) = 35p^2 + 74pq + 8q^2$$
$$(7p + 4q)(5p + 2q) = 35p^2 + 34pq + 8q^2$$

The factorization is $(7p + 4q)(5p + 2q)$.

70. $3(2a + 5b)(5a + 2b)$

71. $18x^2 - 6xy - 24y^2$

(1) Factor out the common factor, 6:

$$6(3x^2 - xy - 4y^2)$$

Then we factor the trinomial $3x^2 - xy - 4y^2$.

(2) Because $3x^2$ can be factored as $3x \cdot x$, we have this possibility:

$$(3x + \quad)(x + \quad)$$

(3) There are 3 pairs of factors of $-4y^2$ and they can be listed two ways:

$$4y, -y \quad -4y, y \quad 2y, -2y$$
$$\text{and} \quad -y, 4y \quad y, -4y \quad -2y, 2y$$

(4) We try some possibilities:

$$(3x + 4y)(x - y) = 3x^2 + xy - 4y^2$$
$$(3x - 4y)(x + y) = 3x^2 - xy - 4y^2$$

Then $3x^2 - xy - 4y^2 = (3x - 4y)(x + y)$, so $18x^2 - 6xy - 24y^2 = 6(3x - 4y)(x + y)$.

72. $5(3a - 4b)(a + b)$

73. $y^2 + 4y + y + 4 = y(y + 4) + 1(y + 4)$
$$= (y + 4)(y + 1)$$

74. $(x + 5)(x + 2)$

75. $x^2 - 4x - x + 4 = x(x - 4) - 1(x - 4)$
$$= (x - 4)(x - 1)$$

76. $(a + 5)(a - 2)$

77. $6x^2 + 4x + 9x + 6 = 2x(3x + 2) + 3(3x + 2)$
$$= (3x + 2)(2x + 3)$$

78. $(3x - 2)(x + 1)$

79. $3x^2 - 4x - 12x + 16 = x(3x - 4) - 4(3x - 4)$
$$= (3x - 4)(x - 4)$$

80. $(4 - 3y)(6 - 5y)$

81. $35x^2 - 40x + 21x - 24 = 5x(7x - 8) + 3(7x - 8)$
$$= (7x - 8)(5x + 3)$$

82. $(4x - 3)(2x - 7)$

83. $4x^2 + 6x - 6x - 9 = 2x(2x + 3) - 3(2x + 3)$
$$= (2x + 3)(2x - 3)$$

84. $(x^2 - 3)(2x^2 - 5)$

85. $2x^2 - 7x - 4$

(a) First look for a common factor. There is none (other than 1).

(b) Multiply the leading coefficient and the constant, 2 and -4: $2(-4) = -8$.

(c) Try to factor -8 so that the sum of the factors is -7.

Pairs of factors	Sums of factors
$-1,\ 8$	7
$1,\ -8$	-7
$-2,\ 4$	2
$2,\ -4$	-2

(d) Split the middle term: $-7x = 1x - 8x$

(e) Factor by grouping:
$$2x^2 - 7x - 4 = 2x^2 + x - 8x - 4$$
$$= x(2x + 1) - 4(2x + 1)$$
$$= (2x + 1)(x - 4)$$

86. $(x + 1)(3x - 4)$

87. $5x^2 + x - 18$

(a) First factor out a common factor. There is none (other than 1).

(b) Multiply the leading coefficient and the constant, 5 and -18: $5(-18) = -90$.

(c) Try to factor -90 so that the sum of the factors is 1.

Pairs of factors	Sums of factors
$-1,\ 90$	89
$1,\ -90$	-89
$-2,\ 45$	43
$2,\ -45$	-43
$-3,\ 30$	27
$3,\ -30$	-27
$-5,\ 18$	13
$5,\ -18$	-13
$-6,\ 15$	9
$6,\ -15$	-9
$-9,\ 10$	1
$9,\ -10$	-1

(d) Split the middle term: $x = -9x + 10x$

(e) Factor by grouping:
$$5x^2 + x - 18 = 5x^2 - 9x + 10x - 18$$
$$= x(5x - 9) + 2(5x - 9)$$
$$= (5x - 9)(x + 2)$$

88. $(x - 3)(3x + 5)$

89. $6x^2 + 23x + 7$

(a) First look for a common factor. There is none (other than 1).

(b) Multiply the leading coefficient and the constant, 6 and 7: $6 \cdot 7 = 42$.

(c) Try to factor 42 so that the sum of the factors is 23. We only need to consider positive factors.

Pairs of factors	Sums of factors
$1,\ \ \ 42$	43
$2,\ \ \ 21$	23
$3,\ \ \ 14$	17
$6,\ \ \ \ 7$	13

(d) Split the middle term: $23x = 2x + 21x$

(e) Factor by grouping:
$$6x^2 + 23x + 7 = 6x^2 + 2x + 21x + 7$$
$$= 2x(3x + 1) + 7(3x + 1)$$
$$= (3x + 1)(2x + 7)$$

90. $(2x + 3)(3x + 2)$

91. $3x^2 + 4x + 1$

(a) First look for a common factor. There is none (other than 1).

(b) Multiply the leading coefficient and the constant, 3 and 1: $3 \cdot 1 = 3$.

(c) Try to factor 3 so that the sum of the factors is 4. The numbers we want are 1 and 3: $1 \cdot 3 = 3$ and $1 + 3 = 4$.

(d) Split the middle term: $4x = 1x + 3x$

(e) Factor by grouping:
$$3x^2 + 4x + 1 = 3x^2 + x + 3x + 1$$
$$= x(3x + 1) + 1(3x + 1)$$
$$= (3x + 1)(x + 1)$$

92. $(x + 2)(7x + 1)$

93. $4x^2 + 4x - 15$

(a) First look for a common factor. There is none (other than 1).

(b) Multiply the leading coefficient and the constant, 4 and -15: $4(-15) = -60$.

(c) Try to factor -60 so that the sum of the factors is 4.

Pairs of factors	Sums of factors
$-1,\quad 60$	59
$1, -60$	-59
$-2,\quad 30$	28
$2, -30$	-28
$-3,\quad 20$	17
$3, -20$	-17
$-4,\quad 15$	11
$4, -15$	-11
$-5,\quad 12$	7
$5, -12$	-7
$-6,\quad 10$	4
$6, -10$	-4

(d) Split the middle term: $4x = -6x + 10x$

(e) Factor by grouping:
$$4x^2 + 4x - 15 = 4x^2 - 6x + 10x - 15$$
$$= 2x(2x - 3) + 5(2x - 3)$$
$$= (2x - 3)(2x + 5)$$

94. $(3x + 4)(3x - 2)$

95. $2x^2 - x - 1$

(a) First look for a common factor, if any. There is none (other than 1).

(b) Multiply the leading coefficient and the constant, 2 and -1: $2(-1) = -2$.

(c) Try to factor -2 so that the sum of the factors is -1. The numbers we want are -2 and 1: $2(-1) = -2$ and $-2 + 1 = -1$.

(d) Split the middle term: $-x = -2x + 1x$

(e) Factor by grouping:
$$2x^2 + x - 1 = 2x^2 - 2x + x - 1$$
$$= 2x(x - 1) + 1(x - 1)$$
$$= (x - 1)(2x + 1)$$

96. $(3x - 5)(5x + 2)$

97. $9x^2 + 18x - 16$

(a) First look for a common factor. There is none (other than 1).

(b) Multiply the leading coefficient and the constant, 9 and -16: $9(-16) = -144$.

(c) Try to factor -144 so that the sum of the factors is 18.

Pairs of factors	Sums of factors
$-1,\quad 144$	143
$1, -144$	-143
$-2,\quad 72$	70
$2, -72$	-70
$-3,\quad 48$	45
$3, -48$	-45
$-4,\quad 36$	32
$4, -36$	-32
$-6,\quad 24$	18
$6, -24$	-18
$-8,\quad 18$	10
$8, -18$	-10
$-9,\quad 16$	7
$9, -16$	-7
$-12,\quad 12$	0

(d) Split the middle term: $18x = -6x + 24x$

(e) Factor by grouping:
$$9x^2 + 18x - 16 = 9x^2 - 6x + 24x - 16$$
$$= 3x(3x - 2) + 8(3x - 2)$$
$$= (3x - 2)(3x + 8)$$

98. $(x + 2)(2x + 1)$

99. $3x^2 - 5x - 2$

(a) First look for a common factor. There is none (other than 1).

(b) Multiply the leading coefficient and the constant, 3 and -2: $3(-2) = -6$.

(c) Try to factor -6 so that the sum of the factors is -5. The number we want are 1 and -6: $1(-6) = -6$ and $1 + (-6) = -5$.

(d) Split the middle term: $-5x = 1x - 6x$

(e) Factor by grouping:
$$3x^2 - 5x - 2 = 3x^2 + x - 6x - 2$$
$$= x(3x + 1) - 2(3x + 1)$$
$$= (3x + 1)(x - 2)$$

100. $(3x + 2)(6x - 5)$

101. $12x^2 + 31x + 20$

(a) First look for a common factor. There is none (other than 1).

(b) Multiply the leading coefficient and the constant, 12 and 20: $12 \cdot 20 = 240$.

(c) Try to factor 240 so that the sum of the factors is 31. We only need to consider positive factors.

Pairs of factors	Sums of factors
1, 240	241
2, 120	122
3, 80	83
4, 60	64
5, 48	53
6, 40	46
8, 30	38
10, 24	34
12, 20	32
15, 16	31

(d) Split the middle term: $31x = 15x + 16x$

(e) Factor by grouping:
$$12x^2 + 31x + 20 = 12x^2 + 15x + 16x + 20$$
$$= 3x(4x + 5) + 4(4x + 5)$$
$$= (4x + 5)(3x + 4)$$

102. $(3x + 5)(5x - 2)$

103. $14x^2 + 19x - 3$

(a) First look for a common factor. There is none (other than 1).

(b) Multiply the leading coefficient and the constant, 14 and -3: $14(-3) = -42$.

(c) Try to factor -42 so that the sum of the factors is 19.

Pairs of factors	Sums of factors
-1, 42	41
1, -42	-41
-2, 21	19
2, -21	-19
-3, 14	11
3, -14	-11
-6, 7	1
6, -7	-1

(d) Split the middle term: $19x = -2x + 21x$

(e) Factor by grouping:
$$14x^2 + 19x - 3 = 14x^2 - 2x + 21x - 3$$
$$= 2x(7x - 1) + 3(7x - 1)$$
$$= (7x - 1)(2x + 3)$$

104. $(7x + 4)(5x + 2)$

105. $9x^2 + 18x + 8$

(a) First look for a common factor. There is none (other than 1).

(b) Multiply the leading coefficient and the constant, 9 and 8: $9 \cdot 8 = 72$.

(c) Try to factor 72 so that the sum of the factors is 18. We only need to consider positive factors.

Pairs of factors	Sums of factors
1, 72	73
2, 36	38
3, 24	27
4, 18	22
6, 12	18
8, 9	17

(d) Split the middle term: $18x = 6x + 12x$

(e) Factor by grouping:
$$9x^2 + 18x + 8 = 9x^2 + 6x + 12x + 8$$
$$= 3x(3x + 2) + 4(3x + 2)$$
$$= (3x + 2)(3x + 4)$$

106. $(2 - 3x)(3 - 2x)$, or $(3x - 2)(2x - 3)$

107. $49 - 42x + 9x^2 = 9x^2 - 42x + 49$

(a) First look for a common factor. There is none (other than 1).

(b) Multiply the leading coefficient and the constant, 9 and 49: $9 \cdot 49 = 441$.

(c) Try to factor 441 so that the sum of the factors is -42. We only need to consider negative factors.

Pairs of factors	Sums of factors
$-1, -441$	-442
$-3, -147$	-150
$-7, -63$	-70
$-9, -49$	-58
$-21, -21$	-42

(d) Split the middle term: $-42x = -21x - 21x$

(e) Factor by grouping:
$$9x^2 - 42x + 49 = 9x^2 - 21x - 21x + 49$$
$$= 3x(3x - 7) - 7(3x - 7)$$
$$= (3x - 7)(3x - 7), \text{ or}$$
$$(3x - 7)^2$$

108. $(5x + 4)^2$

109. *Familiarize.* We will use the formula $C = 2\pi r$, where C is circumference and r is radius, to find the radius in kilometers. Then we will multiply that number by 0.62 to find the radius in miles.

Translate.

$$\underbrace{\text{Circumference}} = \underbrace{2} \cdot \underbrace{\pi} \cdot \underbrace{\text{radius}}$$
$$\downarrow \qquad\qquad \downarrow \quad \downarrow \quad \downarrow$$
$$40,000 \qquad \approx \qquad 2(3.14)r$$

Carry out. First we solve the equation.

$$40,000 \approx 2(3.14)r$$
$$40,000 \approx 6.28r$$
$$6369 \approx r$$

Then we multiply to find the radius in miles:

$6369(0.62) \approx 3949$

Check. If $r = 6369$, then $2\pi r = 2(3.14)(6369) \approx 40,000$. We should also recheck the multiplication we did to find the radius in miles. Both values check.

State. The radius of the earth is about 6369 km or 3949 mi. (These values may differ slightly if a different approximation is used for π.)

110. $40°$

111. Graph: $y = \frac{2}{5}x - 1$

Because the equation is in the form $y = mx + b$, we know the y-intercept is $(0, -1)$. We find two other points on the line, substituting multiples of 5 for x to avoid fractions.

When $x = -5$, $y = \frac{2}{5}(-5) - 1 = -2 - 1 = -3$

When $x = 5$, $y = \frac{2}{5}(5) - 1 = 2 - 1 = 1$.

x	y
0	-1
-5	-3
5	1

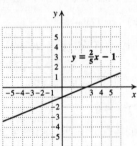

112. y^8

113.

114.

115. $9x^{10} - 12x^5 + 4 = 9(x^5)^2 - 12x^5 + 4$

(a) First look for a common factor. There is none (other than 1).

(b) Multiply the leading coefficient, 9, and the constant, 4: $9 \cdot 4 = 36$.

(c) Look for a factorization of 36 in which the sum of the factors is the coefficient of the middle term, -12. The factors we want are -6 and -6.

(d) Split the middle term: $-12x^5 = -6x^5 - 6x^5$

(e) Factor by grouping:
$$9x^{10} - 12x^5 + 4 = 9x^{10} - 6x^5 - 6x^5 + 4$$
$$= 3x^5(3x^5 - 2) - 2(3x^5 - 2)$$
$$= (3x^5 - 2)(3x^5 - 2), \text{ or}$$
$$= (3x^5 - 2)^2$$

116. $(4x^5 + 1)^2$

117. $20x^{2n} + 16x^n + 3 = 20(x^n)^2 + 16x^n + 3$

(1) There is no common factor (other than 1 or -1).

(2) Because $20x^{2n}$ can be factored as $20x^n \cdot x^n$, $10x^n \cdot 2x^n$, or $5x^n \cdot 4x^n$, we have these possibilities:

$$(20x^n + \ )(x^n + \ ) \text{ and } (10x^n + \ )(2x^n + \ )$$
$$\text{and } (5x^n + \ )(4x^n + \ )$$

(3) Since all the signs are positive, we need consider only the positive factor pair 3,1 when factoring 3. This pair can also be listed as 1,3.

(4) We try some possibilities:

$$(20x^n + 3)(x^n + 1) = 20x^{2n} + 23x^n + 3$$
$$(10x^n + 3)(2x^n + 1) = 20x^{2n} + 16x^n + 3$$

The factorization is $(10x^n + 3)(2x^n + 1)$.

118. $-(3x^m - 4)(5x^m - 2)$

119. $3x^{6a} - 2x^{3a} - 1 = 3(x^{3a})^2 - 2x^{3a} - 1$

(1) There is no common factor (other than 1 or -1).

(2) Because $3x^{6a}$ can be factored as $3x^{3a} \cdot x^{3a}$, we have this possibility:

$$(3x^{3a} + \ )(x^{3a} + \ )$$

(3) There is 1 pair of factors of -1 and it can be listed two ways:

$$-1,1 \quad 1,-1$$

(4) We try these possibilities:

$$(3x^{3a} - 1)(x^{3a} + 1) = 3x^{6a} + 2x^{3a} - 1$$
$$(3x^{3a} + 1)(x^{3a} - 1) = 3x^{6a} - 2x^{3a} - 1$$

The factorization is $(3x^{3a} + 1)(x^{3a} - 1)$.

120. $x(x^n - 1)^2$

121.
$$3(a+1)^{n+1}(a+3)^2 - 5(a+1)^n(a+3)^3$$
$$= (a+1)^n(a+3)^2[3(a+1) - 5(a+3)]$$
$$\qquad\qquad \text{Removing the common factors}$$
$$= (a+1)^n(a+3)^2[3a+3 - 5a - 15] \text{ Simplify-}$$
$$= (a+1)^n(a+3)^2(-2a - 12) \quad \text{ing inside the brackets}$$
$$= (a+1)^n(a+3)^2(-2)(a+6) \text{ Removing the common factor}$$
$$= -2(a+1)^n(a+3)^2(a+6) \quad \text{Rearranging}$$

Exercise Set 5.4

1. $x^2 - 14x + 49$

(a) We know that x^2 and 49 are squares.

(b) There is no minus sign before either x^2 or 49.

(c) If we multiply the square roots, x and 7, and double the product, we get $2 \cdot x \cdot 7 = 14x$. This is the opposite of the remaining term, $-14x$.

Thus, $x^2 - 14x + 49$ is a trinomial square.

2. Yes

3. $x^2 + 16x - 64$

Both x^2 and 64 are squares, but there is a minus sign before 64. Thus, $x^2 + 16x - 64$ is not a trinomial square.

4. No

5. $x^2 - 3x + 9$

a) Both x^2 and 9 are squares.

b) There is no minus sign before either x^2 or 9.

c) If we multiply the square roots, x and 3, and double the product, we get $2 \cdot x \cdot 3 = 6x$. This is neither the remaining term nor its opposite.

Thus, $x^2 - 3x + 9$ is not a trinomial square.

6. No

7. $8x^2 + 40x + 25$

Only one term, 25, is a square. Thus, $8x^2 + 40x + 25$ is not a trinomial square.

8. No

9.
$$x^2 - 14x + 49$$
$$= x^2 - 2 \cdot x \cdot 7 + 7^2 = (x - 7)^2$$
$$\uparrow \quad \uparrow \ \uparrow \ \uparrow \quad \uparrow$$
$$= A^2 - 2 \ \ A \ \ B + B^2 = (A - B)^2$$

10. $(x - 8)^2$

11.
$$x^2 + 16x + 64$$
$$= x^2 + 2 \cdot x \cdot 8 + 8^2 = (x + 8)^2$$
$$\uparrow \quad \uparrow \ \uparrow \ \uparrow \quad \uparrow$$
$$= A^2 + 2 \ \ A \ \ B + B^2 = (A + B)^2$$

12. $(x + 7)^2$

13. $x^2 - 2x + 1 = x^2 - 2 \cdot x \cdot 1 + 1^2 = (x - 1)^2$

14. $(x + 1)^2$

15. $4 + 4x + x^2 = x^2 + 4x + 4$ Changing the order
$$= x^2 + 2 \cdot x \cdot 2 + 2^2$$
$$= (x + 2)^2$$

16. $(x - 2)^2$

17. $9x^2 + 6x + 1 = (3x)^2 + 2 \cdot 3x \cdot 1 + 1^2$
$$= (3x + 1)^2$$

18. $(5x - 1)^2$

19. $49 - 56y + 16y^2 = 16y^2 - 56y + 49$
$$= (4y)^2 - 2 \cdot 4y \cdot 7 + 7^2$$
$$= (4y - 7)^2$$

We could also factor as follows:
$$49 - 56y + 16y^2 = 7^2 - 2 \cdot 7 \cdot 4y + (4y)^2$$
$$= (7 - 4y)^2$$

20. $3(4m + 5)^2$

21. $2x^2 - 4x + 2 = 2(x^2 - 2x + 1)$
$$= 2(x^2 - 2 \cdot x \cdot 1 + 1^2)$$
$$= 2(x - 1)^2$$

22. $2(x - 10)^2$

23. $x^3 - 18x^2 + 81x = x(x^2 - 18x + 81)$
$$= x(x^2 - 2 \cdot x \cdot 9 + 9^2)$$
$$= x(x - 9)^2$$

24. $x(x + 12)^2$

25. $20x^2 + 100x + 125 = 5(4x^2 + 20x + 25)$
$$= 5[(2x)^2 + 2 \cdot 2x \cdot 5 + 5^2]$$
$$= 5(2x + 5)^2$$

26. $3(2x + 3)^2$

27. $49 - 42x + 9x^2 = 7^2 - 2 \cdot 7 \cdot 3x + (3x)^2 = (7 - 3x)^2$

28. $(8 - 7x)^2$, or $(7x - 8)^2$

29. $5y^2 + 10y + 5 = 5(y^2 + 2y + 1)$
$$= 5(y^2 + 2 \cdot y \cdot 1 + 1^2)$$
$$= 5(y + 1)^2$$

30. $2(a + 7)^2$

31. $2 + 20x + 50x^2 = 2(1 + 10x + 25x^2)$
$$= 2[1^2 + 2 \cdot 1 \cdot 5x + (5x)^2]$$
$$= 2(1 + 5x)^2$$

32. $7(1 - a)^2$, or $7(a - 1)^2$

33. $4p^2 + 12pq + 9q^2 = (2p)^2 + 2 \cdot 2p \cdot 3q + (3q)^2$
$$= (2p + 3q)^2$$

34. $(5m + 2n)^2$

35. $a^2 - 14ab + 49b^2 = a^2 - 2 \cdot a \cdot 7b + (7b)^2$
$$= (a - 7b)^2$$

36. $(x - 3y)^2$

37. $64m^2 + 16mn + n^2 = (8m)^2 + 2 \cdot 8m \cdot n + n^2$
$$= (8m + n)^2$$

38. $(9p - q)^2$

39. $16s^2 - 40st + 25t^2 = (4s)^2 - 2 \cdot 4s \cdot 5t + (5t)^2$
$$= (4s - 5t)^2$$

40. $4(3a + 4b)^2$

41. $x^2 - 4$

(a) The first expression is a square: x^2

   The second expression is a square: $4 = 2^2$

(b) The terms have different signs.

   $x^2 - 4$ is a difference of squares.

42. Yes

43. $x^2 + 36$

The terms do not have different signs.

$x^2 + 36$ is not a difference of squares.

44. No

45. $x^2 - 35$

The second expression, 35, is not a square.

$x^2 - 35$ is not a difference of squares.

46. No

47. $16x^2 - 25$

(a) The first expression is a square: $16x^2 = (4x)^2$

The second expression is a square: $25 = 5^2$

(b) The terms have different signs.

$16x^2 - 25$ is a difference of squares.

48. Yes

49. $y^2 - 4 = y^2 - 2^2 = (y + 2)(y - 2)$

50. $(x + 6)(x - 6)$

51. $p^2 - 9 = p^2 - 3^2 = (p + 3)(p - 3)$

52. $(q + 1)(q - 1)$

53. $-49 + t^2 = t^2 - 49 = t^2 - 7^2 = (t + 7)(t - 7)$

54. $(m + 8)(m - 8)$

55. $a^2 - b^2 = (a + b)(a - b)$

56. $(p + q)(p - q)$

57. $25t^2 - m^2 = (5t)^2 - m^2 = (5t + m)(5t - m)$

58. $(w + 7z)(w - 7z)$

59. $100 - k^2 = 10^2 - k^2 = (10 + k)(10 - k)$

60. $(9 + w)(9 - w)$

61. $16a^2 - 9 = (4a)^2 - 3^2 = (4a + 3)(4a - 3)$

62. $(5x + 2)(5x - 2)$

63. $4x^2 - 25y^2 = (2x)^2 - (5y)^2 = (2x + 5y)(2x - 5y)$

64. $(3a + 4b)(3a - 4b)$

65. $8x^2 - 98 = 2(4x^2 - 49) = 2[(2x)^2 - 7^2] =$
$2(2x + 7)(2x - 7)$

66. $6(2x + 3)(2x - 3)$

67. $36x - 49x^3 = x(36 - 49x^2) = x[6^2 - (7x)^2] =$
$x(6 + 7x)(6 - 7x)$

68. $x(4 + 9x)(4 - 9x)$

69. $49a^4 - 81 = (7a^2)^2 - 9^2 = (7a^2 + 9)(7a^2 - 9)$

70. $(5a^2 + 3)(5a^2 - 3)$

71. $x^4 - 1 = (x^2)^2 - 1^2$
$= (x^2 + 1)(x^2 - 1)$
$= (x^2 + 1)(x + 1)(x - 1)$ Factoring
further; $x^2 - 1$ is a
difference of squares

72. $(x^2 + 4)(x + 2)(x - 2)$

73. $4x^4 - 64$
$= 4(x^4 - 16) = 4[(x^2)^2 - 4^2]$
$= 4(x^2 + 4)(x^2 - 4)$
$= 4(x^2 + 4)(x + 2)(x - 2)$ Factoring further;
$x^2 - 4$ is a difference
of squares

74. $5(x^2 + 4)(x + 2)(x - 2)$

75. $1 - y^8$
$= 1^2 - (y^4)^2$
$= (1 + y^4)(1 - y^4)$
$= (1 + y^4)(1 + y^2)(1 - y^2)$ Factoring $1 - y^4$
$= (1 + y^4)(1 + y^2)(1 + y)(1 - y)$
Factoring $1 - y^2$

76. $(x^4 + 1)(x^2 + 1)(x + 1)(x - 1)$

77. $3x^3 - 24x^2 + 48x = 3x(x^2 - 8x + 16)$
$= 3x(x^2 - 2 \cdot x \cdot 4 + 4^2)$
$= 3x(x - 4)^2$

78. $2a^2(a - 9)^2$

79. $x^{12} - 16$
$= (x^6)^2 - 4^2$
$= (x^6 + 4)(x^6 - 4)$
$= (x^6 + 4)(x^3 + 2)(x^3 - 2)$ Factoring $x^6 - 4$

80. $(x^4 + 9)(x^2 + 3)(x^2 - 3)$

81. $y^2 - \dfrac{1}{16} = y^2 - \left(\dfrac{1}{4}\right)^2$
$= \left(y + \dfrac{1}{4}\right)\left(y - \dfrac{1}{4}\right)$

82. $\left(x + \dfrac{1}{5}\right)\left(x - \dfrac{1}{5}\right)$

83. $a^8 - 2a^7 + a^6 = a^6(a^2 - 2a + 1)$
$= a^6(a^2 - 2 \cdot a \cdot 1 + 1^2)$
$= a^6(a - 1)^2$

84. $x^6(x - 4)^2$

85. $25 - \dfrac{1}{49}x^2 = 5^2 - \left(\dfrac{1}{7}x\right)^2$
$= \left(5 + \dfrac{1}{7}x\right)\left(5 - \dfrac{1}{7}x\right)$

86. $\left(2 + \frac{1}{3}y\right)\left(2 - \frac{1}{3}y\right)$

87. $\quad 16m^4 - t^4$

$= (4m^2)^2 - (t^2)^2$

$= (4m^2 + t^2)(4m^2 - t^2)$

$= (4m^2 + t^2)(2m + t)(2m - t)$

$\qquad\qquad$ Factoring $4m^2 - t^2$

88. $(1 + a^2b^2)(1 + ab)(1 - ab)$

89. $\quad m^3 - 7m^2 - 4m + 28$

$= m^2(m - 7) - 4(m - 7)$ Factoring by group-
$\qquad\qquad\qquad\qquad\qquad\qquad$ ing

$= (m - 7)(m^2 - 4)$

$= (m - 7)(m + 2)(m - 2)$ Factoring the differ-
$\qquad\qquad\qquad\qquad\qquad\qquad$ ence of squares

90. $(x + 8)(x + 1)(x - 1)$

91. $\quad a^3 - ab^2 - 2a^2 + 2b^2$

$= a(a^2 - b^2) - 2(a^2 - b^2)$ Factoring by grouping

$= (a^2 - b^2)(a - 2)$

$= (a + b)(a - b)(a - 2)$ Factoring the difference
$\qquad\qquad\qquad\qquad\qquad\qquad$ of squares

92. $(p + 5)(p - 5)(q + 3)$

93. $\quad (a + b)^2 - 100 = (a + b)^2 - 10^2$

$\qquad\qquad\qquad\qquad = (a + b + 10)(a + b - 10)$

94. $(p + 5)(p - 19)$

95. $\quad a^2 + 2ab + b^2 - 9$

$= (a^2 + 2ab + b^2) - 9$ Grouping as a differ-
$\qquad\qquad\qquad\qquad\qquad$ ence of squares

$= (a + b)^2 - 3^2$

$= (a + b + 3)(a + b - 3)$

96. $(x - y + 5)(x - y - 5)$

97. $\quad r^2 - 2r + 1 - 4s^2$

$= (r^2 - 2r + 1) - 4s^2$ Grouping as a differ-
$\qquad\qquad\qquad\qquad\qquad$ ence of squares

$= (r - 1)^2 - (2s)^2$

$= (r - 1 + 2s)(r - 1 - 2s)$

98. $(c + 2d + 3p)(c + 2d - 3p)$

99. $\quad 50a^2 - 2m^2 - 4mn - 2n^2$

$= 2(25a^2 - m^2 - 2mn - n^2)$ Factoring out
$\qquad\qquad\qquad\qquad\qquad\qquad\qquad$ the common factor

$= 2[25a^2 - (m^2 + 2mn + n^2)]$ Factoring out
$\qquad\qquad\qquad\qquad\qquad$ -1 and rewriting as a subtraction

$= 2[25a^2 - (m + n)^2]$ Factoring the trinomial
$\qquad\qquad\qquad\qquad\qquad\qquad$ square

$= 2[5a + (m + n)][5a - (m + n)]$ Factoring a
$\qquad\qquad\qquad\qquad\qquad\qquad$ difference of squares

$= 2(5a + m + n)(5a - m - n)$ Removing
$\qquad\qquad\qquad\qquad\qquad\qquad\qquad$ parentheses

100. $3(x + 2y + 1)(x - 2y - 1)$

101. $\quad = 9 - a^2 + 2ab - b^2$

$= 9 - (a^2 - 2ab + b^2)$ Factoring out
$\qquad\qquad\qquad\qquad$ -1 and rewriting as a subtraction

$= 9 - (a - b)^2$ Factoring the trinomial
$\qquad\qquad\qquad\qquad$ square

$= [3 + (a - b)][3 - (a - b)]$ Factoring a
$\qquad\qquad\qquad\qquad\qquad$ difference of squares

$= (3 + a - b)(3 - a + b)$ Removing
$\qquad\qquad\qquad\qquad\qquad\qquad$ parentheses

102. $(4 + x - y)(4 - x + y)$

103. *Familiarize*. Let $s =$ the score on the fourth test.

Translate.

$$\underbrace{\text{The average score}}_{\dfrac{96 + 98 + 89 + s}{4}} \quad \underbrace{\text{is at least}}_{\geq} \quad \underbrace{90.}_{90}$$

Carry out. We solve the inequality.

$$\frac{96 + 98 + 89 + s}{4} \geq 90$$

$$96 + 98 + 89 + s \geq 360 \qquad \text{Multiplying by 4}$$

$$283 + s \geq 360$$

$$s \geq 77$$

Check. We can obtain a partial check by substituting one number greater than or equal to 77 and another number less than 77 in the inequality. This is left to the student.

State. A score of 77 or better on the last test will earn Bonnie an A in the course. In terms of the inequality we have $s \geq 77$.

104. 3.125 L

105. $(x^3y^5)(x^9y^7) = x^{3+9}y^{5+7} = x^{12}y^{12}$

106. $25a^4b^6$

107.

108.

109. $49x^2 - 216$

There is no common factor. Also, $49x^2$ is a square, but 216 is not so this expression is not a difference of squares. It is not factorable.

110. Prime

111. $\quad 18x^3 + 12x^2 + 2x = 2x(9x^2 + 6x + 1)$

$\qquad\qquad\qquad\qquad = 2x[(3x)^2 + 2 \cdot 3x \cdot 1 + 1^2]$

$\qquad\qquad\qquad\qquad = 2x(3x + 1)^2$

112. $2(81x^2 - 41)$

113. $x^8 - 2^8$
$= (x^4 + 2^4)(x^4 - 2^4)$
$= (x^4 + 2^4)(x^2 + 2^2)(x^2 - 2^2)$
$= (x^4 + 2^4)(x^2 + 2^2)(x + 2)(x - 2)$, or
$= (x^4 + 16)(x^2 + 4)(x + 2)(x - 2)$

114. $4x^2(x + 1)(x - 1)$

115. $3x^5 - 12x^3 = 3x^3(x^2 - 4) = 3x^3(x + 2)(x - 2)$

116. $3\left(x + \dfrac{1}{3}\right)\left(x - \dfrac{1}{3}\right)$, or $\dfrac{1}{3}(3x + 1)(3x - 1)$

117. $18x^3 - \dfrac{8}{25}x = 2x\left(9x^2 - \dfrac{4}{25}\right) =$
$2x\left(3x + \dfrac{2}{5}\right)\left(3x - \dfrac{2}{5}\right)$

118. $(x + 1.5)(x - 1.5)$

119. $0.49p - p^3 = p(0.49 - p^2) = p(0.7 + p)(0.7 - p)$

120. $(0.8x + 1.1)(0.8x - 1.1)$

121. $(x + 3)^2 - 9 = [(x + 3) + 3][(x + 3) - 3] =$
$(x + 6)x$, or $x(x + 6)$

122. $(y - 5 + 6q)(y - 5 - 6q)$

123. $x^2 - \left(\dfrac{1}{x}\right)^2 = \left(x + \dfrac{1}{x}\right)\left(x - \dfrac{1}{x}\right)$

124. $(a^n + 7b^n)(a^n - 7b^n)$

125. $81 - b^{4k} = 9^2 - (b^{2k})^2$
$= (9 + b^{2k})(9 - b^{2k})$
$= (9 + b^{2k})[3^2 - (b^k)^2]$
$= (9 + b^{2k})(3 + b^k)(3 - b^k)$

126. $(x + 3)(x - 3)(x^2 + 1)$

127. $9b^{2n} + 12b^n + 4 = (3b^n)^2 + 2 \cdot 3b^n \cdot 2 + 2^2 =$
$(3b^n + 2)^2$

128. $16(x^2 - 3)^2$

129. $(y + 3)^2 + 2(y + 3) + 1$
$= (y + 3)^2 + 2 \cdot (y + 3) \cdot 1 + 1^2$
$= [(y + 3) + 1]^2$
$= (y + 4)^2$

130. $(7x + 4)^2$

131. $27x^3 - 63x^2 - 147x + 343$
$= 9x^2(3x - 7) - 49(3x - 7)$
$= (3x - 7)(9x^2 - 49)$
$= (3x - 7)(3x + 7)(3x - 7)$, or
$(3x - 7)^2(3x + 7)$

132. $(x - 1)(2x^2 + x + 1)$

133. If $cy^2 + 6y + 1$ is the square of a binomial, then $2 \cdot a \cdot 1 = 6$ where $a^2 = c$. Then $a = 3$, so $c = a^2 = 3^2 = 9$. (The polynomial is $9y^2 + 6y + 1$.)

134. 16

135. See the answer section in the text.

136. 0, 2

Exercise Set 5.5

1. $x^3 + 8 = x^3 + 2^3$
$= (x + 2)(x^2 - 2x + 4)$
$A^3 + B^3 = (A + B)(A^2 - AB + B^2)$

2. $(c + 3)(c^2 - 3c + 9)$

3. $y^3 - 64 = y^3 - 4^3$
$= (y - 4)(y^2 + 4y + 16)$
$A^3 - B^3 = (A - B)(A^2 + AB + B^2)$

4. $(z - 1)(z^2 + z + 1)$

5. $w^3 + 1 = w^3 + 1^3$
$= (w + 1)(w^2 - w + 1)$
$A^3 + B^3 = (A + B)(A^2 - AB + B^2)$

6. $(x + 5)(x^2 - 5x + 25)$

7. $8a^3 + 1 = (2a)^3 + 1^3$
$= (2a + 1)(4a^2 - 2a + 1)$
$A^3 + B^3 = (A + B)(A^2 - AB + B^2)$

8. $(3x + 1)(9x^2 - 3x + 1)$

9. $y^3 - 8 = y^3 - 2^3$
$= (y - 2)(y^2 + 2y + 4)$
$A^3 - B^3 = (A - B)(A^2 + AB + B^2)$

10. $(p - 3)(p^2 + 3p + 9)$

11. $8 - 27b^3 = 2^3 - (3b)^3$
$= (2 - 3b)(4 + 6b + 9b^2)$

12. $(4 - 5x)(16 + 20x + 25x^2)$

13. $64y^3 + 1 = (4y)^3 + 1^3$
$= (4y + 1)(16y^2 - 4y + 1)$

14. $(5x + 1)(25x^2 - 5x + 1)$

15. $8x^3 + 27 = (2x)^3 + 3^3$
$= (2x + 3)(4x^2 - 6x + 9)$

16. $(3y + 4)(9y^2 - 12y + 16)$

17. $a^3 - b^3 = (a - b)(a^2 + ab + b^2)$

18. $(x - y)(x^2 + xy + y^2)$

19. $a^3 + \dfrac{1}{8} = a^3 + \left(\dfrac{1}{2}\right)^3$
$= \left(a + \dfrac{1}{2}\right)\left(a^2 - \dfrac{1}{2}a + \dfrac{1}{4}\right)$

20. $\left(b + \dfrac{1}{3}\right)\left(b^2 - \dfrac{1}{3}b + \dfrac{1}{9}\right)$

21. $\begin{aligned}[t] 2y^3 - 128 &= 2(y^3 - 64) \\ &= 2(y^3 - 4^3) \\ &= 2(y - 4)(y^2 + 4y + 16) \end{aligned}$

22. $3(z - 1)(z^2 + z + 1)$

23. $\begin{aligned}[t] 24a^3 + 3 &= 3(8a^3 + 1) \\ &= 3[(2a)^3 + 1^3] \\ &= 3(2a + 1)(4a^2 - 2a + 1) \end{aligned}$

24. $2(3x + 1)(9x^2 - 3x + 1)$

25. $\begin{aligned}[t] rs^3 + 64r &= r(s^3 + 64) \\ &= r(s^3 + 4^3) \\ &= r(s + 4)(s^2 - 4s + 16) \end{aligned}$

26. $a(b + 5)(b^2 - 5b + 25)$

27. $\begin{aligned}[t] 5x^3 - 40z^3 &= 5(x^3 - 8z^3) \\ &= 5[x^3 - (2z)^3] \\ &= 5(x - 2z)(x^2 + 2xz + 4z^2) \end{aligned}$

28. $2(y - 3z)(y^2 + 3yz + 9z^2)$

29. $\begin{aligned}[t] x^3 + 0.001 &= x^3 + (0.1)^3 \\ &= (x + 0.1)(x^2 - 0.1x + 0.01) \end{aligned}$

30. $(y + 0.5)(y^2 - 0.5y + 0.25)$

31. $\begin{aligned}[t] 64x^6 - 8t^6 &= 8(8x^6 - t^6) \\ &= 8[(2x^2)^3 - (t^2)^3] \\ &= 8(2x^2 - t^2)(4x^4 + 2x^2t^2 + t^4) \end{aligned}$

32. $(5c^2 - 2d^2)(25c^4 + 10c^2d^2 + 4d^4)$

33. $\begin{aligned}[t] 2y^4 - 128y &= 2y(y^3 - 64) \\ &= 2y(y^3 - 4^3) \\ &= 2y(y - 4)(y^2 + 4y + 16) \end{aligned}$

34. $3z^2(z - 1)(z^2 + z + 1)$

35. $\begin{aligned}[t] &z^6 - 1 \\ &= (z^3)^2 - 1^2 \quad \text{Writing as a difference of squares} \\ &= (z^3 + 1)(z^3 - 1) \quad \text{Factoring a difference} \\ &\hspace{4.5cm}\text{of squares} \\ &= (z + 1)(z^2 - z + 1)(z - 1)(z^2 + z + 1) \\ &\hspace{2.5cm}\text{Factoring a sum and} \\ &\hspace{2.5cm}\text{a difference of cubes} \end{aligned}$

36. $(t^2 + 1)(t^4 - t^2 + 1)$

37. $\begin{aligned}[t] t^6 + 64y^6 &= (t^2)^3 + (4y^2)^3 \\ &= (t^2 + 4y^2)(t^4 - 4t^2y^2 + 16y^4) \end{aligned}$

38. $(p + q)(p^2 - pq + q^2)(p - q)(p^2 + pq + q^2)$

39. *Familiarize.* Let $s = $ the score on the fifth quiz.

Translate.

$$\underbrace{\frac{78 + 76 + 82 + 93 + s}{5}}_{\text{The average score}} \quad \underset{\geq}{\underbrace{}_{\text{is at least}}} \quad \underset{80}{\underbrace{}_{80}}$$

Carry out. We solve the inequality.

$$\frac{78 + 76 + 82 + 93 + s}{5} \geq 80$$

$$78 + 76 + 82 + 93 + s \geq 400 \quad \text{Multiplying by 5}$$

$$329 + s \geq 400$$

$$s \geq 71$$

Check. We can obtain a partial check by substituting one number greater than or equal to 71 and another number less than 71 in the inequality. This is left to the student.

State. A score of 71 or better on the last quiz will make your average quiz score at least 80. Using an inequality, we write $s \geq 71$.

40. 56

41. $\begin{aligned}[t] -\frac{2}{3} - \frac{3}{4} + \frac{1}{2} &= -\frac{2}{3} + \left(-\frac{3}{4}\right) + \frac{1}{2} \\ &= -\frac{8}{12} + \left(-\frac{9}{12}\right) + \frac{6}{12} \\ &= -\frac{11}{12} \end{aligned}$

42. $-2y - 2$

43. ◈

44. ◈

45. $\begin{aligned}[t] x^{6a} + y^{3b} &= (x^{2a})^3 + (y^b)^3 \\ &= (x^{2a} + y^b)(x^{4a} - x^{2a}y^b + y^{2b}) \end{aligned}$

46. $(ax - by)(a^2x^2 + axby + b^2y^2)$

47. $\begin{aligned}[t] 3x^{3a} + 24y^{3b} &= 3(x^{3a} + 8y^{3b}) \\ &= 3[(x^a)^3 + (2y^b)^3] \\ &= 3(x^a + 2y^b)(x^{2a} - 2x^ay^b + 4y^{2b}) \end{aligned}$

48. $\left(\dfrac{2}{3}x + \dfrac{1}{4}y\right)\left(\dfrac{4}{9}x^2 - \dfrac{1}{6}xy + \dfrac{1}{16}y^2\right)$

49. $\begin{aligned}[t] &\frac{1}{24}x^3y^3 + \frac{1}{3}z^3 \\ &= \frac{1}{3}\left(\frac{1}{8}x^3y^3 + z^3\right) \\ &= \frac{1}{3}\left[\left(\frac{1}{2}xy\right)^3 + z^3\right] \\ &= \frac{1}{3}\left(\frac{1}{2}xy + z\right)\left(\frac{1}{4}x^2y^2 - \frac{1}{2}xyz + z^2\right) \end{aligned}$

50. $\dfrac{1}{2}\left(\dfrac{1}{2}x^a + y^{2a}z^{3b}\right)\left(\dfrac{1}{4}x^{2a} - \dfrac{1}{2}x^ay^{2a}z^{3b} + y^{4a}z^{6b}\right)$

51.
$$7x^3 + \frac{7}{8} = 7\left(x^3 + \frac{1}{8}\right)$$
$$= 7\left[x^3 + \left(\frac{1}{2}\right)^3\right]$$
$$= 7\left(x + \frac{1}{2}\right)\left(x^2 - \frac{1}{2}x + \frac{1}{4}\right)$$

52. $(c - 2d)^2(c^2 - cd + d^2)^2$

53.
$$(x + y)^3 - x^3$$
$$= [(x + y) - x][(x + y)^2 + x(x + y) + x^2]$$
$$= (x + y - x)(x^2 + 2xy + y^2 + x^2 + xy + x^2)$$
$$= y(3x^2 + 3xy + y^2)$$

54. $(x - 1)^3(x - 2)(x^2 - x + 1)$

55.
$$(a + 2)^3 - (a - 2)^3$$
$$= [(a+2)-(a-2)][(a+2)^2+(a+2)(a-2)+(a-2)^2]$$
$$= (a+2-a+2)(a^2+4a+4+a^2-4+a^2-4a+4)$$
$$= 4(3a^2 + 4)$$

56. $(y - 8)(y - 1)(y^2 + y + 1)$

Exercise Set 5.6

1. $x^2 - 144 = x^2 - 12^2$ Difference of squares
$$= (x + 12)(x - 12)$$

2. $(y + 9)(y - 9)$

3.
$$p^2 + 16p + 64$$
$$= p^2 + 2 \cdot p \cdot 8 + 8^2 \quad \text{Trinomial square}$$
$$= (p + 8)^2$$

4. $(y - 5)^2$

5. $2x^2 - 11x + 12$

There is no common factor (other than 1). This polynomial has three terms, but it is not a trinomial square. Multiply the leading coefficient and the constant, 2 and 12: $2 \cdot 12 = 24$. Try to factor 24 so that the sum of the factors is -11. The numbers we want are -3 and -8: $-3(-8) = 24$ and $-3 + (-8) = -11$. Split the middle term and factor by grouping.
$$2x^2 - 11x + 12 = 2x^2 - 3x - 8x + 12$$
$$= x(2x - 3) - 4(2x - 3)$$
$$= (2x - 3)(x - 4)$$

6. $(2y - 5)(4y + 1)$

7.
$$x^3 + 24x^2 + 144x$$
$$= x(x^2 + 24x + 144) \qquad x \text{ is a common factor}$$
$$= x(x^2 + 2 \cdot x \cdot 12 + 12^2) \quad \text{Trinomial square}$$
$$= x(x + 12)^2$$

8. $x(x - 9)^2$

9.
$$x^3 + 3x^2 - 4x - 12$$
$$= x^2(x + 3) - 4(x + 3) \quad \text{Factoring by grouping}$$
$$= (x + 3)(x^2 - 4)$$
$$= (x + 3)(x + 2)(x - 2) \quad \text{Factoring the difference of squares}$$

10. $(x + 5)(x - 5)^2$

11.
$$9x^2 - 25y^2$$
$$= (3x)^2 - (5y)^2 \qquad \text{Difference of squares}$$
$$= (3x + 5y)(3x - 5y)$$

12. $2(2x + 7y)(2x - 7y)$

13.
$$20x^3 - 4x^2 - 72x$$
$$= 4x(5x^2 - x - 18) \quad 4x \text{ is a common factor}$$
$$= 4x(5x + 9)(x - 2) \quad \text{Factoring the trinomial using trial and error}$$

14. $3x(x + 3)(3x - 5)$

15.
$$a^3b - 8b = b(a^3 - 8) \qquad b \text{ is a common factor}$$
$$= b(a^3 - 2^3) \qquad \text{Difference of cubes}$$
$$= b(a - 2)(a^2 + 2a + 4)$$

16. $5(1 + y)(1 - y + y^2)$

17. $x^4 + 7x^2 - 3x^3 - 21x = x(x^3 + 7x - 3x^2 - 21)$
$$= x[x(x^2 + 7) - 3(x^2 + 7)]$$
$$= x[(x^2 + 7)(x - 3)]$$
$$= x(x^2 + 7)(x - 3)$$

18. $m(m + 8)(m^2 + 8)$

19.
$$x^5 - 14x^4 + 49x^3$$
$$= x^3(x^2 - 14x + 49) \quad x^3 \text{ is a common factor}$$
$$= x^3(x^2 - 2 \cdot x \cdot 7 + 7^2) \quad \text{Trinomial square}$$
$$= x^3(x - 7)^2$$

20. $2x^4(x + 2)^2$

21. $m^6 - 1 = (m^3)^2 - 1^2$ Difference of squares
$$= (m^3 + 1)(m^3 - 1) \quad \text{Sum of cubes; difference of cubes}$$
$$= (m+1)(m^2-m+1)(m-1)(m^2+m+1)$$

22. $(2t + 1)(4t^2 - 2t + 1)(2t - 1)(4t^2 + 2t + 1)$

23. $x^2 + 3x + 1$

There is no common factor (other than 1). This is not a trinomial square, because $2 \cdot x \cdot 1 \neq 3x$. We try factoring by trial and error. We look for two factors whose product is 1 and whose sum is 3. There are none. The polynomial cannot be factored. It is prime.

24. Prime

25. $4x^4 - 64$

$= 4(x^4 - 16)$ 4 is a common factor

$= 4[(x^2)^2 - 4^2]$ Difference of squares

$= 4(x^2 + 4)(x^2 - 4)$ Difference of squares

$= 4(x^2 + 4)(x + 2)(x - 2)$

26. $5x(x^2 + 4)(x + 2)(x - 2)$

27. $t^2 + 25$ is a sum of squares with no common factor (other than 1). It is prime.

28. Prime

29. $x^5 - 4x^4 + 3x^3$

$= x^3(x^2 - 4x + 3)$ x^3 is a common factor

$= x^3(x - 3)(x - 1)$ Factoring the trinomial using trial and error

30. $x^4(x^2 - 2x + 7)$

31. $x^2 + 6x + 9 - 16y^2$

$= (x + 3)^2 - (4y)^2$ Difference of squares

$= (x + 3 + 4y)(x + 3 - 4y)$

32. $(t + 5 + p)(t + 5 - p)$

33. $12n^2 + 24n^3 = 12n^2(1 + 2n)$

34. $a(x^2 + y^2)$

35. $9x^2y^2 - 36xy = 9xy(xy - 4)$

36. $xy(x - y)$

37. $2\pi rh + 2\pi r^2 = 2\pi r(h + r)$

38. $5p^2q^2(2p^2q^2 + 7pq + 2)$

39. $(a + b)(x - 3) + (a + b)(x + 4)$

$= (a + b)[(x - 3) + (x + 4)]$ $(a + b)$ is a common factor

$= (a + b)(2x + 1)$

40. $(a^3 + b)(5c - 1)$

41. $16x^3 + 54y^3$

$= 2(8x^3 + 27y^3)$ 2 is a common factor

$= 2[(2x)^3 + (3y)^3]$ Sum of cubes

$= 2(2x + 3y)(4x^2 - 6xy + 9y^2)$

42. $2(5a - 3b)(25a^2 + 15ab + 9b^2)$

43. $n^2 + 2n + np + 2p$

$= n(n + 2) + p(n + 2)$ Factoring by grouping

$= (n + 2)(n + p)$

44. $(x - 2)(2x + 13)$

45. $ac + cd - ab - bd$

$= c(a + d) - b(a + d)$ Factoring by

$= (a + d)(c - b)$ grouping

46. $(2y - 1)(3y + p)$

47. $x^2 + y^2 - 2xy$

$= x^2 - 2xy + y^2$ Trinomial square

$= (x - y)^2$

48. $(a - 2b)^2$, or $(2b - a)^2$

49. $9c^2 + 6cd + d^2$

$= (3c)^2 + 2 \cdot 3c \cdot d + d^2$ Trinomial square

$= (3c + d)^2$

50. $(4x + 3y)^2$

51. $7p^4 - 7q^4$

$= 7(p^4 - q^4)$ 7 is a common factor

$= 7(p^2 + q^2)(p^2 - q^2)$ Factoring a difference of squares

$= 7(p^2 + q^2)(p + q)(p - q)$ Factoring a difference of squares

52. $(2xy + 3z)^2$

53. $25z^2 + 10zy + y^2$

$= (5z)^2 + 2 \cdot 5z \cdot y + y^2$ Trinomial square

$= (5z + y)^2$

54. $(a^2b^2 + 4)(ab + 2)(ab - 2)$

55. $a^5 + 4a^4b - 5a^3b^2$

$= a^3(a^2 + 4ab - 5b^2)$ a^3 is a common factor

$= a^3(a + 5b)(a - b)$ Factoring the trinomial

56. $p(2p + q)^2$

57. $a^2 - ab - 2b^2 = (a - 2b)(a + b)$ Using trial and error

58. $(3b + a)(b - 6a)$

59. $2mn - 360n^2 + m^2$

$= m^2 + 2mn - 360n^2$ Rewriting

$= (m + 20n)(m - 18n)$ Using trial and error

60. $(xy + 5)(xy + 3)$

61. $m^2n^2 - 4mn - 32$

$= (mn - 8)(mn + 4)$ Using trial and error

62. $(pq + 6)(pq + 1)$

63. $a^5b^2 + 3a^4b - 10a^3$

$= a^3(a^2b^2 + 3ab - 10)$ a^3 is a common factor

$= a^3(ab + 5)(ab - 2)$ Factoring the trinomial

64. $n^4(mn + 8)(mn - 4)$

65. $49m^2 - 112mn + 64n^2$

$= (7m)^2 - 2 \cdot 7m \cdot 8n + (8n)^2$ Trinomial square

$= (7m - 8n)^2$

66. $2t^2(s^3 + 3t)(s^3 + 2t)$

67. $x^6 + x^5 y - 2x^4 y^2$

$= x^4(x^2 + xy - 2y^2)$ x^4 is a common factor

$= x^4(x + 2y)(x - y)$ Factoring the trinomial

68. $a^2(1 + bc)^2$

69. $36a^2 - 15a + \dfrac{25}{16} = (6a)^2 - 2 \cdot 6a \cdot \dfrac{5}{4} + \left(\dfrac{5}{4}\right)^2$

$= \left(6a - \dfrac{5}{4}\right)^2$

70. $\left(\dfrac{1}{9}x - \dfrac{4}{3}\right)^2$, or $\dfrac{1}{9}\left(\dfrac{1}{3}x - 4\right)^2$

71. $\dfrac{1}{4}a^2 + \dfrac{1}{3}ab + \dfrac{1}{9}b^2$

$= \left(\dfrac{1}{2}a\right)^2 + 2 \cdot \dfrac{1}{2}a \cdot \dfrac{1}{3}b + \left(\dfrac{1}{3}b\right)^2$

$= \left(\dfrac{1}{2}a + \dfrac{1}{3}b\right)^2$

72. $(0.1x - 0.5y)^2$, or $0.01(x - 5y)^2$

73. $81a^4 - b^4$

$= (9a^2)^2 - (b^2)^2$ Difference of squares

$= (9a^2 + b^2)(9a^2 - b^2)$ Difference of squares

$= (9a^2 + b^2)(3a + b)(3a - b)$

74. $(1 + n^4)(1 + n^2)(1 + n)(1 - n)$

75. $w^3 - 7w^2 - 4w + 28$

$= w^2(w - 7) - 4(w - 7)$ Factoring by grouping

$= (w - 7)(w^2 - 4)$

$= (w - 7)(w + 2)(w - 2)$ Factoring a difference of squares

76. $(y + 8)(y + 1)(y - 1)$

77. $\dfrac{y = -4x + 7}{11 \; ? \; -4(-1) + 7}$

$\qquad\qquad \begin{array}{c|c} & 4 + 7 \\ 11 & 11 \end{array}$ TRUE

Since $11 = 11$ is true, $(-1, 11)$ is a solution.

$\dfrac{y = -4x + 7}{7 \; ? \; -4 \cdot 0 + 7}$

$\qquad\qquad \begin{array}{c|c} & 0 + 7 \\ 7 & 7 \end{array}$ TRUE

Since $7 = 7$ is true, $(0, 7)$ is a solution.

$\dfrac{y = -4x + 7}{-5 \; ? \; -4 \cdot 3 + 7}$

$\qquad\qquad \begin{array}{c|c} & -12 + 7 \\ -5 & -5 \end{array}$ TRUE

Since $-5 = -5$ is true, $(3, -5)$ is a solution.

78.

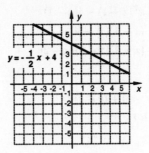

$y = -\dfrac{1}{2}x + 4$

79. $A = aX + bX - 7$

$A + 7 = aX + bX$

$A + 7 = X(a + b)$

$\dfrac{A + 7}{a + b} = X$

80. $\{x \mid x < 32\}$

81.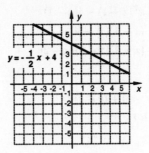

82.

83. $6x^2 - xy - 15y^2$

$= 6 \cdot 1^2 - 1 \cdot 1 - 15 \cdot 1^2$

$= 6 - 1 - 15$

$= -10$

$(2x + 3y)(3x - 5y)$

$= (2 \cdot 1 + 3 \cdot 1)(3 \cdot 1 - 5 \cdot 1)$

$= 5(-2)$

$= -10$

Since the value of both expressions is -10, the factorization is probably correct.

84. 49

85. $18 + y^3 - 9y - 2y^2$

$= y^3 - 2y^2 - 9y + 18$

$= y^2(y - 2) - 9(y - 2)$

$= (y - 2)(y^2 - 9)$

$= (y - 2)(y + 3)(y - 3)$

86. $-(x^2 + 2)(x + 3)(x - 3)$

87. $a^3 + 4a^2 + a + 4 = a^2(a + 4) + 1(a + 4)$

$= (a + 4)(a^2 + 1)$

88. $(x + 1)(x + 2)(x - 2)$

89. $x^4 - 7x^2 - 18 = (x^2 - 9)(x^2 + 2)$

$= (x + 3)(x - 3)(x^2 + 2)$

90. $3(x + 2)(x - 2)(x + 1)(x - 1)$

91. $x^3 - x^2 - 4x + 4 = x^2(x - 1) - 4(x - 1)$
$$= (x - 1)(x^2 - 4)$$
$$= (x - 1)(x + 2)(x - 2)$$

92. $(y + 1)(y - 7)(y + 3)$

93. $y^2(y - 1) - 2y(y - 1) + (y - 1)$
$$= (y - 1)(y^2 - 2y + 1)$$
$$= (y - 1)(y - 1)^2$$
$$= (y - 1)^3$$

94. $(2x + 3y - 2)(3x - y - 3)$

95. $(y + 4)^2 + 2x(y + 4) + x^2$
$$= (y + 4)^2 + 2 \cdot (y + 4) \cdot x + x^2 \qquad \text{Trinomial}$$
$$\text{square}$$
$$= [(y + 4) + x]^2$$

96. $(2a + b + 4)(a - b + 5)$

97. $x^{2k} - 2^{2k} = x^{2 \cdot 4} - 2^{2 \cdot 4} \qquad \text{Substituting 4 for } k$
$$= x^8 - 2^8$$
$$= x^8 - 256$$
$$= (x^4 + 16)(x^4 - 16)$$
$$= (x^4 + 16)(x^2 + 4)(x^2 - 4)$$
$$= (x^4 + 16)(x^2 + 4)(x + 2)(x - 2)$$

98.

Exercise Set 5.7

---

1. $(x + 8)(x + 6) = 0$
$$x + 8 = 0 \quad \text{or} \quad x + 6 = 0$$
$$x = -8 \quad \text{or} \qquad x = -6$$

Check:
For -8:

$$\frac{(x + 8)(x + 6) = 0}{(-8 + 8)(-8 + 6) \;?\; 0}$$
$$\begin{array}{c|c} 0 \cdot (-2) & \\ 0 & 0 \end{array} \quad \text{TRUE}$$

For -6:

$$\frac{(x + 8)(x + 6) = 0}{(-6 + 8)(-6 + 6) \;?\; 0}$$
$$\begin{array}{c|c} 2 \cdot 0 & \\ 0 & 0 \end{array} \quad \text{TRUE}$$

The solutions are -8 and -6.

2. $-3, -2$

3. $(x - 3)(x + 5) = 0$
$$x - 3 = 0 \quad \text{or} \quad x + 5 = 0$$
$$x = 3 \quad \text{or} \qquad x = -5$$

Check:
For 3:

$$\frac{(x - 3)(x + 5) = 0}{(3 - 3)(3 + 5) \;?\; 0}$$
$$\begin{array}{c|c} 0 \cdot 8 & \\ 0 & 0 \end{array} \quad \text{TRUE}$$

For -5:

$$\frac{(x - 3)(x + 5) = 0}{(-5 - 3)(-5 + 5) \;?\; 0}$$
$$\begin{array}{c|c} -8 \cdot 0 & \\ 0 & 0 \end{array} \quad \text{TRUE}$$

The solutions are 3 and -5.

4. $-9, 3$

5. $(x + 12)(x - 11) = 0$
$$x + 12 = 0 \quad \text{or} \quad x - 11 = 0$$
$$x = -12 \quad \text{or} \qquad x = 11$$

The solutions are -12 and 11.

6. $13, -53$

7. $x(x + 5) = 0$
$$x = 0 \quad \text{or} \quad x + 5 = 0$$
$$x = 0 \quad \text{or} \qquad x = -5$$

The solutions are 0 and -5.

8. $0, -7$

9. $0 = y(y + 10)$
$$y = 0 \quad \text{or} \quad y + 10 = 0$$
$$y = 0 \quad \text{or} \qquad y = -10$$

The solutions are 0 and -10.

10. $0, 21$

11. $(2x + 5)(x + 4) = 0$
$$2x + 5 = 0 \quad \text{or} \quad x + 4 = 0$$
$$2x = -5 \quad \text{or} \qquad x = -4$$
$$x = -\frac{5}{2} \quad \text{or} \qquad x = -4$$

The solutions are $-\frac{5}{2}$ and -4.

12. $-\frac{9}{2}, -8$

13. $(5x + 1)(4x - 12) = 0$
$$5x + 1 = 0 \quad \text{or} \quad 4x - 12 = 0$$
$$5x = -1 \quad \text{or} \qquad 4x = 12$$
$$x = -\frac{1}{5} \quad \text{or} \qquad x = 3$$

The solutions are $-\frac{1}{5}$ and 3.

14. $-\frac{9}{4}, \frac{1}{2}$

15. $(7x - 28)(28x - 7) = 0$

$7x - 28 = 0$ or $28x - 7 = 0$

$7x = 28$ or $28x = 7$

$x = 4$ or $x = \dfrac{7}{28} = \dfrac{1}{4}$

The solutions are 4 and $\dfrac{1}{4}$.

16. $\dfrac{11}{12}, \dfrac{5}{8}$

17. $2x(3x - 2) = 0$

$2x = 0$ or $3x - 2 = 0$

$x = 0$ or $3x = 2$

$x = 0$ or $x = \dfrac{2}{3}$

The solutions are 0 and $\dfrac{2}{3}$.

18. $0, \dfrac{9}{8}$

19. $\dfrac{1}{2}x\left(\dfrac{2}{3}x - 12\right) = 0$

$\dfrac{1}{2}x = 0$ or $\dfrac{2}{3}x - 12 = 0$

$x = 0$ or $\dfrac{2}{3}x = 12$

$x = 0$ or $x = \dfrac{3}{2} \cdot 12 = 18$

The solutions are 0 and 18.

20. $0, 8$

21. $\left(\dfrac{1}{5} + 2x\right)\left(\dfrac{1}{9} - 3x\right) = 0$

$\dfrac{1}{5} + 2x = 0$ or $\dfrac{1}{9} - 3x = 0$

$2x = -\dfrac{1}{5}$ or $-3x = -\dfrac{1}{9}$

$x = -\dfrac{1}{10}$ or $x = \dfrac{1}{27}$

The solutions are $-\dfrac{1}{10}$ and $\dfrac{1}{27}$.

22. $\dfrac{1}{21}, \dfrac{18}{11}$

23. $(0.3x - 0.1)(0.05x - 1) = 0$

$0.3x - 0.1 = 0$ or $0.05x - 1 = 0$

$0.3x = 0.1$ or $0.05x = 1$

$x = \dfrac{0.1}{0.3}$ or $x = \dfrac{1}{0.05}$

$x = \dfrac{1}{3}$ or $x = 20$

The solutions are $\dfrac{1}{3}$ and 20.

24. $3, 50$

25. $9x(3x - 2)(2x - 1) = 0$

$9x = 0$ or $3x - 2 = 0$ or $2x - 1 = 0$

$x = 0$ or $3x = 2$ or $2x = 1$

$x = 0$ or $x = \dfrac{2}{3}$ or $x = \dfrac{1}{2}$

The solutions are 0, $\dfrac{2}{3}$, and $\dfrac{1}{2}$.

26. $5, -55, \dfrac{1}{5}$

27. $x^2 + 6x + 5 = 0$

$(x + 5)(x + 1) = 0$ Factoring

$x + 5 = 0$ or $x + 1 = 0$ Using the principle of zero products

$x = -5$ or $x = -1$

The solutions are -5 and -1.

28. $-6, -1$

29. $x^2 + 7x - 18 = 0$

$(x + 9)(x - 2) = 0$ Factoring

$x + 9 = 0$ or $x - 2 = 0$ Using the principle of zero products

$x = -9$ or $x = 2$

The solutions are -9 and 2.

30. $-7, 3$

31. $x^2 - 8x + 15 = 0$

$(x - 5)(x - 3) = 0$

$x - 5 = 0$ or $x - 3 = 0$

$x = 5$ or $x = 3$

The solutions are 5 and 3.

32. $7, 2$

33. $x^2 - 8x = 0$

$x(x - 8) = 0$

$x = 0$ or $x - 8 = 0$

$x = 0$ or $x = 8$

The solutions are 0 and 8.

34. $0, 3$

35. $x^2 + 19x = 0$

$x(x + 19) = 0$

$x = 0$ or $x + 19 = 0$

$x = 0$ or $x = -19$

The solutions are 0 and -19.

36. $0, -12$

37. $x^2 = 16$

$x^2 - 16 = 0$ Subtracting 16

$(x - 4)(x + 4) = 0$

$x - 4 = 0$ or $x + 4 = 0$

$x = 4$ or $x = -4$

The solutions are 4 and -4.

38. -10, 10

39.
$$9x^2 - 4 = 0$$
$$(3x - 2)(3x + 2) = 0$$
$$3x - 2 = 0 \quad \text{or} \quad 3x + 2 = 0$$
$$3x = 2 \quad \text{or} \quad 3x = -2$$
$$x = \frac{2}{3} \quad \text{or} \quad x = -\frac{2}{3}$$
The solutions are $\frac{2}{3}$ and $-\frac{2}{3}$.

40. $-\frac{3}{2}$, $\frac{3}{2}$

41. $0 = 6x + x^2 + 9$
$$0 = x^2 + 6x + 9 \quad \text{Writing in descending order}$$
$$0 = (x + 3)(x + 3)$$
$$x + 3 = 0 \quad \text{or} \quad x + 3 = 0$$
$$x = -3 \quad \text{or} \quad x = -3$$
There is only one solution, -3.

42. -5

43.
$$x^2 + 16 = 8x$$
$$x^2 - 8x + 16 = 0 \quad \text{Subtracting } 8x$$
$$(x - 4)(x - 4) = 0$$
$$x - 4 = 0 \quad \text{or} \quad x - 4 = 0$$
$$x = 4 \quad \text{or} \quad x = 4$$
There is only one solution, 4.

44. 1

45.
$$5x^2 = 6x$$
$$5x^2 - 6x = 0$$
$$x(5x - 6) = 0$$
$$x = 0 \quad \text{or} \quad 5x - 6 = 0$$
$$x = 0 \quad \text{or} \quad 5x = 6$$
$$x = 0 \quad \text{or} \quad x = \frac{6}{5}$$
The solutions are 0 and $\frac{6}{5}$.

46. 0, $\frac{8}{7}$

47.
$$6x^2 - 4x = 10$$
$$6x^2 - 4x - 10 = 0$$
$$2(3x^2 - 2x - 5) = 0$$
$$2(3x - 5)(x + 1) = 0$$
$$3x - 5 = 0 \quad \text{or} \quad x + 1 = 0$$
$$3x = 5 \quad \text{or} \quad x = -1$$
$$x = \frac{5}{3} \quad \text{or} \quad x = -1$$
The solutions are $\frac{5}{3}$ and -1.

48. $-\frac{5}{3}$, 4

49.
$$12y^2 - 5y = 2$$
$$12y^2 - 5y - 2 = 0$$
$$(4y + 1)(3y - 2) = 0$$
$$4y + 1 = 0 \quad \text{or} \quad 3y - 2 = 0$$
$$4y = -1 \quad \text{or} \quad 3y = 2$$
$$y = -\frac{1}{4} \quad \text{or} \quad y = \frac{2}{3}$$
The solutions are $-\frac{1}{4}$ and $\frac{2}{3}$.

50. -5, -1

51.
$$x(x - 5) = 14$$
$$x^2 - 5x = 14 \quad \text{Multiplying on the left side}$$
$$x^2 - 5x - 14 = 0 \quad \text{Adding } -14$$
$$(x - 7)(x + 2) = 0$$

$$x - 7 = 0 \quad \text{or} \quad x + 2 = 0$$
$$x = 7 \quad \text{or} \quad x = -2$$
The solutions are 7 and -2.

52. $\frac{2}{3}$, -1

53.
$$64m^2 - 25 = 56$$
$$64m^2 - 81 = 0$$
$$(8m - 9)(8m + 9) = 0$$
$$8m - 9 = 0 \quad \text{or} \quad 8m + 9 = 0$$
$$8m = 9 \quad \text{or} \quad 8m = -9$$
$$m = \frac{9}{8} \quad \text{or} \quad m = -\frac{9}{8}$$
The solutions are $\frac{9}{8}$ and $-\frac{9}{8}$.

54. $-\frac{7}{10}$, $\frac{7}{10}$

55.
$$3x^2 + 8x = 9 + 2x$$
$$3x^2 + 8x - 2x - 9 = 0 \quad \text{Adding } -2x \text{ and } -9$$
$$3x^2 + 6x - 9 = 0 \quad \text{Collecting like terms}$$
$$3(x^2 + 2x - 3) = 0$$
$$3(x + 3)(x - 1) = 0$$
$$x + 3 = 0 \quad \text{or} \quad x - 1 = 0$$
$$x = -3 \quad \text{or} \quad x = 1$$
The solutions are -3 and 1.

56. 9, -2

57. $(3x + 5)(x + 3) = 7$
$$3x^2 + 14x + 15 = 7 \quad \text{Multiplying on the left}$$
$$3x^2 + 14x + 8 = 0$$
$$(3x + 2)(x + 4) = 0$$

$$3x + 2 = 0 \quad \text{or} \quad x + 4 = 0$$
$$3x = -2 \quad \text{or} \quad x = -4$$
$$x = -\frac{2}{3} \quad \text{or} \quad x = -4$$

The solutions are $-\frac{2}{3}$ and -4.

58. $\frac{6}{5}$, -1

59. We let $y = 0$ and solve for x.
$$0 = x^2 - x - 6$$
$$0 = (x - 3)(x + 2)$$

$$x - 3 = 0 \quad \text{or} \quad x + 2 = 0$$
$$x = 3 \quad \text{or} \quad x = -2$$

The x-intercepts are $(3, 0)$ and $(-2, 0)$.

60. $(-4, 0)$, $(1, 0)$

61. We let $y = 0$ and solve for x.
$$0 = x^2 + 2x - 8$$
$$0 = (x + 4)(x - 2)$$

$$x + 4 = 0 \quad \text{or} \quad x - 2 = 0$$
$$x = -4 \quad \text{or} \quad x = 2$$

The x-intercepts are $(-4, 0)$ and $(2, 0)$.

62. $(5, 0)$, $(-3, 0)$

63. We let $y = 0$ and solve for x.
$$0 = 2x^2 + 3x - 9$$
$$0 = (2x - 3)(x + 3)$$

$$2x - 3 = 0 \quad \text{or} \quad x + 3 = 0$$
$$2x = 3 \quad \text{or} \quad x = -3$$
$$x = \frac{3}{2} \quad \text{or} \quad x = -3$$

The x-intercepts are $\left(\frac{3}{2}, 0\right)$ and $(-3, 0)$.

64. $\left(-\frac{5}{2}, 0\right)$, $(2, 0)$

65. $(a + b)^2$

66. $a^2 + b^2$

67. $2x + 5 < 19$

68. $\frac{1}{2}x - 7 > 24$

69.

70.

71.

72.

73. a)
$$x = -3 \quad \text{or} \quad x = 4$$
$$x + 3 = 0 \quad \text{or} \quad x - 4 = 0$$
$$(x + 3)(x - 4) = 0 \quad \text{Principle of zero products}$$
$$x^2 - x - 12 = 0 \quad \text{Multiplying}$$

b)
$$x = -3 \quad \text{or} \quad x = -4$$
$$x + 3 = 0 \quad \text{or} \quad x + 4 = 0$$
$$(x + 3)(x + 4) = 0$$
$$x^2 + 7x + 12 = 0$$

c)
$$x = \frac{1}{2} \quad \text{or} \quad x = \frac{1}{2}$$
$$x - \frac{1}{2} = 0 \quad \text{or} \quad x - \frac{1}{2} = 0$$
$$\left(x - \frac{1}{2}\right)\left(x - \frac{1}{2}\right) = 0$$
$$x^2 - x + \frac{1}{4} = 0, \quad \text{or}$$
$$4x^2 - 4x + 1 = 0 \quad \text{Multiplying by 4}$$

d)
$$x = 5 \quad \text{or} \quad x = -5$$
$$x - 5 = 0 \quad \text{or} \quad x + 5 = 0$$
$$(x - 5)(x + 5) = 0 \quad \text{Principle of zero products}$$
$$x^2 - 25 = 0 \quad \text{Multiplying}$$

e) $x = 0 \quad \text{or} \quad x = 0.1 \text{ or} \quad x = \frac{1}{4}$
$$x = 0 \quad \text{or} \quad x - 0.1 = 0 \quad \text{or} \quad x - \frac{1}{4} = 0$$
$$x = 0 \quad \text{or} \quad x - \frac{1}{10} = 0 \quad \text{or} \quad x - \frac{1}{4} = 0$$
$$x\left(x - \frac{1}{10}\right)\left(x - \frac{1}{4}\right) = 0$$
$$x\left(x^2 - \frac{7}{20}x + \frac{1}{40}\right) = 0$$
$$x^3 - \frac{7}{20}x^2 + \frac{1}{40}x = 0, \quad \text{or}$$
$$40x^3 - 14x^2 + x = 0 \quad \text{Multiplying by 40}$$

74. -5, 4

75.
$$y(y + 8) = 16(y - 1)$$
$$y^2 + 8y = 16y - 16$$
$$y^2 - 8y + 16 = 0$$
$$(y - 4)(y - 4) = 0$$

$$y - 4 = 0 \quad \text{or} \quad y - 4 = 0$$
$$y = 4 \quad \text{or} \quad y = 4$$

The solution is 4.

76. 5, 3

77.
$$x^2 - \frac{1}{64} = 0$$
$$\left(x - \frac{1}{8}\right)\left(x + \frac{1}{8}\right) = 0$$
$$x - \frac{1}{8} = 0 \quad \text{or} \quad x + \frac{1}{8} = 0$$
$$x = \frac{1}{8} \quad \text{or} \quad x = -\frac{1}{8}$$

The solutions are $\frac{1}{8}$ and $-\frac{1}{8}$.

78. $-\frac{5}{6}, \frac{5}{6}$

79.
$$\frac{5}{16}x^2 = 5$$
$$\frac{5}{16}x^2 - 5 = 0$$
$$5\left(\frac{1}{16}x^2 - 1\right) = 0$$
$$5\left(\frac{1}{4}x - 1\right)\left(\frac{1}{4}x + 1\right) = 0$$
$$\frac{1}{4}x - 1 = 0 \quad \text{or} \quad \frac{1}{4}x + 1 = 0$$
$$\frac{1}{4}x = 1 \quad \text{or} \quad \frac{1}{4}x = -1$$
$$x = 4 \quad \text{or} \quad x = -4$$

The solutions are 4 and -4.

80. $-\frac{5}{9}, \frac{5}{9}$

81. a) $3(3x^2 - 4x + 8) = 3 \cdot 0$ Multiplying (a) by 3
$$9x^2 - 12x + 24 = 0$$

(a) and $9x^2 - 12x + 24 = 0$ are equivalent.

b) $(x - 6)(x + 3) = x^2 - 3x - 18$

(b) and $x^2 - 3x - 18 = 0$ are equivalent.

c) $4(x^2 + 2x + 9) = 4 \cdot 0$ Multiplying (c) by 4
$$4x^2 + 8x + 36 = 0$$

(c) and $4x^2 + 8x + 36 = 0$ are equivalent.

d) $2(2x - 5)(x + 4) = 2 \cdot 0$ Multiplying (d) by 2
$$2(x + 4)(2x - 5) = 0$$
$$(2x + 8)(2x - 5) = 0$$

(d) and $(2x + 8)(2x - 5) = 0$ are equivalent.

e) $5x^2 - 5 = 5(x^2 - 1) = 5(x + 1)(x - 1) =$
$(x + 1)5(x - 1) = (x + 1)(5x - 5)$

(e) and $(x + 1)(5x - 5) = 0$ are equivalent.

f) $2(x^2 + 10x - 2) = 2 \cdot 0$ Multiplying (f) by 2
$$2x^2 + 20x - 4 = 0$$

(f) and $2x^2 + 20x - 4 = 0$ are equivalent.

82.

83.

84. $3.45, -1.65$

85. $-2.33, -6.77$

86. $-0.25, 0.88$

87. $-4.59, -9.15$

88. $4.55, -3.23$

89. $-3.25, -6.75$

Exercise Set 5.8

1. *Familiarize.* Let $x =$ the number (or numbers).

Translate. We reword the problem.

Four times the square of a number   minus   the number   is 3.
$$4x^2 \qquad\qquad - \qquad x \qquad = 3$$

Carry out. We solve the equation.
$$4x^2 - x = 3$$
$$4x^2 - x - 3 = 0$$
$$(4x + 3)(x - 1) = 0$$

$$4x + 3 = 0 \quad \text{or} \quad x - 1 = 0$$
$$4x = -3 \quad \text{or} \qquad x = 1$$
$$x = -\frac{3}{4} \quad \text{or} \qquad x = 1$$

Check. For $-\frac{3}{4}$: Four times the square of $-\frac{3}{4}$ is
$$4\left(-\frac{3}{4}\right)^2 = 4\left(\frac{9}{16}\right) = \frac{9}{4}.$$ If we subtract $-\frac{3}{4}$ from $\frac{9}{4}$
we get $\frac{9}{4} - \left(-\frac{3}{4}\right) = \frac{9}{4} + \frac{3}{4} = \frac{12}{4} = 3$.
For 1: Four times the square of 1 is $4(1)^2 = 4$. If we
subtract 1 from 4 we get $4 - 1 = 3$. Both numbers check.

State. There are two such numbers, $-\frac{3}{4}$ and 1.

2. $5, -5$

3. *Familiarize.* Let $x =$ the number (or numbers).

Translate. We reword the problem.

The square of a number   plus 8 is   six times the number.
$$x^2 \qquad + \quad 8 = \qquad 6x$$

Carry out. We solve the equation.
$$x^2 + 8 = 6x$$
$$x^2 - 6x + 8 = 0$$
$$(x - 4)(x - 2) = 0$$

$x - 4 = 0$ or $\quad x - 2 = 0$

$\qquad x = 4$ or $\qquad x = 2$

Check. The square of 4 is 16, and six times the number 4 is 24. Since $16 + 8 = 24$, the number 4 checks. The square of 2 is 4, and six times the number 2 is 12. Since $4 + 8 = 12$, the number 2 checks.

State. There are two such numbers, 4 and 2.

4. 3, 5

5. Familiarize. The page numbers on facing pages are consecutive integers. Let $x =$ the smaller integer. Then $x + 1 =$ the larger integer.

Translate. We reword the problem.

$$\underbrace{\text{Smaller integer}}_{\displaystyle x} \;\; \underbrace{\text{times}}_{\displaystyle \cdot} \;\; \underbrace{\text{larger integer}}_{\displaystyle (x+1)} \;\; \underbrace{\text{is}}_{\displaystyle =} \;\; \underbrace{210.}_{\displaystyle 210}$$

Carry out. We solve the equation.

$$x(x + 1) = 210$$
$$x^2 + x = 210$$
$$x^2 + x - 210 = 0$$
$$(x + 15)(x - 14) = 0$$

$x + 15 = 0 \qquad$ or $\quad x - 14 = 0$

$\quad x = -15 \quad$ or $\qquad x = 14$

Check. The solutions of the equation are -15 and 14. Since a page number cannot be negative, -15 cannot be a solution of the original problem. We only need to check 14. When $x = 14$, then $x + 1 = 15$, and $14 \cdot 15 = 210$. This checks.

State. The page numbers are 14 and 15.

6. 10 and 11

7. Familiarize. Let $x =$ the smaller even integer. Then $x + 2 =$ the larger even integer.

Translate. We reword the problem.

$$\underbrace{\begin{array}{c}\text{Smaller}\\\text{even integer}\end{array}}_{\displaystyle x} \;\; \underbrace{\text{times}}_{\displaystyle \cdot} \;\; \underbrace{\begin{array}{c}\text{larger}\\\text{even integer}\end{array}}_{\displaystyle (x+2)} \;\; \underbrace{\text{is}}_{\displaystyle =} \;\; \underbrace{168.}_{\displaystyle 168}$$

Carry out.

$$x(x + 2) = 168$$
$$x^2 + 2x = 168$$
$$x^2 + 2x - 168 = 0$$
$$(x + 14)(x - 12) = 0$$

$x + 14 = 0 \qquad$ or $\quad x - 12 = 0$

$\quad x = -14 \quad$ or $\qquad x = 12$

Check. The solutions of the equation are -14 and 12. When x is -14, then $x + 2$ is -12 and $-14(-12) = 168$. The numbers -14 and -12 are consecutive even integers which are solutions of the problem. When x is 12, then $x + 2$ is 14 and $12 \cdot 14 = 168$. The numbers 12 and 14 are also consecutive even integers which are solutions of the problem.

State. We have two solutions, each of which consists of a pair of numbers: -14 and -12, and 12 and 14.

8. 14 and 16, -16 and -14

9. Familiarize. Let $x =$ the smaller odd integer. Then $x + 2 =$ the larger odd integer.

Translate. We reword the problem.

$$\underbrace{\begin{array}{c}\text{Smaller}\\\text{odd integer}\end{array}}_{\displaystyle x} \;\; \underbrace{\text{times}}_{\displaystyle \cdot} \;\; \underbrace{\begin{array}{c}\text{larger}\\\text{odd integer}\end{array}}_{\displaystyle (x+2)} \;\; \underbrace{\text{is}}_{\displaystyle =} \;\; \underbrace{255.}_{\displaystyle 255}$$

Carry out.

$$x(x + 2) = 255$$
$$x^2 + 2x = 255$$
$$x^2 + 2x - 255 = 0$$
$$(x - 15)(x + 17) = 0$$

$x - 15 = 0 \quad$ or $\quad x + 17 = 0$

$\quad x = 15 \quad$ or $\qquad x = -17$

Check. The solutions of the equation are 15 and -17. When x is 15, then $x + 2$ is 17 and $15 \cdot 17 = 255$. The numbers 15 and 17 are consecutive odd integers which are solutions to the problem. When x is -17, then $x + 2$ is -15 and $-17(-15) = 255$. The numbers -17 and -15 are also consecutive odd integers which are solutions to the problem.

State. We have two solutions, each of which consists of a pair of numbers: 15 and 17, and -17 and -15.

10. 11 and 13, -13 and -11

11. Familiarize. Using the labels shown on the drawing in the text, we let $w =$ the width of the rectangle and $w + 4 =$ the length. Recall that the area of a rectangle is length times width.

Translate. We reword the problem.

$$\underbrace{\text{Length}}_{\displaystyle (w+4)} \;\; \underbrace{\text{times}}_{\displaystyle \cdot} \;\; \underbrace{\text{width}}_{\displaystyle w} \;\; \underbrace{\text{is}}_{\displaystyle =} \;\; \underbrace{\text{area.}}_{\displaystyle 96}$$

Carry out.

$$(w + 4) \cdot w = 96$$
$$w^2 + 4w = 96$$
$$w^2 + 4w - 96 = 0$$
$$(w + 12)(w - 8) = 0$$

$w + 12 = 0 \quad$ or $\quad w - 8 = 0$

$\quad w = -12$ or $\qquad w = 8$

Check. The solutions of the equation are -12 and 8. The width of a rectangle cannot have a negative measure, so -12 cannot be a solution. Suppose the width is 8 m. The length is 4 m greater than the width, so the length is 12 m and the area is $12 \cdot 8$, or 96 m^2. The numbers check in the original problem.

State. The length is 12 m, and the width is 8 m.

12. Length: 12 cm, width: 7 cm

13. *Familiarize*. First draw a picture. Let $x =$ the length of a side of the square.

The area of the square is $x \cdot x$, or x^2. The perimeter of the square is $x + x + x + x$, or $4x$.

Translate.

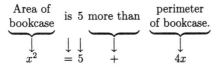

$$\underbrace{\text{Area of bookcase}}_{x^2} \quad \underbrace{\text{is 5 more than}}_{= \quad 5 \quad +} \quad \underbrace{\text{perimeter of bookcase.}}_{4x}$$

Carry out.

$$x^2 = 5 + 4x$$
$$x^2 - 4x - 5 = 0$$
$$(x - 5)(x + 1) = 0$$
$$x - 5 = 0 \quad \text{or} \quad x + 1 = 0$$
$$x = 5 \quad \text{or} \quad x = -1$$

Check. The solutions of the equation are 5 and -1. The length of a side cannot be negative, so we only check 5. The area is $5 \cdot 5$, or 25. The perimeter is $5 + 5 + 5 + 5$, or 20. The area, 25, is 5 more than the perimeter, 20. This checks.

State. The length of a side is 5.

14. 3 or 1

15. *Familiarize*. Using the labels shown on the drawing in the text, we let $h =$ the height and $h + 10 =$ the base. Recall that the formula for the area of a triangle is $\frac{1}{2} \cdot$ (base) $\cdot$ (height).

Translate.

$$\underset{28}{\text{Area}} \; \underset{=}{\text{is}} \; \underset{\frac{1}{2}}{\frac{1}{2}} \; \underset{\cdot}{\text{times}} \; \underset{(h + 10)}{\underbrace{\text{the base}}} \; \underset{\cdot}{\text{times}} \; \underset{h}{\underbrace{\text{the height.}}}$$

Carry out.

$$28 = \frac{1}{2}h(h + 10)$$
$$56 = h(h + 10)$$
$$56 = h^2 + 10h$$
$$0 = h^2 + 10h - 56$$
$$0 = (h + 14)(h - 4)$$

$$h + 14 = 0 \quad \text{or} \quad h - 4 = 0$$
$$h = -14 \text{ or} \quad h = 4$$

Check. The solutions of the equation are -14 and 4. The height of a triangle cannot have a negative length, so -14

cannot be a solution. Suppose the height is 4 cm. The base is 10 cm greater than the height, to the base is 14 cm and the area is $\frac{1}{2} \cdot 14 \cdot 4$, or 28 cm^2. These numbers check.

State. The height is 4 cm and the base is 14 cm.

16. Height: 2 m, base: 10 m

17. *Familiarize*. We make a drawing. Let $x =$ the length of a side of the original square. Then $x + 3 =$ the length of a side of the enlarged square.

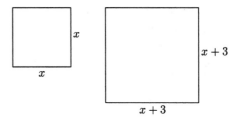

Recall that the area of a square is found by squaring the length of a side.

Translate.

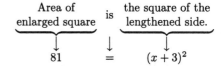

$$\underbrace{\text{Area of enlarged square}}_{81} \quad \underset{=}{\text{is}} \quad \underbrace{\text{the square of the lengthened side.}}_{(x + 3)^2}$$

Carry out.

$$81 = (x + 3)^2$$
$$81 = x^2 + 6x + 9$$
$$0 = x^2 + 6x - 72$$
$$0 = (x + 12)(x - 6)$$

$$x + 12 = 0 \quad \text{or} \quad x - 6 = 0$$
$$x = -12 \quad \text{or} \quad x = 6$$

Check. The solutions of the equation are -12 and 6. The length of a side cannot be negative, so -12 cannot be a solution. Suppose the length of a side of the original square is 6 m. Then the length of a side of the new square is $6 + 3$, or 9 m. Its area is 9^2, or 81 m^2. The numbers check.

State. The length of a side of the original square is 6 m.

18. 4 km

19. *Familiarize*. Let $x =$ the smaller odd whole number. Then $x + 2 =$ the larger odd whole number.

Translate.

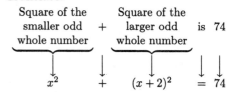

$$\underbrace{\text{Square of the smaller odd whole number}}_{x^2} \quad + \quad \underbrace{\text{Square of the larger odd whole number}}_{(x + 2)^2} \quad \underset{= \quad 74}{\text{is 74}}$$

Carry out.
$$x^2 + (x+2)^2 = 74$$
$$x^2 + x^2 + 4x + 4 = 74$$
$$2x^2 + 4x - 70 = 0$$
$$2(x^2 + 2x - 35) = 0$$
$$2(x+7)(x-5) = 0$$
$$x + 7 = 0 \quad \text{or} \quad x - 5 = 0$$
$$x = -7 \quad \text{or} \quad x = 5$$

Check. The solutions of the equation are -7 and 5. The problem asks for odd whole numbers, so -7 cannot be a solution. When x is 5, $x+2$ is 7. The numbers 5 and 7 are consecutive odd whole numbers. The sum of their squares, $25 + 49$, is 74. The numbers check.

State. The numbers are 5 and 7.

20. 7 and 9

21. *Familiarize.* We will use the formula $n^2 - n = N$.

Translate. Substitute 23 for n.
$$23^2 - 23 = N$$

Carry out. We do the computation of the left.
$$23^2 - 23 = N$$
$$529 - 23 = N$$
$$506 = N$$

Check. We can recheck the computation or we can solve $n^2 - n = 506$. The answer checks.

State. 506 games will be played.

22. 182

23. *Familiarize.* We will use the formula $n^2 - n = N$.

Translate. Substitute 132 for N.
$$n^2 - n = 132$$

Carry out.
$$n^2 - n = 132$$
$$n^2 - n - 132 = 0$$
$$(n-12)(n+11) = 0$$
$$n - 12 = 0 \quad \text{or} \quad n + 11 = 0$$
$$n = 12 \quad \text{or} \quad n = -11$$

Check. The solutions of the equation are 12 and -11. Since the number of teams cannot be negative, -11 cannot be a solution. But 12 checks since $12^2 - 12 = 144 - 12 = 132$.

State. There are 12 teams in the league.

24. 10

25. *Familiarize.* We will use the formula
$$N = \frac{1}{2}(n^2 - n).$$

Translate. Substitute 40 for n.
$$N = \frac{1}{2}(40^2 - 40)$$

Carry out. We do the computation on the right.

$$N = \frac{1}{2}(40^2 - 40)$$
$$N = \frac{1}{2}(1600 - 40)$$
$$N = \frac{1}{2}(1560)$$
$$N = 780$$

Check. We can recheck the computation, or we can solve the equation $780 = \frac{1}{2}(n^2 - n)$. The answer checks.

State. 780 handshakes are possible.

26. 4950

27. *Familiarize.* We will use the formula $N = \frac{1}{2}(n^2 - n)$, since "clicks" can be substituted for handshakes.

Translate. Substitute 190 for N.
$$190 = \frac{1}{2}(n^2 - n)$$

Carry out.
$$190 = \frac{1}{2}(n^2 - n)$$
$$380 = n^2 - n \qquad \text{Multiplying by 2}$$
$$0 = n^2 - n - 380$$
$$0 = (n-20)(n+19)$$
$$n - 20 = 0 \quad \text{or} \quad n + 19 = 0$$
$$n = 20 \quad \text{or} \quad n = -19$$

Check. The solutions of the equation are 20 and -19. Since the number of people cannot be negative, -19 cannot be a solution. However, 20 checks since $\frac{1}{2}(20^2 - 20) = \frac{1}{2}(400 - 20) = \frac{1}{2} \cdot 380 = 190$.

State. 20 people took part in the toast.

28. 25

29. *Familiarize.* We make a drawing. Let $x =$ the length of the unknown leg. Then $x + 2 =$ the length of the hypotenuse.

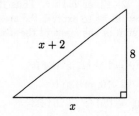

Translate. Use the Pythagorean theorem.
$$a^2 + b^2 = c^2$$
$$8^2 + x^2 = (x+2)^2$$

Carry out. We solve the equation.
$$8^2 + x^2 = (x+2)^2$$
$$64 + x^2 = x^2 + 4x + 4$$
$$60 = 4x \qquad \text{Subtracting } x^2 \text{ and } 4$$
$$15 = x$$

Check. When $x = 15$, then $x + 2 = 17$ and $8^2 + 15^2 = 17^2$. Thus, 15 and 17 check.

State. The lengths of the hypotenuse and the other leg are 17 ft and 15 ft, respectively.

30. Hypotenuse: 26 ft, leg: 10 ft

31. **Familiarize**. We label the drawing. Let $x =$ the length of a side of the dining room.

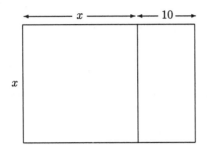

The dimensions of the dining room are x by x, and the dimensions of the kitchen are x by 10. The dimensions of the entire rectangle are $x + 10$ by x. Recall that the area of a rectangle is length × width.

Translate. We reword the problem.

$$\underbrace{\text{Total area}}_{} \quad \text{is} \quad \underbrace{264 \text{ ft}^2}_{}$$
$$\downarrow \qquad\quad \downarrow \qquad \downarrow$$
$$(x + 10)x \quad = \quad 264$$

Carry out. We solve the equation.

$$(x + 10)x = 264$$
$$x^2 + 10x = 264$$
$$x^2 + 10x - 264 = 0$$
$$(x + 22)(x - 12) = 0$$

$$x + 22 = 0 \quad \text{or} \quad x - 12 = 0$$
$$x = -22 \text{ or} \qquad x = 12$$

Check. Since measures cannot be negative, we only consider 12. If $x = 12$, the area of the entire rectangle is $(10 + 12)(12)$, or $22 \cdot 12$, or 264 ft². Thus 12 checks. Another approach would be to express the area of the entire rectangle as the sum of the areas of the dining room and the kitchen:

Dining room: $x \cdot x = 12 \cdot 12$, or 144 ft²

Kitchen: $x \cdot 10 = 12 \cdot 10$, or 120 ft²

Total area: 144 ft² + 120 ft² = 264 ft²

This provides a second check.

State. The dining room is 12 ft by 12 ft, and the kitchen is 12 ft by 10 ft.

32. 4 m

33. Graph: $y = -\dfrac{2}{3}x + 1$

Since the equation is in the form $y = mx + b$, we know that the y-intercept is $(0,1)$. We find two other solutions, using multiples of 3 for x to avoid fractions.

When $x = -3$, $y = -\dfrac{2}{3}(-3) + 1 = 2 + 1 = 3$.

When $x = 3$, $y = -\dfrac{2}{3} \cdot 3 + 1 = -2 + 1 = -1$.

x	y
0	1
-3	3
3	-1

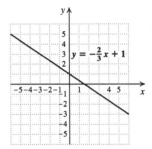

34.

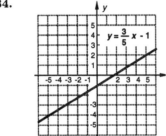

35. $7x^0 = 7 \cdot 4^0 = 7 \cdot 1 = 7$

(Any nonzero number raised to the zero power is 1.)

36. $m^6 n^6$

37.

38. ◈

39. **Familiarize**. Using the labels shown on the drawing in the text, we let $x =$ the width of the walk. Then the length and width of the rectangle formed by the pool and walk together are $40 + 2x$ and $20 + 2x$, respectively.

Translate.

$$\underbrace{\text{Area}}_{} \quad \text{is} \quad \underbrace{\text{length}}_{} \quad \text{times} \quad \underbrace{\text{width}}_{}.$$
$$\downarrow \qquad \downarrow \qquad \downarrow \qquad\quad \downarrow \qquad\quad \downarrow$$
$$1500 \;\; = \;\; (40 + 2x) \quad \cdot \quad (20 + 2x)$$

Carry out. We solve the equation.

$$1500 = (40 + 2x)(20 + 2x)$$

$1500 = 2(20 + x) \cdot 2(10 + x)$ Factoring 2 out of each factor on the right

$$1500 = 4 \cdot (20 + x)(10 + x)$$

$375 = (20 + x)(10 + x)$ Dividing by 4

$$375 = 200 + 30x + x^2$$
$$0 = x^2 + 30x - 175$$
$$0 = (x + 35)(x - 5)$$

$$x + 35 = 0 \quad \text{or} \quad x - 5 = 0$$
$$x = -35 \quad \text{or} \qquad x = 5$$

Check. The solutions of the equation are -35 and 5. Since the width of the walk cannot be negative, -35 is not a solution. When $x = 5$, $40 + 2x = 40 + 2 \cdot 5$, or 50 and $20 + 2x = 20 + 2 \cdot 5$, or 30. The total area of the pool and walk is $50 \cdot 30$, or 1500 ft^2. This checks.

State. The width of the walk is 5 ft.

40. a) 4 sec

b) $3\frac{1}{4}$ sec

41. *Familiarize*. Let $y = $ the ten's digit. Then $y + 4 = $ the one's digit and $10y + y + 4$, or $11y + 4$, represents the number.

Translate.

$$\underbrace{\text{The number}}_{11y + 4} \text{ plus } \underbrace{\begin{array}{c}\text{the product} \\ \text{of the digits}\end{array}}_{y(y + 4)} \underbrace{\text{is } 58.}_{= \ 58}$$

Carry out. We solve the equation.

$$11y + 4 + y(y + 4) = 58$$
$$11y + 4 + y^2 + 4y = 58$$
$$y^2 + 15y + 4 = 58$$
$$y^2 + 15y - 54 = 0$$
$$(y + 18)(y - 3) = 0$$
$$y + 18 = 0 \quad \text{or} \quad y - 3 = 0$$
$$y = -18 \quad \text{or} \quad y = 3$$

Check. Since -18 cannot be a digit of the number, we only need to check 3. When $y = 3$, then $y + 4 = 7$ and the number is 37. We see that $37 + 3 \cdot 7 = 37 + 21$, or 58. The result checks.

State. The number is 37.

42. 7 m

43. *Familiarize*. We make a drawing. Let $w = $ the width of the piece of cardboard. Then $2w = $ the length.

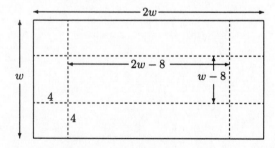

The box will have length $2w - 8$, width $w - 8$, and height 4. Recall that the formula for volume is $V = $ length $\times$ width $\times$ height.

Translate.

$$\underbrace{\text{The volume}}_{(2w - 8)(w - 8)(4)} \underbrace{\text{is}}_{=} \underbrace{616 \text{ cm}^3.}_{616}$$

Carry out. We solve the equation.

$$(2w - 8)(w - 8)(4) = 616$$
$$(2w^2 - 24w + 64)(4) = 616$$
$$8w^2 - 96w + 256 = 616$$
$$8w^2 - 96w - 360 = 0$$
$$8(w^2 - 12w - 45) = 0$$
$$w^2 - 12w - 45 = 0 \qquad \text{Dividing by } 8$$
$$(w - 15)(w + 3) = 0$$
$$w - 15 = 0 \quad \text{or} \quad w + 3 = 0$$
$$w = 15 \quad \text{or} \quad w = -3$$

Check. The width cannot be negative, so we only need to check 15. When $w = 15$, then $2w = 30$ and the dimensions of the box are $30 - 8$ by $15 - 8$ by 4, or 22 by 7 by 4. The volume is $22 \cdot 7 \cdot 4$, or 616.

State. The cardboard is 30 cm by 15 cm.

44. 5 in.

45. *Familiarize*. Let $x = $ the length of a side of the original square. Then $x + 5 = $ the length of a side of the new square.

Translate.

$$\underbrace{\begin{array}{c}\text{The area} \\ \text{of the new} \\ \text{square}\end{array}}_{(x + 5)^2} \underbrace{\text{is}}_{=} \underbrace{2\frac{1}{4}}_{2\frac{1}{4}} \underbrace{\text{times}}_{\cdot} \underbrace{\begin{array}{c}\text{the area of} \\ \text{the original} \\ \text{square.}\end{array}}_{x^2}$$

Carry out. We solve the equation.

$$(x + 5)^2 = 2\frac{1}{4} \cdot x^2$$
$$x^2 + 10x + 25 = \frac{9}{4}x^2 \qquad \text{Writing } 2\frac{1}{4} \text{ as } \frac{9}{4}$$
$$4x^2 + 40x + 100 = 9x^2 \qquad \text{Multiplying by } 4$$
$$0 = 5x^2 - 40x - 100$$
$$0 = 5(x^2 - 8x - 20)$$
$$0 = x^2 - 8x - 20 \qquad \text{Dividing by } 5$$
$$0 = (x - 10)(x + 2)$$

$$x - 10 = 0 \quad \text{or} \quad x + 2 = 0$$
$$x = 10 \quad \text{or} \quad x = -2$$

Check. Since the length of a side of a square cannot be negative, we only need to check 10. If $x = 10$, then $x + 5 = 15$. A square with side 10 has area 10^2, or 100, and a square with side 15 has area 15^2, or 225. Since $225 = 2\frac{1}{4}(100)$, the numbers check.

State. The area of the original square is 100 cm^2, and the area of the new square is 225 cm^2.

Chapter 6

Rational Expressions And Equations

Exercise Set 6.1

---

1. $\dfrac{-5}{2x}$

We find the real number(s) that make the denominator 0. To do so we set the denominator equal to 0 and solve for x:

$$2x = 0$$
$$x = 0$$

The expression is undefined for $x = 0$.

2. 0

3. $\dfrac{a+7}{a-8}$

Set the denominator equal to 0 and solve for a:

$$a - 8 = 0$$
$$a = 8$$

The expression is undefined for $a = 8$.

4. -7

5. $\dfrac{3}{2y+5}$

Set the denominator equal to 0 and solve for y:

$$2y + 5 = 0$$
$$2y = -5$$
$$y = -\frac{5}{2}$$

The expression is undefined for $y = -\dfrac{5}{2}$.

6. 3

7. $\dfrac{x^2+11}{x^2-3x-28}$

Set the denominator equal to 0 and solve for x:

$$x^2 - 3x - 28 = 0$$
$$(x-7)(x+4) = 0$$
$$x - 7 = 0 \quad \text{or} \quad x + 4 = 0$$
$$x = 7 \quad \text{or} \quad x = -4$$

The expression is undefined for $x = 7$ and $x = -4$.

8. 5, 2

9. $\dfrac{m^3-2m}{m^2-25}$

Set the denominator equal to 0 and solve for m:

$$m^2 - 25 = 0$$
$$(m+5)(m-5) = 0$$
$$m + 5 = 0 \quad \text{or} \quad m - 5 = 0$$
$$m = -5 \quad \text{or} \quad m = 5$$

The expression is undefined for $m = -5$ and $m = 5$.

10. $-7, 7$

11. $\dfrac{10a^3b}{30ab^2} = \dfrac{a^2 \cdot 10ab}{3b \cdot 10ab}$ Factoring the numerator and denominator. Note the common factor of $10ab$.

$$= \dfrac{a^2}{3b} \cdot \dfrac{10ab}{10ab} \quad \text{Rewriting as a product of two rational expressions}$$

$$= \dfrac{a^2}{3b} \cdot 1 \qquad \dfrac{10ab}{10ab} = 1$$

$$= \dfrac{a^2}{3b} \qquad \text{Removing a factor of 1}$$

12. $\dfrac{5y}{x^2}$

13. $\dfrac{35x^2y}{14x^3y^5} = \dfrac{5 \cdot 7x^2y}{2xy^4 \cdot 7x^2y}$

$$= \dfrac{5}{2xy^4} \cdot \dfrac{7x^2y}{7x^2y}$$

$$= \dfrac{5}{2xy^4} \cdot 1$$

$$= \dfrac{5}{2xy^4}$$

14. $\dfrac{2a^2b^5}{3}$

15. $\dfrac{9x+15}{6x+10} = \dfrac{3(3x+5)}{2(3x+5)}$

$$= \dfrac{3}{2} \cdot \dfrac{3x+5}{3x+5}$$

$$= \dfrac{3}{2} \cdot 1$$

$$= \dfrac{3}{2}$$

16. $\dfrac{7}{5}$

17. $\dfrac{a^2-25}{a^2+6a+5} = \dfrac{(a+5)(a-5)}{(a+5)(a+1)}$

$$= \dfrac{a+5}{a+5} \cdot \dfrac{a-5}{a+1}$$

$$= 1 \cdot \dfrac{a-5}{a+1}$$

$$= \dfrac{a-5}{a+1}$$

18. $\dfrac{a+2}{a-3}$

19. $\dfrac{48x^4}{18x^6} = \dfrac{8 \cdot 6x^4}{3x^2 \cdot 6x^4}$

$$= \dfrac{8 \cdot \cancel{6x^4}}{3x^2 \cdot \cancel{6x^4}}$$

$$= \dfrac{8}{3x^2}$$

Check: Let $x = 1$.

$$\frac{48x^4}{18x^6} = \frac{48 \cdot 1^4}{18 \cdot 1^6} = \frac{48}{18} = \frac{8}{3}$$

$$\frac{8}{3x^2} = \frac{8}{3 \cdot 1^2} = \frac{8}{3}$$

The answer is probably correct.

20. $\dfrac{19a^2}{6}$

21. $\dfrac{4x - 12}{4x} = \dfrac{4(x - 3)}{4 \cdot x}$

$$= \frac{\cancel{4}(x - 3)}{\cancel{4} \cdot x}$$

$$= \frac{x - 3}{x}$$

Check: Let $x = 2$.

$$\frac{4x - 12}{4x} = \frac{4 \cdot 2 - 12}{4 \cdot 2} = \frac{-4}{8} = -\frac{1}{2}$$

$$\frac{x - 3}{x} = \frac{2 - 3}{2} = \frac{-1}{2} = -\frac{1}{2}$$

The answer is probably correct.

22. $\dfrac{y - 3}{2y}$

23. $\dfrac{3m^2 + 3m}{6m^2 + 9m} = \dfrac{3m(m + 1)}{3m(2m + 3)}$

$$= \frac{3m}{3m} \cdot \frac{m + 1}{2m + 3}$$

$$= 1 \cdot \frac{m + 1}{2m + 3}$$

$$= \frac{m + 1}{2m + 3}$$

Check: Let $m = 1$.

$$\frac{3m^2 + 3m}{6m^2 + 9m} = \frac{3 \cdot 1^2 + 3 \cdot 1}{6 \cdot 1^2 + 9 \cdot 1} = \frac{6}{15} = \frac{2}{5}$$

$$\frac{m + 1}{2m + 3} = \frac{1 + 1}{2 \cdot 1 + 3} = \frac{2}{5}$$

The answer is probably correct.

24. $\dfrac{2(2y - 1)}{5(y - 1)}$

25. $\dfrac{a^2 - 9}{a^2 + 5a + 6} = \dfrac{(a - 3)(a + 3)}{(a + 2)(a + 3)}$

$$= \frac{a - 3}{a + 2} \cdot \frac{a + 3}{a + 3}$$

$$= \frac{a - 3}{a + 2} \cdot 1$$

$$= \frac{a - 3}{a + 2}$$

Check: Let $a = 2$.

$$\frac{a^2 - 9}{a^2 + 5a + 6} = \frac{2^2 - 9}{2^2 + 5 \cdot 2 + 6} = \frac{-5}{20} = -\frac{1}{4}$$

$$\frac{a - 3}{a + 2} = \frac{2 - 3}{2 + 2} = \frac{-1}{4} = -\frac{1}{4}$$

The answer is probably correct.

26. $\dfrac{t - 5}{t - 4}$

27. $\dfrac{2t^2 + 6t + 4}{4t^2 - 12t - 16} = \dfrac{2(t^2 + 3t + 2)}{4(t^2 - 3t - 4)}$

$$= \frac{2(t + 2)(t + 1)}{2 \cdot 2(t - 4)(t + 1)}$$

$$= \frac{2(t + 1)}{2(t + 1)} \cdot \frac{t + 2}{2(t - 4)}$$

$$= 1 \cdot \frac{t + 2}{2(t - 4)}$$

$$= \frac{t + 2}{2(t - 4)}$$

Check: Let $t = 1$.

$$\frac{2t^2 + 6t + 4}{4t^2 - 12t - 16} = \frac{2 \cdot 1^2 + 6 \cdot 1 + 4}{4 \cdot 1^2 - 12 \cdot 1 - 16} = \frac{12}{-24} = -\frac{1}{2}$$

$$\frac{t + 2}{2(t - 4)} = \frac{1 + 2}{2(1 - 4)} = \frac{3}{-6} = -\frac{1}{2}$$

The answer is probably correct.

28. $\dfrac{a - 4}{2(a + 4)}$

29. $\dfrac{x^2 - 25}{x^2 - 10x + 25} = \dfrac{(x - 5)(x + 5)}{(x - 5)(x - 5)}$

$$= \frac{x - 5}{x - 5} \cdot \frac{x + 5}{x - 5}$$

$$= 1 \cdot \frac{x + 5}{x - 5}$$

$$= \frac{x + 5}{x - 5}$$

Check: Let $x = 2$.

$$\frac{x^2 - 25}{x^2 - 10x + 25} = \frac{2^2 - 25}{2^2 - 10 \cdot 2 + 25} = \frac{-21}{9} = -\frac{7}{3}$$

$$\frac{x + 5}{x - 5} = \frac{2 + 5}{2 - 5} = \frac{7}{-3} = -\frac{7}{3}$$

The answer is probably correct.

30. $\dfrac{x + 4}{x - 4}$

31. $\dfrac{a^2 - 1}{a - 1} = \dfrac{(a - 1)(a + 1)}{a - 1}$

$$= \frac{a - 1}{a - 1} \cdot \frac{a + 1}{1}$$

$$= 1 \cdot \frac{a + 1}{1}$$

$$= a + 1$$

Check: Let $a = 2$.

$$\frac{a^2 - 1}{a - 1} = \frac{2^2 - 1}{2 - 1} = \frac{3}{1} = 3$$

$$a + 1 = 2 + 1 = 3$$

The answer is probably correct.

32. $t - 1$

33. $\dfrac{x^2 + 1}{x + 1}$ cannot be simplified.

Neither the numerator nor the denominator can be factored.

34. $\dfrac{y^2 + 4}{y + 2}$

35. $\dfrac{6x^2 - 54}{4x^2 - 36} = \dfrac{2 \cdot 3(x^2 - 9)}{2 \cdot 2(x^2 - 9)}$

$\qquad = \dfrac{2(x^2 - 9)}{2(x^2 - 9)} \cdot \dfrac{3}{2}$

$\qquad = 1 \cdot \dfrac{3}{2}$

$\qquad = \dfrac{3}{2}$

Check: Let $x = 1$.

$\dfrac{6x^2 - 54}{4x^2 - 36} = \dfrac{6 \cdot 1^2 - 54}{4 \cdot 1^2 - 36} = \dfrac{-48}{-32} = \dfrac{3}{2}$

$\dfrac{3}{2} = \dfrac{3}{2}$

The answer is probably correct.

36. 2

37. $\dfrac{6t + 12}{t^2 - t - 6} = \dfrac{6(t + 2)}{(t - 3)(t + 2)}$

$\qquad = \dfrac{6}{t - 3} \cdot \dfrac{t + 2}{t + 2}$

$\qquad = \dfrac{6}{t - 3} \cdot 1$

$\qquad = \dfrac{6}{t - 3}$

Check: Let $t = 1$.

$\dfrac{6t + 12}{t^2 - t - 6} = \dfrac{6 \cdot 1 + 12}{1^2 - 1 - 6} = \dfrac{18}{-6} = -3$

$\dfrac{6}{t - 3} = \dfrac{6}{1 - 3} = \dfrac{6}{-2} = -3$

The answer is probably correct.

38. $\dfrac{5}{y + 6}$

39. $\dfrac{a^2 - 10a + 21}{a^2 - 11a + 28} = \dfrac{(a - 7)(a - 3)}{(a - 7)(a - 4)}$

$\qquad = \dfrac{a - 7}{a - 7} \cdot \dfrac{a - 3}{a - 4}$

$\qquad = 1 \cdot \dfrac{a - 3}{a - 4}$

$\qquad = \dfrac{a - 3}{a - 4}$

Check: Let $a = 2$.

$\dfrac{a^2 - 10a + 21}{a^2 - 11a + 28} = \dfrac{2^2 - 10 \cdot 2 + 21}{2^2 - 11 \cdot 2 + 28} = \dfrac{5}{10} = \dfrac{1}{2}$

$\dfrac{a - 3}{a - 4} = \dfrac{2 - 3}{2 - 4} = \dfrac{-1}{-2} = \dfrac{1}{2}$

The answer is probably correct.

40. $\dfrac{y - 6}{y - 5}$

41. $\dfrac{t^2 - 4}{(t + 2)^2} = \dfrac{(t - 2)(t + 2)}{(t + 2)(t + 2)}$

$\qquad = \dfrac{t - 2}{t + 2} \cdot \dfrac{t + 2}{t + 2}$

$\qquad = \dfrac{t - 2}{t + 2} \cdot 1$

$\qquad = \dfrac{t - 2}{t + 2}$

Check: Let $t = 1$.

$\dfrac{t^2 - 4}{(t + 2)^2} = \dfrac{1^2 - 4}{(1 + 2)^2} = \dfrac{-3}{9} = -\dfrac{1}{3}$

$\dfrac{t - 2}{t + 2} = \dfrac{1 - 2}{1 + 2} = \dfrac{-1}{3} = -\dfrac{1}{3}$

The answer is probably correct.

42. $\dfrac{a - 3}{a + 3}$

43. $\dfrac{6 - x}{x - 6} = \dfrac{-(-6 + x)}{x - 6}$

$\qquad = \dfrac{-1(x - 6)}{x - 6}$

$\qquad = -1 \cdot \dfrac{x - 6}{x - 6}$

$\qquad = -1 \cdot 1$

$\qquad = -1$

Check: Let $x = 3$.

$\dfrac{6 - x}{x - 6} = \dfrac{6 - 3}{3 - 6} = \dfrac{3}{-3} = -1$

$-1 = -1$

The answer is probably correct.

44. -1

45. $\dfrac{a - b}{b - a} = \dfrac{-1(-a + b)}{b - a}$

$\qquad = \dfrac{-1(b - a)}{b - a}$

$\qquad = -1 \cdot \dfrac{b - a}{b - a}$

$\qquad = -1 \cdot 1$

$\qquad = -1$

Check: Let $a = 2$ and $b = 1$.

$\dfrac{a - b}{b - a} = \dfrac{2 - 1}{1 - 2} = \dfrac{1}{-1} = -1$

$-1 = -1$

The answer is probably correct.

46. 1

47. $\dfrac{6t - 12}{2 - t} = \dfrac{-6(-t + 2)}{2 - t}$

$\qquad = \dfrac{-6(2 - t)}{2 - t}$

$\qquad = \dfrac{-6(2 - t)}{2 - t}$

$\qquad = -6$

Check: Let $t = 3$.

$$\frac{6t - 12}{2 - t} = \frac{6 \cdot 3 - 12}{2 - 3} = \frac{6}{-1} = -6$$
$$-6 = -6$$

The answer is probably correct.

48. -5

49.
$$\frac{a^2 - 1}{1 - a} = \frac{(a + 1)(a - 1)}{-1(-1 + a)}$$
$$= \frac{(a + 1)(a - 1)}{-1(a - 1)}$$
$$= \frac{a + 1}{-1} \cdot \frac{a - 1}{a - 1}$$
$$= -(a + 1) \cdot 1$$
$$= -a - 1$$

Check: Let $a = 2$.

$$\frac{a^2 - 1}{1 - a} = \frac{2^2 - 1}{1 - 2} = \frac{3}{-1} = -3$$
$$-a - 1 = -2 - 1 = -3$$

The answer is probably correct.

50. -1

51. $x^2 + 8x + 7$

The factorization is of the form $(x +)(x +)$. We look for two factors of 7 whose sum is 8. The numbers we need are 1 and 7.

$$x^2 + 8x + 7 = (x + 1)(x + 7)$$

52. $(x - 2)(x - 7)$

53. $5x + 2y = 20$

To find the y-intercept, solve:
$$2y = 20$$
$$y = 10$$

The y-intercept is $(0, 10)$.

To find the x-intercept, solve:
$$5x = 20$$
$$x = 4$$

The x-intercept is $(4, 0)$.

We find a third point as a check. Let $x = 2$ and solve for y.
$$5 \cdot 2 + 2y = 20$$
$$10 + 2y = 20$$
$$2y = 10$$
$$y = 5$$

The point $(2, 5)$ appears to line up with the intercepts, so we draw the graph.

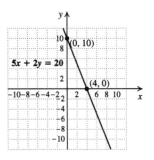

54. y-intercept: $(0, -2)$, x-intercept: $(4, 0)$

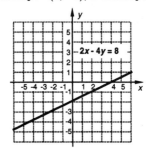

55.

56.

57.
$$\frac{x^4 - 16y^2}{(x^2 + 4y^2)(x - 2y)}$$
$$= \frac{(x^2 + 4y^2)(x + 2y)(x - 2y)}{(x^2 + 4y^2)(x - 2y)}$$
$$= \frac{(x^2 + 4y^2)(x + 2y)(x - 2y)}{(x^2 + 4y^2)(x - 2y)(1)}$$
$$= x + 2y$$

58. $\dfrac{a - b}{-a - b}$, or $\dfrac{b - a}{a + b}$

59.
$$\frac{(t^4 - 1)(t^2 - 9)(t - 9)^2}{(t^4 - 81)(t^2 + 1)(t + 1)^2}$$
$$= \frac{(t^2 + 1)(t + 1)(t - 1)(t + 3)(t - 3)(t - 9)(t - 9)}{(t^2 + 9)(t + 3)(t - 3)(t^2 + 1)(t + 1)(t + 1)}$$
$$= \frac{(t^2 + 1)(t + 1)(t - 1)(t + 3)(t - 3)(t - 9)(t - 9)}{(t^2 + 9)(t + 3)(t - 3)(t^2 + 1)(t + 1)(t + 1)}$$
$$= \frac{(t - 1)(t - 9)(t - 9)}{(t^2 + 9)(t + 1)}, \text{ or } \frac{(t - 1)(t - 9)^2}{(t^2 + 9)(t + 1)}$$

60. 1

61.
$$\frac{(x^2 - y^2)(x^2 - 2xy + y^2)}{(x - y)^2(x^2 - 4xy - 5y^2)}$$
$$= \frac{(x + y)(x - y)(x - y)(x - y)}{(x - y)(x - y)(x - 5y)(x + y)}$$
$$= \frac{(x + y)(x - y)(x - y)(x - y)}{(x - y)(x - y)(x - 5y)(x + y)}$$
$$= \frac{x - y}{x - 5y}$$

62. $\dfrac{1}{x-1}$

63.

Exercise Set 6.2

1. $\dfrac{3x}{2} \cdot \dfrac{x+4}{x-1} = \dfrac{3x(x+4)}{2(x-1)}$

2. $\dfrac{4x(x-3)}{5(x+2)}$

3. $\dfrac{x-1}{x+2} \cdot \dfrac{x+1}{x+2} = \dfrac{(x-1)(x+1)}{(x+2)(x+2)}$

4. $\dfrac{(x-2)(x-2)}{(x-5)(x+5)}$

5. $\dfrac{2x+3}{4} \cdot \dfrac{x+1}{x-5} = \dfrac{(2x+3)(x+1)}{4(x-5)}$

6. $\dfrac{(-5)(-6)}{(3x-4)(5x+6)}$

7. $\dfrac{a-5}{a^2+1} \cdot \dfrac{a+2}{a^2-1} = \dfrac{(a-5)(a+2)}{(a^2+1)(a^2-1)}$

8. $\dfrac{(t+3)(t+3)}{(t^2-2)(t^2-2)}$

9. $\dfrac{x+1}{2+x} \cdot \dfrac{x-1}{x+1} = \dfrac{(x+1)(x-1)}{(2+x)(x+1)}$

10. $\dfrac{(m^2+5)(m^2-4)}{(m+8)(m^2-4)}$

11. $\dfrac{4x^3}{3x} \cdot \dfrac{14}{x}$

$= \dfrac{4x^3 \cdot 14}{3x \cdot x}$ Multiplying the numerators and the denominators

$= \dfrac{4 \cdot x \cdot x \cdot x \cdot 14}{3 \cdot x \cdot x}$ Factoring the numerator and the denominator

$= \dfrac{4 \cdot \cancel{x} \cdot \cancel{x} \cdot x \cdot 14}{3 \cdot \cancel{x} \cdot \cancel{x}}$ Removing a factor of 1

$= \dfrac{56x}{3}$ Simplifying

12. $\dfrac{12}{b^2}$

13. $\dfrac{3c}{d^2} \cdot \dfrac{4d}{6c^3}$

$= \dfrac{3c \cdot 4d}{d^2 \cdot 6c^3}$ Multiplying the numerators and the denominators

$= \dfrac{3 \cdot c \cdot 2 \cdot 2 \cdot d}{d \cdot d \cdot 3 \cdot 2 \cdot c \cdot c \cdot c}$ Factoring the numerator and the denominator

$= \dfrac{\cancel{3} \cdot \cancel{c} \cdot \cancel{2} \cdot 2 \cdot \cancel{d}}{d \cdot d \cdot \cancel{3} \cdot \cancel{2} \cdot \cancel{c} \cdot c \cdot c}$

$= \dfrac{2}{dc^2}$

14. $\dfrac{6x}{y^2}$

15. $\dfrac{x^2-3x-10}{(x-2)^2} \cdot \dfrac{x-2}{x-5} = \dfrac{(x^2-3x-10)(x-2)}{(x-2)^2(x-5)}$

$= \dfrac{(x-5)(x+2)(x-2)}{(x-2)(x-2)(x-5)}$

$= \dfrac{\cancel{(x-5)}(x+2)\cancel{(x-2)}}{\cancel{(x-2)}(x-2)\cancel{(x-5)}}$

$= \dfrac{x+2}{x-2}$

16. $\dfrac{t}{t+2}$

17. $\dfrac{a^2-25}{a^2-4a+3} \cdot \dfrac{2a-5}{2a+5} = \dfrac{(a^2-25)(2a-5)}{(a^2-4a+3)(2a+5)}$

$= \dfrac{(a+5)(a-5)(2a-5)}{(a-3)(a-1)(2a+5)}$

(No simplification is possible.)

18. $\dfrac{(x+3)(x+4)(x+1)}{(x^2+9)(x+9)}$

19. $\dfrac{a^2-9}{a^2} \cdot \dfrac{a^2-3a}{a^2+a-12} = \dfrac{(a-3)(a+3)(a)(a-3)}{a \cdot a(a+4)(a-3)}$

$= \dfrac{(a-3)(a+3)\cancel{(a)}(a-3)}{\cancel{a} \cdot a(a+4)\cancel{(a-3)}}$

$= \dfrac{(a-3)(a+3)}{a(a+4)}$

20. 1

21. $\dfrac{4a^2}{3a^2-12a+12} \cdot \dfrac{3a-6}{2a}$

$= \dfrac{4a^2(3a-6)}{(3a^2-12a+12)2a}$

$= \dfrac{2 \cdot 2 \cdot a \cdot a \cdot 3 \cdot (a-2)}{3 \cdot (a-2) \cdot (a-2) \cdot 2 \cdot a}$

$= \dfrac{\cancel{2} \cdot 2 \cdot \cancel{a} \cdot a \cdot \cancel{3} \cdot \cancel{(a-2)}}{\cancel{3} \cdot \cancel{(a-2)} \cdot (a-2) \cdot \cancel{2} \cdot \cancel{a}}$

$= \dfrac{2a}{a-2}$

22. $\dfrac{5(v-2)}{v-1}$

23. $\dfrac{t^2+2t-3}{t^2+4t-5} \cdot \dfrac{t^2-3t-10}{t^2+5t+6}$

$= \dfrac{(t^2+2t-3)(t^2-3t-10)}{(t^2+4t-5)(t^2+5t+6)}$

$= \dfrac{(t+3)(t-1)(t-5)(t+2)}{(t+5)(t-1)(t+3)(t+2)}$

$= \dfrac{\cancel{(t+3)}\cancel{(t-1)}(t-5)\cancel{(t+2)}}{(t+5)\cancel{(t-1)}\cancel{(t+3)}\cancel{(t+2)}}$

$= \dfrac{t-5}{t+5}$

24. $\dfrac{x+4}{x-4}$

25.
$$\frac{5a^2 - 180}{10a^2 - 10} \cdot \frac{20a + 20}{2a - 12}$$

$$= \frac{(5a^2 - 180)(20a + 20)}{(10a^2 - 10)(2a - 12)}$$

$$= \frac{5(a+6)(a-6)(2)(10)(a+1)}{10(a+1)(a-1)(2)(a-6)}$$

$$= \frac{5(a+6)(a-6)(2)(10)(a+1)}{10(a+1)(a-1)(2)(a-6)}$$

$$= \frac{5(a+6)}{a-1}$$

26. $\dfrac{t+7}{4(t-1)}$

27.
$$\frac{x^2 - 1}{x^2 - 9} \cdot \frac{(x-3)^4}{(x+1)^2}$$

$$= \frac{(x^2 - 1)(x-3)^4}{(x^2 - 9)(x+1)^2}$$

$$= \frac{(x+1)(x-1)(x-3)(x-3)(x-3)(x-3)}{(x+3)(x-3)(x+1)(x+1)}$$

$$= \frac{(x+1)(x-1)(x-3)(x-3)(x-3)(x-3)}{(x+3)(x-3)(x+1)(x+1)}$$

$$= \frac{(x-1)(x-3)(x-3)(x-3)}{(x+3)(x+1)}, \text{ or}$$

$$\frac{(x-1)(x-3)^3}{(x+3)(x+1)}$$

28. $\dfrac{(x+2)^4(x+1)}{(x-1)^2(x+3)}$

29.
$$\frac{a^2 - 4}{a^2 + 2a + 1} \cdot \frac{a-1}{a^4 + 1} = \frac{(a^2 - 4)(a-1)}{(a^2 + 2a + 1)(a^4 + 1)}$$

$$= \frac{(a+2)(a-2)(a-1)}{(a+1)^2(a^4 + 1)}$$

(No simplification is possible.)

30. $\dfrac{(a^2 + 4)(a+3)^2}{(a-3)^2(a^4 + 16)}$

31.
$$\frac{(t-2)^3}{(t-1)^3} \cdot \frac{t^2 - 2t + 1}{t^2 - 4t + 4}$$

$$= \frac{(t-2)^3(t^2 - 2t + 1)}{(t-1)^3(t^2 - 4t + 4)}$$

$$= \frac{(t-2)(t-2)(t-2)(t-1)(t-1)}{(t-1)(t-1)(t-1)(t-2)(t-2)}$$

$$= \frac{(t-2)(t-2)(t-1)(t-1)}{(t-2)(t-2)(t-1)(t-1)} \cdot \frac{t-2}{t-1}$$

$$= \frac{t-2}{t-1}$$

32. $\dfrac{y+4}{y+2}$

33. The reciprocal of $\dfrac{4}{x}$ is $\dfrac{x}{4}$ because $\dfrac{4}{x} \cdot \dfrac{x}{4} = 1.$

34. $\dfrac{a-1}{a+3}$

35. The reciprocal of $x^2 - y^2$ is $\dfrac{1}{x^2 - y^2}$ because
$$\frac{x^2 - y^2}{1} \cdot \frac{1}{x^2 - y^2} = 1.$$

36. $a + b$

37. The reciprocal of $\dfrac{x^2 + 2x - 5}{x^2 - 4x + 7}$ is $\dfrac{x^2 - 4x + 7}{x^2 + 2x - 5}$ because
$$\frac{x^2 + 2x - 5}{x^2 - 4x + 7} \cdot \frac{x^2 - 4x + 7}{x^2 + 2x - 5} = 1.$$

38. $\dfrac{x^2 + 7xy - y^2}{x^2 - 3xy + y^2}$

39.
$$\frac{2}{5} \div \frac{4}{3}$$

$$= \frac{2}{5} \cdot \frac{3}{4} \qquad \text{Multiplying by the reciprocal of the divisor}$$

$$= \frac{2 \cdot 3}{5 \cdot 4}$$

$$= \frac{2 \cdot 3}{5 \cdot 2 \cdot 2} \qquad \text{Factoring the denominator}$$

$$= \frac{2}{2} \cdot \frac{3}{5 \cdot 2} \qquad \text{Factoring the fractional expression}$$

$$= \frac{3}{10} \qquad \text{Simplifying}$$

40. $\dfrac{5}{4}$

41.
$$\frac{2}{x} \div \frac{8}{x} = \frac{2}{x} \cdot \frac{x}{8} \qquad \text{Multiplying by the reciprocal of the divisor}$$

$$= \frac{2 \cdot x}{x \cdot 8}$$

$$= \frac{2 \cdot x \cdot 1}{x \cdot 2 \cdot 4} \qquad \text{Factoring the numerator and the denominator}$$

$$= \frac{2x}{2x} \cdot \frac{1}{9} \qquad \text{Factoring the fractional expression}$$

$$= \frac{1}{4} \qquad \text{Simplifying}$$

42. $\dfrac{x^2}{6}$

43.
$$\frac{x^2}{y} \div \frac{x^3}{y^3} = \frac{x^2}{y} \cdot \frac{y^3}{x^3}$$

$$= \frac{x^2 \cdot y^3}{y \cdot x^3}$$

$$= \frac{x^2 \cdot y \cdot y^2}{y \cdot x^2 \cdot x}$$

$$= \frac{x^2 y}{x^2 y} \cdot \frac{y^2}{x}$$

$$= \frac{y^2}{x}$$

44. $\dfrac{b}{a}$

45. $\dfrac{a+2}{a-3} \div \dfrac{a-1}{a+3} = \dfrac{a+2}{a-3} \cdot \dfrac{a+3}{a-1}$

$\qquad\qquad = \dfrac{(a+2)(a+3)}{(a-3)(a-1)}$

46. $\dfrac{y+2}{2y}$

47. $\dfrac{x^2-1}{x} \div \dfrac{x+1}{x-1} = \dfrac{x^2-1}{x} \cdot \dfrac{x-1}{x+1}$

$\qquad\qquad = \dfrac{(x^2-1)(x-1)}{x(x+1)}$

$\qquad\qquad = \dfrac{(x+1)(x-1)(x-1)}{x(x+1)}$

$\qquad\qquad = \dfrac{x+1}{x+1} \cdot \dfrac{(x-1)(x-1)}{x}$

$\qquad\qquad = \dfrac{(x-1)^2}{x}$

48. $4(y-2)$

49. $\dfrac{x+1}{6} \div \dfrac{x+1}{3} = \dfrac{x+1}{6} \cdot \dfrac{3}{x+1}$

$\qquad\qquad = \dfrac{(x+1)\cdot 3}{6(x+1)}$

$\qquad\qquad = \dfrac{3(x+1)}{2\cdot 3(x+1)}$

$\qquad\qquad = \dfrac{3(x+1)}{3(x+1)} \cdot \dfrac{1}{2}$

$\qquad\qquad = \dfrac{1}{2}$

50. $\dfrac{a}{b}$

51. $(y^2-9) \div \dfrac{y^2-2y-3}{y^2+1} = \dfrac{(y^2-9)}{1} \cdot \dfrac{y^2+1}{y^2-2y-3}$

$\qquad\qquad = \dfrac{(y^2-9)(y^2+1)}{y^2-2y-3}$

$\qquad\qquad = \dfrac{(y+3)(y-3)(y^2+1)}{(y-3)(y+1)}$

$\qquad\qquad = \dfrac{(y+3)(y-3)(y^2+1)}{(y-3)(y+1)}$

$\qquad\qquad = \dfrac{(y+3)(y^2+1)}{y+1}$

52. $\dfrac{(x-6)(x+6)}{x-1}$

53. $\dfrac{5x-5}{16} \div \dfrac{x-1}{6} = \dfrac{5x-5}{16} \cdot \dfrac{6}{x-1}$

$\qquad\qquad = \dfrac{(5x-5)\cdot 6}{16(x-1)}$

$\qquad\qquad = \dfrac{5(x-1)\cdot 2\cdot 3}{2\cdot 8(x-1)}$

$\qquad\qquad = \dfrac{5(x-1)\cdot 2\cdot 3}{2\cdot 8(x-1)}$

$\qquad\qquad = \dfrac{15}{8}$

54. $\dfrac{1}{2}$

55. $\dfrac{-6+3x}{5} \div \dfrac{4x-8}{25} = \dfrac{-6+3x}{5} \cdot \dfrac{25}{4x-8}$

$\qquad\qquad = \dfrac{(-6+3x)\cdot 25}{5(4x-8)}$

$\qquad\qquad = \dfrac{3(x-2)\cdot 5\cdot 5}{5\cdot 4(x-2)}$

$\qquad\qquad = \dfrac{3(x-2)\cdot 5\cdot 5}{5\cdot 4(x-2)}$

$\qquad\qquad = \dfrac{15}{4}$

56. 3

57. $\dfrac{a+2}{a-1} \div \dfrac{3a+6}{a-5} = \dfrac{a+2}{a-1} \cdot \dfrac{a-5}{3a+6}$

$\qquad\qquad = \dfrac{(a+2)(a-5)}{(a-1)(3a+6)}$

$\qquad\qquad = \dfrac{(a+2)(a-5)}{(a-1)\cdot 3\cdot (a+2)}$

$\qquad\qquad = \dfrac{(a+2)(a-5)}{(a-1)\cdot 3\cdot (a+2)}$

$\qquad\qquad = \dfrac{a-5}{3(a-1)}$

58. $\dfrac{t+1}{4(t+2)}$

59. $\quad (x-5) \div \dfrac{2x^2-11x+5}{4x^2-1}$

$\qquad = \dfrac{x-5}{1} \cdot \dfrac{4x^2-1}{2x^2-11x+5}$

$\qquad = \dfrac{(x-5)(4x^2-1)}{1\cdot (2x^2-11x+5)}$

$\qquad = \dfrac{(x-5)(2x+1)(2x-1)}{1\cdot (2x-1)(x-5)}$

$\qquad = \dfrac{(x-5)(2x+1)(2x-1)}{1\cdot (2x-1)(x-5)}$

$\qquad = 2x+1$

60. $\dfrac{(a+7)(a+1)}{3a-7}$

61. $\dfrac{x^2-4}{x} \div \dfrac{x-2}{x+2} = \dfrac{x^2-4}{x} \cdot \dfrac{x+2}{x-2}$

$\qquad\qquad = \dfrac{(x^2-4)(x+2)}{x(x-2)}$

$\qquad\qquad = \dfrac{(x-2)(x+2)(x+2)}{x(x-2)}$

$\qquad\qquad = \dfrac{(x-2)(x+2)(x+2)}{x(x-2)}$

$\qquad\qquad = \dfrac{(x+2)^2}{x}$

62. $\dfrac{(x+y)^2}{x^2+y}$

63. $\dfrac{x^2-9}{4x+12} \div \dfrac{x-3}{6} = \dfrac{x^2-9}{4x+12} \cdot \dfrac{6}{x-3}$

$= \dfrac{(x^2-9)\cdot 6}{(4x+12)(x-3)}$

$= \dfrac{(x-3)(x+3)\cdot 3 \cdot 2}{2\cdot 2(x+3)(x-3)}$

$= \dfrac{\cancel{(x-3)}\cancel{(x+3)}\cdot 3 \cdot \cancel{2}}{\cancel{2}\cdot 2\cancel{(x+3)}\cancel{(x-3)}}$

$= \dfrac{3}{2}$

64. $\dfrac{5x}{2(x+b)}$

65. $\dfrac{c^2+3c}{c^2+2c-3} \div \dfrac{c}{c+1} = \dfrac{c^2+3c}{c^2+2c-3} \cdot \dfrac{c+1}{c}$

$= \dfrac{(c^2+3c)(c+1)}{(c^2+2c-3)c}$

$= \dfrac{c(c+3)(c+1)}{(c+3)(c-1)c}$

$= \dfrac{\cancel{c}(\cancel{c+3})(c+1)}{(\cancel{c+3})(c-1)\cancel{c}}$

$= \dfrac{c+1}{c-1}$

66. $\dfrac{2x}{x+5}$

67. $\dfrac{2y^2-7y+3}{2y^2+3y-2} \div \dfrac{6y^2-5y+1}{3y^2+5y-2}$

$= \dfrac{2y^2-7y+3}{2y^2+3y-2} \cdot \dfrac{3y^2+5y-2}{6y^2-5y+1}$

$= \dfrac{(2y^2-7y+3)(3y^2+5y-2)}{(2y^2+3y-2)(6y^2-5y+1)}$

$= \dfrac{(2y-1)(y-3)(3y-1)(y+2)}{(2y-1)(y+2)(3y-1)(2y-1)}$

$= \dfrac{(\cancel{2y-1})(y-3)(\cancel{3y-1})(\cancel{y+2})}{(\cancel{2y-1})(\cancel{y+2})(\cancel{3y-1})(2y-1)}$

$= \dfrac{y-3}{2y-1}$

68. $\dfrac{x+3}{x-5}$

69. $\dfrac{c^2+10c+21}{c^2-2c-15} \div (c^2+2c-35)$

$= \dfrac{c^2+10c+21}{c^2-2c-25} \cdot \dfrac{1}{c^2+2c-35}$

$= \dfrac{(c^2+10c+21)\cdot 1}{(c^2-2c-15)(c^2+2c-35)}$

$= \dfrac{(c+7)(c+3)}{(c-5)(c+3)(c+7)(c-5)}$

$= \dfrac{(c+7)(c+3)}{(c+7)(c+3)} \cdot \dfrac{1}{(c-5)(c-5)}$

$= \dfrac{1}{(c-5)^2}$

70. $\dfrac{1}{1+2z-z^2}$

71. $\dfrac{(t+5)^3}{(t-5)^3} \div \dfrac{(t+5)^2}{(t-5)^2}$

$= \dfrac{(t+5)^3}{(t-5)^3} \cdot \dfrac{(t-5)^2}{(t+5)^2}$

$= \dfrac{(t+5)^3(t-5)^2}{(t-5)^3(t+5)^2}$

$= \dfrac{(t+5)^2(t-5)^2}{(t+5)^2(t-5)^2} \cdot \dfrac{t+5}{t-5}$

$= \dfrac{t+5}{t-5}$

72. $\dfrac{y-3}{y+3}$

73. **Familiarize**. Let $x =$ the number.

Translate.

$\underbrace{\text{Sixteen}}$	more than	$\overbrace{\text{the square of a number}}$	is	$\overbrace{\text{eight times the number.}}$
$\downarrow$	$\downarrow$	$\downarrow$	$\downarrow$	$\downarrow$
16	+	x^2	=	$8x$

Carry out.

$16+x^2 = 8x$

$x^2-8x+16=0$

$(x-4)(x-4)=0$

$x-4=0 \quad$ or $\quad x-4=0$

$x=4 \quad$ or $\qquad x=4$

Check. The square of 4, which is 16, plus 16 is 32, and eight times 4 is 32. The number checks.

State. The number is 4.

74. $2x^2+16$

75. $(8x^3-3x^2+7)-(8x^2+3x-5) =$
$8x^3-3x^2+7-8x^2-3x+5 =$
$8x^3-11x^2-3x+12$

76. $0.06y^3-0.09y^2+0.01y-1$

77.

78.

79. $\dfrac{2a^2-5ab}{c-3d} \div (4a^2-25b^2)$

$= \dfrac{2a^2-5ab}{c-3d} \cdot \dfrac{1}{4a^2-25b^2}$

$= \dfrac{a(2a-5b)}{(c-3d)(2a+5b)(2a-5b)}$

$= \dfrac{2a-5b}{2a-5b} \cdot \dfrac{a}{(c-3d)(2a+5b)}$

$= \dfrac{a}{(c-3d)(2a+5b)}$

80. 1

81. $\dfrac{3a^2 - 5ab - 12b^2}{3ab + 4b^2} \div (3b^2 - ab)$

$= \dfrac{3a^2 - 5ab - 12b^2}{3ab + 4b^2} \cdot \dfrac{1}{3b^2 - ab}$

$= \dfrac{(3a + 4b)(a - 3b)}{b(3a + 4b) \cdot b(3b - a)}$

$= \dfrac{(3a + 4b)(-1)(3b - a)}{b(3a + 4b) \cdot b(3b - a)}$

$= \dfrac{(3a + 4b)(-1)(3b - a)}{b(3a + 4b) \cdot b(3b - a)}$

$= -\dfrac{1}{b^2}$

82. $\dfrac{1}{(x + y)^2}$

83. $xy \cdot \dfrac{y^2 - 4xy}{y - x} \div \dfrac{16x^2y^2 - y^4}{4x^2 - 3xy - y^2}$

$= \dfrac{xy}{1} \cdot \dfrac{y^2 - 4xy}{y - x} \cdot \dfrac{4x^2 - 3xy - y^2}{16x^2y^2 - y^4}$

$= \dfrac{x \cdot y \cdot y \cdot (y - 4x)(4x + y)(x - y)}{(y - x) \cdot y \cdot y \cdot (4x - y)(4x + y)}$

$= \dfrac{x \cdot y \cdot y \cdot (-1)(4x - y)(4x + y)(-1)(y - x)}{(y - x) \cdot y \cdot y \cdot (4x - y)(4x + y)}$

$= \dfrac{y \cdot y(4x - y)(4x + y)(y - x)}{y \cdot y(4x - y)(4x + y)(y - x)} \cdot \dfrac{x(-1)(-1)}{1}$

$= x$

84. $\dfrac{(z + 4)^3}{(z - 4)^3}$

85. $\dfrac{x^2 - x + xy - y}{x^2 + 6x - 7} \div \dfrac{x^2 + 2xy + y^2}{4x + 4y}$

$= \dfrac{x^2 - x + xy - y}{x^2 + 6x - 7} \cdot \dfrac{4x + 4y}{x^2 + 2xy + y^2}$

$= \dfrac{x(x - 1) + y(x - 1)}{x^2 + 6x - 7} \cdot \dfrac{4x + 4y}{x^2 + 2xy + y^2}$

$= \dfrac{(x - 1)(x + y) \cdot 4(x + y)}{(x + 7)(x - 1)(x + y)(x + y)}$

$= \dfrac{(x - 1)(x + y)(x + y)}{(x - 1)(x + y)(x + y)} \cdot \dfrac{4}{x + 7}$

$= \dfrac{4}{x + 7}$

86. $\dfrac{x(x^2 + 1)}{3(x + y - 1)}$

87. $\dfrac{t^4 - 1}{t^4 - 81} \cdot \dfrac{t^2 - 9}{t^2 + 1} \div \dfrac{(t + 1)^2}{(t - 9)^2}$

$= \dfrac{t^4 - 1}{t^4 - 81} \cdot \dfrac{t^2 - 9}{t^2 + 1} \cdot \dfrac{(t - 9)^2}{(t + 1)^2}$

$= \dfrac{(t^4 - 1)(t^2 - 9)(t - 9)^2}{(t^4 - 81)(t^2 + 1)(t + 1)^2}$

$= \dfrac{(t^2 + 1)(t + 1)(t - 1)(t + 3)(t - 3)(t - 9)(t - 9)}{(t^2 + 9)(t + 3)(t - 3)(t^2 + 1)(t + 1)(t + 1)}$

$= \dfrac{(t^2 + 1)(t + 1)(t - 1)(t + 3)(t - 3)(t - 9)(t - 9)}{(t^2 + 9)(t + 3)(t - 3)(t^2 + 1)(t + 1)(t + 1)}$

$= \dfrac{(t - 1)(t - 9)(t - 9)}{(t^2 + 9)(t + 1)}$, or

$\dfrac{(t - 1)(t - 9)^2}{(t^2 + 9)(t + 1)}$

88. 1

89. $\left(\dfrac{y^2 + 5y + 6}{y^2} \cdot \dfrac{3y^3 + 6y^2}{y^2 - y - 12} \right) \div \dfrac{y^2 - y}{y^2 - 2y - 8}$

$= \dfrac{y^2 + 5y + 6}{y^2} \cdot \dfrac{3y^3 + 6y^2}{y^2 - y - 12} \cdot \dfrac{y^2 - 2y - 8}{y^2 - y}$

$= \dfrac{(y + 3)(y + 2)(3y^2)(y + 2)(y - 4)(y + 2)}{y^2(y - 4)(y + 3)(y)(y - 1)}$

$= \dfrac{y^2(y - 4)(y + 3)}{y^2(y - 4)(y + 3)} \cdot \dfrac{3(y + 2)(y + 2)(y + 2)}{y(y - 1)}$

$= \dfrac{3(y + 2)^3}{y(y - 1)}$

90. $\dfrac{a - 3b}{c}$

Exercise Set 6.3

1. $\dfrac{3}{x} + \dfrac{5}{x} = \dfrac{8}{x}$ Adding numerators

2. $\dfrac{13}{a^2}$

3. $\dfrac{x}{15} + \dfrac{2x + 1}{15} = \dfrac{3x + 1}{15}$ Adding numerators

4. $\dfrac{4a - 4}{7}$

5. $\dfrac{2}{a + 3} + \dfrac{4}{a + 3} = \dfrac{6}{a + 3}$

6. $\dfrac{13}{x + 2}$

7. $\dfrac{9}{a + 6} - \dfrac{5}{a + 6} = \dfrac{4}{a + 6}$ Subtracting numerators

8. $\dfrac{6}{x + 7}$

9. $\dfrac{3y+9}{2y} - \dfrac{y+1}{2y}$

$= \dfrac{3y+9-(y+1)}{2y}$

$= \dfrac{3y+9-y-1}{2y}$ Removing parentheses

$= \dfrac{2y+8}{2y}$

$= \dfrac{2(y+4)}{2y}$

$= \dfrac{\cancel{2}(y+4)}{\cancel{2}\cdot y}$

$= \dfrac{y+4}{y}$

10. $\dfrac{t+4}{4t}$

11. $\dfrac{9x+5}{x+1} + \dfrac{2x+3}{x+1} = \dfrac{11x+8}{x+1}$ Adding numerators

12. $\dfrac{5a+9}{a+4}$

13. $\dfrac{9x+5}{x+1} - \dfrac{2x+3}{x+1} = \dfrac{9x+5-(2x+3)}{x+1}$

$= \dfrac{9x+5-2x-3}{x+1}$

$= \dfrac{7x+2}{x+1}$

14. $\dfrac{a-5}{a+4}$

15. $\dfrac{a^2}{a-4} + \dfrac{a-20}{a-4} = \dfrac{a^2+a-20}{a-4}$

$= \dfrac{(a+5)(a-4)}{a-4}$

$= \dfrac{(a+5)\cancel{(a-4)}}{\cancel{a-4}}$

$= a+5$

16. $x+2$

17. $\dfrac{x^2}{x-2} - \dfrac{6x-8}{x-2} = \dfrac{x^2-(6x-8)}{x-2}$

$= \dfrac{x^2-6x+8}{x-2}$

$= \dfrac{(x-4)(x-2)}{x-2}$

$= \dfrac{(x-4)\cancel{(x-2)}}{\cancel{x-2}}$

$= x-4$

18. $a-5$

19. $\dfrac{t^2+4t}{t-1} + \dfrac{2t-7}{t-1} = \dfrac{t^2+6t-7}{t-1}$

$= \dfrac{(t+7)(t-1)}{t-1}$

$= \dfrac{(t+7)\cancel{(t-1)}}{\cancel{t-1}}$

$= t+7$

20. $y+6$

21. $\dfrac{x+1}{x^2+5x+6} + \dfrac{2}{x^2+5x+6} = \dfrac{x+3}{x^2+5x+6}$

$= \dfrac{x+3}{(x+3)(x+2)}$

$= \dfrac{\cancel{x+3}}{\cancel{(x+3)}(x+2)}$

$= \dfrac{1}{x+2}$

22. $\dfrac{1}{x-1}$

23. $\dfrac{a^2+3}{a^2+5a-6} - \dfrac{4}{a^2+5a-6} = \dfrac{a^2-1}{a^2+5a-6}$

$= \dfrac{(a+1)(a-1)}{(a+6)(a-1)}$

$= \dfrac{(a+1)\cancel{(a-1)}}{(a+6)\cancel{(a-1)}}$

$= \dfrac{a+1}{a+6}$

24. $\dfrac{a+3}{a-4}$

25. $\dfrac{t^2-3t}{t^2+6t+9} + \dfrac{2t-12}{t^2+6t+9} = \dfrac{t^2-t-12}{t^2+6t+9}$

$= \dfrac{(t-4)(t+3)}{(t+3)^2}$

$= \dfrac{(t-4)\cancel{(t+3)}}{(t+3)\cancel{(t+3)}}$

$= \dfrac{t-4}{t+3}$

26. $\dfrac{y-5}{y+4}$

27. $\dfrac{2x^2+3}{x^2-6x+5} - \dfrac{(x^2-5x+9)}{x^2-6x+5}$

$= \dfrac{2x^2+3-(x^2-5x+9)}{x^2-6x+5}$

$= \dfrac{2x^2+3-x^2+5x-9}{x^2-6x+5}$

$= \dfrac{x^2+5x-6}{x^2-6x+5}$

$= \dfrac{(x+6)(x-1)}{(x-5)(x-1)}$

$= \dfrac{(x+6)\cancel{(x-1)}}{(x-5)\cancel{(x-1)}}$

$= \dfrac{x+6}{x-5}$

28. $\dfrac{x+5}{x-6}$

29. $\dfrac{3x}{8} + \dfrac{x}{-8} = \dfrac{3x}{8} + \dfrac{-x}{8}$ Since $\dfrac{a}{-b} = \dfrac{-a}{b}$

$= \dfrac{3x+(-x)}{8}$

$= \dfrac{2x}{8}$

$= \dfrac{2 \cdot x}{2 \cdot 4}$

$= \dfrac{\cancel{2} \cdot x}{\cancel{2} \cdot 4}$

$= \dfrac{x}{4}$

30. $\dfrac{2a}{3}$

31. $\dfrac{3}{t} + \dfrac{4}{-t} = \dfrac{3}{t} + \dfrac{-4}{t}$

$= \dfrac{3+(-4)}{t}$

$= \dfrac{-1}{t}$

$= -\dfrac{1}{t}$

32. $\dfrac{3}{a}$

33. $\dfrac{2x+7}{x-6} + \dfrac{3x}{6-x} = \dfrac{2x+7}{x-6} + \dfrac{3x}{-(x-6)}$

$= \dfrac{2x+7}{x-6} + \dfrac{-3x}{x-6}$

$= \dfrac{(2x+7)+(-3x)}{x-6}$

$= \dfrac{-x+7}{x-6}$

34. $\dfrac{x+3}{4x-3}$

35. $\dfrac{a}{6} - \dfrac{7a}{-6} = \dfrac{a}{6} - \left(-\dfrac{7a}{6}\right)$ Since $\dfrac{a}{-b} = -\dfrac{a}{b}$

$= \dfrac{a}{6} + \dfrac{7a}{6}$

$= \dfrac{8a}{6}$

$= \dfrac{2 \cdot 4a}{2 \cdot 3}$

$= \dfrac{\cancel{2} \cdot 4a}{\cancel{2} \cdot 3}$

$= \dfrac{4a}{3}$

36. $\dfrac{3x}{4}$

37. $\dfrac{5}{a} - \dfrac{8}{-a} = \dfrac{5}{a} - \left(-\dfrac{8}{a}\right)$ Since $\dfrac{a}{-b} = -\dfrac{a}{b}$

$= \dfrac{5}{a} + \dfrac{8}{a}$

$= \dfrac{5+8}{a}$

$= \dfrac{13}{a}$

38. $\dfrac{7}{t}$

39. $\dfrac{x}{4} - \dfrac{3x-5}{-4} = \dfrac{x}{4} - \left(-\dfrac{3x-5}{4}\right)$

$= \dfrac{x}{4} + \dfrac{3x-5}{4}$

$= \dfrac{x+3x-5}{4}$

$= \dfrac{4x-5}{4}$

40. $\dfrac{4}{x-1}$

41. $\dfrac{y^2}{y-3} + \dfrac{9}{3-y} = \dfrac{y^2}{y-3} + \dfrac{9}{-(y-3)}$

$= \dfrac{y^2}{y-3} + \dfrac{-9}{y-3}$

$= \dfrac{y^2+(-9)}{y-3}$

$= \dfrac{y^2-9}{y-3}$

$= \dfrac{(y+3)(y-3)}{y-3}$

$= \dfrac{y+3}{1} \cdot \dfrac{y-3}{y-3}$

$= y+3$

42. $t+2$

43. $\dfrac{b-7}{b^2-16} + \dfrac{7-b}{16-b^2} = \dfrac{b-7}{b^2-16} + \dfrac{7-b}{-(b^2-16)}$

$= \dfrac{b-7}{b^2-16} + \dfrac{-(7-b)}{b^2-16}$

$= \dfrac{b-7}{b^2-16} + \dfrac{b-7}{b^2-16}$

$= \dfrac{(b-7)+(b-7)}{b^2-16}$

$= \dfrac{2b-14}{b^2-16}$

44. 0

45. $\dfrac{3-2t}{t^2-5t+4} + \dfrac{2-3t}{t^2-5t+4} = \dfrac{5-5t}{t^2-5t+4}$

$= \dfrac{-5(-1+t)}{(t-4)(t-1)}$

$= \dfrac{-5(\cancel{t-1})}{(t-4)(\cancel{t-1})}$

$= \dfrac{-5}{t-4}$

46. $\dfrac{-5}{x-4}$

47. $\dfrac{3-x}{x-7} - \dfrac{2x-5}{7-x} = \dfrac{3-x}{x-7} - \dfrac{2x-5}{-(x-7)}$

$= \dfrac{3-x}{x-7} - \left(-\dfrac{2x-5}{x-7}\right)$

$= \dfrac{3-x}{x-7} + \dfrac{2x-5}{x-7}$

$= \dfrac{3-x+2x-5}{x-7}$

$= \dfrac{x-2}{x-7}$

48. $\dfrac{t^2+4}{t-2}$

49. $\dfrac{x-8}{x^2-16} - \dfrac{x-8}{16-x^2} = \dfrac{x-8}{x^2-16} - \dfrac{x-8}{-(x^2-16)}$

$= \dfrac{x-8}{x^2-16} - \left(-\dfrac{x-8}{x^2-16}\right)$

$= \dfrac{x-8}{x^2-16} + \dfrac{x-8}{x^2-16}$

$= \dfrac{x-8+x-8}{x^2-16}$

$= \dfrac{2x-16}{x^2-16}$

50. $\dfrac{4}{x^2-25}$

51. $\dfrac{4-x}{x-9} - \dfrac{3x-8}{9-x} = \dfrac{4-x}{x-9} - \dfrac{3x-8}{-(x-9)}$

$= \dfrac{4-x}{x-9} - \left(-\dfrac{3x-8}{x-9}\right)$

$= \dfrac{4-x}{x-9} + \dfrac{3x-8}{x-9}$

$= \dfrac{4-x+3x-8}{x-9}$

$= \dfrac{2x-4}{x-9}$

52. $\dfrac{x-2}{x-7}$

53. $\dfrac{5-3x}{x^2-2x+1} - \dfrac{x+1}{x^2-2x+1} = \dfrac{5-3x-(x+1)}{x^2-2x+1}$

$= \dfrac{5-3x-x-1}{x^2-2x+1}$

$= \dfrac{4-4x}{x^2-2x+1}$

$= \dfrac{-4(-1+x)}{(x-1)^2}$

$= \dfrac{-4(\cancel{x-1})}{(\cancel{x-1})(x-1)}$

$= \dfrac{-4}{x-1}$

54. $\dfrac{-1}{x-1}$, or $\dfrac{1}{1-x}$

55. Graph: $y = -1$

Any ordered pair, $(x, -1)$ is a solution, so the graph is a line parallel to the x−axis with y−intercept $(0, -1)$.

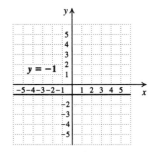

56.

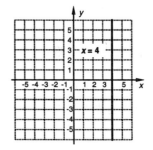

57. Graph: $y = x - 1$

The equation is in the form $y = mx + b$, so we know the y−intercept is $(0, 1)$. We find two other solutions.

When $x = -3$, $y = -3 - 1 = -4$.

When $x = 4$, $y = 4 - 1 = 3$.

x	y
0	-1
-3	-4
4	3

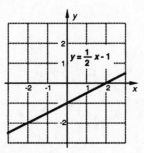

$y = x - 1$

58.

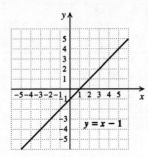

$y = \frac{1}{2}x - 1$

59.

60.

61.

$$\frac{3(2x+5)}{x-1} - \frac{3(2x-3)}{1-x} + \frac{6x+1}{x-1}$$

$$= \frac{3(2x+5)}{x-1} - \frac{3(2x-3)}{-(x-1)} + \frac{6x+1}{x-1}$$

$$= \frac{3(2x+5)}{x-1} - \left(-\frac{3(2x-3)}{x-1}\right) + \frac{6x+1}{x-1}$$

$$= \frac{3(2x+5)}{x-1} + \frac{3(2x-3)}{x-1} + \frac{6x-1}{x-1}$$

$$= \frac{6x+15+6x-9+6x-1}{x-1}$$

$$= \frac{18x+5}{x-1}$$

62. $\dfrac{-2x+4y}{x-y}$

63.

$$\frac{x-y}{x^2-y^2} + \frac{x+y}{x^2-y^2} - \frac{2x}{x^2-y^2}$$

$$= \frac{x-y+x+y-2x}{x^2-y^2}$$

$$= \frac{0}{x^2-y^2}$$

$$= 0$$

64. $\dfrac{-x+3y}{x-y}$

65.

$$\frac{10}{2y-1} - \frac{6}{1-2y} + \frac{y}{2y-1} + \frac{y-4}{1-2y}$$

$$= \frac{10}{2y-1} - \frac{6}{-(2y-1)} + \frac{y}{2y-1} + \frac{y-4}{-(2y-1)}$$

$$= \frac{10}{2y-1} - \left(-\frac{6}{2y-1}\right) + \frac{y}{2y-1} + \frac{-(y-4)}{2y-1}$$

$$= \frac{10}{2y-1} + \frac{6}{2y-1} + \frac{y}{2y-1} + \frac{4-y}{2y-1}$$

$$= \frac{10+6+y+4-y}{2y-1}$$

$$= \frac{20}{2y-1}$$

66. $\dfrac{x^2+3}{(3-2x)(x-3)}$

67.

$$\frac{x}{(x-y)(y-z)} - \frac{y}{(y-x)(z-y)}$$

$$= \frac{x}{(x-y)(y-z)} - \frac{x}{(-1)(x-y)(-1)(y-z)}$$

$$= \frac{x}{(x-y)(y-z)} - \frac{x}{(x-y)(y-z)}$$

$$= \frac{x-x}{(x-y)(y-z)}$$

$$= \frac{0}{(x-y)(y-z)}$$

$$= 0$$

68. $\dfrac{x+y}{x-y}$

69.

$$\frac{x^2}{3x^2-5x-2} - \frac{2x}{3x+1} \cdot \frac{1}{x-2}$$

$$= \frac{x^2}{(3x+1)(x-2)} - \frac{2x}{(3x+1)(x-2)}$$

$$= \frac{x^2-2x}{(3x+1)(x-2)}$$

$$= \frac{x(x-2)}{(3x+1)(x-2)}$$

$$= \frac{x}{3x+1} \cdot \frac{x-2}{x-2}$$

$$= \frac{x}{3x+1}$$

70. $\dfrac{30}{(x+4)(x-3)}$

71.

Exercise Set 6.4

1. $12 = 2 \cdot 2 \cdot 3$

$27 = 3 \cdot 3 \cdot 3$

LCM $= 2 \cdot 2 \cdot 3 \cdot 3 \cdot 3$, or 108

2. 30

3. $8 = 2 \cdot 2 \cdot 2$

$9 = 3 \cdot 3$

LCM $= 2 \cdot 2 \cdot 2 \cdot 3 \cdot 3$, or 72

4. 60

5. $6 = 2 \cdot 3$

$9 = 3 \cdot 3$

$21 = 3 \cdot 7$

LCM $= 2 \cdot 3 \cdot 3 \cdot 7$, or 126

6. 360

7. $24 = 2 \cdot 2 \cdot 2 \cdot 3$

$36 = 2 \cdot 2 \cdot 3 \cdot 3$

$40 = 2 \cdot 2 \cdot 2 \cdot 5$

LCM $= 2 \cdot 2 \cdot 2 \cdot 3 \cdot 3 \cdot 5$, or 360

8. 60

9. $28 = 2 \cdot 2 \cdot 7$

$42 = 2 \cdot 3 \cdot 7$

$60 = 2 \cdot 2 \cdot 3 \cdot 5$

LCM $= 2 \cdot 2 \cdot 3 \cdot 5 \cdot 7$, or 420

10. 500

11. $24 = 2 \cdot 2 \cdot 2 \cdot 3$

$18 = 2 \cdot 3 \cdot 3$

LCD $= 2 \cdot 2 \cdot 2 \cdot 3 \cdot 3$, or 72

$$\frac{7}{24} + \frac{11}{18} = \frac{7}{2 \cdot 2 \cdot 2 \cdot 3} \cdot \frac{3}{3} + \frac{11}{2 \cdot 3 \cdot 3} \cdot \frac{2 \cdot 2}{2 \cdot 2}$$

$$= \frac{21}{2 \cdot 2 \cdot 2 \cdot 3 \cdot 3} + \frac{44}{2 \cdot 2 \cdot 2 \cdot 3 \cdot 3}$$

$$= \frac{65}{72}$$

12. $\dfrac{59}{300}$

13. $\dfrac{1}{6} + \dfrac{3}{40} + \dfrac{2}{75}$

$$= \frac{1}{2 \cdot 3} + \frac{3}{2 \cdot 2 \cdot 2 \cdot 5} + \frac{2}{3 \cdot 5 \cdot 5}$$

LCD is $2 \cdot 2 \cdot 2 \cdot 3 \cdot 5 \cdot 5$, or 600

$$= \frac{1}{2 \cdot 3} \cdot \frac{2 \cdot 2 \cdot 5 \cdot 5}{2 \cdot 2 \cdot 5 \cdot 5} + \frac{3}{2 \cdot 2 \cdot 2 \cdot 5} \cdot \frac{3 \cdot 5}{3 \cdot 5} + \frac{2}{3 \cdot 5 \cdot 5} \cdot \frac{2 \cdot 2 \cdot 2}{2 \cdot 2 \cdot 2}$$

$$= \frac{100 + 45 + 16}{2 \cdot 2 \cdot 2 \cdot 3 \cdot 5 \cdot 5}$$

$$= \frac{161}{600}$$

14. $\dfrac{71}{120}$

15. $\dfrac{2}{15} + \dfrac{5}{9} + \dfrac{3}{20}$

$$= \frac{2}{3 \cdot 5} + \frac{5}{3 \cdot 3} + \frac{3}{2 \cdot 2 \cdot 5}$$

LCD is $2 \cdot 2 \cdot 3 \cdot 3 \cdot 5$, or 180

$$= \frac{2}{3 \cdot 5} \cdot \frac{2 \cdot 2 \cdot 3}{2 \cdot 2 \cdot 3} + \frac{5}{3 \cdot 3} \cdot \frac{2 \cdot 2 \cdot 5}{2 \cdot 2 \cdot 5} + \frac{3}{2 \cdot 2 \cdot 5} \cdot \frac{3 \cdot 3}{3 \cdot 3}$$

$$= \frac{24 + 100 + 27}{2 \cdot 2 \cdot 3 \cdot 3 \cdot 5}$$

$$= \frac{151}{180}$$

16. $\dfrac{23}{180}$

17. $6x^2 = 2 \cdot 3 \cdot x \cdot x$

$12x^3 = 2 \cdot 2 \cdot 3 \cdot x \cdot x \cdot x$

LCM $= 2 \cdot 2 \cdot 3 \cdot x \cdot x \cdot x$, or $12x^3$

18. $8a^2b^2$

19. $2x^2 = 2 \cdot x \cdot x$

$6xy = 2 \cdot 3 \cdot x \cdot y$

$18y^2 = 2 \cdot 3 \cdot 3 \cdot y \cdot y$

LCM $= 2 \cdot 3 \cdot 3 \cdot x \cdot x \cdot y \cdot y$, or $18x^2y^2$

20. c^3d^2

21. $2(y - 3) = 2 \cdot (y - 3)$

$6(y - 3) = 2 \cdot 3 \cdot (y - 3)$

LCM $= 2 \cdot 3 \cdot (y - 3)$, or $6(y - 3)$

22. $8(x - 1)$

23. $t, t + 2, t - 2$

The expressions are not factorable, so the LCM is their product:

LCM $= t(t + 2)(t - 2)$

24. $x(x + 3)(x - 3)$

25. $x^2 - 4 = (x + 2)(x - 2)$

$x^2 + 5x + 6 = (x + 3)(x + 2)$

LCM $= (x + 2)(x - 2)(x + 3)$

26. $(x + 2)(x + 1)(x - 2)$

27. $t^3 + 4t^2 + 4t = t(t^2 + 4t + 4) = t(t + 2)(t + 2)$

$t^2 - 4t = t(t - 4)$

LCM $= t(t + 2)(t + 2)(t - 4) = t(t + 2)^2(t - 4)$

28. $y^2(y + 1)(y - 1)$

29. $9a^5b^2 = 3 \cdot 3 \cdot a \cdot a \cdot a \cdot a \cdot a \cdot b \cdot b$

$6ab^6 = 2 \cdot 3 \cdot a \cdot b \cdot b \cdot b \cdot b \cdot b \cdot b$

LCM $= 2 \cdot 3 \cdot 3 \cdot a \cdot a \cdot a \cdot a \cdot a \cdot b \cdot b \cdot b \cdot b \cdot b \cdot b =$

$\qquad 18a^5b^6$

30. $30a^4b^8$

31. $10x^2y = 2 \cdot 5 \cdot x \cdot x \cdot y$

$6y^2z = 2 \cdot 3 \cdot y \cdot y \cdot z$

$5xz^3 = 5 \cdot x \cdot z \cdot z \cdot z$

LCM $= 2 \cdot 3 \cdot 5 \cdot x \cdot x \cdot y \cdot y \cdot z \cdot z \cdot z = 30x^2y^2z^3$

32. $24x^3y^5z^2$

33. $a + 1 = a + 1$

$(a-1)^2 = (a-1)(a-1)$

$a^2 - 1 = (a+1)(a-1)$

LCM $= (a+1)(a-1)(a-1) = (a+1)(a-1)^2$

34. $2(x+y)^2(x-y)$

35. $m^2 - 5m + 6 = (m-3)(m-2)$

$m^2 - 4m + 4 = (m-2)(m-2)$

LCM $= (m-3)(m-2)(m-2) = (m-3)(m-2)^2$

36. $(2x+1)(x+2)(x-1)$

37. $2 + 3x = 2 + 3x$

$4 - 9x^2 = (2+3x)(2-3x)$

$2 - 3x = 2 - 3x$

LCM $= (2+3x)(2-3x)$

38. $(3+2x)(3-2x)$

39. $10v^2 + 30v = 10v(v+3) = 2 \cdot 5 \cdot v(v+3)$

$5v^2 + 35v + 60 = 5(v^2 + 7v + 12)$

$= 5(v+4)(v+3)$

LCM $= 2 \cdot 5 \cdot v(v+3)(v+4) = 10v(v+3)(v+4)$

40. $12a(a+2)(a+3)$

41. $9x^3 - 9x^2 - 18x = 9x(x^2 - x - 2)$

$= 3 \cdot 3 \cdot x(x-2)(x+1)$

$6x^5 - 24x^4 + 24x^3 = 6x^3(x^2 - 4x + 4)$

$= 2 \cdot 3 \cdot x \cdot x \cdot x(x-2)(x-2)$

LCM $= 2 \cdot 3 \cdot 3 \cdot x \cdot x \cdot x(x-2)(x-2)(x+1) =$

$18x^3(x-2)^2(x+1)$

42. $x^3(x+2)^2(x-2)$

43. $x^5 + 4x^4 + 4x^3 = x^3(x^2 + 4x + 4)$

$= x \cdot x \cdot x(x+2)(x+2)$

$3x^2 - 12 = 3(x^2 - 4) = 3(x+2)(x-2)$

$2x + 4 = 2(x+2)$

LCM $= 2 \cdot 3 \cdot x \cdot x \cdot x(x+2)(x+2)(x-2)$

$= 6x^3(x+2)^2(x-2)$

44. $10x^3(x+1)^2(x-1)$

45. $6x^5 = 2 \cdot 3 \cdot x \cdot x \cdot x \cdot x \cdot x$

$12x^3 = 2 \cdot 2 \cdot 3 \cdot x \cdot x \cdot x$

The LCD is $2 \cdot 2 \cdot 3 \cdot x \cdot x \cdot x \cdot x \cdot x$, or $12x^5$.

The factor of the LCD that is missing from the first denominator is 2. We multiply by 1 using 2/2:

$$\frac{7}{6x^5} \cdot \frac{2}{2} = \frac{14}{12x^5}$$

The second denominator is missing two factors of x, or x^2. We multiply by 1 using x^2/x^2:

$$\frac{y}{12x^3} \cdot \frac{x^2}{x^2} = \frac{x^2y}{12x^5}$$

46. $\dfrac{3a^3}{10a^6}, \dfrac{2b}{10a^6}$

47. $2a^2b = 2 \cdot a \cdot a \cdot b$

$8ab^2 = 2 \cdot 2 \cdot 2 \cdot a \cdot b \cdot b$

The LCD is $2 \cdot 2 \cdot 2 \cdot a \cdot a \cdot b \cdot b$, or $8a^2b^2$.

We multiply the first expression by $\dfrac{4b}{4b}$ to obtain the LCD:

$$\frac{3}{2a^2b} \cdot \frac{4b}{4b} = \frac{12b}{8a^2b^2}$$

We multiply the second expression by a/a to obtain the LCD:

$$\frac{5}{8ab^2} \cdot \frac{a}{a} = \frac{5a}{8a^2b^2}$$

48. $\dfrac{21y}{9x^4y^3}, \dfrac{4x^3}{9x^4y^3}$

49. The LCD is $(x+2)(x-2)(x+3)$. (See Exercise 25.)

$$\frac{x+1}{x^2-4} = \frac{x+1}{(x+2)(x-2)} \cdot \frac{x+3}{x+3}$$

$$= \frac{(x+1)(x+3)}{(x+2)(x-2)(x+3)}$$

$$\frac{x-2}{x^2+5x+6} = \frac{x-2}{(x+3)(x+2)} \cdot \frac{x-2}{x-2}$$

$$= \frac{(x-2)^2}{(x+3)(x+2)(x-2)}$$

50. $\dfrac{(x-4)(x+8)}{(x+3)(x-3)(x+8)}, \dfrac{(x+2)(x-3)}{(x+3)(x+8)(x-3)}$

51. The LCD is $t(t+2)(t-2)$. (See Exercise 23.)

$$\frac{3}{t} = \frac{3}{t} \cdot \frac{(t+2)(t-2)}{(t+2)(t-2)} = \frac{3(t+2)(t-2)}{t(t+2)(t-2)}$$

$$\frac{4}{t+2} = \frac{4}{t+2} \cdot \frac{t(t-2)}{t(t-2)} = \frac{4t(t-2)}{t(t+2)(t-2)}$$

$$\frac{t}{t-2} = \frac{t}{t-2} \cdot \frac{t(t+2)}{t(t+2)} = \frac{t^2(t+2)}{t(t-2)(t+2)}$$

52. $\dfrac{(x+3)(x-3)}{x(x+3)(x-3)}, \dfrac{-2x(x-3)}{x(x+3)(x-3)}, \dfrac{x^3(x+3)}{x(x-3)(x+3)}$

53. $2x - 3 = 2x - 3$

$4x^2 - 9 = (2x+3)(2x-3)$

$2x + 3 = 2x + 3$

LCD $= (2x+3)(2x-3)$

$$\frac{x+1}{2x-3} = \frac{x+1}{2x-3} \cdot \frac{2x+3}{2x+3} = \frac{(x+1)(2x+3)}{(2x-3)(2x+3)}$$

$$\frac{x-2}{4x^2-9} = \frac{x-2}{(2x+3)(2x-3)} \text{ already has the LCD.}$$

$$\frac{x+1}{2x+3} = \frac{x+1}{2x+3} \cdot \frac{2x-3}{2x-3} = \frac{(x+1)(2x-3)}{(2x+3)(2x-3)}$$

54. $\dfrac{x-2}{x^2(x+2)^2(x-2)}, \ \dfrac{x^2(x+2)}{x^2(x+2)^2(x-2)}$

55. $x^2 - 19x + 60$

Since the last term is positive and the middle term is negative, we look for a pair of negative factors of 60 whose sum is -19. The numbers we need are -4 and -15.
$$x^2 - 19x + 60 = (x-4)(x-15)$$

56. $(x+12)(x-3)$

57. The shaded area has dimensions $x-6$ by $x-3$. Then the area is $(x-6)(x-3)$, or $x^2 - 9x + 18$.

58. $s^2 - \pi r^2$

59. ◈

60. ◈

61. $72 = 2 \cdot 2 \cdot 2 \cdot 3 \cdot 3$

$90 = 2 \cdot 3 \cdot 3 \cdot 5$

$96 = 2 \cdot 2 \cdot 2 \cdot 2 \cdot 2 \cdot 3$

LCM $= 2 \cdot 2 \cdot 2 \cdot 2 \cdot 2 \cdot 3 \cdot 3 \cdot 5$, or 1440

62. $120(x+1)(x-1)^2$, or $120(x+1)(x-1)(1-x)$

63. The time it takes the joggers to meet again at the starting place is the LCM of the times it takes them to complete one round of the course.

$6 = 2 \cdot 3$

$8 = 2 \cdot 2 \cdot 2$

LCM $= 2 \cdot 2 \cdot 2 \cdot 3$, or 24

It takes 24 min.

64. ◈

Exercise Set 6.5

1. $\dfrac{2}{x} + \dfrac{5}{x^2} = \dfrac{2}{x} + \dfrac{5}{x \cdot x} \qquad \text{LCD} = x \cdot x, \text{ or } x^2$

$\qquad = \dfrac{2}{x} \cdot \dfrac{x}{x} + \dfrac{5}{x \cdot x}$

$\qquad = \dfrac{2x+5}{x^2}$

2. $\dfrac{4x+8}{x^2}$

3. $\left.\begin{array}{l} 6r = 2 \cdot 3 \cdot r \\ 8r = 2 \cdot 2 \cdot 2 \cdot r \end{array}\right\} \text{LCD} = 2 \cdot 2 \cdot 2 \cdot 3 \cdot r, \text{ or } 24r$

$\dfrac{5}{6r} - \dfrac{7}{8r} = \dfrac{5}{6r} \cdot \dfrac{4}{4} - \dfrac{7}{8r} \cdot \dfrac{3}{3}$

$\qquad = \dfrac{20 - 21}{24r}$

$\qquad = \dfrac{-1}{24r}$

4. $\dfrac{-29}{18t}$

5. $\left.\begin{array}{l} xy^2 = x \cdot y \cdot y \\ x^2y = x \cdot x \cdot y \end{array}\right\} \text{LCD} = x \cdot x \cdot y \cdot y, \text{ or } x^2y^2$

$\dfrac{4}{xy^2} + \dfrac{6}{x^2y} = \dfrac{4}{xy^2} \cdot \dfrac{x}{x} + \dfrac{6}{x^2y} \cdot \dfrac{y}{y}$

$\qquad = \dfrac{4x + 6y}{x^2y^2}$

6. $\dfrac{2d^2 + 7c}{c^2d^3}$

7. $\left.\begin{array}{l} 9t^3 = 3 \cdot 3 \cdot t \cdot t \cdot t \\ 6t^2 = 2 \cdot 3 \cdot t \cdot t \end{array}\right\} \text{LCD} = 2 \cdot 3 \cdot 3 \cdot t \cdot t \cdot t, \text{ or } 18t^3$

$\dfrac{2}{9t^3} - \dfrac{1}{6t^2} = \dfrac{2}{9t^3} \cdot \dfrac{2}{2} - \dfrac{1}{6t^2} \cdot \dfrac{3t}{3t}$

$\qquad = \dfrac{4 - 3t}{18t^3}$

8. $\dfrac{-2xy - 18}{3x^2y^3}$

9. LCD $= 24$ (See Example 1.)

$\dfrac{x+5}{8} + \dfrac{x-3}{12} = \dfrac{x+5}{8} \cdot \dfrac{3}{3} + \dfrac{x-3}{12} \cdot \dfrac{2}{2}$

$\qquad = \dfrac{3(x+5)}{24} + \dfrac{2(x-3)}{24}$

$\qquad = \dfrac{3x + 15}{24} + \dfrac{2x - 6}{24}$

$\qquad = \dfrac{5x + 9}{24}$ Adding numerators

10. $\dfrac{5x+7}{18}$

11. $\left.\begin{array}{l} 6 = 2 \cdot 3 \\ 3 = 3 \end{array}\right\} \text{LCD} = 6$

$\dfrac{x-2}{6} - \dfrac{x+1}{3} = \dfrac{x-2}{6} - \dfrac{x+1}{3} \cdot \dfrac{2}{2}$

$\qquad = \dfrac{x-2}{6} - \dfrac{2x+2}{6}$

$\qquad = \dfrac{x - 2 - (2x+2)}{6}$

$\qquad = \dfrac{x - 2 - 2x - 2}{6}$

$\qquad = \dfrac{-x-4}{6}, \text{ or } \dfrac{-(x+4)}{6}$

12. $\dfrac{a+8}{4}$

13. $\left.\begin{array}{l} 16a = 2 \cdot 2 \cdot 2 \cdot 2 \cdot a \\ 4a^2 = 2 \cdot 2 \cdot a \cdot a \end{array}\right\} \text{LCD} = 2 \cdot 2 \cdot 2 \cdot 2 \cdot a \cdot a, \text{ or } 16a^2$

$$\frac{a+4}{16a} + \frac{3a+4}{4a^2} = \frac{a+4}{16a} \cdot \frac{a}{a} + \frac{3a+4}{4a^2} \cdot \frac{4}{4}$$
$$= \frac{a^2+4a}{16a^2} + \frac{12a+16}{16a^2}$$
$$= \frac{a^2+16a+16}{16a^2}$$

14. $\dfrac{5a^2+7a-3}{9a^2}$

15. $\left.\begin{array}{l} 3z = 3 \cdot z \\ 4z = 2 \cdot 2 \cdot z \end{array}\right\} \text{LCD} = 2 \cdot 2 \cdot 3 \cdot z, \text{ or } 12z$

$$\frac{4z-9}{3z} - \frac{3z-8}{4z} = \frac{4z-9}{3z} \cdot \frac{4}{4} - \frac{3z-8}{4z} \cdot \frac{3}{3}$$
$$= \frac{16z-36}{12z} - \frac{9z-24}{12z}$$
$$= \frac{16z-36-(9z-24)}{12z}$$
$$= \frac{16z-36-9z+24}{12z}$$
$$= \frac{7z-12}{12z}$$

16. $\dfrac{-7x-13}{4x}$

17. LCD $= x^2y^2$ (See Exercise 5.)

$$\frac{x+y}{xy^2} + \frac{3x+y}{x^2y} = \frac{x+y}{xy^2} \cdot \frac{x}{x} + \frac{3x+y}{x^2y} \cdot \frac{y}{y}$$
$$= \frac{x(x+y) + y(3x+y)}{x^2y^2}$$
$$= \frac{x^2+xy+3xy+y^2}{x^2y^2}$$
$$= \frac{x^2+4xy+y^2}{x^2y^2}$$

18. $\dfrac{c^2+3cd-d^2}{c^2d^2}$

19. $\left.\begin{array}{l} 3xt^2 = 3 \cdot x \cdot t \cdot t \\ x^2t = x \cdot x \cdot t \end{array}\right\} \text{LCD} = 3 \cdot x \cdot x \cdot t \cdot t, \text{ or } 3x^2t^2$

$$\frac{4x+2t}{3xt^2} - \frac{5x-3t}{x^2t}$$
$$= \frac{4x+2t}{3xt^2} \cdot \frac{x}{x} - \frac{5x-3t}{x^2t} \cdot \frac{3t}{3t}$$
$$= \frac{4x^2+2tx}{3x^2t^2} - \frac{15xt-9t^2}{3x^2t^2}$$
$$= \frac{4x^2+2tx-(15xt-9t^2)}{3x^2t^2}$$
$$= \frac{4x^2+2tx-15xt+9t^2}{3x^2t^2}$$
$$= \frac{4x^2-13xt+9t^2}{3x^2t^2}$$

(Although $4x^2 - 13xt + 9t^2$ can be factored as $(4x-9t)(x-t)$, doing so will not enable us to simplify the result.)

20. $\dfrac{3y^2-3xy-6x^2}{2x^2y^2}$

(Although $3y^2 - 3xy - 6x^2$ can be factored as $3(y-2x)(y+x)$, doing so will not enable us to simplify the result.)

21. The denominators do not factor, so the LCD is their product, $(x-2)(x+2)$.

$$\frac{3}{x-2} + \frac{3}{x+2} = \frac{3}{x-2} \cdot \frac{x+2}{x+2} + \frac{3}{x+2} \cdot \frac{x-2}{x-2}$$
$$= \frac{3(x+2) + 3(x-2)}{(x-2)(x+2)}$$
$$= \frac{3x+6+3x-6}{(x-2)(x+2)}$$
$$= \frac{6x}{(x-2)(x+2)}$$

22. $\dfrac{4x}{(x-1)(x+1)}$

23. $\dfrac{5}{x+5} - \dfrac{3}{x-5}$ LCD $= (x+5)(x-5)$

$$= \frac{5}{x+5} \cdot \frac{x-5}{x-5} - \frac{3}{x-5} \cdot \frac{x+5}{x+5}$$
$$= \frac{5x-25}{(x+5)(x-5)} - \frac{3x+15}{(x+5)(x-5)}$$
$$= \frac{5x-25-(3x+15)}{(x+5)(x-5)}$$
$$= \frac{5x-25-3x-15}{(x+5)(x-5)}$$
$$= \frac{2x-40}{(x+5)(x-5)}$$

(Although $2x - 40$ can be factored as $2(x - 20)$, doing so will not enable us to simplify the result.)

24. $\dfrac{-z^2+5z}{(z-1)(z+1)}$

(Although $-z^2 + 5z$ can be factored as $-z(z-5)$, doing so will not enable us to simplify the result.)

25. $\left.\begin{array}{l} 3x = 3 \cdot x \\ x+1 = x+1 \end{array}\right\} \text{LCD} = 3x(x+1)$

$$\frac{3}{x+1} + \frac{2}{3x} = \frac{3}{x+1} \cdot \frac{3x}{3x} + \frac{2}{3x} \cdot \frac{x+1}{x+1}$$
$$= \frac{9x+2(x+1)}{3x(x+1)}$$
$$= \frac{9x+2x+2}{3x(x+1)}$$
$$= \frac{11x+2}{3x(x+1)}$$

26. $\dfrac{11x+15}{4x(x+5)}$

27. $\dfrac{3}{2t^2 - 2t} - \dfrac{5}{2t - 2}$

$= \dfrac{3}{2t(t-1)} - \dfrac{5}{2(t-1)}$ LCD $= 2t(t-1)$

$= \dfrac{3}{2t(t-1)} - \dfrac{5}{2(t-1)} \cdot \dfrac{t}{t}$

$= \dfrac{3}{2t(t-1)} - \dfrac{5t}{2t(t-1)}$

$= \dfrac{3 - 5t}{2t(t-1)}$

28. $\dfrac{14 - 3x}{(x+2)(x-2)}$

29. $\left.\begin{array}{l} x^2 - 16 = (x+4)(x-4) \\ x - 4 = x - 4 \end{array}\right\}$ LCD $= (x+4)(x-4)$

$\dfrac{2x}{x^2 - 16} + \dfrac{x}{x-4} = \dfrac{2x}{(x+4)(x-4)} + \dfrac{x}{x-4} \cdot \dfrac{x+4}{x+4}$

$= \dfrac{2x + x(x+4)}{(x+4)(x-4)}$

$= \dfrac{2x + x^2 + 4x}{(x+4)(x-4)}$

$= \dfrac{x^2 + 6x}{(x+4)(x-4)}$

(Although $x^2 + 6x$ can be factored as $x(x+6)$, doing so will not enable us to simplify the result.)

30. $\dfrac{x^2 - x}{(x+5)(x-5)}$

(Although $x^2 - x$ can be factored as $x(x-1)$, doing so will not enable us to simplify the result.)

31. $\dfrac{6}{z+4} - \dfrac{2}{3z+12} = \dfrac{6}{z+4} - \dfrac{2}{3(z+4)}$

$\phantom{= \dfrac{6}{z+4}}$ LCD $= 3(z+4)$

$= \dfrac{6}{z+4} \cdot \dfrac{3}{3} - \dfrac{2}{3(z+4)}$

$= \dfrac{18}{3(z+4)} - \dfrac{2}{3(z+4)}$

$= \dfrac{16}{3(z+4)}$

32. $\dfrac{4t-5}{4(t-3)}$

33. $\dfrac{3}{x-1} + \dfrac{2}{(x-1)^2}$ LCD $= (x-1)^2$

$= \dfrac{3}{x-1} \cdot \dfrac{x-1}{x-1} + \dfrac{2}{(x-1)^2}$

$= \dfrac{3(x-1) + 2}{(x-1)^2}$

$= \dfrac{3x - 3 + 2}{(x-1)^2}$

$= \dfrac{3x - 1}{(x-1)^2}$

34. $\dfrac{2x + 10}{(x+3)^2}$

(Although $2x + 10$ can be factored as $2(x+5)$, doing so will not enable us to simplify the result.)

35. $\dfrac{2t}{t^2 - 9} - \dfrac{3}{t-3} = \dfrac{2t}{(t+3)(t-3)} - \dfrac{3}{t-3}$

 LCD $= (t+3)(t-3)$

$= \dfrac{2t}{(t+3)(t-3)} - \dfrac{3}{t-3} \cdot \dfrac{t+3}{t+3}$

$= \dfrac{2t - 3(t+3)}{(t+3)(t-3)}$

$= \dfrac{2t - 3t - 9}{(t+3)(t-3)}$

$= \dfrac{-t - 9}{(t+3)(t-3)}$

36. $\dfrac{6 - 20x}{15x(x+1)}$

(Although $6 - 20x$ can be factored as $2(3 - 10x)$ or as $-2(10x - 3)$, doing so will not enable us to simplify the result.)

37. $\dfrac{4a}{5a - 10} + \dfrac{3a}{10a - 20} = \dfrac{4a}{5(a-2)} + \dfrac{3a}{2 \cdot 5(a-2)}$

 LCD $= 2 \cdot 5(a-2)$

$= \dfrac{4a}{5(a-2)} \cdot \dfrac{2}{2} + \dfrac{3a}{2 \cdot 5(a-2)}$

$= \dfrac{8a + 3a}{10(a-2)}$

$= \dfrac{11a}{10(a-2)}$

38. $\dfrac{9a}{4(a-5)}$

39. $\dfrac{a}{x+a} - \dfrac{a}{x-a}$ LCD $= (x+a)(x-a)$

$= \dfrac{a}{x+a} \cdot \dfrac{x-a}{x-a} - \dfrac{a}{x-a} \cdot \dfrac{x+a}{x+a}$

$= \dfrac{ax - a^2}{(x+a)(x-a)} - \dfrac{ax + a^2}{(x+a)(x-a)}$

$= \dfrac{ax - a^2 - (ax + a^2)}{(x+a)(x-a)}$

$= \dfrac{ax - a^2 - ax - a^2}{(x+a)(x-a)}$

$= \dfrac{-2a^2}{(x+a)(x-a)}$

40. $\dfrac{t^2 + 2ty - y^2}{(y-t)(y+t)}$

41. $\dfrac{x+4}{x} + \dfrac{x}{x+4}$ $\qquad$ LCD $= x(x+4)$

$\quad = \dfrac{x+4}{x} \cdot \dfrac{x+4}{x+4} + \dfrac{x}{x+4} \cdot \dfrac{x}{x}$

$\quad = \dfrac{(x+4)^2 + x^2}{x(x+4)}$

$\quad = \dfrac{x^2 + 8x + 16 + x^2}{x(x+4)}$

$\quad = \dfrac{2x^2 + 8x + 16}{x(x+4)}$

(Although $2x^2 + 8x + 16$ can be factored as $2(x^2 + 4x + 8)$, doing so will not enable us to simplify the result.)

42. $\dfrac{2x^2 - 10x + 25}{x(x-5)}$

43. $\dfrac{x}{x^2 + 5x + 6} - \dfrac{2}{x^2 + 3x + 2}$

$\quad = \dfrac{x}{(x+3)(x+2)} - \dfrac{2}{(x+2)(x+1)}$

$\qquad\qquad$ LCD $= (x+3)(x+2)(x+1)$

$\quad = \dfrac{x}{(x+3)(x+2)} \cdot \dfrac{x+1}{x+1} - \dfrac{2}{(x+2)(x+1)} \cdot \dfrac{x+3}{x+3}$

$\quad = \dfrac{x^2 + x}{(x+3)(x+2)(x+1)} - \dfrac{2x+6}{(x+3)(x+2)(x+1)}$

$\quad = \dfrac{x^2 + x - (2x+6)}{(x+3)(x+2)(x+1)}$

$\quad = \dfrac{x^2 + x - 2x - 6}{(x+3)(x+2)(x+1)}$

$\quad = \dfrac{x^2 - x - 6}{(x+3)(x+2)(x+1)}$

$\quad = \dfrac{(x-3)(x+2)}{(x+3)(x+2)(x+1)}$

$\quad = \dfrac{(x-3)\cancel{(x+2)}}{(x+3)\cancel{(x+2)}(x+1)}$

$\quad = \dfrac{x-3}{(x+3)(x+1)}$

44. $\dfrac{x-6}{(x+6)(x+4)}$

45. $\dfrac{x}{x^2 + 2x + 1} + \dfrac{1}{x^2 + 5x + 4}$

$\quad = \dfrac{x}{(x+1)(x+1)} + \dfrac{1}{(x+1)(x+4)}$

$\qquad\qquad$ LCD $= (x+1)^2(x+4)$

$\quad = \dfrac{x}{(x+1)(x+1)} \cdot \dfrac{x+4}{x+4} + \dfrac{1}{(x+1)(x+4)} \cdot \dfrac{x+1}{x+1}$

$\quad = \dfrac{x(x+4) + 1 \cdot (x+1)}{(x+1)^2(x+4)} = \dfrac{x^2 + 4x + x + 1}{(x+1)^2(x+4)}$

$\quad = \dfrac{x^2 + 5x + 1}{(x+1)^2(x+4)}$

46. $\dfrac{12a - 11}{(a+2)(a-1)(a-3)}$

47. $\dfrac{x}{x^2 + 15x + 56} - \dfrac{6}{x^2 + 13x + 42}$

$\quad = \dfrac{x}{(x+7)(x+8)} - \dfrac{6}{(x+6)(x+7)}$

$\qquad\qquad$ LCD $= (x+7)(x+8)(x+6)$

$\quad = \dfrac{x}{(x+7)(x+8)} \cdot \dfrac{x+6}{x+6} - \dfrac{6}{(x+6)(x+7)} \cdot \dfrac{x+8}{x+8}$

$\quad = \dfrac{x^2 + 6x}{(x+7)(x+8)(x+6)} - \dfrac{6x + 48}{(x+7)(x+8)(x+6)}$

$\quad = \dfrac{x^2 + 6x - (6x+48)}{(x+7)(x+8)(x+6)}$

$\quad = \dfrac{x^2 + 6x - 6x - 48}{(x+7)(x+8)(x+6)}$

$\quad = \dfrac{x^2 - 48}{(x+7)(x+8)(x+6)}$

48. $\dfrac{-8x - 88}{(x+1)(x+16)(x+8)}$

(Although $-8x - 88$ can be factored as $-8(x+11)$, doing so will not enable us to simplify the result.)

49. $\dfrac{10}{x^2 + x - 6} + \dfrac{3x}{x^2 - 4x + 4}$

$\quad = \dfrac{10}{(x+3)(x-2)} + \dfrac{3x}{(x-2)(x-2)}$

$\qquad\qquad$ LCD $= (x+3)(x-2)^2$

$\quad = \dfrac{10}{(x+3)(x-2)} \cdot \dfrac{x-2}{x-2} + \dfrac{3x}{(x-2)(x-2)} \cdot \dfrac{x+3}{x+3}$

$\quad = \dfrac{10(x-2) + 3x(x+3)}{(x+3)(x-2)^2} = \dfrac{10x - 20 + 3x^2 + 9x}{(x+3)(x-2)^2}$

$\quad = \dfrac{3x^2 + 19x - 20}{(x+3)(x-2)^2}$

50. $\dfrac{5z + 12}{(z-3)(z+2)(z+3)}$

51. $\dfrac{y+2}{y-7} + \dfrac{3-y}{49-y^2} = \dfrac{y+2}{y-7} + \dfrac{3-y}{(7+y)(7-y)}$

$\qquad\qquad = \dfrac{y+2}{y-7} + \dfrac{3-y}{-(y+7)(y-7)}$

$\qquad\qquad\qquad [7+y = y+7;$

$\qquad\qquad\qquad (7-y) = -(y-7)]$

$\qquad\qquad = \dfrac{y+2}{y-7} + \dfrac{-(3-y)}{(y+7)(y-7)}$

$\qquad\qquad\qquad$ LCD $= (y+7)(y-7)$

$\qquad\qquad = \dfrac{y+2}{y-7} \cdot \dfrac{y+7}{y+7} + \dfrac{-(3-y)}{(y+7)(y-7)}$

$\qquad\qquad = \dfrac{(y+2)(y+7) - (3-y)}{(y+7)(y-7)}$

$\qquad\qquad = \dfrac{y^2 + 9y + 14 - 3 + y}{(y+7)(y-7)}$

$\qquad\qquad = \dfrac{y^2 + 10y + 11}{(y+7)(y-7)}$

52. $\dfrac{p^2 + 7p + 1}{(p+5)(p-5)}$

53. $\dfrac{8x}{16 - x^2} - \dfrac{5}{x - 4}$

$= \dfrac{8x}{(4+x)(4-x)} - \dfrac{5}{x-4}$ $\quad 4 - x$ and $x - 4$ are opposites

$= \dfrac{8x}{(4+x)(4-x)} - \dfrac{5}{-(4-x)}$

$= \dfrac{8x}{(4+x)(4-x)} - \dfrac{-5}{4-x}$ $\quad$ LCD $= (4+x)(4-x)$

$= \dfrac{8x}{(4+x)(4-x)} - \dfrac{-5}{4-x} \cdot \dfrac{4+x}{4+x}$

$= \dfrac{8x}{(4+x)(4-x)} - \dfrac{-20 - 5x}{(4-x)(4+x)}$

$= \dfrac{8x - (-20 - 5x)}{(4+x)(4-x)}$

$= \dfrac{8x + 20 + 5x}{(4+x)(4-x)}$

$= \dfrac{13x + 20}{(4+x)(4-x)}$

54. $\dfrac{9x + 12}{(x+3)(x-3)}$

(Although $9x + 12$ can be factored as $3(3x + 4)$, doing so will not enable us to simplify the result.)

55. $\dfrac{a}{a^2 - 1} + \dfrac{2a}{a - a^2}$

$= \dfrac{a}{(a+1)(a-1)} + \dfrac{2a}{-a(a-1)}$

$= \dfrac{a}{(a+1)(a-1)} + \dfrac{-2a}{a(a-1)}$ $\quad$ Since $\dfrac{a}{-b} = \dfrac{-a}{b}$;

$\qquad\qquad\qquad\qquad\qquad$ LCD $= a(a+1)(a-1)$

$= \dfrac{a}{(a+1)(a-1)} \cdot \dfrac{a}{a} + \dfrac{-2a}{a(a-1)} \cdot \dfrac{a+1}{a+1}$

$= \dfrac{a^2 - 2a(a+1)}{a(a+1)(a-1)}$

$= \dfrac{a^2 - 2a^2 - 2a}{a(a+1)(a-1)}$

$= \dfrac{-a^2 - 2a}{a(a+1)(a-1)}$

$= \dfrac{\cancel{a}(-a - 2)}{\cancel{a}(a+1)(a-1)}$

$= \dfrac{-a - 2}{(a+1)(a-1)}$

56. $\dfrac{-3x^2 + 7x + 4}{3(x+2)(2-x)}$, or $\dfrac{3x^2 - 7x - 4}{3(x+2)(x-2)}$

57. $\dfrac{4x}{x^2 - y^2} - \dfrac{6}{y - x}$

$= \dfrac{4x}{(x+y)(x-y)} - \dfrac{6}{-(x-y)}$

$= \dfrac{4x}{(x+y)(x-y)} - \dfrac{-6}{x-y}$ $\quad$ Since $\dfrac{a}{-b} = \dfrac{-a}{b}$;

$\qquad\qquad\qquad\qquad$ LCD $= (x+y)(x-y)$

$= \dfrac{4x}{(x+y)(x-y)} - \dfrac{-6}{x-y} \cdot \dfrac{x+y}{x+y}$

$= \dfrac{4x}{(x+y)(x-y)} - \dfrac{-6x - 6y}{(x+y)(x-y)}$

$= \dfrac{4x - (-6x - 6y)}{(x+y)(x-y)}$

$= \dfrac{4x + 6x + 6y}{(x+y)(x-y)}$

$= \dfrac{10x + 6y}{(x+y)(x-y)}$

(Although $10x + 6y$ can be factored as $2(5x + 3y)$, doing so does not enable us to simplify the result.)

58. $\dfrac{a - 2}{(a+3)(a-3)}$, or $\dfrac{-a + 2}{(3+a)(3-a)}$

59. $\dfrac{4y}{y^2 - 1} - \dfrac{2}{y} - \dfrac{2}{y + 1}$

$= \dfrac{4y}{(y+1)(y-1)} - \dfrac{2}{y} - \dfrac{2}{y+1}$

$\qquad\quad$ LCD $= y(y+1)(y-1)$

$= \dfrac{4y}{(y+1)(y-1)} \cdot \dfrac{y}{y} - \dfrac{2}{y} \cdot \dfrac{(y+1)(y-1)}{(y+1)(y-1)} -$

$\qquad\qquad\qquad\qquad\qquad \dfrac{2}{y+1} \cdot \dfrac{y(y-1)}{y(y-1)}$

$= \dfrac{4y^2 - (2y^2 - 2) - (2y^2 - 2y)}{y(y+1)(y-1)}$

$= \dfrac{4y^2 - 2y^2 + 2 - 2y^2 + 2y}{y(y+1)(y-1)}$

$= \dfrac{2y + 2}{y(y+1)(y-1)}$

$= \dfrac{2(\cancel{y+1})}{y(\cancel{y+1})(y-1)}$

$= \dfrac{2}{y(y-1)}$

60. $\dfrac{2x - 3}{2 - x}$

61. $\dfrac{2z}{1-2z} + \dfrac{3z}{2z+1} - \dfrac{3}{4z^2-1}$

$= \dfrac{2z}{-(2z-1)} + \dfrac{3z}{2z+1} - \dfrac{3}{(2z+1)(2z-1)}$

$= \dfrac{-2z}{2z-1} + \dfrac{3z}{2z+1} - \dfrac{3}{(2z-1)(2z+1)}$

$\qquad$ LCD $= (2z-1)(2z+1)$

$= \dfrac{-2z}{2z-1} \cdot \dfrac{2z+1}{2z+1} + \dfrac{3z}{2z+1} \cdot \dfrac{2z-1}{2z-1} - \dfrac{3}{(2z-1)(2z+1)}$

$= \dfrac{(-4z^2-2z)+(6z^2-3z)-3}{(2z-1)(2z+1)}$

$= \dfrac{2z^2-5z-3}{(2z-1)(2z+1)}$

$= \dfrac{(z-3)(2z+1)}{(2z-1)(2z+1)}$

$= \dfrac{z-3}{2z-1}$

62. 0

63. $\dfrac{5}{3-2x} + \dfrac{3}{2x-3} - \dfrac{x-3}{2x^2-x-3}$

$= \dfrac{5}{-(2x-3)} + \dfrac{3}{2x-3} - \dfrac{x-3}{(2x-3)(x+1)}$

$= \dfrac{-5}{2x-3} + \dfrac{3}{2x-3} - \dfrac{x-3}{(2x-3)(x+1)}$

$\qquad$ LCD $= (2x-3)(x+1)$

$= \dfrac{-5}{2x-3} \cdot \dfrac{x+1}{x+1} + \dfrac{3}{2x-3} \cdot \dfrac{x+1}{x+1} - \dfrac{x-3}{(2x-3)(x+1)}$

$= \dfrac{(-5x-5)+(3x+3)-(x-3)}{(2x-3)(x+1)}$

$= \dfrac{-5x-5+3x+3-x+3}{(2x-3)(x+1)}$

$= \dfrac{-3x+1}{(2x-3)(x+1)}$

64. $\dfrac{2}{r+s}$

65. $\dfrac{3}{2c-1} - \dfrac{1}{c+2} - \dfrac{5}{2c^2+3c-2}$

$= \dfrac{3}{2c-1} - \dfrac{1}{c+2} - \dfrac{5}{(2c-1)(c+2)}$

$\qquad$ LCD $= (2c-1)(c+2)$

$= \dfrac{3}{2c-1} \cdot \dfrac{c+2}{c+2} - \dfrac{1}{c+2} \cdot \dfrac{2c-1}{2c-1} - \dfrac{5}{(2c-1)(c+2)}$

$= \dfrac{(3c+6)-(2c-1)-5}{(2c-1)(c+2)}$

$= \dfrac{3c+6-2c+1-5}{(2c-1)(c+2)}$

$= \dfrac{c+2}{(2c-1)(c+2)}$

$= \dfrac{1}{2c-1}$

66. $\dfrac{2y^2+2y-7}{(2y+3)(y-1)}$

67. $\dfrac{1}{x+y} - \dfrac{1}{x-y} + \dfrac{2x}{x^2-y^2}$

$= \dfrac{1}{x+y} - \dfrac{1}{x-y} + \dfrac{2x}{(x+y)(x-y)}$

$\qquad$ LCD $= (x+y)(x-y)$

$= \dfrac{1}{x+y} \cdot \dfrac{x-y}{x-y} - \dfrac{1}{x-y} \cdot \dfrac{x+y}{x+y} +$

$\qquad\qquad\qquad \dfrac{2x}{(x+y)(x-y)}$

$= \dfrac{x-y-(x+y)+2x}{(x+y)(x-y)}$

$= \dfrac{x-y-x-y+2x}{(x+y)(x-y)}$

$= \dfrac{2x-2y}{(x+y)(x-y)}$

$= \dfrac{2(x-y)}{(x+y)(x-y)}$

$= \dfrac{2(x-y)}{(x+y)(x-y)}$

$= \dfrac{2}{x+y}$

68. $\dfrac{4b}{(a+b)(a-b)}$

69. Graph: $y = \dfrac{1}{2}x - 5$

Since the equation is in the form $y = mx + b$, we know the y-intercept is $(0, -5)$. We find two other solutions, substituting multiples of 2 for x to avoid fractions.

When $x = 2$, $y = \dfrac{1}{2} \cdot 2 - 5 = 1 - 5 = -4$.

When $x = 4$, $y = \dfrac{1}{2} \cdot 4 - 5 = 2 - 5 = -3$.

x	y
0	-5
2	-4
4	-3

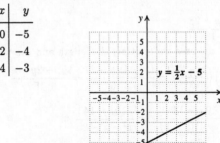

70.

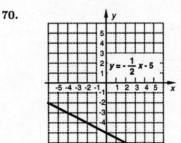

71. Graph: $y = 3$

All solutions are of the form $(x, 3)$. The graph is a line parallel to the x-axis with y-intercept $(0, 3)$.

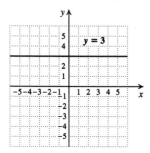

72.

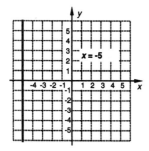

73.

74.

75. To find the perimeter we add the lengths of the sides:

$$\frac{y+4}{3} + \frac{y+4}{3} + \frac{y-2}{5} + \frac{y-2}{5} \quad \text{LCD} = 3 \cdot 5$$

$$= \frac{y+4}{3} \cdot \frac{5}{5} + \frac{y+4}{3} \cdot \frac{5}{5} + \frac{y-2}{5} \cdot \frac{3}{3} + \frac{y-2}{5} \cdot \frac{3}{3}$$

$$= \frac{5y + 20 + 5y + 20 + 3y - 6 + 3y - 6}{3 \cdot 5}$$

$$= \frac{16y + 28}{15}$$

To find the area we multiply the length and the width:

$$\left(\frac{y+4}{3}\right)\left(\frac{y-2}{5}\right) = \frac{(y+4)(y-2)}{3 \cdot 5} = \frac{y^2 + 2y - 8}{15}$$

76. $P = \dfrac{10x - 14}{(x+4)(x-5)}$, or $\dfrac{10x - 14}{x^2 - x - 20}$

$A = \dfrac{6}{x^2 - x - 20}$

77.

$$\frac{5}{z+2} + \frac{4z}{z^2 - 4} + 2$$

$$= \frac{5}{z+2} + \frac{4z}{(z+2)(z-2)} + \frac{2}{1},$$

$$\text{LCD} = (z+2)(z-2)$$

$$= \frac{5}{z+2} \cdot \frac{z-2}{z-2} + \frac{4z}{(z+2)(z-2)} + \frac{2}{1} \cdot \frac{(z+2)(z-2)}{(z+2)(z-2)}$$

$$= \frac{5z - 10 + 4z + 2(z^2 - 4)}{(z+2)(z-2)}$$

$$= \frac{5z - 10 + 4z + 2z^2 - 8}{(z+2)(z-2)} = \frac{2z^2 + 9z - 18}{(z+2)(z-2)}$$

$$= \frac{(2z - 3)(z + 6)}{(z+2)(z-2)}$$

78. $\dfrac{11z^4 - 22z^2 + 6}{(z^2 + 2)(z^2 - 2)(2z^2 - 3)}$

79. $\dfrac{1}{2xy - 6x + ay - 3a} - \dfrac{ay + xy}{(a^2 - 4x^2)(y^2 - 6y + 9)}$

$$= \frac{1}{(2x+a)(y-3)} - \frac{ay+xy}{(a+2x)(a-2x)(y-3)(y-3)}$$

$$\text{LCD} = (a + 2x)(a - 2x)(y - 3)^2$$

$$= \frac{1}{(2x+a)(y-3)} \cdot \frac{(a - 2x)(y - 3)}{(a - 2x)(y - 3)} -$$

$$\frac{ay + xy}{(a+2x)(a-2x)(y-3)(y-3)}$$

$$= \frac{ay - 3a - 2xy + 6x - (ay + xy)}{(a+2x)(a-2x)(y-3)^2}$$

$$= \frac{-3a - 3xy + 6x}{(a+2x)(a-2x)(y-3)^2}$$

80. $\dfrac{x^2 + xy - x^3 + x^2y - xy^2 + y^3}{(x^2 + y^2)(x + y)^2(x - y)}$

81. Answers may vary. $\dfrac{a}{a - b} + \dfrac{3b}{b - a}$

Exercise Set 6.6

1.

$$\frac{1 + \dfrac{9}{16}}{1 - \dfrac{3}{4}} \qquad \text{LCM of the denominators is 16}$$

$$= \frac{1 + \dfrac{9}{16}}{1 - \dfrac{3}{4}} \cdot \frac{16}{16} \qquad \text{Multiplying by 1 using } \frac{16}{16}$$

$$= \frac{\left(1 + \dfrac{9}{16}\right)16}{\left(1 - \dfrac{3}{4}\right)16} \qquad \begin{array}{l}\text{Multiplying numerator and} \\ \text{denominator by 16}\end{array}$$

$$= \frac{1(16) + \dfrac{9}{16}(16)}{1(16) - \dfrac{3}{4}(16)}$$

$$= \frac{16 + 9}{16 - 12}$$

$$= \frac{25}{4}$$

2. $\dfrac{5}{2}$

3. $\dfrac{1 - \dfrac{3}{5}}{1 + \dfrac{1}{5}}$

$= \dfrac{1 \cdot \dfrac{5}{5} - \dfrac{3}{5}}{1 \cdot \dfrac{5}{5} + \dfrac{1}{5}}$ Getting a common denominator in numerator and in denominator

$= \dfrac{\dfrac{5}{5} - \dfrac{3}{5}}{\dfrac{5}{5} + \dfrac{1}{5}}$

$= \dfrac{\dfrac{2}{5}}{\dfrac{6}{5}}$ Subtracting in numerator; adding in denominator

$= \dfrac{2}{5} \cdot \dfrac{5}{6}$ Multiplying by the reciprocal of the divisor

$= \dfrac{2 \cdot 5}{5 \cdot 2 \cdot 3}$

$= \dfrac{2 \cdot 5 \cdot 1}{5 \cdot 2 \cdot 3}$

$= \dfrac{1}{3}$

4. $-\dfrac{65}{18}$

5. $\dfrac{\dfrac{1}{x} + 4}{\dfrac{1}{x} - 3}$ LCM of the denominators is x

$= \dfrac{\dfrac{1}{x} + 4}{\dfrac{1}{x} - 3} \cdot \dfrac{x}{x}$

$= \dfrac{\left(\dfrac{1}{x} + 4\right)x}{\left(\dfrac{1}{x} - 3\right)x}$

$= \dfrac{\dfrac{1}{x} \cdot x + 4 \cdot x}{\dfrac{1}{x} \cdot x - 3 \cdot x} = \dfrac{1 + 4x}{1 - 3x}$

6. $\dfrac{9 + 3s^2}{4s^2}$

7. $\dfrac{\dfrac{1}{2} + \dfrac{3}{4}}{\dfrac{5}{8} - \dfrac{5}{6}}$

$= \dfrac{\dfrac{1}{2} \cdot \dfrac{2}{2} + \dfrac{3}{4}}{\dfrac{5}{8} \cdot \dfrac{3}{3} - \dfrac{5}{6} \cdot \dfrac{4}{4}}$ Getting a common denominator in numerator and denominator

$= \dfrac{\dfrac{2}{4} + \dfrac{3}{4}}{\dfrac{15}{24} - \dfrac{20}{24}}$

$= \dfrac{\dfrac{5}{4}}{\dfrac{-5}{24}}$ Adding in numerator; subtracting in denominator

$= \dfrac{5}{4} \cdot \dfrac{24}{-5}$ Multiplying by the reciprocal of the divisor

$= \dfrac{5 \cdot 4 \cdot 6}{4 \cdot (-1) \cdot 5}$

$= \dfrac{5 \cdot 4 \cdot 6}{4 \cdot (-1) \cdot 5}$

$= -6$

8. $-\dfrac{4}{39}$

9. $\dfrac{\dfrac{2}{y} + \dfrac{1}{2y}}{y + \dfrac{y}{2}}$ LCM of the denominators is $2y$

$= \dfrac{\dfrac{2}{y} + \dfrac{1}{2y}}{y + \dfrac{y}{2}} \cdot \dfrac{2y}{2y}$

$= \dfrac{\left(\dfrac{2}{y} + \dfrac{1}{2y}\right)2y}{\left(y + \dfrac{y}{2}\right)2y}$

$= \dfrac{\dfrac{2}{y}(2y) + \dfrac{1}{2y}(2y)}{y(2y) + \dfrac{y}{2}(2y)} = \dfrac{4 + 1}{2y^2 + y^2} = \dfrac{5}{3y^2}$

10. $\dfrac{2x + 1}{x}$

11. $\dfrac{8 + \dfrac{8}{d}}{1 + \dfrac{1}{d}} = \dfrac{8 \cdot \dfrac{d}{d} + \dfrac{8}{d}}{1 \cdot \dfrac{d}{d} + \dfrac{1}{d}}$

$= \dfrac{\dfrac{8d + 8}{d}}{\dfrac{d + 1}{d}}$

$= \dfrac{8d + 8}{d} \cdot \dfrac{d}{d + 1}$

$= \dfrac{8(d + 1)(d)}{d(d + 1)}$

$= \dfrac{8(d + 1)(d)}{d(d + 1)(1)}$

$= 8$

12. $\dfrac{3(2b - 3)}{b(6 - b)}$

13. $\dfrac{\dfrac{x}{8} - \dfrac{8}{x}}{\dfrac{1}{8} + \dfrac{1}{x}}$ LCM of the denominators is $8x$

$= \dfrac{\dfrac{x}{8} - \dfrac{8}{x}}{\dfrac{1}{8} + \dfrac{1}{x}} \cdot \dfrac{8x}{8x}$

$= \dfrac{\left(\dfrac{x}{8} - \dfrac{8}{x}\right)8x}{\left(\dfrac{1}{8} + \dfrac{1}{x}\right)8x}$

$= \dfrac{\dfrac{x}{8}(8x) - \dfrac{8}{x}(8x)}{\dfrac{1}{8}(8x) + \dfrac{1}{x}(8x)}$

$= \dfrac{x^2 - 64}{x + 8}$

$= \dfrac{(x + 8)(x - 8)}{x + 8}$

$= \dfrac{(x + 8)(x - 8)}{1(x + 8)}$

$= x - 8$

14. $\dfrac{4 + m^2}{m^2 - 4}$

15. $\dfrac{1 + \dfrac{1}{y}}{1 - \dfrac{1}{y^2}} = \dfrac{1 \cdot \dfrac{y}{y} + \dfrac{1}{y}}{1 \cdot \dfrac{y^2}{y^2} - \dfrac{1}{y^2}}$

$= \dfrac{\dfrac{y + 1}{y}}{\dfrac{y^2 - 1}{y^2}}$

$= \dfrac{y + 1}{y} \cdot \dfrac{y^2}{y^2 - 1}$

$= \dfrac{(y + 1)y \cdot y}{y(y + 1)(y - 1)}$

$= \dfrac{(y + 1)y \cdot y}{y(y + 1)(y - 1)}$

$= \dfrac{y}{y - 1}$

16. $\dfrac{1 - q}{q}$

17. $\dfrac{\dfrac{25 - a^2}{5a}}{\dfrac{a + 5}{5}} = \dfrac{25 - a^2}{5a} \cdot \dfrac{5}{a + 5}$ Multiplying by the reciprocal of the divisor

$= \dfrac{(5 + a)(5 - a)}{5a} \cdot \dfrac{5}{a + 5}$

$= \dfrac{(5 + a)(5 - a)(5)}{5 \cdot a(a + 5)}$

$= \dfrac{5 - a}{a}$

18. $\dfrac{x + y}{x}$

19. $\dfrac{\dfrac{x}{x - y}}{\dfrac{x^2}{x^2 - y^2}} = \dfrac{x}{x - y} \cdot \dfrac{x^2 - y^2}{x^2}$ Multiplying by the reciprocal of the divisor

$= \dfrac{x}{x - y} \cdot \dfrac{(x - y)(x + y)}{x \cdot x}$

$= \dfrac{x(x - y)(x + y)}{x \cdot x(x - y)}$

$= \dfrac{x + y}{x}$

20. $x - y$

21.
$$\dfrac{\dfrac{3}{m}+\dfrac{2}{m^3}}{\dfrac{4}{m^2}-\dfrac{3}{m}} \qquad \text{LCM of the denominators is } m^3$$

$$=\dfrac{\dfrac{3}{m}+\dfrac{2}{m^3}}{\dfrac{4}{m^2}-\dfrac{3}{m}}\cdot\dfrac{m^3}{m^3}$$

$$=\dfrac{\dfrac{3}{m}\cdot m^3+\dfrac{2}{m^3}\cdot m^3}{\dfrac{4}{m^2}\cdot m^3-\dfrac{3}{m}\cdot m^3}$$

$$=\dfrac{3m^2+2}{4m-3m^2}$$

22. $\dfrac{1}{a(a-b)}$

23.
$$\dfrac{\dfrac{5}{4x^3}-\dfrac{3}{8x}}{\dfrac{3}{2x}+\dfrac{3}{4x^3}}=\dfrac{\dfrac{5}{4x^3}\cdot\dfrac{2}{2}-\dfrac{3}{8x}\cdot\dfrac{x^2}{x^2}}{\dfrac{3}{2x}\cdot\dfrac{2x^2}{2x^2}+\dfrac{3}{4x^3}}$$

$$=\dfrac{\dfrac{10-3x^2}{8x^3}}{\dfrac{6x^2+3}{4x^3}}$$

$$=\dfrac{10-3x^2}{8x^3}\cdot\dfrac{4x^3}{6x^2+3}$$

$$=\dfrac{\cancel{4x^3}(10-3x^2)}{2\cdot\cancel{4x^3}\cdot 3(2x^2+1)}$$

$$=\dfrac{10-3x^2}{6(2x^2+1)}, \text{ or}$$

$$=\dfrac{10-3x^2}{12x^2+6}$$

24. $\dfrac{15(4-a^3)}{14a^2(9+2a)}, \text{ or } \dfrac{60-15a^3}{126a^2+28a^3}$

25.
$$\dfrac{\dfrac{a}{6b^3}+\dfrac{4}{9b^2}}{\dfrac{5}{6b}-\dfrac{1}{9b^3}} \qquad \text{LCM of the denominators is } 18b^3$$

$$=\dfrac{\dfrac{a}{6b^3}+\dfrac{4}{9b^2}}{\dfrac{5}{6b}-\dfrac{1}{9b^3}}\cdot\dfrac{18b^3}{18b^3}$$

$$=\dfrac{\dfrac{a}{6b^3}\cdot 18b^3+\dfrac{4}{9b^2}\cdot 18b^3}{\dfrac{5}{6b}\cdot 18b^3-\dfrac{1}{9b^3}\cdot 18b^3}$$

$$=\dfrac{3a+8b}{15b^2-2}$$

26. $\dfrac{2xy-3y^3}{xy^3+30}$

27.
$$\dfrac{\dfrac{2}{x^2y}+\dfrac{3}{xy^2}}{\dfrac{2}{xy^3}+\dfrac{1}{x^2y}}=\dfrac{\dfrac{2}{x^2y}\cdot\dfrac{y}{y}+\dfrac{3}{xy^2}\cdot\dfrac{x}{x}}{\dfrac{2}{xy^3}\cdot\dfrac{x}{x}+\dfrac{1}{x^2y}\cdot\dfrac{y^2}{y^2}}$$

$$=\dfrac{\dfrac{2y+3x}{x^2y^2}}{\dfrac{2x+y^2}{x^2y^3}}$$

$$=\dfrac{2y+3x}{x^2y^2}\cdot\dfrac{x^2y^3}{2x+y^2}$$

$$=\dfrac{x^2y^2\cdot y(2y+3x)}{x^2y^2(2x+y^2)}$$

$$=\dfrac{y(2y+3x)}{2x+y^2}, \text{ or } \dfrac{2y^2+3xy}{2x+y^2}$$

28. $\dfrac{5a^2+2b^3}{b^3(5-3a^2)}, \text{ or } \dfrac{5a^2+2b^3}{5b^3-3a^2b^3}$

29.
$$\dfrac{3-\dfrac{2}{a^4}}{2+\dfrac{3}{a^3}}=\dfrac{3-\dfrac{2}{a^4}}{2+\dfrac{3}{a^3}}\cdot\dfrac{a^4}{a^4} \qquad \begin{array}{l}\text{LCM of the denomina-}\\\text{tors is } a^4\end{array}$$

$$=\dfrac{3\cdot a^4-\dfrac{2}{a^4}\cdot a^4}{2\cdot a^4+\dfrac{3}{a^3}\cdot a^4}$$

$$=\dfrac{3a^4-2}{2a^4+3a}$$

30. $\dfrac{2x^4-3x^2}{2x^4+3}$

31.
$$\dfrac{\dfrac{1}{x+h}-\dfrac{1}{x}}{h}$$

$$=\dfrac{\dfrac{1}{x+h}\cdot\dfrac{x}{x}-\dfrac{1}{x}\cdot\dfrac{x+h}{x+h}}{h} \qquad \begin{array}{l}\text{Subtracting in the}\\\text{numerator}\end{array}$$

$$=\dfrac{\dfrac{x-(x+h)}{x(x+h)}}{h}$$

$$=\dfrac{\dfrac{x-x-h}{x(x+h)}}{h}=\dfrac{\dfrac{-h}{x(x+h)}}{h}$$

$$=\dfrac{-h}{x(x+h)}\cdot\dfrac{1}{h} \qquad \begin{array}{l}\text{Multiplying by the recip-}\\\text{rocal of the divisor}\end{array}$$

$$=\dfrac{h}{h}\cdot\dfrac{-1}{x(x+h)} \qquad \text{Removing a factor of 1}$$

$$=\dfrac{-1}{x(x+h)}$$

32. $\dfrac{1}{a(a-h)}$

33. $\dfrac{\dfrac{y^2-y-6}{y^2-5y-14}}{\dfrac{y^2+6y+5}{y^2-6y-7}} = \dfrac{y^2-y-6}{y^2-5y-14} \cdot \dfrac{y^2-6y-7}{y^2+6y+5}$

$\qquad = \dfrac{(y+2)(y-3)}{(y+2)(y-7)} \cdot \dfrac{(y-7)(y+1)}{(y+5)(y+1)}$

$\qquad = \dfrac{y-3}{y+5}$

34. $\dfrac{(x-4)(x-7)}{(x-5)(x+6)}$

35. $\dfrac{\dfrac{x+5}{x^2}}{\dfrac{2}{x}-\dfrac{3}{x^2}} = \dfrac{\dfrac{x+5}{x^2}}{\dfrac{2}{x}\cdot\dfrac{x}{x}-\dfrac{3}{x^2}}$

$\qquad = \dfrac{\dfrac{x+5}{x^2}}{\dfrac{2x-3}{x^2}}$

$\qquad = \dfrac{x+5}{x^2} \cdot \dfrac{x^2}{2x-3}$

$\qquad = \dfrac{x^2(x+5)}{x^2(2x-3)}$

$\qquad = \dfrac{x+5}{2x-3}$

36. $\dfrac{a-7}{a(3+2a)}$, or $\dfrac{a-7}{3a+2a^2}$

37. $\dfrac{x-3+\dfrac{2}{x}}{x-4+\dfrac{3}{x}} = \dfrac{x\cdot\dfrac{x}{x}-3\cdot\dfrac{x}{x}+\dfrac{2}{x}}{x\cdot\dfrac{x}{x}-4\cdot\dfrac{x}{x}+\dfrac{3}{x}}$

$\qquad = \dfrac{\dfrac{x^2-3x+2}{x}}{\dfrac{x^2-4x+3}{x}}$

$\qquad = \dfrac{x^2-3x+2}{x} \cdot \dfrac{x}{x^2-4x+3}$

$\qquad = \dfrac{(x-2)(x-1)}{x} \cdot \dfrac{x}{(x-3)(x-1)}$

$\qquad = \dfrac{x(x-1)}{x(x-1)} \cdot \dfrac{x-2}{x-3}$

$\qquad = \dfrac{x-2}{x-3}$

38. $\dfrac{20x^2-30x}{15x^2+105x-12}$

39. $\dfrac{\dfrac{1}{x-2}+\dfrac{3}{x-1}}{\dfrac{2}{x-1}+\dfrac{5}{x-2}}$ LCM of the denominators is $(x-2)(x-1)$

$= \dfrac{\dfrac{1}{x-2}+\dfrac{3}{x-1}}{\dfrac{2}{x-1}+\dfrac{5}{x-2}} \cdot \dfrac{(x-2)(x-1)}{(x-2)(x-1)}$

$= \dfrac{\dfrac{1}{x-2}\cdot(x-2)(x-1)+\dfrac{3}{x-1}\cdot(x-2)(x-1)}{\dfrac{2}{x-1}\cdot(x-2)(x-1)+\dfrac{5}{x-2}\cdot(x-2)(x-1)}$

$= \dfrac{\dfrac{1}{x-2}\cdot(x-2)(x-1)+\dfrac{3}{x-1}\cdot(x-2)(x-1)}{\dfrac{2}{x-1}\cdot(x-2)(x-1)+\dfrac{5}{x-2}\cdot(x-2)(x-1)}$

$= \dfrac{x-1+3(x-2)}{2(x-2)+5(x-1)}$

$= \dfrac{x-1+3x-6}{2x-4+5x-5}$

$= \dfrac{4x-7}{7x-9}$

40. $\dfrac{3y-1}{7y-5}$

41. $\dfrac{\dfrac{3}{a^2-9}+\dfrac{2}{a+3}}{\dfrac{4}{a^2-9}+\dfrac{1}{a+3}} = \dfrac{\dfrac{3}{(a+3)(a-3)}+\dfrac{2}{a+3}\cdot\dfrac{a-3}{a-3}}{\dfrac{4}{(a+3)(a-3)}+\dfrac{1}{a+3}\cdot\dfrac{a-3}{a-3}}$

$\qquad = \dfrac{\dfrac{3+2(a-3)}{(a+3)(a-3)}}{\dfrac{4+(a-3)}{(a+3)(a-3)}}$

$\qquad = \dfrac{\dfrac{3+2a-6}{(a+3)(a-3)}}{\dfrac{4+a-3}{(a+3)(a-3)}}$

$\qquad = \dfrac{\dfrac{2a-3}{(a+3)(a-3)}}{\dfrac{a+1}{(a+3)(a-3)}}$

$\qquad = \dfrac{2a-3}{(a+3)(a-3)} \cdot \dfrac{(a+3)(a-3)}{a+1}$

$\qquad = \dfrac{2a-3}{a+1}$

42. $\dfrac{a+1}{2a+5}$

43.
$$\frac{\dfrac{1}{x^2-1}+\dfrac{1}{x^2+4x+3}}{\dfrac{1}{x^2-1}+\dfrac{1}{x^2-3x+2}}$$

$$=\frac{\dfrac{1}{(x+1)(x-1)}+\dfrac{1}{(x+1)(x+3)}}{\dfrac{1}{(x+1)(x-1)}+\dfrac{1}{(x-2)(x-1)}}$$

$$=\frac{\dfrac{1}{(x+1)(x-1)}\cdot\dfrac{x+3}{x+3}+\dfrac{1}{(x+1)(x+3)}\cdot\dfrac{x-1}{x-1}}{\dfrac{1}{(x+1)(x-1)}\cdot\dfrac{x-2}{x-2}+\dfrac{1}{(x-2)(x-1)}\cdot\dfrac{x+1}{x+1}}$$

$$=\frac{\dfrac{x+3+x-1}{(x+1)(x-1)(x+3)}}{\dfrac{x-2+x+1}{(x+1)(x-1)(x-2)}}$$

$$=\frac{\dfrac{2x+2}{(x+1)(x-1)(x+3)}}{\dfrac{2x-1}{(x+1)(x-1)(x-2)}}$$

$$=\frac{2(x+1)}{\cancel{(x+1)}\cancel{(x-1)}(x+3)}\cdot\frac{\cancel{(x+1)}\cancel{(x-1)}(x-2)}{2x-1}$$

$$=\frac{2(x+1)(x-2)}{(x+3)(2x-1)}$$

44. $\dfrac{(2x-1)(x-2)}{2x(x+1)}$

45. $(5x^4-6x^3+23x^2-79x+24)-$
$(-18x^4-56x^3+84x-17)=5x^4-6x^3+23x^2-$
$79x+24+18x^4+56x^3-84x+17=$
$23x^4+50x^3+23x^2-163x+41$

46. 14 yd

47.

48.

49.
$$\frac{\dfrac{1}{\dfrac{2}{x-1}-\dfrac{1}{3x-2}}}{}$$

$$=\frac{\dfrac{1}{\dfrac{2}{x-1}-\dfrac{1}{3x-2}}}{}\cdot\frac{(x-1)(3x-2)}{(x-1)(3x-2)}$$

$$=\frac{(x-1)(3x-2)}{\left(\dfrac{2}{x-1}-\dfrac{1}{3x-2}\right)(x-1)(3x-2)}$$

$$=\frac{(x-1)(3x-2)}{\dfrac{2}{x-1}(x-1)(3x-2)-\dfrac{1}{3x-2}(x-1)(3x-2)}$$

$$=\frac{(x-1)(3x-2)}{2(3x-2)-(x-1)}$$

$$=\frac{(x-1)(3x-2)}{6x-4-x+1}$$

$$=\frac{(x-1)(3x-2)}{5x-3}$$

50. $\dfrac{ac}{bd}$

51.
$$\frac{\dfrac{a}{b}-\dfrac{c}{d}}{\dfrac{b}{a}-\dfrac{d}{c}}=\frac{\dfrac{a}{b}\cdot\dfrac{d}{d}-\dfrac{c}{d}\cdot\dfrac{b}{b}}{\dfrac{b}{a}\cdot\dfrac{c}{c}-\dfrac{d}{c}\cdot\dfrac{a}{a}}$$

$$=\frac{\dfrac{ad-bc}{bd}}{\dfrac{bc-ad}{ac}}$$

$$=\frac{ad-bc}{bd}\cdot\frac{ac}{bc-ad}$$

$$=\frac{-1(bc-ad)(ac)}{bd(bc-ad)}$$

$$=\frac{-1\cancel{(bc-ad)}(ac)}{bd\cancel{(bc-ad)}}$$

$$=-\frac{ac}{bd}$$

52. x^5

53. $1 + \dfrac{1}{1 + \dfrac{1}{1 + \dfrac{1}{x}}} = 1 + \dfrac{1}{1 + \dfrac{1}{\frac{x+1}{x}}}$

$\qquad\qquad\qquad\quad = 1 + \dfrac{1}{1 + \dfrac{x}{x+1}}$

$\qquad\qquad\qquad\quad = 1 + \dfrac{1}{\dfrac{x+1+x}{x+1}}$

$\qquad\qquad\qquad\quad = 1 + \dfrac{1}{\dfrac{2x+1}{x+1}}$

$\qquad\qquad\qquad\quad = 1 + \dfrac{x+1}{2x+1}$

$\qquad\qquad\qquad\quad = \dfrac{2x+2+x+1}{2x+1}$

$\qquad\qquad\qquad\quad = \dfrac{3x+2}{2x+1}$

54. $\dfrac{-2z(5z-2)}{(2+z)(-13z+6)}$

55. $\dfrac{a^{-1}+b^{-1}}{\dfrac{a^2-b^2}{ab}} = \dfrac{\dfrac{1}{a}+\dfrac{1}{b}}{\dfrac{a^2-b^2}{ab}}$

$\qquad\qquad = \dfrac{\dfrac{1}{a}+\dfrac{1}{b}}{\dfrac{a^2-b^2}{ab}} \cdot \dfrac{ab}{ab}$

$\qquad\qquad = \dfrac{\dfrac{1}{a}\cdot ab + \dfrac{1}{b}\cdot ab}{\dfrac{a^2-b^2}{ab}\cdot ab}$

$\qquad\qquad = \dfrac{b+a}{a^2-b^2}$

$\qquad\qquad = \dfrac{\cancel{b+a}}{(\cancel{a+b})(a-b)}$

$\qquad\qquad = \dfrac{1}{a-b}$

Exercise Set 6.7

1. $\dfrac{3}{8}+\dfrac{4}{5}=\dfrac{x}{20}$, LCD $= 40$

$40\left(\dfrac{3}{8}+\dfrac{4}{5}\right) = 40\cdot\dfrac{x}{20}$

$40\cdot\dfrac{3}{8}+40\cdot\dfrac{4}{5} = 40\cdot\dfrac{x}{20}$

$15+32 = 2x$

$47 = 2x$

$\dfrac{47}{2} = x$

Check:

$\dfrac{3}{8}+\dfrac{4}{5}=\dfrac{x}{20}$

$\dfrac{3}{8}+\dfrac{4}{5} \;?\; \dfrac{\frac{47}{2}}{20}$

$\dfrac{15}{40}+\dfrac{32}{40} \;\Big|\; \dfrac{47}{2}\cdot\dfrac{1}{20}$

$\dfrac{47}{40} \;\Big|\; \dfrac{47}{40}$ TRUE

This checks, so the solution is $\dfrac{47}{2}$.

2. $\dfrac{57}{5}$

3. $\dfrac{2}{3}-\dfrac{5}{6}=\dfrac{1}{x}$, LCD $= 6x$

$6x\left(\dfrac{2}{3}-\dfrac{5}{6}\right) = 6x\cdot\dfrac{1}{x}$

$6x\cdot\dfrac{2}{3}-6x\cdot\dfrac{5}{6} = 6x\cdot\dfrac{1}{x}$

$4x-5x = 6$

$-x = 6$

$x = -6$

Check:

$\dfrac{1}{x}=\dfrac{2}{3}-\dfrac{5}{6}$

$\dfrac{1}{-6} \;?\; \dfrac{2}{3}-\dfrac{5}{6}$

$-\dfrac{1}{6} \;\Big|\; \dfrac{4}{6}-\dfrac{5}{6}$

$-\dfrac{1}{6} \;\Big|\; -\dfrac{1}{6}$ TRUE

This checks, so the solution is -6.

4. $-\dfrac{40}{19}$

5. $\dfrac{1}{6}+\dfrac{1}{8}=\dfrac{1}{t}$, LCD $= 24t$

$24t\left(\dfrac{1}{6}+\dfrac{1}{8}\right) = 24t\cdot\dfrac{1}{t}$

$24t\cdot\dfrac{1}{6}+24t\cdot\dfrac{1}{8} = 24t\cdot\dfrac{1}{t}$

$4t+3t = 24$

$7t = 24$

$t = \dfrac{24}{7}$

Check:

$$\frac{\frac{1}{6} + \frac{1}{8} = \frac{1}{t}}{\frac{1}{6} + \frac{1}{8} \; ? \; \frac{1}{24/7}}$$

$$\frac{4}{24} + \frac{3}{24} \; \bigg| \; 1 \cdot \frac{7}{24}$$

$$\frac{7}{24} \; \bigg| \; \frac{7}{24} \quad \text{TRUE}$$

This checks, so the solution is $\frac{24}{7}$.

6. $\frac{40}{9}$

7.
$$x + \frac{4}{x} = -5, \; \text{LCD} = x$$

$$x\left(x + \frac{4}{x}\right) = x(-5)$$

$$x \cdot x + x \cdot \frac{4}{x} = x(-5)$$

$$x^2 + 4 = -5x$$

$$x^2 + 5x + 4 = 0$$

$$(x + 4)(x + 1) = 0$$

$$x + 4 = 0 \quad \text{or} \quad x + 1 = 0$$

$$x = -4 \quad \text{or} \quad x = -1$$

Check:
$$x + \frac{4}{x} = -5$$

$$\frac{-4 + \dfrac{4}{-4} \; ? \; -5}{}$$

$$-4 - 1 \; \bigg|$$

$$-5 \; \bigg| \; -5 \quad \text{TRUE}$$

$$x + \frac{4}{x} = -5$$

$$\frac{-1 + \dfrac{4}{-1} \; ? \; -5}{}$$

$$-1 - 4 \; \bigg|$$

$$-5 \; \bigg| \; -5 \quad \text{TRUE}$$

Both of these check, so the two solutions are -4 and -1.

8. $-3, \; -1$

9.
$$\frac{x}{4} - \frac{4}{x} = 0, \; \text{LCD} = 4x$$

$$4x\left(\frac{x}{4} - \frac{4}{x}\right) = 4x \cdot 0$$

$$4x \cdot \frac{x}{4} - 4x \cdot \frac{4}{x} = 4x \cdot 0$$

$$x^2 - 16 = 0$$

$$(x + 4)(x - 4) = 0$$

$$x + 4 = 0 \quad \text{or} \quad x - 4 = 0$$

$$x = -4 \quad \text{or} \quad x = 4$$

Check:

$$\frac{\frac{x}{4} - \frac{4}{x} = 0}{\frac{-4}{4} - \frac{4}{-4} \; ? \; 0}$$

$$-1 - (-1) \; \bigg|$$

$$-1 + 1 \; \bigg|$$

$$0 \; \bigg| \; 0 \quad \text{TRUE}$$

$$\frac{\frac{x}{4} - \frac{4}{x} = 0}{\frac{4}{4} - \frac{4}{4} \; ? \; 0}$$

$$1 - 1 \; \bigg|$$

$$0 \; \bigg| \; 0 \quad \text{TRUE}$$

Both of these check, so the two solutions are -4 and 4.

10. $-5, \; 5$

11.
$$\frac{5}{x} = \frac{6}{x} - \frac{1}{3}, \; \text{LCD} = 3x$$

$$3x \cdot \frac{5}{x} = 3x\left(\frac{6}{x} - \frac{1}{3}\right)$$

$$3x \cdot \frac{5}{x} = 3x \cdot \frac{6}{x} - 3x \cdot \frac{1}{3}$$

$$15 = 18 - x$$

$$-3 = -x$$

$$3 = x$$

Check:
$$\frac{5}{x} = \frac{6}{x} - \frac{1}{3}$$

$$\frac{5}{3} \; ? \; \frac{6}{3} - \frac{1}{3}$$

$$\frac{5}{3} \; \bigg| \; \frac{5}{3} \quad \text{TRUE}$$

This checks, so the solution is 3.

12. 2

13.
$$\frac{5}{3x} + \frac{3}{x} = 1, \; \text{LCD} = 3x$$

$$3x\left(\frac{5}{3x} + \frac{3}{x}\right) = 3x \cdot 1$$

$$3x \cdot \frac{5}{3x} + 3x \cdot \frac{3}{x} = 3x \cdot 1$$

$$5 + 9 = 3x$$

$$14 = 3x$$

$$\frac{14}{3} = x$$

Check:

$$\frac{\frac{5}{3x} + \frac{3}{x} = 1}{\frac{5}{3 \cdot (14/3)} + \frac{3}{(14/3)} \; ? \; 1}$$

$$\frac{5}{14} + \frac{9}{14} \; \bigg|$$

$$\frac{14}{14} \; \bigg|$$

$$1 \; \bigg| \; 1 \quad \text{TRUE}$$

This checks, so the solution is $\frac{14}{3}$.

14. $\dfrac{23}{4}$

15.
$$\dfrac{x-7}{x+2} = \dfrac{1}{4}, \text{ LCD} = 4(x+2)$$
$$4(x+2) \cdot \dfrac{x-7}{x+2} = 4(x+2) \cdot \dfrac{1}{4}$$
$$4(x-7) = x+2$$
$$4x - 28 = x+2$$
$$3x = 30$$
$$x = 10$$

Check:
$$\dfrac{x-7}{x+2} = \dfrac{1}{4}$$

$$\begin{array}{c|c} \dfrac{10-7}{10+2} \; ? \; \dfrac{1}{4} \\[2mm] \dfrac{3}{12} \\[2mm] \dfrac{1}{4} & \dfrac{1}{4} \quad \text{TRUE} \end{array}$$

This checks, so the solution is 10.

16. 5

17.
$$\dfrac{2}{x+1} = \dfrac{1}{x-2},$$
$$\text{LCD} = (x+1)(x-2)$$
$$(x+1)(x-2) \cdot \dfrac{2}{x+1} = (x+1)(x-2) \cdot \dfrac{1}{x-2}$$
$$2(x-2) = x+1$$
$$2x - 4 = x+1$$
$$x = 5$$

This checks, so the solution is 5.

18. $-\dfrac{13}{2}$

19.
$$\dfrac{x}{6} - \dfrac{x}{10} = \dfrac{1}{6}, \text{ LCD} = 30$$
$$30\left(\dfrac{x}{6} - \dfrac{x}{10}\right) = 30 \cdot \dfrac{1}{6}$$
$$30 \cdot \dfrac{x}{6} - 30 \cdot \dfrac{x}{10} = 30 \cdot \dfrac{1}{6}$$
$$5x - 3x = 5$$
$$2x = 5$$
$$x = \dfrac{5}{2}$$

This checks, so the solution is $\dfrac{5}{2}$.

20. 3

21.
$$\dfrac{x+3}{3} - 1 = \dfrac{x-1}{2}, \text{ LCD} = 6$$
$$\dfrac{x+1}{3} - \dfrac{x-1}{2} = 1$$
$$6\left(\dfrac{x+1}{3} - \dfrac{x-1}{2}\right) = 6 \cdot 1$$
$$6 \cdot \dfrac{x+1}{3} - 6 \cdot \dfrac{x-1}{2} = 6 \cdot 1$$
$$2(x+1) - 3(x-1) = 6$$
$$2x + 2 - 3x + 3 = 6$$
$$-x + 5 = 6$$
$$-x = 1$$
$$x = -1$$

This checks, so the solution is -1.

22. -2

23.
$$\dfrac{a-3}{3a+2} = \dfrac{1}{5}, \text{ LCD} = 5(3a+2)$$
$$5(3a+2) \cdot \dfrac{a-3}{3a+2} = 5(3a+2) \cdot \dfrac{1}{5}$$
$$5(a-3) = 3a+2$$
$$5a - 15 = 3a+2$$
$$2a = 17$$
$$a = \dfrac{17}{2}$$

This checks, so the solution is $\dfrac{17}{2}$.

24. $\dfrac{9}{2}$

25.
$$\dfrac{x-1}{x-5} = \dfrac{4}{x-5}, \text{ LCD} = x-5$$
$$(x-5) \cdot \dfrac{x-1}{x-5} = (x-5) \cdot \dfrac{4}{x-5}$$
$$x - 1 = 4$$
$$x = 5$$

Check:
$$\dfrac{x-1}{x-5} = \dfrac{4}{x-5}$$

$$\begin{array}{c|c} \dfrac{5-1}{5-5} \; ? \; \dfrac{4}{5-5} \\[2mm] \dfrac{4}{0} & \dfrac{4}{0} \quad \text{UNDEFINED} \end{array}$$

The number 5 is not a solution because it makes a denominator zero. Thus, there is no solution.

26. No solution

27.
$$\frac{2}{x+3} = \frac{5}{x}, \text{ LCD} = x(x+3)$$
$$x(x+3) \cdot \frac{2}{x+3} = x(x+3) \cdot \frac{5}{x}$$
$$2x = 5(x+3)$$
$$2x = 5x + 15$$
$$-15 = 3x$$
$$-5 = x$$

This checks, so the solution is -5.

28. -16

29.
$$\frac{x-2}{x-3} = \frac{x-1}{x+1},$$
$$\text{LCD} = (x-3)(x+1)$$
$$(x-3)(x+1) \cdot \frac{x-2}{x-3} = (x-3)(x+1) \cdot \frac{x-1}{x+1}$$
$$(x+1)(x-2) = (x-3)(x-1)$$
$$x^2 - x - 2 = x^2 - 4x + 3$$
$$-x - 2 = -4x + 3$$
$$3x = 5$$
$$x = \frac{5}{3}$$

This checks, so the solution is $\frac{5}{3}$.

30. $\frac{1}{5}$

31.
$$\frac{1}{x+3} + \frac{1}{x-3} = \frac{1}{x^2-9},$$
$$\text{LCD} = (x+3)(x-3)$$
$$(x+3)(x-3)\left(\frac{1}{x+3} + \frac{1}{x-3}\right) =$$
$$(x+3)(x-3) \cdot \frac{1}{(x+3)(x-3)}$$
$$(x-3) + (x+3) = 1$$
$$2x = 1$$
$$x = \frac{1}{2}$$

This checks, so the solution is $\frac{1}{2}$.

32. No solution

33.
$$\frac{x}{x+4} - \frac{4}{x-4} = \frac{x^2+16}{x^2-16},$$
$$\text{LCD} = (x+4)(x-4)$$
$$(x+4)(x-4)\left(\frac{x}{x+4} - \frac{x}{x-4}\right) =$$
$$(x+4)(x-4) \cdot \frac{x^2+16}{(x+4)(x-4)}$$
$$x(x-4) - 4(x+4) = x^2 + 16$$
$$x^2 - 4x - 4x - 16 = x^2 + 16$$
$$x^2 - 8x - 16 = x^2 + 16$$
$$-8x - 16 = 16$$
$$-8x = 32$$
$$x = -4$$

The number -4 is not a solution because it makes a denominator zero. Thus, there is no solution.

34. 2

35.
$$\frac{-3}{y-7} = \frac{-10-y}{7-y} \quad \begin{array}{l} y-7 \text{ and } 7-y \\ \text{are opposites} \end{array}$$
$$\frac{-3}{y-7} = \frac{-10-y}{-(y-7)}$$
$$\frac{-3}{y-7} = \frac{-(-10-y)}{y-7} \quad \text{Since } \frac{a}{-b} = \frac{-a}{b}$$
$$\frac{-3}{y-7} = \frac{10+y}{y-7}, \text{ LCD} = y-7$$
$$(y-7)\left(\frac{-3}{y-7}\right) = (y-7)\left(\frac{10+y}{y-7}\right)$$
$$-3 = 10 + y$$
$$-13 = y$$

This checks, so the solution is -13.

36. No solution

37. $(a^2b^5)^{-3} = \frac{1}{(a^2b^5)^3} = \frac{1}{(a^2)^3(b^5)^3} = \frac{1}{a^6b^{15}}$

38. x^8y^{12}

39. $\left(\frac{2x}{t^2}\right)^4 = \frac{(2x)^4}{(t^2)^4} = \frac{2^4x^4}{t^8} \stackrel{.}{=} \frac{16x^4}{t^8}$

40. $\frac{w^4}{y^6}$

41.

42.

43.
$$\frac{4}{y-2} - \frac{2y-3}{y^2-4} = \frac{5}{y+2},$$
$$\text{LCD} = (y+2)(y-2)$$

$$(y+2)(y-2)\left(\frac{4}{y-2} - \frac{2y-3}{(y+2)(y-2)}\right) =$$
$$(y+2)(y-2) \cdot \frac{5}{y+2}$$
$$4(y+2)-(2y-3) = 5(y-2)$$
$$4y+8-2y+3 = 5y-10$$
$$2y+11 = 5y-10$$
$$21 = 3y$$
$$7 = y$$

This checks, so the solution is 7.

44. $-\dfrac{1}{6}$

45.
$$\frac{12-6x}{x^2-4} = \frac{3x}{x+2} - \frac{2x-3}{x-2},$$
$$\text{LCD} = (x+2)(x-2)$$

$$(x+2)(x-2) \cdot \frac{12-6x}{(x+2)(x-2)} =$$
$$(x+2)(x-2)\left(\frac{3x}{x+2} - \frac{2x-3}{x-2}\right)$$
$$12-6x =$$
$$3x(x-2)-(x+2)(2x-3)$$
$$12-6x =$$
$$3x^2-6x-2x^2-x+6$$
$$0 = x^2-x-6$$
$$0 = (x-3)(x+2)$$

$$x-3 = 0 \quad \text{or} \quad x+2 = 0$$
$$x = 3 \quad \text{or} \qquad x = -2$$

The number 3 is a solution, but -2 is not because it makes denominators 0.

46. $0, -1$

47.
$$4a-3 = \frac{a+13}{a+1}, \quad \text{LCD} = a+1$$

$$(a+1)(4a-3) = (a+1)\cdot\frac{a+13}{a+1}$$
$$4a^2+a-3 = a+13$$
$$4a^2-16 = 0$$
$$4(a+2)(a-2) = 0$$
$$a+2 = 0 \quad \text{or} \quad a-2 = 0$$
$$a = -2 \quad \text{or} \qquad a = 2$$

Both of these check, so the two solutions are -2 and 2.

48. 2

49.
$$\frac{y^2-4}{y+3} = 2 - \frac{y-2}{y+3}, \quad \text{LCM} = y+3$$

$$(y+3)\cdot\frac{y^2-4}{y+3} = (y+3)\left(2 - \frac{y-2}{y+3}\right)$$
$$y^2-4 = 2(y+3)-(y-2)$$
$$y^2-4 = 2y+6-y+2$$
$$y^2-4 = y+8$$
$$y^2-y-12 = 0$$
$$(y-4)(y+3) = 0$$
$$y-4 = 0 \quad \text{or} \quad y+3 = 0$$
$$y = 4 \quad \text{or} \qquad y = -3$$

The number 4 is a solution, but -3 is not because it makes a denominator zero.

50. -6

51.

52.

Exercise Set 6.8

1. ***Familiarize***. Let $x =$ the number.

Translate.

A number	minus	twice its reciprocal	is	1.
↓	↓	↓	↓	↓
x	$-$	$2\cdot\dfrac{1}{x}$	$=$	1

Carry out.
$$x - \frac{2}{x} = 1$$
$$x\left(x - \frac{2}{x}\right) = x\cdot1 \quad \text{Multiplying by the LCD}$$
$$x^2-2 = x$$
$$x^2-x-2 = 0$$
$$(x-2)(x+1) = 0$$
$$x-2 = 0 \quad \text{or} \quad x+1 = 0$$
$$x = 2 \quad \text{or} \qquad x = -1$$

Check. Twice the reciprocal of 2 is $2\cdot\dfrac{1}{2}$, or 1. Since $2-1 = 1$, the number 2 is a solution. Twice the reciprocal of -1 is $2(-1)$, or -2. Since $-1-(-2) = -1+2$, or 1, the number -1 is a solution.

State. The solutions are 2 and -1.

2. $4, -1$

3. ***Familiarize***. Let $x =$ the number.

Translate. We reword the problem.

A number	plus	its reciprocal	is	2.
↓	↓	↓	↓	↓
x	$+$	$\dfrac{1}{x}$	$=$	2

Carry out. We solve the equation.

$$x + \frac{1}{x} = 2$$

$$x\left(x + \frac{1}{x}\right) = x \cdot 2 \quad \text{Multiplying by the LCD}$$

$$x^2 + 1 = 2x$$

$$x^2 - 2x + 1 = 0$$

$$(x-1)(x-1) = 0$$

$$x - 1 = 0 \quad \text{or} \quad x - 1 = 0$$

$$x = 1 \quad \text{or} \qquad x = 1$$

Check. The reciprocal of 1 is 1. Since $1 + 1 = 2$, the number 1 is a solution.

State. The solution is 1.

4. 1, 5

5. *Familiarize*. The job takes David 4 hours working alone and Sierra 5 hours working alone. Then in 1 hour David does $\frac{1}{4}$ of the job and Sierra does $\frac{1}{5}$ of the job. Working together, they can do $\frac{1}{4} + \frac{1}{5}$, or $\frac{9}{20}$ of the job in 1 hour. In two hours, David does $2\left(\frac{1}{4}\right)$ of the job and Sierra does $2\left(\frac{1}{5}\right)$ of the job. Working together they can do $2\left(\frac{1}{4}\right) + 2\left(\frac{1}{5}\right)$, or $\frac{9}{10}$ of the job in 2 hours. In 3 hours they can do $3\left(\frac{1}{4}\right) + 3\left(\frac{1}{5}\right)$, or $\frac{27}{20}$ or $1\frac{7}{20}$ of the job which is more of the job than needs to be done. The answer is somewhere between 2 hr and 3 hr.

Translate. If they work together t hours, then David does $t\left(\frac{1}{4}\right)$ of the job and Sierra does $t\left(\frac{1}{5}\right)$ of the job. We want some number t such that

$$t\left(\frac{1}{4}\right) + t\left(\frac{1}{5}\right) = 1.$$

Carry out. We solve the equation.

$$\frac{t}{4} + \frac{t}{5} = 1, \ \text{LCD} = 20$$

$$20\left(\frac{t}{4} + \frac{t}{5}\right) = 20 \cdot 1$$

$$5t + 4t = 20$$

$$9t = 20$$

$$t = \frac{20}{9}, \ \text{or} \ 2\frac{2}{9}$$

Check. The check can be done by repeating the computations. We also have a partial check in that we expected from our familiarization step that the answer would be between 2 hr and 3 hr.

State. Working together, it takes them $2\frac{2}{9}$ hr to complete the job.

6. $6\frac{6}{7}$ hr

7. *Familiarize*. The job takes Vern 45 min working alone and Nina 60 min working alone. Then in 1 minute Vern does $\frac{1}{45}$ of the job and Nina does $\frac{1}{60}$ of the job. Working together, they can do

$\frac{1}{45} + \frac{1}{60}$, or $\frac{7}{180}$ of the job in 1 minute. In 20 minutes, Vern does $\frac{20}{45}$ of the job and Nina does $\frac{20}{60}$ of the job. Working together, they can do $\frac{20}{45} + \frac{20}{60}$, or $\frac{7}{9}$ of the job. In 30 minutes, they can do $\frac{30}{45} + \frac{30}{60}$, or $\frac{7}{6}$ of the job which is more of the job than needs to be done. The answer is somewhere between 20 minutes and 30 minutes.

Translate. If they work together t minutes, then Vern does $t\left(\frac{1}{45}\right)$ of the job and Nina does $t\left(\frac{1}{60}\right)$ of the job. We want some number t such that

$$t\left(\frac{1}{45}\right) + t\left(\frac{1}{60}\right) = 1.$$

Carry out. We solve the equation.

$$\frac{t}{45} + \frac{t}{60} = 1, \ \text{LCD} = 180$$

$$180\left(\frac{t}{45} + \frac{t}{60}\right) = 180 \cdot 1$$

$$4t + 3t = 180$$

$$7t = 180$$

$$t = \frac{180}{7}, \ \text{or} \ 25\frac{5}{7}$$

Check. The check can be done by repeating the computations. We also have a partial check in that we expected from our familiarization step that the answer would be between 20 minutes and 30 minutes.

State. It would take them $25\frac{5}{7}$ minutes to complete the job working together.

8. $1\frac{5}{7}$ hr

9. *Familiarize*. The job takes Rory 12 hours working along and Mira 9 hours working alone. Then in 1 hour Rory does $\frac{1}{12}$ of the job and Mira does $\frac{1}{9}$ of the job. Working together they can do $\frac{1}{12} + \frac{1}{9}$, or $\frac{7}{36}$ of the job in 1 hour. In two hours, Rory does $2\left(\frac{1}{12}\right)$ of the job and Mira does $2\left(\frac{1}{9}\right)$ of the job. Working together they can do $2\left(\frac{1}{12}\right) + 2\left(\frac{1}{9}\right)$, or $\frac{14}{36}$ of the job in two hours. In 3 hours they can do $3\left(\frac{1}{12}\right) + 3\left(\frac{1}{9}\right)$, or $\frac{21}{36}$ of the job. In 5 hours, they can do $\frac{35}{36}$ of the job. In 6 hours, they can do $\frac{42}{36}$, or $1\frac{1}{6}$ of the job which is more of the job than needs to be done. The answer is somewhere between 5 hr and 6 hr.

Translate. If they work together t hours, then Rory does $t\left(\dfrac{1}{12}\right)$ of the job and Mira does $t\left(\dfrac{1}{9}\right)$ of the job. We want some number t such that

$$t\left(\frac{1}{12}\right) + t\left(\frac{1}{9}\right) = 1.$$

Carry out. We solve the equation.

$$\frac{t}{12} + \frac{t}{9} = 1, \; \text{LCD} = 36$$

$$36\left(\frac{t}{12} + \frac{t}{9}\right) = 36 \cdot 1$$

$$3t + 4t = 36$$

$$7t = 36$$

$$t = \frac{36}{7}, \text{ or } 5\frac{1}{7}$$

Check. The check can be done by repeating the computations. We also have a partial check in that we expected from our familiarization step that the answer would be between 5 hr and 6 hr.

State. Working together, it takes them $5\frac{1}{7}$ hr to complete the job.

10. $10\frac{2}{7}$ hr

11. **Familiarize.** Let t = the time it takes Red Bryck and Lotta Mudd to do the job working together.

Translate. We use the work principle.

$$\frac{t}{6} + \frac{t}{8} = 1$$

Carry out. We solve the equation.

$$\frac{t}{6} + \frac{t}{8} = 1, \; \text{LCD} = 24$$

$$24\left(\frac{t}{6} + \frac{t}{8}\right) = 24 \cdot 1$$

$$4t + 3t = 24$$

$$7t = 24$$

$$t = \frac{24}{7}, \text{ or } 3\frac{3}{7}$$

Check. We repeat the computations.

State. It would take them $3\frac{3}{7}$ hr to do the job working together.

12. $11\frac{1}{9}$ min

13. **Familiarize.** We complete the table shown in the text.

$$d \quad = \quad r \quad \cdot \quad t$$

	Distance	Speed	Time
Slow car	150	r	t
Fast car	350	$r+40$	t

Translate. We can replace the t's in the table above using the formula $t = d/r$.

	Distance	Speed	Time
Slow car	150	r	$\dfrac{150}{r}$
Fast car	350	$r+40$	$\dfrac{350}{r+40}$

Since times are the same for both cars, we have the equation

$$\frac{150}{r} = \frac{350}{r+40}.$$

Carry out. We multiply by the LCD, $r(r+40)$.

$$r(r+40) \cdot \frac{150}{r} = r(r+40) \cdot \frac{350}{r+40}$$

$$150(r+40) = 350r$$

$$150r + 6000 = 350r$$

$$6000 = 200r$$

$$30 = r$$

Check. If r is 30 km/h, then $r + 40$ is 70 km/h. The time for the slow car is 150/30, or 5 hr. The time for the fast car is 350/70, or 5 hr. The times are the same. The values check.

State. The speed of the slow car is 30 km/h, and the speed of the fast car is 70 km/h.

14. Slow car: 50 km/h, fast car: 80 km/h

15. **Familiarize.** We complete the table shown in the text.

$$d \quad = \quad r \quad \cdot \quad t$$

	Distance	Speed	Time
Freight	330	$r-14$	t
Passenger	400	r	t

Translate. We can replace the t's in the table above using the formula $t = d/r$.

	Distance	Speed	Time
Freight	330	$r-14$	$\dfrac{330}{r-14}$
Passenger	400	r	$\dfrac{400}{r}$

Since the times are the same for both trains, we have the equation

$$\frac{330}{r-14} = \frac{400}{r}.$$

Carry out. We multiply by the LCD, $r(r-14)$.

$$r(r-14) \cdot \frac{330}{r-14} = r(r-14) \cdot \frac{400}{r}$$

$$330r = 400(r-14)$$

$$330r = 400r - 5600$$

$$-70r = -5600$$

$$r = 80$$

Then substitute 80 for r in either equation to find t:

$$t = \frac{400}{r}$$

$$t = \frac{400}{80} \qquad \text{Substituting 80 for } r$$

$$t = 5$$

Check. If $r = 80$, then $r - 14 = 66$. In 5 hr the freight train travels $66 \cdot 5$, or 330 km/h, and the passenger train travels $80 \cdot 5$, or 400 km/h. The values check.

State. The speed of the passenger train is 80 km/h. The speed of the freight train is 66 km/h.

16. Passenger train: 80 km/h, freight train: 65 km/h

17. **Familiarize**. We let $t =$ the number of hours Dexter rode and organize the given information in a table.

	Distance	Speed	Time
Dexter	16	r	t
Gail	50	r	$t+3$

Translate. We can replace the r's in the table above using the formula $r = d/t$.

	Distance	Speed	Time
Dexter	16	$\dfrac{16}{t}$	t
Gail	50	$\dfrac{50}{t+3}$	$t+3$

Since the speeds are the same for both riders, we have the equation

$$\frac{16}{t} = \frac{50}{t+3}.$$

Carry out. We multiply by the LCD, $t(t+3)$.

$$t(t+3) \cdot \frac{16}{t} = t(t+3) \cdot \frac{50}{t+3}$$

$$16(t+3) = 50t$$

$$16t + 48 = 50t$$

$$48 = 34t$$

$$\frac{24}{17} = t \qquad \text{Dividing by 34 and simplifying}$$

Check. If $t = \frac{24}{17}$, then $t + 3 = \frac{24}{17} + 3$, or $\frac{75}{17}$, and $\frac{16}{t} = \frac{16}{\frac{24}{17}}$, or $\frac{34}{3}$. In $\frac{24}{17}$ hr, Dexter rides $\frac{34}{3} \cdot \frac{24}{17}$, or 16 mi, and in $\frac{75}{17}$ hr Gail rides $\frac{34}{3} \cdot \frac{75}{17}$, or 50 mi. The values check.

State. Dexter rides $\frac{24}{17}$, or $1\frac{7}{17}$ hr.

18. 2 hr

19. **Familiarize**. Let $t =$ the time it takes Caledonia to drive to town and organize the given information in a table.

	Distance	Speed	Time
Caledonia	15	r	t
Manley	20	r	$t+1$

Translate. We can replace the r's in the table above using the formula $r = d/t$.

	Distance	Speed	Time
Caledonia	15	$\dfrac{15}{t}$	t
Manley	20	$\dfrac{20}{t+1}$	$t+1$

Since the speeds are the same for both riders, we have the equation

$$\frac{15}{t} = \frac{20}{t+1}.$$

Carry out. We multiply by the LCD, $t(t+1)$.

$$t(t+1) \cdot \frac{15}{t} = t(t+1) \cdot \frac{20}{t+1}$$

$$15(t+1) = 20t$$

$$15t + 15 = 20t$$

$$15 = 5t$$

$$3 = t$$

Check. If $t = 3$, then $t + 1 = 3 + 1$, or 4, and $\frac{15}{t} = \frac{15}{3}$, or 5. In 3 hr, Caledonia drives $5 \cdot 3$, or 15 mi, and in 4 hr Manley drives $5 \cdot 4$, or 20 mi. The values check.

State. It takes Caledonia 3 hr to drive to town.

20. $1\frac{1}{3}$ hr

21. $\dfrac{54 \text{ days}}{6 \text{ days}} = 9$

22. $16\dfrac{\text{mi}}{\text{gal}}$

23. $\dfrac{4.6 \text{ km}}{2 \text{ hr}} = 2.3$ km/h

24. 186,000 mi/sec

25. **Familiarize**. The coffee beans from 14 trees are required to produce 7.7 kilograms of coffee, and we wish to find how many trees are required to produce 320 kilograms of coffee. We can set up ratios:

$$\frac{T}{320} \qquad \frac{14}{7.7}$$

Translate. Assuming the two ratios are the same, we can translate to a proportion.

$$\begin{array}{r} \text{Trees} \rightarrow \\ \text{Kilograms} \rightarrow \end{array} \frac{T}{320} = \frac{14}{7.7} \begin{array}{l} \leftarrow \text{Trees} \\ \leftarrow \text{Kilograms} \end{array}$$

Carry out. We solve the proportion.

$$320 \cdot \frac{T}{320} = 320 \cdot \frac{14}{7.7}$$

$$T = \frac{4480}{7.7}, \text{ or } \frac{4480}{\frac{77}{10}}$$

$$T = 581\frac{9}{11}$$

Check. $\dfrac{581\frac{9}{11}}{320} = \dfrac{\frac{6400}{11}}{320} = \dfrac{6400}{11} \cdot \dfrac{1}{320} = \dfrac{320 \cdot 20}{11 \cdot 320} =$

$\dfrac{20}{11}$ and $\dfrac{14}{7.7} = \dfrac{14}{\frac{77}{10}} = \dfrac{14}{1} \cdot \dfrac{10}{77} = \dfrac{7 \cdot 2 \cdot 10}{7 \cdot 11} = \dfrac{20}{11}$

The ratios are the same.

State. 582 trees are required to produce 320 kg of coffee. (We round to the nearest whole number.)

26. 200

27. *Familiarize.* Wanda walked 234 kilometers in 14 days, and we wish to find how far she would walk in 42 days. We can set up ratios:

$$\frac{K}{42} \qquad \frac{234}{14}$$

Translate. Assuming the rates are the same, we can translate to a proportion.

$$\begin{array}{c}\text{Kilometers} \rightarrow \\ \text{Days} \rightarrow\end{array} \frac{K}{42} = \frac{234}{14} \begin{array}{c}\leftarrow \text{Kilometers} \\ \leftarrow \text{Days}\end{array}$$

Carry out. We solve the equation.

We multiply by 42 to get K alone.

$$42 \cdot \frac{K}{42} = 42 \cdot \frac{234}{14}$$

$$K = \frac{9828}{14}$$

$$K = 702$$

Check.

$\dfrac{702}{42} \approx 16.7$ and $\dfrac{234}{14} \approx 16.7.$

The ratios are the same.

State. Wanda would travel 702 kilometers in 42 days.

28. $21\frac{2}{3}$ cups

29. *Familiarize.* 10 cm^3 of human blood contains 1.2 grams of hemoglobin, and we wish to find how many grams of hemoglobin are contained in 16 cm^3 of the same blood. We can set up ratios:

$$\frac{H}{16} \qquad \frac{1.2}{10}$$

Translate. Assuming the rates are the same, we can translate to a proportion.

$$\begin{array}{c}\text{Grams} \rightarrow \\ \text{cm}^3 \rightarrow\end{array} \frac{H}{16} = \frac{1.2}{10} \begin{array}{c}\leftarrow \text{Grams} \\ \leftarrow \text{cm}^3\end{array}$$

Carry out. We solve the proportion.

We multiply by 16 to get H alone.

$$16 \cdot \frac{H}{16} = 16 \cdot \frac{1.2}{10}$$

$$H = \frac{19.2}{10}$$

$$H = 1.92$$

Check.

$\dfrac{1.92}{16} = 0.12$ and $\dfrac{1.2}{10} = 0.12.$

The ratios are the same.

State. Thus 16 cm^3 of the same blood would contain 1.92 grams of hemoglobin.

30. 216

31. We write a proportion and then solve it.

$$\frac{b}{6} = \frac{7}{4}$$

$$b = \frac{7}{4} \cdot 6$$

$$b = \frac{42}{4}, \text{ or } 10.5$$

$\left(\text{Note that the proportions } \dfrac{6}{b} = \dfrac{4}{7}, \dfrac{b}{7} = \dfrac{6}{4}, \text{ or } \dfrac{7}{b} = \dfrac{4}{6} \text{ could} \right.$

$\left. \text{also be used.}\right)$

32. 6.75

33. We write a proportion and then solve it.

$$\frac{4}{f} = \frac{6}{4}$$

$$4 \cdot 4 = 6f \quad \text{Cross multiplying}$$

$$\frac{4 \cdot 4}{6} = f \quad \text{Dividing by 6}$$

$$\frac{8}{3} = f \quad \text{Simplifying}$$

$\left(\text{One of the following proportions could also be used:}\right.$

$\left. \dfrac{f}{4} = \dfrac{4}{6}, \dfrac{4}{f} = \dfrac{9}{6}, \dfrac{f}{4} = \dfrac{6}{9}, \dfrac{4}{9} = \dfrac{f}{6}, \dfrac{9}{4} = \dfrac{6}{f}\right)$

34. 7.5

35. We write a proportion and then solve it.

$$\frac{8}{5} = \frac{10}{n}$$

$$8n = 50 \quad \text{Cross multiplying}$$

$$n = \frac{50}{8}, \text{ or } 6.25$$

$\left(\text{One of the following proportions could also be used:}\right.$

$\left. \dfrac{5}{8} = \dfrac{n}{10}, \dfrac{8}{10} = \dfrac{5}{n}, \dfrac{10}{8} = \dfrac{n}{5}\right)$

36. $\dfrac{35}{3}$

37. *Familiarize*. The ratio of trout tagged to the total number of trout in the lake, T, is $\frac{112}{T}$. Of the 82 trout caught later, there were 32 tagged trout. The ratio of tagged trout to trout caught is $\frac{32}{82}$.

***Translate*.** Assuming the two ratios are the same, we can translate to a proportion.

$$\begin{array}{l}\text{Trout tagged} \\ \text{originally} \\ \text{Trout} \\ \text{in lake}\end{array} \begin{array}{l}\longrightarrow \\ \longrightarrow\end{array} \frac{112}{T} = \frac{32}{82} \begin{array}{l}\longleftarrow \\ \longleftarrow\end{array} \begin{array}{l}\text{Tagged trout} \\ \text{caught later} \\ \text{Trout} \\ \text{caught later}\end{array}$$

***Carry out*.** We solve the proportion.

$$82T \cdot \frac{112}{T} = 82T \cdot \frac{32}{82}$$
$$82 \cdot 112 = T \cdot 32$$
$$9184 = 32T$$
$$\frac{9184}{32} = T$$
$$287 = T$$

***Check*.**
$$\frac{112}{287} \approx 0.39 \qquad \frac{32}{82} \approx 0.39$$
The ratios are the same.

***State*.** We estimate that there are 287 trout in the lake.

38. 954

39. *Familiarize*. The ratio of deer tagged to the total number of deer in the forest, D, is $\frac{612}{D}$. Of the 244 deer caught later, 72 are tagged. The ratio of tagged deer to deer caught is $\frac{72}{244}$.

***Translate*.** We translate to a proportion.

$$\begin{array}{l}\text{Deer originally} \\ \text{tagged} \\ \text{Deer} \\ \text{in forest}\end{array} \begin{array}{l}\rightarrow \\ \rightarrow\end{array} \frac{612}{D} = \frac{72}{244} \begin{array}{l}\leftarrow \\ \leftarrow\end{array} \begin{array}{l}\text{Tagged deer} \\ \text{caught later} \\ \text{Deer} \\ \text{caught later}\end{array}$$

***Carry out*.** We solve the proportion. We multiply by the LCD, $244D$.

$$244D \cdot \frac{612}{D} = 244D \cdot \frac{72}{244}$$
$$244 \cdot 612 = D \cdot 72$$
$$149{,}328 = 72D$$
$$\frac{149{,}328}{72} = D$$
$$2074 = D$$

***Check*.**
$$\frac{612}{2074} \approx 0.295 \qquad \frac{72}{244} \approx 0.295$$
The ratios are the same.

***State*.** We estimate that there are 2074 deer in the forest.

40. 42

41. *Familiarize*. Let $D =$ the number of "duds" in a sample of 320 firecrackers. We set up two ratios:

$$\frac{9}{144} \qquad \frac{D}{320}$$

***Translate*.** We write a proportion.

$$\begin{array}{r}\text{"Duds"} \rightarrow \\ \text{Firecrackers} \rightarrow\end{array} \frac{9}{144} = \frac{D}{320} \begin{array}{l}\leftarrow \text{"Duds"} \\ \leftarrow \text{Firecrackers}\end{array}$$

***Carry out*.** We solve the proportion. We multiply by 320 to get D alone.

$$320 \cdot \frac{9}{144} = 320 \cdot \frac{D}{320}$$
$$\frac{2880}{144} = D$$
$$20 = D$$

***Check*.**
$$\frac{9}{144} = 0.0625 \qquad \frac{20}{320} = 0.0625$$
The ratios are the same.

***State*.** You would expect 20 "duds" in a sample of 320 firecrackers.

42. 225

43. *Familiarize*. The ratio of the weight of an object on the moon to the weight of an object on the earth is 0.16 to 1.

a) We wish to find how much a 12-ton rocket would weigh on the moon.

b) We wish to find how much a 180-lb astronaut would weigh on the moon.

We can set up ratios.

$$\frac{0.16}{1} \qquad \frac{T}{12} \qquad \frac{P}{180}$$

***Translate*.** Assuming the ratios are the same, we can translate to proportions.

a)
$$\begin{array}{l}\text{Weight} \\ \text{on moon} \rightarrow \\ \text{Weight} \rightarrow \\ \text{on earth}\end{array} \frac{0.16}{1} = \frac{T}{12} \begin{array}{l}\text{Weight} \\ \leftarrow \text{on moon} \\ \leftarrow \text{Weight} \\ \text{on earth}\end{array}$$

b)
$$\begin{array}{l}\text{Weight} \\ \text{on moon} \rightarrow \\ \text{Weight} \rightarrow \\ \text{on earth}\end{array} \frac{0.16}{1} = \frac{P}{180} \begin{array}{l}\text{Weight} \\ \leftarrow \text{on moon} \\ \leftarrow \text{Weight} \\ \text{on earth}\end{array}$$

***Carry out*.** We solve each proportion.

a) $\quad \dfrac{0.16}{1} = \dfrac{T}{12} \qquad$ b) $\quad \dfrac{0.16}{1} = \dfrac{P}{180}$

$\qquad 12(0.16) = T \qquad\qquad 180(0.16) = P$

$\qquad\quad 1.92 = T \qquad\qquad\quad 28.8 = P$

***Check*.** $\dfrac{0.16}{1} = 0.16$, $\dfrac{1.92}{12} = 0.16$, and $\dfrac{28.8}{180} = 0.16$.
The ratios are the same.

***State*.**

a) A 12-ton rocket would weigh 1.92 tons on the moon.

b) A 180-lb astronaut would weigh 28.8 lb on the moon.

44. a) 4.8 tons

b) 48 lb

45. ***Familiarize***. Let x represent the numerator. Then $104-x$ represents the denominator. The ratio is $\dfrac{x}{104-x}$.

Translate. The ratios are equal.

$$\frac{x}{104-x} = \frac{9}{17}$$

Carry out. We solve the proportion.

We multiply by the LCD, $17(104-x)$.

$$17(104-x) \cdot \frac{x}{104-x} = 17(104-x) \cdot \frac{9}{17}$$
$$17x = 9(104-x)$$
$$17x = 936 - 9x$$
$$26x = 936$$
$$x = \frac{936}{26}$$
$$x = 36$$

Check. If $x = 36$, then $104 - x = 68$. The ratio is $\dfrac{36}{68}$. If we multiply $\dfrac{9}{17}$ by $\dfrac{4}{4}$, a form of 1, we get $\dfrac{36}{68}$. The ratios are equal.

State. The equal ratio is $\dfrac{36}{68}$.

46. 11

47. $(x+2) - (x+1) = x + 2 - x - 1 = 1$

48. $x^2 - 1$

49. $(4y^3 - 5y^2 + 7y - 24) - (-9y^3 + 9y^2 - 5y + 49)$
$= 4y^3 - 5y^2 + 7y - 24 + 9y^3 - 9y^2 + 5y - 49$
$= 13y^3 - 14y^2 + 12y - 73$

50. $25{,}704$ ft^2

51.

52.

53. ***Familiarize***. Let x represent the numerator and $x+1$ represent the denominator of the original fraction. The fraction is $\dfrac{x}{x+1}$. If 2 is subtracted from the numerator and the denominator, the resulting fraction is $\dfrac{x-2}{x+1-2}$, or $\dfrac{x-2}{x-1}$.

Translate.

The resulting fraction is $\dfrac{1}{2}$.

$$\frac{x-2}{x-1} = \frac{1}{2}$$

Carry out. We solve the equation.

$$\frac{x-2}{x-1} = \frac{1}{2}, \quad \text{LCD} = 2(x-1)$$
$$2(x-1) \cdot \frac{x-2}{x-1} = 2(x-1) \cdot \frac{1}{2}$$
$$2(x-2) = x-1$$
$$2x - 4 = x - 1$$
$$x = 3$$

Check. If $x = 3$, then $x + 1 = 4$ and the original fraction is $\dfrac{3}{4}$. If 2 is subtracted from both numerator and denominator, the resulting fraction is $\dfrac{3-2}{4-2}$, or $\dfrac{1}{2}$. The value checks.

State. The original fraction was $\dfrac{3}{4}$.

54. Ann: 6 hr, Betty: 12 hr

55. ***Familiarize***. We organize the information in a table. Let r = the speed of the current and t = the time it takes to travel upstream.

	Distance	Speed	Time
	d	$=$ r	$\cdot$ t
Upstream	24	$10-r$	t
Downstream	24	$10+r$	$5-t$

Translate. From the rows of the table we get two equations:

$$24 = (10-r)t$$
$$24 = (10+r)(5-t)$$

We solve each equation for t and set the results equal:

Solving $24 = (10-r)t$ for t: $t = \dfrac{24}{10-r}$

Solving $24 = (10+r)(5-t)$ for t:

$$\frac{24}{10+r} = 5 - t$$
$$t = 5 - \frac{24}{10+r}$$

Then $\dfrac{24}{10-r} = 5 - \dfrac{24}{10+r}$.

Carry out. We first multiply on both sides of the equation by the LCD, $(10-r)(10+r)$:

$$(10-r)(10+r) \cdot \frac{24}{10-r} =$$
$$(10-r)(10+r)\left(5 - \frac{24}{10+r}\right)$$
$$24(10+r) =$$
$$5(10-r)(10+r) - 24(10-r)$$
$$240 + 24r = 500 - 5r^2 - 240 + 24r$$
$$240 + 24r = 260 - 5r^2 + 24r$$
$$5r^2 - 20 = 0$$
$$5(r^2 - 4) = 0$$
$$5(r+2)(r-2) = 0$$
$$r + 2 = 0 \quad \text{or} \quad r - 2 = 0$$
$$r = -2 \quad \text{or} \quad r = 2$$

Check. We only check 2 since the speed of the current cannot be negative. If $r = 2$, then the speed upstream is $10 - 2$, or 8 mph and the time is $\frac{24}{8}$, or 3 hours. If $r = 2$, then the speed downstream is $10 + 2$, or 12 mph and the time is $\frac{24}{12}$, or 2 hours. The sum of 3 hr and 2 hr is 5 hr. This checks.

State. The speed of the current is 2 mph.

56. $75 + 25$

57. Find a second proportion:

$$\frac{A}{B} = \frac{C}{D} \qquad \text{Given}$$

$$\frac{D}{A} \cdot \frac{A}{B} = \frac{D}{A} \cdot \frac{C}{D} \qquad \text{Multiplying by } \frac{D}{A}$$

$$\frac{D}{B} = \frac{C}{A}$$

Find a third proportion:

$$\frac{A}{B} = \frac{C}{D} \qquad \text{Given}$$

$$\frac{B}{C} \cdot \frac{A}{B} = \frac{B}{C} \cdot \frac{C}{D} \qquad \text{Multiplying by } \frac{B}{C}$$

$$\frac{A}{C} = \frac{B}{D}$$

Find a fourth proportion:

$$\frac{A}{B} = \frac{C}{D} \qquad \text{Given}$$

$$\frac{DB}{AC} \cdot \frac{A}{B} = \frac{DB}{AC} \cdot \frac{C}{D} \qquad \text{Multiplying by } \frac{DB}{AC}$$

$$\frac{D}{C} = \frac{B}{A}$$

58. $27\frac{3}{11}$ minutes after 5:00

59. **Familiarize**. The job takes Rosina 8 days working alone and Ng 10 days working alone. Let x represent the number of days it would take Oscar working alone. Then in 1 day Rosina does $\frac{1}{8}$ of the job, Ng does $\frac{1}{10}$ of the job, and Oscar does $\frac{1}{x}$ of the job. In 1 day they would complete $\frac{1}{8} + \frac{1}{10} + \frac{1}{x}$ of the job, and in 3 days they would complete $3\left(\frac{1}{8} + \frac{1}{10} + \frac{1}{x}\right)$, or $\frac{3}{8} + \frac{3}{10} + \frac{3}{x}$.

Translate. The amount done in 3 days is one entire job, so we have

$$\frac{3}{8} + \frac{3}{10} + \frac{3}{x} = 1.$$

Carry out. We solve the equation.

$$\frac{3}{8} + \frac{3}{10} + \frac{3}{x} = 1, \text{ LCD} = 40x$$

$$40x\left(\frac{3}{8} + \frac{3}{10} + \frac{3}{x}\right) = 40x \cdot 1$$

$$40x \cdot \frac{3}{8} + 40x \cdot \frac{3}{10} + 40x \cdot \frac{3}{x} = 40x$$

$$15x + 12x + 120 = 40x$$

$$120 = 13x$$

$$\frac{120}{13} = x$$

Check. If it takes Oscar $\frac{120}{13}$, or $9\frac{3}{13}$ days, to complete the job, then in one day Oscar does $\frac{1}{\frac{120}{13}}$, or $\frac{13}{120}$, of the job, and in 3 days he does $3\left(\frac{13}{120}\right)$, or $\frac{13}{40}$, of the job. The portion of the job done by Rosina, Ng, and Oscar in 3 days is

$$\frac{3}{8} + \frac{3}{10} + \frac{13}{40} = \frac{15}{40} + \frac{12}{40} + \frac{13}{40} = \frac{40}{40} = 1 \text{ entire job}.$$

The answer checks.

State. It will take Oscar $9\frac{3}{13}$ days to write the program working alone.

60. Michelle: 6 hr, Sal: 3 hr, Kristen: 4 hr

61. **Familiarize**. We organize the information in a table. Let $r =$ the speed on the first part of the trip and $t =$ the time driven at that speed.

$$d = r \cdot t$$

	Distance	Speed	Time
First part	30	r	t
Second part	30	$r + 15$	$1 - t$

Translate. From the rows of the table we obtain two equations:

$$30 = rt$$

$$30 = (r + 15)(1 - t)$$

We solve each equation for t and set the results equal:

Solving $30 = rt$ for t: $t = \frac{30}{r}$

Solving $20 = (r + 15)(1 - t)$ for t:

$$\frac{20}{r + 15} = 1 - t$$

$$t = 1 - \frac{20}{r + 15}$$

Then $\dfrac{30}{r} = 1 - \dfrac{20}{r + 15}$.

Carry out. We first multiply the equation by the LCD, $r(r + 15)$:

$$r(r + 15) \cdot \frac{30}{r} = r(r + 15)\left(1 - \frac{20}{r + 15}\right)$$

$$30(r + 15) = r(r + 15) - 20r$$

$$30r + 450 = r^2 + 15r - 20r$$

$$0 = r^2 - 35r - 450$$

$$0 = (r - 45)(r + 10)$$

$$r - 45 = 0 \quad \text{or} \quad r + 10 = 0$$

$$r = 45 \quad \text{or} \qquad r = -10$$

Check. Since the speed cannot be negative, we only check 45. If $r = 45$, then the time for the first part is $\frac{30}{45}$, or $\frac{2}{3}$ hr. If $r = 45$, then $r + 15 = 60$ and the time for the second part is $\frac{20}{60}$, or $\frac{1}{3}$ hr. The total time is $\frac{2}{3} + \frac{1}{3}$, or 1 hour. The value checks.

State. The speed for the first 30 miles was 45 mph.

62.

63. Let p = the width of the pond. We have similar triangles:

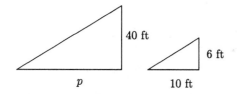

We write a proportion and solve it.

$$\frac{p}{10} = \frac{40}{6}$$

$$p = \frac{40 \cdot 10}{6} \qquad \text{Multiplying by 10}$$

$$p = \frac{400}{6}$$

$$p = \frac{200}{3}, \text{ or } 66\frac{2}{3}$$

The pond is $66\frac{2}{3}$ ft wide.

Exercise Set 6.9

1.
$$S = 2\pi rh$$

$$\frac{S}{2\pi h} = r \qquad \text{Multiplying by } \frac{1}{2\pi h}$$

2. $t = \dfrac{A - P}{Pr}$

3.
$$A = \frac{1}{2}bh$$

$$2A = bh \quad \text{Multiplying by 2}$$

$$\frac{2A}{h} = b \quad \text{Multiplying by } \frac{1}{h}$$

4. $g = \dfrac{2s}{t^2}$

5.
$$\frac{1}{180} = \frac{n - 2}{s}$$

$$\frac{s}{180} = n - 2 \quad \text{Multiplying by } s$$

$$\frac{s}{180} + 2 = n, \qquad \text{Adding 2}$$

$$\text{or } \frac{s + 360}{180} = n$$

6. $a = \dfrac{2S - nl}{n}$

7.
$$V = \frac{1}{3}k(B + b + 4M)$$

$$3V = k(B + b + 4M) \quad \text{Multiplying by 3}$$

$$3V = kB + kb + 4kM \quad \text{Removing parentheses}$$

$$3V - kB - 4kM = kb \qquad \text{Adding } -kb - 4kM$$

$$\frac{3V - kB - 4kM}{k} = b \qquad \text{Multiplying by } \frac{1}{k}$$

8. $P = \dfrac{A}{1 + rt}$

9.
$$rl - rS = L$$

$$r(l - S) = L \quad \text{Factoring out } r$$

$$r = \frac{L}{l - S} \quad \text{Dividing by } l - S$$

10. $m = \dfrac{T}{g - f}$

11.
$$A = \frac{1}{2}h(b_1 + b_2)$$

$$2A = h(b_1 + b_2) \quad \text{Multiplying by 2}$$

$$\frac{2A}{b_1 + b_2} = h \qquad \text{Multiplying by } \frac{1}{b_1 + b_2}$$

12. $h = \dfrac{S}{2\pi r} - r$, or $\dfrac{S - 2\pi r^2}{2\pi r}$

13.
$$ab = ac + d$$

$$ab - ac = d \qquad \text{Subtracting } ac \text{ to get all terms involving } a \text{ on one side}$$

$$a(b - c) = d \qquad \text{Factoring out } a$$

$$a = \frac{d}{b - c} \qquad \text{Dividing by } b - c$$

14. $n = \dfrac{p}{p - m}$

15.
$$\frac{r}{p} = q$$

$$r = pq \quad \text{Multiplying by } p$$

$$\frac{r}{q} = p \quad \text{Dividing by } q$$

16. $v = \dfrac{m}{n}$

17. $\quad a + b = \dfrac{c}{d}$

$\quad d(a+b) = c \qquad$ Multiplying by d

$\quad\quad d = \dfrac{c}{a+b} \qquad$ Dividing by $a+b$

18. $n = \dfrac{m}{p-q}$

19. $\quad \dfrac{x-y}{z} = p + q$

$\quad x - y = z(p+q) \qquad$ Multiplying by z

$\quad \dfrac{x-y}{p+q} = z \qquad$ Dividing by $p+q$

20. $t = \dfrac{M-g}{r+s}$

21. $\qquad \dfrac{1}{p} + \dfrac{1}{q} = \dfrac{1}{f}$

$\quad pqf\left(\dfrac{1}{p} + \dfrac{1}{q}\right) = pqf \cdot \dfrac{1}{f} \qquad$ Multiplying by the LCD, pqf, to clear fractions

$\quad pqf \cdot \dfrac{1}{p} + pqf \cdot \dfrac{1}{q} = pq$

$\qquad\qquad qf + pf = pq$

$\qquad\qquad f(q+p) = pq$

$\qquad\qquad\qquad f = \dfrac{pq}{q+p}$

22. $R = \dfrac{r_1 r_2}{r_2 + r_1}$

23. $\qquad r = \dfrac{v^2 pL}{a}$

$\quad ar = v^2 pL \qquad$ Multiplying by a

$\quad \dfrac{ar}{v^2 L} = p \qquad$ Multiplying by $\dfrac{1}{v^2 L}$

24. $l = \dfrac{P - 2w}{2}$

25. $\qquad \dfrac{a}{c} = n + bn$

$\qquad \dfrac{a}{c} = n(1+b) \qquad$ Factoring out n

$\quad \dfrac{a}{c(1+b)} = n \qquad$ Multiplying by $\dfrac{1}{a+b}$

26. $a = \dfrac{Q}{M(b-c)}$

27. $\qquad S = \dfrac{a + 2b}{3b}$

$\quad 3bS = a + 2b \qquad$ Multiplying by $3b$

$\quad 3bS - 2b = a \qquad$ Adding $-2b$

$\quad b(3S - 2) = a \qquad$ Factoring out b

$\qquad\quad b = \dfrac{a}{3S - 2} \qquad$ Multiplying by $\dfrac{1}{3S - 2}$

28. $a = \dfrac{b}{K - C}$

29. $\qquad C = \dfrac{5}{9}(F - 32)$

$\quad\quad 9C = 5(F - 32) \qquad$ Multiplying by 9

$\quad\quad 9C = 5F - 160 \qquad$ Removing parentheses

$\quad 9C + 160 = 5F \qquad\qquad$ Adding 160

$\quad \dfrac{9C + 160}{5} = F \qquad\qquad$ Multiplying by $\dfrac{1}{5}$

30. $r^3 = \dfrac{3V}{4\pi}$

31. $\qquad \dfrac{W_1}{W_2} = \dfrac{d_1}{d_2}$

$\quad d_2 W_1 = d_1 W_2 \qquad$ Multiplying by $W_2 d_2$

$\quad\quad d_2 = \dfrac{d_1 W_2}{W_1} \qquad$ Dividing by W_1

32. $W_2 = \dfrac{d_2 W_1}{d_1}$

33. $\qquad S = \dfrac{a - ar^n}{1 - r}$

$\quad S(1 - r) = a - ar^n \qquad$ Multiplying by $1 - r$

$\quad S(1 - r) = a(1 - r^n) \qquad$ Factoring out a

$\quad \dfrac{S(1-r)}{1 - r^n} = a \qquad\qquad$ Dividing by $1 - r^n$

34. $r = \dfrac{a - S}{-S}$, or $\dfrac{S - a}{S}$

35. $\qquad f = \dfrac{gm - t}{m}$

$\quad\quad fm = gm - t \qquad$ Multiplying by m

$\quad fm - gm = -t \qquad$ Subtracting gm

$\quad m(f - g) = -t \qquad$ Factoring out m

$\qquad\quad m = \dfrac{-t}{f - g}, \qquad$ Dividing by $f - g$

$\quad$ or $m = \dfrac{t}{g - f} \qquad$ Since $\dfrac{-a}{b} = \dfrac{a}{-b}$

36. $r = \dfrac{Sl - a}{S - l}$

37. ***Familiarize and Translate***. We use the formula given.

Carry out. First solve the formula for R.

$\qquad A = \dfrac{9R}{I}$

$\qquad AI = 9R$

$\qquad \dfrac{AI}{9} = R$

Then substitute 2.4 for A and 45 for I.

$\qquad \dfrac{2.4(45)}{9} = R$

$\qquad\quad 12 = R$

Check. We substitute in the original formula.

$$A = 9R/I$$

2.4 ?	$\dfrac{9 \cdot 12}{45}$
	$\dfrac{108}{45}$
2.4 $\mid$ 2.4	TRUE

The answer checks.

State. 12 earned runs were given up.

38. $5\dfrac{5}{23}$ ohms

39. Familiarize. We want to find r_2 using the formula $\dfrac{1}{R} = \dfrac{1}{r_1} + \dfrac{1}{r_2}$. We know $R = 5$ ohms and $r_1 = 50$ ohms.

Translate. Using a result from Example 5, we have

$$r_2 = \frac{Rr_1}{r_1 - R}.$$

Carry out. We substitute and compute.

$$r_2 = \frac{5 \cdot 50}{50 - 5}$$

$$= \frac{250}{45}$$

$$= \frac{50}{9}, \text{ or } 5\frac{5}{9}$$

Check. We substitute into the original formula.

$$\frac{1}{R} = \frac{1}{r_1} + \frac{1}{r_2}$$

$\dfrac{1}{5}$?	$\dfrac{1}{50} + \dfrac{1}{\frac{50}{9}}$
	$\dfrac{1}{50} + \dfrac{9}{50}$
	$\dfrac{10}{50}$
$\dfrac{1}{5}$ $\mid$ $\dfrac{1}{5}$	TRUE

The answer checks.

State. A resistor with a resistance of $5\dfrac{5}{9}$ ohms should be used.

40. 6 cm

41.

$$P = \frac{A}{1+r}$$

$P(1+r) = A$	Multiplying by $1+r$
$P + Pr = A$	Multiplying
$Pr = A - P$	Subtracting P
$r = \dfrac{A-P}{P}$	Dividing by P

42. 7%

43.

$$-\frac{3}{5}x = \frac{9}{20}$$

$$-\frac{5}{3}\left(-\frac{3}{5}x\right) = -\frac{5}{3} \cdot \frac{9}{20}$$

$$x = -\frac{45}{60}, \text{ or } -\frac{3}{4}$$

The solution is $-\dfrac{3}{4}$.

44. y-intercept: $(0,6)$, x-intercept: $(8,0)$

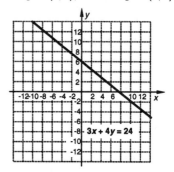

45. $x^2 - 13x - 30$

We look for two factors of -30 whose sum is -13. The numbers we need are -15 and 2.

$$x^2 - 13x - 30 = (x - 15)(x + 2)$$

46. $-3x^3 - 5x^2 + 5$

47. ⬙

48. ⬙

49.

$$u = -F\left(E - \frac{P}{T}\right)$$

$u = -EF + \dfrac{FP}{T}$	Removing parentheses
$T \cdot u = T\left(-EF + \dfrac{FP}{T}\right)$	Multiplying by T
$Tu = -EFT + FP$	
$Tu + EFT = FP$	Adding EFT
$T(u + EF) = FP$	Factoring out T
$T = \dfrac{FP}{u + EF}$	Dividing by $u + EF$

50. $n_2 = \dfrac{n_1 p_2 R + p_1 p_2 n_1}{p_1 p_2 - p_1 R}$

51. When $C = F$, we have

$$C = \frac{5}{9}(C - 32)$$

$9C = 5(C - 32)$ Multiplying by 9

$9C = 5C - 160$ Removing parentheses

$4C = -160$ Subtracting $5C$

$C = -40$ Dividing by 4

At $-40°$ the Fahrenheit and Celsius readings are the same.

52. N decreases when c increases; N increases when c decreases.

53.

$$I_t = \frac{I_f}{I - T}$$

$(1 - T)I_t = I_f$ Multiplying by $(1 - T)$

$1 - T = \dfrac{I_f}{I_t}$ Multiplying by $\dfrac{1}{I_t}$

$-T = \dfrac{I_f}{I_t} - 1$ Adding -1

$-1 \cdot (-T) = -1 \cdot \left(\dfrac{I_f}{I_t} - 1\right)$ Multiplying by -1

$T = -\dfrac{I_f}{I_t} + 1$

54. 28%

Chapter 7

Graphs And Functions

Exercise Set 7.1

1. We can use any two points on the line, such as $(0, 1)$ and $(3, 3)$.

$$m = \frac{\text{change in } y}{\text{change in } x}$$

$$= \frac{3 - 1}{3 - 0} = \frac{2}{3}$$

2. $\frac{3}{4}$

3. We can use any two points on the line, such as $(0, -2)$ and $(2, 0)$.

$$m = \frac{\text{change in } y}{\text{change in } x}$$

$$= \frac{0 - (-2)}{2 - 0} = \frac{2}{2} = 1$$

4. 2

5. We can use any two points on the line, such as $(0, 2)$ and $(3, 3)$.

$$m = \frac{\text{change in } y}{\text{change in } x}$$

$$= \frac{3 - 2}{3 - 0} = \frac{1}{3}$$

6. $\frac{3}{2}$

7. We can use any two points on the line, such as $(-3, 0)$ and $(-2, 3)$.

$$m = \frac{\text{change in } y}{\text{change in } x}$$

$$= \frac{3 - 0}{-2 - (-3)} = \frac{3}{1} = 3$$

8. $\frac{1}{3}$

9. We can use any two points on the line, such as $(0, 3)$ and $(4, 1)$.

$$m = \frac{\text{change in } y}{\text{change in } x}$$

$$= \frac{1 - 3}{4 - 0} = \frac{-2}{4} = -\frac{1}{2}$$

10. -1

11. We can use any two points on the line, such as $(-1, 3)$ and $(3, -3)$.

$$m = \frac{\text{change in } y}{\text{change in } x}$$

$$= \frac{-3 - 3}{3 - (-1)} = \frac{-6}{4} = -\frac{3}{2}$$

12. $-\frac{1}{3}$

13. We can use any two points on the line, such as $(2, 4)$ and $(4, 0)$.

$$m = \frac{\text{change in } y}{\text{change in } x}$$

$$= \frac{0 - 4}{4 - 2} = \frac{-4}{2} = -2$$

14. 0

15. This is a vertical line, so the slope is undefined. If we did not recognize this, we could use any two points on the line and attempt to compute the slope. We use $(2, 0)$ and $(2, 2)$.

$$m = \frac{\text{change in } y}{\text{change in } x}$$

$$= \frac{2 - 0}{2 - 2}$$

$$= \frac{2}{0} \qquad \text{Undefined}$$

16. $-\frac{1}{4}$

17. $(3, 2)$ and $(-1, 5)$

$$m = \frac{5 - 2}{-1 - 3} = \frac{3}{-4} = -\frac{3}{4}$$

18. $\frac{2}{3}$

19. $(-2, 4)$ and $(3, 0)$

$$m = \frac{4 - 0}{-2 - 3} = \frac{4}{-5} = -\frac{4}{5}$$

20. $-\frac{5}{6}$

21. $(4, 0)$ and $(5, 7)$

$$m = \frac{0 - 7}{4 - 5} = \frac{-7}{-1} = 7$$

22. $\frac{2}{3}$

23. $(0, 8)$ and $(-3, 10)$

$$m = \frac{8 - 10}{0 - (-3)} = \frac{8 - 10}{0 + 3} = \frac{-2}{3} = -\frac{2}{3}$$

24. $-\frac{1}{2}$

25. $(3, -2)$ and $(5, -6)$

$$m = \frac{-2 - (-6)}{3 - 5} = \frac{-2 + 6}{3 - 5} = \frac{4}{-2} = -2$$

26. $-\frac{11}{8}$

27. $\left(-2, \dfrac{1}{2}\right)$ and $\left(-5, \dfrac{1}{2}\right)$

$$m = \frac{\dfrac{1}{2} - \dfrac{1}{2}}{-2 - (-5)} = \frac{\dfrac{1}{2} - \dfrac{1}{2}}{-2 + 5} = \frac{0}{3} = 0$$

28. 0

29. $(9, -4)$ and $(9, -7)$

$$m = \frac{-4 - (-7)}{9 - 9} = \frac{-4 + 7}{9 - 9} = \frac{3}{0} \quad \text{(undefined)}$$

The slope is undefined.

30. Undefined

31. $(-1, 5)$ and $(4, 5)$

$$m = \frac{5 - 5}{4 - (-1)} = \frac{0}{5} = 0$$

32. Undefined

33. The line $x = -8$ is a vertical line. The slope is undefined.

34. Undefined

35. The line $y = 2$ is a horizontal line. A horizontal line has slope 0.

36. 0

37. The line $x = 9$ is a vertical line. The slope is undefined.

38. Undefined

39. The line $y = -9$ is a horizontal line. A horizontal line has slope 0.

40. 0

41. Grade $=$ slope $= \dfrac{\text{vertical change}}{\text{horizontal change}} = \dfrac{920.58}{13,740} =$

$0.067 = 6.7\%$

42. $\dfrac{12}{41}$, or about 29%

43. Grade $=$ slope $= \dfrac{\text{vertical change}}{\text{horizontal change}} = \dfrac{0.4}{5} =$

$0.08 = 8\%$

44. $\dfrac{28}{129}$, or about 22%

45. Grade $=$ slope $= \dfrac{\text{vertical change}}{\text{horizontal change}} = \dfrac{158.4}{5280} =$

$0.03 = 3\%$

46. 0.045

47. Grade $= \dfrac{\text{vertical change}}{\text{horizontal change}}$

$0.12 = \dfrac{v}{5}$

$0.6 = v \qquad \text{Multiplying by 5}$

The end of the treadmill is set at 0.6 ft vertically.

48. 30 ft

49. $5x(9x - 3) = 5x \cdot 9x - 5x \cdot 3 = 45x^2 - 15x$

50. $x^3 + x^2 - 11x + 10$

51. $(x - 7)(x + 7) = x^2 - 7^2 = x^2 - 49$

52. $25x^2 + 20x + 4$

53. ◈

54. ◈

55. If the line never enters the second quadrant and is non-vertical, then two points on the line are $(3, 4)$ and $(a, 0)$, $0 \le a < 3$. The slope is of the form $m = \dfrac{4 - 0}{3 - a}$, or $\dfrac{4}{3 - a}$, $0 \le a < 3$, so $m \ge \dfrac{4}{3}$. Then the numbers the line could have for its slope are $\left\{ m \middle| m \ge \dfrac{4}{3} \right\}$.

56. $-\dfrac{6}{5} \le m \le 0$

57. Note that $40 \text{ min} = 40 \text{ min} \cdot \dfrac{1 \text{ hr}}{60 \text{ min}} = \dfrac{40}{60} \text{ hr} = \dfrac{2}{3} \text{ hr}$

Rate $= \dfrac{\text{change in number of candles produced}}{\text{corresponding change in time}}$

$= \dfrac{64 - 46 \text{ candles}}{\dfrac{2}{3} \text{ hr}}$

$= \dfrac{18}{2/3} \dfrac{\text{candles}}{\text{hr}}$

$= \dfrac{18}{1} \cdot \dfrac{3}{2} \text{ candles per hour}$

$= 27 \text{ candles per hour}$

58. 3.6 bushels per hour

59. ◈

60. ◈

Exercise Set 7.2

1. Slope $\dfrac{2}{5}$; y-intercept $(0, 1)$

We plot $(0, 1)$ and from there move up 2 units and right 5 units. This locates the point $(5, 3)$. We plot $(5, 3)$ and draw a line passing through $(0, 1)$ and $(5, 3)$.

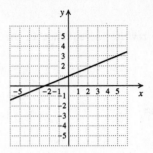

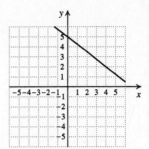

2.

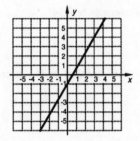

6.

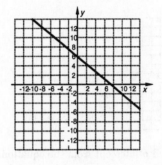

3. Slope $\dfrac{5}{2}$; y-intercept $(0, -3)$

We plot $(0, -3)$ and from there move up 5 units and right 2 units. This locates the point $(2, 2)$. We plot $(2, 2)$ and draw a line passing through $(0, -3)$ and $(2, 2)$.

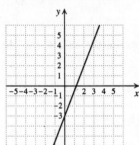

7. Slope 2; y-intercept $(0, -4)$

We plot $(0, -4)$. We can think of the slope as $\dfrac{2}{1}$, so from $(0, -4)$ we move up 2 units and right 1 unit. This locates the point $(1, -2)$. We plot $(1, -2)$ and draw a line passing through $(0, -4)$ and $(1, -2)$.

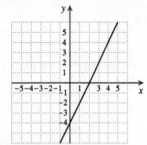

4.

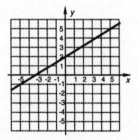

8.

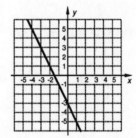

5. Slope $-\dfrac{3}{4}$; y-intercept $(0, 5)$

We plot $(0, 5)$. We can think of the slope as $\dfrac{-3}{4}$, so from $(0, 5)$ we move down 3 units and right 4 units. This locates the point $(4, 2)$. We plot $(4, 2)$ and draw a line passing through $(0, 5)$ and $(4, 2)$.

9. Slope -3; y-intercept $(0, 2)$

We plot $(0, 2)$. We can think of the slope as $\dfrac{-3}{1}$, so from $(0, 2)$ we move down 3 units and right 1 unit. This locates the point $(1, -1)$. We plot $(1, -1)$ and draw a line passing through $(0, 2)$ and $(1, -1)$.

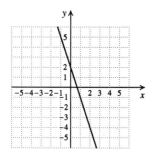

10.

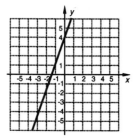

11. We read the slope and y-intercept from the equation.

$$y = 2\,x + 1$$

The slope is 2.   The y-intercept is $(0,1)$.

12. -3, $(0,7)$

13. We read the slope and y-intercept from the equation.

$$y = -\frac{5}{6}\,x + 2$$

The slope is $-\dfrac{5}{6}$.   The y-intercept is $(0,2)$.

14. $\dfrac{7}{2}$, $(0,4)$

15. $y = \dfrac{9}{4}x - 7$

$y = \dfrac{9}{4}x + (-7)$

The slope is $\dfrac{9}{4}$, and the y-intercept is $(0,-7)$.

16. $\dfrac{2}{9}$, $(0,-1)$

17. $y = -\dfrac{2}{5}x$

$y = -\dfrac{2}{5}x + 0$

The slope is $-\dfrac{2}{5}$, and the y-intercept is $(0,0)$.

18. $\dfrac{4}{3}$, $(0,0)$

19. We solve for y to rewrite the equation in the form $y = mx + b$.

$-2x + y = 4$

$y = 2x + 4$

The slope is 2, and the y-intercept is $(0,4)$.

20. 5, $(0,5)$

21. We solve for y to rewrite the equation in the form $y = mx + b$.

$4x - 3y = -12$

$-3y = -4x - 12$

$y = -\dfrac{1}{3}(-4x - 12)$

$y = \dfrac{4}{3}x + 4$

The slope is $\dfrac{4}{3}$, and the y-intercept is $(0,4)$.

22. $\dfrac{1}{2}$, $\left(0, -\dfrac{9}{2}\right)$

23. We solve for y to rewrite the equation in the form $y = mx + b$.

$x - 3y = -2$

$-3y = -x - 2$

$y = -\dfrac{1}{3}(-x - 2)$

$y = \dfrac{1}{3}x + \dfrac{2}{3}$

The slope is $\dfrac{1}{3}$, and the y-intercept is $\left(0, \dfrac{2}{3}\right)$.

24. -1, $(0,7)$

25. We solve for y to rewrite the equation in the form $y = mx + b$.

$-2x + 4y = 8$

$4y = 2x + 8$

$y = \dfrac{1}{4}(2x + 8)$

$y = \dfrac{1}{2}x + 2$

The slope is $\dfrac{1}{2}$, and the y-intercept is $(0,2)$.

26. $\dfrac{5}{7}$, $\left(0, \dfrac{2}{7}\right)$

27. $y = 5$

We can rewrite this as $y = 0 \cdot x + 5$. The slope is 0, and the y-intercept is $(0,5)$.

28. 0, $(0,4)$

29. We use the slope-intercept equation, substituting 3 for m and 1 for b:

$y = mx + b$

$y = 3x + 1$

30. $y = -5x + 3$

31. We use the slope-intercept equation, substituting $\dfrac{3}{5}$ for m and -5 for b:

$y = mx + b$

$y = \dfrac{3}{5}x - 5$

32. $y = -\dfrac{5}{2}x - 1$

33. We use the slope-intercept equation, substituting $-\dfrac{5}{3}$ for m and -8 for b:

$$y = mx + b$$
$$y = -\dfrac{5}{3}x - 8$$

34. $y = \dfrac{3}{4}x + 23$

35. We use the slope-intercept equation, substituting -2 for m and 3 for b:

$$y = mx + b$$
$$y = -2x + 3$$

36. $y = 7x - 6$

37. We use the slope-intercept equation, substituting 1 for m and -2 for b:

$$y = mx + b$$
$$y = 1x - 2$$
$$y = x - 2$$

38. $y = -x + 1$

39. $y = \dfrac{3}{5}x + 2$

First we plot the y-intercept $(0, 2)$. We can start at the y-intercept and use the slope, $\dfrac{3}{5}$, to find another point. We move up 3 units and right 5 units to get a new point $(5, 5)$. Thinking of the slope as $\dfrac{-3}{-5}$ we can start at $(0, 2)$ and move down 3 units and left 5 units to get another point $(-5, -1)$.

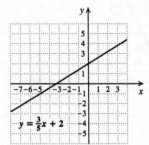

40.

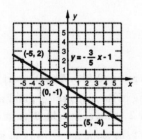

41. $y = -\dfrac{3}{5}x + 4$

First we plot the y-intercept $(0, 4)$. We can start at the y-intercept and, thinking of the slope as $\dfrac{-3}{5}$, find another point by moving down 3 units and right 5 units to the point $(5, 1)$. Thinking of the slope as $\dfrac{3}{-5}$ we can start at $(0, 4)$ and move up 3 units and left 5 units to get another point $(-5, 7)$.

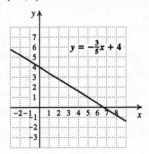

42.

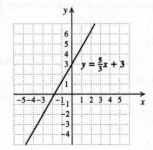

43. $y = \dfrac{5}{3}x + 3$

First we plot the y-intercept $(0, 3)$. We can start at the y-intercept and use the slope, $\dfrac{5}{3}$, to find another point. We move up 5 units and right 3 units to get a new point $(3, 8)$. Thinking of the slope as $\dfrac{-5}{-3}$ we can start at $(0, 3)$ and move down 5 units and left 3 units to get another point $(-3, -2)$.

44.

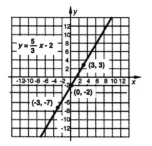

45. $y = -\dfrac{3}{2}x - 2$

First we plot the y-intercept $(0, -2)$. We can start at the y-intercept and, thinking of the slope as $\dfrac{-3}{2}$, find another point by moving down 3 units and right 2 units to the point $(2, -5)$. Thinking of the slope as $\dfrac{3}{-2}$ we can start at $(0, -2)$ and move up 3 units and left 2 units to get another point $(-2, 1)$.

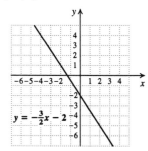

46.

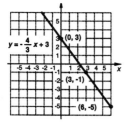

47. We first rewrite the equation in slope-intercept form.
$$2x + y = 1$$
$$y = -2x + 1$$

Now we plot the y-intercept $(0, 1)$. We can start at the y-intercept and, thinking of the slope as $\dfrac{-2}{1}$, find another point by moving down 2 units and right 1 unit to the point $(1, -1)$. In a similar manner, we can move from the point $(1, -1)$ to find a third point $(2, -3)$.

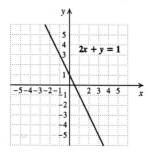

48.

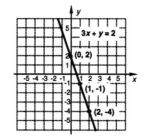

49. We first rewrite the equation in slope-intercept form.
$$3x - y = 4$$
$$-y = -3x + 4$$
$$y = 3x - 4 \quad \text{Multiplying by } -1$$

Now we plot the y-intercept $(0, -4)$. We can start at the y-intercept and, thinking of the slope as $\dfrac{3}{1}$, find another point by moving up 3 units and right 1 unit to the point $(1, -1)$. In a similar manner, we can move from the point $(1, -1)$ to find a third point $(2, 2)$.

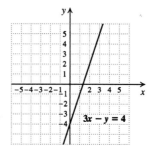

50.

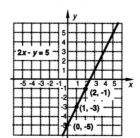

51. We first rewrite the equation in slope-intercept form.
$$2x + 3y = 9$$
$$3y = -2x + 9$$
$$y = \frac{1}{3}(-2x + 9)$$
$$y = -\frac{2}{3}x + 3$$

Now we plot the y-intercept $(0, 3)$. We can start at the y-intercept and, thinking of the slope as $\dfrac{-2}{3}$, find another point by moving down 2 units and right 3 units to the point $(3, 1)$. Thinking of the slope as $\dfrac{2}{-3}$ we can start at $(0, 3)$ and move up 2 units and left 3 units to get another point $(-3, 5)$.

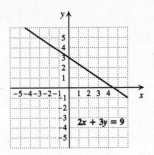

52.

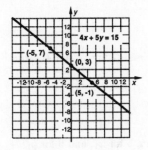

53. We first rewrite the equation in slope-intercept form.

$$x - 4y = 12$$

$$-4y = -x + 12$$

$$y = -\frac{1}{4}(-x + 12)$$

$$y = \frac{1}{4}x - 3$$

Now we plot the y-intercept $(0, -3)$. We can start at the y-intercept and use the slope, $\frac{1}{4}$, to find another point. We move up 1 unit and right 4 units to the point $(4, -2)$. Thinking of the slope as $\frac{-1}{-4}$ we can start at $(0, -3)$ and move down 1 unit and left 4 units to get another point $(-4, -4)$.

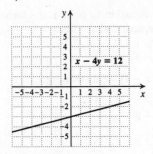

54.

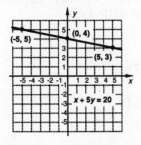

55. We first rewrite the equation in slope-intercept form.

$$5x - 6y = 24$$

$$-6y = -5x + 24$$

$$y = -\frac{1}{6}(-5x + 24)$$

$$y = \frac{5}{6}x - 4$$

Now we plot the y-intercept $(0, -4)$. We can start at the y-intercept and use the slope, $\frac{5}{6}$, to find another point. We move up 5 units and right 6 units to the point $(6, 1)$. Thinking of the slope as $\frac{-5}{-6}$ we can start at $(0, -4)$ and move down 5 units and left 6 units to get another point $(-6, -9)$.

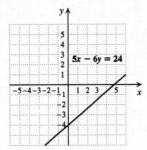

56.

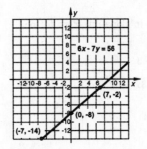

57. $3x^2 - 9x = 0$

$3x(x - 3) = 0$

$3x = 0 \ \text{ or } \ x - 3 = 0$

$x = 0 \ \text{ or } \quad x = 3$

The solutions are 0 and 3.

58. $y(y - 6)(y + 5)$

59. *Familiarize*. Let $y =$ the smaller odd integer. Then $y + 2 =$ the larger odd integer.

Translate. We reword the problem.

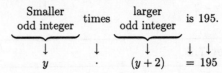

Carry out. We solve the equation.

$$y(y + 2) = 195$$
$$y^2 + 2y = 195$$
$$y^2 + 2y - 195 = 0$$
$$(y + 15)(y - 13) = 0$$
$$y + 15 = 0 \quad \text{or} \quad y - 13 = 0$$
$$y = -15 \quad \text{or} \qquad y = 13$$

Check. If $y = -15$, then $y + 2 = -13$. These are consecutive odd integers and their product is $(-15)(-13)$, or 195. This pair checks. If $y = 13$ then $y + 2 = 15$. These are consecutive odd integers and their product is $13 \cdot 15$, or 195. This pair checks also.

State. The integers are -15 and -13 or 13 and 15.

60. 11, -1

61.

62.

63. See the answer section in the text.

64. $N = \dfrac{3}{2}t + 4$

65. Rewrite each equation in slope-intercept form.

$$3x - 2y = 8$$
$$-2y = -3x + 8$$
$$y = -\frac{1}{2}(-3x + 8)$$
$$y = \frac{3}{2}x - 4$$

The slope is $\dfrac{3}{2}$.

$$2y + 3x = -4$$
$$2y = -3x - 4$$
$$y = \frac{1}{2}(-3x - 4)$$
$$y = -\frac{3}{2}x - 2$$

The y-intercept is $(0, -2)$.

We write an equation of the line with slope $\dfrac{3}{2}$ and y-intercept $(0, -2)$:

$$y = \frac{3}{2}x - 2$$

Exercise Set 7.3

1. $y - y_1 = m(x - x_1)$

We substitute 5 for m, 2 for x_1, and 5 for y_1.

$$y - 5 = 5(x - 2)$$

2. $y - 0 = -2(x - (-3))$

3. $y - y_1 = m(x - x_1)$

We substitute $\dfrac{3}{4}$ for m, 2 for x_1, and 4 for y_1.

$$y - 4 = \frac{3}{4}(x - 2)$$

4. $y - 2 = -1\left(x - \dfrac{1}{2}\right)$

5. $y - y_1 = m(x - x_1)$

We substitute 1 for m, 2 for x_1, and -6 for y_1.

$$y - (-6) = 1(x - 2)$$

6. $y - (-2) = 6(x - 4)$

7. $y - y_1 = m(x - x_1)$

We substitute -3 for m, -7 for x_1, and 0 for y_1.

$$y - 0 = -3(x - (-7))$$

8. $y - 3 = -3(x - 0)$

9. $y - y_1 = m(x - x_1)$

We substitute $\dfrac{2}{3}$ for m, 5 for x_1, and 6 for y_1.

$$y - 6 = \frac{2}{3}(x - 5)$$

10. $y - 7 = \dfrac{5}{6}(x - 2)$

11. First we write the equation in point-slope form.

$$y - y_1 = m(x - x_1)$$
$$y - 7 = 2(x - 3) \quad \text{Substituting}$$

Next we find an equivalent equation of the form $y = mx + b$.

$$y - 7 = 2(x - 3)$$
$$y - 7 = 2x - 6$$
$$y = 2x + 1$$

12. $y = 4x + 1$

13. We first write the equation in point-slope form and then find a equivalent equation of the form $y = mx + b$.

$$y - y_1 = m(x - x_1)$$
$$y - 5 = -1(x - 4) \quad \text{Substituting}$$
$$y - 5 = -x + 4$$
$$y = -x + 9$$

14. $y = x - 5$

15. We first write the equation in point-slope form and then find a equivalent equation of the form $y = mx + b$.

$$y - y_1 = m(x - x_1)$$
$$y - 3 = \frac{1}{2}(x - (-2))$$
$$y - 3 = \frac{1}{2}(x + 2)$$
$$y - 3 = \frac{1}{2}x + 1$$
$$y = \frac{1}{2}x + 4$$

16. $y = -\frac{1}{2}x - 1$

17. We first write the equation in point-slope form and then find a equivalent equation of the form $y = mx + b$.

$$y - y_1 = m(x - x_1)$$
$$y - (-5) = -\frac{1}{3}(x - (-6))$$
$$y + 5 = -\frac{1}{3}(x + 6)$$
$$y + 5 = -\frac{1}{3}x - 2$$
$$y = -\frac{1}{3}x - 7$$

18. $y = \frac{1}{5}x + 1$

19. We first write the equation in point-slope form and then find a equivalent equation of the form $y = mx + b$.

$$y - 0 = \frac{5}{4}(x - 4)$$
$$y = \frac{5}{4}x - 5$$

20. $y = \frac{4}{3}x + 12$

21. $(-6, 1)$ and $(2, 3)$

First we find the slope.
$$m = \frac{1 - 3}{-6 - 2} = \frac{-2}{-8} = \frac{1}{4}$$
Then we use the point-slope equation.
$$y - y_1 = m(x - x_1)$$
We substitute $\frac{1}{4}$ for m, -6 for x_1, and 1 for y_1.
$$y - 1 = \frac{1}{4}(x - (-6))$$
$$y - 1 = \frac{1}{4}(x + 6)$$
$$y - 1 = \frac{1}{4}x + \frac{3}{2}$$
$$y = \frac{1}{4}x + \frac{5}{2}$$
We also could substitute $\frac{1}{4}$ for m, 2 for x_1, and 3 for y_1.
$$y - 3 = \frac{1}{4}(x - 2)$$
$$y - 3 = \frac{1}{4}x - \frac{1}{2}$$
$$y = \frac{1}{4}x + \frac{5}{2}$$

22. $y = x + 4$

23. $(0, 4)$ and $(4, 2)$

First we find the slope.
$$m = \frac{4 - 2}{0 - 4} = \frac{2}{-4} = -\frac{1}{2}$$
Then we use the point-slope equation.
$$y - y_1 = m(x - x_1)$$
We substitute $-\frac{1}{2}$ for m, 0 for x_1, and 4 for y_1.
$$y - 4 = -\frac{1}{2}(x - 0)$$
$$y - 4 = -\frac{1}{2}x$$
$$y = -\frac{1}{2}x + 4$$

24. $y = \frac{1}{2}x$

25. $(3, 2)$ and $(1, 5)$

First we find the slope.
$$m = \frac{2 - 5}{3 - 1} = \frac{-3}{2} = -\frac{3}{2}$$
Then we use the point-slope equation.
$$y - y_1 = m(x - x_1)$$
We substitute $-\frac{3}{2}$ for m, 3 for x_1, and 2 for y_1.
$$y - 2 = -\frac{3}{2}(x - 3)$$
$$y - 2 = -\frac{3}{2}x + \frac{9}{2}$$
$$y = -\frac{3}{2}x + \frac{13}{2}$$

26. $y = x + 5$

27. $(5, 0)$ and $(0, -2)$

First we find the slope.
$$m = \frac{0 - (-2)}{5 - 0} = \frac{2}{5}$$
Then we use the point-slope equation.
$$y - y_1 = m(x - x_1)$$
We substitute $\frac{2}{5}$ for m, 5 for x_1, and 0 for y_1.
$$y - 0 = \frac{2}{5}(x - 5)$$
$$y = \frac{2}{5}x - 2$$

28. $y = \frac{5}{3}x + \frac{4}{3}$

29. $(-2, -4)$ and $(2, -1)$

First we find the slope.
$$m = \frac{-4 - (-1)}{-2 - 2} = \frac{-4 + 1}{-2 - 2} = \frac{-3}{-4} = \frac{3}{4}$$
Then we use the point-slope equation.
$$y - y_1 = m(x - x_1)$$

We substitute $\frac{3}{4}$ for m, -2 for x_1, and -4 for y_1.

$$y - (-4) = \frac{3}{4}(x - (-2))$$

$$y + 4 = \frac{3}{4}(x + 2)$$

$$y + 4 = \frac{3}{4}x + \frac{3}{2}$$

$$y = \frac{3}{4}x - \frac{5}{2}$$

30. $y = -4x - 7$

31. $y - 5 = \frac{1}{2}(x - 3)$ Point-slope form

The line has slope $\frac{1}{2}$ and passes through $(3, 5)$. We plot $(3, 5)$ and then find a second point by moving up 1 unit and right 2 units to $(5, 6)$. We draw the line through these points.

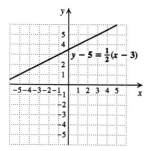

32.

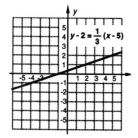

33. $y - 3 = -\frac{1}{2}(x - 5)$ Point-slope form

The line has slope $-\frac{1}{2}$, or $\frac{1}{-2}$ passes through $(5, 3)$. We plot $(5, 3)$ and then find a second point by moving up 1 unit and left 2 units to $(3, 4)$. We draw the line through these points.

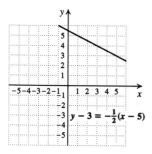

34.

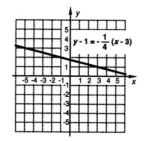

35. $y + 5 = \frac{1}{2}(x - 3)$, or $y - (-5) = \frac{1}{2}(x - 3)$

The line has slope $\frac{1}{2}$ and passes through $(3, -5)$. We plot $(3, -5)$ and then find a second point by moving up 1 unit and right 2 units to $(5, -4)$. We draw the line through these points.

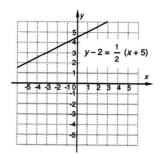

36.

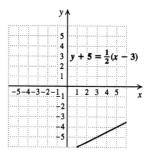

37. $y + 2 = 3(x + 1)$, or $y - (-2) = 3(x - (-1))$

The line has slope 3, or $\frac{3}{1}$, and passes through $(-1, -2)$. We plot $(-1, -2)$ and then find a second point by moving up 3 units and right 1 unit to $(0, 1)$. We draw the line through these points.

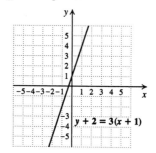

38.

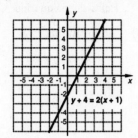

39. $y - 4 = -2(x + 1)$, or $y - 4 = -2(x - (-1))$

The line has slope -2, or $\dfrac{-2}{1}$, and passes through $(-1, 4)$. We plot $(-1, 4)$ and then find a second point by moving down 2 units and right 1 unit to $(0, 2)$. We draw the line through these points.

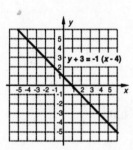

40.

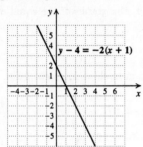

41. $y + 3 = -(x + 2)$, or $y - (-3) = -1(x - (-2))$

The line has slope -1, or $\dfrac{-1}{1}$, and passes through $(-2, -3)$. We plot $(-2, -3)$ and then find a second point by moving down 1 unit and right 1 unit to $(-1, -4)$. We draw the line through these points.

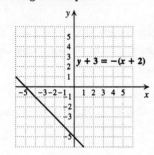

42.

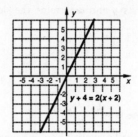

43. $y = \dfrac{2}{3}x + 1$: The slope is $\dfrac{2}{3}$, and the y-intercept is $(0, 1)$.

$y = \dfrac{2}{3}x - 1$: The slope is $\dfrac{2}{3}$, and the y-intercept is $(0, -1)$.

Since both lines have slope $\dfrac{2}{3}$ but different y-intercepts, their graphs are parallel.

44. No

45. The equation $x + 6 = y$, or $y = x + 6$, represents a line with slope 1 and y-intercept $(0, 6)$. We rewrite the second equation in slope-intercept form.

$$y + x = -2$$
$$y = -x - 2$$

The slope is -1 and the y-intercept is $(0, -2)$.

Since the lines have different slopes, their graphs are not parallel.

46. Yes

47. Rewrite each equation in slope-intercept form.

$$3x + 4y = 8$$
$$4y = -3x + 8$$
$$y = \frac{1}{4}(-3x + 8)$$
$$y = -\frac{3}{4}x + 2$$

The slope is $-\dfrac{3}{4}$, and the y-intercept is $(0, 2)$.

$$7 - 12y = 9x$$
$$-12y = 9x - 7$$
$$y = -\frac{1}{12}(9x - 7)$$
$$y = -\frac{3}{4}x + \frac{7}{12}$$

The slope is $-\dfrac{3}{4}$, and the y-intercept is $\left(0, \dfrac{7}{12}\right)$.

Since both lines have slope $-\dfrac{3}{4}$ but different y-intercepts, their graphs are parallel.

48. No

49. First solve the equation for y and determine the slope of the given line.

$$x + 2y = 6 \qquad \text{Given line}$$
$$2y = -x + 6$$
$$y = -\frac{1}{2}x + 3$$

The slope of the given line is $-\frac{1}{2}$.

The slope of every line parallel to the given line must also be $-\frac{1}{2}$. We find the equation of the line with slope $-\frac{1}{2}$ and containing the point $(3, 7)$.

$$y - y_1 = m(x - x_1) \qquad \text{Point-slope equation}$$
$$y - 7 = -\frac{1}{2}(x - 3) \qquad \text{Substituting}$$
$$y - 7 = -\frac{1}{2}x + \frac{3}{2}$$
$$y = -\frac{1}{2}x + \frac{17}{2}$$

50. $y = 3x + 3$

51. First solve the equation for y and determine the slope of the given line.

$$5x - 7y = 8 \qquad \text{Given line}$$
$$5x - 8 = 7y$$
$$\frac{5}{7}x - \frac{8}{7} = y$$
$$y = \frac{5}{7}x - \frac{8}{7}$$

The slope of the given line is $\frac{5}{7}$.

The slope of every line parallel to the given line must also be $\frac{5}{7}$. We find the equation of the line with slope $\frac{5}{7}$ and containing the point $(2, -1)$.

$$y - y_1 = m(x - x_1) \qquad \text{Point-slope equation}$$
$$y - (-1) = \frac{5}{7}(x - 2) \qquad \text{Substituting}$$
$$y + 1 = \frac{5}{7}x - \frac{10}{7}$$
$$y = \frac{5}{7}x - \frac{17}{7}$$

52. $y = -2x - 13$

53. First solve the equation for y and determine the slope of the given line.

$$3x - 9y = 2 \qquad \text{Given line}$$
$$3x - 2 = 9y$$
$$\frac{1}{3}x - \frac{2}{9} = y$$

The slope of the given line is $\frac{1}{3}$.

The slope of every line parallel to the given line must also be $\frac{1}{3}$. We find the equation of the line with slope $\frac{1}{3}$ and containing the point $(-6, 2)$.

$$y - y_1 = m(x - x_1) \qquad \text{Point-slope equation}$$
$$y - 2 = \frac{1}{3}(x - (-6)) \qquad \text{Substituting}$$
$$y - 2 = \frac{1}{3}(x + 6)$$
$$y - 2 = \frac{1}{3}x + 2$$
$$y = \frac{1}{3}x + 4$$

54. $y = -\frac{5}{2}x - \frac{35}{2}$

55. $y = 4x - 5$ represents a line with slope 4.

$y = -\frac{1}{4}x + 8$ represents a line with slope $-\frac{1}{4}$.

Since $4\left(-\frac{1}{4}\right) = -1$, the graphs of the equations are perpendicular.

56. No

57. Rewrite each equation in slope-intercept form.

$$3x + 5y = 10$$
$$5y = -3x + 10$$
$$y = \frac{1}{5}(-3x + 10)$$
$$y = -\frac{3}{5}x + 2$$

The slope is $-\frac{3}{5}$.

$$15x + 9y = 18$$
$$9y = -15x + 18$$
$$y = \frac{1}{9}(-15x + 18)$$
$$y = -\frac{5}{3}x + 2$$

The slope is $-\frac{5}{3}$.

Since $-\frac{3}{5}\left(-\frac{5}{3}\right) = 1 \neq -1$, the graphs of the equations are not perpendicular.

58. Yes

59. Since $x = 5$ represents a vertical line and $y = \frac{1}{2}$ represents a horizontal line, the graphs of the equations are perpendicular.

60. Yes

61. First solve the equation for y and determine the slope of the given line.

$$2x + y = -3 \qquad \text{Given line}$$
$$y = -2x - 3$$

The slope of the given line is -2.

The slope of a perpendicular line is given by the opposite of the reciprocal of -2, $\frac{1}{2}$.

We find the equation of the line with slope $\frac{1}{2}$ containing the point $(2, 5)$.

$$y - y_1 = m(x - x_1) \qquad \text{Point-slope equation}$$

$$y - 5 = \frac{1}{2}(x - 2) \qquad \text{Substituting}$$

$$y - 5 = \frac{1}{2}x - 1$$

$$y = \frac{1}{2}x + 4$$

62. $y = -3x + 12$

63. First solve the equation for y and determine the slope of the given line.

$$3x + 4y = 5 \qquad \text{Given line}$$

$$4y = -3x + 5$$

$$y = -\frac{3}{4}x + \frac{5}{4}$$

The slope of the given line is $-\frac{3}{4}$.

The slope of a perpendicular line is given by the opposite of the reciprocal of $-\frac{3}{4}$, $\frac{4}{3}$.

We find the equation of the line with slope $\frac{4}{3}$ and containing the point $(3, -2)$.

$$y - y_1 = m(x - x_1) \quad \text{Point-slope equation}$$

$$y - (-2) = \frac{4}{3}(x - 3) \qquad \text{Substituting}$$

$$y + 2 = \frac{4}{3}x - 4$$

$$y = \frac{4}{3}x - 6$$

64. $y = -\frac{2}{5}x - \frac{31}{5}$

65. First solve the equation for y and determine the slope of the given line.

$$2x + 5y = 7 \qquad \text{Given line}$$

$$5y = -2x + 7$$

$$y = -\frac{2}{5}x + \frac{7}{5}$$

The slope of the given line is $-\frac{2}{5}$.

The slope of a perpendicular line is given by the opposite of the reciprocal of $-\frac{2}{5}$, $\frac{5}{2}$.

We find the equation of the line with slope $\frac{5}{2}$ and containing the point $(0, 9)$.

$$y - y_1 = m(x - x_1) \quad \text{Point-slope equation}$$

$$y - 9 = \frac{5}{2}(x - 0) \qquad \text{Substituting}$$

$$y - 9 = \frac{5}{2}x$$

$$y = \frac{5}{2}x + 9$$

66. $y = -2x - 10$

67. $7x^3y^2 + 35x^2y^6 = 7x^2y^2 \cdot x + 7x^2y^2 \cdot 5y^4$
$$= 7x^2y^2(x + 5y^4)$$

68. $(3x - 1)(5x^2 + 1)$

69. $\dfrac{5x^2 + 5x}{10x^3 - 10x^2} = \dfrac{5x(x + 1)}{10x^2(x - 1)}$

$$= \dfrac{\cancel{5} \cdot \cancel{x}(x + 1)}{\cancel{5} \cdot 2 \cdot \cancel{x} \cdot x(x - 1)}$$

$$= \dfrac{x + 1}{2x(x - 1)}$$

70. $\dfrac{x - 5}{x + 2}$

71.

72.

73. First find the slope of $3x - y + 4 = 0$.

$$3x - y + 4 = 0$$

$$3x + 4 = y$$

The slope is 3.

Then find an equation of the line containing $(2, -3)$ and having slope 3.

$$y - y_1 = m(x - x_1)$$

We substitute 3 for m, 2 for x_1, and -3 for y_1.

$$y - (-3) = 3(x - 2)$$

$$y + 3 = 3x - 6$$

$$y = 3x - 9$$

74. $y = \frac{1}{5}x - 2$

75. First we find the slope of the given line:

$$4x - 8y = 12$$

$$-8y = -4x + 12$$

$$y = \frac{1}{2}x - \frac{3}{2}$$

The slope is $\frac{1}{2}$.

Then we use the point-slope equation to find the equation of a line with slope $\frac{1}{2}$ containing the point $(-2, 0)$:

$$y - y_1 = m(x - x_1)$$

$$y - 0 = \frac{1}{2}(x - (-2))$$

$$y = \frac{1}{2}(x + 2)$$

$$y = \frac{1}{2}x + 1$$

76.

Exercise Set 7.4

1. The correspondence is not a function, because a member of the domain (a) corresponds to more than one member of the range.

2. Yes

3. The correspondence is a function, because each member of the domain corresponds to just one member of the range.

4. Yes

5. The correspondence is a function, because each member of the domain corresponds to just one member of the range.

6. No

7. The correspondence is not a function, because a member of the domain (Viola) corresponds to more than one member of the range.

8. No

9. The correspondence is a function, because each member of the domain corresponds to just one member of the range.

10. Yes

11. This correspondence is a function, because each class member has only one seat number.

12. Function

13. This correspondence is a function, because each shape has only one number for its area.

14. Function

15. The correspondence is not a function, since it is reasonable to assume that at least one person in the town has more than one aunt.

The correspondence is a relation, since it is reasonable to assume that each person in the town has at least one aunt.

16. A relation but not a function

17. $g(x) = x + 1$

  a) $g(0) = 0 + 1 = 1$

  b) $g(-4) = -4 + 1 = -3$

  c) $g(-7) = -7 + 1 = -6$

  d) $g(8) = 8 + 1 = 9$

  e) $g(a + 2) = a + 2 + 1 = a + 3$

18. a) 0,   b) 4,   c) -7,   d) -8,   e) $a - 5$

19. $f(n) = 5n^2 + 4$

  a) $f(0) = 5(0)^2 + 4 = 0 + 4 = 4$

  b) $f(-1) = 5(-1)^2 + 4 = 5 + 4 = 9$

  c) $f(3) = 5(3)^2 + 4 = 45 + 4 = 49$

  d) $f(t) = 5(t)^2 + 4 = 5t^2 + 4$

  e) $f(2a) = 5(2a)^2 + 4 = 5 \cdot 4a^2 + 4 = 20a^2 + 4$

20. a) -2,   b) 1,   c) 25,   d) $3t^2 - 2$,   e) $12a^2 - 2$

21. $g(r) = 3r^2 + 2r - 1$

  a) $g(2) = 3(2)^2 + 2(2) - 1 = 12 + 4 - 1 = 15$

  b) $g(3) = 3(3)^2 + 2(3) - 1 = 27 + 6 - 1 = 32$

  c) $g(-3) = 3(-3)^2 + 2(-3) - 1 = 27 - 6 - 1 = 20$

  d) $g(1) = 3(1)^2 + 2(1) - 1 = 3 + 2 - 1 = 4$

  e) $g(3r) = 3(3r)^2 + 2(3r) - 1 = 3 \cdot 9r^2 + 6r - 1 =$
  $$27r^2 + 6r - 1$$

22. a) 35,   b) 2,   c) 7,   d) 20,   e) $36r^2 - 3r + 2$

23. $f(x) = \dfrac{x - 3}{2x - 5}$

  a) $f(0) = \dfrac{0 - 3}{2 \cdot 0 - 5} = \dfrac{-3}{0 - 5} = \dfrac{-3}{-5} = \dfrac{3}{5}$

  b) $f(4) = \dfrac{4 - 3}{2 \cdot 4 - 5} = \dfrac{1}{8 - 5} = \dfrac{1}{3}$

  c) $f(-2) = \dfrac{-1 - 3}{2(-1) - 5} = \dfrac{-4}{-2 - 5} = \dfrac{-4}{-7} = \dfrac{4}{7}$

  d) $f(3) = \dfrac{3 - 3}{2 \cdot 3 - 5} = \dfrac{0}{6 - 5} = \dfrac{0}{1} = 0$

  e) $f(x + 2) = \dfrac{x + 2 - 3}{2(x + 2) - 5} = \dfrac{x - 1}{2x + 4 - 5} = \dfrac{x - 1}{2x - 1}$

24. a) $\dfrac{26}{25}$,   b) $\dfrac{2}{9}$,   c) undefined,   d) $-\dfrac{7}{3}$,   e) $\dfrac{3x + 5}{2x + 11}$

25. $A(s) = s^2 \dfrac{\sqrt{3}}{4}$

  $A(4) = 4^2 \dfrac{\sqrt{3}}{4} = 4\sqrt{3}$

  The area is $4\sqrt{3}$ cm^2.

26. $9\sqrt{3}$ in^2

27. $V(r) = 4\pi r^2$

  $V(3) = 4\pi(3)^2 = 36\pi$

  The area is 36π in$^2 \approx 113.04$ in^2.

28. 314 cm^2

29. $$F(C) = \frac{9}{5}C + 32$$

  $$F(-10) = \frac{9}{5}(-10) + 32 = -18 + 32 = 14$$

  The equivalent temperature is 14° F.

30. 41° F

31. $H(x) = 2.75x + 71.48$

  $H(32) = 2.75(32) + 71.48 = 159.48$

  The predicted height is 159.48 cm.

32. 167.73 cm

33. Locate the point that is directly above 225. Then estimate its second coordinate by moving horizontally from the point to the vertical axis. The rate is about 75 per 10,000 men.

34. 125 per 10,000 men

35. Locate the point that is directly above 1989. Then estimate its second coordinate by moving horizontally from the point to the vertical axis. The number of wood bats sold in 1989 was about 1.4 million.

36. 0.7 million

37. Plot and connect the points, using body weight as the first coordinate and the corresponding number of drinks as the second coordinate.

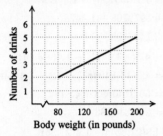

To estimate the number of drinks that a 140-lb person would have to drink to be considered intoxicated, first locate the point that is directly above 140. Then estimate its second coordinate by moving horizontally from the point to the vertical axis. Read the approximate function value there. The estimated number of drinks is 3.5.

38. 3 drinks

39. Plot and connect the points, using the counter reading as the first coordinate and the time of the tape as the second coordinate.

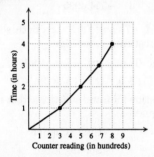

To estimate the time elapsed when the counter has reached 600, first locate the point that is directly above 600. Then estimate its second coordinate by moving horizontally from the point to the vertical axis. Read the approximate function value there. The time elapsed is about 2.5 hr.

40. About 0.5 hr

41. Plot and connect the points, using the year as the first coordinate and the population as the second.

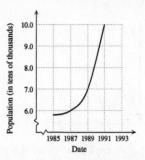

To estimate what the population was in 1988, first locate the point that is directly above 1988. Then estimate its second coordinate by moving horizontally from the point to the vertical axis. Read the approximate function value there. The population was about 64,000.

42. About 150,000

43. Plot and connect the points, using the year as the first coordinate and the sales total as the second coordinate.

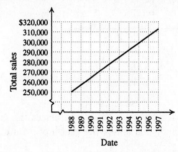

To predict the total sales for 1997, first locate the point directly above 1997. Then estimate its second coordinate by moving horizontally to the vertical axis. Read the approximate function value there. The predicted 1997 sales total is about $310,000.

44. About $270,000

45. We can use the vertical line test:

If it is possible for a vertical line to intersect a graph more than once, the graph is not the graph of a function.

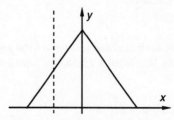

Visualize moving this vertical line across the graph. Ask yourself the question:

Will this line ever intersect the graph more than once?

If the answer is yes, the graph is not a graph of a function. If the answer is no, the graph is a graph of a function.

In this problem the vertical line will not intersect the graph more than once. Thus, the graph is a graph of a function.

46. No

47. We can use the vertical line test:

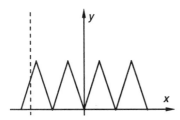

Visualize moving this vertical line across the graph. The vertical line will not intersect the graph more than once. Thus, the graph is a graph of a function.

48. No

49. We can use the vertical line test:

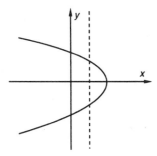

It is possible for a vertical line to intersect the graph more than once. Thus this is not the graph of a function.

50. Yes

51. We can use the vertical line test.

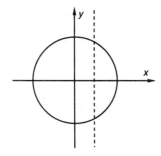

It is possible for a vertical line to intersect the graph more than once. Thus this is not a graph of a function.

52. Yes

53. $3x^4 - 9x^3 = 3x^3 \cdot x - 3x^3 \cdot 3$
$\quad\quad\quad\quad = 3x^3(x - 3)$

54. $7(y + 3z - 2)$

55. $x^3 - 7x^2 + 3x - 21 = x^2(x - 7) + 3(x - 7)$
$\quad\quad\quad\quad\quad\quad\quad\quad = (x - 7)(x^2 + 3)$

56. $(y - 1)(y^2 - 2)$

57. ◈

58. ◈

59. $\quad\quad\quad f(x) = 4.3x^2 - 1.4x$

a) $f(1.034) = 4.3(1.034)^2 - 1.4(1.034)$
$\quad\quad\quad\quad = 4.3(1.069156) - 1.4(1.034)$
$\quad\quad\quad\quad = 4.5973708 - 1.4476$
$\quad\quad\quad\quad = 3.1497708$

b) $f(-3.441) = 4.3(-3.441)^2 - 1.4(-3.441)$
$\quad\quad\quad\quad = 4.3(11.840481) - 1.4(-3.441)$
$\quad\quad\quad\quad = 50.9140683 + 4.8174$
$\quad\quad\quad\quad = 55.7314683$

c) $f(27.35) = 4.3(27.35)^2 - 1.4(27.35)$
$\quad\quad\quad\quad = 4.3(748.0225) - 1.4(27.35)$
$\quad\quad\quad\quad = 3216.49675 - 38.29$
$\quad\quad\quad\quad = 3178.20675$

d) $f(-16.31) = 4.3(-16.31)^2 - 1.4(-16.31)$
$\quad\quad\quad\quad = 4.3(266.0161) - 1.4(-16.31)$
$\quad\quad\quad\quad = 1143.86923 + 22.834$
$\quad\quad\quad\quad = 1166.70323$

60. a) $11,394.477$,   b) $-582,136.93$
c) 3.5018862,   d) -554.4995

61. We know that $(-1, -7)$ and $(3, 8)$ are both solutions of $g(x) = mx + b$. Substituting, we have
$$-7 = m(-1) + b, \text{ or } -7 = -m + b,$$
$$\text{and } 8 = m(3) + b, \text{ or } 8 = 3m + b.$$
Solve the first equation for b and substitute that expression into the second equation.

$\quad -7 = -m + b$ First equation
$\quad m - 7 = b$ Solving for b
$\quad 8 = 3m + b$ Second equation
$\quad 8 = 3m + (m - 7)$ Substituting
$\quad 8 = 3m + m - 7$
$\quad 8 = 4m - 7$
$\quad 15 = 4m$
$\quad \dfrac{15}{4} = m$

We know that $m - 7 = b$, so $\dfrac{15}{4} - 7 = b$, or $-\dfrac{13}{4} = b$.

Then we have $m = \dfrac{15}{4}$, $b = -\dfrac{13}{4}$. We can express $g(x)$ as
$$g(x) = \frac{15}{4}x - \frac{13}{4}.$$

62. 35

63. ◈

64.

65. The correspondence in the chart is a function, because each member of the domain, or activity, corresponds to just one member of the range.

66. 22 mm

67. Locate the highest point on the graph. For this graph, there are two equally high points that are higher than all the other points of the graph. Then move down a vertical line through each point to the horizontal axis and read the corresponding times.

The times are 2 min, 40 sec and 5 min, 40 sec.

68.

69. From Exercise 67 we know that the first largest contraction occurred at 2 min, 40 sec, and the second occurred at 5 min, 40 sec. Find the time between the contractions:

$$\begin{array}{r} 5 \text{ min } 40 \text{ sec} \\ -2 \text{ min } 40 \text{ sec} \\ \hline 3 \text{ min} \end{array}$$

Then the frequency of the largest contractions is 1 contraction every 3 min.

70.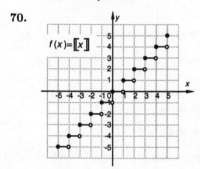

Exercise Set 7.5

1. Since $f(1) = -3 \cdot 1 + 1 = -2$ and $g(1) = 1^2 + 2 = 3$, we have $f(1) + g(1) = -2 + 3 = 1$.

2. 1

3. Since $f(-1) = -3(-1) + 1 = 4$ and $g(-1) = (-1)^2 + 2 = 3$ we have $f(-1) + g(-1) = 4 + 3 = 7$.

4. 13

5. Since $f(-7) = -3(-7) + 1 = 22$ and $g(-7) = (-7)^2 + 2 = 51$, we have $f(-7) - g(-7) = 22 - 51 = -29$.

6. −11

7. Since $f(5) = -3(5) + 1 = -14$ and $g(5) = 5^2 + 2 = 27$, we have $f(5) - g(5) = -14 - 27 = -41$.

8. −29

9. Since $f(2) = -3 \cdot 2 + 1 = -5$ and $g(2) = 2^2 + 2 = 6$, we have $f(2) \cdot g(2) = -5 \cdot 6 = -30$.

10. −88

11. Since $f(-3) = -3(-3) + 1 = 10$ and $g(-3) = (-3)^2 + 2 = 11$, we have $f(-3) \cdot g(-3) = 10 \cdot 11 = 110$.

12. 234

13. Since $f(0) = -3 \cdot 0 + 1 = 1$ and $g(0) = 0^2 + 2 = 2$, we have $f(0)/g(0) = 1/2$.

14. −2/3

15. Using our work in Exercise 11, we have
$$f(-3)/g(-3) = 10/11.$$

16. 11/10

17. $(F + G)(x) = F(x) + G(x)$
$$= x^2 - 3 + 4 - x$$
$$= x^2 - x + 1 \quad \text{Combining like terms}$$
$(F + G)(-3) = (-3)^2 - (-3) + 1 = 9 + 3 + 1 = 13$

18. 7

19. Using our work in Exercise 17, we have
$$(F + G)(x) = x^2 - x + 1.$$

20. $a^2 - a + 1$

21. $(F - G)(x) = F(x) - G(x)$
$$= x^2 - 3 - (4 - x)$$
$$= x^2 - 3 - 4 + x$$
$$= x^2 + x - 7$$
$(F - G)(-4) = (-4)^2 + (-4) - 7 = 16 - 4 - 7 = 5$

22. 13

23. $(F \cdot G)(x) = F(x) \cdot G(x)$
$$= (x^2 - 3)(4 - x)$$
$$= 4x^2 - x^3 - 12 + 3x$$
$(F \cdot G)(2) = 4 \cdot 2^2 - 2^3 - 12 + 3 \cdot 2 =$
$$16 - 8 - 12 + 6 = 2$$

24. 6

25. Using our work in Exercise 23, we have:
$(F \cdot G)(-3) = 4(-3)^2 - (-3)^3 - 12 + 3(-3)$
$$= 4 \cdot 9 - (-27) - 12 - 9$$
$$= 36 + 27 - 12 - 9$$
$$= 42$$

26. 104

27. $(F/G)(x) = F(x)/G(x)$
$$= \frac{x^2 - 3}{4 - x}$$
$(F/G)(0) = \frac{0^2 - 3}{4 - 0} = -\frac{3}{4}$

28. $-\dfrac{2}{3}$

29. Using our work in Exercise 27, we have
$$(F/G)(-2) = \frac{(-2)^2 - 3}{4 - (-2)} = \frac{4 - 3}{4 + 2} = \frac{1}{6}.$$

30. 6

31. Since division by 0 is not defined we set the denominator equal to 0 and solve for x.
$$3 - x = 0$$
$$3 = x \quad \text{Adding } x \text{ on both sides}$$
Domain of $f = \{x | x$ is a real number and $x \neq 3\}$

32. $\left\{ x \Big| x \text{ is a real number and } x \neq -\dfrac{1}{2} \right\}$

33. Since division by 0 is not defined we set the denominator equal to 0 and solve for x.
$$x^2 + 4x = 0$$
$$x(x + 4) = 0$$
$$x = 0 \quad \text{or} \quad x + 4 = 0 \quad \text{Principle of zero products}$$
$$x = 0 \quad \text{or} \qquad x = -4$$
Domain of $v = \{x | x$ is a real number and $x \neq 0$ and $x \neq -4\}$

34. $\{x | x$ is a real number and $x \neq -5$ and $x \neq 5\}$

35. Any real number can be an input of
$$f(x) = x^2 + x - 30.$$
Domain of $f = \{x | x$ is a real number$\}$

36. $\{x | x$ is a real number$\}$

37. Since division by 0 is not defined we set the denominator equal to 0 and solve for x.
$$x^2 - 8x + 12 = 0$$
$$(x - 2)(x - 6) = 0$$
$$x - 2 = 0 \quad \text{or} \quad x - 6 = 0$$
$$x = 2 \quad \text{or} \qquad x = 6$$
Domain of $g = \{x | x$ is a real number and $x \neq 2$ and $x \neq 6\}$

38. $\{x | x$ is a real number and $x \neq 2$ and $x \neq 4\}$

39. The domain of f and of g is all real numbers. Thus, Domain of $f + g =$ Domain of $f - g =$ Domain of $f \cdot g = \{x | x$ is a real number$\}$.

40. $\{x | x$ is a real number$\}$.

41. Because division by 0 is undefined, we have Domain of $f = \{x | x$ is a real number and $x \neq 2\}$.

Because any real number can be an input of $g(x)$, Domain of $g = \{x | x$ is a real number $\}$.

Thus, Domain of $f + g =$ Domain of $f - g =$ Domain of $f \cdot g = \{x | x$ is a real number and $x \neq 2\}$.

42. $\{x | x$ is a real number and $x \neq -4\}$

43. Because division by 0 is undefined, we have Domain of $f = \{x | x$ is a real number and $x \neq 1\}$.

Because any real number can be an input of $g(x)$, Domain of $g = \{x | x$ is a real number $\}$.

Thus, Domain of $f + g =$ Domain of $f - g =$ Domain of $f \cdot g = \{x | x$ is a real number and $x \neq 1\}$.

44. $\{x | x$ is a real number and $x \neq 5\}$

45. Because division by 0 is undefined, we have Domain of $f = \{x | x$ is a real number and $x \neq 2\}$,

and

Domain of $g = \{x | x$ is a real number and $x \neq 4\}$.

Thus, Domain of $f + g =$ Domain of $f - g =$ Domain of $f \cdot g = \{x | x$ is a real number and $x \neq 2$ and $x \neq 4\}$.

46. $\{x | x$ is a real number and $x \neq -3$ and $x \neq 2\}$

47. Because division by 0 is undefined, we have Domain of $f = \{x | x$ is a real number and $x \neq -2\}$,

and

Domain of $g = \left\{ x \Big| x \text{ is a real number and } x \neq \dfrac{4}{3} \right\}$.

Thus, Domain of $f + g =$ Domain of $f - g =$ Domain of $f \cdot g = \Big\{ x \Big| x$ is a real number and

$x \neq -2$ and $x \neq \dfrac{4}{3} \Big\}$.

48. $\left\{ x \Big| x \text{ is a real number and } x \neq 3 \text{ and } x \neq \dfrac{5}{2} \right\}$

49. Domain of $f =$ Domain of $g = \{x | x$ is a real number$\}$.

Since $g(x) = 0$ when $x - 3 = 0$, we have $g(x) = 0$ when $x = 3$. We conclude that Domain of $f/g = \{x | x$ is a real number and $x \neq 3\}$.

50. $\{x | x$ is a real number and $x \neq 5\}$

51. Domain of $f =$ Domain of $g = \{x | x$ is a real number$\}$.

Since $g(x) = 0$ when $2x - 8 = 0$, we have $g(x) = 0$ when $x = 4$. We conclude that Domain of $f/g = \{x | x$ is a real number and $x \neq 4\}$.

52. $\{x | x$ is a real number and $x \neq 3\}$

53. Domain of $f = \{x | x$ is a real number and $x \neq 4\}$.

Domain of $g = \{x | x$ is a real number$\}$.

Since $g(x) = 0$ when $5 - x = 0$, we have $g(x) = 0$ when $x = 5$. We conclude that Domain of $f/g = \{x | x$ is a real number and $x \neq 4$ and $x \neq 5\}$.

54. $\{x | x$ is a real number and $x \neq 2$ and $x \neq 7\}$

55. Domain of $f = \{x | x$ is a real number and $x \neq 4\}$. Domain of $g = \{x | x$ is a real number$\}$. Since $g(x) = 0$ when $3x^2 = 0$, we have $g(x) = 0$ when $x = 0$. We conclude that Domain of $f/g = \{x | x$ is a real number and $x \neq 4$ and $x \neq 0\}$.

56. $\{x | x$ is a real number and $x \neq -3$ and $x \neq 0\}$

57. Domain of $f = \{x | x$ is a real number$\}$. Domain of $g = \left\{x \middle| x \text{ is a real number and } x \neq \frac{4}{3}\right\}$. Since $g(x) = 0$ when $\dfrac{x-1}{3x-4} = 0$, we have $g(x) = 0$ when $x = 1$. We conclude that Domain of $f/g = \left\{x \middle| x \text{ is a real number and } x \neq \frac{4}{3} \text{ and } x \neq 1\right\}$.

58. $\{x | x$ is a real number and $x \neq -3$ and $x \neq -2\}$

59. Domain of $f = \{x | x$ is a real number and $x \neq 2\}$. Domain of $g = \{x | x$ is a real number and $x \neq 4\}$. Since $g(x) = 0$ when $\dfrac{x-3}{x-4} = 0$, we have $g(x) = 0$ when $x = 3$. We conclude that Domain of $f/g = \{x | x$ is a real number and $x \neq 2$, $x \neq 4$, and $x \neq 3\}$.

60. $\left\{x \middle| x \text{ is a real number and } x \neq -3 \text{ and } x \neq -\frac{1}{2} \text{ and } x \neq -2\right\}$

61. From the graph we see that
Domain of $F = \{x | 0 \leq x \leq 9\}$ and
Domain of $G = \{x | 3 \leq x \leq 10\}$. Then
Domain of $F + G = \{x | 3 \leq x \leq 9\}$.
Since $G(x)$ is never 0,
Domain of $F/G = \{x | 3 \leq x \leq 9\}$.

62. $\{x | 3 \leq x \leq 9\}$; $\{x | 3 \leq x \leq 9\}$;
$\{x | 3 \leq x \leq 9$ and $x \neq 6$ and $x \neq 8\}$

63. We use $(F + G)(x) = F(x) + G(x)$.

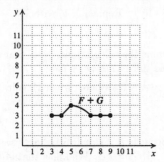

64.

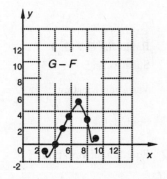

65.
$$(2x^2 - x - 1)(x + 3)$$
$$= (2x^2 - x - 1)(x) + (2x^2 - x - 1)(3)$$
$$= 2x^3 - x^2 - x + 6x^2 - 3x - 3$$
$$= 2x^3 + 5x^2 - 4x - 3$$

66. $9x^2 - 25$

67. $3a^3 + 18a^2 - 4a - 24 = 3a^2(a + 6) - 4(a + 6)$
$$= (a + 6)(3a^2 - 4)$$

68. $\dfrac{x - 7}{x + 5}$

69. ◈

70. ◈

71. Domain of $p = \{x | x$ is a real number and $x \neq 1\}$
(since p is not defined for $x = 1$).
Domain of $q = \{x | x$ is a real number and $x \neq 2\}$.
Since $q(x) = 0$ when $\dfrac{x-3}{x-2} = 0$, we have $q(x) = 0$ when $x = 3$. We conclude that Domain of $p/q = \{x | x$ is a real number and $x \neq 1$, $x \neq 2$, and $x \neq 3\}$.

72. $\left\{x \middle| x \text{ is a real number and } x \neq \frac{3}{2} \text{ and } -1 < x < 5\right\}$

73. The domain of each function is the set of first coordinates for that function.

Domain of $f = \{-2, -1, 0, 1, 2, \}$, and

Domain of $g = \{-4, -3, -2, -1, 0, 1\}$.

Domain of $f + g =$ Domain of $f - g =$ Domain of $f \cdot g = \{-2, -1, 0, 1\}$.

Since $g(-1) = 0$, we conclude that Domain of $f/g = \{-2, 0, 1\}$.

74. 5; 15; 2/3

75. Domain of $F = \{x | x$ is a real number and $x \neq -1$ and $x \neq 1\}$. Domain of $G = \{x | x$ is a real number and $x \neq 3\}$. Since $G(x) = 0$ when $\dfrac{x^2 - 4}{x - 3} = 0$, we have $G(x) = 0$ when $x^2 - 4 = 0$, or when $x = 2$ or $x = -2$. We conclude that Domain of $F/G = \{x | x$ is a real number and $x \neq -1$, $x \neq 1$, $x \neq 3$, $x \neq 2$, and $x \neq -2\}$.

76. $\left\{x \middle| x \text{ is a real number and } x \neq -\frac{5}{2}, x \neq -3, x \neq 1, \text{ and } x \neq -1\right\}$

77. Answers may vary. $f(x) = \dfrac{1}{x + 2}$, $g(x) = \dfrac{1}{x - 5}$

78. Answers may vary.

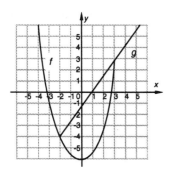

79. a) $(k + l + n)(1983) = k(1983) + l(1983) + n(1983) \approx 67$ million

b) $(l+n)(1986) = l(1986) + n(1986) \approx 50$ million

c) $(l + k)(1986) = (k + l + n)(1986) - n(1986) \approx 80 - 30 \approx 50$ million

---

Exercise Set 7.6

1. $y = kx$

$24 = k \cdot 3$ Substituting

$8 = k$

The variation constant is 8.
The equation of variation is $y = 8x$.

2. $k = \dfrac{5}{12};\ y = \dfrac{5}{12}x$

3. $y = kx$

$3.6 = k \cdot 1$ Substituting

$3.6 = k$

The variation constant is 3.6.
The equation of variation is $y = 3.6x$.

4. $k = \dfrac{2}{5};\ y = \dfrac{2}{5}x$

5. $y = kx$

$30 = k \cdot 8$ Substituting

$\dfrac{30}{8} = k$

$\dfrac{15}{4} = k$

The variation constant is $\dfrac{15}{4}$.

The equation of variation is $y = \dfrac{15}{4}x$.

6. $k = 3;\ y = 3x$

7. $y = kx$

$0.8 = k(0.5)$ Substituting

$8 = k \cdot 5$ Clearing decimals

$\dfrac{8}{5} = k$

$1.6 = k$

The variation constant is 1.6.
The equation of variation is $y = 1.6x$.

8. $k = 1.5;\ y = 1.5x$

9. **Familiarize**. Because of the phrase "$I \ldots$ varies directly as $\ldots V$," we express the current as a function of the voltage. Thus we have $I(V) = kV$. We know that $I(12) = 4$.

Translate. We find the variation constant and then find the equation of variation.

$I(V) = kV$

$I(12) = k \cdot 12$ Replacing V with 12

$4 = k \cdot 12$ Substituting

$\dfrac{4}{12} = k$

$\dfrac{1}{3} = k$

The equation of variation is $I(V) = \dfrac{1}{3}V$.

Carry out. We compute $I(18)$.

$I(V) = \dfrac{1}{3}V$

$I(18) = \dfrac{1}{3} \cdot 18$ Replacing V with 18

$= 6$

Check. Reexamine the calculations. Note that the answer seems reasonable since $12/4 = 18/6$.

State. The current is 6 amperes when 18 volts is applied.

10. $66\dfrac{2}{3}$ cm

11. **Familiarize**. Because A varies directly as G, we write A as a function of G: $A(G) = kG$. We know that $A(9) = 9.66$.

Translate.

$A(G) = kG$

$A(9) = k \cdot 9$ Replacing G with 9

$9.66 = k \cdot 9$ Substituting

$\dfrac{9.66}{9} = k$ Variation constant

$\dfrac{3.22}{3} = k$ Simplifying

$A(G) = \dfrac{3.22}{3}G$ Equation of variation

Carry out. Find $A(4)$.

$A(G) = \dfrac{3.22}{3}G$

$A(4) = \dfrac{3.22}{3} \cdot 4$

≈ 4.29

Check. Reexamine the calculations.

State. The average weekly allowance of a 4th-grade student is about $4.29.

12. 72.5 g

13. **Familiarize**. Because W varies directly as the total mass, we write $W(m) = km$. We know that $W(96) = 64$.

Translate.
$$W(m) = km$$
$$W(96) = k \cdot 96 \quad \text{Replacing } m \text{ with } 96$$
$$64 = k \cdot 96 \quad \text{Substituting}$$
$$\frac{2}{3} = k \quad \text{Variation constant}$$
$$W(m) = \frac{2}{3}m \quad \text{Equation of variation}$$

Carry out. Find $W(75)$.
$$W(m) = \frac{2}{3}m$$
$$W(75) = \frac{2}{3} \cdot 75$$
$$= 50$$

Check. Reexamine the calculations.

State. There are 50 kg of water in a 75 kg person.

14. 40 lb

15. **Familiarize**. Because the f-stop varies directly as F, we write $f(F) = kF$. We know that $F(150) = 6.3$.

Translate.
$$f(F) = kF$$
$$f(150) = k \cdot 150 \quad \text{Replacing } F \text{ with } 150$$
$$6.3 = k \cdot 150 \quad \text{Substituting}$$
$$0.042 = k \quad \text{Variation constant}$$
$$f(F) = 0.042F \quad \text{Equation of variation}$$

Carry out. Find $f(80)$.
$$f(F) = 0.042F$$
$$f(80) = 0.042(80)$$
$$= 3.36$$

Check. Reexamine the calculations.

State. An 80 mm focal length has an f-stop of 3.36.

16. 3,200,000

17.
$$y = \frac{k}{x}$$
$$6 = \frac{k}{10} \quad \text{Substituting}$$
$$60 = k$$

The variation constant is 60.

The equation of variation is $y = \frac{60}{x}$.

18. $k = 64$; $y = \frac{64}{x}$

19.
$$y = \frac{k}{x}$$
$$4 = \frac{k}{3} \quad \text{Substituting}$$
$$12 = k$$

The variation constant is 12.

The equation of variation is $y = \frac{12}{x}$.

20. $k = 36$; $y = \frac{36}{x}$

21.
$$y = \frac{k}{x}$$
$$12 = \frac{k}{3} \quad \text{Substituting}$$
$$36 = k$$

The variation constant is 36.

The equation of variation is $y = \frac{36}{x}$.

22. $k = 45$; $y = \frac{45}{x}$

23.
$$y = \frac{k}{x}$$
$$27 = \frac{k}{\frac{1}{3}} \quad \text{Substituting}$$
$$9 = k$$

The variation constant is 9.

The equation of variation is $y = \frac{9}{x}$.

24. $k = 9$; $y = \frac{9}{x}$

25. **Familiarize**. Because I varies inversely as R, we express I as a function of R. Thus we write $I(R) = k/R$. We know that $I(960) = 2$.

Translate.
$$I(R) = \frac{k}{R}$$
$$I(960) = \frac{k}{960} \quad \text{Replacing } R \text{ with } 960$$
$$2 = \frac{k}{960} \quad \text{Substituting}$$
$$1920 = k \quad \text{Variation constant}$$
$$I(R) = \frac{1920}{R} \quad \text{Equation of variation}$$

Carry out. Find $I(540)$.
$$I(540) = \frac{1920}{540}$$
$$= \frac{32}{9}$$

Check. Reexamine the calculations. Note that, as expected, when the resistance decreases the current increases.

State. The current is $\frac{32}{9}$ amperes.

26. 27 min

27. *Familiarize*. Because V varies inversely as P, we write $V(P) = k/P$. We know that $V(32) = 200$.

Translate.

$$V(P) = \frac{k}{P}$$

$$V(32) = \frac{k}{32} \qquad \text{Replacing } P \text{ with } 32$$

$$200 = \frac{k}{32} \qquad \text{Substituting}$$

$$6400 = k \qquad \text{Variation constant}$$

$$V(P) = \frac{6400}{P} \qquad \text{Equation of variation}$$

Carry out. Find $V(40)$.

$$V(40) = \frac{6400}{40}$$
$$= 160$$

Check. Reexamine the calculations.

State. The volume will be 160 cm^3.

28. 3.5 hr

29. *Familiarize*. Because t varies inversely as r, we write $t(r) = k/r$. We know that $t(80) = 5$.

Translate.

$$t(r) = \frac{k}{r}$$

$$t(80) = \frac{k}{80} \qquad \text{Replacing } r \text{ with } 80$$

$$5 = \frac{k}{80} \qquad \text{Substituting}$$

$$400 = k \qquad \text{Variation constant}$$

$$t(r) = \frac{400}{r} \qquad \text{Equation of variation}$$

Carry out. Find $t(60)$.

$$t(60) = \frac{400}{60}$$
$$= 6\frac{2}{3}$$

Check. Reexamine the calculations.

State. It will take $6\frac{2}{3}$ hr.

30. 2.4 ft

31.
$$y = kx^2$$
$$0.15 = k(0.1)^2 \quad \text{Substituting}$$
$$0.15 = 0.01k$$
$$\frac{0.15}{0.01} = k$$
$$15 = k \qquad \text{Variation constant}$$

The equation of variation is $y = 15x^2$.

32. $y = \frac{2}{3}x^2$

33.
$$y = \frac{k}{x^2}$$
$$0.15 = \frac{k}{(0.1)^2} \qquad \text{Substituting}$$
$$0.15 = \frac{k}{0.01}$$
$$0.15(0.01) = k$$
$$0.0015 = k \qquad \text{Variation constant}$$

The equation of variation is $y = \frac{0.0015}{x^2}$.

34. $y = \frac{54}{x^2}$

35.
$$y = kxz$$
$$56 = k \cdot 7 \cdot 8 \quad \text{Substituting 56 for } y, \text{ 7 for } x, \text{ and}$$
$$\qquad\qquad\qquad\text{8 for } z$$
$$56 = 56k$$
$$1 = k \qquad \text{Variation constant}$$

The equation of variation is $y = xz$.

36. $y = \frac{5x}{z}$

37.
$$y = kxz^2$$
$$105 = k \cdot 14 \cdot 5^2 \quad \text{Substituting 105 for } y,$$
$$\qquad\qquad\qquad\text{14 for } x, \text{ and 5 for } z$$
$$105 = 350k$$
$$\frac{105}{350} = k$$
$$0.3 = k$$

The equation of variation is $y = 0.3xz^2$.

38. $y = \frac{xz}{w}$

39.
$$y = k \cdot \frac{wx^2}{z}$$
$$49 = k \cdot \frac{3 \cdot 7^2}{12} \qquad \text{Substituting}$$
$$4 = k \qquad \text{Variation constant}$$

The equation of variation is $y = \frac{4wx^2}{z}$.

40. $y = \frac{6x}{wz^2}$

41.
$$y = k \cdot \frac{xz}{wp}$$

$$\frac{3}{28} = k \cdot \frac{3 \cdot 10}{7 \cdot 8} \quad \text{Substituting}$$

$$\frac{3}{28} = k \cdot \frac{30}{56}$$

$$\frac{3}{28} \cdot \frac{56}{30} = k$$

$$\frac{1}{5} = k \quad \text{Variation constant}$$

The equation of variation is $y = \dfrac{xz}{5wp}$.

42. $y = \dfrac{5xz}{4w^2}$

43. Familiarize. Because d varies directly as the square of r, we write $d = kr^2$. We know that $d = 200$ when $r = 60$.

Translate.
$$d = kr^2$$
$$200 = k(60)^2$$
$$\frac{1}{18} = k$$

$$d = \frac{1}{18}r^2 \quad \text{Equation of variation}$$

Carry out. Substitute 72 for d and solve for r.
$$72 = \frac{1}{18}r^2$$
$$1296 = r^2$$
$$36 = r$$

Check. Recheck the calculations and perhaps make an estimate to see if the answer seems reasonable.

State. The car can go 36 mph.

44. 220 cm^3

45. Familiarize. I varies inversely as d^2, so we write $I = k/d^2$. We know that $I = 25$ when $d = 2$.

Translate. First we find k.
$$I = \frac{k}{d^2}$$
$$25 = \frac{k}{2^2}$$
$$100 = k$$

$$I = \frac{100}{d^2} \quad \text{Equation of variation}$$

Carry out. Substitute 2.56 for I and solve for d.
$$2.56 = \frac{100}{d^2}$$
$$d^2 = \frac{100}{2.56}$$
$$d = 6.25$$

Check. Recheck the calculations.

State. You are 6.25 km from the transmitter.

46. 5 sec

47. Familiarize. W varies inversely as d^2, so we write $W = k/d^2$. We know that $W = 100$ when $d = 6400$.

Translate. Find k.
$$W = \frac{k}{d^2}$$
$$100 = \frac{k}{(6400)^2}$$
$$4,096,000,000 = k$$

$$W = \frac{4,096,000,000}{d^2} \quad \text{Equation of variation}$$

Carry out. Substitute 64 for w and solve for d.
$$64 = \frac{4,096,000,000}{d^2}$$
$$d^2 = \frac{4,096,000,000}{64}$$
$$d = 8000$$

Note that a distance of 8000 km from the center of the earth is $8000 - 6400$, or 1600 km, above the earth.

Check. Recheck the calculations.

State. The astronaut must be 1600 km above the earth in order to weigh 64 lb.

48. 4.8 cm

49. Familiarize. R varies directly as l and inversely as d^2, so we write $R = kl/d^2$. We know that $R = 0.1$ when $l = 50$ and $d = 1$.

Translate. Find k.
$$R = \frac{kl}{d^2}$$
$$0.1 = \frac{k \cdot 50}{1^2}$$
$$0.002 = k$$

$$R = \frac{0.002l}{d^2} \quad \text{Equation of variation}$$

Carry out. Substitute 1 for R and 2000 for l and solve for d.
$$1 = \frac{0.002(2000)}{d^2}$$
$$d^2 = 4$$
$$d = 2$$

Check. Recheck the calculations.

State. The diameter is 2 mm.

50. 73 kilowatt-hours

51. $ac + bc - a - b = c(a + b) - (a + b)$
$$= (a + b)(c - 1)$$

52. $\dfrac{x(x-1)}{x+4}$

53. $(2x + 0.1)^2 = (2x)^2 + 2(2x)(0.1) + (0.1)^2$
$$(A + B)^2 = A^2 + 2AB + B^2$$
$$= 4x^2 + 0.4x + 0.01$$

54. $x^3 + 1$

55.

56.

57.

58.

59. $P = kS$, where k = the number of sides

In this case $k = 8$, so we have $P = 8S$.

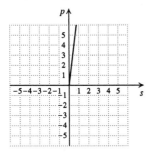

60. $C = 2\pi r$

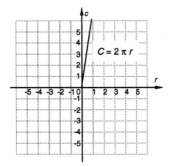

61. $B = kN$

62. $C = kA$

63. a) ***Familiarize.*** We write $N = \dfrac{kP_1P_2}{d^2}$. We let P_1 = the population of Indianapolis and P_2 = the population of Cincinnati. We know that $N = 11,153$ when $P_1 = 744,624$, $P_2 = 452,524$, and $d = 174$.

Translate. We substitute.
$$11,153 = \frac{k(744,624)(452,524)}{(174)^2}$$

Carry out. We solve for k.
$$11,153 = \frac{k(744,624)(452,524)}{(174)^2}$$
$$11,153 = \frac{k(744,624)(452,524)}{30,276}$$
$$337,668,228 = k(744,624)(452,524)$$
$$0.001 \approx k$$

Check. Reexamine the calculations.

State. The value of k is approximately 0.001.

The equation of variation is $N = \dfrac{0.001P_1P_2}{d^2}$.

b) ***Familiarize.*** We will use the equation of variation found in part (a): $N = \dfrac{0.001P_1P_2}{d^2}$. We let P_1 = the population of Indianapolis and P_2 = the population of New York. We know that $N = 4270$ when $P_1 = 744,624$ and $P_2 = 7,895,563$.

Translate. We substitute.
$$4270 = \frac{0.001(744,624)(7,895,563)}{d^2}$$

Carry out. We solve for d.
$$4270 = \frac{0.001(744,624)(7,895,563)}{d^2}$$
$$d^2 = \frac{0.001(744,624)(7,895,563)}{4270}$$
$$d^2 \approx 1,376,868$$
$$d \approx 1173$$

Check. Reexamine the calculations.

State. The distance between Indianapolis and New York is approximately 1173 km.

64. $d = \dfrac{28}{s}$; 70 yd

65. $Q = \dfrac{kp^2}{q^3}$

Q varies directly as the square of p and inversely as the cube of q.

66. W varies jointly as m_1 and M_1 and inversely as the square of d.

67.
$$p = kq$$
$$\frac{1}{k} \cdot p = \frac{1}{k} \cdot kq$$
$$\frac{1}{k} \cdot p = q, \text{ or } q = \frac{1}{k}p$$

Since k is a constant, so is $\dfrac{1}{k}$, and q varies directly as p.

68.

Chapter 8

Solving Equations and Inequalities

Exercise Set 8.1

---

1. *Familiarize*. Let x = the first number and y = the second number.

***Translate*.**

The sum of two numbers is -42.

Rewording:

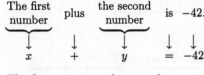

$$x \quad + \quad y \quad = \quad -42$$

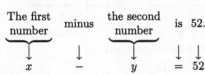

$$x \quad - \quad y \quad = \quad 52$$

We have a system of equations:

$$x + y = -42,$$
$$x - y = 52$$

2. $\quad x - y = 11,$
$\quad 3x + 2y = 123$

3. *Familiarize*. Let x = the number of white scarves sold and y = the number of printed scarves sold.

***Translate*.** Organize the information in a table.

Kind of scarf	White	Printed	Total
Number sold	x	y	40
Price	$4.95	$7.95	
Amount taken in	$4.95x$	$7.95y$	282

The "Number sold" row of the table gives us one equation:

$$x + y = 40$$

The "Amount taken in" row gives us a second equation:

$$4.95x + 7.95y = 282$$

We have a system of equations:

$$x + y = 40,$$
$$4.95x + 7.95y = 282$$

We can multiply the second equation on both sides by 100 to clear the decimals:

$$x + y = 40,$$
$$495x + 795y = 28,200$$

4. $\quad x + y = 45,$
$\quad 8.50x + 9.75y = 398.75$

5. *Familiarize*. Let x = the measure of the first angle and y = the measure of the second angle.

***Translate*.**

Two angles are complementary.

Rewording:

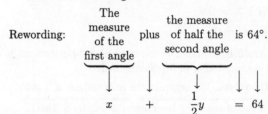

The sum of the measures of the first angle and half the second angle is $64°$.

Rewording:

$$x \quad + \quad \frac{1}{2}y \quad = \quad 64$$

We have a system of equations:

$$x + y = 90,$$
$$x + \frac{1}{2}y = 64$$

6. $\quad x + y = 180,$
$\quad x = 2y - 3$

7. *Familiarize*. Let x = the number of children's plates and y = the number of adult's plates served.

***Translate*.** We organize the information in a table.

Kind of plate	Children's	Adult's	Total
Number sold	x	y	250
Price	$3.50	$7.00	
Amount collected	$3.50x$	$7.00y$	1347.50

The "Number sold" row of the table gives us one equation:

$$x + y = 250$$

The "Amount collected" row gives us a second equation:

$$3.50x + 7.00y = 1347.50$$

We have a system of equations:

$$x + y = 250,$$
$$3.50x + 7.00y = 1347.50$$

We can multiply the both sides of the second equation by 10 to clear the decimals:

$$x + y = 250,$$
$$35x + 70y = 13,475$$

8. $x + y = 18,$

$2x + y = 30$

9. *Familiarize*. The tennis court is a rectangle with perimeter 228 ft. Let l = the length and w = width. Recall that for a rectangle with length l and width w, the perimeter P is given by $P = 2l + 2w$.

***Translate*.** The formula for perimeter gives us one equation:

$$2l + 2w = 228$$

The statement relating width and length gives us a second equation:

Width is 42 ft less than length.

$$w = l - 42$$

We have a system of equations:

$$2w + 2l = 228,$$

$$w = l - 42$$

10. $2l + 2w = 288,$

$l = 44 + w$

11. *Familiarize*. Let x = the number of 2-pointers made and y = the number of 3-pointers made.

***Translate*.** We organize the information in a table.

Type of score	2-pointer	3-pointer	Total
Number scored	x	y	40
Points scored	$2x$	$3y$	89

The "Number scored" row of the table gives us one equation:

$$x + y = 40$$

The "Points scored" row gives us a second equation:

$$2x + 3y = 89$$

We have a system of equations:

$$x + y = 40,$$

$$2x + 3y = 89$$

12. $x + y = 77,$

$3.00x + 1.50y = 213$

13. *Familiarize*. Let x = the number of units of lumber produced and y = the number of units of plywood produced in a day.

***Translate*.** We organize the information in a table.

Product	Lumber	Plywood	Total
Number of units produced	x	y	400
Profit per unit	$20	$30	
Amount of profit	$20x$	$30y$	11,000

The "Number of units produced" row of the table gives us one equation:

$$x + y = 400$$

The "Amount of profit" row gives us a second equation:

$$20x + 30y = 11,000$$

We have a system of equations:

$$x + y = 400,$$

$$20x + 30y = 11,000$$

14. $2x + y = 60,$

$x = 9 + y$

15. *Familiarize*. Let x = number of 30-sec commercials and y = number of 60-sec commercials. The total time used by x 30-sec commercials is $30x$; the total time used by y 60-sec commercials is $60y$. Also note that 10 min = 10×60, or 600 sec.

***Translate*.**

Total number of commercials is 12.

$$x + y \qquad = 12$$

Total commercial time is 10 min, or 600 sec.

$$30x + 60y \qquad = \qquad 600$$

We have a system of equations:

$$x + y = 12,$$

$$30x + 60y = 600$$

16. $x + y = 152,$

$x = 5 + 6y$

17. We use alphabetical order for the variables. We replace x by 1 and y by 2.

$4x - y = 2$		$10x - 3y = 4$	
$4 \cdot 1 - 2 \ ? \ 2$		$10 \cdot 1 - 3 \cdot 2 \ ? \ 4$	
$4 - 2$		$10 - 6$	
2	2 TRUE	4	4 TRUE

The pair $(1, 2)$ makes both equations true, so it is a solution of the system.

18. Yes

19. We use alphabetical order for the variables. We replace x by 2 and y by 5.

$y = 3x - 1$		$2x + y = 4$	
$5 \ ? \ 3 \cdot 2 - 1$		$2 \cdot 2 + 5 \ ? \ 4$	
$6 - 1$		$4 + 5$	
5	5　　TRUE	9	4 FALSE

The pair $(2, 5)$ is not a solution of $2x + y = 4$. Therefore, it is not a solution of the system of equations.

20. No

21. We replace x by 1 and y by 5.

$$\frac{x+y=6}{\begin{array}{c|c} 1+5 \ ? \ 6 & \\ 6 & 6 \quad \text{TRUE} \end{array}}$$

$$\frac{y=2x+3}{\begin{array}{c|c} 5 \ ? \ 2\cdot 1+3 \\ & 2+3 \\ 5 & 5 \qquad \text{TRUE} \end{array}}$$

The pair $(1,5)$ makes both equations true, so it is a solution of the system.

22. Yes

23. We replace a by 2 and b by -7.

$$\frac{3a+b=-1}{\begin{array}{c|c} 3\cdot 2+(-7) \ ? \ -1 & \\ 6-7 & \\ -1 & -1 \quad \text{TRUE} \end{array}}$$

$$\frac{2a-3b=-8}{\begin{array}{c|c} 2\cdot 2-3\cdot(-7) \ ? \ -8 & \\ 4+21 & \\ 25 & -8 \quad \text{FALSE} \end{array}}$$

The pair $(2,-7)$ is not a solution of $2a-3b=-8$. Therefore, it is not a solution of the system of equations.

24. No

25. We replace x by 3 and y by 1.

$$\frac{3x+4y=13}{\begin{array}{c|c} 3\cdot 3+4\cdot 1 \ ? \ 13 & \\ 9+4 & \\ 13 & 13 \quad \text{TRUE} \end{array}}$$

$$\frac{5x-4y=11}{\begin{array}{c|c} 5\cdot 3-4\cdot 1 \ ? \ 11 & \\ 15-4 & \\ 11 & 11 \quad \text{TRUE} \end{array}}$$

The pair $(3,1)$ makes both equations true, so it is a solution of the system.

26. No

27. Graph both lines on the same set of axes.

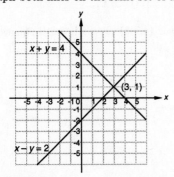

The solution (point of intersection) seems to be the point $(3,1)$.

Check:

$$\frac{x+y=4}{\begin{array}{c|c} 3+1 \ ? \ 4 & \\ 4 & 4 \quad \text{TRUE} \end{array}}$$

$$\frac{x-y=2}{\begin{array}{c|c} 3-1 \ ? \ 2 & \\ 2 & 2 \quad \text{TRUE} \end{array}}$$

The solution is $(3,1)$.

28. $(4,1)$

29. Graph both lines on the same set of axes.

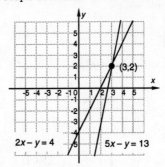

The solution (point of intersection) seems to be the point $(3,2)$.

Check:

$$\frac{2x-y=4}{\begin{array}{c|c} 2\cdot 3-2 \ ? \ 4 & \\ 6-2 & \\ 4 & 4 \quad \text{TRUE} \end{array}}$$

$$\frac{5x-y=13}{\begin{array}{c|c} 5\cdot 3-2 \ ? \ 13 & \\ 15-2 & \\ 13 & 13 \quad \text{TRUE} \end{array}}$$

The solution is $(3,2)$.

30. $(2,-1)$

31. Graph both lines on the same set of axes.

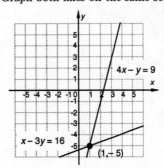

The solution seems to be the point $(1,-5)$.

Check:

$$\frac{4x-y=9}{\begin{array}{c|c} 4\cdot 1-(-5) \ ? \ 9 & \\ 4+5 & \\ 9 & 9 \quad \text{TRUE} \end{array}}$$

$$\frac{x-3y=16}{\begin{array}{c|c} 1-3(-5) \ ? \ 16 & \\ 1+15 & \\ 16 & 16 \quad \text{TRUE} \end{array}}$$

The solution is $(1,-5)$.

32. $(4,3)$

33. Graph both lines on the same set of axes.

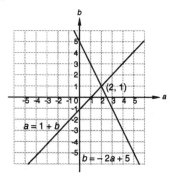

The solution seems to be the point $(2, 1)$.

Check:

$a = 1 + b$		$b = -2a + 5$	
$2 ? 1 + 1$		$1 ? -2 \cdot 2 + 5$	
$2 \mid 2$	TRUE	$-4 + 5$	
		$1 \mid 1$	TRUE

The solution is $(2, 1)$.

34. $(-3, -2)$

35. Graph both lines on the same set of axes.

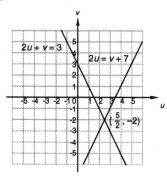

The solution seems to be $\left(\dfrac{5}{2}, -2\right)$.

Check:

$2u + v = 3$		$2u = v + 7$	
$2 \cdot \dfrac{5}{2} + (-2) ? 3$		$2 \cdot \dfrac{5}{2} ? -2 + 7$	
$5 - 2$		$5 \mid 5$	TRUE
	$3 \mid 3$ TRUE		

The solution is $\left(\dfrac{5}{2}, -2\right)$.

36. $(7, 2)$

37. Graph both lines on the same set of axes.

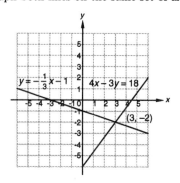

The ordered pair $(3, -2)$ checks in both equations. It is the solution.

38. $(4, 0)$

39. Graph both lines on the same set of axes.

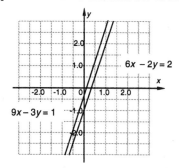

The lines are parallel. The solution set is $\emptyset$.

40. $\emptyset$

41. Graph both lines on the same set of axes.

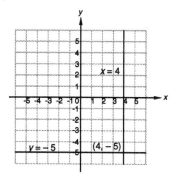

The ordered pair $(4, -5)$ checks in both equations. It is the solution.

42. $(-3, 2)$

43. Graph both lines on the same set of axes.

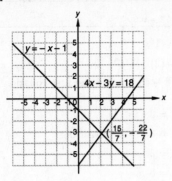

The ordered pair $(3, -4)$ checks in both equations. It is the solution.

44. $\left(\dfrac{1}{2}, 3\right)$

45. Graph both lines on the same set of axes.

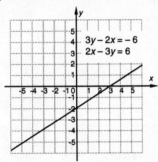

The graphs are the same. Any solution of one equation is a solution of the other. Each equation has infinitely many solutions. The solution set is the set of all pairs (x, y) for which $2x - 3y = 6$, or $\{(x, y) | 2x - 3y = 6\}$. (In place of $2x - 3y = 6$ we could have used $3y - 2x = -6$ since the two equations are equivalent.)

46. $\{(x, y) | y = 3 - x\}$

47. A system of equations is consistent if it has at least one solution. Of the systems under consideration, only the one in Exercise 39 has no solution. Therefore, all except the system in Exercise 39 are consistent.

48. All except 40

49. A system of two equations in two variables is dependent if it has infinitely many solutions. Only the system in Exercise 45 is dependent.

50. 46

51. $3x + 4 = x - 2$

$\quad 2x + 4 = -2 \quad$ Adding $-x$ on both sides

$\quad\quad 2x = -6 \quad$ Adding -4 on both sides

$\quad\quad\ x = -3 \quad$ Multiplying by $\dfrac{1}{2}$ on both sides

The solution is -3.

52. -35

53. $4x - 5x = 8x - 9 + 11x$

$\quad -x = 19x - 9 \quad$ Collecting like terms

$\quad -20x = -9 \quad$ Adding $-19x$ on both sides

$\quad\quad x = \dfrac{9}{20} \quad$ Multiplying by $-\dfrac{1}{20}$ on both sides

The solution is $\dfrac{9}{20}$.

54. $b = a - 4Q$

55. ◈

56. ◈

57. 1983 and 1990

58. 1989

59. 1987

60. $A = -\dfrac{17}{4}, \ B = -\dfrac{12}{5}$

61. a) There are many correct answers. One can be found by expressing the sum and difference of the two numbers:

$$x + y = 6,$$
$$x - y = 4$$

b) There are many correct answers. For example, write an equation in two variables. Then write a second equation by multiplying the left side of the first equation by one constant and multiplying the right side by another constant.

$$x + y = 1,$$
$$2x + 2y = 3$$

c) There are many correct answers. One can be found by writing an equation in two variables and then writing a constant multiple of that equation:

$$x + y = 1,$$
$$2x + 2y = 2$$

62. a) $(4, -5)$; answers may vary.

b) Infinitely many

63. *Familiarize*. Let b = Burt's age and s = his son's age. Ten years ago, Burt's age was $b - 10$ and his son's age was $s - 10$.

Translate.

Burt is twice as old as his son.

$b \ = 2s$

Ten years ago

Burt was three times as old as his son.

$b - 10 \ = \ 3(s - 10)$

We have a system of equations:

$$b = 2s,$$
$$b - 10 = 3(s - 10)$$

64. $x + y = 46,$

$\quad x - 2 = 2.5(y - 2)$

65. *Familiarize*. Let l = the length and w = the width. If you cut 6 in. off the width, the new width is $w - 6$.

***Translate*.**

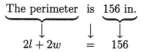

The perimeter is 156 in.

$$2l + 2w \quad = \quad 156$$

If you cut 6 in. off the width

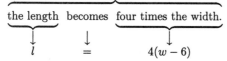

the length becomes four times the width.

$$l \qquad = \qquad 4(w - 6)$$

We have a system of equations:

$$2l + 2w = 156,$$
$$l = 4(w - 6)$$

66. $w + l = 104,$

$\quad w - 8 = 3l$

67. Graph both equations on the same set of axes.

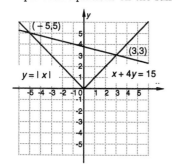

The solutions seem to be $(-5, 5)$ and $(3, 3)$. Both pairs check.

68. $(0, 0)$, $(1, 1)$

69. $(-0.39, -1.10)$

70. $(0.07, -7.95)$

71. $(-0.13, 0.67)$

72. $(0.02, 1.25)$

Exercise Set 8.2

1. $3x + 5y = 3,$ (1)

$\quad x = 8 - 4y$ (2)

We substitute $8 - 4y$ for x in the first equation and solve for y.

$$3x + 5y = 3 \qquad (1)$$
$$3(8 - 4y) + 5y = 3 \qquad \text{Substituting}$$
$$24 - 12y + 5y = 3$$
$$24 - 7y = 3$$
$$-7y = -21$$
$$y = 3$$

Next we substitute 3 for y in either equation of the original system and solve for x.

$$x = 8 - 4y \qquad (2)$$
$$x = 8 - 4 \cdot 3 \quad \text{Substituting}$$
$$x = 8 - 12$$
$$x = -4$$

We check the ordered pair $(-4, 3)$.

$$\frac{3x + 5y = 3}{3(-4) + 5 \cdot 3 \ ? \ 3}$$
$$\begin{array}{c|c} -12 + 15 & \\ 3 & 3 \quad \text{TRUE} \end{array}$$

$$\frac{x = 8 - 4y}{-4 \ ? \ 8 - 4 \cdot 3}$$
$$\begin{array}{c|c} & 8 - 12 \\ -4 & -4 \qquad \text{TRUE} \end{array}$$

Since $(-4, 3)$ checks, it is the solution.

2. $(2, -3)$

3. $9x - 2y = 3,$ (1)

$\quad 3x - 6 = y$ (2)

We substitute $3x - 6$ for y in the first equation and solve for x.

$$9x - 2y = 3 \qquad (1)$$
$$9x - 2(3x - 6) = 3 \qquad \text{Substituting}$$
$$9x - 6x + 12 = 3$$
$$3x + 12 = 3$$
$$3x = -9$$
$$x = -3$$

Next we substitute -3 for x in either equation of the original system and solve for y.

$$3x - 6 = y \qquad (2)$$
$$3(-3) - 6 = y \quad \text{Substituting}$$
$$-9 - 6 = y$$
$$-15 = y$$

We check the ordered pair $(-3, -15)$.

$$\frac{9x - 2y = 3}{\begin{array}{c|c} 9(-3) - 2(-15) \ ? \ 3 & \\ -27 + 30 & \\ \hline 3 & 3 \quad \text{TRUE} \end{array}}$$

$$\frac{3x - 6 = y}{\begin{array}{c|c} 3(-3) - 6 \ ? \ -15 & \\ -9 - 6 & \\ \hline -15 & -15 \quad \text{TRUE} \end{array}}$$

Since $(-3, -15)$ checks, it is the solution.

4. $\left(\dfrac{21}{5}, \dfrac{12}{5}\right)$

5. $5m + n = 8,$ (1)

$3m - 4n = 14$ (2)

We solve the first equation for n.

$5m + n = 8$ (1)

$n = 8 - 5m$ (3)

We substitute $8 - 5m$ for n in the second equation and solve for m.

$$3m - 4n = 14 \quad (2)$$
$$3m - 4(8 - 5m) = 14 \quad \text{Substituting}$$
$$3m - 32 + 20m = 14$$
$$23m - 32 = 14$$
$$23m = 46$$
$$m = 2$$

Now we substitute 2 for m in Equation (1), (2), or (3). It is easiest to use Equation (3) since it is already solved for n.

$$n = 8 - 5m = 8 - 5(2) = 8 - 10 = -2$$

We check the ordered pair $(2, -2)$.

$$\frac{5m + n = 8}{\begin{array}{c|c} 5 \cdot 2 + (-2) \ ? \ 8 & \\ 10 - 2 & \\ \hline 8 & 8 \quad \text{TRUE} \end{array}}$$

$$\frac{3m - 4n = 14}{\begin{array}{c|c} 3 \cdot 2 - 4(-2) \ ? \ 14 & \\ 6 + 8 & \\ \hline 14 & 14 \quad \text{TRUE} \end{array}}$$

Since $(2, -2)$ checks, it is the solution.

6. $(2, -7)$

7. $4x + 12y = 4,$ (1)

$-5x + y = 11$ (2)

We solve the second equation for y.

$-5x + y = 11$ (2)

$y = 5x + 11$ (3)

We substitute $5x + 11$ for y in the first equation and solve for x.

$$4x + 12y = 4 \quad (1)$$
$$4x + 12(5x + 11) = 4 \quad \text{Substituting}$$
$$4x + 60x + 132 = 4$$
$$64x + 132 = 4$$
$$64x = -128$$
$$x = -2$$

Now substitute -2 for x in Equation (3).

$$y = 5x + 11 = 5(-2) + 11 = -10 + 11 = 1$$

We check the ordered pair $(-2, 1)$.

$$\frac{4x + 12y = 4}{\begin{array}{c|c} 4(-2) + 12 \cdot 1 \ ? \ 4 & \\ -8 + 12 & \\ \hline 4 & 4 \quad \text{TRUE} \end{array}}$$

$$\frac{-5x + y = 11}{\begin{array}{c|c} -5(-2) + 1 \ ? \ 11 & \\ 10 + 1 & \\ \hline 11 & 11 \quad \text{TRUE} \end{array}}$$

Since $(-2, 1)$ checks, it is the solution.

8. $(4, -1)$

9. $3x - y = 1,$ (1)

$2x + 2y = 2$ (2)

We solve the first equation for y.

$3x - y = 1$ (1)

$3x - 1 = y$ (3)

We substitute $3x - 1$ for y in the second equation and solve for x.

$$2x + 2y = 2 \quad (2)$$
$$2x + 2(3x - 1) = 2 \quad \text{Substituting}$$
$$2x + 6x - 2 = 2$$
$$8x - 2 = 2$$
$$8x = 4$$
$$x = \frac{1}{2}$$

Next we substitute $\dfrac{1}{2}$ for x in Equation (3).

$$y = 3x - 1 = 3 \cdot \frac{1}{2} - 1 = \frac{3}{2} - 1 = \frac{1}{2}$$

We check the ordered pair $\left(\dfrac{1}{2}, \dfrac{1}{2}\right)$.

$$\frac{3x - y = 1}{\begin{array}{c|c} 3\cdot\frac{1}{2} - \frac{1}{2} \ ? \ 1 \\ \frac{3}{2} - \frac{1}{2} \\ 1 & 1 \ \text{TRUE} \end{array}} \qquad \frac{2x + 2y = 2}{\begin{array}{c|c} 2\cdot\frac{1}{2} + 2\cdot\frac{1}{2} \ ? \ 2 \\ 1 + 1 \\ 2 & 2 \ \text{TRUE} \end{array}}$$

Since $\left(\frac{1}{2}, \frac{1}{2}\right)$ checks, it is the solution.

10. $(3, -2)$

11. $3x - y = 7$,   (1)

  $2x + 2y = 5$ (2)

We solve the first equation for y.

$3x - y = 7$ (1)

$3x - 7 = y$ (3)

We substitute $3x - 7$ for y in the second equation and solve for x.

$$2x + 2y = 5 \qquad (2)$$
$$2x + 2(3x - 7) = 5 \qquad \text{Substituting}$$
$$2x + 6x - 14 = 5$$
$$8x - 14 = 5$$
$$8x = 19$$
$$x = \frac{19}{8}$$

Now we substitute $\frac{19}{8}$ for x in Equation (3).

$$y = 3x - 7 = 3\cdot\frac{19}{8} - 7 = \frac{57}{8} - 7 = \frac{1}{8}$$

The ordered pair $\left(\frac{19}{8}, \frac{1}{8}\right)$ checks in both equations. It is the solution.

12. $\left(\frac{25}{23}, -\frac{11}{23}\right)$

13. $x + 2y = 6$,   (1)

  $x = 4 - 2y$ (2)

We substitute $4 - 2y$ for x in the first equation and solve for y.

$$x + 2y = 6 \quad (1)$$
$$(4 - 2y) + 2y = 6 \quad \text{Substituting}$$
$$4 - 2y + 2y = 6$$
$$4 = 6 \quad \text{Collecting like terms}$$

We have a false equation. Therefore, there is no solution. The solution set is $\emptyset$.

14. $\emptyset$

15. $x - 3 = y$,   (1)

  $2x - 2y = 6$ (2)

We substitute $x - 3$ for y in the second equation and solve for x.

$$2x - 2y = 6 \quad (2)$$
$$2x - 2(x - 3) = 6 \quad \text{Substituting}$$
$$2x - 2x + 6 = 6$$
$$6 = 6 \quad \text{Collecting like terms}$$

We have an equation that is true for all numbers x and y. The system is dependent and has an infinite number of solutions. Using (1), we can express the solution set as $\{(x, y) | x - 3 = y\}$.

16. $\{(x, y) | 3y = x - 2\}$

17. $x + 3y = 7$

  $\dfrac{-x + 4y = 7}{0 + 7y = 14} \quad \text{Adding}$

$$7y = 14$$
$$y = 2$$

Substitute 2 for y in one of the original equations and solve for x.

$$x + 3y = 7$$
$$x + 3\cdot 2 = 7 \quad \text{Substituting}$$
$$x + 6 = 7$$
$$x = 1$$

Check:

$$\frac{x + 3y = 7}{\begin{array}{c|c} 1 + 3\cdot 2 \ ? \ 7 \\ 1 + 6 \\ 7 & 7 \ \text{TRUE} \end{array}} \qquad \frac{-x + 4y = 7}{\begin{array}{c|c} -1 + 4\cdot 2 \ ? \ 7 \\ -1 + 8 \\ 7 & 7 \ \text{TRUE} \end{array}}$$

Since $(1, 2)$ checks, it is the solution.

18. $(2, 7)$

19. $2x + y = 6$

  $\dfrac{x - y = 3}{3x + 0 = 9} \quad \text{Adding}$

$$3x = 9$$
$$x = 3$$

Substitute 3 for x in one of the original equations and solve for y.

$$2x + y = 6$$
$$2\cdot 3 + y = 6 \quad \text{Substituting}$$
$$6 + y = 6$$
$$y = 0$$

We obtain $(3, 0)$. This checks, so it is the solution.

20. $(10, 2)$

21. $9x + 3y = -3$

  $\dfrac{2x - 3y = -8}{11x + 0 = -11} \quad \text{Adding}$

$$11x = -11$$
$$x = -1$$

Substitute -1 for x in one of the original equations and solve for y.

$$9x + 3y = -3$$
$$9(-1) + 3y = -3 \quad \text{Substituting}$$
$$-9 + 3y = -3$$
$$3y = 6$$
$$y = 2$$

We obtain $(-1, 2)$. This checks, so it is the solution.

22. $\left(\dfrac{1}{2}, -5\right)$

23. $5x + 3y = 19, \quad (1)$
$2x - 5y = 11 \quad (2)$

We multiply twice to make two terms become opposites.

From (1): $\quad 25x + 15y = 95 \quad$ Multiplying by 5
From (2): $\quad \underline{6x - 15y = 33} \quad$ Multiplying by 3
$ 31x = 128 \quad$ Adding
$$x = \frac{128}{31}$$

Substitute $\dfrac{128}{31}$ for x in one of the original equations and solve for y.

$$5x + 3y = 19$$
$$5 \cdot \frac{128}{31} + 3y = 19 \quad \text{Substituting}$$
$$\frac{640}{31} + 3y = \frac{589}{31}$$
$$3y = -\frac{51}{31}$$
$$\frac{1}{3} \cdot 3y = \frac{1}{3} \cdot \left(-\frac{51}{31}\right)$$
$$y = -\frac{17}{31}$$

We obtain $\left(\dfrac{128}{31}, -\dfrac{17}{31}\right)$. This checks, so it is the solution.

24. $\left(\dfrac{10}{21}, \dfrac{11}{14}\right)$

25. $5r - 3s = 24, \quad (1)$
$3r + 5s = 28 \quad (2)$

We multiply twice to make two terms become additive inverses.

From (1): $\quad 25r - 15s = 120 \quad$ Multiplying by 5
From (2): $\quad \underline{9r + 15s = 84} \quad$ Multiplying by 3
$ 34r + 0 = 204 \quad$ Adding
$$r = 6$$

Substitute 6 for r in one of the original equations and solve for s.

$$3r + 5s = 28$$
$$3 \cdot 6 + 5s = 28 \quad \text{Substituting}$$
$$18 + 5s = 28$$
$$5s = 10$$
$$s = 2$$

We obtain $(6, 2)$. This checks, so it is the solution.

26. $(1, 3)$

27. $0.3x - 0.2y = 4,$
$0.2x + 0.3y = 1$

We first multiply each equation by 10 to clear decimals.

$$3x - 2y = 40 \quad (1)$$
$$2x + 3y = 10 \quad (2)$$

We use the multiplication principle with both equations of the resulting system.

From (1): $\quad 9x - 6y = 120 \quad$ Multiplying by 3
From (2): $\quad \underline{4x + 6y = 20} \quad$ Multiplying by 2
$13x + 0 = 140 \quad$ Adding
$$x = \frac{140}{13}$$

Substitute $\dfrac{140}{13}$ for x in one of the equations in which the decimals were cleared and solve for y.

$$2x + 3y = 10$$
$$2 \cdot \frac{140}{13} + 3y = 10 \quad \text{Substituting}$$
$$\frac{280}{13} + 3y = \frac{130}{13}$$
$$3y = -\frac{150}{13}$$
$$y = -\frac{50}{13}$$

We obtain $\left(\dfrac{140}{13}, -\dfrac{50}{13}\right)$. This checks, so it is the solution.

28. $(2, 3)$

29. $\dfrac{1}{2}x + \dfrac{1}{3}y = 4, \quad (1)$
$\phantom{\dfrac{1}{2}x +} \dfrac{1}{4}x + \dfrac{1}{3}y = 3 \quad (2)$

We first multiply each equation by the LCM of the denominators to clear fractions.

From (1): $3x + 2y = 24 \quad$ Multiplying by 6
From (2): $3x + 4y = 36 \quad$ Multiplying by 12

We multiply by -1 on both sides of the first equation and then add.

$\quad -3x - 2y = -24 \quad$ Multiplying by -1
$\quad \underline{3x + 4y = 36}$
$ 2y = 12 \quad$ Adding
$$y = 6$$

Substitute 6 for y in one of the equations in which the fractions were cleared and solve for x.

$$3x + 2y = 24$$
$$3x + 2 \cdot 6 = 24 \quad \text{Substituting}$$
$$3x + 12 = 24$$
$$3x = 12$$
$$x = 4$$

We obtain $(4, 6)$. This checks, so it is the solution.

30. $(-21, 21)$

31. $\dfrac{2}{5}x + \dfrac{1}{2}y = 2,$

$\dfrac{1}{2}x - \dfrac{1}{6}y = 3$

We first multiply each equation by the LCM of the denominators to clear fractions.

From (1): $4x + 5y = 20$ Multiplying by 10

From (2): $3x - y = 18$ Multiplying by 6

We multiply by 5 on both sides of the second equation and then add.

$$4x + 5y = 20$$
$$\underline{15x - 5y = 90} \quad \text{Multiplying by 5}$$
$$19x + 0 = 110 \quad \text{Adding}$$
$$19x = 110$$
$$x = \frac{110}{19}$$

Substitute $\dfrac{110}{19}$ for x in one of the equations in which fractions were cleared and solve for y.

$$3x - y = 18$$
$$3\left(\frac{110}{19}\right) - y = 18 \quad \text{Substituting}$$
$$\frac{330}{19} - y = \frac{342}{19}$$
$$-y = \frac{12}{19}$$
$$y = -\frac{12}{19}$$

We obtain $\left(\dfrac{110}{19}, -\dfrac{12}{19}\right)$. This checks, so it is the solution.

32. $(12, 15)$

33. $2x + 3y = 1,$

$4x + 6y = 2$

Multiply the first equation by -2 and then add.

$$-4x - 6y = -2$$
$$\underline{4x + 6y = 2}$$
$$0 = 0 \quad \text{Adding}$$

We have an equation that is true for all numbers x and y. The system is dependent and has an infinite number of solutions. The solution set can be expressed using either equation. Using the first equation, we have $\{(x, y) | 2x + 3y = 1\}$.

34. $\{(x, y) | 3x - 2y = 1\}$

35. $2x - 4y = 5,$

$2x - 4y = 6$

Multiply the first equation by -1 and then add.

$$-2x + 4y = -5$$
$$\underline{2x - 4y = 6}$$
$$0 = 1$$

We have a false equation. The system has no solution. The solution set is $\emptyset$.

36. $\emptyset$

37. $5x - 9y = 7,$

$7y - 3x = -5$

We first write the second equation in the form $Ax + By = C$.

$$5x - 9y = 7 \quad (1)$$
$$-3x + 7y = -5 \quad (2)$$

We use the multiplication principle with both equations and then add.

From (1): $15x - 27y = 21$ Multiplying by 3

From (2): $\underline{-15x + 35y = -25}$ Multiplying by 5

$$8y = -4 \quad \text{Adding}$$
$$y = -\frac{1}{2}$$

Substitute $-\dfrac{1}{2}$ for y in one of the original equations and solve for x.

$$5x - 9y = 7 \quad (1)$$
$$5x - 9\left(-\frac{1}{2}\right) = 7 \quad \text{Substituting}$$
$$5x + \frac{9}{2} = \frac{14}{2}$$
$$5x = \frac{5}{2}$$
$$x = \frac{1}{2}$$

We obtain $\left(\dfrac{1}{2}, -\dfrac{1}{2}\right)$. This checks, so it is the solution.

38. $(-2, -9)$

39. $3(a - b) = 15,$

$4a = b + 1$

We first write each equation in the form $Ax + By = C$.

$$3a - 3b = 15$$
$$4a - b = 1$$

We multiply by -3 on both sides of the second equation and then add.

$$3a - 3b = 15$$

$$\underline{-12a + 3b = -3} \quad \text{Multiplying by } -3$$

$$-9a \qquad = 12$$

$$a = -\frac{12}{9}$$

$$a = -\frac{4}{3}$$

Substitute $-\frac{4}{3}$ for a in one of the equations and solve for b.

$$4a - b = 1$$

$$4\left(-\frac{4}{3}\right) - b = 1 \qquad \text{Substituting}$$

$$-\frac{16}{3} - b = \frac{3}{3}$$

$$-b = \frac{19}{3}$$

$$b = -\frac{19}{3}$$

We obtain $\left(-\frac{4}{3}, -\frac{19}{3}\right)$. This checks, so it is the solution.

40. $(30, 6)$

41. $x - \frac{1}{10}y = 100,$

$$y - \frac{1}{10}x = -100$$

We first write the second equation in the form $Ax + By = C$.

$$x - \frac{1}{10}y = 100$$

$$-\frac{1}{10}x + \quad y = -100$$

Next we multiply each equation by 10 to clear fractions.

$$10x - \quad y = 1000$$

$$-x + 10y = -1000$$

We multiply by 10 on both sides of the first equation and then add.

$$100x - 10y = 10,000 \quad \text{Multiplying by 10}$$

$$\underline{-x + 10y = -1000}$$

$$99x \qquad = 9000$$

$$x = \frac{9000}{99}$$

$$x = \frac{1000}{11}$$

Substitute $\frac{1000}{11}$ for x in one of the equations in which the fractions were cleared and solve for y.

$$10x - y = 1000$$

$$10\left(\frac{1000}{11}\right) - y = 1000 \qquad \text{Substituting}$$

$$\frac{10,000}{11} - y = \frac{11,000}{11}$$

$$-y = \frac{1000}{11}$$

$$y = -\frac{1000}{11}$$

We obtain $\left(\frac{1000}{11}, -\frac{1000}{11}\right)$. This checks, so it is the solution.

42. $(-20, 20)$

43. $0.05x + 0.25y = 22,$

$$0.15x + 0.05y = 24$$

We first multiply each equation by 100 to clear decimals.

$$5x + 25y = 2200$$

$$15x + \quad 5y = 2400$$

We multiply by -5 on both sides of the second equation and add.

$$5x + 25y = \qquad 2200$$

$$\underline{-75x - 25y = -12,000} \quad \text{Multiplying by } -5$$

$$-70x \qquad = -9800 \quad \text{Adding}$$

$$x = \frac{-9800}{-70}$$

$$x = 140$$

Substitute 140 for x in one of the equations in which the decimals were cleared and solve for y.

$$5x + 25y = 2200$$

$$5 \cdot 140 + 25y = 2200 \quad \text{Substituting}$$

$$700 + 25y = 2200$$

$$25y = 1500$$

$$y = 60$$

We obtain $(140, 60)$. This checks, so it is the solution.

44. $(10, 5)$

45. $9a^5b^2 = 3 \cdot 3 \cdot a \cdot a \cdot a \cdot a \cdot a \cdot b \cdot b$

$$6ab^6 = 2 \cdot 3 \cdot a \cdot b \cdot b \cdot b \cdot b \cdot b \cdot b$$

The LCM includes each factor the greatest number of times that it occurs in any one factorization.

LCM $= 2 \cdot 3 \cdot 3 \cdot a \cdot a \cdot a \cdot a \cdot a \cdot b \cdot b \cdot b \cdot b \cdot b \cdot b$, or $18a^5b^6$

46. $t(t + 2)(t - 2)$

47. ◈

48. ◈

49. $3.5x - 2.1y = 106.2$,

    $4.1x + 16.7y = -106.28$

Since this is a calculator exercise, you may choose not to clear the decimals. We will do so here, however.

    $35x - 21y = 1062$ Multiplying by 10

  $410x + 1670y = -10,628$ Multiplying by 100

Multiply twice to make two terms become opposites.

$58,450x - 35,070y = 1,773,540$ Multiplying by 1670

$\underline{\quad 8610x + 35,070y = -223,188\quad}$ Multiplying by 21

$67,060x \qquad\qquad = 1,550,352$ Adding

$$x \approx 23.118879$$

Substitute 23.118879 for x in one of the equations in which the decimals were cleared and solve for y.

$$35x - 21y = 1062$$

$$35(23.118879) - 21y = 1062 \qquad \text{Substituting}$$

$$809.160765 - 21y = 1062$$

$$-21y = 252.839235$$

$$y \approx -12.039964$$

The numbers check, so the solution is $(23.118879, -12.039964)$.

50. $\left(-\dfrac{32}{17}, \dfrac{38}{17}\right)$

51. $5x + 2y = a$,

    $x - y = b$

We multiply by 2 on both sides of the second equation and then add.

$5x + 2y = a$

$\underline{2x - 2y = 2b\qquad}$ Multiplying by 2

$7x \qquad = a + 2b$ Adding

$$x = \frac{a+2b}{7}$$

Next we multiply by -5 on both sides of the second equation and then add.

$5x + 2y = a$

$\underline{-5x + 5y = -5b\qquad}$ Multiplying by -5

$7y = a - 5b$

$$y = \frac{a-5b}{7}$$

We obtain $\left(\dfrac{a+2b}{7}, \dfrac{a-5b}{7}\right)$. This checks, so it is the solution.

52. $a = 5$, $b = 2$

53. $(0, -3)$ and $\left(-\dfrac{3}{2}, 6\right)$ are two solutions of $px - qy = -1$.

Substitute 0 for x and -3 for y.

$$p \cdot 0 - q \cdot (-3) = -1$$

$$3q = -1$$

$$q = -\frac{1}{3}$$

Substitute $-\dfrac{3}{2}$ for x and 6 for y.

$$p \cdot \left(-\frac{3}{2}\right) - q \cdot 6 = -1$$

$$-\frac{3}{2}p - 6q = -1$$

Substitute $-\dfrac{1}{3}$ for q and solve for p.

$$-\frac{3}{2}p - 6 \cdot \left(-\frac{1}{3}\right) = -1$$

$$-\frac{3}{2}p + 2 = -1$$

$$-\frac{3}{2}p = -3$$

$$-\frac{2}{3} \cdot \left(-\frac{3}{2}p\right) = -\frac{2}{3} \cdot (-3)$$

$$p = 2$$

Thus, $p = 2$ and $q = -\dfrac{1}{3}$.

54. $m = -\dfrac{1}{2}$, $b = \dfrac{5}{2}$

55. $\dfrac{1}{x} - \dfrac{3}{y} = 2$, $\dfrac{1}{x} - 3 \cdot \dfrac{1}{y} = 2$,

                         or

$\dfrac{6}{x} + \dfrac{5}{y} = -34$ $6 \cdot \dfrac{1}{x} + 5 \cdot \dfrac{1}{y} = -34$

Substitute u for $\dfrac{1}{x}$ and v for $\dfrac{1}{y}$.

$u - 3v = 2$,   (1)

$6u + 5v = -34$ (2)

$-6u + 18v = -12$ Multiplying (1) by -6

$\underline{\quad 6u + 5v = -34\quad}$ (2)

$23v = -46$ Adding

$v = -2$

Substitute -2 for v in (1).

$u - 3v = 2$

$u - 3(-2) = 2$

$u = -4$

If $u = -4$, then $\dfrac{1}{x} = -4$, so $x = -\dfrac{1}{4}$.

If $v = -2$, then $\dfrac{1}{y} = -2$, so $y = -\dfrac{1}{2}$.

The solution is $\left(-\dfrac{1}{4}, -\dfrac{1}{2}\right)$.

56. $\left(-\dfrac{1}{5}, \dfrac{1}{10}\right)$

Exercise Set 8.3

1. The Familiarize and Translate steps were done in Exercise 1 of Exercise Set 8.1

Carry out. We solve the system of equations

$$x + y = -42,$$
$$x - y = 52$$

where $x =$ the first number and $y =$ the second number. We use the elimination method.

$$
\begin{array}{r}
x + y = -42 \\
\underline{x - y = 52} \\
2x = 10 \quad \text{Adding} \\
x = 5
\end{array}
$$

Substitute 5 for x in one of the equations and solve for y.

$$x + y = -42$$
$$5 + y = -42$$
$$y = -47$$

Check. The sum of the numbers is $5 + (-47)$, or -42. The difference is $5 - (-47)$, or 52. The numbers check.

State. The numbers are 5 and -47.

2. 29 and 18

3. The Familiarize and Translate steps were done in Exercise 3 of Exercise Set 8.1

Carry out. We solve the system of equations

$$x + y = 40,$$
$$495x + 795y = 28,200$$

where $x =$ the number of white scarves sold and $y =$ the number of printed scarves sold. We use the elimination method. We eliminate x by multiplying the first equation by -495 and then adding it to the second.

$$
\begin{array}{r}
-495x - 495y = -19,800 \quad \text{Multiplying by } -495 \\
\underline{495x + 795y = 28,200} \\
300y = 8400 \quad \text{Adding} \\
y = 28
\end{array}
$$

Substitute 28 for y in one of the original equations and solve for x.

$$x + 28 = 40$$
$$x = 12$$

Check. The number of scarves sold is $12 + 28$, or 40.

Money from white scarves $= \$4.95 \times 12 = \$59.40$

Money from printed scarves $= \$7.95 \times 28 = \$222.60$

$$\text{Total} = \overline{\$282.00}$$

The numbers check.

State. 12 white scarves and 28 printed scarves were sold.

4. 32 of the \$8.50 pens, 13 of the \$9.75 pens

5. The Familiarize and Translate steps were done in Exercise 5 of Exercise Set 8.1.

Carry out. We solve the system of equations

$$x + y = 90, \quad (1)$$
$$x + \frac{1}{2}y = 64 \quad (2)$$

where $x =$ the measure of the first angle and $y =$ the measure of the second angle. We use elimination.

$$
\begin{array}{r}
x + \phantom{\frac{1}{2}}y = 90 \quad (1) \\
\underline{-x - \frac{1}{2}y = -64} \quad \text{Multiplying (2) by } -1 \\
\frac{1}{2}y = 26 \\
y = 52
\end{array}
$$

Substitute 52 for y in (1) and solve for x.

$$x + 52 = 90$$
$$x = 38$$

Check. The sum of the angle measures is $38° + 52°$, or $90°$, so the angles are complementary. The sum of the measures of the first angle and half the second angle is $38° + \frac{1}{2} \cdot 52°$, or $38° + 26°$, or $64°$. The numbers check.

State. The measures of the angles are $38°$ and $52°$.

6. $119°$, $61°$

7. The Familiarize and Translate steps were done in Exercise 7 of Exercise Set 8.1.

Carry out. We solve the system of equations

$$x + y = 250, \quad (1)$$
$$35x + 70y = 13,475 \quad (2)$$

where $x =$ the number of children's plates and and $y =$ the number of adult's plates served. We use elimination.

$$
\begin{array}{r}
-35x - 35y = -8750 \quad \text{Multiplying (1) by } -35 \\
\underline{35x + 70y = 13,475} \\
35y = 4725 \quad \text{Adding} \\
y = 135
\end{array}
$$

Substitute 135 for y in (1) and solve for x.

$$x + 135 = 250$$
$$x = 115$$

Check. Number of dinners: $115 + 135 = 250$

Money from children's plates: $\$3.50(115) = \$402.50$

Money from adult's plates: $\$7.00(135) = \$945.00$

Total money: $\$402.50 + \$945.00 = \$1347.50$

The numbers check.

State. 115 children's plates and 135 adult's plates were served.

8. 12 field goals, 6 free throws

9. The Familiarize and Translate steps were done in Exercise 9 of Exercise Set 8.1.

Carry out. We solve the system of equations

$$2w + 2l = 228, \quad (1)$$
$$w = l - 42 \quad (2)$$

where $w =$ the width and $l =$ the length. We use elimination.

$$2(l - 42) + 2l = 228 \quad \text{Substituting } l - 42$$
$$\text{for } w \text{ in } (1)$$
$$2l - 84 + 2l = 228$$
$$4l - 84 = 228$$
$$4l = 312$$
$$l = 78$$

$$w = 78 - 42 \quad \text{Substituting 78 for } l \text{ in } (2)$$
$$w = 36$$

Check. The perimeter is $2 \cdot 36 + 2 \cdot 78$, or $72 + 156$, or 228. Since $78 - 42 = 36$, the width is 42 ft less than the length. The numbers check.

State. The width is 36 ft, and the length is 78 ft.

10. Length: 94 ft, width: 50 ft

11. The Familiarize and Translate steps were done in Exercise 11 of Exercise Set 8.1.

Carry out. We solve the system of equations
$$x + y = 40, \quad (1)$$
$$2x + 3y = 89 \quad (2)$$
where $x =$ the number of 2-pointers and $y =$ the number of 3-pointers. We use elimination.
$$-2x - 2y = -80 \quad \text{Multiplying } (1) \text{ by } -2$$
$$\underline{2x + 3y = \quad 89}$$
$$y = \quad 9 \quad \text{Adding}$$

Substitute 9 for y in (1) and solve for x.
$$x + 9 = 40$$
$$x = 31$$

Check. Total field goals: $31 + 9 = 40$

Points from 2-pointers: $2 \cdot 31 = 62$

Points from 3-pointers: $3 \cdot 9 = 27$

Total points: $62 + 27 = 89$

The numbers check.

State. The Cougars made 31 2-pointers and 9 3-pointers.

12. 65 general interest films, 12 children's films

13. The Familiarize and Translate steps were done in Exercise 13 of Exercise Set 8.1.

Carry out. We solve the system of equations
$$x + y = 400, \quad (1)$$
$$20x + 30y = 11,000 \quad (2)$$
where $x =$ the number of units of lumber produced and $y =$ the number of units of plywood produced. We use elimination.
$$-20x - 20y = -8000 \quad \text{Multiplying } (1) \text{ by } -20$$
$$\underline{20x + 30y = 11,000}$$
$$10y = \quad 3000 \quad \text{Adding}$$
$$y = \quad 300$$

Substitute 300 for y in (1) and solve for x.
$$x + 300 = 400$$
$$x = 100$$

Check. Total output: $100 + 300 = 400$ units

Profit from lumber: $\$20 \cdot 100 = \$2000$

Profit from plywood: $\$30 \cdot 300 = \$9000$

Total profit: $\$2000 + \$9000 = \$11,000$

State. The lumber company must produce 100 units of lumber and 300 units of plywood.

14. 23 wins, 14 ties

15. The Familiarize and Translate steps were done in Exercise 15 of Exercise Set 8.1.

Carry out. We solve the system of equations
$$x + y = 12, \quad (1)$$
$$30x + 60y = 600 \quad (2)$$
where $x =$ the number of 30-sec commercials and $y =$ the number of 60-sec commercials. We use elimination.
$$-30x - 30y = -360 \quad \text{Multiplying } (1) \text{ by } -30$$
$$\underline{30x + 60y = \quad 600}$$
$$30y = \quad 240 \quad \text{Adding}$$
$$y = \quad 8$$

Substitute 8 for y in (1) and solve for x.
$$x + 8 = 12$$
$$x = 4$$

Check. The total number of commercials: $4 + 8 = 12$

Commercial time for 30-sec commercials:
$$30 \cdot 4 = 120 \text{ sec}$$

Commercial time for 60-sec commercials:
$$60 \cdot 8 = 480 \text{ sec}$$

Total commercial time: $120 \text{ sec} + 480 \text{ sec} =$
$$600 \text{ sec, or 10 min.}$$

The numbers check.

State. There were 4 30-sec commercials and 8 60-sec commercials.

16. 131 coach-class seats, 21 first-class seats

17. **Familiarize**. Let $x =$ the number of pounds of soybean meal and $y =$ the number of pounds of corn meal in the mixture.

Translate. We organize the information in a table.

Let $x =$ the number of pounds of soybean meal and $y =$ the number of pounds of corn meal.

	Soybean	Corn	Mixture
Pounds of meal	x	y	350
Percent of protein	16%	9%	12%
Amount of protein	$0.16x$	$0.09y$	0.12×350 or 42

The "Pounds of meal" row gives us one equation: $x + y = 350$

The last row gives us a second equation:
$0.16x + 0.09y = 42$

After clearing decimals, we have this system:

$$x + y = 350, \quad (1)$$
$$16x + 9y = 4200 \quad (2)$$

Carry out. We use the elimination method to solve the system of equations.

$$-9x - 9y = -3150 \quad \text{Multiplying (1) by } -9$$
$$\underline{16x + 9y = 4200}$$
$$7x = 1050$$
$$x = 150$$

$$150 + y = 350 \quad \text{Substituting 150 for } x \text{ in (1)}$$
$$y = 200$$

Check. The total number of pounds is $150 + 200$, or 350. Also, 16% of 150 is 24, and 9% of 200 is 18. Their total is 42. The numbers check.

State. 150 lb of soybean meal and 200 lb of corn meal should be mixed.

18. 4 L of 25% solution, 6 L of 50% solution

19. *Familiarize*. Let $x =$ the number of liters of Artic Antifreeze and $y =$ the number of liters of Frost-No-More in the mixture.

Translate. We organize the information in a table.

	Artic Anti-freeze	Frost-No-more	Mixture
Number of liters	x	y	20
Percent of alcohol	18%	10%	15%
Amount of alcohol	$0.18x$	$0.10y$	0.15×20 or 3

The "Number of liters" row gives us one equation: $x + y = 20$

The last row gives us a second equation:
$0.18x + 0.10y = 3$

After clearing decimals, we have this system:

$$x + y = 20, \quad (1)$$
$$18x + 10y = 300 \quad (2)$$

Carry out. We use the elimination method to solve the system of equations.

$$-10x - 10y = -200 \quad \text{Multiplying (1) by } -10$$
$$\underline{18x + 10y = 300}$$
$$8x = 100$$
$$x = 12\frac{1}{2}$$

$$12\frac{1}{2} + y = 20 \quad \text{Substituting } 12\frac{1}{2} \text{ for } x \text{ in (1)}$$
$$y = 7\frac{1}{2}$$

Check. Total liters: $12\frac{1}{2} + 7\frac{1}{2} = 20$

Total amount of alcohol: $18\% \times 12\frac{1}{2} + 10\% \times 7\frac{1}{2} = 2.25 + 0.75 = 3$

Percentage of alcohol in mixture:

$$\frac{\text{Total amount of alcohol}}{\text{Total liters in mixture}} = \frac{3}{20} = 0.15 = 15\%$$

The numbers check.

State. $12\frac{1}{2}$ L of Arctic Antifreeze and $7\frac{1}{2}$ L of Frost-No-More should be used.

20. 5 liters of each

21. *Familiarize*. Let $x =$ the amount invested at 9% and $y =$ the amount invested at 10%. Recall the formula for simple interest:

Interest = Principal $\cdot$ Rate $\cdot$ Time

Translate. We organize the information in a table.

	First Investment	Second Investment	Total
Principal	x	y	\$15,000
Interest Rate	9%	10%	
Time	1 year	1 year	
Interest	$0.09x$	$0.10y$	\$1432

The "Principal" row gives us one equation:
$x + y = 15,000$

The last row gives us a second equation:
$0.09x + 0.10y = 1432$

After clearing the decimals we have this system:

$$x + y = 15,000, \quad (1)$$
$$9x + 10y = 143,200 \quad (2)$$

Carry out. We use the elimination method to solve the system of equations.

$$-9x - 9y = -135,000 \quad \text{Multiplying (1) by } -9$$
$$\underline{9x + 10y = 143,200}$$
$$y = 8200$$

$$x + 8200 = 15,000 \quad \text{Substituting 8200 for } y \text{ in (1)}$$
$$x = 6800$$

Check. Total investment: $\$6800 + \$8200 = \$15,000$

Total interest: $9\% \times \$6800 + 10\% \times \$8200 = \$612 + \$820 = \$1432$

The numbers check.

State. $6800 is invested at 9%, and $8200 is invested at 10%.

22. $4100 at 14%, $4700 at 16%

23. ***Familiarize***. Let x = the amount invested at 10% and y = the amount invested at 12%. Recall the formula for simple interest:

Interest = Principal · Rate · Time

Translate. We organize the information in a table.

	First Investment	Second Investment	Total
Principal	x	y	$27,000
Interest Rate	10%	12%	
Time	1 year	1 year	
Interest	$0.10x$	$0.12y$	$2990

The "Principal" row gives us one equation:
$x + y = 27,000$

The last row gives us a second equation:
$0.10x + 0.12y = 2990$

After clearing the decimals we have this system:
$$x + y = 27,000 \qquad (1)$$
$$10x + 12y = 299,000 \quad (2)$$

Carry out. We use the elimination method to solve the system of equations.

$$-10x - 10y = -270,000 \quad \text{Multiplying (1) by } -10$$
$$\underline{10x + 12y = 299,000}$$
$$2y = 29,000$$
$$y = 14,500$$

$x + 14,500 = 27,000$ Substituting 14,500 for y in (1)
$$x = 12,500$$

Check. Total investment: $12,500 + $14,500 = $27,000

Total interest: $10\% \times \$12,500 + 12\% \times \$14,500 =$
$$\$1250 + \$1740 = \$2990$$

The numbers check.

State. $12,500 is invested at 10%, and $14,500 is invested at 12%.

24. $725 at 12%, $425 at 11%

25. ***Familiarize***. Let a = the number of adult's tickets sold and c = the number of children's tickets sold.

Translate. We organize the information in a table.

	Adult's tickets	Children's tickets	Total
Number Sold	a	c	117
Price	$1.25	$0.75	
Amount taken in	$1.25a$	$0.75c$	129.75

The "Number sold" row gives us one equation:
$a + c = 117$

The last row gives us a second equation:
$1.25a + 0.75c = 129.75$

After clearing the decimals we have this system:
$$a + c = 117 \qquad (1)$$
$$125a + 75c = 12,975 \quad (2)$$

Carry out. We use the elimination method to solve the system of equations.

$$-75a - 75c = -8775 \quad \text{Multiplying (1) by } -75$$
$$\underline{125a + 75c = 12,975}$$
$$50a = 4200$$
$$a = 84$$

$84 + c = 117$ Substituting 84 for a in (1)
$$c = 33$$

Check. Total tickets sold: $84 + 33 = 117$

Total taken in: $\$1.25 \times 84 + \$0.75 \times 33 =$
$$\$105 + \$24.75 = \$129.75$$

The numbers check.

State. 84 adult's tickets and 33 children's tickets were sold.

26. 8 white, 22 yellow

27. ***Familiarize***. Let x = Paula's age now and y = Bob's age now. Four years from now, Paula's age will be $x + 4$ and Bob's age will be $y + 4$.

Translate.

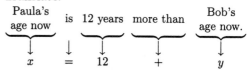

We have a system of equations:
$$x = 12 + y, \qquad (1)$$
$$y + 4 = \frac{2}{3}(x + 4) \quad (2)$$

Carry out. We use the substitution method.

$$y + 4 = \frac{2}{3}(12 + y + 4) \quad \text{Substituting } 12 + y \text{ for } x \text{ in (2)}$$
$$y + 4 = \frac{2}{3}(y + 16)$$
$$3(y + 4) = 3 \cdot \frac{2}{3}(y + 16) \quad \text{Clearing the fraction}$$
$$3y + 12 = 2(y + 16)$$
$$3y + 12 = 2y + 32$$
$$y = 20$$

$x = 12 + 20$ Substituting 20 for y in (1)

$x = 32$

Check. Bob's age plus 12 years is $20 + 12$, or 32, which is Paula's age. Four years from now Bob will be 24 and Paula will be 36, and 24 is $\frac{2}{3}$ of 36. The numbers check.

State. Now Paula is 32 years old and Bob is 20 years old.

28. Carlos: 28, Maria: 20

29. Familiarize. We first make a drawing.

We let $l = $ length and $w = $ width.

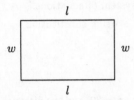

The formula for perimeter is $P = 2l + 2w$.

Translate.

The perimeter is 190 m.

$$2l + 2w = 190$$

The width is one fourth of the length.

$$w = \frac{1}{4} \cdot l$$

We have a system of equations:

$$2l + 2w = 190, \quad (1)$$

$$w = \frac{1}{4}l \qquad (2)$$

Carry out. We use the substitution method.

$2l + 2 \cdot \frac{1}{4}l = 190$ Substituting $\frac{1}{4}l$ for w in (1)

$$2l + \frac{1}{2}l = 190$$

$$\frac{5}{2}l = 190$$

$$l = 76$$

$w = \frac{1}{4} \cdot 76$ Substituting 76 for l in (2)

$w = 19$

Check. The perimeter of a rectangle with width 19 m and length 76 m is $2 \cdot 76 + 2 \cdot 19 = 152 + 38 = 190$ m. Also, 19 is one fourth of 76. The numbers check.

State. The width is 19 m and the length is 76 m.

30. Length: 78 yd, width: 19 yd

31. Familiarize. Let $x = $ the number of $5 bills and $y = $ the number of $1 bills. The total value of the $5 bills is $5x$, and the total value of the $1 bills is $1 \cdot y$, or y.

Translate.

The total number of bills is 22.

$$x + y = 22$$

The total value of the bills is $50.

$$5x + y = 50$$

We have a system of equations:

$$x + y = 22, \quad (1)$$

$$5x + y = 50 \quad (2)$$

Carry out. We use the elimination method.

$$\begin{array}{rl} -x - y = -22 & \text{Multiplying (1) by } -1 \\ \underline{5x + y = 50} & \\ 4x = 28 & \\ x = 7 & \end{array}$$

$7 + y = 22$ Substituting 7 for x in (1)

$y = 15$

Check. Total number of bills: $7 + 15 = 22$

Total value of bills: $\$5 \cdot 7 + \$1 \cdot 15 = \$35 + \$15 = \$50$.

The numbers check.

State. There are 7 $5 bills and 15 $1 bills.

32. 17 quarters, 13 fifty-cent pieces

33. Familiarize. We first make a drawing.

Slow train		
d kilometers	75 km/h	$(t + 2)$ hr

Fast train		
d kilometers	125 km/h	t hr

From the drawing we see that the distances are the same. Now complete the chart.

$$d = r \cdot t$$

	Distance	Rate	Time	
Slow train	d	75	$t + 2$	$\rightarrow d = 75(t+2)$
Fast train	d	125	t	$\rightarrow d = 125t$

Translate. Using $d = rt$ in each row of the table, we get a system of equations:

$$d = 75(t + 2),$$

$$d = 125t$$

Carry out. We solve the system of equations.

$125t = 75(t + 2)$ Using substitution

$$125t = 75t + 150$$

$$50t = 150$$

$$t = 3$$

Then $d = 125t = 125 \cdot 3 = 375$

Check. At 125 km/h, in 3 hr the fast train will travel $125 \cdot 3 = 375$ km. At 75 km/h, in $3 + 2$, or 5 hr the slow train will travel $75 \cdot 5 = 375$ km. The numbers check.

State. The trains will meet 375 km from the station.

34. 3 hr

35. **Familiarize**. We first make a drawing.

Chicago   Indianapolis
110 km/h t hours t hours   90 km/h

|——————— 350 km ———————|

The sum of the distances is 350 km. The times are the same. We let t = the time, d = the distance from Chicago, and $350 - d$ = the distance from Indianapolis. We organize the information in a table.

	Distance	Rate	Time	
From Chicago	d	110	t	$\rightarrow d = 110t$
From Indpls.	$350 - d$	90	t	$\rightarrow 350 - d = 90t$

Translate. Using $d = rt$ in each row of the table, we get a system of equations:

$$d = 110t, \qquad (1)$$
$$350 - d = 90t \quad (2)$$

Carry out. We use the substitution method.

$$350 - 110t = 90t \quad \text{Substituting } 110t \text{ for } d$$
$$\text{in (2)}$$
$$350 = 200t$$
$$\frac{350}{200} = t$$
$$\frac{7}{4} = t$$

Check. The motorcycle from Chicago will travel $110 \cdot \frac{7}{4}$, or 192.5 km. The motorcycle from Indianapolis will travel $90 \cdot \frac{7}{4}$, or 157.5 km. The sum of the two distances is $192.5 + 157.5$, or 350 km. The value checks.

State. In $1\frac{3}{4}$ hours the motorcycles will meet.

36. 2 hr

37. **Familiarize**. We first make a drawing. Let d = the distance and r = the speed of the boat in still water. Then when the boat travels downstream its speed is $r + 6$, and its speed upstream is $r - 6$. From the drawing we see that the distances are the same.

Downstream, 6 mph current

d mi, $r + 6$, 3 hr

Upstream, 6 mph current

d mi, $r - 6$, 5 hr

Organize the information in a table.

$$d \quad = \quad r \quad \cdot \quad t$$

	Distance	Rate	Time	
Down-stream	d	$r + 6$	3	$\rightarrow d = (r+6)3$
Up-stream	d	$r - 6$	5	$\rightarrow d = (r-6)5$

Translate. Using $d = rt$ in each row of the table, we get a system of equations:

$$d = 3r + 18,$$
$$d = 5r - 30$$

Carry out. Solve the system of equations.

$$3r + 18 = 5r - 30 \quad \text{Using substitution}$$
$$18 = 2r - 30$$
$$48 = 2r$$
$$24 = r$$

Check. When $r = 24$, $r + 6 = 30$, and the distance traveled in 3 hr is $30 \cdot 3 = 90$ mi. Also, $r - 6 = 18$, and the distance traveled in 5 hr is $18 \cdot 5 = 90$ mi. The value checks.

State. The speed of the boat in still water is 24 mph.

38. 14 km/h

39. $x^2 + 6x + 5$

We look for a pair of numbers whose product is 5 and whose sum is 6. They are 1 and 5.

$$x^2 + 6x + 5 = (x + 1)(x + 5)$$

40. $(2x + 1)(x - 5)$

41. We use the point-slope equation.

$$y - y_1 = m(x - x_1)$$
$$y - (-5) = -\frac{3}{4}(x - 2) \quad \text{Substituting}$$
$$y + 5 = -\frac{3}{4}x + \frac{3}{2}$$
$$y = -\frac{3}{4}x - \frac{7}{2}$$

42.

$y = -\frac{1}{2}x + 5$

43. The Familiarize and Translate steps were done in Exercise 63 of Exercise Set 8.1.

Carry out. We solve the system of equations

$$x = 2y, \qquad (1)$$
$$x + 20 = 3y \quad (2)$$

where x = Burl's age now and y = his son's age now.

$2y + 20 = 3y$ Substituting $2y$ for x in (2)

$20 = y$

$x = 2 \cdot 20$ Substituting 20 for y in (1)

$x = 40$

Check. Burl's age now, 40, is twice his son's age now, 20. Ten years ago Burl was 30 and his son was 10, and $30 = 3 \cdot 10$. The numbers check.

State. Now Burl is 40 and his son is 20.

44. Lou: 32 years, Juanita: 14 years

45. The Familiarize and Translate steps were done in Exercise 65 of Exercise Set 8.1.

Carry out. We solve the system of equations

$2l + 2w = 156$,   (1)

$l = 4(w - 6)$ (2)

where l = length and w = width.

$2 \cdot 4(w - 6) + 2w = 156$ Substituting $4(w - 6)$ for l in (1)

$8w - 48 + 2w = 156$

$10w - 48 = 156$

$10w = 204$

$w = \dfrac{204}{10}$, or $\dfrac{102}{5}$

$l = 4\left(\dfrac{102}{5} - 6\right)$ Substituting $\dfrac{102}{5}$ for w in (2)

$l = 4\left(\dfrac{102}{5} - \dfrac{30}{5}\right)$

$l = 4\left(\dfrac{72}{5}\right)$

$l = \dfrac{288}{5}$

Check. The perimeter of a rectangle with width $\dfrac{102}{5}$ in. and length $\dfrac{288}{5}$ in. is

$2\left(\dfrac{288}{5}\right) + 2\left(\dfrac{102}{5}\right) = \dfrac{576}{5} + \dfrac{204}{5} = \dfrac{780}{5} = 156$ in.

If 6 in. is cut off the width, the new width is

$\dfrac{102}{5} - 6 = \dfrac{102}{5} - \dfrac{30}{5} = \dfrac{72}{5}$. The length, $\dfrac{288}{5}$, is $4\left(\dfrac{72}{5}\right)$. The numbers check.

State. The original piece of posterboard had width $\dfrac{102}{5}$ in. and length $\dfrac{288}{5}$ in.

46. 80 wins, 24 losses

47. Familiarize. Let x = the amount of the original solution that remains after some of the original solution is drained and replaced with pure antifreeze. Let y = the amount of the original solution that is drained and replaced with pure antifreeze.

Translate. We organize the information in a table. Keep in mind that the table contains information regarding the solution *after* some of the original solution is drained and replaced with pure antifreeze.

	Original Solution	Pure Anti-freeze	New Mixture
Amount of solution	x	y	16 L
Percent of antifreeze	30%	100%	50%
Amount of antifreeze in solution	$0.3x$	$1 \cdot y$, or y	$0.5(16)$, or 8

The "Amount of solution" row gives us one equation:

$x + y = 16$

The last row gives us a second equation:

$0.3x + y = 8$

After clearing the decimal we have the following system of equations:

$x + y = 16$,   (1)

$3x + 10y = 80$ (2)

Carry out. We use the elimination method.

$-3x - 3y = -48$ Multiplying (1) by -10

$\underline{3x + 10y = 80}$

$7y = 32$

$y = \dfrac{32}{7}$, or $4\dfrac{4}{7}$

Although the problem only asks for the amount of pure antifreeze added, we will also find x in order to check.

$x + 4\dfrac{4}{7} = 16$ Substituting $4\dfrac{4}{7}$ for y in (1)

$x = 11\dfrac{3}{7}$

Check. Total amount of new mixture: $11\dfrac{3}{7} + 4\dfrac{4}{7} =$

16 L

Amount of antifreeze in new mixture:

$0.3\left(11\dfrac{3}{7}\right) + 4\dfrac{4}{7} = \dfrac{3}{10} \cdot \dfrac{80}{7} + \dfrac{32}{7} = \dfrac{56}{7} = 8$ L

The numbers check.

State. Michelle should drain $4\dfrac{4}{7}$ L of the original solution and replace it with pure antifreeze.

48. 4 km

49. Familiarize. Let x = the ten's digit and y = the unit's digit. Then the number is $10x + y$. If the digits are interchanged, the new number is $10y + x$.

Translate.

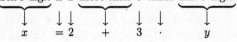

Ten's digit is 2 more than 3 times unit's digit.

$x \quad = 2 \quad + \quad 3 \quad \cdot \quad y$

If the digits are interchanged,

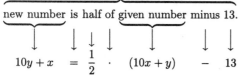

new number is half of given number minus 13.

$$10y + x = \frac{1}{2} \cdot (10x + y) - 13$$

The system of equations is

$$x = 2 + 3y, \quad (1)$$

$$10y + x = \frac{1}{2}(10x + y) - 13 \quad (2)$$

Carry out. We use the substitution method. Substitute $2 + 3y$ for x in (2).

$$10y + (2 + 3y) = \frac{1}{2}[10(2 + 3y) + y] - 13$$

$$13y + 2 = \frac{1}{2}[20 + 30y + y] - 13$$

$$13y + 2 = \frac{1}{2}[20 + 31y] - 13$$

$$13y + 2 = 10 + \frac{31}{2}y - 13$$

$$13y + 2 = \frac{31}{2}y - 3$$

$$5 = \frac{5}{2}y$$

$$2 = y$$

$$x = 2 + 3 \cdot 2 \quad \text{Substituting 2 for } y \text{ in (1)}$$

$$x = 2 + 6$$

$$x = 8$$

Check. If $x = 8$ and $y = 2$, the given number is 82 and the new number is 28. In the given number the ten's digit, 8, is two more than three times the unit's digit, 2. The new number is 13 less than one-half the given number: $28 = \frac{1}{2}(82) - 13$. The values check.

State. The given integer is 82.

50. 180

51. **Familiarize**. We first make a drawing. Let $r_1 =$ the speed of the first train and $r_2 =$ the speed of the second train. If the first train leaves at 9 A.M. and the second at 10 A.M., we have:

Train 1   Train 2
r_1, 3 hr r_2, 2 hr
Union ————————————— Central
Station   Station
├———— 216 km ————┤

If the second train leaves at 9 A.M. and the first at 10:30 A.M. we have:

Train 1   Train 2
$r_1, \frac{3}{2}$ hr r_2, 3 hr
Union ————————————— Central
Station   Station
├———— 216 km ————┤

The total distance traveled in each case is 216 km and is equal to the sum of the distances traveled by each train.

Translate. We will use the formula $d = rt$. For each situation we have:

Total distance	is	Train 1's distance	plus	Train 2's distance.
216	=	$3r_1$	+	$2r_2$

$$\text{and} \quad 216 = \frac{3}{2}r_1 + 3r_2$$

Clearing the fraction, we have this system:

$$216 = 3r_1 + 2r_2, \quad (1)$$

$$432 = 3r_1 + 6r_2 \quad (2)$$

Carry out. Solve the system of equations.

$$-216 = -3r_1 - 2r_2 \quad \text{Multiplying (1) by } -1$$

$$\underline{432 = 3r_1 + 6r_2}$$

$$216 = 4r_2$$

$$54 = r_2$$

$$216 = 3r_1 + 2(54) \quad \text{Substituting 54 for } r_2 \text{ in (1)}$$

$$216 = 3r_1 + 108$$

$$108 = 3r_1$$

$$36 = r_1$$

Check. If Train 1 travels for 3 hr at 36 km/h and Train 2 travels for 2 hr at 54 km/h, the total distance traveled is $3 \cdot 36 + 2 \cdot 54 = 108 + 108 = 216$ km. If Train 1 travels for $\frac{3}{2}$ hr at 36 km/h and Train 2 travels for 3 hr at 54 km/h, then the total distance traveled is $\frac{3}{2} \cdot 36 + 3 \cdot 54 = 54 + 162 = 216$ km. The numbers check.

State. The speed of the first train is 36 km/h, and the speed of the second train is 54 km/h.

52. City: 261 mi, highway: 204 mi

53. **Familiarize**. Let $g =$ the number of girls and $b =$ the number of boys. Then Phyllis has b brothers and $g - 1$ sisters, and Phil has $b - 1$ brothers and g sisters.

Translate.

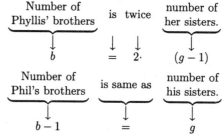

Number of Phyllis' brothers	is twice	number of her sisters.
b	$= 2 \cdot$	$(g - 1)$

Number of Phil's brothers	is same as	number of his sisters.
$b - 1$	$=$	g

We have a system of equations:

$$b = 2(g - 1), \quad (1)$$

$$b - 1 = g \quad (2)$$

Carry out. We use the substitution method.

$$b = 2[(b - 1) - 1] \quad \text{Substituting } b - 1 \text{ for } g \text{ in (1)}$$

$$b = 2(b - 2)$$

$$b = 2b - 4$$

$$4 = b$$

$4 - 1 = g$ Substituting 4 for b in (2)
$ 3 = g$

Check. If there are 3 girls and 4 boys in the family, then Phyllis has 4 brothers and 2 sisters. She has twice as many brothers as sisters. Also, Phil has 3 brothers and 3 sisters, the same number of each. The numbers check.

State. There are 3 girls and 4 boys in the family.

54.

Exercise Set 8.4

1. Substitute $(1, -2, 3)$ into the three equations, using alphabetical order.

$$\frac{x + y + z = 2}{1 + (-2) + 3 \ ? \ 2}$$
$$ 2 \ \Big| \ 2 \quad \text{TRUE}$$

$$\frac{x - 2y - z = 2}{1 - 2(-2) - 3 \ ? \ 2}$$
$$ 1 + 4 - 3 \ \Big|$$
$$ 2 \ \Big| \ 2 \quad \text{TRUE}$$

$$\frac{3x + 2y + z = 2}{3 \cdot 1 + 2(-2) + 3 \ ? \ 2}$$
$$ 3 - 4 + 3 \ \Big|$$
$$ 2 \ \Big| \ 2 \quad \text{TRUE}$$

The triple $(1, -2, 3)$ makes all three equations true, so it is a solution.

2. No

3. $x + y + z = 6,$ (1)
$ 2x - y + 3z = 9,$ (2)
$ -x + 2y + 2z = 9$ (3)

1., 2. The equations are already in standard form with no fractions or decimals.

3. Add equations (1) and (2) to eliminate y:

$$\begin{array}{l} x + y + z = 6 \quad (1) \\ \underline{2x - y + 3z = 9} \quad (2) \\ 3x + 4z = 15 \quad (4) \quad \text{Adding} \end{array}$$

4. Use a different pair of equations and eliminate y:

$$\begin{array}{l} 4x - 2y + 6z = 18 \quad \text{Multiplying (2) by 2} \\ \underline{-x + 2y + 2z = 9} \quad (3) \\ 3x + 8z = 27 \quad (5) \end{array}$$

5. Now solve the system of equations (4) and (5).

$$3x + 4z = 15 \quad (4)$$
$$3x + 8z = 27 \quad (5)$$

$-3x - 4z = -15$ Multiplying (4) by -1
$\underline{3x + 8z = 27}$
$ 4z = 12$
$ z = 3$

$3x + 4 \cdot 3 = 15$ Substituting 3 for z in (4)
$ 3x + 12 = 15$
$ 3x = 3$
$ x = 1$

6. Substitute in one of the original equations to find y.

$1 + y + 3 = 6$ Substituting 1 for x and 3 for z in (1)
$ y + 4 = 6$
$ y = 2$

We obtain $(1, 2, 3)$. This checks, so it is the solution.

4. $(4, 0, 2)$

5. $2x - y - 3z = -1,$ (1)
$ 2x - y + z = -9,$ (2)
$ x + 2y - 4z = 17$ (3)

1., 2. The equations are already in standard form with no fractions or decimals.

3., 4. We eliminate z from two different pairs of equations.

$$\begin{array}{l} 2x - y - 3z = -1 \quad (1) \\ \underline{6x - 3y + 3z = -27} \quad \text{Multiplying (2) by 3} \\ 8x - 4y = -28 \quad (4) \quad \text{Adding} \end{array}$$

$$\begin{array}{l} 8x - 4y + 4z = -36 \quad \text{Multiplying (2) by 4} \\ \underline{x + 2y - 4z = 17} \\ 9x - 2y = -19 \quad (5) \quad \text{Adding} \end{array}$$

5. Now solve the system of equations (4) and (5).

$$8x - 4y = -28 \quad (4)$$
$$9x - 2y = -19 \quad (5)$$

$$\begin{array}{l} 8x - 4y = -28 \quad (4) \\ \underline{-18x + 4y = 38} \quad \text{Multiplying (5) by } -2 \\ -10x = 10 \quad \text{Adding} \\ x = -1 \end{array}$$

$8(-1) - 4y = -28$ Substituting -1 for x in (4)
$ -8 - 4y = -28$
$ -4y = -20$
$ y = 5$

6. Substitute in one of the original equations to find z.

$$2(-1) - 5 + z = -9 \quad \text{Substituting } -1 \text{ for } x$$
$$\text{and } 5 \text{ for } y \text{ in (2)}$$
$$-2 - 5 + z = -9$$
$$-7 + z = -9$$
$$z = -2$$

We obtain $(-1, 5, -2)$. This checks, so it is the solution.

6. $(2, -2, 2)$

7. $2x - 3y + z = 5, \quad (1)$
$x + 3y + 8z = 22, \quad (2)$
$3x - y + 2z = 12 \quad (3)$

1., 2. The equations are already in standard form with no fractions or decimals.

3., 4. We eliminate y from two different pairs of equations.

$$\begin{array}{rl} 2x - 3y + z = 5 & (1) \\ \underline{x + 3y + 8z = 22} & (2) \\ 3x \phantom{{}+3y} + 9z = 27 & (4) \quad \text{Adding} \end{array}$$

$$\begin{array}{rl} x + 3y + 8z = 22 & (2) \\ \underline{9x - 3y + 6z = 36} & \text{Multiplying (3) by 3} \\ 10x \phantom{{}+3y} + 14z = 58 & (5) \quad \text{Adding} \end{array}$$

5. Solve the system of Equations (4) and (5).

$$3x + 9z = 27 \quad (4)$$
$$10x + 14z = 58 \quad (5)$$

$$\begin{array}{rl} 30x + 90z = 270 & \text{Multiplying (4) by 10} \\ \underline{-30x - 42z = -174} & \text{Multiplying (5) by } -3 \\ 48z = 96 & \text{Adding} \\ z = 2 \end{array}$$

$$3x + 9 \cdot 2 = 27 \quad \text{Substituting 2 for } z \text{ in (4)}$$
$$3x + 18 = 27$$
$$3x = 9$$
$$x = 3$$

6. Substitute in one of the original equations to find y.

$$2 \cdot 3 - 3y + 2 = 5 \quad \text{Substituting 3 for } x \text{ and}$$
$$2 \text{ for } z \text{ in (1)}$$
$$-3y + 8 = 5$$
$$-3y = -3$$
$$y = 1$$

We obtain $(3, 1, 2)$. This checks, so it is the solution.

8. $(3, -2, 1)$

9. $3a - 2b + 7c = 13, \quad (1)$
$a + 8b - 6c = -47, \quad (2)$
$7a - 9b - 9c = -3 \quad (3)$

1., 2. The equations are already in standard form with no fractions or decimals.

3., 4. We eliminate a from two different pairs of equations.

$$\begin{array}{rl} 3a - 2b + 7c = 13 & (1) \\ \underline{-3a - 24b + 18c = 141} & \text{Multiplying (2) by } -3 \\ - 26b + 25c = 154 & (4) \quad \text{Adding} \end{array}$$

$$\begin{array}{rl} -7a - 56b + 42c = 329 & \text{Multiplying (2) by } -7 \\ \underline{7a - 9b - 9c = -3} & (3) \\ - 65b + 33c = 326 & (5) \quad \text{Adding} \end{array}$$

5. Now solve the system of Equations (4) and (5).

$$-26b + 25c = 154 \quad (4)$$
$$-65b + 33c = 326 \quad (5)$$

$$\begin{array}{rl} -130b + 125c = 770 & \text{Multiplying (4) by 5} \\ \underline{130b - 66c = -652} & \text{Multiplying (5) by } -2 \\ 59c = 118 \\ c = 2 \end{array}$$

$$-26b + 25 \cdot 2 = 154 \quad \text{Substituting 2 for } c$$
$$\text{in (4)}$$
$$-26b + 50 = 154$$
$$-26b = 104$$
$$b = -4$$

6. Substitute in one of the original equations to find a.

$$a + 8(-4) - 6(2) = -47 \quad \text{Substituting } -4$$
$$\text{for } b \text{ and 2 for } c$$
$$\text{in (2)}$$
$$a - 32 - 12 = -47$$
$$a - 44 = -47$$
$$a = -3$$

We obtain $(-3, -4, 2)$. This checks, so it is the solution.

10. $(7, -3, -4)$

11. $2x + 3y + z = 17, \quad (1)$
$x - 3y + 2z = -8, \quad (2)$
$5x - 2y + 3z = 5 \quad (3)$

1., 2. The equations are already in standard form with no fractions or decimals.

3., 4. We eliminate y from two different pairs of equations.

$$\begin{array}{rl} 2x + 3y + z = 17 & (1) \\ \underline{x - 3y + 2z = -8} & (2) \\ 3x \phantom{{}+3y} + 3z = 9 & (4) \quad \text{Adding} \end{array}$$

$$\begin{array}{rl} 4x + 6y + 2z = 34 & \text{Multiplying (1) by 2} \\ \underline{15x - 6y + 9z = 15} & \text{Multiplying (3) by 3} \\ 19x \phantom{{}+6y} + 11z = 49 & (5) \quad \text{Adding} \end{array}$$

5. Now solve the system of Equations (4) and (5).

$$3x + 3z = 9 \quad (4)$$
$$19x + 11z = 49 \quad (5)$$

$$33x + 33z = 99 \quad \text{Multiplying (4) by 11}$$
$$\underline{-57x - 33z = -147} \quad \text{Multiplying (5) by } -3$$
$$-24x = -48$$
$$x = 2$$

$$3 \cdot 2 + 3z = 9 \quad \text{Substituting 2 for } x \text{ in (4)}$$
$$6 + 3z = 9$$
$$3z = 3$$
$$z = 1$$

6. Substitute in one of the original equations to find y.
$$2 \cdot 2 + 3y + 1 = 17 \quad \text{Substituting 2 for } x \text{ and}$$
$$ 1 \text{ for } z \text{ in (1)}$$
$$3y + 5 = 17$$
$$3y = 12$$
$$y = 4$$

We obtain $(2, 4, 1)$. This checks, so it is the solution.

12. $(2, 1, 3)$

13.
$$2x + y + z = -2, \quad (1)$$
$$2x - y + 3z = 6, \quad (2)$$
$$3x - 5y + 4z = 7 \quad (3)$$

1., 2. The equations are already in standard form with no fractions or decimals.

3., 4. We eliminate y from two different pairs of equations.
$$2x + y + z = -2 \quad (1)$$
$$\underline{2x - y + 3z = 6} \quad (2)$$
$$4x + 4z = 4 \quad (4) \quad \text{Adding}$$

$$10x + 5y + 5z = -10 \quad \text{Multiplying (1) by 5}$$
$$\underline{3x - 5y + 4z = 7} \quad (3)$$
$$13x + 9z = -3 \quad (5) \quad \text{Adding}$$

5. Now solve the system of Equations (4) and (5).
$$4x + 4z = 4 \quad (4)$$
$$13x + 9z = -3 \quad (5)$$

$$36x + 36z = 36 \quad \text{Multiplying (4) by 9}$$
$$\underline{-52x - 36z = 12} \quad \text{Multiplying (5) by } -4$$
$$-16x = 48 \quad \text{Adding}$$
$$x = -3$$

$$4(-3) + 4z = 4 \quad \text{Substituting } -3 \text{ for } x \text{ in (4)}$$
$$-12 + 4z = 4$$
$$4z = 16$$
$$z = 4$$

6. Substitute in one of the original equations to find y.
$$2(-3) + y + 4 = -2 \quad \text{Substituting } -3 \text{ for}$$
$$ x \text{ and 4 for } z \text{ in (1)}$$
$$y - 2 = -2$$
$$y = 0$$

We obtain $(-3, 0, 4)$. This checks, so it is the solution.

14. $(2, -5, 6)$

15.
$$x - y + z = 4, \quad (1)$$
$$5x + 2y - 3z = 2, \quad (2)$$
$$4x + 3y - 4z = -2 \quad (3)$$

1., 2. The equations are already in standard form with no fractions or decimals.

3., 4. We eliminate z from two different pairs of equations.
$$3x - 3y + 3z = 12 \quad \text{Multiplying (1) by 3}$$
$$\underline{5x + 2y - 3z = 2} \quad (2)$$
$$8x - y = 14 \quad (4) \quad \text{Adding}$$

$$4x - 4y + 4z = 16 \quad \text{Multiplying (1) by 4}$$
$$\underline{4x + 3y - 4z = -2} \quad (3)$$
$$8x - y = 14 \quad (5) \quad \text{Adding}$$

5. Now solve the system of Equations (4) and (5).
$$8x - y = 14 \quad (4)$$
$$8x - y = 14 \quad (5)$$

$$8x - y = 14 \quad (4)$$
$$\underline{-8x + y = -14} \quad \text{Multiplying (5) by } -1$$
$$0 = 0 \quad (6)$$

Equation (6) indicates the equations (1), (2), and (3) are dependent. (Note that if equation (1) is subtracted from equation (2), the result is equation (3).)

16. The system is dependent.

17.
$$4x - y - z = 4, \quad (1)$$
$$2x + y + z = -1, \quad (2)$$
$$6x - 3y - 2z = 3 \quad (3)$$

1., 2. The equations are already in standard form with no fractions or decimals.

3. Add equations (1) and (2) to eliminate y.
$$4x - y - z = 4 \quad (1)$$
$$\underline{2x + y + z = -1} \quad (2)$$
$$6x = 3 \quad (4) \quad \text{Adding}$$

4. At this point we can either continue by eliminating y from a second pair of equations or we can solve (4) for x and substitute that value in a different pair of the original equations to obtain a system of two equations in two variables. We take the second option.

$$6x = 3 \quad (4)$$
$$x = \frac{1}{2}$$

Substitute $\frac{1}{2}$ for x in (1):

$$4\left(\frac{1}{2}\right) - y - z = 4$$
$$2 - y - z = 4$$
$$-y - z = 2 \quad (5)$$

Substitute $\frac{1}{2}$ for x in (3):

$$6\left(\frac{1}{2}\right) - 3y - 2z = 3$$
$$3 - 3y - 2z = 3$$
$$-3y - 2z = 0 \quad (6)$$

5. Solve the system of Equations (5) and (6).

$$\begin{array}{rl} 2y + 2z = -4 & \text{Multiplying (5) by } -2 \\ \underline{-3y - 2z = 0} & (6) \\ -y = -4 \\ y = 4 \end{array}$$

6. Substitute to find z.

$$-4 - z = 2 \qquad \text{Substituting 4 for } y \text{ in (5)}$$
$$-z = 6$$
$$z = -6$$

We obtain $\left(\frac{1}{2}, 4, -6\right)$. This checks, so it is the solution.

18. $(3, -5, 8)$

19. $\quad 2r + 3s + 12t = 4, \quad (1)$
    $\quad 4r - 6s + 6t = 1, \quad (2)$
    $\quad r + s + t = 1 \quad (3)$

1., 2. The equations are already in standard form with no fractions or decimals.

3., 4. We eliminate s from two different pairs of equations.

$$\begin{array}{rl} 4r + 6s + 24t = 8 & \text{Multiplying (1) by 2} \\ \underline{4r - 6s + 6t = 1} & (2) \\ 8r + 30t = 9 & (4) \quad \text{Adding} \end{array}$$

$$\begin{array}{rl} 4r - 6s + 6t = 1 & (2) \\ \underline{6r + 6s + 6t = 6} & \text{Multiplying (3) by 6} \\ 10r + 12t = 7 & (5) \quad \text{Adding} \end{array}$$

5. Solve the system of Equations (4) and (5).

$$\begin{array}{rl} 40r + 150t = 45 & \text{Multiplying (4) by 5} \\ \underline{-40r - 48t = -28} & \text{Multiplying (5) by } -4 \\ 102t = 17 \end{array}$$
$$t = \frac{17}{102}$$
$$t = \frac{1}{6}$$

$$8r + 30\left(\frac{1}{6}\right) = 9 \qquad \text{Substituting } \frac{1}{6} \text{ for } t$$
$$8r + 5 = 9 \qquad\qquad \text{in (4)}$$
$$8r = 4$$
$$r = \frac{1}{2}$$

6. Substitute in one of the original equations to find s.

$$\frac{1}{2} + s + \frac{1}{6} = 1 \qquad \text{Substituting } \frac{1}{2} \text{ for } r \text{ and}$$
$$\frac{1}{6} \text{ for } t \text{ in (3)}$$
$$s + \frac{2}{3} = 1$$
$$s = \frac{1}{3}$$

We obtain $\left(\frac{1}{2}, \frac{1}{3}, \frac{1}{6}\right)$. This checks, so it is the solution.

20. $\left(\frac{3}{5}, \frac{2}{3}, -3\right)$

21. $\quad 4a + 9b = 8, \quad (1)$
    $\quad 8a + 6c = -1, \quad (2)$
    $\quad 6b + 6c = -1 \quad (3)$

1., 2. The equations are already in standard form with no fractions or decimals.

3., 4. Note that there is no c in Equation (1). We will use equations (2) and (3) to obtain another equation with no c term.

$$\begin{array}{rl} 8a + 6c = -1 & (2) \\ \underline{- 6b - 6c = 1} & \text{Multiplying (3)} \\ & \text{by } -1 \\ 8a - 6b = 0 & (4) \text{ Adding} \end{array}$$

5. Now solve the system of equations (1) and (4).

$$\begin{array}{rl} -8a - 18b = -16 & \text{Multiplying (1) by } -2 \\ \underline{8a - 6b = 0} \\ - 24b = -16 \end{array}$$
$$b = \frac{2}{3}$$

$$8a - 6\left(\frac{2}{3}\right) = 0 \qquad \text{Substituting } \frac{2}{3} \text{ for } b$$
$$8a - 4 = 0 \qquad\qquad \text{in (4)}$$
$$8a = 4$$
$$a = \frac{1}{2}$$

6. Substitute in equation (2) or (3) to find c.

$$8\left(\frac{1}{2}\right) + 6c = -1 \qquad \text{Substituting } \frac{1}{2} \text{ for } a$$
$$4 + 6c = -1 \qquad\qquad \text{in (2)}$$
$$6c = -5$$
$$c = -\frac{5}{6}$$

We obtain $\left(\frac{1}{2}, \frac{2}{3}, -\frac{5}{6}\right)$. This checks, so it is the solution.

22. $\left(4, \frac{1}{2}, -\frac{1}{2}\right)$

23. $\quad x + y + z = 57, \quad (1)$
$$-2x + y \quad\quad = 3, \quad (2)$$
$$x \quad\quad - z = 6 \quad (3)$$

1., 2. The equations are already in standard form with no fractions or decimals.

3., 4. Note that there is no z in equation (2). We will use equations (1) and (3) to obtain another equation with no z term.

$$x + y + z = 57 \quad (1)$$
$$\underline{x \quad\quad - z = 6} \quad (3)$$
$$2x + y \quad\quad = 63 \quad (4)$$

5. Now solve the system of equations (2) and (4).

$$-2x + y = 3 \quad (2)$$
$$\underline{2x + y = 63} \quad (4)$$
$$2y = 66$$
$$y = 33$$

$$2x + 33 = 63 \quad \text{Substituting 33 for y in (4)}$$
$$2x = 30$$
$$x = 15$$

6. Substitute in equation (1) or (3) to find z.

$$15 - z = 6 \quad \text{Substituting 15 for } x \text{ in (3)}$$
$$9 = z$$

We obtain $(15, 33, 9)$. This checks, so it is the solution.

24. $(17, 9, 79)$

25. $\quad 2a - 3b \quad\quad = 2, \quad (1)$
$$7a \quad\quad + 4c = \frac{3}{4}, \quad (2)$$
$$-3b + 2c = 1 \quad (3)$$

1. The equations are already in standard form.

2. Multiply equation (2) by 4 to clear the fraction. The resulting system is

$$2a - 3b \quad\quad = 2, \quad (1)$$
$$28a \quad\quad + 16c = 3, \quad (4)$$
$$-3b + 2c = 1 \quad (3)$$

3. Note that there is no b in equation (2). We will use equations (1) and (3) to obtain another equation with no b term.

$$2a - 3b \quad\quad = 2 \quad (1)$$
$$\underline{\quad\quad 3b - 2c = -1} \quad \text{Multiplying (3) by } -1$$
$$2a \quad\quad - 2c = 1 \quad (5)$$

5. Now solve the system of equations (4) and (5).

$$28a + 16c = 3 \quad (4)$$
$$\underline{16a - 16c = 8} \quad \text{Multiplying (5) by 8}$$
$$44a \quad\quad = 11$$
$$a = \frac{1}{4}$$

$$2 \cdot \frac{1}{4} - 2c = 1 \quad \text{Substituting } \frac{1}{4} \text{ for } a \text{ in (5)}$$
$$\frac{1}{2} - 2c = 1$$
$$-2c = \frac{1}{2}$$
$$c = -\frac{1}{4}$$

6. $2\left(\frac{1}{4}\right) - 3b = 2 \quad \text{Substituting } \frac{1}{4} \text{ for } a \text{ in (1)}$
$$\frac{1}{2} - 3b = 2$$
$$-3b = \frac{3}{2}$$
$$b = -\frac{1}{2}$$

We obtain $\left(\frac{1}{4}, -\frac{1}{2}, -\frac{1}{4}\right)$. This checks, so it is the solution.

26. $(3, 4, -1)$

27. $\quad x + y + z = 180, \quad (1)$
$$y = 2 + 3x, \quad (2)$$
$$z = 80 + x \quad (3)$$

1. Only Equation (1) is in standard form. Rewrite the system with all equations in standard form.

$$x + y + z = 180, \quad (1)$$
$$-3x + y \quad\quad = 2, \quad (4)$$
$$-x \quad\quad + z = 80 \quad (5)$$

2. There are no fractions or decimals.

3., 4. Note that there is no z in equation (4). We will use equations (1) and (5) to obtain another equation with no z term.

$$x + y + z = 180 \quad (1)$$
$$\underline{x \quad\quad - z = -80} \quad \text{Multiplying (5) by } -1$$
$$2x + y \quad\quad = 100 \quad (6)$$

5. Now solve the system of equations (4) and (6).

$$-3x + y = 2 \quad (4)$$
$$\underline{-2x - y = -100} \quad \text{Multiplying (6) by } -1$$
$$-5x \quad\quad = -98$$
$$x = \frac{98}{5}$$

$$-3 \cdot \frac{98}{5} + y = 2 \quad \text{Substituting } \frac{98}{5} \text{ for}$$
$$x \text{ in (4)}$$
$$-\frac{294}{5} + y = 2$$
$$y = \frac{304}{5}$$

6. Substitute in equation (1) or (5) to find z.

$$-\frac{98}{5} + z = 80 \quad \text{Substituting } \frac{98}{5} \text{ for } x \text{ in (5)}$$
$$z = \frac{498}{5}$$

We obtain $\left(\frac{98}{5}, \frac{304}{5}, \frac{498}{5}\right)$. This checks, so it is the solution.

28. $(2, 5, -3)$

29.
$$\begin{aligned} x \quad\;\;\; + \;z &= 0, \quad (1) \\ x + y + 2z &= 3, \quad (2) \\ y + \;z &= 2 \quad (3) \end{aligned}$$

1., 2. The equations are already in standard form with no fractions or decimals.

3., 4. The variable y is missing in equation (1). We use equations (2) and (3) to obtain another equation with no y term.

$$\begin{array}{ll} x + y + 2z = \;\;\;3 & (2) \\ \underline{\;\;\;\; - y - \;\;z = -2} & \text{Multiplying (3) by } -1 \\ x \quad\quad + \;\;z = \;\;\;1 & (4) \quad \text{Adding} \end{array}$$

5. Now solve the system of equations (1) and (4).

$$\begin{array}{ll} x + z = \;\;\;0 & (1) \\ \underline{-x - z = -1} & \text{Multiplying (4) by } -1 \\ \;\;\;\;\;\; 0 = -1 & \text{Adding} \end{array}$$

We get a false equation. There is no solution. The solution set is $\emptyset$.

30. $\emptyset$

31.
$$\begin{aligned} x + \;\;y + z &= \;\;\;1, \quad (1) \\ -x + 2y + z &= \;\;\;2, \quad (2) \\ 2x - \;\;y \quad\;\; &= -1 \quad (3) \end{aligned}$$

1., 2. The equations are already in standard form with no fractions or decimals.

3. Use equations (1) and (2) to eliminate z:

$$\begin{array}{ll} x + \;\;y + z = \;\;\;1 & (1) \\ \underline{x - 2y - z = -2} & \text{Multiplying (2) by } -1 \\ 2x - \;\;y \quad\;\; = -1 & \text{Adding} \end{array}$$

Equations (3) and (4) are identical, so the system is dependent. (We have seen that if equation (2) is multiplied by -1 and added to equation (1), the result is equation (3).)

32. The system is dependent.

33. $g(x) = 2x - 7$

$g(-1) = 2(-1) - 7 = -2 - 7 = -9$

34. $2a + 2h - 7$

35.

36.

37.
$$\begin{aligned} \frac{x+2}{3} - \frac{y+4}{2} + \frac{z+1}{6} &= 0, \\ \frac{x-4}{3} + \frac{y+1}{4} - \frac{z-2}{2} &= -1, \\ \frac{x+1}{2} + \frac{y}{2} + \frac{z-1}{4} &= \frac{3}{4} \end{aligned}$$

1., 2. We clear fractions and write each equation in standard form.

To clear fractions, we multiply both sides of each equation by the LCM of its denominators. The LCM's are 6, 12, and 4, respectively.

$$6\left(\frac{x+2}{3} - \frac{y+4}{2} + \frac{z+1}{6}\right) = 6 \cdot 0$$
$$2(x+2) - 3(y+4) + (z+1) = 0$$
$$2x + 4 - 3y - 12 + z + 1 = 0$$
$$2x - 3y + z = 7$$

$$12\left(\frac{x-4}{3} + \frac{y+1}{4} - \frac{z-2}{2}\right) = 12 \cdot (-1)$$
$$4(x-4) + 3(y+1) - 6(z-2) = -12$$
$$4x - 16 + 3y + 3 - 6z + 12 = -12$$
$$4x + 3y - 6z = -11$$

$$4\left(\frac{x+1}{2} + \frac{y}{2} + \frac{z-1}{4}\right) = 4 \cdot \frac{3}{4}$$
$$2(x+1) + 2(y) + (z-1) = 3$$
$$2x + 2 + 2y + z - 1 = 3$$
$$2x + 2y + z = 2$$

The resulting system is

$$\begin{aligned} 2x - 3y + \;\;z &= \;\;\;7, \quad (1) \\ 4x + 3y - 6z &= -11, \quad (2) \\ 2x + 2y + \;\;z &= \;\;\;2 \quad (3) \end{aligned}$$

3., 4. We eliminate z from two different pairs of equations.

$$\begin{array}{ll} 12x - 18y + 6z = \;\;\;42 & \text{Multiplying (1) by 6} \\ \underline{4x + \;\;3y - 6z = -11} & (2) \\ 16x - 15y \quad\quad\; = \;\;\;31 & (4) \quad \text{Adding} \end{array}$$

$$\begin{array}{ll} 2x - 3y + z = \;\;\;7 & (1) \\ \underline{-2x - 2y - z = -2} & \text{Multiplying (3) by } -1 \\ -5y \quad\quad\;\; = \;\;\;5 & (5) \quad \text{Adding} \end{array}$$

5. Solve (5) for y: $\quad -5y = 5$
$$y = -1$$

Substitute -1 for y in (4):

$$16x - 15(-1) = 31$$
$$16x + 15 = 31$$
$$16x = 16$$
$$x = 1$$

6. Substitute 1 for x and -1 for y in (1):

$$2 \cdot 1 - 3(-1) + z = 7$$
$$5 + z = 7$$
$$z = 2$$

We obtain $(1, -1, 2)$. This checks, so it is the solution.

38. $(1, 1, 1)$

39.
$$w + x + y + z = 2, \quad (1)$$
$$w + 2x + 2y + 4z = 1, \quad (2)$$
$$w - x + y + z = 6, \quad (3)$$
$$w - 3x - y + z = 2 \quad (4)$$

The equations are already in standard form with no fractions or decimals.

Start by eliminating w from three different pairs of equations.

$$\begin{array}{ll} w + x + y + z = 2 & (1) \\ \underline{-w - 2x - 2y - 4z = -1} & \text{Multiplying (2) by } -1 \\ - x - y - 3z = 1 & (5) \quad \text{Adding} \end{array}$$

$$\begin{array}{ll} w + x + y + z = 2 & (1) \\ \underline{-w + x - y - z = -6} & \text{Multiplying (3) by } -1 \\ 2x = -4 & (6) \quad \text{Adding} \end{array}$$

$$\begin{array}{ll} w + x + y + z = 2 & (1) \\ \underline{-w + 3x + y - z = -2} & \text{Multiplying (4) by } -1 \\ 4x + 2y = 0 & (7) \quad \text{Adding} \end{array}$$

We can solve (6) for x:

$$2x = -4$$
$$x = -2$$

Substitute -2 for x in (7):

$$4(-2) + 2y = 0$$
$$-8 + 2y = 0$$
$$2y = 8$$
$$y = 4$$

Substitute -2 for x and 4 for y in (5):

$$-(-2) - 4 - 3z = 1$$
$$-2 - 3z = 1$$
$$-3z = 3$$
$$z = -1$$

Substitute -2 for x, 4 for y, and -1 for z in (1):

$$w - 2 + 4 - 1 = 2$$
$$w + 1 = 2$$
$$w = 1$$

We obtain $(1, -2, 4, -1)$. This checks, so it is the solution.

40. $(-3, -1, 0, 4)$

41.
$$\frac{2}{x} - \frac{1}{y} - \frac{3}{z} = -1,$$
$$\frac{2}{x} - \frac{1}{y} + \frac{1}{z} = -9,$$
$$\frac{1}{x} + \frac{2}{y} - \frac{4}{z} = 17$$

Let u represent $\frac{1}{x}$, v represent $\frac{1}{y}$, and w represent $\frac{1}{z}$. Substituting, we have

$$2u - v - 3w = -1, \quad (1)$$
$$2u - v + w = -9, \quad (2)$$
$$u + 2v - 4w = 17 \quad (3)$$

1., 2. The equations in u, v, and w are in standard form with no fractions or decimals.

3., 4. We eliminate v from two different pairs of equations.

$$\begin{array}{ll} 2u - v - 3w = -1 & (1) \\ \underline{-2u + v - w = 9} & \text{Multiplying (2) by } -1 \\ -4w = 8 & (4) \quad \text{Adding} \end{array}$$

$$\begin{array}{ll} 4u - 2v - 6w = -2 & \text{Multiplying (1) by 2} \\ \underline{u + 2v - 4w = 17} & (3) \\ 5u - 10w = 15 & (5) \quad \text{Adding} \end{array}$$

5. We can solve (4) for w:

$$-4w = 8$$
$$w = -2$$

Substitute -2 for w in (5):

$$5u - 10(-2) = 15$$
$$5u + 20 = 15$$
$$5u = -5$$
$$u = -1$$

6. Substitute -1 for u and -2 for w in (1):

$$2(-1) - v - 3(-2) = -1$$
$$-v + 4 = -1$$
$$-v = -5$$
$$v = 5$$

Solve for x, y, and z. We substitute -1 for u, 5 for v and -2 for w.

$$u = \frac{1}{x} \qquad v = \frac{1}{y} \qquad w = \frac{1}{z}$$
$$-1 = \frac{1}{x} \qquad 5 = \frac{1}{y} \qquad -2 = \frac{1}{z}$$
$$x = -1 \qquad y = \frac{1}{5} \qquad z = -\frac{1}{2}$$

We obtain $\left(-1, \frac{1}{5}, -\frac{1}{2}\right)$. This checks, so it is the solution.

42. $\left(-\frac{1}{2}, -1, -\frac{1}{3}\right)$

43. $x - 3y + 2z = 1,$ (1)

$2x + y - z = 3,$ (2)

$9x - 6y + 3z = k$ (3)

Eliminate z from two different pairs of equations.

$x - 3y + 2z = 1$ (1)

$\underline{4x + 2y - 2z = 6}$ Multiplying (2) by 2

$5x - y \quad\quad = 7$ (4)

$6x + 3y - 3z = 9$ Multiplying (2) by 3

$\underline{9x - 6y + 3z = k}$ (3)

$15x - 3y \quad\quad = 9 + k$ (5)

Solve the system of equations (4) and (5).

$-15x + 3y = -21$ Multiplying (4) by -3

$\underline{15x - 3y = 9 + k}$

$\quad\quad\quad 0 = -12 + k$ (6)

The system is dependent for the value of k that makes equation (6) true. We solve for k:

$0 = -12 + k$

$12 = k$

44. 14

45. $Ax + By + Cz = 12$

Three solutions are $\left(1, \frac{3}{4}, 3\right)$, $\left(\frac{4}{3}, 1, 2\right)$, and $(2, 1, 1)$. We substitute for x, y, and z and solve for A, B, and C.

$A + \frac{3}{4}B + 3C = 12,$

$\frac{4}{3}A + B + 2C = 12,$

$2A + B + C = 12$

1., 2. The equations are in standard form. We clear fractions.

$4A + 3B + 12C = 48,$ (1)

$4A + 3B + 6C = 36,$ (2)

$2A + B + C = 12$ (3)

3. Use equations (1) and (2) to eliminate A.

$4A + 3B + 12C = 48$ (1)

$\underline{-4A - 3B - 6C = -36}$ Multiplying (2)

$6C = 12$ by -1

$C = 2$

4. At this point we can use another point of equations to eliminate A, or we can substitute 2 for C in either equation (1) or (2) and in equation (3) to get a system of equations in B and C. We choose the latter option.

$4A + 3B = 24$ (4)   Substituting in (1)

$2A + B = 10$ (5)   Substituting in (3)

5., 6. Solve the system of equations (4) and (5).

$4A + 3B = 24$ (4)

$\underline{-4A - 2B = -20}$ Multiplying (5) by -2

$B = 4$

$2A + 4 = 10$ Substituting 4 for B in (5)

$2A = 6$

$A = 3$

The solution is $(3, 4, 2)$, so the equation is $3x + 4y + 2z = 12$.

46. $z = 8 - 2x - 4y$

Exercise Set 8.5

1. *Familiarize.* Let $x =$ the first number, $y =$ the second number, and $z =$ the third number.

Translate.

The sum of three numbers is 105.

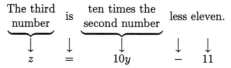

$$x + y + z = 105$$

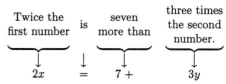

The third number is ten times the second number less eleven.

$$z = 10y - 11$$

Twice the first number is seven more than three times the second number.

$$2x = 7 + 3y$$

We now have a system of equations.

$x + y + z = 105$ or $x + y + z = 105$

$z = 10y - 11$ $-10y + z = -11$

$2x = 7 + 3y$ $2x - 3y = 7$

Carry out. Solving the system we get $(17, 9, 79)$.

Check. The sum of the three numbers is 105. Ten times the second number is 90, and 11 less than 90 is 79, the third number. Three times the second number is 27, and 7 more than 27 is 34, which is twice the first number. The numbers check.

State. The numbers are 17, 9, and 79.

2. 16, 19, 22

3. *Familiarize.* Let $x =$ the first number, $y =$ the second number, and $z =$ the third number.

Translate.

The sum of the three numbers is 5.

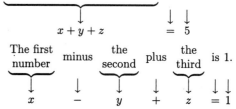

$$x + y + z = 5$$

The first number minus the second plus the third is 1.

$$x - y + z = 1$$

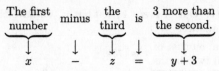

We now have a system of equations.

$$x + y + z = 5 \quad \text{or} \quad x + y + z = 5$$
$$x - y + z = 1 \qquad\qquad x - y + z = 1$$
$$x - z = y + 3 \qquad\quad x - y - z = 3$$

Carry out. Solving the system we get $(4, 2, -1)$.

Check. The sum of the numbers is 5. The first minus the second plus the third is $4 - 2 + (-1)$, or 1. The first minus the third is 5, which is three more than the second. The numbers check.

State. The numbers are 4, 2, and -1.

4. 8, 21, -3

5. Familiarize. We first make a drawing.

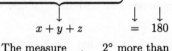

We let x, y, and z represent the measures of angles A, B, and C, respectively. The measures of the angles of a triangle add up to $180°$.

Translate.

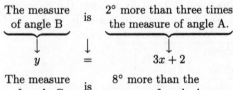

We now have a system of equations.

$$x + y + z = 180,$$
$$y = 3x + 2,$$
$$z = x + 8$$

Carry out. Solving the system we get $(34, 104, 42)$.

Check. The sum of the numbers is 180, so that checks. Three times the measure of angle A is $3 \cdot 34°$, or $102°$, and $2°$ added to $102°$ is $104°$, the measure of angle B. The measure of angle C, $42°$, is $8°$ more than $34°$, the measure of angle A. These values check.

State. Angles A, B, and C measure $34°$, $104°$, and $42°$, respectively.

6. $30°$, $90°$, $60°$

7. Familiarize. We first make a drawing.

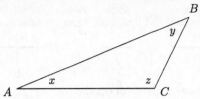

We let x, y, and z represent the measures of angles, A, B, and C, respectively. The measures of the angles of a triangle add up to $180°$.

Translate.

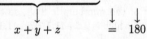

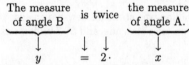

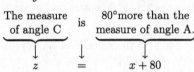

We now have a system of equations.

$$x + y + z = 180,$$
$$y = 2x,$$
$$z = x + 80$$

Carry out. Solving the system we get $(25, 50, 105)$.

Check. The sum of the numbers is 180, so that checks. The measure of angle B, $50°$, is twice $25°$, the measure of angle A. The measure of angle C, $105°$, is $80°$ more than $25°$, the measure of angle A. The values check.

State. Angles A, B, and C measure $25°$, $50°$, and $105°$, respectively.

8. $32°$, $96°$, $52°$

9. Familiarize. Let x, y, and z represent the amount spent on newspaper, television and radio ads, respectively, in billions of dollars.

Translate.

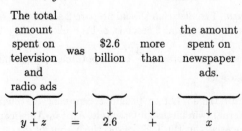

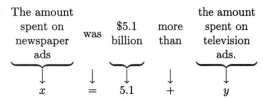

We now have a system of equations.

$$x + y + z = 84.8,$$
$$y + z = 2.6 + x,$$
$$x = 5.1 + y$$

Carry out. Solving the system we get
$(41.1., 36, 7.7)$.

Check. The sum of the numbers is 84.8, so that checks. Also, $36 + 7.7 = 43.7$ which is $2.6 + 41.1$, and $41.1 = 5.1 + 36$. These values check.

State. \$41.1 billion was spent on newspaper ads, \$36 billion was spent on television ads, and \$7.7 billion was spent on radio ads.

10. Automatic transmission: \$865, power door locks: \$520, air conditioning: \$375

11. **Familiarize**. Let s = the number of servings of steak, p = the number of baked potatoes, and b = the number of servings of broccoli. Then s servings of steak contain $300s$ Calories, $20s$ g of protein, and no vitamin C. In p baked potatoes there are $100p$ Calories, $5p$ g of protein, and $20p$ mg of vitamin C. And b servings of broccoli contain $50b$ Calories, $5b$ g of protein, and $100b$ mg of vitamin C. The patient requires 800 Calories, 55 g of protein, and 200 mg of vitamin C.

Translate. Write equations for the total number of calories, the total amount of protein, and the total amount of vitamin C.

$$300s + 100p + 50b = 800 \quad \text{(Calories)}$$
$$20s + 5p + 5b = 55 \quad \text{(protein)}$$
$$20p + 100b = 220 \quad \text{(vitamin C)}$$

We now have a system of equations.

Carry out. Solving the system we get $(2, 1, 2)$.

Check. Two servings of steak provide 600 Calories, 40 g of protein, and no vitamin C. One baked potato provides 100 Calories, 5 g of protein, and 20 mg of vitamin C. And 2 servings of broccoli provide 100 Calories, 10 g of protein, and 200 mg of vitamin C. Together, then, they provide 800 Calories, 55 g of protein, and 220 mg of vitamin C. The values check.

State. The dietician should prepare 2 servings of steak, 1 baked potato, and 2 servings of broccoli.

12. $1\frac{1}{8}$ servings of steak, $2\frac{3}{4}$ baked potatoes, $3\frac{3}{4}$ servings of asparagus

13. **Familiarize**. Let x, y, and z represent the number of fraternal twin births for Orientals, African-Americans, and Caucasians in the U.S., respectively, out of every 15,400 births.

Translate. Out of every 15,400 births, we have the following statistics:

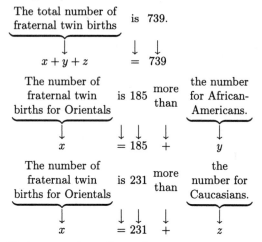

We have a system of equations.

$$x + y + z = 739,$$
$$x = 185 + y,$$
$$x = 231 + y$$

Carry out. Solving the system we get
$(385, 200, 154)$.

Check. The total of the numbers is 739. Also 385 is 185 more than 200, and it is 231 more than 154.

State. Out of every 15,400 births, there are 385 births of fraternal twins for Orientals, 200 for African-Americans, and 154 for Caucasians.

14. A man: 3.6 times, a woman: 18.1 times, a one-year old child: 50 times

15. **Familiarize**. It helps to organize the information in a table. We let x, y, and z represent the weekly productions of the individual machines.

Machines Working	A	B	C
Weekly Production	x	y	z

Machines Working	A + B	B + C	A, B, & C
Weekly Production	3400	4200	5700

Translate. From the table, we obtain three equations.

$$x + y + z = 5700 \quad \text{(All three machines working)}$$
$$x + y = 3400 \quad \text{(A and B working)}$$
$$y + z = 4200 \quad \text{(B and C working)}$$

Carry out. Solving the system we get
$(1500, 1900, 2300)$.

Check. The sum of the weekly productions of machines A, B & C is $1500 + 1900 + 2300$, or 5700. The sum of the weekly productions of machines A and B is $1500 + 1900$, or 3400. The sum of the weekly productions of machines B and C is $1900 + 2300$, or 4200. The numbers check.

State. In a week Machine A can polish 1500 lenses, Machine B can polish 1900 lenses, and Machine C can polish 2300 lenses.

16. $A: 2200$, $B: 2500$, $C: 2700$

17. *Familiarize*. It helps to organize the information in a table. We let x, y, and z represent the number of gallons per hour which can be pumped by pumps A, B, and C, respectively.

Pumps Working	A	B	C
Gallons Per hour	x	y	z

Pumps Working	A + B	A + C	A, B, & C
Gallons Per hour	2200	2400	3700

Translate. From the table, we obtain three equations.

$$x + y + z = 3700 \quad \text{(All three pumps working)}$$
$$x + y = 2200 \quad \text{(A and B working)}$$
$$x + z = 2400 \quad \text{(A and C working)}$$

Carry out. Solving the system we get $(900, 1300, 1500)$.

Check. The sum of the gallons per hour pumped when all three are pumping is $900 + 1300 + 1500$, or 3700. The sum of the gallons per hour pumped when only pump A and pump B are pumping is $900 + 1300$, or 2200. The sum of the gallons per hour pumped when only pump A and pump C are pumping is $900 + 1500$, or 2400. The numbers check.

State. The pumping capacities of pumps A, B, and C are respectively 900, 1300, and 1500 gallons per hour.

18. Pat: 10, Chris: 12, Jean: 15

19. *Familiarize*. Let x, y, and z represent the amount invested at 8%, 6%, and 9%, respectively. The interest earned is $0.08x$, $0.06y$, and $0.09z$.

Translate.

The total invested was $80,000.
$$x + y + z = 80,000$$

The total interest was $6300.
$$0.08x + 0.06y + 0.09z = 6300$$

The interest at 8% was 4 times the interest at 6%.
$$0.08x = 4 \cdot 0.06y$$

We have a system of equations.

$$x + y + z = 80,000,$$
$$0.08x + 0.06y + 0.09z = 6300,$$
$$0.08x = 4(0.06y)$$

Carry out. Solving the system we get $(45,000, 15,000, 20,000)$.

Check. The numbers add up to 80,000. Also, $0.08(45,000) + 0.06(15,000) + 0.09(20,000) = 6300$. In addition, $0.08(45,000) = 3600$ which is $4[0.06(15,000)]$. The values check.

State. $45,000 was invested at 8%, $15,000 at 6%, and $20,000 at 9%.

20. 1869

21. $\dfrac{2}{t+1} + \dfrac{t}{t+1} = \dfrac{2+t}{t+1}$ \quad Adding the numerators

22. 1

23.
$$\frac{2}{t} + \frac{t}{t+1} \quad \text{LCD is } t(t+1)$$
$$= \frac{2}{t} \cdot \frac{t+1}{t+1} + \frac{t}{t+1} \cdot \frac{t}{t}$$
$$= \frac{2(t+1)}{t(t+1)} + \frac{t^2}{t(t+1)}$$
$$= \frac{t^2 + 2t + 2}{t(t+1)}$$

24. $\dfrac{3t+1}{(t+1)(t-1)}$

25. We let w, x, y, and z represent the ages of Tammy, Carmen, Dennis, and Mark, respectively.

Translate.

Tammy's age is Carmen's age plus Dennis's age.
$$w = x + y$$

Carmen's age is 2 plus Dennis's age plus Mark's age.
$$x = 2 + y + z$$

Dennis's age is four times Mark's age.
$$y = 4 \cdot z$$

The sum of all ages is 42.
$$w + x + y + z = 42$$

We now have a system of equations.

$$w = x + y,$$
$$x = 2 + y + z,$$
$$y = 4z,$$
$$w + x + y + z = 42$$

Carry out. Solving the system we get $(20, 12, 8, 2)$.

Check. The sum of all four numbers is 42. Tammy's age, 20, is the sum of the ages of Carmen and Dennis, $12 + 8$. Carmen's age, 12, is 2 more than the sum of the ages of Dennis and Mark, $8 + 2$. Dennis's age, 8, is four times Mark's age, 2. The numbers check.

State. Tammy is 20 years old.

26. 464

27. ***Familiarize***. We first make a drawing with additional labels.

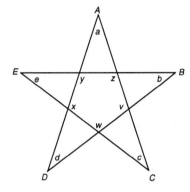

We let a, b, c, d, and e represent the angle measures at the tips of the star. We also label the interior angles of the pentagon v, w, x, y, and z. We recall the following geometric fact:

The sum of the measures of the interior angles of a polygon of n sides is given by $(n - 2)180°$.

Using this fact we know:

1. The sum of the angle measures of a triangle is $(3 - 2)180°$, or $180°$.

2. The sum of the angle measures of a pentagon is $(5 - 2)180°$, or $3(180°)$.

Translate. Using fact (1) listed above we obtain a system of 5 equations.

$$a + v + d = 180$$
$$b + w + e = 180$$
$$c + x + a = 180$$
$$d + y + b = 180$$
$$e + z + c = 180$$

Carry out. Adding we obtain

$$2a + 2b + 2c + 2d + 2e + v + w + x + y + z = $$
$$5(180)$$
$$2(a + b + c + d + e) + (v + w + x + y + z) = $$
$$5(180)$$

Using fact (2) listed above we substitute $3(180)$ for $(v + w + x + y + z)$ and solve for $(a + b + c + d + e)$.

$$2(a + b + c + d + e) + 3(180) = 5(180)$$
$$2(a + b + c + d + e) = 2(180)$$
$$a + b + c + d + e = 180$$

Check. We should repeat the above calculations.

State. The sum of the angle measures at the tips of the star is $180°$.

28. 5 adults, 1 student, 94 children

29. ***Familiarize***. Let T, G, and H represent the number of tickets Tom, Gary, and Hal begin with, respectively. After Hal gives tickets to Tom and Gary, each has the following number of tickets:

  Tom: $T + T$, or $2T$,

  Gary: $G + G$, or $2G$,

  Hal: $H - T - G$.

After Tom gives tickets to Gary and Hal, each has the following number of tickets:

  Gary: $2G + 2G$, or $4G$,

  Hal: $(H - T - G) + (H - T - G)$, or

           $2(H - T - G)$,

  Tom: $2T - 2G - (H - T - G)$, or

           $3T - H - G$

After Gary gives tickets to Hal and Tom, each has the following number of tickets:

  Hal: $2(H - T - G) + 2(H - T - G)$, or

          $4(H - T - G)$

  Tom: $(3T - H - G) + (3T - H - G)$, or

          $2(3T - H - G)$,

  Gary: $4G - 2(H - T - G) - (3T - H - G)$, or

          $7G - H - T$.

Translate. Since Hal, Tom, and Gary each finish with 40 tickets, we write the following system of equations:

$$4(H - T - G) = 40,$$
$$2(3T - H - G) = 40,$$
$$7G - H - T = 40$$

Carry out. Solving the system we find that $T = 35$.

Check. The check is left to the student.

State. Tom started with 35 tickets.

Exercise Set 8.6

1. $4x + 2y = 11,$

   $3x - y = 2$

Write a matrix using only the constants.

$$\begin{bmatrix} 4 & 2 & \vdots & 11 \\ 3 & -1 & \vdots & 2 \end{bmatrix}$$

Multiply row 2 by 4 to make the first number in row 2 a multiple of 4.

$$\begin{bmatrix} 4 & 2 & \vdots & 11 \\ 12 & -4 & \vdots & 8 \end{bmatrix} \text{ New Row 2 = 4(Row 2)}$$

Multiply row 1 by -3 and add it to row 2.

$$\begin{bmatrix} 4 & 2 & \vdots & 11 \\ 0 & -10 & \vdots & -25 \end{bmatrix} \begin{array}{l} \text{New Row } 2 = -3(\text{Row } 1) + \\ \text{Row } 2 \end{array}$$

Reinserting the variables, we have

$$4x + 2y = 11, \quad (1)$$
$$-10y = -25. \quad (2)$$

Solve Equation (2) for y.

$$-10y = -25$$
$$y = \frac{5}{2}$$

Back-substitute $\frac{5}{2}$ for y in Equation (1) and solve for x.

$$4x + 2y = 11$$
$$4x + 2 \cdot \frac{5}{2} = 11$$
$$4x + 5 = 11$$
$$4x = 6$$
$$x = \frac{3}{2}$$

The solution is $\left(\frac{3}{2}, \frac{5}{2} \right)$.

2. $\left(-\frac{1}{3}, -4 \right)$

3. $x + 4y = 8,$
$3x + 5y = 3$

We first write a matrix using only the constants.

$$\begin{bmatrix} 1 & 4 & \vdots & 8 \\ 3 & 5 & \vdots & 3 \end{bmatrix}$$

Multiply the first row by -3 and add it to the second row.

$$\begin{bmatrix} 1 & 4 & \vdots & 8 \\ 0 & -7 & \vdots & -21 \end{bmatrix} \begin{array}{l} \text{New Row } 2 = -3(\text{Row } 1) + \\ \text{Row } 2 \end{array}$$

Reinserting the variables, we have

$$x + 4y = 8, \quad (1)$$
$$-7y = -21. \quad (2)$$

Solve Equation (2) for y.

$$-7y = -21$$
$$y = 3$$

Back-substitute 3 for y in Equation (1) and solve for x.

$$x + 4 \cdot 3 = 8$$
$$x + 12 = 8$$
$$x = -4$$

The solution is $(-4, 3)$.

4. $(-3, 2)$

5. $5x - 3y = -2,$
$4x + 2y = 5$

Write a matrix using only the constants.

$$\begin{bmatrix} 5 & -3 & \vdots & -2 \\ 4 & 2 & \vdots & 5 \end{bmatrix}$$

Multiply the second row by 5 to make the first number in row 2 a multiple of 5.

$$\begin{bmatrix} 5 & -3 & \vdots & -2 \\ 20 & 10 & \vdots & 25 \end{bmatrix} \text{New Row } 2 = 5(\text{Row } 2)$$

Now multiply the first row by -4 and add it to the second row.

$$\begin{bmatrix} 5 & -3 & \vdots & -2 \\ 0 & 22 & \vdots & 33 \end{bmatrix} \begin{array}{l} \text{New Row } 2 = -4(\text{Row } 1) + \\ \text{Row } 2 \end{array}$$

Reinserting the variables, we have

$$5x - 3y = -2, \quad (1)$$
$$22y = 33. \quad (2)$$

Solve Equation (2) for y.

$$22y = 33$$
$$y = \frac{3}{2}$$

Back-substitute $\frac{3}{2}$ for y in Equation (1) and solve for x.

$$5x - 3y = -2$$
$$5x - 3 \cdot \frac{3}{2} = -2$$
$$5x - \frac{9}{2} = -\frac{4}{2}$$
$$5x = \frac{5}{2}$$
$$x = \frac{1}{2}$$

The solution is $\left(\frac{1}{2}, \frac{3}{2} \right)$.

6. $\left(-1, \frac{5}{2} \right)$

7. $4x - y - 3z = 1,$
$8x + y - z = 5,$
$2x + y + 2z = 5$

Write a matrix using only the constants.

$$\begin{bmatrix} 4 & -1 & -3 & \vdots & 1 \\ 8 & 1 & -1 & \vdots & 5 \\ 2 & 1 & 2 & \vdots & 5 \end{bmatrix}$$

First interchange rows 1 and 3 so that each number below the first number in the first row is a multiple of that number.

$$\begin{bmatrix} 2 & 1 & 2 & \vdots & 5 \\ 8 & 1 & -1 & \vdots & 5 \\ 4 & -1 & -3 & \vdots & 1 \end{bmatrix}$$

Multiply row 1 by -4 and add it to row 2.

Multiply row 1 by -2 and add it to row 3.

$$\begin{bmatrix} 2 & 1 & 2 & | & 5 \\ 0 & -3 & -9 & | & -15 \\ 0 & -3 & -7 & | & -9 \end{bmatrix}$$

Multiply row 2 by -1 and add it to row 3.

$$\begin{bmatrix} 2 & 1 & 2 & | & 5 \\ 0 & -3 & -9 & | & -15 \\ 0 & 0 & 2 & | & 6 \end{bmatrix}$$

Reinserting the variables, we have

$$\begin{aligned} 2x + y + 2z &= 5, &(1) \\ -3y - 9z &= -15, &(2) \\ 2z &= 6. &(3) \end{aligned}$$

Solve (3) for z.

$$2z = 6$$
$$z = 3$$

Back-substitute 3 for z in (2) and solve for y.

$$-3y - 9z = -15$$
$$-3y - 9(3) = -15$$
$$-3y - 27 = -15$$
$$-3y = 12$$
$$y = -4$$

Back-substitute 3 for z and -4 for y in (1) and solve for x.

$$2x + y + 2z = 5$$
$$2x + (-4) + 2(3) = 5$$
$$2x - 4 + 6 = 5$$
$$2x = 3$$
$$x = \frac{3}{2}$$

The solution is $\left(\frac{3}{2}, -4, 3\right)$.

8. $\left(2, \frac{1}{2}, -2\right)$

9. $\quad p + q + r = 1,$
$\quad p - 2q - 3r = 3,$
$4p + 5q + 6r = 4$

We first write a matrix using only the constants.

$$\begin{bmatrix} 1 & 1 & 1 & | & 1 \\ 1 & -2 & -3 & | & 3 \\ 4 & 5 & 6 & | & 4 \end{bmatrix}$$

$$\begin{bmatrix} 1 & 1 & 1 & | & 1 \\ 0 & -3 & -4 & | & 2 \\ 0 & 1 & 2 & | & 0 \end{bmatrix}$$
New Row 2 = -1(Row 1) + Row 2
New Row 3 = -4(Row 1) + Row 3

$$\begin{bmatrix} 1 & 1 & 1 & | & 1 \\ 0 & -3 & -4 & | & 2 \\ 0 & 3 & 6 & | & 0 \end{bmatrix}$$
New Row 3 = 3(Row 3)

$$\begin{bmatrix} 1 & 1 & 1 & | & 1 \\ 0 & -3 & -4 & | & 2 \\ 0 & 0 & 2 & | & 2 \end{bmatrix}$$
New Row 3 = Row 2 + Row 3

Reinstating the variables, we have

$$\begin{aligned} p + q + r &= 1, &(1) \\ -3q - 4r &= 2, &(2) \\ 2r &= 2 &(3) \end{aligned}$$

Solve (3) for r.

$$2r = 2$$
$$r = 1$$

Back-substitute 1 for r in (2) and solve for q.

$$-3q - 4 \cdot 1 = 2$$
$$-3q - 4 = 2$$
$$-3q = 6$$
$$q = -2$$

Back-substitute -2 for q and 1 for r in (1) and solve for p.

$$p + (-2) + 1 = 1$$
$$p - 1 = 1$$
$$p = 2$$

The solution is $(2, -2, 1)$.

10. $(-1, 2, -2)$

11. $\quad 3p \qquad + 2r = 11,$
$\qquad\quad q - 7r = 4,$
$\quad p - 6q \qquad = 1$

We first write a matrix using only the constants.

$$\begin{bmatrix} 3 & 0 & 2 & | & 11 \\ 0 & 1 & -7 & | & 4 \\ 1 & -6 & 0 & | & 1 \end{bmatrix}$$

$$\begin{bmatrix} 1 & -6 & 0 & | & 1 \\ 0 & 1 & -7 & | & 4 \\ 3 & 0 & 2 & | & 11 \end{bmatrix}$$
Interchange Row 1 and Row 3

$$\begin{bmatrix} 1 & -6 & 0 & | & 1 \\ 0 & 1 & -7 & | & 4 \\ 0 & 18 & 2 & | & 8 \end{bmatrix}$$
New Row 3 = -3(Row 1) + Row 3

$$\begin{bmatrix} 1 & -6 & 0 & | & 1 \\ 0 & 1 & -7 & | & 4 \\ 0 & 0 & 128 & | & -64 \end{bmatrix}$$
New Row 3 = -18(Row 2) + Row 3

Reinserting the variables, we have

$$\begin{aligned} p - 6q &= 1, &(1) \\ q - 7r &= 4, &(2) \\ 128r &= -64. &(3) \end{aligned}$$

Solve (3) for r.

$$128r = -64$$
$$r = -\frac{1}{2}$$

Back-substitute $-\frac{1}{2}$ for r in (2) and solve for q.

$$q - 7r = 4$$
$$q - 7\left(-\frac{1}{2}\right) = 4$$
$$q + \frac{7}{2} = 4$$
$$q = \frac{1}{2}$$

Back-substitute $\frac{1}{2}$ for q in (1) and solve for p.

$$p - 6 \cdot \frac{1}{2} = 1$$
$$p - 3 = 1$$
$$p = 4$$

The solution is $\left(4, \frac{1}{2}, -\frac{1}{2}\right)$.

12. $\left(\frac{1}{2}, \frac{2}{3}, -\frac{5}{6}\right)$

13. We will rewrite the equations with the variables in alphabetical order:

$$-2w + 2x + 2y - 2z = -10,$$
$$w + x + y + z = -5,$$
$$3w + x - y + 4z = -2,$$
$$w + 3x - 2y + 2z = -6$$

Write a matrix using only the constants.

$$\begin{bmatrix} -2 & 2 & 2 & -2 & | & -10 \\ 1 & 1 & 1 & 1 & | & -5 \\ 3 & 1 & -1 & 4 & | & -2 \\ 1 & 3 & -2 & 2 & | & -6 \end{bmatrix}$$

$$\begin{bmatrix} -1 & 1 & 1 & -1 & | & -5 \\ 1 & 1 & 1 & 1 & | & -5 \\ 3 & 1 & -1 & 4 & | & -2 \\ 1 & 3 & -2 & 2 & | & -6 \end{bmatrix} \quad \begin{matrix} \text{New Row 1} = \\ \frac{1}{2}(\text{Row 1}) \end{matrix}$$

$$\begin{bmatrix} -1 & 1 & 1 & -1 & | & -5 \\ 0 & 2 & 2 & 0 & | & -10 \\ 0 & 4 & 2 & 1 & | & -17 \\ 0 & 4 & -1 & 1 & | & -11 \end{bmatrix} \quad \begin{matrix} \text{New Row 2} = \\ \text{Row 1 + Row 2} \\ \text{New Row 3} = \\ 3(\text{Row 1}) + \text{Row 3} \\ \text{New Row 4} = \\ \text{Row 1 + Row 4} \end{matrix}$$

$$\begin{bmatrix} -1 & 1 & 1 & -1 & | & -5 \\ 0 & 2 & 2 & 0 & | & -10 \\ 0 & 0 & -2 & 1 & | & 3 \\ 0 & 0 & -5 & 1 & | & 9 \end{bmatrix} \quad \begin{matrix} \text{New Row 3} = \\ -2(\text{Row 2}) + \text{Row 3} \\ \text{New Row 4} = \\ -2(\text{Row 2}) + \text{Row 4} \end{matrix}$$

$$\begin{bmatrix} -1 & 1 & 1 & -1 & | & -5 \\ 0 & 2 & 2 & 0 & | & -10 \\ 0 & 0 & -2 & 1 & | & 3 \\ 0 & 0 & -10 & 2 & | & 18 \end{bmatrix} \quad \begin{matrix} \text{New Row 4} = \\ 2(\text{Row 4}) \end{matrix}$$

$$\begin{bmatrix} -1 & 1 & 1 & -1 & | & -5 \\ 0 & 2 & 2 & 0 & | & -10 \\ 0 & 0 & -2 & 1 & | & 3 \\ 0 & 0 & 0 & -3 & | & 3 \end{bmatrix} \quad \begin{matrix} \text{New Row 4} = \\ -5(\text{Row 3}) + \text{Row 4} \end{matrix}$$

Reinserting the variables, we have

$$-w + x + y - z = -5, \quad (1)$$
$$2x + 2y = -10, \quad (2)$$
$$-2y + z = 3, \quad (3)$$
$$-3z = 3. \quad (4)$$

Solve (4) for z.

$$-3z = 3$$
$$z = -1$$

Back-substitute -1 for z in (3) and solve for y.

$$-2y + (-1) = 3$$
$$-2y = 4$$
$$y = -2$$

Back-substitute -2 for y in (2) and solve for x.

$$2x + 2(-2) = -10$$
$$2x - 4 = -10$$
$$2x = -6$$
$$x = -3$$

Back-substitute -3 for x, -2 for y, and -1 for z in (1) and solve for w.

$$-w + (-3) + (-2) - (-1) = -5$$
$$-w - 3 - 2 + 1 = -5$$
$$-w - 4 = -5$$
$$-w = -1$$
$$w = 1$$

The solution is $(1, -3, -2, -1)$.

14. $(7, 4, 5, 6)$

15. *Familiarize*. Let d = the number of dimes and n = the number of nickels. The value of d dimes is \$0.10d, and the value of n nickels is \$0.05n.

Translate.

Total number of coins is 34.

$$d + n = 34$$

Total value of coins is \$1.90.

$$0.10d + 0.05n = 1.90$$

After clearing decimals, we have this system.

$$d + \ n = \ 34,$$
$$10d + 5n = 190$$

Carry out. Solve using matrices.

$$\begin{bmatrix} 1 & 1 & \vdots & 34 \\ 10 & 5 & \vdots & 190 \end{bmatrix}$$

$$\begin{bmatrix} 1 & 1 & \vdots & 34 \\ 0 & -5 & \vdots & -150 \end{bmatrix} \text{New Row 2} = -10(\text{Row 1}) + \text{Row 2}$$

Reinserting the variables, we have

$$d + n = \ 34, \quad (1)$$
$$-5n = -150 \quad (2)$$

Solve (2) for n.

$$-5n = -150$$
$$n = 30$$

$$d + 30 = 34 \quad \text{Back-substituting}$$
$$d = 4$$

Check. The sum of the two numbers is 34. The total value is $\$0.10(4) + \$0.50(30) = \$0.40 + \$1.50 = \$1.90$. The numbers check.

State. There are 4 dimes and 30 nickels.

16. 21 dimes, 22 quarters

17. *Familiarize.* We let x represent the number of pounds of the \$4.05 kind and y represent the number of pounds of the \$2.70 kind of granola. We organize the information in a table.

Granola	Number of pounds	Price per pound	Value
\$4.05 kind	x	\$4.05	\$4.05x
\$2.70 kind	y	\$2.70	\$2.70y
Mixture	15	\$3.15	\$3.15 × 15 or \$47.25

Translate.

Total number of pounds is 15.

$$x + y = 15$$

Total value of mixture is \$47.25.

$$4.05x + 2.70y = 47.25$$

After clearing decimals, we have this system:

$$x + \ y = \ 15,$$
$$405x + 270y = 4725$$

Carry out. Solve using matrices.

$$\begin{bmatrix} 1 & 1 & \vdots & 15 \\ 405 & 270 & \vdots & 4725 \end{bmatrix}$$

$$\begin{bmatrix} 1 & 1 & \vdots & 15 \\ 0 & -135 & \vdots & -1350 \end{bmatrix} \text{New Row 2} = -405(\text{Row 1}) + \text{Row 2}$$

Reinserting the variables, we have

$$x + y = 15, \quad (1)$$
$$-135y = -1350 \quad (2)$$

Solve (2) for y.

$$-135y = -1350$$
$$y = 10$$

Back-substitute 10 for y in (1) and solve for x.

$$x + 10 = 15$$
$$x = 5$$

Check. The sum of the numbers is 15. The total value is $\$4.05(5) + \$2.70(10)$, or $\$20.25 + \$27.00$, or \$47.25. The numbers check.

State. 5 pounds of the \$4.05 per lb granola and 10 pounds of the \$2.70 per lb granola should be used.

18. Candy: 14 lb, nuts: 6 lb

19. *Familiarize.* We let x, y, and z represent the amounts invested at 7%, 8%, and 9%, respectively. Recall the formula for simple interest:

$$\text{Interest} = \text{Principal} \ \times \ \text{Rate} \ \times \ \text{Time}$$

Translate. We organize the imformation in a table.

	First Investment	Second Investment	Third Investment	Total
P	x	y	z	\$2500
R	7%	8%	9%	
T	1 yr	1 yr	1 yr	
I	0.07x	0.08y	0.09z	\$212

The first row gives us one equation:

$$x + y + z = 2500$$

The last row gives a second equation:

$$0.07x + 0.08y + 0.09z = 212$$

Amount invested at 9% is \$1100 more than amount invested at 8%.

$$z = \$1100 + y$$

After clearing decimals, we have this system:

$$x + \ y + \ z = \ 2500,$$
$$7x + \ 8y + 9z = 21,200,$$
$$-y + \ z = \ 1100$$

Carry out. Solve using matrices.

$$\begin{bmatrix} 1 & 1 & 1 & \vdots & 2500 \\ 7 & 8 & 9 & \vdots & 21,200 \\ 0 & -1 & 1 & \vdots & 1100 \end{bmatrix}$$

$$\begin{bmatrix} 1 & 1 & 1 & \vdots & 2500 \\ 0 & 1 & 2 & \vdots & 3700 \\ 0 & -1 & 1 & \vdots & 1100 \end{bmatrix} \quad \begin{matrix} \text{New Row 2} = \\ -7(\text{Row 1}) + \text{Row 2} \end{matrix}$$

$$\begin{bmatrix} 1 & 1 & 1 & \vdots & 2500 \\ 0 & 1 & 2 & \vdots & 3700 \\ 0 & 0 & 3 & \vdots & 4800 \end{bmatrix} \quad \begin{matrix} \text{New Row 3} = \\ \text{Row 2} + \text{Row 3} \end{matrix}$$

Reinserting the variables, we have

$$\begin{aligned} x + y + z &= 2500, \quad (1) \\ y + 2z &= 3700, \quad (2) \\ 3z &= 4800 \quad (3) \end{aligned}$$

Solve (3) for z.

$$3z = 4800$$

$$z = 1600$$

Back-substitute 1600 for z in (2) and solve for y.

$$y + 2 \cdot 1600 = 3700$$

$$y + 3200 = 3700$$

$$y = 500$$

Back-substitute 500 for y and 1600 for z in (1) and solve for x.

$$x + 500 + 1600 = 2500$$

$$x + 2100 = 2500$$

$$x = 400$$

Check. The total investment is \$400 + \$500 + \$1600, or \$2500. The total interest is 0.07(\$400) + 0.08(\$500) + 0.09(\$1600) = \$28 + \$40 + \$144 = \$212. The amount invested at 9%, \$1600, is \$1100 more than the amount invested at 8%, \$500. The numbers check.

State. \$400 is invested at 7%, \$500 is invested at 8%, and \$1600 is invested at 9%.

20. \$500 at 8%, \$400 at 9%, \$2300 at 10%

21. $x^2 + x - 6$

We look for a pair of numbers whose product is -6 and whose sum is 1. They are 3 and -2.

$$x^2 + x - 6 = (x + 3)(x - 2)$$

22. $(2x + 1)(2x - 3)$

23. $x^2 + \dfrac{3}{4}x + \dfrac{1}{8}$

We look for a pair of numbers whose product is $\dfrac{1}{8}$ and whose sum is $\dfrac{3}{4}$. They are $\dfrac{1}{2}$ and $\dfrac{1}{4}$.

$$x^2 + \frac{3}{4}x + \frac{1}{8} = \left(x + \frac{1}{2}\right)\left(x + \frac{1}{4}\right)$$

24. $(6x + 1)(2x + 3)$

25.

26.

27. *Familiarize*. Let w, x, y, and z represent the thousand's, hundred's, ten's, and one's digits, respectively.

Translate.

The sum of the digits is 10.
$$w + x + y + z = 10$$

Twice the sum of the thousand's and ten's digits is the sum of the hundred's and one's digits less one.
$$2(w + y) = x + z - 1$$

The ten's digit is twice the thousand's digit.
$$y = 2 \cdot w$$

The one's digit equals the sum of the thousand's and hundred's digits.
$$z = w + x$$

We have a system of equations which can be written as

$$\begin{aligned} w + x + y + z &= 10, \\ 2w - x + 2y - z &= -1, \\ -2w + y &= 0, \\ w + x - z &= 0. \end{aligned}$$

Carry out. We can use matrices to solve the system. We get $(1, 3, 2, 4)$.

Check. The sum of the digits is 10. Twice the sum of 1 and 2 is 6. This is one less than the sum of 3 and 4. The ten's digit, 2, is twice the thousand's digit, 1. The one's digit, 4, equals $1 + 3$. The numbers check.

State. The number is 1324.

28. $x = \dfrac{ce - bf}{ae - bd}$, $y = \dfrac{af - cd}{ae - bd}$

Exercise Set 8.7

1. $\begin{vmatrix} 2 & 7 \\ 1 & 5 \end{vmatrix} = 2 \cdot 5 - 1 \cdot 7 = 10 - 7 = 3$

2. -13

3. $\begin{vmatrix} 6 & -9 \\ 2 & 3 \end{vmatrix} = 6 \cdot 3 - 2 \cdot (-9) = 18 + 18 = 36$

4. 29

5.
$$\begin{vmatrix} 0 & 2 & 0 \\ 3 & -1 & 1 \\ 1 & -2 & 2 \end{vmatrix}$$

$$= 0\begin{vmatrix} -1 & 1 \\ -2 & 2 \end{vmatrix} - 3\begin{vmatrix} 2 & 0 \\ -2 & 2 \end{vmatrix} + 1\begin{vmatrix} 2 & 0 \\ -1 & 1 \end{vmatrix}$$

$$= 0 - 3[2 \cdot 2 - (-2) \cdot 0] + 1[2 \cdot 1 - (-1) \cdot 0]$$

$$= 0 - 3 \cdot 4 + 1 \cdot 2$$

$$= 0 - 12 + 2$$

$$= -10$$

6. 1

7.
$$\begin{vmatrix} -1 & -2 & -3 \\ 3 & 4 & 2 \\ 0 & 1 & 2 \end{vmatrix}$$

$$= -1\begin{vmatrix} 4 & 2 \\ 1 & 2 \end{vmatrix} - 3\begin{vmatrix} -2 & -3 \\ 1 & 2 \end{vmatrix} + 0\begin{vmatrix} -2 & -3 \\ 4 & 2 \end{vmatrix}$$

$$= -1[4 \cdot 2 - 1 \cdot 2] - 3[-2 \cdot 2 - 1(-3)] + 0$$

$$= -1 \cdot 6 - 3 \cdot (-1) + 0$$

$$= -6 + 3 + 0$$

$$= -3$$

8. 3

9.
$$\begin{vmatrix} 3 & 2 & -2 \\ -2 & 1 & 4 \\ -4 & -3 & 3 \end{vmatrix}$$

$$= 3\begin{vmatrix} 1 & 4 \\ -3 & 3 \end{vmatrix} - (-2)\begin{vmatrix} 2 & -2 \\ -3 & 3 \end{vmatrix} + (-4)\begin{vmatrix} 2 & -2 \\ 1 & 4 \end{vmatrix}$$

$$= 3[1 \cdot 3 - (-3) \cdot 4] + 2[2 \cdot 3 - (-3)(-2)] - 4[2 \cdot 4 - 1(-2)]$$

$$= 3 \cdot 15 + 2 \cdot 0 - 4 \cdot 10$$

$$= 45 + 0 - 40$$

$$= 5$$

10. −6

11. $3x - 4y = 6,$
$\quad 5x + 9y = 10$

We compute D, D_x, and D_y.

$$D = \begin{vmatrix} 3 & -4 \\ 5 & 9 \end{vmatrix} = 27 - (-20) = 47$$

$$D_x = \begin{vmatrix} 6 & -4 \\ 10 & 9 \end{vmatrix} = 54 - (-40) = 94$$

$$D_y = \begin{vmatrix} 3 & 6 \\ 5 & 10 \end{vmatrix} = 30 - 30 = 0$$

Then,
$$x = \frac{D_x}{D} = \frac{94}{47} = 2$$
and
$$y = \frac{D_y}{D} = \frac{0}{47} = 0.$$
The solution is $(2, 0)$.

12. $(-3, 2)$

13. $-2x + 4y = 3,$
$\quad 3x - 7y = 1$

We compute D, D_x, and D_y.

$$D = \begin{vmatrix} -2 & 4 \\ 3 & -7 \end{vmatrix} = 14 - 12 = 2$$

$$D_x = \begin{vmatrix} 3 & 4 \\ 1 & -7 \end{vmatrix} = -21 - 4 = -25$$

$$D_y = \begin{vmatrix} -2 & 3 \\ 3 & 1 \end{vmatrix} = -2 - 9 = -11$$

Then,
$$x = \frac{D_x}{D} = \frac{-25}{2} = -\frac{25}{2}$$
and
$$y = \frac{D_y}{D} = \frac{-11}{2} = -\frac{11}{2}.$$
The solution is $\left(-\dfrac{25}{2}, -\dfrac{11}{2}\right)$.

14. $\left(\dfrac{9}{19}, \dfrac{51}{38}\right)$

15. $3x + 2y - z = 4$
$\quad 3x - 2y + z = 5$
$\quad 4x - 5y - z = -1$

We compute D, D_x, and D_y.

$$D = \begin{vmatrix} 3 & 2 & -1 \\ 3 & -2 & 1 \\ 4 & -5 & -1 \end{vmatrix}$$

$$= 3\begin{vmatrix} -2 & 1 \\ -5 & -1 \end{vmatrix} - 3\begin{vmatrix} 2 & -1 \\ -5 & -1 \end{vmatrix} + 4\begin{vmatrix} 2 & -1 \\ -2 & 1 \end{vmatrix}$$

$$= 3(7) - 3(-7) + 4(0)$$

$$= 21 + 21 + 0$$

$$= 42$$

$$D_x = \begin{vmatrix} 4 & 2 & -1 \\ 5 & -2 & 1 \\ -1 & -5 & -1 \end{vmatrix}$$

$$= 4\begin{vmatrix} -2 & 1 \\ -5 & -1 \end{vmatrix} - 5\begin{vmatrix} 2 & -1 \\ -5 & -1 \end{vmatrix} + (-1)\begin{vmatrix} 2 & -1 \\ -2 & 1 \end{vmatrix}$$

$$= 4(7) - 5(-7) - 1(0)$$

$$= 28 + 35 - 0$$

$$= 63$$

$$D_y = \begin{vmatrix} 3 & 4 & -1 \\ 3 & 5 & 1 \\ 4 & -1 & -1 \end{vmatrix}$$

$$= 3 \begin{vmatrix} 5 & 1 \\ -1 & -1 \end{vmatrix} - 3 \begin{vmatrix} 4 & -1 \\ -1 & -1 \end{vmatrix} + 4 \begin{vmatrix} 4 & -1 \\ 5 & 1 \end{vmatrix}$$

$$= 3(-4) - 3(-5) + 4(9)$$

$$= -12 + 15 + 36$$

$$= 39$$

Then,

$$x = \frac{D_x}{D} = \frac{63}{42} = \frac{3}{2}$$

and

$$y = \frac{D_y}{D} = \frac{39}{42} = \frac{13}{14}.$$

We substitute in the second equation to find z.

$$3 \cdot \frac{3}{2} - 2 \cdot \frac{13}{14} + z = 5$$

$$\frac{9}{2} - \frac{13}{7} + z = 5$$

$$\frac{37}{14} + z = 5$$

$$z = \frac{33}{14}$$

The solution is $\left(\frac{3}{2}, \frac{13}{14}, \frac{33}{14} \right)$.

16. $\left(-1, -\frac{6}{7}, \frac{11}{7} \right)$

17. $2x - 3y + 5z = 27,$
$\quad x + 2y - z = -4,$
$\quad 5x - y + 4z = 27$

We compute D, D_x, and D_y.

$$D = \begin{vmatrix} 2 & -3 & 5 \\ 1 & 2 & -1 \\ 5 & -1 & 4 \end{vmatrix}$$

$$= 2 \begin{vmatrix} 2 & -1 \\ -1 & 4 \end{vmatrix} - 1 \begin{vmatrix} -3 & 5 \\ -1 & 4 \end{vmatrix} + 5 \begin{vmatrix} -3 & 5 \\ 2 & -1 \end{vmatrix}$$

$$= 2(7) - 1(-7) + 5(-7)$$

$$= 14 + 7 - 35$$

$$= -14$$

$$D_x = \begin{vmatrix} 27 & -3 & 5 \\ -4 & 2 & -1 \\ 27 & -1 & 4 \end{vmatrix}$$

$$= 27 \begin{vmatrix} 2 & -1 \\ -1 & 4 \end{vmatrix} - (-4) \begin{vmatrix} -3 & 5 \\ -1 & 4 \end{vmatrix} + 27 \begin{vmatrix} -3 & 5 \\ 2 & -1 \end{vmatrix}$$

$$= 27(7) + 4(-7) + 27(-7)$$

$$= 189 - 28 - 189$$

$$= -28$$

$$D_y = \begin{vmatrix} 2 & 27 & 5 \\ 1 & -4 & -1 \\ 5 & 27 & 4 \end{vmatrix}$$

$$= 2 \begin{vmatrix} -4 & -1 \\ 27 & 4 \end{vmatrix} - 1 \begin{vmatrix} 27 & 5 \\ 27 & 4 \end{vmatrix} + 5 \begin{vmatrix} 27 & 5 \\ -4 & -1 \end{vmatrix}$$

$$= 2(11) - 1(-27) + 5(-7)$$

$$= 22 + 27 - 35$$

$$= 14$$

Then,

$$x = \frac{D_x}{D} = \frac{-28}{-14} = 2,$$

and

$$y = \frac{D_y}{D} = \frac{14}{-14} = -1.$$

We substitute in the second equation to find z.

$$2 + 2(-1) - z = -4$$

$$2 - 2 - z = -4$$

$$-z = -4$$

$$z = 4$$

The solution is $(2, -1, 4)$.

18. $(-3, 2, 1)$

19. $r - 2s + 3t = 6,$
$\quad 2r - s - t = -3,$
$\quad r + s + t = 6$

We compute D, D_r, and D_s.

$$D = \begin{vmatrix} 1 & -2 & 3 \\ 2 & -1 & -1 \\ 1 & 1 & 1 \end{vmatrix}$$

$$= 1 \begin{vmatrix} -1 & -1 \\ 1 & 1 \end{vmatrix} - 2 \begin{vmatrix} -2 & 3 \\ 1 & 1 \end{vmatrix} + 1 \begin{vmatrix} -2 & 3 \\ -1 & -1 \end{vmatrix}$$

$$= 1(0) - 2(-5) + 1(5)$$

$$= 0 + 10 + 5$$

$$= 15$$

$$D_r = \begin{vmatrix} 6 & -2 & 3 \\ -3 & -1 & -1 \\ 6 & 1 & 1 \end{vmatrix}$$

$$= 6 \begin{vmatrix} -1 & -1 \\ 1 & 1 \end{vmatrix} - (-3) \begin{vmatrix} -2 & 3 \\ 1 & 1 \end{vmatrix} + 6 \begin{vmatrix} -2 & 3 \\ -1 & -1 \end{vmatrix}$$

$$= 6(0) + 3(-5) + 6(5)$$

$$= 0 - 15 + 30$$

$$= 15$$

$$D_s = \begin{vmatrix} 1 & 6 & 3 \\ 2 & -3 & -1 \\ 1 & 6 & 1 \end{vmatrix}$$

$$= 1 \begin{vmatrix} -3 & -1 \\ 6 & 1 \end{vmatrix} - 2 \begin{vmatrix} 6 & 3 \\ 6 & 1 \end{vmatrix} + 1 \begin{vmatrix} 6 & 3 \\ -3 & -1 \end{vmatrix}$$

$$= 1(3) - 2(-12) + 1(3)$$

$$= 3 + 24 + 3$$

$$= 30$$

Then,

$$r = \frac{D_r}{D} = \frac{15}{15} = 1,$$

and

$$s = \frac{D_s}{D} = \frac{30}{15} = 2.$$

Substitute in the third equation to find t.

$$1 + 2 + t = 6$$

$$3 + t = 6$$

$$t = 3$$

The solution is $(1, 2, 3)$.

20. $(3, 4, -1)$

21. $f(x) = 2x + 5, \ g(x) = x^2$

$f(x) + g(x) = (2x + 5) + (x^2) = x^2 + 2x + 5$

22. $x^2 + 2x + 5$

23. Domain of $f(x)$ = Domain of $g(x)$ = $\{x | x$ is a real number$\}$

Since $f(x) = 0$ when $2x + 5 = 0$, we have $f(x) = 0$ when x is $-\frac{5}{2}$. Then Domain of $(g/f)(x)$ = $\left\{ x | x$ is a real number and $x \neq -\frac{5}{2} \right\}$.

24. 12

25. $\begin{vmatrix} 2 & x & -1 \\ -1 & 3 & 2 \\ -2 & 1 & 1 \end{vmatrix} = -12$

$$2 \begin{vmatrix} 3 & 2 \\ 1 & 1 \end{vmatrix} - (-1) \begin{vmatrix} x & -1 \\ 1 & 1 \end{vmatrix} + (-2) \begin{vmatrix} x & -1 \\ 3 & 2 \end{vmatrix} = -12$$

$$2(1) + 1(x + 1) - 2(2x + 3) = -12$$

$$2 + x + 1 - 4x - 6 = -12$$

$$-3x - 3 = -12$$

$$-3x = -9$$

$$x = 3$$

26. 10

27. $\begin{vmatrix} x & y & 1 \\ x_1 & y_1 & 1 \\ x_2 & y_2 & 1 \end{vmatrix} = 0$

is equivalent to

$$x \begin{vmatrix} y_1 & 1 \\ y_2 & 1 \end{vmatrix} - x_1 \begin{vmatrix} y & 1 \\ y_2 & 1 \end{vmatrix} + x_2 \begin{vmatrix} y & 1 \\ y_1 & 1 \end{vmatrix} = 0$$

or

$$x(y_1 - y_2) - x_1(y - y_2) + x_2(y - y_1) = 0$$

or

$$xy_1 - xy_2 - x_1 y + x_1 y_2 + x_2 y - x_2 y_1 = 0. \qquad (1)$$

Since the slope of the line through (x_1, y_1) and (x_2, y_2) is $\frac{y_2 - y_1}{x_2 - x_1}$, an equation of the line through (x_1, y_1) and (x_2, y_2) is

$$y - y_1 = \frac{y_2 - y_1}{x_2 - x_1}(x - x_1)$$

which is equivalent to

$$(x_2 - x_1)(y - y_1) = (y_2 - y_1)(x - x_1)$$

or

$$x_2 y - x_2 y_1 - x_1 y + x_1 y_1 = y_2 x - y_2 x_1 - y_1 x + y_1 x_1$$

or

$$x_2 y - x_2 y_1 - x_1 y - x y_2 + x_1 y_2 + x y_1 = 0. \qquad (2)$$

Equations (1) and (2) are equivalent.

28.

Exercise Set 8.8

1. $C(x) = 45x + 600,000 \qquad R(x) = 65x$

a) $P(x) = R(x) - C(x)$

$$= 65x - (45x + 600,000)$$

$$= 65x - 45x - 600,000$$

$$= 20x - 600,000$$

b) To find the break-even point we solve the system

$$R(x) = 65x,$$
$$C(x) = 45x + 600,000.$$

Since $R(x) = C(x)$ at the break-even point, we can rewrite the system:

$$R(x) = 65x, \qquad (1)$$
$$R(x) = 45x + 600,000 \quad (2)$$

We solve using substitution.

$65x = 45x + 600,000$ Substituting $65x$ for $R(x)$ in (2)

$$20x = 600,000$$
$$x = 30,000$$

Thus, 30,000 units must be produced and sold in order to break even.

2. a) $P(x) = 45x - 360,000$

    b) 8000 units

3. $C(x) = 10x + 120,000$ $R(x) = 60x$

   a) $P(x) = R(x) - C(x)$
$$= 60x - (10x + 120,000)$$
$$= 60x - 10x - 120,000$$
$$= 50x - 120,000$$

   b) Solve the system

$$R(x) = 60x,$$
$$C(x) = 10x + 120,000.$$

Since $R(x) = C(x)$ at the break-even point, we can rewrite the system:

$$R(x) = 60x, \qquad (1)$$
$$R(x) = 10x + 120,000 \quad (2)$$

We solve using substitution.

$60x = 10x + 120,000$ Substituting $60x$ for $R(x)$ in (2)

$$50x = 120,000$$
$$x = 2400$$

Thus, 2400 units must be produced and sold in order to break even.

4. a) $P(x) = 55x - 49,500$

    b) 900 units

5. $C(x) = 20x + 10,000$ $R(x) = 100x$

   a) $P(x) = R(x) - C(x)$
$$= 100x - (20x + 10,000)$$
$$= 100x - 20x - 10,000$$
$$= 80x - 10,000$$

   b) Solve the system

$$R(x) = 100x,$$
$$C(x) = 20x + 10,000.$$

Since $R(x) = C(x)$ at the break-even point, we can rewrite the system:

$$R(x) = 100x, \qquad (1)$$
$$R(x) = 20x + 10,000 \quad (2)$$

We solve using substitution.

$100x = 20x + 10,000$ Substituting $100x$ for $R(x)$ in (2)

$$80x = 10,000$$
$$x = 125$$

Thus, 125 units must be produced and sold in order to break even.

6. a) $P(x) = 45x - 22,500$

    b) 500 units

7. $C(x) = 15x + 75,000$ $R(x) = 55x$

   a) $P(x) = R(x) - C(x)$
$$= 55x - (15x + 75,000)$$
$$= 55x - 15x - 75,000$$
$$= 40x - 75,000$$

   b) Solve the system

$$R(x) = 55x,$$
$$C(x) = 15x + 75,000.$$

Since $R(x) = C(x)$ at the break-even point, we can rewrite the system:

$$R(x) = 55x, \qquad (1)$$
$$R(x) = 15x + 75,000 \quad (2)$$

We solve using substitution.

$55x = 15x + 75,000$ Substituting $55x$ for $R(x)$ in (2)

$$40x = 75,000$$
$$x = 1875$$

To break even 1875 units must be produced and sold.

8. a) $P(x) = 18x - 16,000$

    b) 889 units

9. $C(x) = 50x + 195,000$ $R(x) = 125x$

   a) $P(x) = R(x) - C(x)$
$$= 125x - (50x + 195,000)$$
$$= 125x - 50x - 195,000$$
$$= 75x - 195,000$$

   b) Solve the system

$$R(x) = 125x,$$
$$C(x) = 50x + 195,000.$$

Since $R(x) = C(x)$ at the break-even point, we can rewrite the system:

$$R(x) = 125x, \qquad (1)$$
$$R(x) = 50x + 195,000 \quad (2)$$

We solve using substitution.

$$125x = 50x + 195,000 \quad \text{Substituting } 125x$$
$$\text{for } R(x) \text{ in (2)}$$
$$75x = 195,000$$
$$x = 2600$$

To break even 2600 units must be produced and sold.

10. a) $P(x) = 94x - 928,000$

b) 9873 units

11. $D(p) = 2000 - 60p,$

$S(p) = 460 + 94p$

Since both demand and supply are quantities, the system can be rewritten:

$$q = 2000 - 60p, \quad (1)$$
$$q = 460 + 94p \quad (2)$$

Substitute $2000 - 60p$ for q in (2) and solve.

$$2000 - 60p = 460 + 94p$$
$$1540 = 154p$$
$$10 = p$$

The equilibrium price is $10 per unit. To find the equilibrium quantity we substitute $10 into either $D(p)$ or $S(p)$.

$$D(10) = 2000 - 60(10) = 2000 - 600 = 1400$$

The equilibrium quantity is 1400 units.

The equilibrium point is ($10, 1400$).

12. ($50, 500$)

13. $D(p) = 760 - 13p,$

$S(p) = 430 + 2p$

Rewrite the system:

$$q = 760 - 13p, \quad (1)$$
$$q = 430 + 2p \quad (2)$$

Substitute $760 - 13p$ for q in (2) and solve.

$$760 - 13p = 430 + 2p$$
$$330 = 15p$$
$$22 = p$$

The equilibrium price is $22 per unit.

To find the equilibrium quantity we substitute $22 into either $D(p)$ or $S(p)$.

$$S(22) = 430 + 2(22) = 430 + 44 = 474$$

The equilibrium quantity is 474 units.

The equilibrium point is ($22, 474$).

14. ($10, 370$)

15. $D(p) = 7500 - 25p,$

$S(p) = 6000 + 5p$

Rewrite the system:

$$q = 7500 - 25p, \quad (1)$$
$$q = 6000 + 5p \quad (2)$$

Substitute $7500 - 25p$ for q in (2) and solve.

$$7500 - 25p = 6000 + 5p$$
$$1500 = 30p$$
$$50 = p$$

The equilibrium price is $50 per unit.

To find the equilibrium quantity we substitute $50 into either $D(p)$ or $S(p)$.

$$D(50) = 7500 - 25(50) = 7500 - 1250 = 6250$$

The equilibrium quantity is 6250 units.

The equilibrium point is ($50, 6250$).

16. ($40, 7600$)

17. $D(p) = 1600 - 53p,$

$S(p) = 320 + 75p$

Rewrite the system:

$$q = 1600 - 53p, \quad (1)$$
$$q = 320 + 75p \quad (2)$$

Substitute $1600 - 53p$ for q in (2) and solve.

$$1600 - 53p = 320 + 75p$$
$$1280 = 128p$$
$$10 = p$$

The equilibrium price is $10 per unit.

To find the equilibrium quantity we substitute $10 into either $D(p)$ or $S(p)$.

$$S(10) = 320 + 75(10) = 320 + 750 = 1070$$

The equilibrium quantity is 1070 units.

The equilibrium point is ($10, 1070$).

18. ($36, 4060$)

19. a) $C(x) = \text{Fixed costs} + \text{Variable costs}$

$C(x) = 125,000 + 750x,$

where x is the number of computers produced.

b) Each computer sells for $1050. The total revenue is 1050 times the number of computers sold. We assume that all computers produced are sold.

$R(x) = 1050x$

c) $P(x) = R(x) - C(x)$

$$P(x) = 1050x - (125,000 + 750x)$$
$$= 1050x - 125,000 - 750x$$
$$= 300x - 125,000$$

d) $P(x) = 300x - 125,000$

$$P(400) = 300(400) - 125,000$$
$$= 120,000 - 125,000$$
$$= -5000$$

The company will realize a $5000 loss when 400 computers are produced and sold.

$$P(700) = 300(700) - 125,000$$
$$= 210,000 - 125,000$$
$$= 85,000$$

The company will realize a profit of $85,000 from the production and sale of 700 computers.

e) Solve the system

$$R(x) = 1050x,$$
$$C(x) = 125,000 + 750x.$$

Since $R(x) = C(x)$ at the break-even point, we can rewrite the system:

$$R(x) = 1050x, \qquad (1)$$
$$R(x) = 125,000 + 750x \quad (2)$$

We solve using substitution.

$1050x = 125,000 + 750x$ Substituting $1050x$ for $R(x)$ in (2)

$$300x = 125,000$$
$$x \approx 417$$

To break even 417 units must be produced and sold.

20. a) $C(x) = 22,500 + 40x$

b) $R(x) = 85x$

c) $P(x) = 45x - 22,500$

d) A profit of $112,500, a loss of $4500)

e) 500 units

21. a) $C(x) =$ Fixed costs + Variable costs

$$C(x) = 10,000 + 20x,$$

where x is the number of sport coats produced.

b) Each sport coat sells for $100. The total revenue is 100 times the number of coats sold. We assume that all coats produced are sold.

$$R(x) = 100x$$

c) $P(x) = R(x) - C(x)$

$$P(x) = 100x - (10,000 + 20x)$$
$$= 100x - 10,000 - 20x$$
$$= 80x - 10,000$$

d) $P(x) = 80x - 10,000$

$$P(2000) = 80(2000) - 10,000$$
$$= 160,000 - 10,000$$
$$= 150,000$$

The clothing firm will realize a $150,000 profit when 2000 sport coats are produced and sold.

$$P(50) = 80(50) - 10,000$$
$$= 4000 - 10,000$$
$$= -6000$$

The company will realize a $6000 loss when 50 coats are produced and sold.

e) Solve the system

$$R(x) = 100x,$$
$$C(x) = 10,000 + 20x.$$

Since $R(x) = C(x)$ at the break-even point, we can rewrite the system:

$$R(x) = 100x, \qquad (1)$$
$$R(x) = 10,000 + 20x \quad (2)$$

We solve using substitution.

$100x = 10,000 + 20x$ Substituting $100x$ for $R(x)$ in (2)

$$80x = 10,000$$
$$x = 125$$

To break even 125 coats must be produced and sold.

22. a) $C(x) = 16,400 + 6x$

b) $R(x) = 18x$

c) $P(x) = 12x - 16,400$

d) A profit of $19,600, a loss of $4400)

e) 1367 units

23. $x^2 + 7x + 12$

We look for a pair of numbers whose product is 12 and whose sum is 7. They are 3 and 4.

$$x^2 + 7x + 12 = (x + 3)(x + 4)$$

24. $(3x - 2)(2x + 1)$

25. $\dfrac{2}{x} + \dfrac{x+1}{x^2}$ LCD is x^2

$$= \frac{2}{x} \cdot \frac{x}{x} + \frac{x+1}{x^2}$$
$$= \frac{2x}{x^2} + \frac{x+1}{x^2}$$
$$= \frac{2x + x + 1}{x^2}$$
$$= \frac{3x + 1}{x^2}$$

26. $x^3 + x^2 + 3x + 3$

27. The supply function contains the points ($2, 100) and ($8, 500). We find its equation:

$$m = \frac{500 - 100}{8 - 2} = \frac{400}{6} = \frac{200}{3}$$

$y - y_1 = m(x - x_1)$ Point-slope form

$$y - 100 = \frac{200}{3}(x - 2)$$
$$y - 100 = \frac{200}{3}x - \frac{400}{3}$$
$$y = \frac{200}{3}x - \frac{100}{3}$$

We can equivalently express supply S as a function of price p:

$$S(p) = \frac{200}{3}p - \frac{100}{3}$$

The demand function contains the points ($1, 500$) and ($9, 100$). We find its equation:

$$m = \frac{100 - 500}{9 - 1} = \frac{-400}{8} = -50$$

$$y - y_1 = m(x - x_1)$$

$$y - 500 = -50(x - 1)$$

$$y - 500 = -50x + 50$$

$$y = -50x + 550$$

We can equivalently express demand D as a function of price p:

$$D(p) = -50p + 550$$

We have a system of equations

$$S(p) = \frac{200}{3}p - \frac{100}{3},$$

$$D(p) = -50p + 550.$$

Rewrite the system:

$$q = \frac{200}{3}p - \frac{100}{3}, \quad (1)$$

$$q = -50p + 550 \quad (2)$$

Substitute $\frac{200}{3}p - \frac{100}{3}$ for q in (2) and solve.

$$\frac{200}{3}p - \frac{100}{3} = -50p + 550$$

$$200p - 100 = -150p + 1650 \quad \text{Multiplying by 3 to clear fractions}$$

$$350p - 100 = 1650$$

$$350p = 1750$$

$$p = 5$$

The equilibrium price is $5 per unit.

To find the equilibrium quantity, we substitute $5 into either $S(p)$ or $D(p)$.

$$D(5) = -50(5) + 550 = -250 + 550 = 300$$

The equilibrium quantity is 300 units.

The equilibrium point is ($5, 300$).

28. 308 pairs

29.

30.

31. a) 4526 units

b) $870

32. a) $8.74

b) 24,509 units

Chapter 9

Inequalities and Linear Programming

Exercise Set 9.1

1. $x > 4$

Graph: The solutions consist of all real numbers greater than 4, so we shade all numbers to the right of 4 and use an open circle at 4 to indicate that it is not a solution.

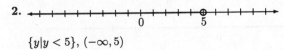

Set builder notation: $\{x | x > 4\}$

Interval notation: $(4, \infty)$

2.
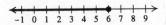

$\{y | y < 5\}$, $(-\infty, 5)$

3. $t \leq 6$

Graph: We shade all numbers to the left of 6 and use a solid endpoint at 6 to indicate that it is also a solution.

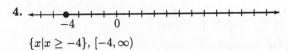

Set builder notation: $\{t | t \leq 6\}$

Interval notation: $(-\infty, 6]$

4.
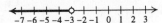

$\{x | x \geq -4\}$, $[-4, \infty)$

5. $y < -3$

Graph: We shade all numbers to the left of -3 and use an open circle at -3 to indicate that it is not a solution.

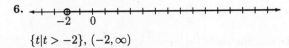

Set builder notation: $\{y | y < -3\}$

Interval notation: $(-\infty, 3)$

6.
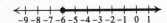

$\{t | t > -2\}$, $(-2, \infty)$

7. $x \geq -6$

Graph: We shade all numbers to the right of -6 and use a solid endpoint at -6 to indicate that it is also a solution.

Set builder notation: $\{x | x \geq -6\}$

Interval notation: $[-6, \infty)$

8.
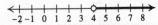

$\{x | x \leq -5\}$, $(-\infty, -5]$

9.
$$x + 8 > 3$$
$$x + 8 + (-8) > 3 + (-8) \qquad \text{Adding } -8$$
$$x > -5$$

The solution set is $\{x | x > -5\}$, or $(-5, \infty)$.

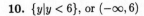

10. $\{y | y < 6\}$, or $(-\infty, 6)$

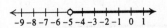

11.
$$a + 9 \leq -12$$
$$a + 9 + (-9) \leq -12 + (-9) \qquad \text{Adding } -9$$
$$y \leq -21$$

The solution set is $\{a | a \leq -21\}$, or $(-\infty, -21]$.

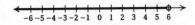

12. $\{a | a \leq -20\}$, or $(-\infty, -20]$

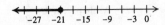

13.
$$8x \geq 24$$
$$\frac{1}{8} \cdot 8x \geq \frac{1}{8} \cdot 24 \qquad \text{Multiplying by } \frac{1}{8}$$
$$x \geq 3$$

The solution set is $\{x | x \geq 3\}$, or $[3, \infty)$.

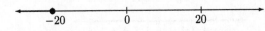

14. $\{t | t < -9\}$, or $(-\infty, -9)$

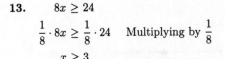

15.
$$-9x \geq -8.1$$
$$-\frac{1}{9}(-9x) \leq -\frac{1}{9}(-8.1) \qquad \begin{array}{l}\text{Multiplying by } -\frac{1}{9} \\ \text{and reversing the in-} \\ \text{equality sign}\end{array}$$
$$x \leq \frac{8.1}{9}$$
$$x \leq 0.9$$

The solution set is $\{x | x \leq 0.9\}$, or $(-\infty, 0.9]$.

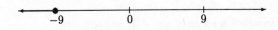

16. $\{y | y \geq -0.4\}$, or $[-0.4, \infty)$

17.
$$-\frac{3}{4}x \geq -\frac{5}{8}$$

$-\frac{4}{3}\left(-\frac{3}{4}x\right) \leq -\frac{4}{3}\left(-\frac{5}{8}\right)$ Multiplying by $-\frac{4}{3}$ and reversing the inequality sign

$$x \leq \frac{20}{24}$$

$$x \leq \frac{5}{6}$$

The solution set is $\left\{x \big| x \leq \frac{5}{6}\right\}$, or $\left(-\infty, \frac{5}{6}\right]$.

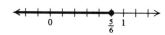

18. $\left\{y \big| y \geq \frac{9}{10}\right\}$, or $\left[\frac{9}{10}, \infty\right)$

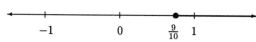

19.
$$2x + 7 < 19$$

$2x + 7 + (-7) < 19 + (-7)$ Adding -7

$$2x < 12$$

$\frac{1}{2} \cdot 2x < \frac{1}{2} \cdot 12$ Multiplying by $\frac{1}{2}$

$$x < 6$$

The solution set is $\{x | x < 6\}$, or $(-\infty, 6)$.

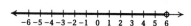

20. $\{y | y > 3\}$, or $(3, \infty)$

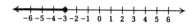

21.
$$5y + 2y \leq -21$$

$7y \leq -21$ Collecting like terms

$\frac{1}{7}(7y) \leq \frac{1}{7}(-21)$ Multiplying by $\frac{1}{7}$

$$y \leq -3$$

The solution set is $\{y | y \leq -3\}$, or $(-\infty, -3]$.

22. $\{x | x \leq 4\}$, or $(-\infty, 4]$

23.
$$3x - \frac{1}{8} \leq \frac{3}{8} + 2x$$

$-2x + 3x - \frac{1}{8} \leq -2x + \frac{3}{8} + 2x$ Adding $-2x$

$$x - \frac{1}{8} \leq \frac{3}{8}$$

$x - \frac{1}{8} + \frac{1}{8} \leq \frac{3}{8} + \frac{1}{8}$ Adding $\frac{1}{8}$

$$x \leq \frac{4}{8}$$

$$x \leq \frac{1}{2}$$

The solution set is $\left\{x \big| x \leq \frac{1}{2}\right\}$, or $\left(-\infty, \frac{1}{2}\right]$.

24. $\left\{x \big| x < \frac{13}{3}\right\}$, or $\left(-\infty, \frac{13}{3}\right)$

25.
$$3(2x + 1) \geq 4(3x - 2)$$

$$6x + 3 \geq 12x - 8$$

$$-6x + 3 \geq -8$$

$$-6x \geq -11$$

$$x \leq \frac{11}{6}$$

The solution set is $\left\{x \big| x \leq \frac{11}{6}\right\}$, or $\left(-\infty, \frac{11}{6}\right]$.

26. $\left\{y \big| y > -\frac{11}{7}\right\}$, or $\left(-\frac{11}{7}, \infty\right)$

27.
$$5[3m - (m + 4)] > -2(m - 4)$$

$$5(3m - m - 4) > -2(m - 4)$$

$$5(2m - 4) > -2(m - 4)$$

$$10m - 20 > -2m + 8$$

$$12m - 20 > 8$$

$$12m > 28$$

$$m > \frac{28}{12}$$

$$m > \frac{7}{3}$$

The solution set is $\left\{m \big| m > \frac{7}{3}\right\}$, or $\left(\frac{7}{3}, \infty\right)$.

28. $\left\{x \big| x \leq -\frac{23}{2}\right\}$, or $\left(-\infty, -\frac{23}{2}\right]$

29.
$$2[4 - 2(3 - x)] - 1 \geq 4[2(4x - 3) + 7] - 25$$
$$2[4 - 6 + 2x] - 1 \geq 4[8x - 6 + 7] - 25$$
$$2[-2 + 2x] - 1 \geq 4[8x + 1] - 25$$
$$-4 + 4x - 1 \geq 32x + 4 - 25$$
$$4x - 5 \geq 32x - 21$$
$$-28x - 5 \geq -21$$
$$-28x \geq -16$$
$$x \leq \frac{-16}{-28}, \text{ or } \frac{4}{7}$$

The solution set is $\left\{x \middle| x \leq \frac{4}{7}\right\}$, or $\left(-\infty, \frac{4}{7}\right]$.

30. $\left\{t \middle| t \geq -\frac{27}{19}\right\}$, or $\left[-\frac{27}{19}, \infty\right)$

31. *Familiarize*. Let m represent the daily mileage. Then the total daily rental cost is $\$42.95 + \$0.46m$ for a Ridem truck and $\$75$ for an Atlas truck.

Translate. We write an inequality stating that the cost of a Ridem truck is less than the cost of an Atlas truck.

$$42.95 + 0.46m < 75$$

Carry out.
$$42.95 + 0.46m < 75$$
$$0.46m < 32.05$$
$$m < 69.7$$

Check. We can do a partial check by substituting a value for m less than 69.7. When $m = 69$, the cost of a Ridem truck is $\$42.95 + \$0.46(69) = \$74.69$. This is less than $\$75$, the cost of an Atlas truck. We cannot check all possible values for m, so we stop here.

State. It would be less expensive to rent a Ridem truck for daily mileages less than 69.7 miles.

32. More than 36.8 miles per day

33. *Familiarize*. Let m = the length of a telephone call, in minutes. Then the number of minutes after the first minute of the call is $m - 1$. Using Down East Calling, the cost of the call, in cents, is $20 + 16(m - 1)$. Using Long Call Systems, the cost of the call, in cents, is $19 + 18(m - 1)$.

Translate.

Cost using Down East Calling	is less than	cost using Long Call Systems.
$20 + 16(m-1)$	$<$	$19 + 18(m-1)$

Carry out.
$$20 + 16(m - 1) < 19 + 18(m - 1)$$
$$20 + 16m - 16 < 19 + 18m - 18$$
$$4 + 16m < 1 + 18m$$
$$3 < 2m$$
$$1.5 < m$$

Check. We substitute a value of m greater than 1.5. We choose 2. For $m = 2$, the cost of the Down East call is $20 + 16(2 - 1)$, or 36¢. The cost of the Long Call call

is $19 + 18(2 - 1)$, or 37¢, so the Down East call is less expensive. Since we cannot check all possible values, we stop here.

State. For calls longer than 1.5 minutes, Down East calling is less expensive.

34. For times greater than 6 hours

35. *Familiarize*. We make a table of information.

Plan A: Monthly Income	Plan B: Monthly Income
$500 salary	$750 salary
4% of sales	5% of sales over $8000
Total: 500 + 4% of sales	Total: 750 + 5% of sales over 8000

Suppose your gross sales were $14,000 in one month. Compute the income from each plan.

Plan A: Plan B:

$500 + 4\%(14,000)$ $750 + 5\%(14,000 - 8000)$

$500 + 0.04(14,000)$ $750 + 0.05(6000)$

$500 + 560$ $750 + 300$

$1060 $1050

When gross sales are $14,000, Plan A is better.

Compute the income from each plan for gross monthly sales of $16,000.

Plan A: Plan B:

$500 + 4\%(16,000)$ $750 + 5\%(16,000 - 8000)$

$500 + 0.04(16,000)$ $750 + 0.05(8000)$

$500 + 640$ $750 + 400$

$1140 $1150

When gross sales are $16,000, Plan B is better.

To determine all values for which Plan B is better, we solve an inequality.

Translate. We write an inequality stating that the income from Plan B is greater than the income from Plan A. We let S represent gross sales. Then $S - 8000$ represents gross sales over 8000.

$$750 + 5\%(S - 8000) > 500 + 4\%S$$

Carry out.
$$750 + 0.05S - 400 > 500 + 0.04S$$
$$350 + 0.05S > 500 + 0.04S \quad \text{Collecting like terms}$$
$$0.01S > 150 \quad \text{Adding } -350 \text{ and } -0.04S$$
$$S > \frac{150}{0.01} \quad \text{Multiplying by } \frac{1}{0.01}$$
$$S > 15,000$$

Check. We calculate for $S = \$15,000$ and for some amount greater than $15,000 and some amount less than $15,000.

Plan A:	Plan B:
$500 + 4\%(15,000)$	$750 + 5\%(15,000 - 8000)$
$500 + 0.04(15,000)$	$750 + 0.05(7000)$
$500 + 600$	$750 + 350$
$\$1100$	$\$1100$

When $S = \$15,000$, the income from Plan A is equal to the income from Plan B.

In the Familiarize step we saw that Plan A is better for sales of \$14,000 and Plan B is better for sales of \$16,000. We cannot check all possible values for S, so we will stop here.

State. Plan B is better than Plan A when gross sales are greater than \$15,000.

36. Less than \$116,666.67 per year

37. Familiarize. We let n represent the number of hours it will take the painter to complete the job. Plan A will pay the painter $500 + 6n$, while Plan B will pay the painter $11n$.

Suppose the job takes 90 hr. Compute the income from each plan.

Plan A:	Plan B:
$500 + 6(90)$	$11(90)$
$500 + 540$	$\$990$
$\$1040$	

If the job takes 90 hr, Plan A is better.

Compute the income from each plan if the job takes 120 hr.

Plan A:	Plan B:
$500 + 6(120)$	$11(120)$
$500 + 720$	$\$1320$
$\$1220$	

If the job takes 120 hr, Plan B is better.

To determine all values for which Plan A is better, we solve an inequality.

Translate. We write an inequality stating that the income from Plan A is greater than the income from Plan B.

Carry out. $500 + 6n > 11n$

$$500 > 5n \quad \text{Adding } -6n$$

$$100 > n \quad \text{Multiplying by } \frac{1}{5}$$

Check.

Plan A:	Plan B:
$500 + 6(100)$	$11(100)$
$500 + 600$	$\$1100$
$\$1100$	

When $n = 100$, the income from Plan A is equal to the income from Plan B.

In the Familiarize Step we saw that Plan A is better for a 90 hr job and Plan B is better for a 120 hr job. We cannot check all possible values for n, so we stop here.

State. Plan A is better for the painter for $n < 100$.

38. $n > 85.7$

39. Familiarize. Organize the information in a table. Let $x =$ the amount invested at 7%. Then $25,000 - x =$ the amount invested at 8%.

Amount invested	Rate of interest	Time
x	7%	1 yr
$25,000 - x$	8%	1 yr
Total $\$25,000$		

Amount invested	Interest $(I = Prt)$
x	$0.07x$
$25,000 - x$	$0.08(25,000 - x)$
Total $\$25,000$	$\$1800$ or more

Translate. Use the information in the table to write an inequality:

$$0.07x + 0.08(25,000 - x) \geq 1800$$

Carry out.

$0.07x + 2000 - 0.08x \geq 1800$

$$2000 - 0.01x \geq 1800 \quad \text{Collecting like terms}$$

$$-0.01x \geq -200 \quad \text{Adding } -2000$$

$$x \leq \frac{200}{0.01}$$

$$\text{Multiplying by } -\frac{1}{0.01}$$
$$\text{and reversing the}$$
$$\text{inequality sign}$$

$$x \leq 20,000$$

Check. For $x = \$20,000$, $7\%(20,000) = 0.07(20,000)$, or \$1400, and $8\%(25,000 - 20,000) = 0.08(5000)$, or \$400. The total interest earned is $1400 + 400$, or \$1800. We also calculate for some amount less than \$20,000 and for some amount greater than \$20,000. For $x = \$15,000$, $7\%(15,000) = 0.07(15,000)$, or \$1050, and $8\%(25,000 - 15,000) = 0.08(10,000)$, or \$800. The total interest earned is $1050 + 800$, or \$1850. For $x = \$22,000$, $7\%(22,000) = 0.07(22,000)$, or \$1540, and $8\%(25,000 - 22,000) = 0.08(3000)$, or \$240. The total interest earned is $1540 + 240$, or \$1780. For these values the inequality, $x \leq 20,000$, gives correct results.

State. To make at least \$1800 interest per year, \$20,000 is the most that can be invested at 7%.

40. \$5000

41. Familiarize. We let x represent the ticket price. Then $300x$ represents the total receipts from the ticket sales assuming 300 people will attend. The first band will play for $250 + 50\%(300x)$. The second band will play for \$550.

Translate. For school profit to be greater when the first band plays, the amount the first band charges must be less than the amount the second band charges. We now have an inequality.

$$250 + 50\%(300x) < 550$$

Carry out.

$$250 + 0.5(300x) < 550$$
$$250 + 150x < 550$$
$$150x < 300 \quad \text{Adding } -250$$
$$x < 2 \quad \text{Multiplying by } \frac{1}{150}$$

Check. For $x = \$2$, the total receipts are $300(2)$, or $600. The first band charges $250 + 50\%(600)$, or $550. The second band also charges $550. The school profit is the same using either band. For $x = \$1.99$, the total receipts are $300(1.99)$, or $597. The first band charges $250 + 50\%(597)$, or $548.50. Using the first band, the school profit would be $597 - 548.50$, or $48.50. Using the second band, the profit would be $597 - 590$, or $47. Thus, the first band produces more profit. For $x = \$2.01$, the total receipts are $300(2.01)$, or $603. The first band charges $250 + 50\%(603)$, or $551.50. Using the first band, the school profit would be $603 - 551.50$, or $51.50. Using the second band, the school profit would be $603 - 550$, or $53. Thus, the second band produces more profit. For these values, the inequality $x < 2$ gives correct results.

State. The ticket price must be less than $2. The highest price, rounded to the nearest cent, less than $2 is $1.99.

42. For more than 6 crossings

43. *Familiarize*. We make a table listing the information. We let x represent the total medical bill.

	Plan A	Plan B
You pay the first	100	250
Insurance pays	$80\%(x-100)$	$90\%(x-250)$
You also pay	$20\%(x-100)$	$10\%(x-250)$
Total you pay	$100+20\%(x-100)$	$250+10\%(x-250)$

Translate. We write an inequality stating that the amount you pay is less when you choose Plan B.

$$250 + 10\%(x - 250) < 100 + 20\%(x - 100)$$

Carry out.

$$250 + 0.1x - 25 < 100 + 0.2x - 20$$
$$225 + 0.1x < 80 + 0.2x \quad \text{Collecting like terms}$$
$$145 < 0.1x \quad \text{Adding } -80 \text{ and } -0.1x$$
$$1450 < x \quad \text{Multiplying by } \frac{1}{0.1}$$

or

$$x > 1450$$

Check. We calculate for $x = \$1450$ and also for some amount greater than $1450 and some amount less than $1450.

Plan A:	Plan B:
$100 + 20\%(1450 - 100)$	$250 + 10\%(1450 - 250)$
$100 + 0.2(1350)$	$250 + 0.1(1200)$
$100 + 270$	$250 + 120$
$370	$370

When $x = \$1450$, you pay the same with either plan.

Plan A:	Plan B:
$100 + 20\%(1500 - 100)$	$250 + 10\%(1500 - 250)$
$100 + 0.2(1400)$	$250 + 0.1(1250)$
$100 + 280$	$250 + 125$
$380	$375

When $x = \$1500$, you pay less with Plan B.

Plan A:	Plan B:
$100 + 20\%(1400 - 100)$	$250 + 10\%(1400 - 100)$
$100 + 0.2(1300)$	$250 + 0.1(1300)$
$100 + 260$	$250 + 130$
$360	$380

When $x = \$1400$, you pay more with Plan B.

For these values, the inequality $x > \$1450$ gives correct results.

State. For Plan B to save you money, the total of the medical bills must be greater than $1450.

44. For more than 8.75 pounds

45. a) *Familiarize*. Find the values of F for which C is less than $1063°$.

Translate.

$$\frac{5}{9}(F - 32) < 1063$$

Carry out.

$$5(F - 32) < 9567$$
$$5F - 160 < 9567$$
$$5F < 9727$$
$$F < 1945.4$$

Check. When $F = 1945.4$, $C = \frac{5}{9}(1945.4 - 32) = 1063$. Then values of F less than 1945.4 will give values of C less than 1063.

State. Gold is solid at temperatures less than $1945.4°$ F.

b) *Familiarize*. Find the values of F for which C is less than 960.8.

Translate.

$$\frac{5}{9}(F - 32) < 960.8$$

Carry out.

$$5(F - 32) < 8647.2$$
$$5F - 160 < 8647.2$$
$$5F < 8807.2$$
$$F < 1761.44$$

Check. When $F=1761.44$, $C=\dfrac{5}{9}(1761.44-32)=$ 960.8. Then values of F less than 1761.44 will give values of C less than 960.8.

State. Silver is solid at temperatures less than $1761.44°$ F.

46. a) 44,000, 165,978, 275,758

b) From 1987 on

47. a) **Familiarize.** Find the values of x for which $R(x) < C(x)$.

Translate.

$$26x < 90,000 + 15x$$

Carry out.

$$11x < 90,000$$

$$x < 8181\frac{9}{11}$$

Check. $R\left(8181\frac{9}{11}\right)=\$212,727.27=C\left(8181\frac{9}{11}\right).$

Calculate $R(x)$ and $C(x)$ for some x greater than $8181\frac{9}{11}$ and for some x less than $8181\frac{9}{11}$.

Suppose $x = 8200$:

$$R(x) = 26(8200) = 213,200 \quad \text{and}$$

$$C(x) = 90,000 + 15(8200) = 213,000.$$

In this case $R(x) > C(x)$.

Suppose $x = 8000$:

$$R(x) = 26(8000) = 208,000 \quad \text{and}$$

$$C(x) = 90,000 + 15(8000) = 210,000.$$

In this case $R(x) < C(x)$.

Then for $x < 8181\frac{9}{11}$, $R(x) < C(x)$.

State. We will state the result in terms of integers, since the company cannot sell a fraction of a radio. For 8181 or fewer radios the company loses money.

b) Our check in part a) shows that for $x > 8181\frac{9}{11}$, $R(x) > C(x)$ and the company makes a profit. Again, we will state the result in terms of an integer. For more than 8182 radios the company makes money.

48. a) $p < 10$

b) $p > 10$

49. $5y - 10 = 2x$

x	y
0	2
-5	0
5	4

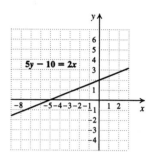

50. -4

51. $|-7| = 7$ -7 is 7 units from 0

52. -7

53.

54.

55. $3ax + 2x \geq 5ax - 4$

$$2x - 2ax \geq -4$$

$$2x(1 - a) \geq -4$$

$$x(1 - a) \geq -2$$

$$x \leq -\frac{2}{1 - a}, \text{ or } \frac{2}{a - 1}$$

We reversed the inequality sign when we divided because when $a > 1$, then $1 - a < 0$.

The solution set is $\left\{x \mid x \leq \dfrac{2}{a - 1}\right\}$.

56. $\left\{y \mid y \geq -\dfrac{10}{b + 4}\right\}$

57. $a(by - 2) \geq b(2y + 5)$

$$aby - 2a \geq 2by + 5b$$

$$aby - 2by \geq 2a + 5b$$

$$y(ab - 2b) \geq 2a + 5b$$

$$y \geq \frac{2a + 5b}{ab - 2b}, \text{ or } \frac{2a + 5b}{b(a - 2)}$$

The inequality sign remained unchanged when we divided because when $a > 2$, then $ab - 2b > 0$.

The solution set is $\left\{y \mid y \geq \dfrac{2a + 5b}{b(a - 2)}\right\}$.

58. $\left\{x \mid x < \dfrac{4c + 3d}{6c - 2d}\right\}$

59.
$$c(2 - 5x) + dx > m(4 + 2x)$$
$$2c - 5cx + dx > 4m + 2mx$$
$$-5cx + dx - 2mx > 4m - 2c$$
$$x(-5c + d - 2m) > 4m - 2c$$
$$x[d - (5c + 2m)] > 4m - 2c$$
$$x > \frac{4m - 2c}{d - (5c + 2m)}$$

The inequality sign remained unchanged when we divided because when $5c + 2m < d$, then $d - (5c + 2m) > 0$.

The solution set is $\left\{x \middle| x > \dfrac{4m - 2c}{d - (5c + 2m)}\right\}$.

60. $\left\{x \middle| x < \dfrac{-3a + 2d}{c - (4a + 5d)}\right\}$

61. False. If $a = 2$, $b = 3$, $c = 4$, $d = 5$, then $2 - 4 = 3 - 5$.

62. False, because $-3 < -2$, but $9 > 4$.

63.

64.

65. $x + 5 \le 5 + x$

  $5 \le 5$ Subtracting x

We get an inequality that is true for all real numbers x. Thus the solution is all real numbers.

66. No solution

67. $0^2 = 0$, $x^2 > 0$ for $x \ne 0$

The solution is all real numbers except 0.

68. All real numbers

Exercise Set 9.2

1. $\{5, 6, 7, 8\} \cap \{4, 6, 8, 10\}$

The numbers 6 and 8 are common to the two sets, so the intersection is $\{6, 8\}$.

2. $\{10\}$

3. $\{2, 4, 6, 8\} \cap \{1, 3, 5\}$

There are no numbers common to the two sets, so the intersection is the empty set, $\emptyset$.

4. $\{0\}$

5. $\{1, 2, 3, 4\} \cap \{1, 2, 3, 4\}$

The numbers 1, 2, 3, and 4 are common to the two sets, so the intersection is $\{1, 2, 3, 4\}$.

6. $\emptyset$

7. $1 < x < 6$; $(1, 6)$

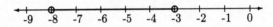

8. $0 \le y \le 3$; $[0, 3]$

9. $-7 \le y \le -3$; $[-7, -3]$

10. $-9 \le x < -5$; $[-9, -5)$

11. $-4 \le -x < 3$

  $4 \ge x > -3$ Multiplying by -1

  $(-3, 4]$

12. $x > -8$ and $x < -3$; $(-8, -3)$

13. $6 > -x \ge -2$

  $-6 < x \le 2$ Multiplying by -1

  $(-6, 2]$

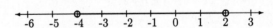

14. $x > -4$ and $x < 2$; $(-4, 2)$

15. $5 > x \ge -2$; $[-2, 5)$

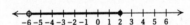

16. $3 > x \ge 0$; $[0, 3)$

17. $x < 5$ and $x \ge 1$; $[1, 5)$

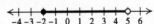

18. $x \ge -2$ and $x < 2$; $[-2, 2)$

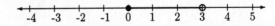

19. $-2 < x + 2 < 8$

  $-2 + (-2) < x + 2 + (-2) < 8 + (-2)$

                    Adding -2

      $-4 < x < 6$

The solution set is $\{x \mid -4 < x < 6\}$, or $(-4, 6)$.

20. $\{x|-2 < x \le 5\}$, or $(-2, 5]$

21.
$$1 < 2y + 5 \le 9$$
$$1 + (-5) < 2y + 5 + (-5) \le 9 + (-5)$$
$$\qquad\qquad\qquad\qquad\qquad \text{Adding } -5$$
$$-4 < 2y \le 4$$
$$\frac{1}{2} \cdot (-4) < \frac{1}{2} \cdot 2y \le \frac{1}{2} \cdot 4 \quad \text{Multiplying by } \frac{1}{2}$$
$$-2 < y \le 2$$

The solution set is $\{y|-2 < y \le 2\}$, or $(-2, 2]$.

22. $\{x|0 \le x \le 1\}$, or $[0, 1]$

23.
$$-10 \le 3x - 5 \le -1$$
$$-10 + 5 \le 3x - 5 + 5 \le -1 + 5 \quad \text{Adding } 5$$
$$-5 \le 3x \le 4$$
$$\frac{1}{3} \cdot (-5) \le \frac{1}{3} \cdot 3x \le \frac{1}{3} \cdot 4 \quad \text{Multiplying by } \frac{1}{3}$$
$$-\frac{5}{3} \le x \le \frac{4}{3}$$

The solution set is $\left\{x\middle|-\frac{5}{3} \le x \le \frac{4}{3}\right\}$, or $\left[-\frac{5}{3}, \frac{4}{3}\right]$.

24. $\left\{x\middle|-\frac{7}{2} < x \le \frac{11}{2}\right\}$, or $\left(-\frac{7}{2}, \frac{11}{2}\right]$.

25.
$$2 < x + 3 \le 9$$
$$2 + (-3) < x + 3 + (-3) \le 9 + (-3) \quad \text{Adding } -3$$
$$-1 < x \le 6$$

The solution set is $\{x|-1 < x \le 6\}$, or $(-1, 6]$.

26. $\{x|-7 \le x < 8\}$, or $[-7, 8)$.

27.
$$-6 \le 2x - 3 < 6$$
$$-6 + 3 \le 2x - 3 + 3 < 6 + 3$$
$$-3 \le 2x < 9$$
$$\frac{1}{2} \cdot (-3) \le \frac{1}{2} \cdot 2x < \frac{1}{2} \cdot 9$$
$$-\frac{3}{2} \le x < \frac{9}{2}$$

The solution set is $\left\{x\middle|-\frac{3}{2} \le x < \frac{9}{2}\right\}$, or $\left[-\frac{3}{2}, \frac{9}{2}\right)$.

28. $\left\{m\middle|-\frac{11}{3} < m \le -3\right\}$, or $\left(-\frac{11}{3}, -3\right]$.

29.
$$-\frac{1}{2} < \frac{1}{4}x - 3 \le \frac{1}{2}$$
$$-\frac{1}{2} + 3 < \frac{1}{4}x - 3 + 3 \le \frac{1}{2} + 3$$
$$\frac{5}{2} < \frac{1}{4}x \le \frac{7}{2}$$
$$4 \cdot \frac{5}{2} < 4 \cdot \frac{1}{4}x \le 4 \cdot \frac{7}{2}$$
$$10 < x \le 14$$

The solution set is $\{x|10 < x \le 14\}$ or $(10, 14]$.

30. $\left\{x\middle|\frac{40}{3} < x \le \frac{56}{3}\right\}$, or $\left(\frac{40}{3}, \frac{56}{3}\right]$

31. $\{4, 5, 6, 7, 8\} \cup \{1, 4, 6, 11\}$

The numbers in either or both sets are $1, 4, 5, 6, 7, 8$, and 11, so the union is $\{1, 4, 5, 6, 7, 8, 11\}$.

32. $\{2, 8, 9, 27\}$

33. $\{2, 4, 6, 8\} \cup \{1, 3, 5\}$

The numbers in either or both sets are $1, 2, 3, 4, 5, 6$, and 8, so the union is $\{1, 2, 3, 4, 5, 6, 8\}$.

34. $\{8, 9, 10\}$

35. $\{4, 8, 11\} \cup \emptyset$

The numbers in either or both sets are $4, 8$, and 11, so the union is $\{4, 8, 11\}$.

36. $\emptyset$

37. $x < -1 \; or \; x > 2$

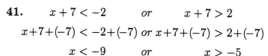

38. $x < -2 \; or \; x > 0$

39. $x \le -3 \; or \; x > 1$

40. $x \le -1 \; or \; x > 3$

41.
$$x + 7 < -2 \qquad or \qquad x + 7 > 2$$
$$x + 7 + (-7) < -2 + (-7) \; or \; x + 7 + (-7) > 2 + (-7)$$
$$x < -9 \qquad or \qquad x > -5$$

The solution set is $\{x|x < -9 \; or \; x > -5\}$, or $(-\infty, -9) \cup (-5, \infty)$.

42. $\{x|x < -13 \; or \; x > -5\}$, or $(-\infty, -13) \cup (-5, \infty)$.

43.
$$2x - 8 \le -3 \qquad or \qquad x - 8 \ge 3$$
$$2x - 8 + 8 \le -3 + 8 \; or \; x - 8 + 8 \ge 3 + 8$$
$$2x \le 5 \qquad or \qquad x \ge 11$$
$$\frac{1}{2} \cdot 2x \le \frac{1}{2} \cdot 5 \quad or \qquad x \ge 11$$
$$x \le \frac{5}{2} \qquad or \qquad x \ge 11$$

The solution set is $\left\{x\middle|x \le \frac{5}{2} \; or \; x \ge 11\right\}$, or $\left(-\infty, \frac{5}{2}\right] \cup [11, \infty)$.

44. $\{x|x \le -9 \; or \; x \ge 3\}$, or $(-\infty, -9] \cup [3, \infty)$.

45.
$$7x + 4 \ge -17 \; or \; 6x + 5 \ge -7$$
$$7x \ge -21 \; or \qquad 6x \ge -12$$
$$x \ge -3 \; or \qquad x \ge -2$$

The solution set is $\{x|x \ge -3\} \cup \{x|x \ge -2\}$. This is $\{x|x \ge -3\}$, or $[-3, \infty)$.

46. $\{x|x < 4\}$, or $(-\infty, 4)$

47.
$$7 > -4x + 5 \qquad or \qquad 10 \le -4x + 5$$
$$7 + (-5) > -4x + 5 + (-5) \quad or \quad 10 + (-5) \le -4x + 5 + (-5)$$
$$2 > -4x \qquad or \qquad 5 \le -4x$$
$$-\frac{1}{4} \cdot 2 < -\frac{1}{4}(-4x) \qquad or \qquad -\frac{1}{4} \cdot 5 \ge -\frac{1}{4}(-4x)$$
$$-\frac{1}{2} < x \qquad or \qquad -\frac{5}{4} \ge x$$

The solution set is $\left\{x \middle| x \le -\frac{5}{4} \ or \ x > -\frac{1}{2}\right\}$, or

$\left(-\infty, -\frac{5}{4}\right] \cup \left(-\frac{1}{2}, \infty\right)$.

48. The set of all real numbers

49.
$$3x - 7 > -10 \quad or \quad 5x + 2 \le 22$$
$$3x > -3 \quad or \quad 5x \le 20$$
$$x > -1 \quad or \quad x \le 4$$

Since all real numbers are greater than -1 or less than or equal to 4, the two sets fill up the entire number line. Thus the solution set is the set of all real numbers.

50. The set of all real numbers

51.
$$-2x - 2 < -6 \qquad or \qquad -2x - 2 > 6$$
$$-2x - 2 + 2 < -6 + 2 \quad or \quad -2x - 2 + 2 > 6 + 2$$
$$-2x < -4 \qquad or \qquad -2x > 8$$
$$-\frac{1}{2}(-2x) > -\frac{1}{2}(-4) \quad or \quad -\frac{1}{2}(-2x) < -\frac{1}{2} \cdot 8$$
$$x > 2 \qquad or \qquad x < -4$$

The solution set is $\{x | x < -4 \ or \ x > 2\}$, or $(-\infty, -4) \cup (2, \infty)$.

52. $\left\{m \middle| m < -4 \ or \ m > -\frac{2}{3}\right\}$, or $(-\infty, -4) \cup \left(-\frac{2}{3}, \infty\right)$

53.
$$\frac{2}{3}x - 14 < -\frac{5}{6} \qquad or \qquad \frac{2}{3}x - 14 > \frac{5}{6}$$
$$6\left(\frac{2}{3}x - 14\right) < 6\left(-\frac{5}{6}\right) \quad or \quad 6\left(\frac{2}{3}x - 14\right) > 6 \cdot \frac{5}{6}$$
$$4x - 84 < -5 \qquad or \qquad 4x - 84 > 5$$
$$4x - 84 + 84 < -5 + 84 \quad or \quad 4x - 84 + 84 > 5 + 84$$
$$4x < 79 \qquad or \qquad 4x > 89$$
$$\frac{1}{4} \cdot 4x < \frac{1}{4} \cdot 79 \qquad or \qquad \frac{1}{4} \cdot 4x > \frac{1}{4} \cdot 89$$
$$x < \frac{79}{4} \qquad or \qquad x > \frac{89}{4}$$

The solution set is $\left\{x \middle| x < \frac{79}{4} \ or \ x > \frac{89}{4}\right\}$, or

$\left(-\infty, \frac{79}{4}\right) \cup \left(\frac{89}{4}, \infty\right)$.

54. $\left\{x \middle| x \le -\frac{91}{100} \ or \ x \ge \frac{79}{60}\right\}$, or $\left(-\infty, -\frac{91}{100}\right] \cup \left[\frac{79}{60}, \infty\right)$

55.
$$\frac{2x - 5}{6} \le -3 \qquad or \qquad \frac{2x - 5}{6} \ge 4$$
$$6\left(\frac{2x - 5}{6}\right) \le 6(-3) \quad or \quad 6\left(\frac{2x - 5}{6}\right) \ge 6 \cdot 4$$
$$2x - 5 \le -18 \qquad or \qquad 2x - 5 \ge 24$$
$$2x - 5 + 5 \le -18 + 5 \quad or \quad 2x - 5 + 5 \ge 24 + 5$$
$$2x \le -13 \qquad or \qquad 2x \ge 29$$
$$\frac{1}{2} \cdot 2x \le \frac{1}{2}(-13) \qquad or \qquad \frac{1}{2} \cdot 2x \ge \frac{1}{2} \cdot 29$$
$$x \le -\frac{13}{2} \qquad or \qquad x \ge \frac{29}{2}$$

The solution set is $\left\{x \middle| x \le -\frac{13}{2} \ or \ x \ge \frac{29}{2}\right\}$, or

$\left(-\infty, -\frac{13}{2}\right] \cup \left[\frac{29}{2}, \infty\right)$.

56. $\left\{x \middle| x < -\frac{13}{3} \ or \ x > 9\right\}$, or $\left(-\infty, -\frac{13}{3}\right) \cup (9, \infty)$

57.
$$5x - 7 \le 13 \quad or \quad 2x - 1 \ge -7$$
$$5x \le 20 \ or \qquad 2x \ge -6$$
$$x \le 4 \ or \qquad x \ge -3$$

Since all real numbers are greater than or equal to -3 or less than or equal to 4, the two sets fill up the entire number line. Thus, the solution set is the set of all real numbers.

58. $\{x | x \le 2\}$, or $(-\infty, 2]$

59.
$$2x - 3y = 7, \qquad (1)$$
$$3x + 2y = -10 \quad (2)$$

We will use the elimination method.
$$\begin{array}{ll} 4x - 6y = 14 & \text{Multiplying (1) by 2} \\ 9x + 6y = -30 & \text{Multiplying (2) by 3} \\ \hline 13x = -16 & \text{Adding} \\ x = -\dfrac{16}{13} \end{array}$$

Substitute $-\dfrac{16}{13}$ for x in (1) and solve for y.

$$2\left(-\frac{16}{13}\right) - 3y = 7$$
$$-\frac{32}{13} - 3y = 7$$
$$\frac{32}{13} - \frac{32}{13} - 3y = \frac{32}{13} + \frac{91}{13}$$
$$-3y = \frac{123}{13}$$
$$y = -\frac{41}{13}$$

These numbers check, so the solution is

$\left(-\dfrac{16}{13}, -\dfrac{41}{13}\right)$.

60. $-\dfrac{8}{3}$

61. $5(2x + 3) = 3(x - 4)$

$10x + 15 = 3x - 12$

$7x + 15 = -12$

$7x = -27$

$x = -\dfrac{27}{7}$

The solution is $-\dfrac{27}{7}$.

62.

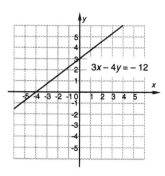

$3x - 4y = -12$

63.

64.

65. a) Substitute $\dfrac{5}{9}(F - 32)$ for C in the given inequality.

$1063 \le \dfrac{5}{9}(F - 32) < 2660$

$9 \cdot 1063 \le 9 \cdot \dfrac{5}{9}(F - 32) < 9 \cdot 2660$

$9567 \le 5(F - 32) < 23{,}940$

$9567 \le 5F - 160 < 23{,}940$

$9727 \le 5F < 24{,}100$

$1945.4 \le F < 4820$

The inequality for Fahrenheit temperatures is $1945.4° \le F < 4820°$.

b) Substitute $\dfrac{5}{9}(F - 32)$ for C in the given inequality.

$960.8 \le \dfrac{5}{9}(F - 32) < 2180$

$9(960.8) \le 9 \cdot \dfrac{5}{9}(F - 32) < 9 \cdot 2180$

$8647.2 \le 5(F - 32) < 19{,}620$

$8647.2 \le 5F - 160 < 19{,}620$

$8807.2 \le 5F < 19{,}780$

$1761.44 \le F < 3956$

The inequality for Fahrenheit temperatures is $1761.44° \le F < 3956°$.

66. All the numbers between -5 and 11

67. Solve $32 < f(x) < 46$, or $32 < 2(x + 10) < 46$.

$32 < 2(x + 10) < 46$

$32 < 2x + 20 < 46$

$12 < 2x < 26$

$6 < x < 13$

For U.S. dress sizes between 6 and 13, dress sizes in Italy will be between 32 and 46.

68. $0 \text{ ft} \le d \le 198 \text{ ft}$

69. Solve $50{,}000 \le N \le 250{,}000$ where $N = 12{,}197.8t + 44{,}000$. Keep in mind that t is the number of years since 1971.

$50{,}000 \le 12{,}197.8t + 44{,}000 \le 250{,}000$

$6000 \le 12{,}197.8t \le 206{,}000$

$0.49 \le t \le 16.89$

0.49 years after the end of 1971 is in 1972 and 16.89 years after the end of 1971 is in 1988. We give the result in terms of *entire* years that satisfy the inequality. Therefore, there will be at least 50,000 and at most 250,000 women in the active duty military force from 1973 to 1987.

70. $1966 \le y \le 1980$

71. $4a - 2 \le a + 1 \le 3a + 4$

$4a - 2 \le a + 1 \quad and \quad a + 1 \le 3a + 4$

$3a \le 3 \qquad and \qquad -3 \le 2a$

$a \le 1 \qquad and \qquad -\dfrac{3}{2} \le a$

The solution set is $\left\{ a \middle| -\dfrac{3}{2} \le a \le 1 \right\}$, or $\left[-\dfrac{3}{2}, 1 \right]$.

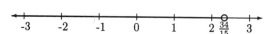

72. $\left\{ m \middle| m < \dfrac{6}{5} \right\}$, or $\left(-\infty, \dfrac{6}{5} \right)$

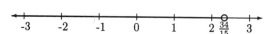

73. $x - 10 < 5x + 6 \le x + 10$

$-10 < 4x + 6 \le 10$

$-16 < 4x \le 4$

$-4 < x \le 1$

The solution set is $\{x | -4 < x \le 1\}$, or $(-4, 1]$.

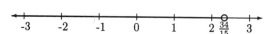

74. $\left\{ y \middle| y < \dfrac{34}{15} \right\}$, or $\left(-\infty, \dfrac{34}{15} \right)$

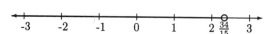

75. $3x < 4 - 5x < 5 + 3x$

$0 < 4 - 8x < 5$

$-4 < -8x < 1$

$\frac{1}{2} > x > -\frac{1}{8}$

The solution set is $\left\{x \middle| -\frac{1}{8} < x < \frac{1}{2}\right\}$, or $\left(-\frac{1}{8}, \frac{1}{2}\right)$.

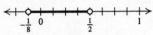

76. $\left\{x \middle| x > \frac{21}{4}\right\}$, or $\left(\frac{21}{4}, \infty\right)$

77. Given any two numbers b and c either $b < c$, $b = c$, or $b > c$. If $b > c$ is given, we know that $b \not< c$ and $b \neq c$. Thus, $b \not\leq c$ is true.

78. True

79. Let $c = 6$ and $a = 10$. Then $c \neq a$, but $a \not< c$. The given statement is false.

80. False

81. Let $a = 5$, $c = 12$, and $b = 2$. Then $a < c$ and $b < c$, but $a \not< b$. The given statement is false.

82. False

83. $[4x - 2 < 8 \quad or \quad 3(x-1) < -2] \ and \ -2 \leq 5x \leq 10$

$\left[4x - 2 < 8 \quad or \quad 3x - 3 < -2\right] \ and \ -\frac{2}{5} \leq x \leq 2$

$\left[4x < 10 \ or \quad 3x < 1\right] \ and \ -\frac{2}{5} \leq x \leq 2$

$\left[x < \frac{5}{2} \ or \quad x < \frac{1}{3}\right] \ and \ -\frac{2}{5} \leq x \leq 2$

$\left[x < \frac{5}{2}\right] \ and \ -\frac{2}{5} \leq x \leq 2$

The solution set is $\left\{x \middle| -\frac{2}{5} \leq x \leq 2\right\}$, or $\left[-\frac{2}{5}, 2\right]$.

84. $\emptyset$

Exercise Set 9.3

1. $|x| = 3$

$x = -3 \ or \ x = 3$ Using the absolute-value principle

The solution set is $\{-3, 3\}$.

2. $\{-5, 5\}$

3. $|x| = -3$

The absolute value of a number is always nonnegative. Therefore, the solution set is $\emptyset$.

4. $\emptyset$

5. $|p| = 0$

The only number whose absolute value is 0 is 0. The solution set is $\{0\}$.

6. $\{-8.6, 8.6\}$

7. $|t| = 5.5$

$t = -5.5 \ or \ t = 5.5$ Using the absolute-value principle

The solution set is $\{-5.5, 5.5\}$.

8. $\{0\}$

9. $|x - 3| = 12$

$x - 3 = -12 \ or \ x - 3 = 12$ Absolute-value principle

$x = -9 \ or \quad x = 15$

The solution set is $\{-9, 15\}$.

10. $\left\{-\frac{4}{3}, \frac{8}{3}\right\}$

11. $|2x - 3| = 4$

$2x - 3 = -4 \ or \ 2x - 3 = 4$ Absolute-value principle

$2x = -1 \ or \quad 2x = 7$

$x = -\frac{1}{2} \ or \quad x = \frac{7}{2}$

The solution set is $\left\{-\frac{1}{2}, \frac{7}{2}\right\}$.

12. $\left\{-1, \frac{1}{5}\right\}$

13. $|2y - 7| = 10$

$2y - 7 = -10 \ or \ 2y - 7 = 10$

$2y = -3 \ or \quad 2y = 17$

$y = -\frac{3}{2} \ or \quad x = \frac{17}{2}$

The solution set is $\left\{-\frac{3}{2}, \frac{17}{2}\right\}$.

14. $\left\{-\frac{4}{3}, 4\right\}$

15. $|3x - 10| = -8$

Absolute value is always nonnegative, so the equation has no solution. The solution set is $\emptyset$.

16. $\emptyset$

17. $|x| + 7 = 18$

$|x| + 7 - 7 = 18 - 7$ Adding -7

$|x| = 11$

$x = -11 \ or \ x = 11$ Absolute-value principle

The solution set is $\{-11, 11\}$.

18. $\{-8.3, 8.3\}$

19. $|5x| - 3 = 37$

$|5x| = 40$ Adding 3

$5x = -40$ *or* $5x = 40$

$x = -8$ *or* $x = 8$

The solution set is $\{-8, 8\}$.

20. $\{-9, 9\}$

21. $5|q| - 2 = 9$

$5|q| = 11$ Adding 2

$|q| = \dfrac{11}{5}$ Multiplying by $\dfrac{1}{5}$

$q = -\dfrac{11}{5}$ *or* $q = \dfrac{11}{5}$

The solution set is $\left\{-\dfrac{11}{5}, \dfrac{11}{5}\right\}$.

22. $\{-2, 2\}$

23. $\left|\dfrac{2x-1}{3}\right| = 5$

$\dfrac{2x-1}{3} = -5$ *or* $\dfrac{2x-1}{3} = 5$

$2x - 1 = -15$ *or* $2x - 1 = 15$

$2x = -14$ *or* $2x = 16$

$x = -7$ *or* $x = 8$

The solution set is $\{-7, 8\}$.

24. $\left\{-\dfrac{38}{5}, \dfrac{46}{5}\right\}$

25. $|m + 5| + 9 = 16$

$|m + 5| = 7$ Adding -9

$m + 5 = -7$ *or* $m + 5 = 7$

$m = -12$ *or* $m = 2$

The solution set is $\{-12, 2\}$.

26. $\{6, 8\}$

27. $\left|\dfrac{2x-1}{3}\right| = 1$

$\dfrac{2x-1}{3} = -1$ *or* $\dfrac{2x-1}{3} = 1$

$2x - 1 = -3$ *or* $2x - 1 = 3$

$2x = -2$ *or* $2x = 4$

$x = -1$ *or* $x = 2$

The solution set is $\{-1, 2\}$.

28. $\left\{-\dfrac{8}{3}, 4\right\}$

29. $5 - 2|3x - 4| = -5$

$-2|3x - 4| = -10$

$|3x - 4| = 5$

$3x - 4 = -5$ *or* $3x - 4 = 5$

$3x = -1$ *or* $3x = 9$

$x = -\dfrac{1}{3}$ *or* $x = 3$

The solution set is $\left\{-\dfrac{1}{3}, 3\right\}$.

30. $\left\{\dfrac{3}{2}, \dfrac{7}{2}\right\}$

31. The distance between 13 and 17 is $|17 - 13| = |4| = 4$. The distance is also given by $|13 - 17| = |-4| = 4$.

32. 6

33. The distance between 25 and 14 is $|25 - 14| = |11| = 11$, or $|14 - 25| = |-11| = 11$.

34. 15

35. The distance between -9 and 24 is $|24 - (-9)| = |24 + 9| = |33| = 33$, or $|-9 - 24| = |-33| = 33$.

36. 19

37. The distance between -8 and -42 is $|-8 - (-42)| = |-8 + 42| = |34| = 34$, or $|-42 - (-8)| = |-42 + 8| = |-34| = 34$.

38. 27

39. $|3x + 4| = |x - 7|$

$3x + 4 = x - 7$ *or* $3x + 4 = -(x - 7)$

$2x + 4 = -7$ *or* $3x + 4 = -x + 7$

$2x = -11$ *or* $4x + 4 = 7$

$x = -\dfrac{11}{2}$ *or* $4x = 3$

$x = \dfrac{3}{4}$

The solution set is $\left\{-\dfrac{11}{2}, \dfrac{3}{4}\right\}$.

40. $\left\{11, \dfrac{5}{3}\right\}$

41. $|x + 5| = |x - 2|$

$x + 5 = x - 2$ *or* $x + 5 = -(x - 2)$

$5 = -2$ *or* $x + 5 = -x + 2$

False — $2x + 5 = 2$

yields no $2x = -3$

solution $x = -\dfrac{3}{2}$

The solution set is $\left\{-\dfrac{3}{2}\right\}$.

42. $\left\{-\dfrac{1}{2}\right\}$

43. $|2a + 4| = |3a - 1|$

$2a + 4 = 3a - 1$ *or* $2a + 4 = -(3a - 1)$

$-a + 4 = -1$ *or* $2a + 4 = -3a + 1$

$-a = -5$ *or* $5a + 4 = 1$

$a = 5$ *or* $5a = -3$

$a = -\dfrac{3}{5}$

The solution set is $\left\{5, -\dfrac{3}{5}\right\}$.

44. $\left\{-4, -\dfrac{10}{9}\right\}$

45. $|y - 3| = |3 - y|$

$\quad y - 3 = 3 - y \quad or \quad y - 3 = -(3 - y)$

$\quad 2y - 3 = 3 \qquad or \quad y - 3 = -3 + y$

$\quad\quad 2y = 6 \qquad or \quad -3 = -3$

$\quad\quad\quad y = 3 \qquad$ True for all real values of y

The solution set is $\{y | y \text{ is a real number}\}$.

46. $\{m | m \text{ is a real number}\}$

47. $|5 - p| = |p + 8|$

$\quad 5 - p = p + 8 \quad or \quad 5 - p = -(p + 8)$

$\quad 5 - 2p = 8 \qquad or \quad 5 - p = -p - 8$

$\quad\quad -2p = 3 \qquad or \qquad 5 = -8$

$\quad\quad\quad p = -\dfrac{3}{2} \qquad$ False

The solution set is $\left\{ -\dfrac{3}{2} \right\}$.

48. $\left\{ -\dfrac{11}{2} \right\}$

49. $\left| \dfrac{2x - 3}{6} \right| = \left| \dfrac{4 - 5x}{8} \right|$

$\quad \dfrac{2x - 3}{6} = \dfrac{4 - 5x}{8} \quad or \quad \dfrac{2x - 3}{6} = -\left(\dfrac{4 - 5x}{8} \right)$

$\quad 24\left(\dfrac{2x - 3}{6} \right) = 24\left(\dfrac{4 - 5x}{8} \right) or \quad \dfrac{2x - 3}{6} = \dfrac{-4 + 5x}{8}$

$\quad 8x - 12 = 12 - 15x \quad or \, 24\left(\dfrac{2x - 3}{6} \right) = 24\left(\dfrac{-4 + 5x}{8} \right)$

$\quad 23x - 12 = 12 \qquad or \quad 8x - 12 = -12 + 15x$

$\quad\quad 23x = 24 \qquad or \quad -7x - 12 = -12$

$\quad\quad\quad x = \dfrac{24}{23} \qquad or \qquad -7x = 0$

$\quad\quad\quad\quad\quad\quad\quad\quad\quad\quad x = 0$

The solution set is $\left\{ \dfrac{24}{23}, 0 \right\}$.

50. $\left\{ -\dfrac{23}{31}, 47 \right\}$

51. $\left| \dfrac{1}{2}x - 5 \right| = \left| \dfrac{1}{4}x + 3 \right|$

$\quad \dfrac{1}{2}x - 5 = \dfrac{1}{4}x + 3 \quad or \quad \dfrac{1}{2}x - 5 = -\left(\dfrac{1}{4}x + 3 \right)$

$\quad \dfrac{1}{4}x - 5 = 3 \qquad or \quad \dfrac{1}{2}x - 5 = -\dfrac{1}{4}x - 3$

$\quad\quad \dfrac{1}{4}x = 8 \qquad or \quad \dfrac{3}{4}x - 5 = -3$

$\quad\quad\quad x = 32 \qquad or \qquad \dfrac{3}{4}x = 2$

$\quad\quad\quad\quad\quad\quad\quad\quad\quad\quad x = \dfrac{8}{3}$

The solution set is $\left\{ 32, \dfrac{8}{3} \right\}$.

52. $\left\{ -\dfrac{48}{37}, -\dfrac{144}{5} \right\}$

53. $|x| < 3$

$\quad -3 < x < 3 \qquad$ Part (b)

The solution set is $\{x | -3 < x < 3\}$, or $(-3, 3)$.

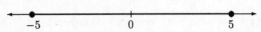

54. $\{x | -5 \leq x \leq 5\}$, or $[-5, 5]$

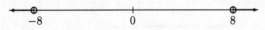

55. $|x| \geq 2$

$\quad x \leq -2 \text{ or } 2 \leq x \qquad$ Part (c)

The solution set is $\{x | x \leq -2 \text{ or } x \geq 2\}$, or $(-\infty, -2] \cup [2, \infty)$.

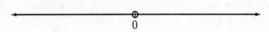

56. $\{y | y < -8 \text{ or } y > 8\}$, or $(-\infty, -8) \cup (8, \infty)$

57. $|t| \geq 5.5$

$\quad t \leq -5.5 \text{ or } 5.5 \leq t \qquad$ Part (c)

The solution set is $\{t | t \leq -5.5 \text{ or } t \geq 5.5\}$, or $(-\infty, -5.5] \cup [5.5, \infty)$.

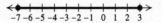

58. $\{m | m \neq 0\}$, or $(-\infty, 0) \cup (0, \infty)$

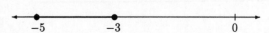

59. $|x - 3| < 1$

$\quad -1 < x - 3 < 1 \qquad$ Part (b)

$\quad 2 < x < 4 \qquad$ Adding 3

The solution set is $\{x | 2 < x < 4\}$, or $(2, 4)$.

60. $\{x | -4 < x < 8\}$, or $(-4, 8)$

61. $|x + 2| \leq 5$

$\quad -5 \leq x + 2 \leq 5 \qquad$ Part (b)

$\quad -7 \leq x \leq 3 \qquad$ Adding -2

The solution set is $\{x | -7 \leq x \leq 3\}$, or $[-7, 3]$.

62. $\{x | -5 \leq x \leq -3\}$, or $[-5, -3]$

63. $|x - 3| > 1$

$x - 3 < -1 \quad or \quad 1 < x - 3 \quad$ Part (c)

$x < 2 \quad or \quad 4 < x \qquad$ Adding 3

The solution set is $\{x | x < 2 \ or \ x > 4\}$, or $(-\infty, 2) \cup (4, \infty)$.

64. $\{x | x < -4 \ or \ x > 8\}$, or $(-\infty, -4) \cup (8, \infty)$

65. $|2x - 3| \leq 4$

$-4 \leq 2x - 3 \leq 4 \quad$ Part (b)

$-1 \leq 2x \leq 7 \qquad$ Adding 3

$-\dfrac{1}{2} \leq x \leq \dfrac{7}{2} \qquad$ Multiplying by $\dfrac{1}{2}$

The solution set is $\left\{x \middle| -\dfrac{1}{2} \leq x \leq \dfrac{7}{2}\right\}$, or $\left[-\dfrac{1}{2}, \dfrac{7}{2}\right]$.

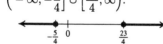

66. $\left\{x \middle| -1 \leq x \leq \dfrac{1}{5}\right\}$, or $\left[-1, \dfrac{1}{5}\right]$

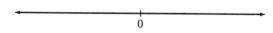

67. $|2y - 7| > -1$

Since absolute value is never negative, any value of $2y - 7$, and hence any value of y, will satisfy the inequality. The solution set is the set of all real numbers.

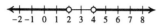

68. $\left\{y \middle| y < -\dfrac{4}{3} \ or \ y > 4\right\}$, or $\left(-\infty, -\dfrac{4}{3}\right) \cup (4, \infty)$

69. $|4x - 9| \geq 14$

$4x - 9 \leq -14 \quad or \quad 14 \leq 4x - 9 \quad$ Part (c)

$4x \leq -5 \quad or \quad 23 \leq 4x$

$x \leq -\dfrac{5}{4} \quad or \quad \dfrac{23}{4} \leq x$

The solution set is $\left\{x \middle| x \leq -\dfrac{5}{4} \ or \ x \geq \dfrac{23}{4}\right\}$, or $\left(-\infty, -\dfrac{5}{4}\right] \cup \left[\dfrac{23}{4}, \infty\right)$.

70. The set of all real numbers

71. $|y - 3| < 12$

$-12 < y - 3 < 12 \quad$ Part (b)

$-9 < y < 15 \qquad$ Adding 3

The solution set is $\{y | -9 < y < 15\}$, or $(-9, 15)$.

72. $\{p | -1 < p < 5\}$ or $(-1, 5)$

73. $|2x + 3| \leq 4$

$-4 \leq 2x + 3 \leq 4 \quad$ Part (b)

$-7 \leq 2x \leq 1 \qquad$ Adding -3

$-\dfrac{7}{2} \leq x \leq \dfrac{1}{2} \qquad$ Multiplying by $\dfrac{1}{2}$

The solution set is $\left\{x \middle| -\dfrac{7}{2} \leq x \leq \dfrac{1}{2}\right\}$, or $\left[-\dfrac{7}{2}, \dfrac{1}{2}\right]$.

74. $\left\{x \middle| -\dfrac{1}{5} \leq x \leq 1\right\}$, or $\left[-\dfrac{1}{5}, 1\right]$

75. $|4 - 3y| > 8$

$4 - 3y < -8 \quad or \quad 8 < 4 - 3y \quad$ Part (c)

$-3y < -12 \quad or \quad 4 < -3y \qquad$ Adding -4

$y > 4 \quad or \quad -\dfrac{4}{3} > y \quad$ Multiplying by $-\dfrac{1}{3}$

The solution set is $\left\{y \middle| y < -\dfrac{4}{3} \ or \ y > 4\right\}$, or $\left(-\infty, -\dfrac{4}{3}\right) \cup (4, \infty)$.

76. $\emptyset$

77. $|9 - 4x| \leq 14$

$-14 \leq 9 - 4x \leq 14 \quad$ Part (b)

$-23 \leq -4x \leq 5 \qquad$ Adding -9

$\dfrac{23}{4} \geq x \geq -\dfrac{5}{4} \qquad$ Multiplying by $-\dfrac{1}{4}$

The solution set is $\left\{x \middle| -\dfrac{5}{4} \leq x \leq \dfrac{23}{4}\right\}$, or $\left[-\dfrac{5}{4}, \dfrac{23}{4}\right]$.

78. $\left\{p \middle| p \leq -\dfrac{5}{3} \ or \ p \geq \dfrac{19}{9}\right\}$, or $\left(-\infty, -\dfrac{5}{3}\right] \cup \left[\dfrac{19}{9}, \infty\right]$

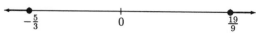

79. $|3 - 4x| < -5$

Absolute value is always nonnegative, so the inequality has no solution. The solution set is $\emptyset$.

80. $\left\{ x \middle| -5 \le x \le \dfrac{25}{7} \right\}$, or $\left[-5, \dfrac{25}{7} \right]$

81. $7 + |2x - 1| > 16$

$\qquad |2x - 1| > 9$

$\qquad 2x - 1 < -9 \quad or \quad 9 < 2x - 1 \quad$ Part (c)

$\qquad 2x < -8 \quad or \quad 10 < 2x \qquad$ Adding 1

$\qquad x < -4 \quad or \quad 5 < x \qquad$ Multiplying by $\dfrac{1}{2}$

The solution set is $\{ x | x < -4 \ or \ x > 5 \}$, or $(-\infty, -4) \cup (5, \infty)$.

82. $\left\{ x \middle| x < -\dfrac{16}{3} \ or \ x > 4 \right\}$, or $\left(-\infty, -\dfrac{16}{3} \right) \cup (4, \infty)$

83. $\left| \dfrac{x - 7}{3} \right| < 4$

$\qquad -4 < \dfrac{x - 7}{3} < 4 \qquad$ Part (b)

$\qquad -12 < x - 7 < 12 \qquad$ Multiplying by 3

$\qquad -5 < x < 19 \qquad$ Adding 7

The solution set is $\{ x | -5 < x < 19 \}$, or $(-5, 19)$.

84. $\{ x | -13 \le x \le 3 \}$, or $[-13, 3]$

85. $\left| \dfrac{2 - 5x}{4} \right| \ge \dfrac{2}{3}$

$\qquad \dfrac{2 - 5x}{4} \le -\dfrac{2}{3} \quad or \quad \dfrac{2}{3} \le \dfrac{2 - 5x}{4} \quad$ Part (c)

$\qquad 2 - 5x \le -\dfrac{8}{3} \quad or \quad \dfrac{8}{3} \le 2 - 5x \quad$ Multiplying by 4

$\qquad -5x \le -\dfrac{14}{3} \quad or \quad \dfrac{2}{3} \le -5x \quad$ Adding -2

$\qquad x \ge \dfrac{14}{15} \quad or \quad -\dfrac{2}{15} \ge x \quad$ Multiplying by $-\dfrac{1}{5}$

The solution set is $\left\{ x \middle| x \le -\dfrac{2}{15} \ or \ x \ge \dfrac{14}{15} \right\}$, or $\left(-\infty, -\dfrac{2}{15} \right] \cup \left[\dfrac{14}{15}, \infty \right)$.

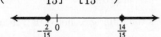

86. $\left\{ x \middle| x < -\dfrac{43}{24} \ or \ x > \dfrac{9}{8} \right\}$, or $\left(-\infty, -\dfrac{43}{24} \right) \cup \left(\dfrac{9}{8}, \infty \right)$

87. $|m + 5| + 9 \le 16$

$\qquad |m + 5| \le 7 \quad$ Adding -9

$\qquad -7 \le m + 5 \le 7$

$\qquad -12 \le m \le 2$

The solution set is $\{ m | -12 \le m \le 2 \}$, or $[-12, 2]$.

88. $\{ t | t \le 6 \ or \ t \ge 8 \}$, or $(-\infty, 6] \cup [8, \infty)$

89. $|g + 7| + 13 > 9$

$\qquad |g + 7| > -4 \quad$ Adding -13

Since absolute value is never negative, any value of $g + 7$, and hence any value of g, will satisfy the inequality. The solution set is the set of all real numbers.

90. The set of all real numbers

91. $\left| \dfrac{2x - 1}{3} \right| \le 1$

$\qquad -1 \le \dfrac{2x - 1}{3} \le 1$

$\qquad -3 \le 2x - 1 \le 3$

$\qquad -2 \le 2x \le 4$

$\qquad -1 \le x \le 2$

The solution set is $\{ x | -1 \le x \le 2 \}$, or $[-1, 2]$.

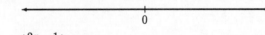

92. $\left\{ x \middle| -\dfrac{8}{3} \le x \le 4 \right\}$, or $\left[-\dfrac{8}{3}, 4 \right]$

93. *Familiarize.* Let l represent the length and w represent the width. Recall that perimeter $P = 2l + 2w$, and area $A = lw$.

Translate.

The perimeter is 628 m.

$\qquad 2l + 2w \quad = \quad 628$

The length is 6 m more than the width.

$\qquad l \quad = \quad 6 \quad + \quad w$

We have a system of equations:

$\qquad 2l + 2w = 628, \quad (1)$

$\qquad l = 6 + w \qquad (2)$

Carry out. Use the substitution method to solve the system of equations and then compute the area. Substitute $6 + w$ for l in (1).

$$2(6 + w) + 2w = 628$$
$$12 + 2w + 2w = 628$$
$$12 + 4w = 628$$
$$4w = 616$$
$$w = 154$$

Substitute 154 for w in (2).

$$l = 6 + 154 = 160$$
$$A = lw = 160(154) = 24{,}640$$

Check. $2l + 2w = 2 \cdot 160 + 2 \cdot 154 = 320 + 308 = 628$. 160 is 6 more than 154. The numbers check.

State. The area is 24,640 m².

94. 132 adult plates, 118 children's plates

95. ◈

96. ◈

97. From the definition of absolute value, $|2x - 5| = 2x - 5$ only when $2x - 5 \ge 0$. Solve $2x - 5 \ge 0$.

$$2x - 5 \ge 0$$
$$2x \ge 5$$
$$x \ge \frac{5}{2}$$

The solution set is $\left\{ x \middle| x \ge \frac{5}{2} \right\}$.

98. $\{-31, -33\}$

99. $|x + 5| = x + 5$

From the definition of absolute value, $|x + 5| = x + 5$ only when $x + 5 \ge 0$, or $x \ge -5$. The solution set is $\{x | x \ge -5\}$.

100. $\{x | x \ge 1\}$

101. $|7x - 2| = x + 4$

From the definition of absolute value, we know $x + 4 \ge 0$, or $x \ge -4$. So we have $x \ge -4$ and

$$7x - 2 = x + 4 \quad or \quad 7x - 2 = -(x + 4)$$
$$6x = 6 \quad\quad or \quad 7x - 2 = -x - 4$$
$$x = 1 \quad\quad or \quad\quad 8x = -2$$
$$x = -\frac{1}{4}$$

The solution set is $\left\{ x \middle| x \ge -4 \text{ and } x = 1 \text{ or } x = -\frac{1}{4} \right\}$, or $\left\{ 1, -\frac{1}{4} \right\}$.

102. $\{x | x \text{ is a real number}\}$

103. $\left| 5.2x - \dfrac{6}{7} \right| \le -8$

From the definition of absolute value we know that $\left| 5.2x - \dfrac{6}{7} \right| \ge 0$. Thus, $\left| 5.2x - \dfrac{6}{7} \right| \le -8$ is false for all x. The solution set is $\emptyset$.

104. $\emptyset$

105. $|x + 5| > x$

The inequality is true for all $x < 0$ (because absolute value must be nonnegative). The solution set in this case is $\{x | x < 0\}$. If $x = 0$, we have $|0 + 5| > 0$, which is true. The solution set in this case is $\{0\}$. If $x > 0$, we have the following:

$$x + 5 < -x \quad or \quad x < x + 5$$
$$2x < -5 \quad or \quad 0 < 5$$
$$x < -\frac{5}{2}$$

Although $x > 0$ and $x < -\dfrac{5}{2}$ yields no solution, $x > 0$ and $5 > 0$ (true for all x) yields the solution set $\{x | x > 0\}$ in this case. The solution set for the inequality is $\{x | x < 0\} \cup \{0\} \cup \{x | x > 0\}$, or $\{x | x \text{ is a real number}\}$.

106. $\{x | -4 \le x \le -1 \ or \ 3 \le x \le 6\}$

107. Using part (b), we find that $-3 < x < 3$ is equivalent to $|x| < 3$.

108. $|y| \le 5$

109. $x \le -6 \ or \ 6 \le x$
$|x| \ge 6$ Using part (c)

110. $|x| > 4$

111. $x < -8 \ or \ 2 < x$
$x + 3 < -5 \ or \ 5 < x + 3$ Adding 3
$|x + 3| > 5$ Using part (c)

112. $|x + 2| < 3$

113. We first write an inequality in terms of feet.

Convert $\dfrac{1}{8}$ in. to feet:

$$\frac{1}{8} \text{ in.} = \frac{1}{8} \text{ in.} \times \frac{1 \text{ ft}}{12 \text{ in.}} = \frac{1}{96} \text{ ft}$$

Then we have:

$$5 - \frac{1}{96} \le p \le 5 + \frac{1}{96}$$
$$-\frac{1}{96} \le p - 5 \le \frac{1}{96} \quad \text{Adding } -5$$
$$|p - 5| \le \frac{1}{96} \quad\quad \text{Part (b)}$$

We could also express the difference between 5 ft and p as $|5 - p|$, so the result can also be stated as $|5 - p| \le \dfrac{1}{96}$.

We could also write an inequality in terms of inches.

Convert 5 ft to inches:

$$5 \text{ ft} = 5 \text{ ft} \times \frac{12 \text{ in.}}{1 \text{ ft}} = 60 \text{ in.}$$

Then we have:

$$60 - \frac{1}{8} \le p \le 60 + \frac{1}{8}$$

$$-\frac{1}{8} \le p - 60 \le \frac{1}{8} \quad \text{Adding } -60$$

$$|p - 60| \le \frac{1}{8} \qquad \text{Part (b)}$$

We could also express the difference between 60 in. and p as $|60 - p|$, so the result can also be stated as $|60 - p| \le \frac{1}{8}$.

114. $\left\{ d \middle| 5\frac{1}{2} \text{ ft} \le d \le 6\frac{1}{2} \text{ ft} \right\}$, or $\left[5\frac{1}{2} \text{ ft}, 6\frac{1}{2} \text{ ft} \right]$

115. ▨

Exercise Set 9.4

1. We replace x by -4 and y by 2.

$$\begin{array}{c|c} \multicolumn{2}{c}{2x + y < -5} \\ \hline 2(-4) + 2 \; ? \; -5 & \\ -8 + 2 & \\ \hline -6 & -5 \qquad \text{TRUE} \end{array}$$

Since $-6 < -5$ is true, $(-4, 2)$ is a solution.

2. Yes

3. We replace x by 8 and y by 14.

$$\begin{array}{c|c} \multicolumn{2}{c}{2y - 3x > 5} \\ \hline 2 \cdot 14 - 3 \cdot 8 \; ? \; 5 & \\ 28 - 24 & \\ \hline 4 & 5 \qquad \text{FALSE} \end{array}$$

Since $4 > 5$ is false, $(8, 14)$ is not a solution.

4. Yes

5. Graph: $y > 2x$

We first graph the line $y = 2x$. We draw the line dashed since the inequality symbol is $>$. To determine which half-plane to shade, test a point not on the line. We try $(1, 1)$ and substitute:

$$\begin{array}{c|c} \multicolumn{2}{c}{y > 2x} \\ \hline 1 \; ? \; 2 \cdot 1 & \\ \hline 1 & 2 \qquad \text{FALSE} \end{array}$$

Since $1 > 2$ is false, $(1, 1)$ is not a solution, nor are any points in the half-plane containing $(1, 1)$. The points in the other half-plane are solutions, so we shade that half-plane and obtain the graph.

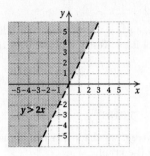

6.

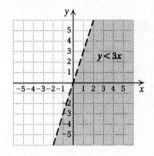

7. Graph: $y < x + 1$

First graph the line $y = x + 1$. Draw it dashed since the inequality symbol is $<$. Test the point $(0, 0)$ to determine if it is a solution.

$$\begin{array}{c|c} \multicolumn{2}{c}{y < x + 1} \\ \hline 0 \; ? \; 0 + 1 & \\ \hline 0 & 1 \qquad \text{TRUE} \end{array}$$

Since $0 < 1$ is true, we shade the half-plane containing $(0, 0)$ and obtain the graph.

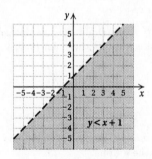

8.

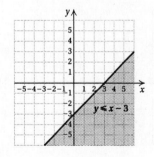

9. Graph: $y > x - 2$

First graph the line $y = x - 2$. Draw a dashed line since the inequality symbol is $>$. Test the point $(0, 0)$ to determine if it is a solution.

$$\frac{y > x - 2}{\begin{array}{c|c} 0 \ ? \ 0 - 2 \\ \hline 0 & -2 \quad \text{TRUE} \end{array}}$$

Since $0 > -2$ is true, we shade the half-plane containing $(0, 0)$ and obtain the graph.

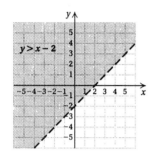

10.

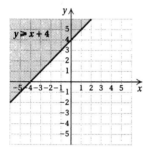

11. Graph: $x + y < 4$

First graph $x + y = 4$. Draw the line dashed since the inequality symbol is $<$. Test the point $(0, 0)$ to determine if it is a solution.

$$\frac{x + y < 4}{\begin{array}{c|c} 0 + 0 \ ? \ 4 \\ \hline 0 & 4 \quad \text{TRUE} \end{array}}$$

Since $0 < 4$ is true, we shade the half-plane containing $(0, 0)$ and obtain the graph.

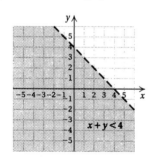

12.

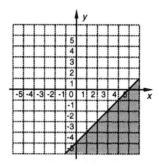

13. Graph: $3x + 4y \leq 12$

We first graph $3x + 4y = 12$. Draw the line solid since the inequality symbol is $\leq$. Test the point $(0, 0)$ to determine if it is a solution.

$$\frac{3x + 4y \leq 12}{\begin{array}{c|c} 3 \cdot 0 + 4 \cdot 0 \ ? \ 12 \\ \hline 0 & 12 \quad \text{TRUE} \end{array}}$$

Since $0 \leq 12$ is true, we shade the half-plane containing $(0, 0)$ and obtain the graph.

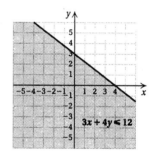

14.

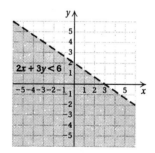

15. Graph: $2y - 3x > 6$

We first graph $2y - 3x = 6$. Draw the line dashed since the inequality symbol is $>$. Test the point $(0, 0)$ to determine if it is a solution.

$$\frac{2y - 3x > 6}{\begin{array}{c|c} 2 \cdot 0 - 3 \cdot 0 \ ? \ 6 \\ \hline 0 & 6 \quad \text{FALSE} \end{array}}$$

Since $0 > 6$ is false, we shade the half-plane that does not contain $(0, 0)$ and obtain the graph.

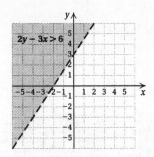

16.

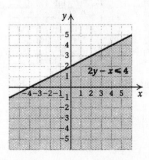

17. Graph: $3x - 2 \leq 5x + y$

$$-2 \leq 2x + y$$

We first graph $-2 = 2x + y$. Draw the line solid since the inequality symbol is $\leq$. Test the point $(0,0)$ to determine if it is a solution.

$$\frac{-2 \leq 2x + y}{-2 ~?~ 2 \cdot 0 + 0}$$
$$-2 ~\Big|~ 0 \qquad \text{TRUE}$$

Since $-2 \leq 0$ is true, we shade the half-plane containing $(0,0)$ and obtain the graph.

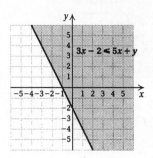

18.

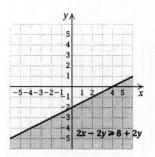

19. Graph: $x < -4$

We first graph $x = -4$. Draw the line dashed since the inequality symbol is $<$. Test the point $(0,0)$ to determine if it is a solution.

$$\frac{x < -4}{0 ~?~ -4} \qquad \text{FALSE}$$

Since $0 < -4$ is false, we shade the half-plane that does not contain $(0,0)$ and obtain the graph.

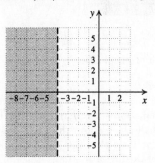

20.

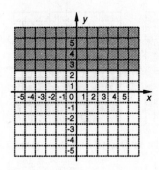

21. Graph: $y > -2$

We first graph $y = -2$. We draw the line dashed since the inequality symbol is $>$. Test the point $(0,0)$ to determine if it is a solution.

$$\frac{y > -2}{0 ~?~ -2} \qquad \text{TRUE}$$

Since $0 > -2$ is true, we shade the half-plane containing $(0,0)$ and obtain the graph.

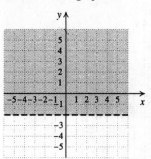

22.

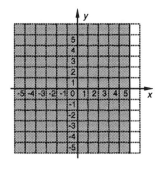

23. Graph: $-4 < y < -1$

This is a system of inequalities:

$$-4 < y,$$
$$y < -1$$

We graph the equation $-4 = y$ and see that the graph of $-4 < y$ is the half-plane above the line $-4 = y$. We also graph $y = -1$ and see that the graph of $y < -1$ is the half-plane below the line $y = -1$.

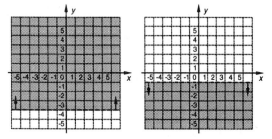

Finally, we shade the intersection of these graphs.

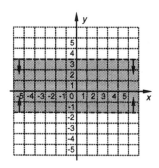

24.

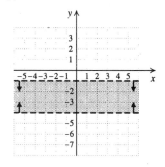

25. Graph: $-3 \le x \le 3$

This is a system of inequalities:

$$-3 \le x,$$
$$x \le 3$$

Graph $-3 \le x$ and $x \le 3$.

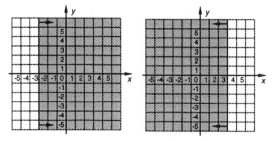

Then we shade the intersection of these graphs.

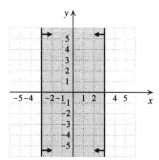

26.

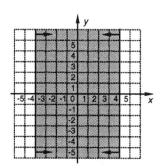

27. Graph: $0 \le x \le 5$

This is a system of inequalities:

$$0 \le x,$$
$$x \le 5$$

Graph $0 \le x$ and $x \le 5$.

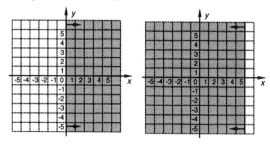

Then we shade the intersection of these graphs.

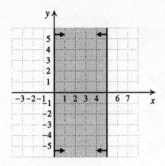

28.

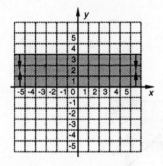

29. Graph: $y < x$,   (1)

$y > -x + 3$ (2)

We graph the lines $y = x$ and $y = -x + 3$, using dashed lines. We indicate the region for each inequality by the arrows at the ends of the lines. Note where the regions overlap and shade the region of solutions.

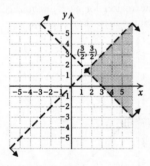

To find the vertex we solve the system of related equations:

$y = x$,

$y = -x + 3$

Solving, we obtain the vertex $\left(\dfrac{3}{2}, \dfrac{3}{2}\right)$.

30.

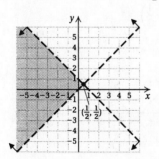

31. Graph: $y \geq x$,   (1)

$y \leq -x + 4$ (2)

We graph the lines $y = x$ and $y = -x+4$, using solid lines. We indicate the region for each inequality by the arrows at the ends of the lines. Note where the regions overlap and shade the region of solutions.

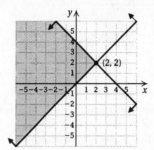

To find the vertex we solve the system of related equations:

$y = x$,

$y = -x + 4$

Solving, we obtain the vertex $(2, 2)$.

32.

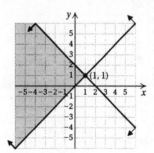

33. Graph: $y \geq -2$,

$x \geq 1$

We graph the lines $y = -2$ and $x = 1$, using solid lines. We indicate the region for each inequality by arrows. Shade the region where they overlap.

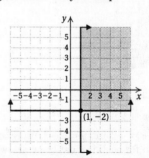

To find the vertex, we solve the system of related equations:

$y = -2$,

$x = 1$

Solving, we obtain the vertex $(1, -2)$.

34.

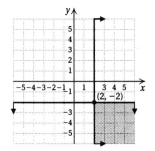

35. Graph: $x < 3$,

$\qquad y > -3x + 2$

We graph the lines $x = 3$ and $y = -3x + 2$, using dashed lines. Indicate the region for each inequality by arrows, and shade the region where they overlap.

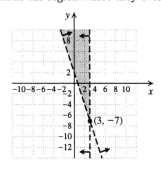

To find the vertex we solve the system of related equations:

$\qquad x = 3$,

$\qquad y = -3x + 2$

Solving, we obtain the vertex $(3, -7)$.

36.

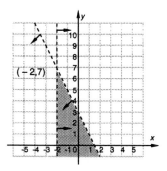

37. Graph: $y \geq -2$,

$\qquad y \geq x + 3$

Graph the lines $y = -2$ and $y = x+3$, using solid lines. Indicate the region for each inequality by arrows, and shade the region where they overlap.

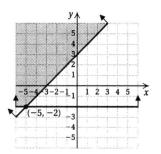

To find the vertex we solve the system of related equations:

$\qquad y = -2$,

$\qquad y = x + 3$

The vertex is $(-5, -2)$.

38.

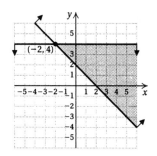

39. Graph: $x + y < 1$,

$\qquad x - y < 2$

Graph the lines $x + y = 1$ and $x - y = 2$, using dashed lines. Indicate the region for each inequality by arrows, and shade the region where they overlap.

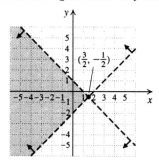

To find the vertex we solve the system of related equations:

$\qquad x + y = 1$,

$\qquad x - y = 2$

The vertex is $\left(\dfrac{3}{2}, -\dfrac{1}{2}\right)$.

40.

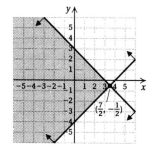

41. Graph: $y - 2x \geq 1$,

$y - 2x \leq 3$

Graph the lines $y - 2x = 1$ and $y - 2x = 3$, using solid lines. Indicate the region for each inequality by arrows, and shade the region where they overlap.

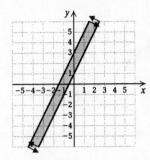

We can see from the graph that the lines are parallel. Hence there are no vertices.

42.

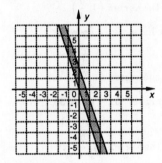

There are no vertices.

43. Graph: $2y - x \leq 2$,

$y - 3x \geq -1$

Graph the lines $2y - x = 2$ and $y - 3x = -1$, using solid lines. Indicate the region for each inequality by arrows, and shade the region where they overlap.

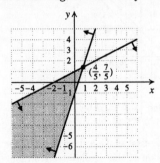

To find the vertex we solve the system of related equations:

$2y - x = 2$,

$y - 3x = -1$

The vertex is $\left(\dfrac{4}{5}, \dfrac{7}{5}\right)$.

44.

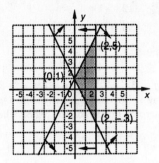

45. Graph: $x - y \leq 2$, (1)

$x + 2y \geq 8$, (2)

$y \leq 4$ (3)

Graph the lines $x - y = 2$, $x + 2y = 8$, and $y = 4$, using solid lines. Indicate the region for each inequality by arrows, and shade the region where they overlap.

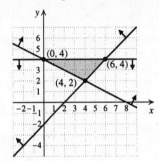

To find the vertices we solve three different systems of equations.

From (1) and (2) we have $x - y = 2$,

$x + 2y = 8$.

Solving, we obtain the vertex $(4, 2)$.

From (1) and (3) we have $x - y = 2$,

$y = 4$.

Solving, we obtain the vertex $(6, 4)$.

From (2) and (3) we have $x + 2y = 8$,

$y = 4$.

Solving, we obtain the vertex $(0, 4)$.

46.

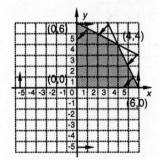

47. Graph: $4y - 3x \geq -12$, (1)

$4y + 3x \geq -36$ (2)

$y \leq 0$, (3)

$x \leq 0$ (4)

Graph the lines $4y - 3x = -12$, $4y + 3x = -36$, $y = 0$, and $x = 0$, using solid lines. Indicate the region for each inequality by arrows, and shade the region where they overlap.

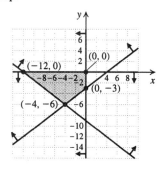

To find the vertices we solve four different systems of equations.

From (1) and (2) we have $4y - 3x = -12$,

$4y + 3x = -36$.

Solving, we obtain the vertex $(-4, -6)$.

From (1) and (4) we have $4y - 3x = -12$,

$x = 0$.

Solving, we obtain the vertex $(0, -3)$.

From (2) and (3) we have $4y + 3x = -36$,

$y = 0$.

Solving, we obtain the vertex $(-12, 0)$.

From (3) and (4) we have $y = 0$,

$x = 0$.

Solving, we obtain the vertex $(0, 0)$.

48.

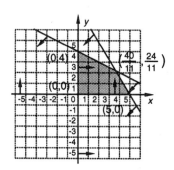

49. Graph: $3x + 4y \geq 12$, (1)

$5x + 6y \leq 30$, (2)

$1 \leq x \leq 3$ (3)

Think of (3) as two inequalities:

$1 \leq x$, (4)

$x \leq 3$ (5)

Graph the lines $3x + 4y = 12$, $5x + 6y = 30$, $x = 1$, and $x = 3$, using solid lines. Indicate the region for each inequality by arrows, and shade the region where they overlap.

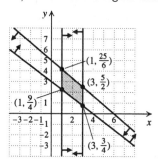

To find the vertices we solve four different systems of equations.

From (1) and (4) we have $3x + 4y = 12$,

$x = 1$.

Solving, we obtain the vertex $\left(1, \dfrac{9}{4}\right)$.

From (1) and (5) we have $3x + 4y = 12$,

$x = 3$.

Solving, we obtain the vertex $\left(3, \dfrac{3}{4}\right)$.

From (2) and (4) we have $5x + 6y = 30$,

$x = 1$.

Solving, we obtain the vertex $\left(1, \dfrac{25}{6}\right)$.

From (2) and (5) we have $5x + 6y = 30$,

$x = 3$.

Solving, we obtain the vertex $\left(3, \dfrac{5}{2}\right)$.

50.

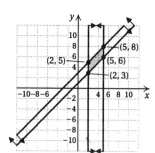

51. Familiarize. We first make a drawing. We let x represent the length of a side of the equilateral triangle. Then $x - 5$ represents the length of a side of the square.

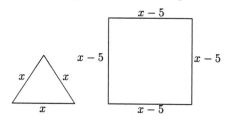

Translate.

Perimeter of triangle = Perimeter of square

$$3x \qquad = \qquad 4(x - 5)$$

Carry out.

$$3x = 4(x - 5)$$

$$3x = 4x - 20$$

$$20 = x$$

Then $x - 5 = 20 - 5 = 15$.

Check. If the length of a side of the triangle is 20 and the length of a side of the square is 15, the perimeter of the triangle is $3 \cdot 20$, or 60 and the perimeter of the square is $4 \cdot 15$, or 60. The values check.

State. The length of a side of the square is 15, and the length of a side of the triangle is 20.

52. $\left(-\dfrac{44}{23}, \dfrac{13}{23}\right)$

53. $5(3x - 4) = -2(x + 5)$

$$15x - 20 = -2x - 10$$

$$17x - 20 = -10$$

$$17x = 10$$

$$x = \dfrac{10}{17}$$

The solution is $\dfrac{10}{17}$.

54. $-\dfrac{14}{13}$

55.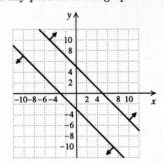

56.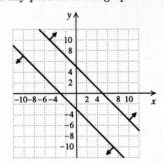

57. Graph: $x + y \geq 5$,

$\qquad\qquad x + y \leq -3$

Graph the lines $x + y = 5$ and $x + y = -3$, using solid lines, and indicate the region for each inequality by arrows. The regions do not overlap (the solution set is $\emptyset$), so we do not shade any portion of the graph.

58.

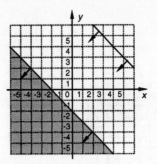

59. Graph: $x - 2y \leq 0$,

$\qquad\qquad -2x + y \leq 2$,

$\qquad\qquad x \leq 2$,

$\qquad\qquad y \leq 2$,

$\qquad\qquad x + y \leq 4$

Graph the five inequalities above, and shade the region where they overlap.

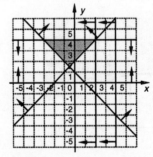

60.

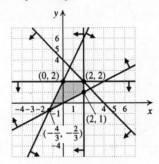

61. We graph the following system of inequalities:

$$0 \leq L \leq 94,$$

$$0 \leq W \leq 50$$

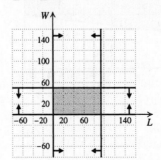

62. $2w + t \geq 60,$

$\quad\quad w \geq 0,$

$\quad\quad t \geq 0$

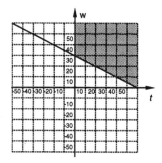

63. Let $c =$ the number of children and $a =$ the number of adults.

Graph: $35c + 75a > 1000,$

$\quad\quad\quad c \geq 0,$

$\quad\quad\quad a \geq 0$

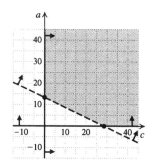

64. a) $3x + 6y > 2$

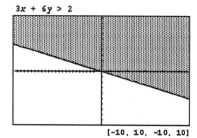

b) $x - 5y \leq 10$

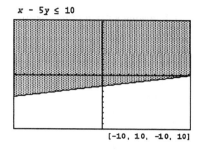

c) $13x - 25y + 10 \leq 0$

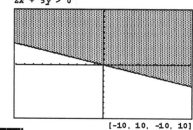

$[-10, 10, -10, 10]$

d) $2x + 5y > 0$

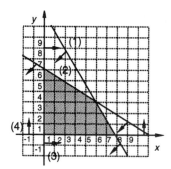

$[-10, 10, -10, 10]$

65.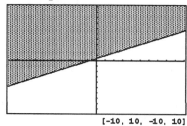

Exercise Set 9.5

---

1. Find the maximum and minimum values of
$F = 4x + 28y,$

subject to

$\quad\quad 5x + 3y \leq 34, \quad (1)$

$\quad\quad 3x + 5y \leq 30, \quad (2)$

$\quad\quad\quad\quad x \geq 0, \quad (3)$

$\quad\quad\quad\quad y \geq 0. \quad (4)$

Graph the system of inequalities and find the coordinates of the vertices.

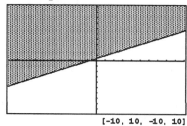

To find one vertex we solve the system

$\quad x = 0,$

$\quad y = 0.$

This vertex is $(0, 0).$

To find a second vertex we solve the system

$\quad\quad 5x + 3y = 34,$

$\quad\quad\quad\quad y = 0.$

This vertex is $\left(\dfrac{34}{5}, 0\right).$

To find a third vertex we solve the system

$$5x + 3y = 34,$$
$$3x + 5y = 30.$$

This vertex is $(5, 3)$.

To find the fourth vertex we solve the system

$$3x + 5y = 30,$$
$$x = 0.$$

This vertex is $(0, 6)$.

Now find the value of F at each of these points.

Vertex (x, y)	$F = 4x + 28y$	
$(0, 0)$	$4 \cdot 0 + 28 \cdot 0 = 0 + 0 = 0$	← Minimum
$\left(\frac{34}{5}, 0\right)$	$4 \cdot \frac{34}{5} + 28 \cdot 0 = \frac{136}{5} + 0 = 27\frac{1}{5}$	
$(5, 3)$	$4 \cdot 5 + 28 \cdot 3 = 20 + 84 = 104$	
$(0, 6)$	$4 \cdot 0 + 28 \cdot 6 = 0 + 168 = 168$	← Maximum

The maximum value of F is 168 when $x = 0$ and $y = 6$.
The minimum value of F is 0 when $x = 0$ and $y = 0$.

2. The maximum is 92.8 when $x = 0$ and $y = 5.8$. The minimum is 0 when $x = 0$ and $y = 0$.

3. Find the maximum and minimum values of
$P = 16x - 2y + 40$,

subject to

$$6x + 8y \le 48, \quad (1)$$
$$0 \le y \le 4, \quad (2)$$
$$0 \le x \le 7. \quad (3)$$

Think of (2) as $0 \le y$, (4)

  $y \le 4$. (5)

Think of (3) as $0 \le x$, (6)

  $x \le 7$. (7)

Graph the system of inequalities.

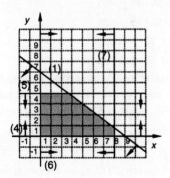

To determine the coordinates of the vertices, we solve the following systems:

$x = 0,$ $x = 7,$ $6x + 8y = 48,$
$y = 0;$ $y = 0;$ $x = 7;$

$6x + 8y = 48,$ $x = 0,$
$y = 4;$ $y = 4$

The vertices are $(0,0)$, $(7,0)$, $\left(7, \frac{3}{4}\right)$, $\left(\frac{8}{3}, 4\right)$, and $(0, 4)$, respectively. Compute the value of P at each of these points.

Vertex (x, y)	$P = 16x - 2y + 40$	
$(0, 0)$	$16 \cdot 0 - 2 \cdot 0 + 40 =$ $0 - 0 + 40 = 40$	
$(7, 0)$	$16 \cdot 7 - 2 \cdot 0 + 40 =$ $112 - 0 + 40 =$ 152	← Maximum
$\left(7, \frac{3}{4}\right)$	$16 \cdot 7 - 2 \cdot \frac{3}{4} + 40 =$ $112 - \frac{3}{2} + 40 = 150\frac{1}{2}$	
$\left(\frac{8}{3}, 4\right)$	$16 \cdot \frac{8}{3} - 2 \cdot 4 + 40 =$ $\frac{128}{3} - 8 + 40 = 74\frac{2}{3}$	
$(0, 4)$	$16 \cdot 0 - 2 \cdot 4 + 40 =$ $0 - 8 + 40 =$ 32	← Minimum

The maximum is 152 when $x = 7$ and $y = 0$. The minimum is 32 when $x = 0$ and $y = 4$.

4. The maximum is 124 when $x = 3$ and $y = 0$. The minimum is 40 when $x = 0$ and $y = 4$.

5. Find the maximum and minimum values of
$F = 5x + 2y + 3$,

subject to

$$y \le 2x + 1, \quad (1)$$
$$x \le 5, \quad (2)$$
$$y \ge 1 \quad (3)$$

Graph the system of inequalities and find the coordinates of the vertices.

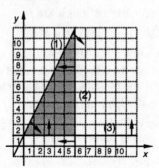

To determine the coordinates of the vertices, we solve the following systems:

$y = 2x + 1,$ $x = 5,$ $y = 2x + 1,$
$y = 1;$ $y = 1;$ $x = 5$

The solutions of the systems are $(0, 1)$, $(5, 1)$, and $(5, 11)$, respectively. Now find the value of F at each of these points.

Vertex (x, y)	$F = 5x + 2y + 3$	
$(0, 1)$	$5 \cdot 0 + 2 \cdot 1 + 3 = 5$	←—Minimum
$(5, 1)$	$5 \cdot 5 + 2 \cdot 1 + 3 = 30$	
$(5, 11)$	$5 \cdot 5 + 2 \cdot 11 + 3 = 50$	←Maximum

The maximum is 50 when $x = 5$ and $y = 11$. The minimum value is 5 when $x = 0$ and $y = 1$.

6. The maximum is 4 when $x = 2$ and $y = 5$. The minimum is -12 when $x = 2$ and $y = -3$.

7. *Familiarize*. Let $x =$ the number of Biscuit Jumbos and $y =$ the number of Mitimite Biscuits to be made per day.

***Translate*.** The income I is given by

$$I = \$0.10x + \$0.08y.$$

We wish to maximize I subject to these facts (constraints) about x and y:

$$x + y \leq 200,$$
$$2x + y \leq 300,$$
$$x \geq 0,$$
$$y \geq 0.$$

***Carry out*.** We graph the system of inequalities, determine the vertices, and evaluate I at each vertex.

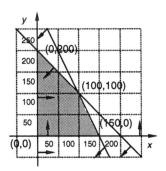

Vertex	$I = \$0.10x + \$0.08y$
$(0, 0)$	$\$0.10(0) + \$0.08(0) = \$0$
$(0, 200)$	$\$0.10(0) + \$0.08(200) = \$16$
$(100, 100)$	$\$0.10(100) + \$0.08(100) = \$18$
$(150, 0)$	$\$0.10(150) + \$0.08(0) = \$15$

The greatest income in the table is $18, obtained when 100 Biscuit Jumbos and 100 Mitimite Biscuits are made.

***Check*.** Go over the algebra and arithmetic.

***State*.** The company will have a maximum income of $18 when 100 of each type of biscuit is made.

8. The maximum number of miles is 480 when the car uses 9 gal and the moped uses 3 gal.

9. *Familiarize*. We organize the information in a table. Let $x =$ the number of matching questions and $y =$ the number of essay questions you answer.

Type	Number of points for each	Number answered	Total points
Matching	10	$3 \leq x \leq 12$	$10x$
Essay	25	$4 \leq y \leq 15$	$25y$
Total		$x + y \leq 20$	$10x + 25y$

Since no more than 20 questions can be answered, we have the inequality $x + y \leq 20$ in the "Number answered" column. The expression $10x + 25y$ in the "Total points" column gives the total score on the test.

***Translate*.** The score S is given by

$$S = 10x + 25y.$$

We wish to maximize S subject to these facts (constraints) about x and y.

$$3 \leq x \leq 12,$$
$$4 \leq y \leq 15,$$
$$x + y \leq 20.$$

***Carry out*.** We graph the system of inequalities, determine the vertices, and evaluate S at each vertex.

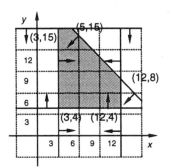

Vertex	$S = 10x + 25y$
$(3, 4)$	$10 \cdot 3 + 25 \cdot 4 = 130$
$(3, 15)$	$10 \cdot 3 + 25 \cdot 15 = 405$
$(5, 15)$	$10 \cdot 5 + 25 \cdot 15 = 425$
$(12, 8)$	$10 \cdot 12 + 25 \cdot 8 = 320$
$(12, 4)$	$10 \cdot 12 + 25 \cdot 4 = 220$

The greatest score in the table is 425, obtained when 5 matching and 15 essay questions are answered correctly.

***Check*.** Go over the algebra and arithmetic.

***State*.** The maximum score is 425 points when 5 matching questions and 15 essay questions are answered correctly.

10. The maximum test score is 102 when 8 short-answer questions and 10 word problems are answered correctly.

11. *Familiarize*. Let $x =$ the number of motorcycles manufactured and $y =$ the number of bicycles manufactured each month.

***Translate*.** The profit P is given by

$$P = \$134x + \$20y.$$

We wish to maximize P subject to these facts (constraints) about x and y.

$10 \leq x \leq 60$ From 10 to 60 motorcycles will be produced.

$0 \leq y \leq 120$ No more than 120 bicycles can be produced.

$x + y \leq 160$ Total production cannot exceed 160.

Carry out. We graph the system of inequalities, determine the vertices, and evaluate P at each vertex.

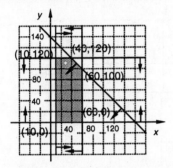

Vertex	$P = \$134x + \$20y$
$(10, 0)$	$\$134 \cdot 10 + \$20 \cdot 0 = \$1340$
$(60, 0)$	$\$134 \cdot 60 + \$20 \cdot 0 = \$8040$
$(60, 100)$	$\$134 \cdot 60 + \$20 \cdot 100 = \$10{,}040$
$(40, 120)$	$\$134 \cdot 40 + \$20 \cdot 120 = \$7760$
$(10, 120)$	$\$134 \cdot 10 + \$20 \cdot 120 = \$3740$

The greatest profit in the table is $10,040, obtained when 60 motorcycles and 100 bicycles are manufactured.

Check. Go over the algebra and arithmetic.

State. The maximum profit occurs when 60 motorcycles and 100 bicycles are manufactured.

12. The maximum profit occurs when 40 hamburgers and 50 hot dogs are sold.

13. Familiarize. Let $x =$ the number of units of lumber and $y =$ the number of units of plywood produced per week.

Translate. The profit P is given by

$P = \$20x + \$30y.$

We wish to maximize P subject to these facts (constraints) about x and y.

$x + y \leq 400,$

$x \geq 100,$

$y \geq 150.$

Carry out. We graph the system of inequalities, determine the vertices, and evaluate P at each vertex.

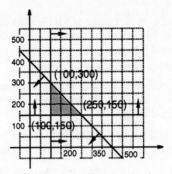

Vertex	$P = \$20x + \$30y$
$(100, 150)$	$\$20 \cdot 100 + \$30 \cdot 150 = \$6500$
$(100, 300)$	$\$20 \cdot 100 + \$30 \cdot 300 = \$11{,}000$
$(250, 150)$	$\$20 \cdot 250 + \$30 \cdot 150 = \$9500$

The greatest profit in the table is $11,000, obtained when 100 units of lumber and 300 units of plywood are produced.

Check. Go over the algebra and arithmetic.

State. The maximum profit is achieved by producing 100 units of lumber and 300 units of plywood.

14. The maximum profit occurs by planting 80 acres of corn and 160 acres of oats.

15. Familiarize. Let $x =$ the amount invested in corporate bonds and $y =$ the amount invested in municipal bonds.

Translate. The income I is given by

$I = 0.08x + 0.075y.$

We wish to maximize I subject to these facts (constraints) about x and y:

$x + y \leq \$40{,}000,$

$\$6000 \leq x \leq \$22{,}000$

$0 \leq y \leq \$30{,}000.$

Carry out. We graph the system of inequalities, determine the vertices, and evaluate I at each vertex.

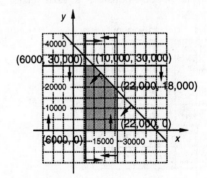

Vertex	$I = 0.08x + 0.075y$
$(\$6000, \$0)$	$\$480$
$(\$6000, \$30{,}000)$	$\$2730$
$(\$10{,}000, \$30{,}000)$	$\$3050$
$(\$22{,}000, \$18{,}000)$	$\$3110$
$(\$22{,}000, \$0)$	$\$1760$

The greatest profit in the table is $3110, obtained when $22,000 is invested in corporate bonds and $18,000 is invested in municipal bonds.

Check. Go over the algebra and arithmetic.

State. The maximum income of $3110 occurs when $22,000 is invested in corporate bonds and $18,000 is invested in municipal bonds.

16. The maximum interest income is $1395 when $7000 is invested in City Bank and $15,000 is invested in State Bank.

17. **Familiarize**. Let x = the number of batches of Smello and y = the number of batches of Roppo to be made. We organize the information in a table.

	Composition		Number of lb
	Smello	Roppo	available
Number of batches	x	y	
English tobacco	12	8	3000
Virginia tobacco	0	8	2000
Latakia tobacco	4	0	500
Profit per batch	$10.56	$6.40	

Translate. The profit P is given by

$$P = \$10.56x + \$6.40y.$$

We wish to maximize P subject to these facts (constraints) about x and y.

$$12x + 8y \leq 3000,$$
$$8y \leq 2000,$$
$$4x \leq 500,$$
$$x \geq 0,$$
$$y \geq 0.$$

Carry out. We graph the system of inequalities, determine the vertices, and evaluate P at each vertex.

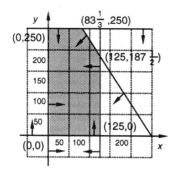

Vertex	$P = \$10.56x + \$6.40y$
$(0,0)$	$0
$(125,0)$	$1320
$(125,187.5)$	$2520
$\left(83\frac{1}{3}, 250\right)$	$2480
$(0,250)$	$1600

The greatest profit in the table is $2520, obtained when 125 batches of Smello and 187.5 batches of Roppo are produced.

Check. Go over the algebra and arithmetic.

State. The maximum profit or $2520 occurs when 125 batches of Smello and 187.5 batches of Roppo are made.

18. The maximum profit per day is $192 when 2 knit suits and 4 worsted suits are made.

19. **Familiarize**. Let x represent the number of P-1 planes and y represent the number of P-2 planes. Organize the information in a table.

Plane	Number of planes	Passengers		
		First	Tourist	Economy
P-1	x	$40x$	$40x$	$120x$
P-2	y	$80y$	$30y$	$40y$

Plane	Cost per mile
P-1	$12,000x$
P-2	$10,000y$

Translate. Suppose C is the total cost per mile. Then $C = 12,000x + 10,000y$. We wish to minimize C subject to these facts (constraints) about x and y.

$$40x + 80y \geq 2000,$$
$$40x + 30y \geq 1500,$$
$$120x + 40y \geq 2400,$$
$$x \geq 0,$$
$$y \geq 0$$

Carry out. Graph the system of inequalities, determine the vertices, and evaluate C at each vertex.

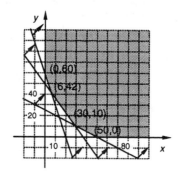

Vertex	$C = 12,000x + 10,000y$
$(0,60)$	$12,000(0) + 10,000(60) = 600,000$
$(6,42)$	$12,000(6) + 10,000(42) = 492,000$
$(30,10)$	$12,000(30) + 10,000(10) = 460,000$
$(50,0)$	$12,000(50) + 10,000(0) = 600,000$

The smallest value of C in the table is 460,000, obtained when $x = 30$ and $y = 10$.

Check. Go over the algebra and arithmetic.

State. In order to minimize the operating cost, 30 P-1 planes and 10 P-2 planes should be used.

20. The airline should use 30 P-2's and 15 P-3's.

21. *Familiarize*. Let x = the number of chairs and y = the number of sofas produced.

Translate. Find the maximum value of

$$I = \$20x + \$300y$$

subject to

$$20x + 100y \leq 1900,$$
$$x + 50y \leq 500,$$
$$2x + 20y \leq 240,$$
$$x \geq 0,$$
$$y \geq 0.$$

Carry out. Graph the system of inequalities, determine the vertices, and evaluate I at each vertex.

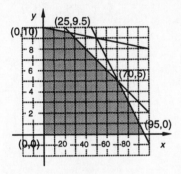

Vertex	$I = \$20x + \$300y$
$(0, 0)$	$\$0$
$(0, 10)$	$\$3000$
$(25, 9.5)$	$\$3350$
$(70, 5)$	$\$2900$
$(95, 0)$	$\$1900$

The greatest income in the table is $3350, obtained when 25 chairs and 9.5 sofas are made.

Check. Go over the algebra and arithmetic.

State. The maximum income of $3350 occurs when 25 chairs and 9.5 sofas are made. A more practical answer is that the maximum income of $3200 is achieved when 25 chairs and 9 sofas are made.

Chapter 10

Rational Exponents and Radicals

Exercise Set 10.1

1. The square roots of 16 are 4 and -4, because $4^2 = 16$ and $(-4)^2 = 16$.

2. 15, -15

3. The square roots of 144 are 12 and -12, because $12^2 = 144$ and $(-12)^2 = 144$.

4. 3, -3

5. The square roots of 400 are 20 and -20, because $20^2 = 400$ and $(-20)^2 = 400$.

6. 9, -9

7. The square roots of 49 are 7 and -7, because $7^2 = 49$ and $(-7)^2 = 49$.

8. 30, -30

9. $-\sqrt{\dfrac{49}{36}} = -\dfrac{7}{6}$ Since $\sqrt{\dfrac{49}{36}} = \dfrac{7}{6}$, $-\sqrt{\dfrac{49}{36}} = -\dfrac{7}{6}$.

10. $-\dfrac{19}{3}$

11. $\sqrt{196} = 14$ Remember, $\sqrt{}$ indicates the principle square root.

12. 21

13. $-\sqrt{\dfrac{16}{81}} = -\dfrac{4}{9}$ Since $\sqrt{\dfrac{16}{81}} = \dfrac{4}{9}$, $-\sqrt{\dfrac{16}{81}} = -\dfrac{4}{9}$.

14. $-\dfrac{9}{12}$, or $-\dfrac{3}{4}$

15. $\sqrt{0.09} = 0.3$

16. 0.6

17. $-\sqrt{0.0049} = -0.07$

18. 0.12

19. $5\sqrt{p^2 + 4}$

The radicand is the expression written under the radical sign, $p^2 + 4$.

20. $y^2 - 8$

21. $x^2 y^2 \sqrt{\dfrac{x}{y+4}}$

The radicand is the expression written under the radical sign, $\dfrac{x}{y+4}$.

22. $\dfrac{a}{a^2 - b}$

23. $f(y) = \sqrt{5y - 10}$

$f(6) = \sqrt{5 \cdot 6 - 10} = \sqrt{20}$

$f(2) = \sqrt{5 \cdot 2 - 10} = \sqrt{0} = 0$

$f(0) = \sqrt{5 \cdot 0 - 10} = \sqrt{-10}$

Since negative numbers do not have real-number square roots, 0 is not in the domain of f.

$f(-3) = \sqrt{5(-3) - 10} = \sqrt{-25}$

Since negative numbers do not have real-number square roots, -3 is not in the domain of f.

24. 0, 0, 0 is not in the domain of g, $\sqrt{11}$, $\sqrt{11}$

25. $t(x) = -\sqrt{2x + 1}$

$t(4) = -\sqrt{2 \cdot 4 + 1} = -\sqrt{9} = -3$

$t(-4) = -\sqrt{2(-4) + 1} = -\sqrt{-7};$

-4 is not in the domain of t

$t(0) = -\sqrt{2 \cdot 0 + 1} = -\sqrt{1} = -1$

$t(12) = -\sqrt{2 \cdot 12 + 1} = -\sqrt{25} = -5$

26. 0 is not in the domain of p, $\sqrt{30}$, $\sqrt{30}$, $\sqrt{180}$, $\sqrt{180}$

27. $f(t) = \sqrt{t^2 + 1}$

$f(5) = \sqrt{5^2 + 1} = \sqrt{26}$

$f(-5) = \sqrt{(-5)^2 + 1} = \sqrt{26}$

$f(0) = \sqrt{0^2 + 1} = \sqrt{1} = 1$

$f(10) = \sqrt{10^2 + 1} = \sqrt{101}$

$f(-10) = \sqrt{(-10)^2 + 1} = \sqrt{101}$

28. -2, 0, -4, -2, -6, -4

29. $g(x) = \sqrt{x^3 + 9}$

$g(2) = \sqrt{2^3 + 9} = \sqrt{17}$

$g(-2) = \sqrt{(-2)^3 + 9} = \sqrt{1} = 1$

$g(3) = \sqrt{3^3 + 9} = \sqrt{36} = 6$

$g(-3) = \sqrt{(-3)^3 + 9} = \sqrt{-18};$

-3 is not in the domain of g

30. 2 is not in the domain of f, -2 is not in the domain of f, $\sqrt{17}$, -3 is not in the domain of f

31. $\sqrt{16x^2} = \sqrt{(4x)^2} = |4x| = 4|x|$

Since x might be negative, absolute-value notation is necessary.

32. $5|t|$

33. $\sqrt{(-7c)^2} = |-7c| = |-7| \cdot |c| = 7|c|$

Since c might be negative, absolute-value notation is necessary.

34. $6|b|$

35. $\sqrt{(a+1)^2} = |a+1|$

Since $a + 1$ might be negative, absolute-value notation is necessary.

36. $|5 - b|$

37. $\sqrt{x^2 - 4x + 4} = \sqrt{(x-2)^2} = |x - 2|$

Since $x - 2$ might be negative, absolute-value notation is necessary.

38. $|y + 8|$

39. $\sqrt{4x^2 + 28x + 49} = \sqrt{(2x+7)^2} = |2x + 7|$

Since $2x + 7$ might be negative, absolute-value notation is necessary.

40. $|3x - 5|$

41. $\sqrt[4]{625} = 5$ Since $5^4 = 625$

42. -4

43. $\sqrt[5]{-1} = -1$ Since $(-1)^5 = -1$

44. 2

45. $\sqrt[5]{-\dfrac{32}{243}} = -\dfrac{2}{3}$ Since $\left(-\dfrac{2}{3}\right)^5 = -\dfrac{32}{243}$

46. $-\dfrac{1}{2}$

47. $\sqrt[6]{x^6} = |x|$

The index is even. Use absolute-value notation since x could have a negative value.

48. $|y|$

49. $\sqrt[4]{(5a)^4} = |5a| = 5|a|$

The index is even. Use absolute-value notation since a could have a negative value.

50. $7|b|$

51. $\sqrt[10]{(-6)^{10}} = |-6| = 6$

52. 10

53. $\sqrt[414]{(a+b)^{414}} = |a + b|$

The index is even. Use absolute-value notation since $a + b$ could have a negative value.

54. $|2a + b|$

55. $\sqrt[7]{y^7} = y$

We do not use absolute-value notation when the index is odd.

56. -6

57. $\sqrt[5]{(x-2)^5} = x - 2$

We do not use absolute-value notation when the index is odd.

58. $2xy$

59. $\sqrt{16x^2} = \sqrt{(4x)^2} = 4x$ Assuming x is nonnegative

60. $5t$

61. $\sqrt{(-6b)^2} = \sqrt{36b^2} = 6b$ Assuming b is nonnegative

62. $7c$

63. $\sqrt{(a+1)^2} = a + 1$ Assuming $a + 1$ is nonnegative

64. $5 + b$

65. $\sqrt{4x^2 + 8x + 4} = \sqrt{4(x^2 + 2x + 1)} =$
$\sqrt{[2(x+1)]^2} = 2(x+1)$, or $\boxed{2x + 2}$

66. $3(x + 2)$, or $3x + 6$

67. $\sqrt{9t^2 - 12t + 4} = \sqrt{(3t-2)^2} = 3t - 2$

68. $5t - 2$

69. $\sqrt[3]{27} = 3$ $[3^3 = 27]$

70. -4

71. $\sqrt[4]{16x^4} = \sqrt[4]{(2x)^4} = 2x$

72. $3x$

73. $\sqrt[3]{-216} = -6$ $[(-6)^3 = -216]$

74. 10

75. $-\sqrt[3]{-125y^3} = -(-5y)$ $[(-5y)^3 = -125y^3]$
$\qquad\qquad = 5y$

76. $4x$

77. $\sqrt[5]{0.00032(x+1)^5} = 0.2(x + 1)$
$\qquad\qquad [0.00032(x+1)^5 = (0.2(x+1))^5]$

78. $0.02(y - 2)$

79. $\sqrt[6]{64x^{12}} = 2x^2$ $[64x^{12} = (2x^2)^6]$

80. $3y^3$

81. $\quad f(x) = \sqrt[3]{x + 1}$
$\quad f(7) = \sqrt[3]{7 + 1} = \sqrt[3]{8} = 2$
$\quad f(26) = \sqrt[3]{26 + 1} = \sqrt[3]{27} = 3$
$\quad f(-9) = \sqrt[3]{-9 + 1} = \sqrt[3]{-8} = -2$
$\quad f(-65) = \sqrt[3]{-65 + 1} = \sqrt[3]{-64} = -4$

82. $1, 5, 3, -5$

83. $g(t) = \sqrt[4]{t-3}$

$g(19) = \sqrt[4]{19-3} = \sqrt[4]{16} = 2$

$g(-13) = \sqrt[4]{-13-3} = \sqrt[4]{-16};$

$\qquad -13$ is not in the domain of g

$g(1) = \sqrt[4]{1-3} = \sqrt[4]{-2};$

$\qquad 1$ is not in the domain of g

$g(84) = \sqrt[4]{84-3} = \sqrt[4]{81} = 3$

84. $1, 2, -82$ is not in the domain of f, 3

85. $f(x) = \sqrt{x+7}$

We need to find all x-values such that $x+7$ is nonnegative. We solve the inequality:

$x + 7 \geq 0$

$\qquad x \geq -7 \quad$ Adding -7 on both sides

Domain of $f = \{x | x \geq -7\}$

86. $\{x | x \geq 10\}$

87. $g(t) = \sqrt[4]{2t-9}$

Since the index is even, the radicand must be nonnegative. We solve the inequality:

$2t - 9 \geq 0$

$\qquad 2t \geq 9 \quad$ Adding 9 on both sides

$\qquad t \geq \dfrac{9}{2} \quad$ Multiply by $\dfrac{1}{2}$ on both sides

Domain of $g = \left\{t | t \geq \dfrac{9}{2}\right\}$

88. $\{x | x \geq -1\}$

89. $g(x) = \sqrt[4]{5-x}$

Since the index is even, the radicand must be nonnegative. We solve the inequality:

$5 - x \geq 0$

$\qquad 5 \geq x$

Domain of $g = \{x | x \leq 5\}$

90. $\{t | t$ is a real number$\}$

91. $f(t) = \sqrt[5]{3t+7}$

Since the index is odd, the radicand can be any real number.

Domain of $f = \{t | t$ is a real number$\}$

92. $\left\{t | t \geq -\dfrac{5}{2}\right\}$

93. $h(z) = -\sqrt[6]{5z+3}$

Since the index is even, the radicand must be nonnegative. We solve the inequality:

$5z + 3 \geq 0$

$\qquad 5z \geq -3$

$\qquad z \geq -\dfrac{3}{5}$

Domain of $h = \left\{z | z \geq -\dfrac{3}{5}\right\}$

94. $\left\{x | x \geq \dfrac{5}{7}\right\}$

95. $f(t) = 7 + 2\sqrt[8]{3t-5}$

Since the index is even, the radicand must be nonnegative. We solve the inequality:

$3t - 5 \geq 0$

$\qquad 3t \geq 5$

$\qquad t \geq \dfrac{5}{3}$

Domain of $f = \left\{t | t \geq \dfrac{5}{3}\right\}$

96. $\left\{t | t \geq \dfrac{4}{5}\right\}$

97. $(a^3 b^2 c^5)^3 = a^{3 \cdot 3} b^{2 \cdot 3} c^{5 \cdot 3} = a^9 b^6 c^{15}$

98. $10 a^{10} b^9$

99. $(x-3)(x+3) = x^2 - 3^2 = x^2 - 9$

100. $a^2 - b^2 x^2$

101.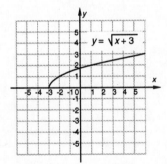

102.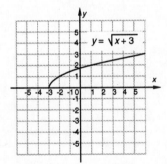

103. $N = 2.5\sqrt{A}$

a) $N = 2.5\sqrt{25} = 2.5(5) = 12.5 \approx 13$

b) $N = 2.5\sqrt{36} = 2.5(6) = 15$

c) $N = 2.5\sqrt{49} = 2.5(7) = 17.5 \approx 18$

d) $N = 2.5\sqrt{64} = 2.5(8) = 20$

104.

105. $y = \sqrt{x} + 3$

Make a table of values. Note that x must be nonnegative. Plot these points and draw the graph.

x	y
0	3
1	4
2	4.4
3	4.7
4	5
5	5.2

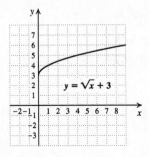

106.

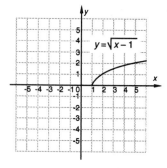

107. $y = \sqrt{x} - 2$

Make a table of values. Note that x must be nonnegative. Plot these points and draw the graph.

x	y
0	-2
1	-1
3	-0.3
4	0
5	0.2

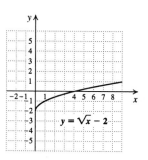

Exercise Set 10.2

1. $x^{1/4} = \sqrt[4]{x}$

2. $\sqrt[5]{y}$

3. $(8)^{1/3} = \sqrt[3]{8} = 2$

4. 4

5. $81^{1/4} = \sqrt[4]{81} = 3$

6. 4

7. $9^{1/2} = \sqrt{9} = 3$

8. 5

9. $(xyz)^{1/3} = \sqrt[3]{xyz}$

10. $\sqrt[4]{ab}$

11. $(a^2b^2)^{1/5} = \sqrt[5]{a^2b^2}$

12. $\sqrt[4]{x^3y^3}$

13. $a^{2/3} = \sqrt[3]{a^2}$

14. $\sqrt{b^3}$, or $b\sqrt{b}$

15. $16^{3/4} = \sqrt[4]{16^3} = (\sqrt[4]{16})^3 = 2^3 = 8$

16. 128

17. $49^{3/2} = \sqrt{49^3} = (\sqrt{49})^3 = 7^3 = 343$

18. 81

19. $9^{5/2} = \sqrt{9^5} = (\sqrt{9})^5 = 3^5 = 243$

20. 729

21. $(81x)^{3/4} = \sqrt[4]{(81x)^3} = \sqrt[4]{81^3x^3}$, or $\sqrt[4]{81^3} \cdot \sqrt[4]{x^3} = (\sqrt[4]{81})^3 \cdot \sqrt[4]{x^3} = 3^3\sqrt[4]{x^3} = 27\sqrt[4]{x^3}$

22. $25\sqrt[3]{a^2}$

23. $(25x^4)^{3/2} = \sqrt{(25x^4)^3} = \sqrt{25^3 \cdot x^{12}} = \sqrt{25^3} \cdot \sqrt{x^{12}} = (\sqrt{25})^3x^6 = 5^3x^6 = 125x^6$

24. $27y^9$

25. $(8a^2b^4)^{2/3} = \sqrt[3]{(8a^2b^4)^2} = \sqrt[3]{64a^4b^8}$, or $\sqrt[3]{64a^3b^6 \cdot ab^2} = 4ab^2\sqrt[3]{ab^2}$

26. $9a^3\sqrt[3]{ab^2}$

27. $\sqrt[3]{20} = 20^{1/3}$

28. $19^{1/3}$

29. $\sqrt{17} = 17^{1/2}$

30. $6^{1/2}$

31. $\sqrt{x^3} = x^{3/2}$

32. $a^{5/2}$

33. $\sqrt[5]{m^2} = m^{2/5}$

34. $n^{4/5}$

35. $\sqrt[4]{cd} = (cd)^{1/4}$ Parentheses are required.

36. $(xy)^{1/5}$

37. $\sqrt[5]{xy^2z} = (xy^2z)^{1/5}$

38. $(x^3y^2z^2)^{1/7}$

39. $(\sqrt{3mn})^3 = (3mn)^{3/2}$

40. $(7xy)^{4/3}$

41. $(\sqrt[7]{8x^2y})^5 = (8x^2y)^{5/7}$

42. $(2a^5b)^{7/6}$

43. $x^{-1/3} = \dfrac{1}{x^{1/3}}$

44. $\dfrac{1}{y^{1/4}}$

45. $(2rs)^{-3/4} = \dfrac{1}{(2rs)^{3/4}}$

46. $\dfrac{1}{(5xy)^{5/6}}$

47. $\left(\dfrac{1}{10}\right)^{-2/3} = \dfrac{1}{\left(\dfrac{1}{10}\right)^{2/3}} = 1 \cdot \left(\dfrac{10}{1}\right)^{2/3} = 10^{2/3}$

48. $8^{3/4}$

49. $\dfrac{1}{x^{-2/3}} = x^{2/3}$

50. $x^{5/6}$

51. $5x^{-1/4} = 5 \cdot x^{-1/4} = 5 \cdot \dfrac{1}{x^{1/4}} = \dfrac{5}{x^{1/4}}$

52. $\dfrac{8}{a^{1/3}}$

53. $\dfrac{3}{5m^{-1/2}} = \dfrac{3}{5} \cdot \dfrac{1}{m^{-1/2}} = \dfrac{3}{5} \cdot m^{1/2} = \dfrac{3m^{1/2}}{5}$

54. $\dfrac{2x^{1/3}}{7}$

55. $5^{3/4} \cdot 5^{1/8} = 5^{3/4+1/8} = 5^{6/8+1/8} = 5^{7/8}$

We added exponents after finding a common denominator.

56. $11^{7/6}$

57. $\dfrac{7^{5/8}}{7^{3/8}} = 7^{5/8-3/8} = 7^{2/8} = 7^{1/4}$

We subtracted exponents and simplified using arithmetic.

58. $9^{2/11}$

59. $\dfrac{8.3^{3/4}}{8.3^{2/5}} = 8.3^{3/4-2/5} = 8.3^{15/20-8/20} = 8.3^{7/20}$

We subtracted exponents after finding a common denominator.

60. $3.9^{7/20}$

61. $(10^{3/5})^{2/5} = 10^{3/5 \cdot 2/5} = 10^{6/25}$

We multiplied exponents.

62. $5^{15/28}$

63. $a^{2/3} \cdot a^{5/4} = a^{2/3+5/4} = a^{8/12+15/12} = a^{23/12}$

We added exponents after finding a common denominator.

64. $x^{17/12}$

65. $(x^{2/3})^{3/7} = x^{2/3 \cdot 3/7} = x^{6/21} = x^{2/7}$

We multiplied exponents and simplified using arithmetic.

66. $a^{3/5}$

67. $(m^{2/3}n^{1/2})^{1/4} = m^{2/3 \cdot 1/4} \cdot n^{1/2 \cdot 1/4} = m^{2/12}n^{1/8} = m^{1/6}n^{1/8}$

We raised each factor to the power $1/4$, multiplied exponents, and simplified using arithmetic.

68. $x^{1/12}y^{1/10}$

69. $(a^{-2/3}b^{-1/4})^{-6} = a^{-2/3 \cdot (-6)} \cdot b^{-1/4 \cdot (-6)} = a^{12/3}b^{6/4} = a^4 b^{3/2}$

We raised each factor to the power -6, multiplied exponents, and simplified using arithmetic.

70. $m^2 n^{25/3}$

71. $\sqrt[6]{a^4} = a^{4/6}$ Converting to an exponential expression

$\quad = a^{2/3}$ Using arithmetic to simplify the exponent

$\quad = \sqrt[3]{a^2}$ Converting back to radical notation

72. $\sqrt[3]{y}$

73. $\sqrt[3]{8y^6} = \sqrt[3]{2^3 y^6}$ Recognizing that 8 is 2^3

$\quad = (2^3 y^6)^{1/3}$ Converting to an exponential expression

$\quad = 2^{3/3} y^{6/3}$ Using the product and power rules

$\quad = 2y^2$ Using arithmetic to simpify the exponents

74. $x^2 y^3$

75. $\sqrt[4]{32} = \sqrt[4]{2^5} = 2^{5/4} = 2^{4/4+1/4} = 2^{4/4} \cdot 2^{1/4} = 2\sqrt[4]{2}$

76. $\sqrt{3}$

77. $\sqrt[6]{4x^2} = \sqrt[6]{2^2 x^2} = \sqrt[6]{(2x)^2} = [(2x)^2]^{1/6} = (2x)^{2/6} = (2x)^{1/3} = \sqrt[3]{2x}$

78. $2x\sqrt{y}$

79. $\sqrt[5]{32c^{10}d^{15}} = \sqrt[5]{2^5 c^{10} d^{15}} = (2^5 c^{10} d^{15})^{1/5} = 2^{5/5} c^{10/5} d^{15/5} = 2c^2 d^3$

80. $2x^3 y^4$

81. $\sqrt[6]{\dfrac{m^{12}n^{24}}{64}} = \sqrt[6]{\dfrac{m^{12}n^{24}}{2^6}} = \left(\dfrac{m^{12}n^{24}}{2^6}\right)^{1/6} = \dfrac{m^{12/6}n^{24/6}}{2^{6/6}} = \dfrac{m^2 n^4}{2}$

82. $\dfrac{x^3 y^4}{2}$

83. $\sqrt[8]{r^4 s^2} = (r^4 s^2)^{1/8} = r^{4/8} s^{2/8} = r^{2/4} s^{1/4} = (r^2 s)^{1/4} = \sqrt[4]{r^2 s}$

84. $\sqrt{2ts}$

85. $\sqrt[3]{27a^3 b^9} = \sqrt[3]{3^3 a^3 b^9} = (3^3 a^3 b^9)^{1/3} = 3^{3/3} a^{3/3} b^{9/3} = 3ab^3$

86. $3x^2 y^2$

87. $\qquad x^2 - 1 = 8$

$\qquad x^2 - 9 = 0$

$\quad (x+3)(x-3) = 0$

$\quad x + 3 = 0 \quad \text{or} \quad x - 3 = 0$

$\qquad x = -3 \quad \text{or} \qquad x = 3$

Both values check. The solutions are -3 and 3.

88. $-\dfrac{11}{2}$

89. $\dfrac{1}{x} + 2 = 5$

$\qquad \dfrac{1}{x} = 3$

$\quad x \cdot \dfrac{1}{x} = x \cdot 3$

$\qquad 1 = 3x$

$\qquad \dfrac{1}{3} = x$

This value checks. The solution is $\dfrac{1}{3}$.

90. $93,500.

91.

92.

93. $\sqrt[5]{x^2y\sqrt{xy}} = \sqrt[5]{x^2y(xy)^{1/2}} = \sqrt[5]{x^2yx^{1/2}y^{1/2}} = \sqrt[5]{x^{5/2}y^{3/2}} = (x^{5/2}y^{3/2})^{1/5} = x^{5/10}y^{3/10} = (x^5y^3)^{1/10} = \sqrt[10]{x^5y^3}$

94. $x^3\sqrt[6]{x}$

95. $\sqrt[4]{\sqrt[3]{8x^3y^6}} = \sqrt[4]{(2^3x^3y^6)^{1/3}} = \sqrt[4]{2^{3/3}x^{3/3}y^{6/3}} = \sqrt[4]{2xy^2}$

96. $\sqrt[6]{p+q}$

97. a) $L = \dfrac{(0.000169)60^{2.27}}{1} \approx 1.8$ m

b) $L = \dfrac{(0.000169)75^{2.27}}{0.9906} \approx 3.1$ m

c) $L = \dfrac{(0.000169)80^{2.27}}{2.4} \approx 1.5$ m

d) $L = \dfrac{(0.000169)100^{2.27}}{1.1} \approx 5.3$ m

98.

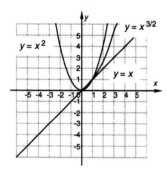

99. $r(1900) = 10^{-12}2^{-1900/5700} \approx 10^{-12}(0.7937)$, or 7.937×10^{-13}. The ratio is about 7.937×10^{-13} to 1.

Exercise Set 10.3

1. $\sqrt{3}\sqrt{2} = \sqrt{3 \cdot 2} = \sqrt{6}$

2. $\sqrt{35}$

3. $\sqrt[3]{2}\sqrt[3]{5} = \sqrt[3]{2 \cdot 5} = \sqrt[3]{10}$

4. $\sqrt[3]{14}$

5. $\sqrt[4]{8}\sqrt[4]{9} = \sqrt[4]{8 \cdot 9} = \sqrt[4]{72}$

6. $\sqrt[4]{18}$

7. $\sqrt{3a}\sqrt{10b} = \sqrt{3a \cdot 10b} = \sqrt{30ab}$

8. $\sqrt{26xy}$

9. $\sqrt[5]{9t^2}\sqrt[5]{2t} = \sqrt[5]{9t^2 \cdot 2t} = \sqrt[5]{18t^3}$

10. $\sqrt[5]{80y^4}$

11. $\sqrt{x-a}\sqrt{x+a} = \sqrt{(x-a)(x+a)} = \sqrt{x^2-a^2}$

12. $\sqrt{y^2-b^2}$

13. $\sqrt[3]{0.3x}\sqrt[3]{0.2x} = \sqrt[3]{0.3x \cdot 0.2x} = \sqrt[3]{0.06x^2}$

14. $\sqrt[3]{0.21y^2}$

15. $\sqrt[4]{x-1}\sqrt[4]{x^2+x+1} = \sqrt[4]{(x-1)(x^2+x+1)} = \sqrt[4]{x^3-1}$

16. $\sqrt[5]{(x-2)^3}$

17. $\sqrt{\dfrac{6}{x}}\sqrt{\dfrac{y}{5}} = \sqrt{\dfrac{6}{x} \cdot \dfrac{y}{5}} = \sqrt{\dfrac{6y}{5x}}$

18. $\sqrt{\dfrac{7s}{11t}}$

19. $\sqrt[7]{\dfrac{x-3}{4}}\sqrt[7]{\dfrac{5}{x+2}} = \sqrt[7]{\dfrac{x-3}{4} \cdot \dfrac{5}{x+2}} = \sqrt[7]{\dfrac{5x-15}{4x+8}}$

20. $\sqrt[6]{\dfrac{3a}{b^2-4}}$

21. $\quad \sqrt[3]{7} \cdot \sqrt{2}$

$= 7^{1/3} \cdot 2^{1/2}$ Converting to exponential notation

$= 7^{2/6} \cdot 2^{3/6}$ Rewriting so that exponents have a common denominator

$= (7^2 \cdot 2^3)^{1/6}$ Using the laws of exponents

$= \sqrt[6]{7^2 \cdot 2^3}$ Converting back to radical notation

$= \sqrt[6]{392}$ Multiplying under the radical

22. $\sqrt[12]{300,125}$

23. $\sqrt{x}\sqrt[3]{2x} = x^{1/2} \cdot (2x)^{1/3} = x^{1/2} \cdot 2^{1/3} \cdot x^{1/3} = 2^{1/3}x^{1/2+1/3} = 2^{1/3}x^{3/6+2/6} = 2^{1/3} \cdot x^{5/6} = 2^{2/6}x^{5/6} = (2^2 \cdot x^5)^{1/6} = \sqrt[6]{2^2 \cdot x^5} = \sqrt[6]{4x^5}$

24. $\sqrt[15]{27y^8}$

25. $\sqrt{x}\sqrt[3]{x-2} = x^{1/2} \cdot (x-2)^{1/3} = x^{3/6} \cdot (x-2)^{2/6} = [x^3(x-2)^2]^{1/6} = \sqrt[6]{x^3(x-2)^2} = \sqrt[6]{x^3(x^2-4x+4)} = \sqrt[6]{x^5-4x^4+4x^3}$

26. $\sqrt[4]{3xy^2+24xy+48x}$

27. $\sqrt[5]{yx^2}\sqrt{xy} = (yx^2)^{1/5}(xy)^{1/2} = y^{1/5}x^{2/5}x^{1/2}y^{1/2} = x^{2/5+1/2}y^{1/5+1/2} = x^{4/10+5/10}y^{2/10+5/10} = x^{9/10}y^{7/10} = (x^9y^7)^{1/10} = \sqrt[10]{x^9y^7}$

28. $\sqrt[10]{2^2a^9b^9}$, or $\sqrt[10]{4a^9b^9}$

29. $\sqrt[4]{x(y+1)^2}\sqrt[3]{x^2(y+1)} = [x(y+1)^2]^{1/4}[x^2(y+1)]^{1/3} = x^{1/4}(y+1)^{2/4}x^{2/3}(y+1)^{1/3} = x^{11/12}(y+1)^{10/12} = [x^{11}(y+1)^{10}]^{1/12} = \sqrt[12]{x^{11}(y+1)^{10}}$

30. $2\sqrt[10]{2^2a^9b^7}$, or $2\sqrt[10]{4a^9b^7}$

31. $\sqrt[5]{a^2bc^3}\sqrt[3]{ab^2c} = (a^2bc^3)^{1/5}(ab^2c)^{1/3} =$
$a^{2/5}b^{1/5}c^{3/5}a^{1/3}b^{2/3}c^{1/3} = a^{11/15}b^{13/15}c^{14/15} =$
$(a^{11}b^{13}c^{14})^{1/15} = \sqrt[15]{a^{11}b^{13}c^{14}}$

32. $\sqrt[12]{x^{11}(y-1)^{10}}$

33. $\qquad\sqrt{27}$
$= \sqrt{9 \cdot 3}$ 9 is the largest perfect square factor of 27
$= \sqrt{9} \cdot \sqrt{3}$
$= 3\sqrt{3}$

34. $2\sqrt{7}$

35. $\qquad\sqrt{45}$
$= \sqrt{9 \cdot 5}$ 9 is the largest perfect square factor of 45
$= \sqrt{9} \cdot \sqrt{5}$
$= 3\sqrt{5}$

36. $2\sqrt{3}$

37. $\sqrt{8} = \sqrt{4 \cdot 2} = \sqrt{4} \cdot \sqrt{2} = 2\sqrt{2}$

38. $3\sqrt{2}$

39. $\sqrt{24} = \sqrt{4 \cdot 6} = \sqrt{4} \cdot \sqrt{6} = 2\sqrt{6}$

40. $2\sqrt{5}$

41. $\qquad\sqrt{180x^4}$
$= \sqrt{36 \cdot 5 \cdot x^4}$ Factoring the radicand
$= \sqrt{36 \cdot x^4 \cdot 5}$
$= \sqrt{36}\sqrt{x^4}\sqrt{5}$ Factoring into several radicals
$= 6x^2\sqrt{5}$

42. $5y^3\sqrt{7}$

43. $\qquad\sqrt[3]{800}$
$= \sqrt[3]{8 \cdot 100}$ 8 is the largest perfect cube factor of 800
$= \sqrt[3]{8} \cdot \sqrt[3]{100}$
$= 2\sqrt[3]{100}$

44. $3\sqrt[3]{10}$

45. $\qquad\sqrt[3]{-16x^6}$
$= \sqrt[3]{-8 \cdot 2 \cdot x^6}$ Factoring the radicand
$= \sqrt[3]{-8 \cdot x^6 \cdot 2}$
$= \sqrt[3]{-8}\sqrt[3]{x^6}\sqrt[3]{2}$ Factoring into several radicals
$= -2x^2\sqrt[3]{2}$ Taking cube roots

46. $-2a^2\sqrt[3]{4}$

47. $\sqrt[3]{54x^8} = \sqrt[3]{27 \cdot 2 \cdot x^6 \cdot x^2} = \sqrt[3]{27 \cdot x^6 \cdot 2 \cdot x^2} =$
$\sqrt[3]{27}\sqrt[3]{x^6}\sqrt[3]{2x^2} = 3x^2\sqrt[3]{2x^2}$

48. $2y\sqrt[3]{5}$

49. $\sqrt[3]{80x^8} = \sqrt[3]{8 \cdot 10 \cdot x^6 \cdot x^2} = \sqrt[3]{8 \cdot x^6 \cdot 10 \cdot x^2} =$
$\sqrt[3]{8}\sqrt[3]{x^6}\sqrt[3]{10x^2} = 2x^2\sqrt[3]{10x^2}$

50. $3m\sqrt[3]{4m^2}$

51. $\sqrt[4]{32} = \sqrt[4]{16 \cdot 2} = \sqrt[4]{16} \cdot \sqrt[4]{2} = 2\sqrt[4]{2}$

52. $2\sqrt[4]{5}$

53. $\sqrt[4]{810} = \sqrt[4]{81 \cdot 10} = \sqrt[4]{81} \cdot \sqrt[4]{10} = 3\sqrt[4]{10}$

54. $2\sqrt[4]{10}$

55. $\sqrt[4]{96a^8} = \sqrt[4]{16 \cdot 6 \cdot a^8} = \sqrt[4]{16 \cdot a^8 \cdot 6} =$
$\sqrt[4]{16}\sqrt[4]{a^8}\sqrt[4]{6} = 2a^2\sqrt[4]{6}$

56. $2x^2\sqrt[4]{15}$

57. $\sqrt[4]{162c^4d^6} = \sqrt[4]{81 \cdot 2 \cdot c^4 \cdot d^4 \cdot d^2} =$
$\sqrt[4]{81 \cdot c^4 \cdot d^4 \cdot 2 \cdot d^2} = \sqrt[4]{81}\sqrt[4]{c^4}\sqrt[4]{d^4}\sqrt[4]{2d^2} =$
$3cd\sqrt[4]{2d^2}$

58. $3x^2y^2\sqrt[4]{3y^2}$

59. $\sqrt[3]{(x+y)^4} = \sqrt[3]{(x+y)^3(x+y)} =$
$\sqrt[3]{(x+y)^3}\sqrt[3]{x+y} = (x+y)\sqrt[3]{x+y}$

60. $(a-b)\sqrt[3]{(a-b)^2}$

61. $\sqrt[3]{8000(m+n)^8} = \sqrt[3]{8000(m+n)^6(m+n)^2} =$
$\sqrt[3]{8000}\sqrt[3]{(m+n)^6}\sqrt[3]{(m+n)^2} =$
$20(m+n)^2\sqrt[3]{(m+n)^2}$

62. $-10(x+y)^3\sqrt[3]{x+y}$

63. $\sqrt[5]{-a^6b^{11}c^{17}} = \sqrt[5]{-1 \cdot a^5 \cdot a \cdot b^{10} \cdot b \cdot c^{15} \cdot c^2} =$
$\sqrt[5]{-1 \cdot a^5 \cdot b^{10} \cdot c^{15} \cdot a \cdot b \cdot c^2} =$
$\sqrt[5]{-1}\sqrt[5]{a^5}\sqrt[5]{b^{10}}\sqrt[5]{c^{15}}\sqrt[5]{abc^2} = -ab^2c^3\sqrt[5]{abc^2}$

64. $x^2yz^4\sqrt[5]{x^3y^3z^2}$

65. $\sqrt{3}\sqrt{6} = \sqrt{3 \cdot 6} = \sqrt{18} = \sqrt{9 \cdot 2} = 3\sqrt{2}$

66. $5\sqrt{2}$

67. $\sqrt{15}\sqrt{12} = \sqrt{15 \cdot 12} = \sqrt{180} = \sqrt{36 \cdot 5} = 6\sqrt{5}$

68. 8

69. $\sqrt{6}\sqrt{8} = \sqrt{6 \cdot 8} = \sqrt{48} = \sqrt{16 \cdot 3} = 4\sqrt{3}$

70. $6\sqrt{7}$

71. $\sqrt[3]{3}\sqrt[3]{18} = \sqrt[3]{3 \cdot 18} = \sqrt[3]{54}$
$= \sqrt[3]{27 \cdot 2} = \sqrt[3]{27}\sqrt[3]{2} = 3\sqrt[3]{2}$

72. $30\sqrt{3}$

73. $\sqrt{5b^3}\,\sqrt{10c^4}$

$= \sqrt{5b^3 \cdot 10c^4}$ Multiplying radicands

$= \sqrt{50b^3c^4}$

$= \sqrt{25 \cdot 2 \cdot b^2 \cdot b \cdot c^4}$ Factoring the radicand

$= \sqrt{25b^2c^4}\,\sqrt{2b}$ Factoring into radicals

$= 5bc^2\,\sqrt{2b}$ Taking the square root

74. $-2a\sqrt[3]{15a^2}$

75. $\sqrt[3]{10x^5}\,\sqrt[3]{-75x^2}$

$= \sqrt[3]{10x^5 \cdot (-75x^2)}$ Multiplying radicands

$= \sqrt[3]{-750x^7}$

$= \sqrt[3]{-125x^6 \cdot 6x}$ Factoring the radicand

$= \sqrt[3]{-125}\,\sqrt[3]{x^6}\,\sqrt[3]{6x}$ Factoring into radicals

$= -5x^2\sqrt[3]{6x}$ Taking the cube root

76. $2x^2y\sqrt{6}$

77. $\sqrt[3]{y^4}\,\sqrt[3]{16y^5} = \sqrt[3]{y^4 \cdot 16y^5}$

$= \sqrt[3]{16y^9}$

$= \sqrt[3]{8 \cdot 2 \cdot y^9}$

$= \sqrt[3]{8y^9}\,\sqrt[3]{2}$

$= 2y^3\,\sqrt[3]{2}$

78. $5^2t^3\sqrt[3]{t}$, or $25t^3\sqrt[3]{t}$

79. $\sqrt[3]{(b+3)^4}\,\sqrt[3]{(b+3)^2} = \sqrt[3]{(b+3)^4(b+3)^2} =$

$\sqrt[3]{(b+3)^6} = (b+3)^2$

80. $(x+y)^2\sqrt[3]{(x+y)^2}$

81. $\sqrt{12a^3b}\,\sqrt{8a^4b^2} = \sqrt{12a^3b \cdot 8a^4b^2}$

$= \sqrt{96a^7b^3}$

$= \sqrt{16a^6b^2 \cdot 6ab}$

$= \sqrt{16}\sqrt{a^6}\sqrt{b^2}\sqrt{6ab}$

$= 4a^3b\sqrt{6ab}$

82. $6a^2b^4\sqrt{15ab}$

83. $\sqrt[5]{a^2(b+c)^4}\,\sqrt[5]{a^4(b+c)^7} =$

$\sqrt[5]{a^2(b+c)^4 \cdot a^4(b+c)^7} = \sqrt[5]{a^6(b+c)^{11}} =$

$\sqrt[5]{a^5(b+c)^{10} \cdot a(b+c)} =$

$\sqrt[5]{a^5}\,\sqrt[5]{(b+c)^{10}}\,\sqrt[5]{a(b+c)} = a(b+c)^2\sqrt[5]{a(b+c)}$

84. $x(y-z)^3\sqrt[5]{x^4(y-z)}$

85. $\sqrt[5]{a^3b}\,\sqrt{ab}$

$= (a^3b)^{1/5}(ab)^{1/2}$ Converting to exponential notation

$= a^{3/5}b^{1/5}a^{1/2}b^{1/2}$ Using the laws

$= a^{3/5+1/2}b^{1/5+1/2}$ of exponents

$= a^{11/10}b^{7/10}$ Adding exponents

$= a^{1\frac{1}{10}}b^{\frac{7}{10}}$

$= a \cdot a^{1/10} \cdot b^{7/10}$ Factoring $a^{11/10}$

$= a(ab^7)^{1/10}$ Adding exponents

$= a\sqrt[10]{ab^7}$ Converting back to radical notation

86. $xy\sqrt[6]{xy^5}$

87. $\sqrt[3]{4xy^2}\,\sqrt{2x^3y^3}$

$= (4xy^2)^{1/3}(2x^3y^3)^{1/2}$

$= (2^2xy^2)^{1/3}(2x^3y^3)^{1/2}$ Writing 4 as 2^2

$= 2^{2/3}x^{1/3}y^{2/3}2^{1/2}x^{3/2}y^{3/2}$

$= 2^{7/6}x^{11/6}y^{13/6}$

$= 2^{1\frac{1}{6}}x^{1\frac{5}{6}}y^{2\frac{1}{6}}$

$= 2 \cdot 2^{1/6} \cdot x \cdot x^{5/6} \cdot y^2 \cdot y^{1/6}$

$= 2xy^2(2x^5y)^{1/6}$

$= 2xy^2\sqrt[6]{2x^5y}$

88. $3a^2b\sqrt[4]{ab}$

89. $\sqrt[4]{x^3y^5}\,\sqrt{xy} = (x^3y^5)^{1/4}(xy)^{1/2}$

$= x^{3/4}y^{5/4}x^{1/2}y^{1/2}$

$= x^{5/4}y^{7/4}$

$= x^{1\frac{1}{4}}y^{1\frac{3}{4}}$

$= x \cdot x^{1/4} \cdot y \cdot y^{3/4}$

$= xy(xy^3)^{1/4}$

$= xy\sqrt[4]{xy^3}$

90. $a\sqrt[10]{ab^7}$

91. $\sqrt{a^4b^3c^4}\,\sqrt[3]{ab^2c} = (a^4b^3c^4)^{1/2}(ab^2c)^{1/3}$

$= a^{4/2}b^{3/2}c^{4/2}a^{1/3}b^{2/3}c^{1/3}$

$= a^{14/6}b^{13/6}c^{14/6}$

$= a^{2\frac{2}{6}}b^{2\frac{1}{6}}c^{2\frac{2}{6}}$

$= a^2 \cdot a^{2/6} \cdot b^2 \cdot b^{1/6} \cdot c^2 \cdot c^{2/6}$

$= a^2b^2c^2(a^2bc^2)^{1/6}$

$= a^2b^2c^2\sqrt[6]{a^2bc^2}$

92. $xyz\sqrt[6]{x^5yz^2}$

93. $\sqrt[3]{x^2yz^2}\sqrt[4]{xy^3z^3} = (x^2yz^2)^{1/3}(xy^3z^3)^{1/4}$

$= x^{2/3}y^{1/3}z^{2/3}x^{1/4}y^{3/4}z^{3/4}$

$= x^{11/12}y^{13/12}z^{17/12}$

$= x^{1\frac{11}{12}}y^{1\frac{1}{12}}z^{1\frac{5}{12}}$

$= x^{11/12} \cdot y \cdot y^{1/12} \cdot z \cdot z^{5/12}$

$= yz(x^{11}yz^5)^{1/12}$

$= yz\sqrt[12]{x^{11}yz^5}$

94. $ac\sqrt[12]{a^2b^{11}c^5}$

95. $\sqrt[3]{4a^2(b-5)^2}\sqrt{8a(b-5)^3}$

$= [4a^2(b-5)^2]^{1/3}[8a(b-5)^3]^{1/2}$

$= (2^2)^{1/3}a^{2/3}(b-5)^{2/3}(2^3)^{1/2}a^{1/2}(b-5)^{3/2}$

$\qquad\qquad (4 = 2^2,\ 8 = 2^3)$

$= 2^{2/3}a^{2/3}(b-5)^{2/3}2^{3/2}a^{1/2}(b-5)^{3/2}$

$= 2^{13/6}a^{7/6}(b-5)^{13/6}$

$= 2^{2\frac{1}{6}}a^{1\frac{1}{6}}(b-5)^{2\frac{1}{6}}$

$= 2^2 \cdot 2^{1/6} \cdot a \cdot a^{1/6} \cdot (b-5)^2(b-5)^{1/6}$

$= 2^2a(b-5)^2[2a(b-5)]^{1/6}$

$= 2^2a(b-5)^2\sqrt[6]{2a(b-5)}$

96. $9x(y-2)^2\sqrt[6]{3^5x^5(y-2)}$

97. Using a calculator,

$\sqrt{180} \approx 13.41640787 \approx 13.416$

Using Table 1,

$\sqrt{180} = \sqrt{36 \cdot 5} = \sqrt{36}\sqrt{5} = 6\sqrt{5} \approx 6(2.236) \approx 13.416.$

98. 11.136

99. Using a calculator,

$\dfrac{8+\sqrt{480}}{4} \approx \dfrac{8+21.9089023}{4} \approx \dfrac{29.9089023}{4}$

$\approx 7.477225575 \approx 7.477$

Using Table 1,

$\dfrac{8+\sqrt{480}}{4} = \dfrac{8+\sqrt{16\cdot30}}{4} = \dfrac{8+4\sqrt{30}}{4} =$

$\dfrac{4(2+\sqrt{30})}{4} = 2+\sqrt{30} \approx 2+5.477 = 7.477$

100. $\left(-3.071\right)$

101. Using a calculator,

$\dfrac{16-\sqrt{48}}{20} \approx \dfrac{16-6.92820323}{20} \approx \dfrac{9.07179677}{20}$

$\approx 0.453589838 \approx 0.454$

Using Table 1,

$\dfrac{16-\sqrt{48}}{20} \approx \dfrac{16-6.928}{20} = \dfrac{9.072}{20} = 0.4536 \approx 0.454$

102. 0.919

103. Using a calculator,

$\dfrac{24+\sqrt{128}}{8} \approx \dfrac{24+11.3137085}{8} \approx \dfrac{35.3137085}{8}$

$\approx 4.414213562 \approx 4.414$

Using Table 1,

$\dfrac{24+\sqrt{128}}{8} = \dfrac{24+\sqrt{64\cdot2}}{8} = \dfrac{24+8\sqrt{2}}{8} =$

$\dfrac{8(3+\sqrt{2})}{8} = 3+\sqrt{2} \approx 3+1.414 = 4.414$

104. 7.209

105. $\sqrt{24,500,000,000}$

$= \sqrt{245 \cdot 10^8}$ Factoring the radicand

$= \sqrt{245} \cdot \sqrt{10^8}$ Factoring the expression

$\approx 15.65247584 \cdot 10^4$ Approximating $\sqrt{245}$ with a calculator and finding $\sqrt{10^8}$

$\approx 156,524.7584$ Multiplying

106. 128,452.3258

107. $\sqrt{468,200,000,000} = \sqrt{4682 \cdot 10^8}$

$= \sqrt{4682} \cdot \sqrt{10^8}$

$\approx 68.42514158 \cdot 10^4$

$\approx 684,251.4158$

108. 315,277.6554

109. $\sqrt{0.0000000395} = \sqrt{3.95 \cdot 10^{-8}}$

$= \sqrt{3.95} \cdot \sqrt{10^{-8}}$

$\approx 1.987460691 \cdot 10^{-4}$

≈ 0.0001987460691

110. 0.0003928103868

111. $\sqrt{0.0000005001} = \sqrt{50.01 \cdot 10^{-8}}$

$= \sqrt{50.01} \cdot \sqrt{10^{-8}}$

$\approx 7.071774883 \cdot 10^{-4}$

≈ 0.0007071774883

112. 0.003178207042

113. *Familiarize*. Let x and y represent the number of 30-sec and 60-sec commercials, respectively. Then the total number of minutes of commercial time during the show is $\dfrac{30x+60y}{60}$, or $\dfrac{x}{2}+y$. (We divide by 60 to convert seconds to minutes.)

Translate. Rewording when necessary, we write two equations.

$\underbrace{\text{Total number of commercials}}_{\displaystyle x+y} \quad \overset{\displaystyle\downarrow}{\text{is}} \quad \overset{\displaystyle\downarrow}{\underset{\displaystyle 12}{=}} \ 12.$

Number of   total minutes
30-sec   is   of commercial   less 6.
commercials   time

$\underbrace{\qquad\qquad}$ $\downarrow$ $\underbrace{\qquad}$ $\downarrow\ \downarrow$

x $=$ $\dfrac{x}{2}+y$ $-$ 6

Carry out. Solving the system of equations, we get (4,8).

Check. If there are 4 30-sec and 8 60-sec commercials, the total number of commercials is 12. The total amount of commercial time is $4 \cdot 30$ sec $+$ $8 \cdot 60$ sec $= 600$ sec, or 10 min. Then the number of 30-sec commercials is 6 less than the total number of minutes of commercial time. The values check.

State. 8 60-sec commercials were used.

114. $4x^2 - 9$

115. $4x^2 - 49 = (2x)^2 - 7^2 = (2x+7)(2x-7)$

116. $2(x-9)(x-4)$

117.

118.

119. $\qquad r = 2\sqrt{5L}$

a) $r = 2\sqrt{5 \cdot 20}$
$= 2\sqrt{100}$
$= 2 \cdot 10 = 20$ mph

b) $r = 2\sqrt{5 \cdot 70}$
$= 2\sqrt{350}$
$\approx 2 \times 18.708$ Using Table 1 or a calculator

≈ 37.4 mph   Multiplying and rounding

c) $r = 2\sqrt{5 \cdot 90}$
$= 2\sqrt{450}$
$\approx 2 \times 21.213$ Using Table 1 or a calculator

≈ 42.4 mph   Multiplying and rounding

120. a) $-3.3°$ C
b) $-16.6°$ C
c) $-25.5°$ C
d) $-54.0°$ C

121. $\sqrt[3]{5x^{k+1}}\,\sqrt[3]{25x^k} = 5x^7$

$\sqrt[3]{5x^{k+1} \cdot 25x^k} = 5x^7$

$\sqrt[3]{125x^{2k+1}} = 5x^7$

$\sqrt[3]{125}\,\sqrt[3]{x^{2k+1}} = 5x^7$

$5\sqrt[3]{x^{2k+1}} = 5x^7$

$\sqrt[3]{x^{2k+1}} = x^7$

$(x^{2k+1})^{1/3} = x^7$

$x^{\frac{2k+1}{3}} = x^7$

Since the base is the same, the exponents must be equal. We have:

$\dfrac{2k+1}{3} = 7$

$2k+1 = 21$

$2k = 20$

$k = 10$

122. 6

123. We must assume $2x + 3$ is nonnegative.

$2x + 3 \geq 0$

$2x \geq -3$

$x \geq -\dfrac{3}{2}$

Thus we must assume $x \geq -\dfrac{3}{2}$.

124.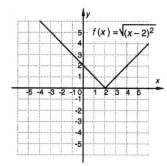

$\{x | x$ is a real number$\}$

Exercise Set 10.4

1. $\dfrac{\sqrt{21a}}{\sqrt{3a}} = \sqrt{\dfrac{21a}{3a}} = \sqrt{7}$

2. $= \sqrt{7}$

3. $\dfrac{\sqrt[3]{54}}{\sqrt[3]{2}} = \sqrt[3]{\dfrac{54}{2}} = \sqrt[3]{27} = 3$

4. 2

5. $\dfrac{\sqrt{40xy^3}}{\sqrt{8x}} = \sqrt{\dfrac{40xy^3}{8x}} = \sqrt{5y^3} = \sqrt{y^2 \cdot 5y} =$
$\sqrt{y^2}\,\sqrt{5y} = y\sqrt{5y}$

6. $2b\sqrt{2b}$

7. $\dfrac{\sqrt[3]{96a^4b^2}}{\sqrt[3]{12a^2b}} = \sqrt[3]{\dfrac{96a^4b^2}{12a^2b}} = \sqrt[3]{8a^2b} = \sqrt[3]{8}\,\sqrt[3]{a^2b} =$
$2\sqrt[3]{a^2b}$

8. $3xy\sqrt[3]{y^2}$

9. $\dfrac{\sqrt{144xy}}{2\sqrt{2}} = \dfrac{1}{2}\sqrt{\dfrac{144xy}{2}} = \dfrac{1}{2}\sqrt{72xy} =$
$\dfrac{1}{2}\sqrt{36\cdot 2\cdot x\cdot y} = \dfrac{1}{2}\sqrt{36}\sqrt{2xy} = \dfrac{1}{2}\cdot 6\sqrt{2xy} =$
$3\sqrt{2xy}$

10. $\dfrac{5}{3}\sqrt{ab}$

11. $\dfrac{\sqrt[4]{48x^9y^{13}}}{\sqrt[4]{3xy^5}} = \sqrt[4]{\dfrac{48x^9y^{13}}{3xy^5}} = \sqrt[4]{16x^8y^8} = 2x^2y^2$

12. $2a^2b^5\sqrt[5]{b}$

13. $\dfrac{\sqrt{x^3-y^3}}{\sqrt{x-y}} = \sqrt{\dfrac{x^3-y^3}{x-y}} =$

$\sqrt{\dfrac{(x-y)(x^2+xy+y^2)}{x-y}} =$

$\sqrt{\dfrac{(x\!\!\!\!\diagdown y)(x^2+xy+y^2)}{x\!\!\!\!\diagdown y}} = \sqrt{x^2+xy+y^2}$

14. $\sqrt{r^2-rs+s^2}$

15. $\dfrac{\sqrt[3]{a^2}}{\sqrt[4]{a}}$

$= \dfrac{a^{2/3}}{a^{1/4}}$ Converting to exponential notation

$= a^{2/3-1/4}$ Subtracting exponents

$= a^{5/12}$ Converting back

$= \sqrt[12]{a^5}$ to radical notation

16. $\sqrt[15]{x^7}$

17. $\dfrac{\sqrt[4]{x^2y^3}}{\sqrt[3]{xy^2}}$

$= \dfrac{(x^2y^3)^{1/4}}{(xy^2)^{1/3}}$ Converting to exponential

$= \dfrac{x^{2/4}y^{3/4}}{x^{1/3}y^{2/3}}$ notation; using the product

       and power rules

$= x^{2/4-1/3}y^{3/4-2/3}$ Subtracting exponents

$= x^{2/12}y^{1/12}$ Converting back

$= (x^2y)^{1/12}$ to radical

$= \sqrt[12]{x^2y}$ notation

18. $\sqrt[15]{a^4b^2}$

19. $\dfrac{\sqrt{ab^3c}}{\sqrt[5]{a^2b^3c}} = \dfrac{(ab^3c)^{1/2}}{(a^2b^3c)^{1/5}}$

$= \dfrac{a^{1/2}b^{3/2}c^{1/2}}{a^{2/5}b^{3/5}c^{1/5}}$

$= a^{1/10}b^{9/10}c^{3/10}$ Subtracting exponents

$= (ab^9c^3)^{1/10}$

$= \sqrt[10]{ab^9c^3}$

20. $\sqrt[10]{xy^3z^3}$

21. $\dfrac{\sqrt[4]{(3x-1)^3}}{\sqrt[5]{(3x-1)^3}}$

$= \dfrac{(3x-1)^{3/4}}{(3x-1)^{3/5}}$ Converting to exponential notation

$= (3x-1)^{3/4-3/5}$ Subtracting exponents

$= (3x-1)^{3/20}$ Converting back

$= \sqrt[20]{(3x-1)^3}$ to radical notation

22. $\sqrt[12]{5-3x}$

23. $\sqrt{\dfrac{16}{25}} = \dfrac{\sqrt{16}}{\sqrt{25}} = \dfrac{4}{5}$

24. $\dfrac{10}{9}$

25. $\sqrt[3]{\dfrac{64}{27}} = \dfrac{\sqrt[3]{64}}{\sqrt[3]{27}} = \dfrac{4}{3}$

26. $\dfrac{7}{8}$

27. $\sqrt{\dfrac{49}{y^2}} = \dfrac{\sqrt{49}}{\sqrt{y^2}} = \dfrac{7}{y}$

28. $\dfrac{11}{x}$

29. $\sqrt{\dfrac{25y^3}{x^4}} = \dfrac{\sqrt{25y^3}}{\sqrt{x^4}} = \dfrac{\sqrt{25y^2\cdot y}}{\sqrt{x^4}} = \dfrac{\sqrt{25y^2}\,\sqrt{y}}{\sqrt{x^4}} =$
$\dfrac{5y\sqrt{y}}{x^2}$

30. $\dfrac{6a^2\sqrt{a}}{b^3}$

31. $\sqrt[3]{\dfrac{8x^5}{27y^3}} = \dfrac{\sqrt[3]{8x^5}}{\sqrt[3]{27y^3}} = \dfrac{\sqrt[3]{8x^3\cdot x^2}}{\sqrt[3]{27y^3}} = \dfrac{\sqrt[3]{8x^3}\,\sqrt[3]{x^2}}{\sqrt[3]{27y^3}} =$
$\dfrac{2x\sqrt[3]{x^2}}{3y}$

32. $\dfrac{2x^2\sqrt[3]{x}}{3y^2}$

33. $\sqrt[4]{\dfrac{16a^4}{81}} = \dfrac{\sqrt[4]{16a^4}}{\sqrt[4]{81}} = \dfrac{2a}{3}$

34. $\dfrac{3x}{y^2}$

35. $\sqrt[4]{\dfrac{a^5b^8}{c^{10}}} = \dfrac{\sqrt[4]{a^5b^8}}{\sqrt[4]{c^{10}}} = \dfrac{\sqrt[4]{a^4b^8 \cdot a}}{\sqrt[4]{c^8 \cdot c^2}} = \dfrac{\sqrt[4]{a^4b^8}\,\sqrt[4]{a}}{\sqrt[4]{c^8}\,\sqrt[4]{c^2}} = \dfrac{ab^2\,\sqrt[4]{a}}{c^2\,\sqrt[4]{c^2}}$, or $\dfrac{ab^2}{c^2}\sqrt[4]{\dfrac{a}{c^2}}$

36. $\dfrac{x^2y^3}{z}\sqrt[4]{\dfrac{x}{z^2}}$

37. $\sqrt[5]{\dfrac{32x^6}{y^{11}}} = \dfrac{\sqrt[5]{32x^6}}{\sqrt[5]{y^{11}}} = \dfrac{\sqrt[5]{32x^5 \cdot x}}{\sqrt[5]{y^{10} \cdot y}} = \dfrac{\sqrt[5]{32x^5}\,\sqrt[5]{x}}{\sqrt[5]{y^{10}}\,\sqrt[5]{y}} = \dfrac{2x\,\sqrt[5]{x}}{y^2\,\sqrt[5]{y}}$, or $\dfrac{2x}{y^2}\sqrt[5]{\dfrac{x}{y}}$

38. $\dfrac{3a}{b^2}\sqrt[5]{\dfrac{a^4}{b^3}}$

39. $\sqrt[6]{\dfrac{x^6y^8}{z^{15}}} = \dfrac{\sqrt[6]{x^6y^8}}{\sqrt[6]{z^{15}}} = \dfrac{\sqrt[6]{x^6y^6 \cdot y^2}}{\sqrt[6]{z^{12} \cdot z^3}} = \dfrac{\sqrt[6]{x^6y^6}\,\sqrt[6]{y^2}}{\sqrt[6]{z^{12}}\,\sqrt[6]{z^3}} = \dfrac{xy\,\sqrt[6]{y^2}}{z^2\,\sqrt[6]{z^3}}$, or $\dfrac{xy}{z^2}\sqrt[6]{\dfrac{y^2}{z^3}}$

40. $\dfrac{ab^2}{c^2}\sqrt[6]{\dfrac{a^3}{c}}$

41. a) $\sqrt{(6a)^3} = \sqrt{6^3a^2} = \sqrt{6^2a^2}\,\sqrt{6a} = 6a\sqrt{6a}$

b) $(\sqrt{6a})^3 = \sqrt{6a}\,\sqrt{6a}\,\sqrt{6a} = 6a\sqrt{6a}$

42. $7y\sqrt{7y}$

43. a) $(\sqrt{16b^2})^3 = \sqrt{16b^2}\,\sqrt{16b^2}\,\sqrt{16b^2} = $
$16b^2\sqrt{16b^2} = 16b^2 \cdot 4b = 64b^3$

b) $\sqrt{(16b^2)^3} = \sqrt{(2^4b^2)^3} = \sqrt{2^{12}b^6} = 2^6b^3 = $
$64b^3$

44. $125r^3$

45. a) $\sqrt{(18a^2b)^3} = \sqrt{5832a^6b^3} = \sqrt{2916a^6b^2 \cdot 2b} = $
$54a^3b\sqrt{2b}$

b) $(\sqrt{18a^2b})^3 = \sqrt{18a^2b}\,\sqrt{18a^2b}\,\sqrt{18a^2b} = $
$18a^2b\sqrt{18a^2b} = 18a^2b\sqrt{9a^2 \cdot 2b} = $
$18a^2b \cdot 3a\sqrt{2b} = 54a^3b\sqrt{2b}$

46. $24x^3y\sqrt{3y}$

47. a) $(\sqrt[3]{3c^2d})^4 = \sqrt[3]{3c^2d}\,\sqrt[3]{3c^2d}\,\sqrt[3]{3c^2d}\,\sqrt[3]{3c^2d} = $
$3c^2d\sqrt[3]{3c^2d}$

b) $\sqrt[3]{(3c^2d)^4} = \sqrt[3]{81c^8d^4} = \sqrt[3]{27c^6d^3 \cdot 3c^2d} = $
$3c^2d\sqrt[3]{3c^2d}$

48. $2x^2y\sqrt[3]{2x^2y}$

49. a) $\sqrt[3]{(5x^2y)^2} = \sqrt[3]{25x^4y^2} = \sqrt[3]{x^3 \cdot 25xy^2} = $
$x\sqrt[3]{25xy^2}$

b) $(\sqrt[3]{5x^2y})^2 = \sqrt[3]{5x^2y}\,\sqrt[3]{5x^2y} = \sqrt[3]{25x^4y^2} = $
$x\sqrt[3]{25xy^2}$, as in part (a)

50. $b\sqrt[3]{36a^2b}$

51. a) $\sqrt[4]{(x^2y)^3} = \sqrt[4]{x^6y^3} = \sqrt[4]{x^4 \cdot x^2y^3} = x\sqrt[4]{x^2y^3}$

b) $(\sqrt[4]{x^2y})^3 = \sqrt[4]{x^2y}\,\sqrt[4]{x^2y}\,\sqrt[4]{x^2y} = \sqrt[4]{x^6y^3} = $
$x\sqrt[4]{x^2y^3}$, as in part (a)

52. $a^2\sqrt[4]{8a}$

53. a) $(\sqrt[3]{8a^4b})^2 = (\sqrt[3]{8a^3 \cdot ab})^2 = (2a\sqrt[3]{ab})^2 = $
$2a\sqrt[3]{ab} \cdot 2a\sqrt[3]{ab} = 4a^2\sqrt[3]{a^2b^2}$

b) $\sqrt[3]{(8a^4b)^2} = \sqrt[3]{64a^8b^2} = \sqrt[3]{64a^6 \cdot a^2b^2} = $
$4a^2\sqrt[3]{a^2b^2}$

54. $9y^3\sqrt[3]{x^2y}$

55. a) $(\sqrt[4]{16x^2y^3})^2 = (2\sqrt[4]{x^2y^3})^2 = $
$2\sqrt[4]{x^2y^3} \cdot 2\sqrt[4]{x^2y^3} = 4\sqrt[4]{x^4y^6} = 4\sqrt[4]{x^4y^4 \cdot y^2} = $
$4xy\sqrt[4]{y^2}$, or $4xy(y^{2/4}) = 4xy(y^{1/2}) = 4xy\sqrt{y}$

b) $\sqrt[4]{(16x^2y^3)^2} = \sqrt[4]{256x^4y^6} = \sqrt[4]{256x^4y^4 \cdot y^2} = $
$4xy\sqrt[4]{y^2} = 4xy\sqrt{y}$, as in part (a)

56. $8y^3\sqrt[4]{x^3y^3}$

57.
$$\dfrac{12x}{x-4} - \dfrac{3x^2}{x+4} = \dfrac{384}{x^2-16}$$
$$\dfrac{12x}{x-4} - \dfrac{3x^2}{x+4} = \dfrac{384}{(x+4)(x-4)},$$
$$\text{LCM is } (x+4)(x-4).$$
$$(x+4)(x-4)\left[\dfrac{12x}{x-4} - \dfrac{3x^2}{x+4}\right] = $$
$$(x+4)(x-4) \cdot \dfrac{384}{(x+4)(x-4)}$$
$$12x(x+4) - 3x^2(x-4) = 384$$
$$12x^2 + 48x - 3x^3 + 12x^2 = 384$$
$$-3x^3 + 24x^2 + 48x - 384 = 0$$
$$-3(x^3 - 8x^2 - 16x + 128) = 0$$
$$-3[x^2(x-8) - 16(x-8)] = 0$$
$$-3(x-8)(x^2-16) = 0$$
$$-3(x-8)(x+4)(x-4) = 0$$
$$x-8 = 0 \quad \text{or} \quad x+4 = 0 \quad \text{or} \quad x-4 = 0$$
$$x = 8 \quad \text{or} \quad x = -4 \quad \text{or} \quad x = 4$$

Check: For 8:

$$\dfrac{\dfrac{12x}{x-4} - \dfrac{3x^2}{x+4}}{} = \dfrac{\dfrac{384}{x^2-16}}{}$$

$$\dfrac{12 \cdot 8}{8-4} - \dfrac{3 \cdot 8^2}{8+4} \quad \Bigg| \quad \dfrac{384}{8^2-16}$$

$$\dfrac{96}{4} - \dfrac{192}{12} \quad \Bigg| \quad \dfrac{384}{48}$$

$$24 - 16 \quad \Bigg| \quad 8$$

$$8 \quad \Bigg| \quad \text{TRUE}$$

8 is a solution.

For -4:

$$\frac{12x}{x-4} - \frac{3x^2}{x+4} = \frac{384}{x^2-16}$$

$$\begin{array}{c|c} \dfrac{12(-4)}{-4-4} - \dfrac{3(-4)^2}{-4+4} & \dfrac{384}{(-4)^2-16} \\[3mm] \dfrac{-48}{-8} - \dfrac{48}{0} & \dfrac{384}{16-16} \quad \text{UNDEFINED} \end{array}$$

-4 is not a solution.

For 4:

$$\frac{12x}{x-4} - \frac{3x^2}{x+4} = \frac{384}{x^2-16}$$

$$\begin{array}{c|c} \dfrac{12\cdot 4}{4-4} - \dfrac{3\cdot 4^2}{4+4} & \dfrac{384}{4^2-16} \\[3mm] \dfrac{48}{0} - \dfrac{48}{8} & \dfrac{384}{16-16} \quad \text{UNDEFINED} \end{array}$$

4 is not a solution.

The solution is 8.

58. $\dfrac{15}{2}$

59. *Familiarize*. Let x and y represent the width and length of the rectangle, respectively.

Translate. We write two equations.

The width is one-fourth the length.

$$x = \frac{1}{4}\cdot y$$

The area is twice the perimeter.

$$xy = 2\cdot (2x+2y)$$

Carry out. Solving the system of equations we get $(5,20)$.

Check. The width, 5, is one-fourth the length, 20. The area is $5\cdot 20$, or 100. The perimeter is $2\cdot 5 + 2\cdot 20$, or 50. Since $100 = 2\cdot 50$, the area is twice the perimeter. The values check.

State. The width is 5, and the length is 20.

60.

61.

62. a) 1.62 sec

b) 1.99 sec

c) 2.20 sec

63. $\dfrac{7\sqrt{a^2b}\sqrt{25xy}}{5\sqrt{a^{-4}b^{-1}}\sqrt{49x^{-1}y^{-3}}} = \dfrac{7\sqrt{25a^2bxy}}{5\sqrt{49a^{-4}b^{-1}x^{-1}y^{-3}}} =$

$\dfrac{7}{5}\sqrt{\dfrac{25a^2bxy}{49a^{-4}b^{-1}x^{-1}y^{-3}}} = \dfrac{7}{5}\sqrt{\dfrac{25}{49}\cdot a^6b^2x^2y^4} =$

$\dfrac{7}{5}\cdot\dfrac{5}{7}\cdot a^3bxy^2 = a^3bxy^2$

64. $9\sqrt[3]{9n^2}$

65. $\dfrac{\sqrt{44x^2y^9z}\sqrt{22y^9z^6}}{(\sqrt{11xy^8z^2})^2} = \dfrac{\sqrt{44\cdot 22x^2y^{18}z^7}}{\sqrt{11\cdot 11x^2y^{16}z^4}} =$

$\sqrt{\dfrac{44\cdot 22x^2y^{18}z^7}{11\cdot 11x^2y^{16}z^4}} = \sqrt{4\cdot 2y^2z^3} = \sqrt{4y^2z^2\cdot 2z} = 2yz\sqrt{2z}$

Exercise Set 10.5

1. $6\sqrt{3} + 2\sqrt{3} = (6+2)\sqrt{3} = 8\sqrt{3}$

2. $17\sqrt{5}$

3. $9\sqrt[3]{5} - 6\sqrt[3]{5} = (9-6)\sqrt[3]{5} = 3\sqrt[3]{5}$

4. $8\sqrt[5]{2}$

5. $4\sqrt[3]{y} + 9\sqrt[3]{y} = (4+9)\sqrt[3]{y} = 13\sqrt[3]{y}$

6. $3\sqrt[4]{t}$

7. $8\sqrt{2} - 6\sqrt{2} + 5\sqrt{2} = (8-6+5)\sqrt{2} = 7\sqrt{2}$

8. $7\sqrt{6}$

9. $4\sqrt[3]{3} - \sqrt{5} + 2\sqrt[3]{3} + \sqrt{5} =$
$(4+2)\sqrt[3]{3} + (-1+1)\sqrt{5} = 6\sqrt[3]{3}$

10. $6\sqrt{7} + \sqrt[4]{11}$

11. $\quad 8\sqrt{27} - 3\sqrt{3}$

$= 8\sqrt{9\cdot 3} - 3\sqrt{3}$ Factoring the

$= 8\sqrt{9}\cdot\sqrt{3} - 3\sqrt{3}$ first radical

$= 8\cdot 3\sqrt{3} - 3\sqrt{3}$ Taking the square root of 9

$= 24\sqrt{3} - 3\sqrt{3}$

$= (24-3)\sqrt{3}$ Factoring out $\sqrt{3}$ to collect like radical terms

$= 21\sqrt{3}$

12. $41\sqrt{2}$

13. $\quad 8\sqrt{45} + 7\sqrt{20}$

$= 8\sqrt{9\cdot 5} + 7\sqrt{4\cdot 5}$ Factoring the

$= 8\sqrt{9}\cdot\sqrt{5} + 7\sqrt{4}\cdot\sqrt{5}$ radicals

$= 8\cdot 3\sqrt{5} + 7\cdot 2\sqrt{5}$ Taking the square roots

$= 24\sqrt{5} + 14\sqrt{5}$

$= (24+14)\sqrt{5}$ Factoring out $\sqrt{5}$ to collect like radical terms

$= 38\sqrt{5}$

14. $66\sqrt{3}$

15. $\quad 18\sqrt{72} + 2\sqrt{98} = 18\sqrt{36\cdot 2} + 2\sqrt{49\cdot 2} =$
$18\sqrt{36}\cdot\sqrt{2} + 2\sqrt{49}\cdot\sqrt{2} = 18\cdot 6\sqrt{2} + 2\cdot 7\sqrt{2} =$
$108\sqrt{2} + 14\sqrt{2} = (108+14)\sqrt{2} = 122\sqrt{2}$

16. $4\sqrt{5}$

17. $3\sqrt[3]{16} + \sqrt[3]{54} = 3\sqrt[3]{8 \cdot 2} + \sqrt[3]{27 \cdot 2} =$
$3\sqrt[3]{8} \cdot \sqrt[3]{2} + \sqrt[3]{27} \cdot \sqrt[3]{2} = 3 \cdot 2\sqrt[3]{2} + 3\sqrt[3]{2} =$
$6\sqrt[3]{2} + 3\sqrt[3]{2} = (6+3)\sqrt[3]{2} = 9\sqrt[3]{2}$

18. -7

19. $\sqrt{5a} + 2\sqrt{45a^3} = \sqrt{5a} + 2\sqrt{9a^2 \cdot 5a} =$
$\sqrt{5a} + 2\sqrt{9a^2} \cdot \sqrt{5a} = \sqrt{5a} + 2 \cdot 3a\sqrt{5a} =$
$\sqrt{5a} + 6a\sqrt{5a} = (1 + 6a)\sqrt{5a}$

20. $(4x - 2)\sqrt{3x}$

21. $\sqrt[3]{24x} - \sqrt[3]{3x^4} = \sqrt[3]{8 \cdot 3x} - \sqrt[3]{x^3 \cdot 3x} =$
$\sqrt[3]{8} \cdot \sqrt[3]{3x} - \sqrt[3]{x^3} \cdot \sqrt[3]{3x} = 2\sqrt[3]{3x} - x\sqrt[3]{3x} =$
$(2 - x)\sqrt[3]{3x}$

22. $(3 - x)\sqrt[3]{2x}$

23. $\sqrt{8y - 8} + \sqrt{2y - 2} = \sqrt{4(2y - 2)} + \sqrt{2y - 2} =$
$\sqrt{4} \cdot \sqrt{2y - 2} + \sqrt{2y - 2} = 2\sqrt{2y - 2} + \sqrt{2y - 2} =$
$(2 + 1)\sqrt{2y - 2} = 3\sqrt{2y - 2}$

24. $3\sqrt{3t + 3}$

25. $\sqrt{x^3 - x^2} + \sqrt{9x - 9} = \sqrt{x^2(x - 1)} + \sqrt{9(x - 1)} =$
$\sqrt{x^2} \cdot \sqrt{x - 1} + \sqrt{9} \cdot \sqrt{x - 1} =$
$x\sqrt{x - 1} + 3\sqrt{x - 1} = (x + 3)\sqrt{x - 1}$

26. $(2 - x)\sqrt{x - 1}$

27. $5\sqrt[3]{32} - \sqrt[3]{108} + 2\sqrt[3]{256} =$
$5\sqrt[3]{8 \cdot 4} - \sqrt[3]{27 \cdot 4} + 2\sqrt[3]{64 \cdot 4} =$
$5\sqrt[3]{8} \cdot \sqrt[3]{4} - \sqrt[3]{27} \cdot \sqrt[3]{4} + 2\sqrt[3]{64} \cdot \sqrt[3]{4} =$
$5 \cdot 2\sqrt[3]{4} - 3\sqrt[3]{4} + 2 \cdot 4\sqrt[3]{4} = 10\sqrt[3]{4} - 3\sqrt[3]{4} + 8\sqrt[3]{4} =$
$(10 - 3 + 8)\sqrt[3]{4} = 15\sqrt[3]{4}$

28. $2\sqrt[3]{x}$

29. $\sqrt{x^3 + x^2} + \sqrt{4x^3 + 4x^2} - \sqrt{9x^3 + 9x^2} =$
$\sqrt{x^2(x + 1)} + \sqrt{4x^2(x + 1)} - \sqrt{9x^2(x + 1)} =$
$\sqrt{x^2} \cdot \sqrt{x + 1} + \sqrt{4x^2} \cdot \sqrt{x + 1} - \sqrt{9x^2} \cdot \sqrt{x + 1} =$
$x\sqrt{x + 1} + 2x\sqrt{x + 1} - 3x\sqrt{x + 1} =$
$(x + 2x - 3x)\sqrt{x + 1} = 0$

30. $-8\sqrt{5x^2 + 4}$

31. $\sqrt[4]{x^5 - x^4} + 3\sqrt[4]{x^9 - x^8} =$
$\sqrt[4]{x^4(x - 1)} + 3\sqrt[4]{x^8(x - 1)} =$
$\sqrt[4]{x^4} \cdot \sqrt[4]{x - 1} + 3\sqrt[4]{x^8}\sqrt[4]{x - 1} =$
$x\sqrt[4]{x - 1} + 3x^2\sqrt[4]{x - 1} = (x + 3x^2)\sqrt[4]{x - 1}$, or
$(3x^2 + x)\sqrt[4]{x - 1}$

32. $(2a - 2a^2)\sqrt[4]{1 + a}$

33. $\sqrt{6}(2 - 3\sqrt{6}) = \sqrt{6} \cdot 2 - \sqrt{6} \cdot 3\sqrt{6} = 2\sqrt{6} - 3 \cdot 6 =$
$2\sqrt{6} - 18$

34. $4\sqrt{3} + 3$

35. $\sqrt{2}(\sqrt{3} - \sqrt{5}) = \sqrt{2} \cdot \sqrt{3} - \sqrt{2} \cdot \sqrt{5} = \sqrt{6} - \sqrt{10}$

36. $5 - \sqrt{10}$

37. $\sqrt{3}(2\sqrt{5} - 3\sqrt{4}) = \sqrt{3}(2\sqrt{5} - 3 \cdot 2) =$
$\sqrt{3} \cdot 2\sqrt{5} - \sqrt{3} \cdot 6 = 2\sqrt{15} - 6\sqrt{3}$

38. $6\sqrt{5} - 4$

39. $\sqrt[3]{2}(\sqrt[3]{4} - 2\sqrt[3]{32}) = \sqrt[3]{2} \cdot \sqrt[3]{4} - \sqrt[3]{2} \cdot 2\sqrt[3]{32} =$
$\sqrt[3]{8} - 2\sqrt[3]{64} = 2 - 2 \cdot 4 = 2 - 8 = -6$

40. $3 - 4\sqrt[3]{63}$

41. $\sqrt[3]{a}(\sqrt[3]{2a^2} + \sqrt[3]{16a^2}) = \sqrt[3]{a} \cdot \sqrt[3]{2a^2} + \sqrt[3]{a} \cdot \sqrt[3]{16a^2} =$
$\sqrt[3]{2a^3} + \sqrt[3]{16a^3} = \sqrt[3]{a^3 \cdot 2} + \sqrt[3]{8a^3 \cdot 2} =$
$a\sqrt[3]{2} + 2a\sqrt[3]{2} = 3a\sqrt[3]{2}$

42. $-2x\sqrt[3]{3}$

43. $\sqrt[4]{x}(\sqrt[4]{x^7} + \sqrt[4]{3x^2}) = \sqrt[4]{x} \cdot \sqrt[4]{x^7} + \sqrt[4]{x} \cdot \sqrt[4]{3x^2} =$
$\sqrt[4]{x^8} + \sqrt[4]{3x^3} = x^2 + \sqrt[4]{3x^3}$

44. $\sqrt[4]{2a^2} - a^3$

45. $(5 - \sqrt{7})(5 + \sqrt{7}) = 5^2 - (\sqrt{7})^2 = 25 - 7 = 18$

46. 4

47. $(\sqrt{5} + \sqrt{8})(\sqrt{5} - \sqrt{8}) = (\sqrt{5})^2 - (\sqrt{8})^2 =$
$5 - 8 = -3$

48. -2

49. $(3 - 2\sqrt{7})(3 + 2\sqrt{7}) = 3^2 - (2\sqrt{7})^2 = 9 - 4 \cdot 7 =$
$9 - 28 = -19$

50. -2

51. $\quad (\sqrt{a} + \sqrt{2})(\sqrt{a} + \sqrt{3})$
$= \sqrt{a}\sqrt{a} + \sqrt{a}\sqrt{3} + \sqrt{2}\sqrt{a} + \sqrt{2}\sqrt{3}$
$\qquad\qquad\qquad\qquad\qquad$ Using FOIL
$= a + \sqrt{3a} + \sqrt{2a} + \sqrt{6}\qquad$ Multiplying radicals

52. $2 - 3\sqrt{x} + x$

53. $\quad (3 - \sqrt[3]{5})(2 + \sqrt[3]{5})$
$= 3 \cdot 2 + 3\sqrt[3]{5} - 2\sqrt[3]{5} - \sqrt[3]{5}\sqrt[3]{5}\quad$ Using FOIL
$= 6 + 3\sqrt[3]{5} - 2\sqrt[3]{5} - \sqrt[3]{25}$
$= 6 + \sqrt[3]{5} - \sqrt[3]{25}\qquad\qquad\quad$ Simplifying

54. $8 + 2\sqrt[3]{6} - \sqrt[3]{36}$

55. $(2\sqrt{7} - 4\sqrt{2})(3\sqrt{7} + 6\sqrt{2}) =$
$2\sqrt{7} \cdot 3\sqrt{7} + 2\sqrt{7} \cdot 6\sqrt{2} - 4\sqrt{2} \cdot 3\sqrt{7} - 4\sqrt{2} \cdot 6\sqrt{2} =$
$6 \cdot 7 + 12\sqrt{14} - 12\sqrt{14} - 24 \cdot 2 =$
$42 + 12\sqrt{14} - 12\sqrt{14} - 48 = -6$

56. $24 - 7\sqrt{15}$

57. $(2\sqrt[3]{3} + \sqrt[3]{2})(\sqrt[3]{3} - 2\sqrt[3]{2}) =$
$2\sqrt[3]{3} \cdot \sqrt[3]{3} - 2\sqrt[3]{3} \cdot 2\sqrt[3]{2} + \sqrt[3]{2} \cdot \sqrt[3]{3} - \sqrt[3]{2} \cdot 2\sqrt[3]{2} =$
$2\sqrt[3]{9} - 4\sqrt[3]{6} + \sqrt[3]{6} - 2\sqrt[3]{4} = 2\sqrt[3]{9} - 3\sqrt[3]{6} - 2\sqrt[3]{4}$

58. $6\sqrt[4]{63} - 9\sqrt[4]{42} + 2\sqrt[4]{54} - 3\sqrt[4]{36}$

59. $(1 + \sqrt{3})^2$
$= 1^2 + 2\sqrt{3} + (\sqrt{3})^2$ Squaring a binomial
$= 1 + 2\sqrt{3} + 3$
$= 4 + 2\sqrt{3}$

60. $6 + 2\sqrt{5}$

61. $(a + \sqrt{b})^2$
$= a^2 + 2a\sqrt{b} + (\sqrt{b})^2$ Squaring a binomial
$= a^2 + 2a\sqrt{b} + b$

62. $x^2 - 2x\sqrt{y} + y$

63. $(2x - \sqrt[4]{y})^2 = (2x)^2 - 2 \cdot 2x\sqrt[4]{y} + (\sqrt[4]{y})^2$
$= 4x^2 - 4x\sqrt[4]{y} + y^{2/4}$
$= 4x^2 - 4x\sqrt[4]{y} + y^{1/2}$
$= 4x^2 - 4x\sqrt[4]{y} + \sqrt{y}$

64. $9a^2 + 6a\sqrt[4]{b} + \sqrt{b}$

65. $(\sqrt{m} + \sqrt{n})^2 = (\sqrt{m})^2 + 2\sqrt{m} \cdot \sqrt{n} + (\sqrt{n})^2 =$
$m + 2\sqrt{mn} + n$

66. $r - 2\sqrt{rs} + s$

67. $\sqrt{a}(\sqrt[3]{a} + \sqrt[4]{ab})$
$= a^{1/2}[a^{1/3} + (ab)^{1/4}]$ Converting to exponential notation
$= a^{1/2}[(a^{1/3} + a^{1/4}b^{1/4})$ Using the laws of exponents
$= a^{1/2}a^{1/3} + a^{1/2}a^{1/4}b^{1/4}$ Using the distributive law
$= a^{1/2+1/3} + a^{1/2+1/4}b^{1/4}$ Adding exponents
$= a^{5/6} + a^{3/4}b^{1/4}$
$= a^{5/6} + (a^3b)^{1/4}$ Using the laws of exponents
$= \sqrt[6]{a^5} + \sqrt[4]{a^3b}$ Converting back to radical notation

68. $\sqrt[6]{x^5y^2} + x\sqrt[4]{y}$

69. $\sqrt[3]{x^2y}(\sqrt{xy} - \sqrt[5]{xy^3})$
$= (x^2y)^{1/3}[(xy)^{1/2} - (xy^3)^{1/5}]$
$= x^{2/3}y^{1/3}(x^{1/2}y^{1/2} - x^{1/5}y^{3/5})$
$= x^{2/3}y^{1/3}x^{1/2}y^{1/2} - x^{2/3}y^{1/3}x^{1/5}y^{3/5}$
$= x^{2/3+1/2}y^{1/3+1/2} - x^{2/3+1/5}y^{1/3+3/5}$
$= x^{7/6}y^{5/6} - x^{13/15}y^{14/15}$
$= x^{1\frac{1}{6}}y^{\frac{5}{6}} - x^{13/15}y^{14/15}$
Writing a mixed numeral
$= x \cdot x^{1/6}y^{5/6} - x^{13/15}y^{14/15}$
$= x(xy^5)^{1/6} - (x^{13}y^{14})^{1/15}$
$= x\sqrt[6]{xy^5} - \sqrt[15]{x^{13}y^{14}}$

70. $a\sqrt[12]{a^2b^7} - \sqrt[20]{a^{18}b^{13}}$

71. $(m + \sqrt[3]{n^2})(2m + \sqrt[4]{n})$
$= (m + n^{2/3})(2m + n^{1/4})$ Converting to exponential notation
$= 2m^2 + mn^{1/4} + 2mn^{2/3} + n^{2/3}n^{1/4}$ Using FOIL
$= 2m^2 + mn^{1/4} + 2mn^{2/3} + n^{2/3+1/4}$ Adding exponents
$= 2m^2 + mn^{1/4} + 2mn^{2/3} + n^{11/12}$
$= 2m^2 + m\sqrt[4]{n} + 2m\sqrt[3]{n^2} + \sqrt[12]{n^{11}}$ Converting back to radical notation

72. $3r^2 - r\sqrt[5]{s} - 3r\sqrt[4]{s^3} + \sqrt[20]{s^{19}}$

73. $(a + \sqrt[4]{b})(a - \sqrt[3]{c})$
$= (a + b^{1/4})(a - c^{1/3})$
$= a^2 - ac^{1/3} + ab^{1/4} - b^{1/4}c^{1/3}$
$= a^2 - ac^{1/3} + ab^{1/4} - b^{3/12}c^{4/12}$ Finding a common denominator
$= a^2 - ac^{1/3} + ab^{1/4} - (b^3c^4)^{1/12}$
$= a^2 - a\sqrt[3]{c} + a\sqrt[4]{b} - \sqrt[12]{b^3c^4}$ Converting back to rational notation

74. $x^2 + x\sqrt[4]{z} - x\sqrt[3]{y} - \sqrt[12]{y^4z^3}$

75. $(\sqrt{x} - \sqrt[3]{yz})(\sqrt[4]{x} + \sqrt[5]{xz})$
$= [x^{1/2} - (yz)^{1/3}][x^{1/4} + (xz)^{1/5}]$
$= (x^{1/2} - y^{1/3}z^{1/3})(x^{1/4} + x^{1/5}z^{1/5})$
$= x^{1/2}x^{1/4} + x^{1/2}x^{1/5}z^{1/5} - x^{1/4}y^{1/3}z^{1/3} - x^{1/5}y^{1/3}z^{1/3}z^{1/5}$
$= x^{3/4} + x^{7/10}z^{1/5} - x^{1/4}y^{1/3}z^{1/3} - x^{1/5}y^{1/3}z^{8/15}$
Adding exponents
$= x^{3/4} + x^{7/10}z^{2/10} - x^{3/12}y^{4/12}z^{4/12} - x^{3/15}y^{5/15}z^{8/15}$
Finding common denominators
$= x^{3/4} + (x^7z^2)^{1/10} - (x^3y^4z^4)^{1/12} - (x^3y^5z^8)^{1/15}$
$= \sqrt[4]{x^3} + \sqrt[10]{x^7z^2} - \sqrt[12]{x^3y^4z^4} - \sqrt[15]{x^3y^5z^8}$

76. $\sqrt[6]{4a^5} - \sqrt[10]{a^5b^2c^2} + \sqrt[12]{432a^4b^3} - \sqrt[20]{243b^9c^4}$

77.

$$\frac{5}{x-1} + \frac{9}{x^2+x+1} = \frac{15}{x^3-1}$$

$$\frac{5}{x-1} + \frac{9}{x^2+x+1} =$$

$$\frac{15}{(x-1)(x^2+x+1)}$$

The LCD is $(x-1)(x^2+x+1)$.

$$(x-1)(x^2+x+1)\left(\frac{5}{x-1} + \frac{9}{x^2+x+1}\right) =$$

$$(x-1)(x^2+x+1) \cdot \frac{15}{(x-1)(x^2+x+1)}$$

$$5(x^2+x+1) + 9(x-1) = 15$$

$$5x^2 + 5x + 5 + 9x - 9 = 15$$

$$5x^2 + 14x - 4 = 15$$

$$5x^2 + 14x - 19 = 0$$

$$(5x+19)(x-1) = 0$$

$$5x + 19 = 0 \quad \text{or} \quad x - 1 = 0$$

$$5x = -19 \quad \text{or} \quad x = 1$$

$$x = -\frac{19}{5} \quad \text{or} \quad x = 1$$

Only $-\dfrac{19}{5}$ checks. The solution is $-\dfrac{19}{5}$.

78. $\dfrac{x-2}{x+3}$

79. $\sqrt{432} - \sqrt{6125} + \sqrt{845} - \sqrt{4800} =$

$$\sqrt{144 \cdot 3} - \sqrt{1225 \cdot 5} + \sqrt{169 \cdot 5} - \sqrt{1600 \cdot 3} =$$

$$12\sqrt{3} - 35\sqrt{5} + 13\sqrt{5} - 40\sqrt{3} = -28\sqrt{3} - 22\sqrt{5}$$

80. $(25x - 30y - 9xy)\sqrt{2xy}$

81. $\dfrac{1}{2}\sqrt{36a^5bc^4} - \dfrac{1}{2}\sqrt[3]{64a^4bc^6} + \dfrac{1}{6}\sqrt{144a^3bc^2} =$

$$\frac{1}{2}\sqrt{36a^4c^4 \cdot ab} - \frac{1}{2}\sqrt[3]{64a^3c^6 \cdot ab} + \frac{1}{6}\sqrt{144a^2c^2 \cdot ab} =$$

$$\frac{1}{2}(6a^2c^2)\sqrt{ab} - \frac{1}{2}(4ac^2)\sqrt[3]{ab} + \frac{1}{6}(12ac)\sqrt{ab} =$$

$$3a^2c^2\sqrt{ab} - 2ac^2\sqrt[3]{ab} + 2ac\sqrt{ab}$$

$$(3a^2c^2 + 2ac)\sqrt{ab} - 2ac^2\sqrt[3]{ab}, \text{ or}$$

$$ac[(3ac+2)\sqrt{ab} - 2c\sqrt[3]{ab}]$$

82. $(7x^2 - 2y^2)\sqrt{x+y}$

83. $\sqrt{9+3\sqrt{5}}\sqrt{9-3\sqrt{5}} = \sqrt{(9+3\sqrt{5})(9-3\sqrt{5})} =$

$$\sqrt{81 - 9 \cdot 5} = \sqrt{81 - 45} = \sqrt{36} = 6$$

84. $2x - 2\sqrt{x^2-4}$

85. $(\sqrt{3}+\sqrt{5}-\sqrt{6})^2 = [(\sqrt{3}+\sqrt{5}) - \sqrt{6}]^2 =$

$$(\sqrt{3}+\sqrt{5})^2 - 2(\sqrt{3}+\sqrt{5})(\sqrt{6}) + (\sqrt{6})^2 =$$

$$3 + 2\sqrt{15} + 5 - 2\sqrt{18} - 2\sqrt{30} + 6 =$$

$$14 + 2\sqrt{15} - 2\sqrt{9 \cdot 2} - 2\sqrt{30} =$$

$$14 + 2\sqrt{15} - 6\sqrt{2} - 2\sqrt{30}$$

86. $\sqrt[3]{y} - y$

87. $(\sqrt[3]{9} - 2)(\sqrt[3]{9} + 4)$

$$= \sqrt[3]{81} + 2\sqrt[3]{9} - 8$$

$$= \sqrt[3]{27 \cdot 3} + 2\sqrt[3]{9} - 8$$

$$= 3\sqrt[3]{3} + 2\sqrt[3]{9} - 8$$

88. $12 + 6\sqrt{3}$

Exercise Set 10.6

1. $\sqrt{\dfrac{6}{5}} = \sqrt{\dfrac{6}{5} \cdot \dfrac{5}{5}} = \sqrt{\dfrac{30}{25}} = \dfrac{\sqrt{30}}{\sqrt{25}} = \dfrac{\sqrt{30}}{5}$

2. $\dfrac{\sqrt{66}}{6}$

3. $\sqrt{\dfrac{10}{7}} = \sqrt{\dfrac{10}{7} \cdot \dfrac{7}{7}} = \sqrt{\dfrac{70}{49}} = \dfrac{\sqrt{70}}{\sqrt{49}} = \dfrac{\sqrt{70}}{7}$

4. $\dfrac{\sqrt{66}}{3}$

5. $\dfrac{6\sqrt{5}}{5\sqrt{3}} = \dfrac{6\sqrt{5}}{5\sqrt{3}} \cdot \dfrac{\sqrt{3}}{\sqrt{3}} = \dfrac{6\sqrt{15}}{5 \cdot 3} = \dfrac{2\sqrt{15}}{5}$

6. $\dfrac{\sqrt{6}}{5}$

7. $\sqrt[3]{\dfrac{16}{9}} = \sqrt[3]{\dfrac{16}{9} \cdot \dfrac{3}{3}} = \sqrt[3]{\dfrac{48}{27}} = \dfrac{\sqrt[3]{8 \cdot 6}}{\sqrt[3]{27}} = \dfrac{2\sqrt[3]{6}}{3}$

8. $\dfrac{\sqrt[3]{6}}{3}$

9. $\dfrac{\sqrt[3]{3a}}{\sqrt[3]{5c}} = \dfrac{\sqrt[3]{3a}}{\sqrt[3]{5c}} \cdot \dfrac{\sqrt[3]{5^2c^2}}{\sqrt[3]{5^2c^2}} = \dfrac{\sqrt[3]{75ac^2}}{\sqrt[3]{5^3c^3}} = \dfrac{\sqrt[3]{75ac^2}}{5c}$

10. $\dfrac{\sqrt[3]{63xy^2}}{3y}$

11. $\dfrac{\sqrt[3]{5y^4}}{\sqrt[3]{6x^4}} = \dfrac{\sqrt[3]{5y^4}}{\sqrt[3]{6x^4}} \cdot \dfrac{\sqrt[3]{36x^2}}{\sqrt[3]{36x^2}} = \dfrac{\sqrt[3]{y^3 \cdot 180x^2y}}{\sqrt[3]{216x^6}} =$

$$\dfrac{y\sqrt[3]{180x^2y}}{6x^2}$$

12. $\dfrac{a\sqrt[3]{147ab}}{7b}$

13. $\dfrac{1}{\sqrt[3]{xy}} = \dfrac{1}{\sqrt[3]{xy}} \cdot \dfrac{\sqrt[3]{(xy)^2}}{\sqrt[3]{(xy)^2}} = \dfrac{\sqrt[3]{(xy)^2}}{\sqrt[3]{(xy)^3}} = \dfrac{\sqrt[3]{x^2y^2}}{xy}$

14. $\dfrac{\sqrt[3]{a^2b^2}}{ab}$

15. $\sqrt{\dfrac{7a}{18}} = \sqrt{\dfrac{7a}{18} \cdot \dfrac{2}{2}} = \sqrt{\dfrac{14a}{36}} = \dfrac{\sqrt{14a}}{\sqrt{36}} = \dfrac{\sqrt{14a}}{6}$

16. $\dfrac{\sqrt{30x}}{10}$

17. $\sqrt{\dfrac{9}{20x^2y}} = \sqrt{\dfrac{9}{20x^2y} \cdot \dfrac{5y}{5y}} = \sqrt{\dfrac{9 \cdot 5y}{100x^2y^2}} =$

$\dfrac{\sqrt{9 \cdot 5y}}{\sqrt{100x^2y^2}} = \dfrac{3\sqrt{5y}}{10xy}$

18. $\dfrac{\sqrt{10a}}{8ab}$

19. $\sqrt[3]{\dfrac{9}{100x^2y^5}} = \sqrt[3]{\dfrac{9}{100x^2y^5} \cdot \dfrac{10xy}{10xy}} = \sqrt[3]{\dfrac{90xy}{1000x^3y^6}} =$

$\dfrac{\sqrt[3]{90xy}}{\sqrt[3]{1000x^3y^6}} = \dfrac{\sqrt[3]{90xy}}{10xy^2}$

20. $\dfrac{\sqrt[3]{42a^2b^2}}{6a^2b}$

21. $\dfrac{\sqrt{7}}{\sqrt{3x}} = \dfrac{\sqrt{7}}{\sqrt{3x}} \cdot \dfrac{\sqrt{7}}{\sqrt{7}} = \dfrac{7}{\sqrt{21x}}$

22. $\dfrac{6}{\sqrt{30x}}$

23. $\sqrt{\dfrac{14}{21}} = \sqrt{\dfrac{2}{3}} = \sqrt{\dfrac{2}{3} \cdot \dfrac{2}{2}} = \sqrt{\dfrac{4}{6}} = \dfrac{\sqrt{4}}{\sqrt{6}} = \dfrac{2}{\sqrt{6}}$

24. $\dfrac{2}{\sqrt{5}}$

25. $\dfrac{4\sqrt{13}}{3\sqrt{7}} = \dfrac{4\sqrt{13}}{3\sqrt{7}} \cdot \dfrac{\sqrt{13}}{\sqrt{13}} = \dfrac{4\sqrt{169}}{3\sqrt{91}} = \dfrac{4 \cdot 13}{3\sqrt{91}} = \dfrac{52}{3\sqrt{91}}$

26. $\dfrac{105}{2\sqrt{105}}$

27. $\dfrac{\sqrt[3]{7}}{\sqrt[3]{2}} = \dfrac{\sqrt[3]{7}}{\sqrt[3]{2}} \cdot \dfrac{\sqrt[3]{7^2}}{\sqrt[3]{7^2}} = \dfrac{\sqrt[3]{7^3}}{\sqrt[3]{98}} = \dfrac{7}{\sqrt[3]{98}}$

28. $\dfrac{5}{\sqrt[3]{100}}$

29. $\sqrt{\dfrac{7x}{3y}} = \sqrt{\dfrac{7x}{3y} \cdot \dfrac{7x}{7x}} = \dfrac{\sqrt{(7x)^2}}{\sqrt{21xy}} = \dfrac{7x}{\sqrt{21xy}}$

30. $\dfrac{6a}{\sqrt{30ab}}$

31. $\dfrac{\sqrt[3]{5y^4}}{\sqrt[3]{6x^5}} = \dfrac{\sqrt[3]{5y^4}}{\sqrt[3]{6x^5}} \cdot \dfrac{\sqrt[3]{5^2y^2}}{\sqrt[3]{5^2y^2}} = \dfrac{\sqrt[3]{5^3y^6}}{\sqrt[3]{150x^5y^2}} =$

$\dfrac{5y^2}{x\sqrt[3]{150x^2y^2}}$

32. $\dfrac{3a^2}{\sqrt[3]{63ab^2}}$

33. $\dfrac{\sqrt{ab}}{3} = \dfrac{\sqrt{ab}}{3} \cdot \dfrac{\sqrt{ab}}{\sqrt{ab}} = \dfrac{ab}{3\sqrt{ab}}$

34. $\dfrac{xy}{5\sqrt{xy}}$

35. $\sqrt{\dfrac{x^3y}{2}} = \sqrt{\dfrac{x^3y}{2} \cdot \dfrac{xy}{xy}} = \sqrt{\dfrac{x^4y^2}{2xy}} = \dfrac{\sqrt{x^4y^2}}{\sqrt{2xy}} = \dfrac{x^2y}{\sqrt{2xy}}$

36. $\dfrac{ab^3}{\sqrt{3ab}}$

37. $\dfrac{\sqrt[3]{a^2b}}{\sqrt[3]{5}} = \dfrac{\sqrt[3]{a^2b}}{\sqrt[3]{5}} \cdot \dfrac{\sqrt[3]{ab^2}}{\sqrt[3]{ab^2}} = \dfrac{ab}{\sqrt[3]{5ab^2}}$

38. $\dfrac{xy}{\sqrt[3]{7x^2y}}$

39. $\sqrt[3]{\dfrac{x^4y^2}{3}} = \sqrt[3]{\dfrac{x^4y^2}{3} \cdot \dfrac{x^2y}{x^2y}} = \sqrt[3]{\dfrac{x^6y^3}{3x^2y}} = \dfrac{\sqrt[3]{x^6y^3}}{\sqrt[3]{3x^2y}} = \dfrac{x^2y}{\sqrt[3]{3x^2y}}$

40. $\dfrac{a^2b}{\sqrt[3]{2ab^2}}$

41. $\dfrac{5}{8-\sqrt{6}} = \dfrac{5}{8-\sqrt{6}} \cdot \dfrac{8+\sqrt{6}}{8+\sqrt{6}} = \dfrac{5(8+\sqrt{6})}{8^2 - (\sqrt{6})^2} =$

$\dfrac{5(8+\sqrt{6})}{64-6} = \dfrac{5(8+\sqrt{6})}{58}$

42. $\dfrac{7(9-\sqrt{10})}{71}$

43. $\dfrac{-4\sqrt{7}}{\sqrt{5}-\sqrt{3}} = \dfrac{-4\sqrt{7}}{\sqrt{5}-\sqrt{3}} \cdot \dfrac{\sqrt{5}+\sqrt{3}}{\sqrt{5}+\sqrt{3}} =$

$\dfrac{-4\sqrt{7}(\sqrt{5}+\sqrt{3})}{(\sqrt{5})^2 - (\sqrt{3})^2} = \dfrac{-4\sqrt{7}(\sqrt{5}+\sqrt{3})}{5-3} =$

$\dfrac{-4\sqrt{7}(\sqrt{5}+\sqrt{3})}{2} = -2\sqrt{7}(\sqrt{5}+\sqrt{3}) =$

$-2\sqrt{35} - 2\sqrt{21}$

44. $\dfrac{3\sqrt{2}(\sqrt{3}+\sqrt{5})}{2}$

45. $\dfrac{\sqrt{5}-2\sqrt{6}}{\sqrt{3}-4\sqrt{5}} = \dfrac{\sqrt{5}-2\sqrt{6}}{\sqrt{3}-4\sqrt{5}} \cdot \dfrac{\sqrt{3}+4\sqrt{5}}{\sqrt{3}+4\sqrt{5}} =$

$\dfrac{\sqrt{15}+4\cdot 5 - 2\sqrt{18} - 8\sqrt{30}}{(\sqrt{3})^2 - (4\sqrt{5})^2} =$

$\dfrac{\sqrt{15}+20 - 2\sqrt{9\cdot 2} - 8\sqrt{30}}{3 - 16\cdot 5} =$

$\dfrac{\sqrt{15}+20 - 6\sqrt{2} - 8\sqrt{30}}{-77}$

46. $\dfrac{3\sqrt{2} + 2\sqrt{42} - 3\sqrt{15} - 6\sqrt{35}}{-25}$

47. $\dfrac{\sqrt{x}-\sqrt{y}}{\sqrt{x}+\sqrt{y}} = \dfrac{\sqrt{x}-\sqrt{y}}{\sqrt{x}+\sqrt{y}} \cdot \dfrac{\sqrt{x}-\sqrt{y}}{\sqrt{x}-\sqrt{y}} =$

$\dfrac{x - \sqrt{xy} - \sqrt{xy} + y}{x-y} = \dfrac{x - 2\sqrt{xy} + y}{x-y}$

48. $\dfrac{a + 2\sqrt{ab} + b}{a-b}$

49. $\dfrac{5\sqrt{3}-3\sqrt{2}}{3\sqrt{2}-2\sqrt{3}} = \dfrac{5\sqrt{3}-3\sqrt{2}}{3\sqrt{2}-2\sqrt{3}} \cdot \dfrac{3\sqrt{2}+2\sqrt{3}}{3\sqrt{2}+2\sqrt{3}} =$

$\dfrac{15\sqrt{6}+10\cdot 3-9\cdot 2-6\sqrt{6}}{9\cdot 2-4\cdot 3} = \dfrac{12+9\sqrt{6}}{6} =$

$\dfrac{3(4+3\sqrt{6})}{3\cdot 2} = \dfrac{4+3\sqrt{6}}{2}$

50. $\dfrac{4\sqrt{6}+9}{3}$

51. $\dfrac{3\sqrt{x}+\sqrt{y}}{2\sqrt{x}+3\sqrt{y}} = \dfrac{3\sqrt{x}+\sqrt{y}}{2\sqrt{x}+3\sqrt{y}} \cdot \dfrac{2\sqrt{x}-3\sqrt{y}}{2\sqrt{x}-3\sqrt{y}} =$

$\dfrac{6x-9\sqrt{xy}+2\sqrt{xy}-3y}{4x-9y} = \dfrac{6x-7\sqrt{xy}-3y}{4x-9y}$

52. $\dfrac{6a-7\sqrt{ab}+2b}{9a-4b}$

53. $\dfrac{\sqrt{3}+5}{8} = \dfrac{\sqrt{3}+5}{8} \cdot \dfrac{\sqrt{3}-5}{\sqrt{3}-5} =$

$\dfrac{3-25}{8(\sqrt{3}-5)} = \dfrac{-22}{8(\sqrt{3}-5)} = \dfrac{2(-11)}{2\cdot 4(\sqrt{3}-5)} =$

$\dfrac{2}{2}\cdot\dfrac{-11}{4(\sqrt{3}-5)} = \dfrac{-11}{4(\sqrt{3}-5)}$

54. $\dfrac{7}{5(3+\sqrt{2})}$

55. $\dfrac{\sqrt{3}-5}{\sqrt{2}+5} = \dfrac{\sqrt{3}-5}{\sqrt{2}+5} \cdot \dfrac{\sqrt{3}+5}{\sqrt{3}+5} =$

$\dfrac{3-25}{\sqrt{6}+5\sqrt{2}+5\sqrt{3}+25} = \dfrac{-22}{\sqrt{6}+5\sqrt{2}+5\sqrt{3}+25}$

56. $\dfrac{-3}{3\sqrt{2}+3\sqrt{3}+7\sqrt{6}+21}$

57. $\dfrac{\sqrt{x}-\sqrt{y}}{\sqrt{x}+\sqrt{y}} = \dfrac{\sqrt{x}-\sqrt{y}}{\sqrt{x}+\sqrt{y}} \cdot \dfrac{\sqrt{x}+\sqrt{y}}{\sqrt{x}+\sqrt{y}} =$

$\dfrac{x-y}{x+\sqrt{xy}+\sqrt{xy}+y} = \dfrac{x-y}{x+2\sqrt{xy}+y}$

58. $\dfrac{x-y}{x-2\sqrt{xy}+y}$

59. $\dfrac{4\sqrt{6}-5\sqrt{3}}{2\sqrt{3}+7\sqrt{6}} = \dfrac{4\sqrt{6}-5\sqrt{3}}{2\sqrt{3}+7\sqrt{6}} \cdot \dfrac{4\sqrt{6}+5\sqrt{3}}{4\sqrt{6}+5\sqrt{3}} =$

$\dfrac{16\cdot 6-25\cdot 3}{8\sqrt{18}+10\cdot 3+28\cdot 6+35\sqrt{18}} =$

$\dfrac{96-75}{43\sqrt{18}+30+168} = \dfrac{21}{43\sqrt{9\cdot 2}+198} =$

$\dfrac{21}{43\cdot 3\sqrt{2}+198} = \dfrac{3\cdot 7}{3(43\sqrt{2}+66)} = \dfrac{7}{43\sqrt{2}+66}$

60. $\dfrac{53}{75\sqrt{6}-187}$

61. $\dfrac{\sqrt{3}+2\sqrt{x}}{\sqrt{3}-\sqrt{x}} = \dfrac{\sqrt{3}+2\sqrt{x}}{\sqrt{3}-\sqrt{x}} \cdot \dfrac{\sqrt{3}-2\sqrt{x}}{\sqrt{3}-2\sqrt{x}} =$

$\dfrac{3-4x}{3-2\sqrt{3x}-\sqrt{3x}+2x} = \dfrac{3-4x}{3-3\sqrt{3x}+2x}$

62. $\dfrac{5-9x}{5+4\sqrt{5x}+3x}$

63. $\dfrac{a+b\sqrt{c}}{a+\sqrt{c}} = \dfrac{a+b\sqrt{c}}{a+\sqrt{c}} \cdot \dfrac{a-b\sqrt{c}}{a-b\sqrt{c}} =$

$\dfrac{a^2-b^2c}{a^2-ab\sqrt{c}+a\sqrt{c}-bc} =$

$\dfrac{a^2-b^2c}{a^2+(-ab+a)\sqrt{c}-bc},$ or

$\dfrac{a^2-b^2c}{a^2+(a-ab)\sqrt{c}-bc}$

64. $\dfrac{a^2b-c^2}{ab+(1-a)c\sqrt{b}-c^2}$

65. $\dfrac{1}{2}-\dfrac{1}{3}=\dfrac{1}{t},$ LCD is $6t$

$6t\left(\dfrac{1}{2}-\dfrac{1}{3}\right) = 6t\left(\dfrac{1}{t}\right)$

$3t-2t = 6$

$t = 6$

Check:

$\dfrac{1}{2}-\dfrac{1}{3}=\dfrac{1}{t}$

$\begin{array}{c|c} \dfrac{1}{2}-\dfrac{1}{3} & \dfrac{1}{6} \\[2mm] \dfrac{3}{6}-\dfrac{2}{6} & \\[2mm] \dfrac{1}{6} & \text{TRUE} \end{array}$

The solution is 6.

66. 1

67.

68.

69. $\dfrac{a-\sqrt{a+b}}{\sqrt{a+b}-b} = \dfrac{a-\sqrt{a+b}}{\sqrt{a+b}-b} \cdot \dfrac{\sqrt{a+b}+b}{\sqrt{a+b}+b} =$

$\dfrac{a\sqrt{a+b}+ab-(a+b)-b\sqrt{a+b}}{a+b-b^2} =$

$\dfrac{ab+(a-b)\sqrt{a+b}-a-b}{a+b-b^2}$

70. $\dfrac{15y+6\sqrt{y(z+y)}+20y\sqrt{z}+8\sqrt{yz(z+y)}}{21y-4z}$

71. $\dfrac{b+\sqrt{b}}{1+b+\sqrt{b}} = \dfrac{b+\sqrt{b}}{(1+b)+\sqrt{b}} \cdot \dfrac{(1+b)-\sqrt{b}}{(1+b)-\sqrt{b}}$

$= \dfrac{(b+\sqrt{b})(1+b-\sqrt{b})}{(1+b)^2-(\sqrt{b})^2}$

$= \dfrac{b+b^2-b\sqrt{b}+\sqrt{b}+b\sqrt{b}-b}{1+2b+b^2-b}$

$= \dfrac{b^2+\sqrt{b}}{1+b+b^2}$

72. $\dfrac{1}{\sqrt{y+18}+\sqrt{y}}$

73. $\dfrac{\sqrt{x+6}-5}{\sqrt{x+6}+5} = \dfrac{\sqrt{x+6}-5}{\sqrt{x+6}+5} \cdot \dfrac{\sqrt{x+6}+5}{\sqrt{x+6}+5}$

$\qquad = \dfrac{(x+6)-25}{(x+6)+10\sqrt{x+6}+25}$

$\qquad = \dfrac{x-19}{x+10\sqrt{x+6}+31}$

74. $\dfrac{-3\sqrt{a^2-3}}{a^2-3}$

75. $5\sqrt{\dfrac{x}{y}} + 4\sqrt{\dfrac{y}{x}} - \dfrac{3}{\sqrt{xy}} = \dfrac{5\sqrt{x}}{\sqrt{y}} + \dfrac{4\sqrt{y}}{\sqrt{x}} - \dfrac{3}{\sqrt{xy}} =$

$\dfrac{5\sqrt{x}}{\sqrt{y}} \cdot \dfrac{\sqrt{x}}{\sqrt{x}} + \dfrac{4\sqrt{y}}{\sqrt{x}} \cdot \dfrac{\sqrt{y}}{\sqrt{y}} - \dfrac{3}{\sqrt{xy}} = \dfrac{5x}{\sqrt{xy}} + \dfrac{4y}{\sqrt{xy}} - \dfrac{3}{\sqrt{xy}} =$

$\dfrac{5x+4y-3}{\sqrt{xy}} = \dfrac{5x+4y-3}{\sqrt{xy}} \cdot \dfrac{\sqrt{xy}}{\sqrt{xy}} =$

$\dfrac{(5x+4y-3)\sqrt{xy}}{xy}$

76. $1 - \sqrt{w}$

77. $\dfrac{1}{4+\sqrt{3}} + \dfrac{1}{\sqrt{3}} + \dfrac{1}{\sqrt{3}-4} =$

$\dfrac{1}{4+\sqrt{3}} \cdot \dfrac{\sqrt{3}(\sqrt{3}-4)}{\sqrt{3}(\sqrt{3}-4)} + \dfrac{1}{\sqrt{3}} \cdot \dfrac{(4+\sqrt{3})(\sqrt{3}-4)}{(4+\sqrt{3})(\sqrt{3}-4)} + \dfrac{1}{\sqrt{3}-4} \cdot$

$\dfrac{\sqrt{3}(4+\sqrt{3})}{\sqrt{3}(4+\sqrt{3})} =$

$\dfrac{3-4\sqrt{3}-16+3+4\sqrt{3}+3}{\sqrt{3}(4+\sqrt{3})(\sqrt{3}-4)} = \dfrac{-7}{\sqrt{3}(-16+3)} =$

$\dfrac{-7}{-13\sqrt{3}} \cdot \dfrac{\sqrt{3}}{\sqrt{3}} = \dfrac{7\sqrt{3}}{39}$

Exercise Set 10.7

1. $\sqrt{2x-3} = 1$

$\quad (\sqrt{2x-3})^2 = 1^2 \quad$ Principle of powers (squaring)

$\qquad 2x-3 = 1$

$\qquad 2x = 4$

$\qquad x = 2$

Check: $\quad \dfrac{\sqrt{2x-3} = 1}{\sqrt{2 \cdot 2 - 3} \; ? \; 1}$

$\qquad\qquad\quad \dfrac{\sqrt{1}}{1} \; \Big| \; 1 \qquad$ TRUE

The solution is 2.

2. 33

3. $\sqrt{3x}+1 = 7$

$\quad \sqrt{3x} = 6 \qquad$ Adding to isolate the radical

$\quad (\sqrt{3x})^2 = 6^2 \qquad$ Principle of powers (squaring)

$\qquad 3x = 36$

$\qquad x = 12$

Check: $\quad \dfrac{\sqrt{3x}+1 = 7}{\sqrt{3 \cdot 12}+1 \; ? \; 7}$

$\qquad\qquad\quad \begin{array}{c|c} 6+1 & \\ 7 & 7 \end{array} \qquad$ TRUE

The solution is 12.

4. 32

5. $\sqrt{y+1}-5 = 8$

$\quad \sqrt{y+1} = 13 \qquad$ Adding to isolate the radical

$\quad (\sqrt{y+1})^2 = 13^2 \qquad$ Principle of powers (squaring)

$\qquad y+1 = 169$

$\qquad y = 168$

Check: $\quad \dfrac{\sqrt{y+1}-5 = 8}{\sqrt{168+1}-5 \; ? \; 8}$

$\qquad\qquad\quad \begin{array}{c|c} 13-5 & \\ 8 & 8 \end{array} \qquad$ TRUE

The solution is 168.

6. 11

7. $\sqrt{y-3}+4 = 2$

$\quad \sqrt{y-3} = -2$

At this point we might observe that this equation has no real-number solution, because the principle square root of a number is never negative. However, we will continue with the solution.

$\quad (\sqrt{y-3})^2 = (-2)^2$

$\qquad y-3 = 4$

$\qquad y = 7$

Check: $\quad \dfrac{\sqrt{y-3}+4 = 2}{\sqrt{7-3}+4 \; ? \; 2}$

$\qquad\qquad\quad \begin{array}{c|c} \sqrt{4}+4 & \\ 6 & 2 \end{array} \qquad$ FALSE

The number 7 does not check. There is no solution.

8. -3

9. $\sqrt[3]{x+5} = 2$

$\quad (\sqrt[3]{x+5})^3 = 2^3$

$\qquad x+5 = 8$

$\qquad x = 3$

Check: $\quad \dfrac{\sqrt[3]{x+5} = 2}{\sqrt[3]{3+5} \; ? \; 2}$

$\qquad\qquad\quad \begin{array}{c|c} \sqrt[3]{8} & \\ 2 & 2 \end{array} \qquad$ TRUE

The solution is 3.

10. 29

11.
$$\sqrt[4]{y-3} = 2$$
$$(\sqrt[4]{y-3})^4 = 2^4$$
$$y - 3 = 16$$
$$y = 19$$

Check:

$\sqrt[4]{y-3} = 2$	
$\sqrt[4]{19-3}$? 2	
$\sqrt[4]{16}$	
2	2 TRUE

The solution is 19.

12. 78

13.
$$\sqrt{3y+1} = 9$$
$$(\sqrt{3y+1})^2 = 9^2$$
$$3y + 1 = 81$$
$$3y = 80$$
$$y = \frac{80}{3}$$

Check:

$\sqrt{3y+1} = 9$	
$\sqrt{3 \cdot \dfrac{80}{3} + 1}$? 9	
$\sqrt{81}$	
9	9 TRUE

The solution is $\frac{80}{3}$.

14. 84

15.
$$3\sqrt{x} = 6$$
$$\sqrt{x} = 2 \quad \text{Multiplying by } \frac{1}{3}$$
$$(\sqrt{x})^2 = 2^2$$
$$x = 4$$

Check:

$3\sqrt{x} = 6$	
$3\sqrt{4}$? 6	
$3 \cdot 2$	
6	6 TRUE

The solution is 4.

16. $\dfrac{1}{16}$

17.
$$2y^{1/2} - 7 = 9$$
$$2\sqrt{y} - 7 = 9$$
$$2\sqrt{y} = 16$$
$$\sqrt{y} = 8$$
$$(\sqrt{y})^2 = 8^2$$
$$y = 64$$

Check:

$2y^{1/2} - 7 = 9$	
$2 \cdot 64^{1/2} - 7$? 9	
$2 \cdot 8 - 7$	
9	9 TRUE

The solution is 64.

18. No solution

19.
$$\sqrt[3]{x} = -3$$
$$(\sqrt[3]{x})^3 = (-3)^3$$
$$x = -27$$

Check:

$\sqrt[3]{x} = -3$	
$\sqrt[3]{-27}$? -3	
-3	-3 TRUE

The solution is -27.

20. -64

21.
$$\sqrt{y+3} - 20 = 0$$
$$\sqrt{y+3} = 20$$
$$(\sqrt{y+3})^2 = 20^2$$
$$y + 3 = 400$$
$$y = 397$$

Check:

$\sqrt{y+3} - 20 = 0$	
$\sqrt{397+3} - 20$? 0	
$\sqrt{400} - 20$	
0	0 TRUE

The solution is 397.

22. 117

23.
$$(x+2)^{1/2} = -4$$
$$\sqrt{x+2} = -4$$

We might observe that this equation has no real-number solution, since the principal square root of a number is never negative. However, we will go through the solution process.

$$(\sqrt{x+2})^2 = (-4)^2$$
$$x + 2 = 16$$
$$x = 14$$

Check:

$(x+2)^{1/2} = -4$	
$(14+2)^{1/2}$? -4	
$16^{1/2}$	
4	-4 FALSE

The number 14 does not check. The equation has no solution.

24. No solution

25.
$$\sqrt{2x+3} - 5 = -2$$
$$\sqrt{2x+3} = 3$$
$$(\sqrt{2x+3})^2 = 3^2$$
$$2x + 3 = 9$$
$$2x = 6$$
$$x = 3$$

Check: $\dfrac{\sqrt{2x+3}-5=-2}{}$

$$\sqrt{2\cdot 3+3}-5 \ ? \ -2$$
$$\sqrt{9}-5 \ \Big|$$
$$-2 \ \Big| \ -2 \qquad \text{TRUE}$$

The solution is 3.

26. $\dfrac{8}{3}$

27. $8 = x^{-1/2}$

$$8 = \frac{1}{x^{1/2}}$$
$$8 = \frac{1}{\sqrt{x}}$$
$$8 \cdot \sqrt{x} = \frac{1}{\sqrt{x}} \cdot \sqrt{x}$$
$$8\sqrt{x} = 1$$
$$(8\sqrt{x})^2 = 1^2$$
$$64x = 1$$
$$x = \frac{1}{64}$$

Check: $\dfrac{8 = x^{-1/2}}{}$

$$8 \ ? \ \left(\frac{1}{64}\right)^{-1/2}$$
$$\Big| \ 64^{1/2}$$
$$8 \ \Big| \ 8 \qquad \text{TRUE}$$

The solution is $\dfrac{1}{64}$.

28. $\dfrac{1}{9}$

29. $\sqrt[3]{6x+9}+8 = 5$

$$\sqrt[3]{6x+9} = -3$$
$$(\sqrt[3]{6x+9})^3 = (-3)^3$$
$$6x+9 = -27$$
$$6x = -36$$
$$x = -6$$

Check: $\dfrac{\sqrt[3]{6x+9}+8 = 5}{}$

$$\sqrt[3]{6(-6)+9}+8 \ ? \ 5$$
$$\sqrt[3]{-27}+8 \ \Big|$$
$$5 \ \Big| \ 5 \qquad \text{TRUE}$$

The solution is -6.

30. $-\dfrac{5}{3}$

31. $\sqrt{3y+1} = \sqrt{2y+6}$ One radical is already isolated.

$(\sqrt{3y+1})^2 = (\sqrt{2y+6})^2$ Squaring both sides

$$3y+1 = 2y+6$$
$$y = 5$$

The number 5 checks and is the solution.

32. 2

33. $2\sqrt{1-x} = \sqrt{5}$ One radical is already isolated

$(2\sqrt{1-x})^2 = (\sqrt{5})^2$ Squaring both sides

$$4(1-x) = 5$$
$$4 - 4x = 5$$
$$-4x = 1$$
$$x = -\frac{1}{4}$$

The number $-\dfrac{1}{4}$ checks and is the solution.

34. 3

35. $2\sqrt{t-1} = \sqrt{3t-1}$

$$(2\sqrt{t-1})^2 = (\sqrt{3t-1})^2$$
$$4(t-1) = 3t-1$$
$$4t-4 = 3t-1$$
$$t = 3$$

The number 3 checks and is the solution.

36. -1

37. $\sqrt{y-5}+\sqrt{y} = 5$

$\sqrt{y-5} = 5 - \sqrt{y}$ Adding $-\sqrt{y}$; this isolates one of the radical terms

$(\sqrt{y-5})^2 = (5-\sqrt{y})^2$ Squaring both sides

$$y-5 = 25 - 10\sqrt{y} + y$$

$-30 = -10\sqrt{y}$ Isolating the remaining radical term

$3 = \sqrt{y}$ Multiplying by $-\dfrac{1}{10}$

$$3^2 = (\sqrt{y})^2$$
$$9 = y$$

The number 9 checks and is the solution.

38. No solution

39. $3 + \sqrt{z-6} = \sqrt{z+9}$ One radical is already isolated.

$(3+\sqrt{z-6})^2 = (\sqrt{z+9})^2$ Squaring both sides

$$9 + 6\sqrt{z-6} + z - 6 = z+9$$
$$6\sqrt{z-6} = 6$$

$\sqrt{z-6} = 1$ Multiplying by $\dfrac{1}{6}$

$$(\sqrt{z-6})^2 = 1^2$$
$$z-6 = 1$$
$$z = 7$$

The number 7 checks and is the solution.

40. 7, 3

41. $\sqrt{20-x}+8 = \sqrt{9-x}+11$

$\qquad \sqrt{20-x} = \sqrt{9-x}+3 \quad$ Isolating one radical

$\qquad (\sqrt{20-x})^2 = (\sqrt{9-x}+3)^2 \quad$ Squaring both sides

$\qquad 20-x = 9-x+6\sqrt{9-x}+9$

$\qquad 2 = 6\sqrt{9-x} \qquad$ Isolating the remaining radical

$\qquad 1 = 3\sqrt{9-x} \qquad$ Multiplying by $\dfrac{1}{2}$

$\qquad 1^2 = (3\sqrt{9-x})^2 \quad$ Squaring both sides

$\qquad 1 = 9(9-x)$

$\qquad 1 = 81-9x$

$\qquad -80 = -9x$

$\qquad \dfrac{80}{9} = x$

The number $\dfrac{80}{9}$ checks and is the solution.

42. $\dfrac{15}{4}$

43. $\sqrt{x+2}+\sqrt{3x+4} = 2$

$\qquad \sqrt{x+2} = 2-\sqrt{3x+4} \quad$ Isolating one radical

$\qquad (\sqrt{x+2})^2 = (2-\sqrt{3x+4})^2$

$\qquad x+2 = 4-4\sqrt{3x+4}+3x+4$

$\qquad -2x-6 = -4\sqrt{3x+4} \quad$ Isolating the remaining radical

$\qquad x+3 = 2\sqrt{3x+4} \quad$ Multiplying by $-\dfrac{1}{2}$

$\qquad (x+3)^2 = (2\sqrt{3x+4})^2$

$\qquad x^2+6x+9 = 4(3x+4)$

$\qquad x^2+6x+9 = 12x+16$

$\qquad x^2-6x-7 = 0$

$\qquad (x-7)(x+1) = 0$

$\qquad x-7=0 \quad \text{or} \quad x+1=0$

$\qquad x=7 \quad \text{or} \quad x=-1$

Check:

For 7:

$$\begin{array}{c|c} \sqrt{x+2}+\sqrt{3x+4} = 2 \\ \hline \sqrt{7+2}+\sqrt{3\cdot 7+4} \ ? \ 2 \\ \sqrt{9}+\sqrt{25} \\ 8 \ \big| \ 2 \qquad \text{FALSE} \end{array}$$

For -1:

$$\begin{array}{c|c} \sqrt{x+2}+\sqrt{3x+4} = 2 \\ \hline \sqrt{-1+2}+\sqrt{3\cdot(-1)+4} \ ? \ 2 \\ \sqrt{1}+\sqrt{1} \\ 2 \ \big| \ 2 \qquad \text{TRUE} \end{array}$$

Since -1 checks but 7 does not, the solution is -1.

44. $\dfrac{1}{3}, \ -1$

45. $\sqrt{4y+1}-\sqrt{y-2} = 3$

$\qquad \sqrt{4y+1} = 3+\sqrt{y-2} \quad$ Isolating one radical

$\qquad (\sqrt{4y+1})^2 = (3+\sqrt{y-2})^2$

$\qquad 4y+1 = 9+6\sqrt{y-2}+y-2$

$\qquad 3y-6 = 6\sqrt{y-2} \quad$ Isolating the remaining radical

$\qquad y-2 = 2\sqrt{y-2} \quad$ Multiplying by $\dfrac{1}{3}$

$\qquad (y-2)^2 = (2\sqrt{y-2})^2$

$\qquad y^2-4y+4 = 4(y-2)$

$\qquad y^2-4y+4 = 4y-8$

$\qquad y^2-8y+12 = 0$

$\qquad (y-6)(y-2) = 0$

$\qquad y-6=0 \quad \text{or} \quad y-2=0$

$\qquad y=6 \quad \text{or} \qquad y=2$

The numbers 6 and 2 check, so they are the solutions.

46. 1

47. $\sqrt{3x-5}+\sqrt{2x+3}+1 = 0$

$\qquad \sqrt{3x-5}+1 = -\sqrt{2x+3}$

$\qquad (\sqrt{3x-5}+1)^2 = (-\sqrt{2x+3})^2$

$\qquad 3x-5+2\sqrt{3x-5}+1 = 2x+3$

$\qquad 2\sqrt{3x-5} = -x+7$

$\qquad (2\sqrt{3x-5})^2 = (-x+7)^2$

$\qquad 4(3x-5) = x^2-14x+49$

$\qquad 12x-20 = x^2-14x+49$

$\qquad 0 = x^2-26x+69$

$\qquad 0 = (x-23)(x-3)$

$\qquad x-23=0 \quad \text{or} \quad x-3=0$

$\qquad x=23 \quad \text{or} \qquad x=3$

Neither number checks. There is no solution. (At the outset we might have observed that there is no solution since the sum on the left side of the equation must be at least 1.)

48. 2

49.
$$2\sqrt{3x+6} - \sqrt{4x+9} = 5$$
$$2\sqrt{3x+6} = 5 + \sqrt{4x+9}$$
$$(2\sqrt{3x+6})^2 = (5+\sqrt{4x+9})^2$$
$$4(3x+6) = 25 + 10\sqrt{4x+9} + 4x + 9$$
$$12x + 24 = 34 + 10\sqrt{4x+9} + 4x$$
$$8x - 10 = 10\sqrt{4x+9}$$
$$4x - 5 = 5\sqrt{4x+9}$$
$$(4x-5)^2 = (5\sqrt{4x+9})^2$$
$$16x^2 - 40x + 25 = 25(4x+9)$$
$$16x^2 - 40x + 25 = 100x + 225$$
$$16x^2 - 140x - 200 = 0$$
$$4(4x+5)(x-10) = 0$$
$$4x + 5 = 0 \quad \text{or} \quad x - 10 = 0$$
$$x = -\frac{5}{4} \quad \text{or} \quad x = 10$$

Since 10 checks and $-\frac{5}{4}$ does not, 10 is the solution.

50. 7

51.
$$3\sqrt{t+1} - \sqrt{2t-5} = 7$$
$$3\sqrt{t+1} = 7 + \sqrt{2t-5}$$
$$(3\sqrt{t+1})^2 = (7+\sqrt{2t-5})^2$$
$$9(t+1) = 49 + 14\sqrt{2t-5} + 2t - 5$$
$$9t + 9 = 44 + 14\sqrt{2t-5} + 2t$$
$$7t - 35 = 14\sqrt{2t-5}$$
$$t - 5 = 2\sqrt{2t-5}$$
$$(t-5)^2 = (2\sqrt{2t-5})^2$$
$$t^2 - 10t + 25 = 4(2t-5)$$
$$t^2 - 10t + 25 = 8t - 20$$
$$t^2 - 18t + 45 = 0$$
$$(t-15)(t-3) = 0$$
$$t - 15 = 0 \quad \text{or} \quad t - 3 = 0$$
$$t = 15 \quad \text{or} \quad t = 3$$

Since 15 checks and 3 does not, the solution is 15.

52. 5

53.
$$\frac{3}{2x} + \frac{1}{x} = \frac{2x+3.5}{3x} \qquad \text{LCD is } 6x$$
$$6x\left(\frac{3}{2x} + \frac{1}{x}\right) = 6x\left(\frac{2x+3.5}{3x}\right)$$
$$9 + 6 = 4x + 7$$
$$8 = 4x$$
$$2 = x$$

The number 2 checks and is the solution.

54. Height: 7 in., base: 9 in.

55.

56.

57.
$$V = 1.2\sqrt{h}$$
$$V = 1.2\sqrt{30,000}$$
$$V \approx 208 \text{ mi} \qquad \text{Using a calculator to compute}$$

58. 72.25 ft

59.
$$\frac{x+\sqrt{x+1}}{x-\sqrt{x+1}} = \frac{5}{11}$$
$$11(x+\sqrt{x+1}) = 5(x-\sqrt{x+1})$$
$$11x + 11\sqrt{x+1} = 5x - 5\sqrt{x+1}$$
$$16\sqrt{x+1} = -6x$$
$$8\sqrt{x+1} = -3x$$
$$(8\sqrt{x+1})^2 = (-3x)^2$$
$$64(x+1) = 9x^2$$
$$64x + 64 = 9x^2$$
$$0 = 9x^2 - 64x - 64$$
$$0 = (9x+8)(x-8)$$

$$9x + 8 = 0 \quad \text{or} \quad x - 8 = 0$$
$$9x = -8 \quad \text{or} \qquad x = 8$$
$$x = -\frac{8}{9} \quad \text{or} \qquad x = 8$$

Since $-\frac{8}{9}$ checks but 8 does not, the solution is $-\frac{8}{9}$.

60. 6912

61.
$$\sqrt[4]{z^2+17} = 3$$
$$(\sqrt[4]{z^2+17})^4 = 3^4$$
$$z^2 + 17 = 81$$
$$z^2 - 64 = 0$$
$$(z+8)(z-8) = 0$$

$$z + 8 = 0 \quad \text{or} \quad z - 8 = 0$$
$$z = -8 \quad \text{or} \qquad z = 8$$

The numbers -8 and 8 check and are the solutions.

62. 0

63.
$$\sqrt[3]{x^2+x+15} - 3 = 0$$
$$\sqrt[3]{x^2+x+15} = 3$$
$$(\sqrt[3]{x^2+x+15})^3 = 3^3$$
$$x^2 + x + 15 = 27$$
$$x^2 + x - 12 = 0$$
$$(x+4)(x-3) = 0$$

$$x + 4 = 0 \quad \text{or} \quad x - 3 = 0$$
$$x = -4 \quad \text{or} \qquad x = 3$$

The numbers -4 and 3 check and are the solutions.

64. 6, −1

65.
$$\sqrt{8-b} = b\sqrt{8-b}$$
$$(\sqrt{8-b})^2 = (b\sqrt{8-b})^2$$
$$(8-b) = b^2(8-b)$$
$$0 = b^2(8-b) - (8-b)$$
$$0 = (8-b)(b^2-1)$$
$$0 = (8-b)(b+1)(b-1)$$
$$8-b=0 \ \text{ or } \ b+1=0 \ \text{or} \ b-1=0$$
$$8=b \ \text{ or } \qquad b=-1 \ \text{or} \quad b=1$$

Since the numbers 8 and 1 check but −1 does not, 8 and 1 are the solutions.

66. 2

67.
$$6\sqrt{y} + 6y^{-1/2} = 37$$
$$6\sqrt{y} + \frac{6}{\sqrt{y}} = 37$$
$$\sqrt{y}\left(6\sqrt{y} + \frac{6}{\sqrt{y}}\right) = \sqrt{y}\cdot 37$$
$$6y + 6 = 37\sqrt{y}$$
$$(6y+6)^2 = (37\sqrt{y})^2$$
$$36y^2 + 72y + 36 = 1369y$$
$$36y^2 - 1297y + 36 = 0$$
$$(36y-1)(y-36) = 0$$

$$36y-1=0 \quad \text{or} \quad y-36=0$$
$$36y=1 \quad \text{or} \qquad y=36$$
$$y=\frac{1}{36} \quad \text{or} \qquad y=36$$

The numbers $\frac{1}{36}$ and 36 check and are the solutions.

68. 0, $\dfrac{125}{4}$

Exercise Set 10.8

1. $a=3, \quad b=5$
Find c.
$$c^2 = a^2 + b^2 \qquad \text{Pythagorean equation}$$
$$c^2 = 3^2 + 5^2 \qquad \text{Substituting}$$
$$c^2 = 9 + 25$$
$$c^2 = 34$$
$$c = \sqrt{34} \qquad \text{Exact answer}$$
$$c \approx 5.831 \qquad \text{Approximation}$$

2. $\sqrt{164}; \ 12.806$

3. $a=12, \quad b=12$
Find c.
$$c^2 = a^2 + b^2 \qquad \text{Pythagorean equation}$$
$$c^2 = 12^2 + 12^2 \qquad \text{Substituting}$$
$$c^2 = 144 + 144$$
$$c^2 = 288$$
$$c = \sqrt{288} \qquad \text{Exact answer}$$
$$c \approx 16.971 \qquad \text{Approximation}$$

4. $\sqrt{200}; \ 14.142$

5. $b=12, \quad c=13$
Find a.
$$a^2 + b^2 = c^2 \qquad \text{Pythagorean equation}$$
$$a^2 + 12^2 = 13^2 \qquad \text{Substituting}$$
$$a^2 + 144 = 169$$
$$a^2 = 25$$
$$a = 5$$

6. $\sqrt{119}; \ 10.909$

7. $c=6, \quad a=\sqrt{5}$
Find b.
$$c^2 = a^2 + b^2$$
$$(\sqrt{5})^2 + b^2 = 6^2$$
$$5 + b^2 = 36$$
$$b^2 = 31$$
$$b = \sqrt{31} \qquad \text{Exact answer}$$
$$b \approx 5.568 \qquad \text{Approximation}$$

8. 4

9. $b=1, \quad c=\sqrt{13}$
Find a.
$$a^2 + b^2 = c^2 \qquad \text{Pythagorean equation}$$
$$a^2 + 1^2 = (\sqrt{13})^2 \qquad \text{Substituting}$$
$$a^2 + 1 = 13$$
$$a^2 = 12$$
$$a = \sqrt{12} \qquad \text{Exact answer}$$
$$a \approx 3.464 \qquad \text{Approximation}$$

10. $\sqrt{19}; \ 4.359$

11. $a=1, \quad c=\sqrt{n}$
Find b.
$$a^2 + b^2 = c^2$$
$$1^2 + b^2 = (\sqrt{n})^2$$
$$1 + b^2 = n$$
$$b^2 = n - 1$$
$$b = \sqrt{n-1}$$

12. $\sqrt{4-n}$

13. We make a drawing and let d = the length of the guy wire.

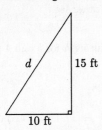

We use the Pythagorean equation to find d.

$d^2 = 10^2 + 15^2$

$d^2 = 100 + 225$

$d^2 = 325$

$d = \sqrt{325}$

$d \approx 18.028$

The wire is $\sqrt{325}$, or about 18.028 ft long.

14. $\sqrt{8450} \approx 91.924$ ft

15. We first make a drawing and let d = the distance, in feet, to second base. A right triangle is formed in which the length of the leg from second base to third base is 90 ft. The length of the leg from third base to where the catcher fields the ball is $90 - 10$, or 80 ft.

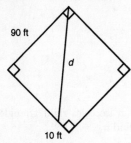

We substitute these values into the Pythagorean equation to find d.

$d^2 = 90^2 + 80^2$

$d^2 = 8100 + 6400$

$d^2 = 14,500$

$d = \sqrt{14,500}$

Exact answer: $d = \sqrt{14,500}$ ft

Approximation: $d \approx 120.416$ ft

16. $\sqrt{340} + 8 \approx 26.439$ ft

17. We use the drawing in the text, replacing w with 16 in.

We use the Pythagorean equation to find h.

$h^2 + 16^2 = 20^2$

$h^2 + 256 = 400$

$h^2 = 144$

$h = 12$

The height is 12 in.

18. 20 in.

19.

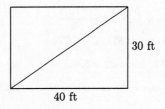

$d^2 = 30^2 + 40^2$

$d^2 = 900 + 1600$

$d^2 = 2500$

$d = 50$ ft   Exact answer

20. $\sqrt{13,000} \approx 114.018$ m

21. Since one acute angle is 45°, this is an isosceles right triangle with $a = 9$. Then $b = 9$ also. We substitute to find c.

$c = a\sqrt{2}$

$c = 9\sqrt{2}$

Exact answer: $b = 9$, $c = 9\sqrt{2}$

Approximation: $c \approx 12.728$

22. $a = 12,;\ c = 12\sqrt{2} \approx 16.971$

23. This is a 30-60-90 right triangle with $c = 14$. We substitute to find a and b.

$c = 2a$

$14 = 2a$

$7 = a$

$b = a\sqrt{3}$

$b = 7\sqrt{3}$

Exact answer: $a = 7$, $b = 7\sqrt{3}$

Approximation: $b \approx 12.124$

24. $a = 9$; $b = 9\sqrt{3} \approx 15.588$

25. This is a 30-60-90 right triangle with $b = 5$. We substitute to find a and c.

$$b = a\sqrt{3}$$
$$5 = a\sqrt{3}$$
$$\frac{5}{\sqrt{3}} = a$$
$$\frac{5\sqrt{3}}{3} = a \qquad \text{Rationalizing the denominator}$$

$$c = 2a$$
$$c = 2 \cdot \frac{5\sqrt{3}}{3}$$
$$c = \frac{10\sqrt{3}}{3}$$

Exact answer: $a = \dfrac{5\sqrt{3}}{3}$, $c = \dfrac{10\sqrt{3}}{3}$

Approximation: $a \approx 2.887$, $c \approx 5.774$

26. $a = 4\sqrt{2} \approx 5.657$; $b = 4\sqrt{2} \approx 5.657$

27. This is an isosceles right triangle with $c = 13$. We substitute to find a.

$$a = \frac{c\sqrt{2}}{2}$$
$$a = \frac{13\sqrt{2}}{2}$$

Since $a = b$, we have $b = \dfrac{13\sqrt{2}}{2}$ also.

Exact answer: $a = \dfrac{13\sqrt{2}}{2}$, $b = \dfrac{13\sqrt{2}}{2}$

Approximation: $a \approx 9.192$, $b \approx 9.192$

28. $a = \dfrac{7}{\sqrt{3}} \approx 4.041$; $c = \dfrac{14\sqrt{3}}{3} \approx 8.083$

29. This is a 30-60-90 triangle with $a = 12$. We substitute to find b and c.

$$b = a\sqrt{3} \qquad\qquad c = 2a$$
$$b = 12\sqrt{3} \qquad\qquad c = 2 \cdot 12$$
$$\qquad\qquad\qquad c = 24$$

Exact answer: $b = 12\sqrt{3}$, $c = 24$

Approximations: $b \approx 20.785$

30. $b = 9\sqrt{3} \approx 15.588$; $c = 18$

31.

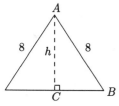

This is an equilateral triangle, so all the angles are 60°. The altitude bisects one angle and one side. Then triangle ABC is a 30-60-90 right triangle with the shorter leg of

length 8/2, or 4, and hypotenuse of length 8. We substitute to find the length of the other leg.

$$b = a\sqrt{3}$$
$$h = 4\sqrt{3} \qquad \text{Substituting } h \text{ for } b \text{ and 4 for } a$$

Exact answer: $h = 4\sqrt{3}$

Approximation: $h \approx 6.928$

32. $5\sqrt{3} \approx 8.660$

33.

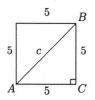

Triangle ABC is an isosceles right triangle with $a = 5$. We substitute to find c.

$$c = a\sqrt{2}$$
$$c = 5\sqrt{2}$$

Exact answer: $c = 5\sqrt{2}$

Approximation: $c \approx 7.071$

34. $7\sqrt{2} \approx 9.899$

35.

Triangle ABC is an isosceles right triangle with $c = 19$. We substitute to find a.

$$a = \frac{c\sqrt{2}}{2}$$
$$a = \frac{19\sqrt{2}}{2}$$

Since $a = b$, we have $b = \dfrac{19\sqrt{2}}{2}$ also.

Exact answer: $a = \dfrac{19\sqrt{2}}{2}$, $b = \dfrac{19\sqrt{2}}{2}$

Approximation: $a \approx 13.435$, $b \approx 13.435$

36. $\dfrac{15\sqrt{2}}{2} \approx 10.607$

37.

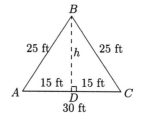

Triangle ABC is an isosceles triangle. The height from B to $\overline{AC}$ bisects $\overline{AC}$. Thus, $\overline{DC}$ measures 15 ft.

$$h^2 + 15^2 = 25^2$$
$$h^2 + 225 = 625$$
$$h^2 = 400$$
$$h = \sqrt{400}$$
$$h = 20$$

If the height of triangle ABC is 20 ft and the base is 30 ft, the area is $\frac{1}{2} \cdot 30 \cdot 20$, or 300 ft^2.

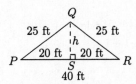

Triangle PQR is an isosceles triangle. The height from Q to $\overline{PR}$ bisects $\overline{PR}$. Thus, $\overline{SR}$ measures 20 ft.

$$h^2 + 20^2 = 25^2$$
$$h^2 + 400 = 625$$
$$h^2 = 225$$
$$h = \sqrt{225}$$
$$h = 15$$

If the height of triangle PQR is 15 ft and the base is 40 ft, the area is $\frac{1}{2} \cdot 40 \cdot 15$, or 300 ft^2.

The areas of the two triangles are the same.

38. $\sqrt{10,561} \approx 102.767$ ft

39.

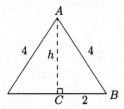

The entrance is an equilateral triangle, so all the angles are 60°. The altitude bisects one angle and one side. Then triangle ABC is a 30-60-90 right triangle with the shorter leg of length 4/2, or 2, and hypotenuse of length 4. We substitute to find h, the height of the tent.

$$b = a\sqrt{3}$$
$$h = 2\sqrt{3} \qquad \text{Substituting } h \text{ for } b \text{ and } 2 \text{ for } a$$

Exact answer: $h = 2\sqrt{3}$ ft

Approximation: $h \approx 3.464$ ft

40. $d = s + s\sqrt{2}$

41.

Triangle ABC is an isosceles right triangle with $c = 8\sqrt{2}$. We substitute to find a.

$$a = \frac{c\sqrt{2}}{2} = \frac{8\sqrt{2} \cdot \sqrt{2}}{2} = \frac{8 \cdot 2}{2} = 8$$

The length of a side of the square is 8 ft.

42. $\sqrt{181} \approx 13.454$ cm

43.

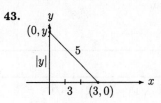

$$|y|^2 + 3^2 = 5^2$$
$$y^2 + 9 = 25$$
$$y^2 = 16$$
$$y = \pm 4$$

The points are $(0, -4)$ and $(0, 4)$.

44. $(3, 0), (-3, 0)$

45.
$$x^2 - 11x + 24 = 0$$
$$(x - 8)(x - 3) = 0$$
$$x - 8 = 0 \quad \text{or} \quad x - 3 = 0$$
$$x = 8 \quad \text{or} \quad x = 3$$

The solutions are 8 and 3.

46. $\frac{3}{2}, -7$

47. ▨

48. ▨

49.

First find the length of a diagonal of the base of the cube. It is the hypotenuse of an isosceles right triangle with $a = 5$ cm. Then $c = a\sqrt{2} = 5\sqrt{2}$ cm.

Triangle ABC is a right triangle with legs of $5\sqrt{2}$ cm and 5 cm and hypotenuse d. Use the Pythagorean equation to find d, the length of the diagonal that connects two opposite corners of the cube.

$$d^2 = (5\sqrt{2})^2 + 5^2$$
$$d^2 = 25 \cdot 2 + 25$$
$$d^2 = 50 + 25$$
$$d^2 = 75$$
$$d = \sqrt{75}$$

Exact answer: $d = \sqrt{75}$ cm

50. 9

Exercise Set 10.9

1. $\sqrt{-15} = \sqrt{-1 \cdot 15} = \sqrt{-1} \cdot \sqrt{15} = i\sqrt{15}$, or $\sqrt{15}i$

2. $i\sqrt{17}$, or $\sqrt{17}i$

3. $\sqrt{-16} = \sqrt{-1 \cdot 16} = \sqrt{-1} \cdot \sqrt{16} = i \cdot 4 = 4i$

4. $5i$

5. $-\sqrt{-12} = -\sqrt{-1 \cdot 12} = -\sqrt{-1} \cdot \sqrt{12} = -i \cdot 2\sqrt{3} =$
$-2i\sqrt{3}$, or $-2\sqrt{3}i$

6. $-2i\sqrt{5}$, or $-2\sqrt{5}i$

7. $\sqrt{-3} = \sqrt{-1 \cdot 3} = \sqrt{-1} \cdot \sqrt{3} = i\sqrt{3}$, or $\sqrt{3}i$

8. $2i$

9. $\sqrt{-81} = \sqrt{-1 \cdot 81} = \sqrt{-1} \cdot \sqrt{81} = i \cdot 9 = 9i$

10. $3i\sqrt{3}$, or $3\sqrt{3}i$

11. $\sqrt{-98} = \sqrt{-1 \cdot 98} = \sqrt{-1} \cdot \sqrt{98} = i \cdot 7\sqrt{2} = 7i\sqrt{2}$, or $7\sqrt{2}i$

12. $-3i\sqrt{2}$, or $= -3\sqrt{2}i$

13. $-\sqrt{-49} = -\sqrt{-1 \cdot 49} = -\sqrt{-1} \cdot \sqrt{49} = -i \cdot 7 = -7i$

14. $-5i\sqrt{5}$, or $-5\sqrt{5}i$

15. $4 - \sqrt{-60} = 4 - \sqrt{-1 \cdot 60} = 4 - \sqrt{-1} \cdot \sqrt{60} =$
$4 - i \cdot 2\sqrt{15} = 4 - 2\sqrt{15}i$, or $4 - 2i\sqrt{15}$

16. $6 - 2i\sqrt{21}$, or $6 - 2\sqrt{21}i$

17. $\sqrt{-4} + \sqrt{-12} = \sqrt{-1 \cdot 4} + \sqrt{-1 \cdot 12} =$
$\sqrt{-1} \cdot \sqrt{4} + \sqrt{-1} \cdot \sqrt{12} = i \cdot 2 + i \cdot 2\sqrt{3} =$
$(2 + 2\sqrt{3})i$

18. $(-2\sqrt{19} + 5\sqrt{5})i$

19. $(3 + 2i) + (5 - i)$
$= (3 + 5) + (2 - 1)i$ Collecting the real and the
                         imaginary parts
$= 8 + i$

20. $5 + 11i$

21. $(4 - 3i) + (5 - 2i)$
$= (4 + 5) + (-3 - 2)i$ Collecting the real and
                          the imaginary parts
$= 9 - 5i$

22. $-1 - 8i$

23. $(9 - i) + (-2 + 5i) = (9 - 2) + (-1 + 5)i$
$= 7 + 4i$

24. $8 + i$

25. $(3 - i) - (5 + 2i) = (3 - 5) + (-1 - 2)i$
$= -2 - 3i$

26. $-9 + 5i$

27. $(4 - 2i) - (5 - 3i) = (4 - 5) + [-2 - (-3)]i$
$= -1 + i$

28. $-3 + 2i$

29. $(9 + 5i) - (-2 - i) = [9 - (-2)] + [5 - (-1)]i$
$= 11 + 6i$

30. $4 - 7i$

31. $\sqrt{-25}\sqrt{-36} = \sqrt{-1} \cdot \sqrt{25} \cdot \sqrt{-1} \cdot \sqrt{36}$
$= i \cdot 5 \cdot i \cdot 6$
$= i^2 \cdot 30$
$= -1 \cdot 30 \qquad i^2 = -1$
$= -30$

32. -63

33. $\sqrt{-6}\sqrt{-5}$
$= \sqrt{-1} \cdot \sqrt{6} \cdot \sqrt{-1} \cdot \sqrt{5} = i \cdot \sqrt{6} \cdot i \cdot \sqrt{5}$
$= i^2 \cdot \sqrt{30} = -1 \cdot \sqrt{30} = -\sqrt{30}$

34. $-\sqrt{70}$

35. $\sqrt{-50}\sqrt{-3} = \sqrt{-1 \cdot 25 \cdot 2} \cdot \sqrt{-1 \cdot 3} =$
$\sqrt{-1} \cdot \sqrt{25} \cdot \sqrt{2} \cdot \sqrt{-1} \cdot \sqrt{3} = i \cdot 5 \cdot \sqrt{2} \cdot i \cdot \sqrt{3} =$
$i^2 \cdot 5 \cdot \sqrt{6} = -1 \cdot 5 \cdot \sqrt{6} = -5\sqrt{6}$

36. $-6\sqrt{6}$

37. $\sqrt{-48}\sqrt{-6} = \sqrt{-1 \cdot 16 \cdot 3} \cdot \sqrt{-1 \cdot 3 \cdot 2}$
$= \sqrt{-1} \cdot \sqrt{16} \cdot \sqrt{3} \cdot \sqrt{-1} \cdot \sqrt{3} \cdot \sqrt{2}$
$= i \cdot 4 \cdot 3 \cdot i \cdot \sqrt{2} \qquad \sqrt{3} \cdot \sqrt{3} = 3$
$= i^2 \cdot 12 \cdot \sqrt{2}$
$= -1 \cdot 12 \cdot \sqrt{2}$
$= -12\sqrt{2}$

38. $-15\sqrt{5}$

39. $5i \cdot 8i = 40 \cdot i^2$
$= 40 \cdot (-1) \qquad i^2 = -1$
$= -40$

40. -54

41. $5i \cdot (-7i) = -35 \cdot i^2$
$= -35 \cdot (-1) \qquad i^2 = -1$
$= 35$

42. 28

43.
$$5i(3 - 2i)$$
$$= 5i \cdot 3 + 5i(-2i) \quad \text{Using the distributive law}$$
$$= 15i - 10i^2$$
$$= 15i + 10 \qquad i^2 = -1$$
$$= 10 + 15i$$

44. $24 + 20i$

45. $-3i(7 - 4i) = -3i \cdot 7 + (-3i)(-4i) = -21i + 12i^2 = -21i - 12 = -12 - 21i$

46. $-21 - 63i$

47.
$$(3 + 2i)(1 + i)$$
$$= 3 + 3i + 2i + 2i^2 \quad \text{Using FOIL}$$
$$= 3 + 3i + 2i - 2 \quad i^2 = -1$$
$$= 1 + 5i$$

48. $-7 + 26i$

49.
$$(2 + 3i)(6 - 2i)$$
$$= 12 - 4i + 18i - 6i^2 \quad \text{Using FOIL}$$
$$= 12 - 4i + 18i + 6 \quad i^2 = -1$$
$$= 18 + 14i$$

50. $16 + 7i$

51. $(6 - 5i)(3 + 4i) = 18 + 24i - 15i - 20i^2 =$
$$= 18 + 24i - 15i + 20$$
$$= 38 + 9i$$

52. $40 + 13i$

53. $(7 - 2i)(2 - 6i) = 14 - 42i - 4i + 12i^2 =$
$$= 14 - 42i - 4i - 12$$
$$= 2 - 46i$$

54. $8 + 31i$

55. $(5 - 3i)(4 - 5i) = 20 - 25i - 12i + 15i^2 =$
$20 - 25i - 12i - 15 = 5 - 37i$

56. $7 - 61i$

57. $(-2 + 3i)(-2 + 5i) = 4 - 10i - 6i + 15i^2 =$
$4 - 10i - 6i - 15 = -11 - 16i$

58. $-15 - 30i$

59. $(-5 - 4i)(3 + 7i) = -15 - 35i - 12i - 28i^2 =$
$-15 - 35i - 12i + 28 = 13 - 47i$

60. $39 - 37i$

61.
$$(3 - 2i)^2$$
$$= 3^2 - 2 \cdot 3 \cdot 2i + (2i)^2 \quad \text{Squaring a binomial}$$
$$= 9 - 12i + 4i^2$$
$$= 9 - 12i - 4 \qquad i^2 = -1$$
$$= 5 - 12i$$

62. $21 - 20i$

63.
$$(2 + 3i)^2$$
$$= 2^2 + 2 \cdot 2 \cdot 3i + (3i)^2 \quad \text{Squaring a binomial}$$
$$= 4 + 12i + 9i^2$$
$$= 4 + 12i - 9$$
$$= -5 + 12i$$

64. $12 + 16i$

65. $(-2 + 3i)^2 = 4 - 12i + 9i^2 = 4 - 12i - 9 =$
$-5 - 12i$

66. $21 + 20i$

67. $i^7 = i^6 \cdot i = (i^2)^3 \cdot i = (-1)^3 \cdot i = -1 \cdot i = -i$

68. $-i$

69. $i^{24} = (i^2)^{12} = (-1)^{12} = 1$

70. $-i$

71. $i^{42} = (i^2)^{21} = (-1)^{21} = -1$

72. 1

73. $i^9 = (i^2)^4 \cdot i = (-1)^4 \cdot i = 1 \cdot i = i$

74. i

75. $i^6 = (i^2)^3 = (-1)^3 = -1$

76. 1

77. $(5i)^3 = 5^3 \cdot i^3 = 125 \cdot i^2 \cdot i = 125(-1)(i) = -125i$

78. $-243i$

79. $7 + i^4 = 7 + (i^2)^2 = 7 + (-1)^2 = 7 + 1 = 8$

80. $-18 - i$

81. $i^4 - 26i = (i^2)^2 - 26i = (-1)^2 - 26i = 1 - 26i$

82. $38i$

83. $i^2 + i^4 = -1 + (i^2)^2 = -1 + (-1)^2 = -1 + 1 = 0$

84. i

85. $i^5 + i^7 = i^4 \cdot i + i^6 \cdot i = (i^2)^2 \cdot i + (i^2)^3 \cdot i =$
$(-1)^2 \cdot i + (-1)^3 \cdot i = 1 \cdot i + (-1)i = i - i = 0$

86. 0

87.
$$1 + i + i^2 + i^3 + i^4$$
$$= 1 + i + i^2 + i^2 \cdot i + (i^2)^2$$
$$= 1 + i + (-1) + (-1) \cdot i + (-1)^2$$
$$= 1 + i - 1 - i + 1$$
$$= 1$$

88. i

89. $5 - \sqrt{-64} = 5 - \sqrt{-1} \cdot \sqrt{64} = 5 - i \cdot 8 = 5 - 8i$

90. $(2\sqrt{3} + 36)i$

91.

$$\frac{5}{3-i}$$

$$= \frac{5}{3-i} \cdot \frac{3+i}{3+i} \qquad \text{Multiplying by 1, using the conjugate}$$

$$= \frac{15+5i}{10} \qquad \text{Performing the multiplication}$$

$$= \frac{15}{10} + \frac{5}{10}i, \text{ or } \frac{3}{2} + \frac{1}{2}i \quad \text{Writing in the form } a+bi$$

92. $\dfrac{15}{26} - \dfrac{3}{26}i$

93.

$$\frac{2i}{7+3i}$$

$$= \frac{2i}{7+3i} \cdot \frac{7-3i}{7-3i} \qquad \text{Multiplying by 1, using the conjugate}$$

$$= \frac{14i+6}{58} \qquad \text{Performing the multiplication}$$

$$= \frac{6}{58} + \frac{14}{58}i, \text{ or } \frac{3}{29} + \frac{7}{29}i \quad \text{Writing in the form } a+bi$$

94. $-\dfrac{20}{29} + \dfrac{8}{29}i$

95. $\dfrac{7}{6i} = \dfrac{7}{6i} \cdot \dfrac{-6i}{-6i} = \dfrac{-42i}{36} = -\dfrac{7}{6}i$

96. $-\dfrac{3}{10}i$

97.

$$\frac{8-3i}{7i} = \frac{8-3i}{7i} \cdot \frac{-7i}{-7i}$$

$$= \frac{-56i-21}{49}$$

$$= -\frac{21}{49} - \frac{56}{49}i$$

$$= -\frac{3}{7} - \frac{8}{7}i$$

98. $\dfrac{8}{5} - \dfrac{3}{5}i$

99. $\dfrac{3+2i}{2+i} = \dfrac{3+2i}{2+i} \cdot \dfrac{2-i}{2-i} = \dfrac{8+i}{5} = \dfrac{8}{5} + \dfrac{1}{5}i$

100. $\dfrac{15}{26} + \dfrac{29}{26}i$

101. $\dfrac{5-2i}{2+5i} = \dfrac{5-2i}{2+5i} \cdot \dfrac{2-5i}{2-5i} = \dfrac{-29i}{29} = -i$

102. $\dfrac{6}{25} - \dfrac{17}{25}i$

103. $\dfrac{3-5i}{3-2i} = \dfrac{3-5i}{3-2i} \cdot \dfrac{3+2i}{3+2i} = \dfrac{19-9i}{13} = \dfrac{19}{13} - \dfrac{9}{13}i$

104. $\dfrac{38}{41} - \dfrac{27}{41}i$

105. Substitute $1+2i$ for x in the equation.

$$\begin{array}{c|c}
\multicolumn{2}{c}{x^2 - 2x + 5 = 0} \\
\hline
(1+2i)^2 - 2(1+2i) + 5 \ ? \ 0 & \\
1 + 4i + 4i^2 - 2 - 4i + 5 & \\
1 + 4i - 4 - 2 - 4i + 5 & \\
0 & 0 \qquad \text{TRUE}
\end{array}$$

$1+2i$ is a solution.

106. Yes

107. Substitute $1-i$ for x in the equation.

$$\begin{array}{c|c}
\multicolumn{2}{c}{x^2 + 2x + 2 = 0} \\
\hline
(1-i)^2 + 2(1-i) + 2 \ ? \ 0 & \\
1 - 2i + i^2 + 2 - 2i + 2 & \\
1 - 2i - 1 + 2 - 2i + 2 & \\
4 - 4i & 0 \qquad \text{FALSE}
\end{array}$$

$1-i$ is not a solution.

108. No

109. $\dfrac{196}{x^2 - 7x + 49} - \dfrac{2x}{x+7} = \dfrac{2058}{x^3 + 343}$

Note: $x^3 + 343 = (x+7)(x^2 - 7x + 49)$.

The LCD $= (x+7)(x^2 - 7x + 49)$.

$$(x+7)(x^2-7x+49)\left(\frac{196}{x^2-7x+49} - \frac{2x}{x+7}\right) =$$
$$(x+7)(x^2-7x+49) \cdot \frac{2058}{x^3+343}$$

$$196(x+7) - 2x(x^2 - 7x + 49) = 2058$$
$$196x + 1372 - 2x^3 + 14x^2 - 98x = 2058$$
$$98x - 686 - 2x^3 + 14x^2 = 0$$
$$49x - 343 - x^3 + 7x^2 = 0 \quad \begin{array}{l}\text{Dividing} \\ \text{by 2}\end{array}$$
$$49(x-7) - x^2(x-7) = 0$$
$$(49 - x^2)(x-7) = 0$$
$$(7-x)(7+x)(x-7) = 0$$

$$7 - x = 0 \quad \text{or} \quad 7 + x = 0 \quad \text{or} \quad x - 7 = 0$$
$$7 = x \quad \text{or} \qquad x = -7 \quad \text{or} \qquad x = 7$$

Only 7 checks. It is the solution.

110. $\dfrac{70}{29}$

111. $28 = 3x^2 - 17x$

$$0 = 3x^2 - 17x - 28$$
$$0 = (3x+4)(x-7)$$

$$3x + 4 = 0 \quad \text{or} \quad x - 7 = 0$$

$$3x = -4 \quad \text{or} \qquad x = 7$$

$$x = -\frac{4}{3} \quad \text{or} \qquad x = 7$$

Both values check. The solutions are $-\dfrac{4}{3}$ and 7.

112. $-4 - 8i;\ -2 + 4i;\ 8 - 6i$

113. $\dfrac{1}{\dfrac{1-i}{10} - \left(\dfrac{1-i}{10}\right)^2} = \dfrac{1}{\dfrac{1-i}{10} - \left(\dfrac{-2i}{100}\right)} =$

$\dfrac{1}{\dfrac{1-i}{10} + \dfrac{i}{50}} = \dfrac{1}{\dfrac{1-i}{10} + \dfrac{i}{50}} \cdot \dfrac{50}{50} = \dfrac{50}{5 - 5i + i} =$

$\dfrac{50}{5 - 4i} = \dfrac{50}{5 - 4i} \cdot \dfrac{5 + 4i}{5 + 4i} = \dfrac{250 + 200i}{41} = \dfrac{250}{41} + \dfrac{200}{41}i$

114. 0

115. $(1 - i)^3 (1 + i)^3 =$

$(1 - i)(1 + i) \cdot (1 - i)(1 + i) \cdot (1 - i)(1 + i) =$

$(1 - i^2)(1 - i^2)(1 - i^2) = (1 + 1)(1 + 1)(1 + 1) =$

$2 \cdot 2 \cdot 2 = 8$

116. $-1 - \sqrt{5}i$

117. $\dfrac{6}{1 + \dfrac{3}{i}} = \dfrac{6}{\dfrac{i + 3}{i}} = \dfrac{6i}{i + 3} = \dfrac{6i}{i + 3} \cdot \dfrac{-i + 3}{-i + 3} =$

$\dfrac{-6i^2 + 18i}{-i^2 + 9} = \dfrac{6 + 18i}{10} = \dfrac{6}{10} + \dfrac{18}{10}i = \dfrac{3}{5} + \dfrac{9}{5}i$

118. $-\dfrac{2}{3}i$

119. $\dfrac{i - i^{38}}{1 + i} = \dfrac{i - (i^2)^{19}}{1 + i} = \dfrac{i - (-1)^{19}}{1 + i} = \dfrac{i - (-1)}{1 + i} =$

$\dfrac{i + 1}{1 + i} = 1$

Chapter 11

Quadratic Equations And Functions

Exercise Set 11.1

1.
$$x^2 = 25$$
$x = \sqrt{25}$ or $x = -\sqrt{25}$ Using the principle of square roots

$x = 5$ or $x = -5$

$x^2 = 25$		$x^2 = 25$	
5^2 ? 25		$(-5)^2$? 25	
25	25 TRUE	25	25 TRUE

The solutions are 5 and -5.

2. 6, -6

3.
$$a^2 = 81$$
$a = \sqrt{81}$ or $a = -\sqrt{81}$ Using the principle of square roots

$a = 9$ or $a = -9$

Both numbers check. The solutions are 9 and -9

4. 11, -11

5.
$$m^2 = 15$$
$m = \sqrt{15}$ or $m = -\sqrt{15}$

Both numbers check. The solutions are $\sqrt{15}$ and $-\sqrt{15}$

6. $\sqrt{10}$, $-\sqrt{10}$

7.
$$x^2 = 19$$
$x = \sqrt{19}$ or $x = -\sqrt{19}$

Both numbers check. The solutions are $\sqrt{19}$ and $-\sqrt{19}$

8. $\sqrt{29}$, $-\sqrt{29}$

9.
$$5a^2 = 35$$
$a^2 = 7$ Dividing by 5

$a = \sqrt{7}$ or $a = -\sqrt{7}$ Principle of square roots

Both numbers check. The solutions are $\sqrt{7}$ and $-\sqrt{7}$.

10. $2\sqrt{7}$, $-2\sqrt{7}$

11.
$$7x^2 = 140$$
$$x^2 = 20$$
$x = \sqrt{20}$ or $x = -\sqrt{20}$

$x = \sqrt{4 \cdot 5}$ or $x = -\sqrt{4 \cdot 5}$ $\Big\}$ Simplifying

$x = 2\sqrt{5}$ or $x = -2\sqrt{5}$

Both numbers check. The solutions are $2\sqrt{5}$ and $-2\sqrt{5}$.

12. $2\sqrt{2}$, $-2\sqrt{2}$

13.
$$4t^2 - 25 = 0$$
$$4t^2 = 25$$
$$t^2 = \frac{25}{4}$$
$t = \sqrt{\dfrac{25}{4}}$ or $t = -\sqrt{\dfrac{25}{4}}$

$t = \dfrac{\sqrt{25}}{\sqrt{4}}$ or $t = -\dfrac{\sqrt{25}}{\sqrt{4}}$

$t = \dfrac{5}{2}$ or $t = -\dfrac{5}{2}$

Both numbers check. The solutions are $\dfrac{5}{2}$ and $-\dfrac{5}{2}$.

14. $\dfrac{2}{3}$, $-\dfrac{2}{3}$

15. $25x^2 + 4 = 0$
$$x^2 = -\frac{4}{25}$$
$x = \sqrt{-\dfrac{4}{25}}$ or $x = \sqrt{-\dfrac{4}{25}}$

$x = \dfrac{2}{5}i$ or $x = -\dfrac{2}{5}i$

Both numbers check. The solutions are $\dfrac{2}{5}i$ and $-\dfrac{2}{5}i$.

16. $\dfrac{4}{3}i$, $-\dfrac{4}{3}i$

17. $9y^2 + 2 = 7$
$$9y^2 = 5$$
$$y^2 = \frac{5}{9}$$
$y = \sqrt{\dfrac{5}{9}}$ or $y = -\sqrt{\dfrac{5}{9}}$

$y = \dfrac{\sqrt{5}}{3}$ or $y = -\dfrac{\sqrt{5}}{3}$

Both numbers check. The solutions are $\dfrac{\sqrt{5}}{3}$ and $-\dfrac{\sqrt{5}}{3}$.

18. $\dfrac{\sqrt{21}}{2}$, $-\dfrac{\sqrt{21}}{2}$

19. $3x^2 - 7 = 0$
$$x^2 = \frac{7}{3}$$
$x = \sqrt{\dfrac{7}{3}}$ or $x = -\sqrt{\dfrac{7}{3}}$

$x = \dfrac{\sqrt{7}}{\sqrt{3}} \cdot \dfrac{\sqrt{3}}{\sqrt{3}}$ or $x = -\dfrac{\sqrt{7}}{\sqrt{3}} \cdot \dfrac{\sqrt{3}}{\sqrt{3}}$ Rationalizing the denominators

$x = \dfrac{\sqrt{21}}{3}$ or $x = -\dfrac{\sqrt{21}}{3}$

Both numbers check. The solutions are $\dfrac{\sqrt{21}}{3}$ and $-\dfrac{\sqrt{21}}{3}$.

20. $\dfrac{\sqrt{6}}{2},\ -\dfrac{\sqrt{6}}{2}$

21. $(x-2)^2 = 49$

$x - 2 = 7$ or $x - 2 = -7$ Using the principle
of square roots

$x = 9$ or $\quad x = -5$

The solutions are 9 and -5.

22. $4,\ -6$

23. $\qquad (x+3)^2 = 36$

$x + 3 = 6$ or $x + 3 = -6$ Using the principle
of square roots

$x = 3$ or $\quad x = -9$

The solutions are 3 and -9.

24. $13,\ -5$

25. $(m+3)^2 = 21$

$m + 3 = \sqrt{21} \qquad$ or $\ m + 3 = -\sqrt{21}$
$\qquad\qquad\qquad$ Principle of square roots

$m = -3 + \sqrt{21}$ or $\quad m = -3 - \sqrt{21}$

The solutions are $-3 + \sqrt{21}$ and $-3 - \sqrt{21}$, or $-3 \pm \sqrt{21}$.

26. $3 \pm \sqrt{6}$

27. $(a+13)^2 = 8$

$a + 13 = \sqrt{8} \qquad$ or $\quad a + 13 = -\sqrt{8}$
$a + 13 = \sqrt{4 \cdot 2} \qquad$ or $\quad a + 13 = -\sqrt{4 \cdot 2}$
$a + 13 = 2\sqrt{2} \qquad$ or $\quad a + 13 = -2\sqrt{2}$
$\qquad a = -13 + 2\sqrt{2}$ or $\qquad a = -13 - 2\sqrt{2}$

The solutions are $-13 + 2\sqrt{2}$ and $-13 - 2\sqrt{2}$, or $-13 \pm 2\sqrt{2}$.

28. $21,\ 5$

29. $(x-7)^2 = 12$

$x - 7 = \sqrt{12} \qquad$ or $\quad x - 7 = -\sqrt{12}$
$x - 7 = 2\sqrt{3} \qquad$ or $\quad x - 7 = -2\sqrt{3}$
$x = 7 + 2\sqrt{3}$ or $\qquad x = 7 - 2\sqrt{3}$

The solutions are $7 + 2\sqrt{3}$ and $7 - 2\sqrt{3}$, or $7 \pm 2\sqrt{3}$.

30. $-1 \pm \sqrt{14}$

31. $(x+9)^2 = 34$

$x + 9 = \sqrt{34} \qquad$ or $\quad x + 9 = -\sqrt{34}$
$\quad x = -9 + \sqrt{34}$ or $\qquad x = -9 - \sqrt{34}$

The solutions are $-9 + \sqrt{34}$ and $-9 - \sqrt{34}$, or $-9 \pm \sqrt{34}$.

32. $3,\ -7$

33. $\left(x + \dfrac{3}{2}\right)^2 = \dfrac{7}{2}$

$x + \dfrac{3}{2} = \sqrt{\dfrac{7}{2}} \qquad$ or $\quad x + \dfrac{3}{2} = -\sqrt{\dfrac{7}{2}}$

$x = -\dfrac{3}{2} + \sqrt{\dfrac{7}{2}} \quad$ or $\qquad x = -\dfrac{3}{2} - \sqrt{\dfrac{7}{2}}$

$x = -\dfrac{3}{2} + \dfrac{\sqrt{7}}{\sqrt{2}} \quad$ or $\qquad x = -\dfrac{3}{2} - \dfrac{\sqrt{7}}{\sqrt{2}}$

$x = -\dfrac{3}{2} + \dfrac{\sqrt{7}}{\sqrt{2}} \cdot \dfrac{\sqrt{2}}{\sqrt{2}} \quad$ or $\quad x = -\dfrac{3}{2} - \dfrac{\sqrt{7}}{\sqrt{2}} \cdot \dfrac{\sqrt{2}}{\sqrt{2}}$

$x = -\dfrac{3}{2} + \dfrac{\sqrt{14}}{2} \quad$ or $\qquad x = -\dfrac{3}{2} - \dfrac{\sqrt{14}}{2}$

$x = \dfrac{-3 + \sqrt{14}}{2} \quad$ or $\qquad x = \dfrac{-3 - \sqrt{14}}{2}$

The solutions are $\dfrac{-3 \pm \sqrt{14}}{2}$.

34. $\dfrac{3 \pm \sqrt{17}}{4}$

35. $x^2 - 6x + 9 = 64$

$(x-3)^2 = 64$ Factoring the left side

$x - 3 = 8$ or $\quad x - 3 = -8$ Principle of
square roots

$x = 11$ or $\qquad x = -5$

The solutions are 11 and -5.

36. $15,\ -5$

37. $y^2 + 14y + 49 = 4$

$(y+7)^2 = 4$ Factoring the left side

$y + 7 = 2$ or $\quad y + 7 = -2$ Principle of
square roots

$y = -5$ or $\qquad y = -9$

The solutions are -5 and -9.

38. $-3,\ -5$

39. $m^2 - 2m + 1 = 5$

$(m-1)^2 = 5$

$m - 1 = \sqrt{5} \qquad$ or $\quad m - 1 = -\sqrt{5}$
$\quad m = 1 + \sqrt{5}$ or $\qquad m = 1 - \sqrt{5}$

The solutions are $1 \pm \sqrt{5}$.

40. $-3 \pm \sqrt{13}$

41. $x^2 + 4x + 4 = 12$

$(x+2)^2 = 12$

$x + 2 = \sqrt{12} \qquad$ or $\quad x + 2 = -\sqrt{12}$
$\quad x = -2 + \sqrt{12} \qquad$ or $\qquad x = -2 - \sqrt{12}$
$\quad x = -2 + \sqrt{4 \cdot 3} \qquad$ or $\qquad x = -2 - \sqrt{4 \cdot 3}$
$\quad x = -2 + 2\sqrt{3} \qquad$ or $\qquad x = -2 - 2\sqrt{3}$

The solutions are $-2 \pm 2\sqrt{3}$.

42. $6 \pm 3\sqrt{2}$

43. $\sqrt{6}\,\sqrt{10} = \sqrt{6\cdot10} = \sqrt{60} = \sqrt{4\cdot15} = 2\sqrt{15}$

44. $3x^2\sqrt{2}$

45. $2x + y = 5,$ (1)

     $y = 4 - x$ (2)

Substitute $4 - x$ for y in Equation (1) and solve for x.

     $2x + (4 - x) = 5$ Substituting

         $x + 4 = 5$ Collecting like terms

             $x = 1$ Subtracting 4 on both sides

Substitute 1 for x in Equation (2) and simplify.

     $y = 4 - 1 = 3$

The solution is $(1, 3)$.

46. $\left(\dfrac{13}{5}, \dfrac{1}{5}\right)$

47.

48.

49. $x^2 + 5x + \dfrac{25}{4} = \dfrac{13}{4}$

     $\left(x + \dfrac{5}{2}\right)^2 = \dfrac{13}{4}$

   $x + \dfrac{5}{2} = \dfrac{\sqrt{13}}{2}$ or $x + \dfrac{5}{2} = -\dfrac{\sqrt{13}}{2}$

     $x = -\dfrac{5}{2} + \dfrac{\sqrt{13}}{2}$ or $x = -\dfrac{5}{2} - \dfrac{\sqrt{13}}{2}$

     $x = \dfrac{-5 + \sqrt{13}}{2}$ or $x = \dfrac{-5 - \sqrt{13}}{2}$

The solutions are $\dfrac{-5 \pm \sqrt{13}}{2}$.

50. $x = \dfrac{7 \pm \sqrt{7}}{6}$

51. $m^2 - \dfrac{3}{2}m + \dfrac{9}{16} = \dfrac{17}{16}$

     $\left(m - \dfrac{3}{4}\right)^2 = \dfrac{17}{16}$

   $m - \dfrac{3}{4} = \dfrac{\sqrt{17}}{4}$ or $m - \dfrac{3}{4} = -\dfrac{\sqrt{17}}{4}$

     $m = \dfrac{3 + \sqrt{17}}{4}$ or $m = \dfrac{3 - \sqrt{17}}{4}$

The solutions are $\dfrac{3 \pm \sqrt{17}}{4}$.

52. $2, -5$

53. $x^2 + 0.5x + 0.0625 = 13.69$

         $(x + 0.25)^2 = 13.69$

  $x + 0.25 = 3.7$ or $x + 0.25 = -3.7$

     $x = 3.45$ or $x = -3.95$

The solutions are 3.45 and -3.95.

54. $1.85, -4.35$

55. $a^2 - 3.8a + 3.61 = 27.04$

         $(a - 1.9)^2 = 27.04$

  $a - 1.9 = 5.2$ or $a - 1.9 = -5.2$

     $a = 7.1$ or $a = -3.3$

The solutions are 7.1 and -3.3.

56. $9.9, -4.7$

Exercise Set 11.2

1. $x^2 - 2x$

     $\left(\dfrac{-2}{2}\right)^2 = (-1)^2 = 1$ Taking half the x-coefficient and squaring

        $x^2 - 2x + 1$

The trinomial $x^2 - 2x + 1$ is the square of $x - 1$.

2. $x^2 - 4x + 4$

3. $x^2 + 18x$ $\left(\dfrac{18}{2}\right)^2 = 9^2 = 81$

The trinomial $x^2 + 18x + 81$ is the square of $x + 9$.

4. $x^2 + 22x + 121$

5. $x^2 - x$ $\left(\dfrac{-1}{2}\right)^2 = \left(-\dfrac{1}{2}\right)^2 = \dfrac{1}{4}$

The trinomial $x^2 - x + \dfrac{1}{4}$ is the square of $x - \dfrac{1}{2}$.

6. $x^2 + x + \dfrac{1}{4}$

7. $t^2 + 5t$ $\left(\dfrac{5}{2}\right)^2 = \dfrac{25}{4}$

The trinomial $t^2 + 5t + \dfrac{25}{4}$ is the square of $t + \dfrac{5}{2}$.

8. $y^2 - 9y + \dfrac{81}{4}$

9. $x^2 - \dfrac{3}{2}x$ $\left(\dfrac{-\frac{3}{2}}{2}\right)^2 = \left(-\dfrac{3}{4}\right)^2 = \dfrac{9}{16}$

The trinomial $x^2 - \dfrac{3}{2}x + \dfrac{9}{16}$ is the square of $x - \dfrac{3}{4}$.

10. $x^2 + \dfrac{4}{3}x + \dfrac{4}{9}$

11. $m^2 + \dfrac{9}{2}m$ $\left(\dfrac{\frac{9}{2}}{2}\right)^2 = \left(\dfrac{9}{4}\right)^2 = \dfrac{81}{16}$

The trinomial $m^2 + \dfrac{9}{2}m + \dfrac{81}{16}$ is the square of $m + \dfrac{9}{4}$.

12. $r^2 - \dfrac{2}{5}r + \dfrac{1}{25}$

13. $x^2 - 6x - 16 = 0$

$x^2 - 6x \qquad = 16 \qquad$ Adding 16

$x^2 - 6x + \;\; 9 = 16 + 9 \qquad$ Adding 9: $\left(\dfrac{-6}{2}\right)^2 =$
$\qquad\qquad\qquad\qquad\qquad\qquad (-3)^2 = 9$

$\qquad (x - 3)^2 = 25$

$x - 3 = 5 \quad$ or $\quad x - 3 = -5 \quad$ Principle of
$\qquad\qquad\qquad\qquad\qquad\qquad$ square roots

$\qquad x = 8 \quad$ or $\qquad x = -2$

The solutions are 8 and -2.

14. $-3,\; -5$

15. $x^2 + 22x + \;\; 21 = 0$

$x^2 + 22x \qquad\quad = -21 \qquad\qquad$ Subtracting 21

$x^2 + 22x + 121 = -21 + 121 \qquad$ Adding 121:

$\qquad\qquad\qquad \left(\dfrac{22}{2}\right)^2 = 11^2 = 121$

$\qquad\qquad (x + 11)^2 = 100$

$x + 11 = 10 \quad$ or $\quad x + 11 = -10 \quad$ Principle of
$\qquad\qquad\qquad\qquad\qquad\qquad\qquad$ square roots

$\qquad\quad x = -1 \quad$ or $\qquad\quad x = -21$

The solutions are -1 and -21.

16. $1,\; -15$

17. $\qquad 3x^2 - 6x - 15 = 0$

$\dfrac{1}{3}(3x^2 - 6x - 15) = \dfrac{1}{3} \cdot 0 \quad$ Multiplying by $\dfrac{1}{3}$ to
$\qquad\qquad\qquad\qquad\qquad\qquad$ make the x^2−coefficient 1

$\qquad x^2 - 2x - \;\; 5 = 0$

$\qquad x^2 - 2x \qquad = 5$

$\qquad x^2 - 2x + \;\; 1 = 5 + 1 \quad$ Adding 1: $\left(\dfrac{-2}{2}\right)^2 =$
$\qquad\qquad\qquad\qquad\qquad\qquad\qquad (-1)^2 = 1$

$\qquad\qquad (x - 1)^2 = 6$

$x - 1 = \sqrt{6} \qquad$ or $\quad x - 1 = -\sqrt{6}$

$\qquad x = 1 + \sqrt{6} \quad$ or $\qquad x = 1 - \sqrt{6}$

The solutions are $1 \pm \sqrt{6}$.

18. $2 \pm \sqrt{15}$

19. $\qquad x^2 - 10x = 22$

$x^2 - 10x + 25 = 22 + 25 \quad$ Adding 25:

$\qquad\qquad\qquad \left(\dfrac{-10}{2}\right)^2 = (-5)^2 = 25$

$\qquad (x - 5)^2 = 47$

$x - 5 = \sqrt{47} \qquad$ or $\quad x - 5 = -\sqrt{47}$

$\qquad x = 5 + \sqrt{47} \quad$ or $\qquad x = 5 - \sqrt{47}$

The solutions are $5 \pm \sqrt{47}$.

20. $-4 \pm 3\sqrt{10}$

21. $x^2 + 6x + 13 = 0$

$x^2 + 6x \qquad = -13$

$x^2 + 6x + 9 = -13 + 9 \quad$ Adding 9: $\left(\dfrac{6}{2}\right)^2 =$
$\qquad\qquad\qquad\qquad\qquad\qquad 3^2 = 9$

$\qquad (x + 3)^2 = -4$

$x + 3 = \sqrt{-4} \qquad$ or $\quad x + 3 = -\sqrt{-4}$

$x + 3 = 2i \qquad\quad$ or $\quad x + 3 = -2i$

$\qquad x = -3 + 2i \quad$ or $\qquad x = -3 - 2i$

The solutions are $-3 \pm 2i$.

22. $-4 \pm 3i$

23. $x^2 - 7x - 2 \;\; = 0$

$x^2 - 7x \qquad\quad = 2$

$x^2 - 7x + \dfrac{49}{4} = 2 + \dfrac{49}{4} \qquad$ Adding $\dfrac{49}{4}$:

$\qquad\qquad\qquad\qquad \left(\dfrac{-7}{2}\right)^2 = \dfrac{49}{4}$

$\left(x - \dfrac{7}{2}\right)^2 = \dfrac{8}{4} + \dfrac{49}{4} = \dfrac{57}{4}$

$x - \dfrac{7}{2} = \dfrac{\sqrt{57}}{2} \qquad$ or $\quad x - \dfrac{7}{2} = -\dfrac{\sqrt{57}}{2}$

$x = \dfrac{7}{2} + \dfrac{\sqrt{57}}{2} \quad$ or $\qquad x = \dfrac{7}{2} - \dfrac{\sqrt{57}}{2}$

$x = \dfrac{7 + \sqrt{57}}{2} \quad$ or $\qquad x = \dfrac{7 - \sqrt{57}}{2}$

The solutions are $\dfrac{7 \pm \sqrt{57}}{2}$.

24. $\dfrac{-7 \pm \sqrt{57}}{2}$

25. $\qquad 2x^2 + 6x - 56 = 0$

$\dfrac{1}{2}(2x^2 + 6x - 56) = \dfrac{1}{2} \cdot 0 \quad$ Multiplying by $\dfrac{1}{2}$
$\qquad\qquad\qquad\qquad\qquad\qquad$ to make the
$\qquad\qquad\qquad\qquad\qquad\qquad x^2$−coefficient 1

$\qquad x^2 + 3x - 28 = 0$

$\qquad x^2 + 3x \qquad = 28$

$\qquad x^2 + 3x + \dfrac{9}{4} = 28 + \dfrac{9}{4} \qquad$ Adding $\dfrac{9}{4}$:

$\qquad\qquad\qquad\qquad\qquad \left(\dfrac{3}{2}\right)^2 = \dfrac{9}{4}$

$\qquad \left(x + \dfrac{3}{2}\right)^2 = \dfrac{121}{4}$

$x + \dfrac{3}{2} = \dfrac{11}{2} \quad$ or $\quad x + \dfrac{3}{2} = -\dfrac{11}{2}$

$\qquad x = \dfrac{8}{2} \quad$ or $\qquad x = -\dfrac{14}{2}$

$\qquad x = 4 \quad$ or $\qquad x = -7$

The solutions are 4 and -7.

26. $7,\; -4$

27. $x^2 + \frac{3}{2}x - \frac{1}{2} = 0$

$x^2 + \frac{3}{2}x \qquad = \frac{1}{2}$

$x^2 + \frac{3}{2}x + \frac{9}{16} = \frac{1}{2} + \frac{9}{16}$ Adding $\frac{9}{16}$:

$$\left(\frac{3/2}{2}\right)^2 = \left(\frac{3}{4}\right)^2 = \frac{9}{16}$$

$$\left(x + \frac{3}{4}\right)^2 = \frac{17}{16}$$

$x + \frac{3}{4} = \frac{\sqrt{17}}{4}$ or $x + \frac{3}{4} = -\frac{\sqrt{17}}{4}$

$x = -\frac{3}{4} + \frac{\sqrt{17}}{4}$ or $x = -\frac{3}{4} - \frac{\sqrt{17}}{4}$

$x = \frac{-3 + \sqrt{17}}{4}$ or $x = \frac{-3 - \sqrt{17}}{4}$

The solutions are $\dfrac{-3 \pm \sqrt{17}}{4}$.

28. $\dfrac{3 \pm \sqrt{41}}{4}$

29. $2x^2 - 5x - 3 = 0$

$\frac{1}{2}(2x^2 - 5x - 3) = \frac{1}{2} \cdot 0$ Multiplying by $\frac{1}{2}$ to make the x^2-coefficient 1

$x^2 - \frac{5}{2}x - \frac{3}{2} = 0$

$x^2 - \frac{5}{2}x \qquad = \frac{3}{2}$

$x^2 - \frac{5}{2}x + \frac{25}{16} = \frac{3}{2} + \frac{25}{16}$ Adding $\frac{25}{16}$:

$$\left[\frac{1}{2}\left(-\frac{5}{2}\right)\right]^2 =$$
$$\left[-\frac{5}{4}\right]^2 = \frac{25}{16}$$

$$\left(x - \frac{5}{4}\right)^2 = \frac{49}{16}$$

$x - \frac{5}{4} = \frac{7}{4}$ or $x - \frac{5}{4} = -\frac{7}{4}$

$x = \frac{12}{4}$ or $x = -\frac{2}{4}$

$x = 3$ or $x = -\frac{1}{2}$

The solutions are 3 and $-\frac{1}{2}$.

30. $\dfrac{-2 \pm \sqrt{2}}{2}$

31. $3x^2 + 4x - 1 = 0$

$\frac{1}{3}(3x^2 + 4x - 1) = \frac{1}{3} \cdot 0$

$x^2 + \frac{4}{3}x - \frac{1}{3} = 0$

$x^2 + \frac{4}{3}x \qquad = \frac{1}{3}$

$x^2 + \frac{4}{3}x + \frac{4}{9} = \frac{1}{3} + \frac{4}{9}$

$$\left(x + \frac{2}{3}\right)^2 = \frac{7}{9}$$

$x + \frac{2}{3} = \frac{\sqrt{7}}{3}$ or $x + \frac{2}{3} = -\frac{\sqrt{7}}{3}$

$x = \frac{-2 + \sqrt{7}}{3}$ or $x = -\frac{-2 - \sqrt{7}}{3}$

The solutions are $\dfrac{-2 \pm \sqrt{7}}{3}$.

32. $\dfrac{2 \pm \sqrt{13}}{3}$

33. $\qquad 2x^2 = 9x + 5$

$2x^2 - 9x - 5 = 0$ Standard form

$\frac{1}{2}(2x^2 - 9x - 5) = \frac{1}{2} \cdot 0$

$x^2 - \frac{9}{2}x - \frac{5}{2} = 0$

$x^2 - \frac{9}{2}x \qquad = \frac{5}{2}$

$x^2 - \frac{9}{2}x + \frac{81}{16} = \frac{5}{2} + \frac{81}{16}$

$$\left(x - \frac{9}{4}\right)^2 = \frac{121}{16}$$

$x - \frac{9}{4} = \frac{11}{4}$ or $x - \frac{9}{4} = -\frac{11}{4}$

$x = \frac{20}{4}$ or $x = -\frac{2}{4}$

$x = 5$ or $x = -\frac{1}{2}$

The solutions are 5 and $-\frac{1}{2}$.

34. $4, -\dfrac{3}{2}$

35. $\qquad 4x^2 + 12x = 7$

$4x^2 + 12x - 7 = 0$

$\frac{1}{4}(4x^2 + 12x - 7) = \frac{1}{4} \cdot 0$

$x^2 + 3x - \frac{7}{4} = 0$

$x^2 + 3x \qquad = \frac{7}{4}$

$x^2 + 3x + \frac{9}{4} = \frac{7}{4} + \frac{9}{4}$

$$\left(x + \frac{3}{2}\right)^2 = \frac{16}{4} = 4$$

$$x + \frac{3}{2} = 2 \quad \text{or} \quad x + \frac{3}{2} = -2$$

$$x = \frac{1}{2} \quad \text{or} \qquad x = -\frac{7}{2}$$

The solutions are $\frac{1}{2}$ and $-\frac{7}{2}$.

36. $\frac{2}{3}, \ -\frac{5}{2}$

37. *Familiarize.* Let $x =$ the number of pounds of Kenyan coffee and $y =$ the number of pounds of Peruvian coffee in the mixture. We organize the information in a table.

Type of Coffee	Kenyan	Peruvian	Mixture
Price per pound	$4.50	$7.50	$5.70
Number of pounds	x	y	50
Total cost	$4.50x	$7.50y	$5.70 × 50, or $285

Translate. From the last two rows of the table we get a system of equations.

$$x + y = 50,$$
$$4.50x + 7.50y = 285$$

Solve. Solving the system of equations, we get $(30, 20)$.

Check. The total number of pounds in the mixture is $30 + 20$, or 50. The total cost of the mixture is $\$4.50(30) + \$7.50(20)$, or $\$135 + \$150$, or $\$285$. The values check.

State. The mixture should consist of 30 lb of Kenyan coffee and 20 lb of Peruvian coffee.

38. $\frac{1}{16}$

39. $\quad \sqrt[5]{x} = -2$

$\quad (\sqrt[5]{x})^5 = (-2)^5 \quad$ Principle of powers

$\qquad x = -32$

The number -32 checks and is the solution.

40. $\frac{3}{2}$

41. ◈

42. ◈

43. $x^2 + bx + 36$

The trinomial is a square if the square of one-half the x-coefficient is equal to 36. Thus we have:

$$\left(\frac{b}{2}\right)^2 = 36$$

$$\frac{b^2}{4} = 36$$

$$b^2 = 144$$

$b = 12$ or $b = -12 \qquad$ Principle of square roots

44. $2\sqrt{55}, \ -2\sqrt{55}$

45. $x^2 + bx + 128$

The trinomial is a square if the square of one-half the x-coefficient is equal to 128. Thus we have:

$$\left(\frac{b}{2}\right)^2 = 128$$

$$\frac{b^2}{4} = 128$$

$$b^2 = 512$$

$$b = \sqrt{512} \quad \text{or} \quad b = -\sqrt{512}$$

$$b = 16\sqrt{2} \quad \text{or} \quad b = -16\sqrt{2}$$

46. $16, \ -16$

47. $x^2 + bx + c$

The trinomial is a square if the square of one-half the x-coefficient is equal to c. Thus we have:

$$\left(\frac{b}{2}\right)^2 = c$$

$$\frac{b^2}{4} = c$$

$$b^2 = 4c$$

$$b = \sqrt{4c} \quad \text{or} \quad b = -\sqrt{4c}$$

$$b = 2\sqrt{c} \quad \text{or} \quad b = -2\sqrt{c}$$

48. $2\sqrt{ac}, \ -2\sqrt{ac}$

49. $8.00, 2.00$

50. $2.00, -8.00$

51. $-0.39, -7.61$

52. $7.41, 4.59$

53. $7.27, -0.27$

54. $0.27, -7.27$

55. $-0.50, 5.00$

56. $4.00, -1.50$

Exercise Set 11.3

1.
$$x^2 - 4x = 21$$
$$x^2 - 4x - 21 = 0 \qquad \text{Standard form}$$

We can factor.
$$x^2 - 4x - 21 = 0$$
$$(x - 7)(x + 3) = 0$$

$$x - 7 = 0 \quad \text{or} \quad x + 3 = 0$$
$$x = 7 \quad \text{or} \quad x = -3$$

The solutions are 7 and -3.

2. $-9, 2$

3.
$$x^2 = 6x - 9$$
$$x^2 - 6x + 9 = 0 \qquad \text{Standard form}$$

We can factor.
$$x^2 - 6x + 9 = 0$$
$$(x - 3)(x - 3) = 0$$

$$x - 3 = 0 \quad \text{or} \quad x - 3 = 0$$
$$x = 3 \quad \text{or} \quad x = 3$$

The solution is 3.

4. 4

5. $x^2 + 6x + 4 = 0$

We use the quadratic formula.

$a = 1, \ b = 6, \ c = 4$
$$x = \frac{-6 \pm \sqrt{6^2 - 4 \cdot 1 \cdot 4}}{2 \cdot 1}$$
$$x = \frac{-6 \pm \sqrt{20}}{2} = \frac{-6 \pm \sqrt{4 \cdot 5}}{2}$$
$$x = \frac{-6 \pm 2\sqrt{5}}{2} = \frac{2(-3 \pm \sqrt{5})}{2}$$
$$x = -3 \pm \sqrt{5}$$

6. $3 \pm \sqrt{13}$

7. $y^2 + 16 = 0$

We use the quadratic formula. (Note that we could also have written $y^2 = -16$ and used the principle of square roots.)

$a = 1, \ b = 0, \ c = 16$
$$y = \frac{0 \pm \sqrt{0^2 - 4 \cdot 1 \cdot 16}}{2 \cdot 1}$$
$$y = \frac{\pm\sqrt{-64}}{2} = \frac{\pm 8i}{2}$$
$$y = \pm 4i$$

8. $\pm i$

9.
$$x^2 - 9 = 0 \qquad \text{Difference of squares}$$
$$(x + 3)(x - 3) = 0$$

$$x + 3 = 0 \quad \text{or} \quad x - 3 = 0$$
$$x = -3 \quad \text{or} \quad x = 3$$

The solutions are -3 and 3.

10. $-2, 2$

11. $x^2 - 2x - 2 = 0$

We use the quadratic formula.

$a = 1, \ b = -2, \ c = -2$
$$x = \frac{-(-2) \pm \sqrt{(-2)^2 - 4 \cdot 1 \cdot (-2)}}{2 \cdot 1}$$
$$x = \frac{2 \pm \sqrt{12}}{2} = \frac{2 \pm \sqrt{4 \cdot 3}}{2}$$
$$x = \frac{2 \pm 2\sqrt{3}}{2} = \frac{2(1 \pm \sqrt{3})}{2}$$
$$x = 1 \pm \sqrt{3}$$

12. $2 \pm \sqrt{11}$

13. $3u^2 - 5u + 4 = 0$

We use the quadratic formula.

$a = 3, \ b = -5, \ c = 4$
$$u = \frac{-(-5) \pm \sqrt{(-5)^2 - 4 \cdot 3 \cdot 4}}{2 \cdot 3}$$
$$u = \frac{5 \pm \sqrt{-23}}{6}$$
$$u = \frac{5 \pm i\sqrt{23}}{6}$$

14. $\dfrac{1 \pm i\sqrt{23}}{3}$

15.
$$x^2 + 4x + 4 = 7$$
$$x^2 + 4x - 3 = 0 \qquad \text{Adding } -7 \text{ to get standard form}$$

We use the quadratic formula.

$a = 1, \ b = 4, \ c = -3$
$$x = \frac{-4 \pm \sqrt{4^2 - 4 \cdot 1 \cdot (-3)}}{2 \cdot 1}$$
$$x = \frac{-4 \pm \sqrt{28}}{2} = \frac{-4 \pm \sqrt{4 \cdot 7}}{2}$$
$$x = \frac{-4 \pm 2\sqrt{7}}{2} = \frac{2(-2 \pm \sqrt{7})}{2}$$
$$x = -2 \pm \sqrt{7}$$

16. $1 \pm \sqrt{5}$

17. $3x^2 + 8x + 2 = 0$

We use the quadratic formula.

$a = 3, \ b = 8, \ c = 2$
$$x = \frac{-8 \pm \sqrt{8^2 - 4 \cdot 3 \cdot 2}}{2 \cdot 3}$$
$$x = \frac{-8 \pm \sqrt{40}}{6} = \frac{-8 \pm \sqrt{4 \cdot 10}}{6}$$
$$x = \frac{-8 \pm 2\sqrt{10}}{6} = \frac{2(-4 \pm \sqrt{10})}{2 \cdot 3}$$
$$x = \frac{-4 \pm \sqrt{10}}{3}$$

18. $\dfrac{2 \pm \sqrt{10}}{3}$

19. $x^2 + x + 2 = 0$

We use the quadratic formula.

$a = 1,\ b = 1,\ c = 2$

$x = \dfrac{-1 \pm \sqrt{1^2 - 4 \cdot 1 \cdot 2}}{2 \cdot 1}$

$x = \dfrac{-1 \pm \sqrt{-7}}{2}$

$x = \dfrac{-1 \pm i\sqrt{7}}{2}$

20. $1 \pm i\sqrt{2}$

21. $4y^2 - 4y - 1 = 0$

We use the quadratic formula.

$a = 4,\ b = -4,\ c = -1$

$y = \dfrac{-(-4) \pm \sqrt{(-4)^2 - 4 \cdot 4 \cdot (-1)}}{2 \cdot 4}$

$y = \dfrac{4 \pm \sqrt{32}}{8} = \dfrac{4 \pm \sqrt{16 \cdot 2}}{8}$

$y = \dfrac{4 \pm 4\sqrt{2}}{8} = \dfrac{4(1 \pm \sqrt{2})}{4 \cdot 2}$

$y = \dfrac{1 \pm \sqrt{2}}{2}$

22. $\dfrac{-1 \pm \sqrt{2}}{2}$

23. $2t^2 + 6t + 5 = 0$

We use the quadratic formula.

$a = 2,\ b = 6,\ c = 5$

$t = \dfrac{-6 \pm \sqrt{6^2 - 4 \cdot 2 \cdot 5}}{2 \cdot 2}$

$t = \dfrac{-6 \pm \sqrt{-4}}{4} = \dfrac{-6 \pm 2i}{4}$

$t = \dfrac{2(-3 \pm i)}{2 \cdot 2} = \dfrac{-3 \pm i}{2}$

24. $\dfrac{-3 \pm i\sqrt{23}}{8}$

25. $\qquad 3x^2 = 5x + 4$

$3x^2 - 5x - 4 = 0$

We use the quadratic formula.

$a = 3,\ b = -5,\ c = -4$

$x = \dfrac{-(-5) \pm \sqrt{(-5)^2 - 4 \cdot 3 \cdot (-4)}}{2 \cdot 3}$

$x = \dfrac{5 \pm \sqrt{73}}{6}$

26. $\dfrac{-3 \pm \sqrt{17}}{4}$

27. $25x^2 - 20x + 4 = 0 \qquad$ We can factor.

$(5x - 2)^2 = 0$

$5x - 2 = 0 \ \ \text{or} \ \ 5x - 2 = 0$

$x = \dfrac{2}{5} \ \ \text{or} \qquad x = \dfrac{2}{5}$

The solution is $\dfrac{2}{5}$.

28. $-\dfrac{7}{6}$

29. $\quad 6x^2 - 9x = 0 \qquad$ We can factor.

$3x(2x - 3) = 0$

$3x = 0 \ \ \text{or} \ \ 2x - 3 = 0$

$x = 0 \ \ \text{or} \qquad 2x = 3$

$x = 0 \ \ \text{or} \qquad x = \dfrac{3}{2}$

The solutions are 0 and $\dfrac{3}{2}$.

30. $\dfrac{3 \pm i\sqrt{5}}{7}$

31. $\qquad 5t^2 - 7t = -4$

$5t^2 - 7t + 4 = 0$

We use the quadratic formula.

$a = 5,\ b = -7,\ c = 4$

$t = \dfrac{-(-7) \pm \sqrt{(-7)^2 - 4 \cdot 5 \cdot 4}}{2 \cdot 5}$

$t = \dfrac{7 \pm \sqrt{-31}}{10}$

$t = \dfrac{7 \pm i\sqrt{31}}{10}$

32. $0,\ -\dfrac{2}{3}$

33. $\qquad 5t^2 = 100$

$5t^2 - 100 = 0$

$t^2 - 20 = 0 \qquad$ Dividing by 5 on both sides

We use the quadratic formula. (Note that this equation could also be solved using the principle of square roots.)

$a = 1,\ b = 0,\ c = -20$

$t = \dfrac{0 \pm \sqrt{0^2 - 4 \cdot 1 \cdot (-20)}}{2 \cdot 1}$

$t = \dfrac{\pm\sqrt{80}}{2} = \dfrac{\pm\sqrt{16 \cdot 5}}{2}$

$t = \dfrac{\pm 4\sqrt{5}}{2}$

$t = \pm 2\sqrt{5}$

34. $\pm\dfrac{3\sqrt{10}}{2}$

35. $x^2 - 4x - 7 = 0$

$a = 1, \ b = -4, \ c = -7$

$$x = \frac{-(-4) \pm \sqrt{(-4)^2 - 4 \cdot 1 \cdot (-7)}}{2 \cdot 1}$$

$$x = \frac{4 \pm \sqrt{16 + 28}}{2} = \frac{4 \pm \sqrt{44}}{2}$$

$$x = \frac{4 \pm \sqrt{4 \cdot 11}}{2} = \frac{4 \pm 2\sqrt{11}}{2}$$

$$x = \frac{2(2 \pm \sqrt{11})}{2} = 2 \pm \sqrt{11}$$

Using a calculator or Table 1, we see that $\sqrt{11} \approx 3.317$:

$$\begin{aligned} 2 + \sqrt{11} &\approx 2 + 3.317 \quad \text{or} \quad 2 - \sqrt{11} \approx 2 - 3.317 \\ &\approx 5.317 \quad \text{or} \quad \approx -1.317 \end{aligned}$$

The approximate solutions, to the nearest thousandth, are 5.317 and -1.317.

36. $0.732, \ -2.732$

37. $y^2 - 6y - 1 = 0$

$a = 1, \ b = -6, \ c = -1$

$$y = \frac{-(-6) \pm \sqrt{(-6)^2 - 4 \cdot 1 \cdot (-1)}}{2 \cdot 1}$$

$$y = \frac{6 \pm \sqrt{36 + 4}}{2} = \frac{6 \pm \sqrt{40}}{2}$$

$$y = \frac{6 \pm \sqrt{4 \cdot 10}}{2} = \frac{6 \pm 2\sqrt{10}}{2}$$

$$y = \frac{2(3 \pm \sqrt{10})}{2} = 3 \pm \sqrt{10}$$

Using a calculator or Table 1, we see that $\sqrt{10} \approx 3.162$:

$$\begin{aligned} 3 + \sqrt{10} &\approx 3 + 3.162 \quad \text{or} \quad 3 - \sqrt{10} \approx 3 - 3.162 \\ &\approx 6.162 \quad \text{or} \quad \approx -0.162 \end{aligned}$$

The approximate solutions, to the nearest thousandth, are 6.162 and -0.162.

38. $-3.268, \ -6.732$

39. $\qquad 4x^2 + 4x = 1$

$4x^2 + 4x - 1 = 0 \qquad$ Standard form

$a = 4, \ b = 4, \ c = -1$

$$x = \frac{-4 \pm \sqrt{4^2 - 4 \cdot 4 \cdot (-1)}}{2 \cdot 4}$$

$$x = \frac{-4 \pm \sqrt{16 + 16}}{8} = \frac{-4 \pm \sqrt{32}}{8}$$

$$x = \frac{-4 \pm \sqrt{16 \cdot 2}}{8} = \frac{-4 \pm 4\sqrt{2}}{8}$$

$$x = \frac{4(-1 \pm \sqrt{2})}{4 \cdot 2} = \frac{-1 \pm \sqrt{2}}{2}$$

Using a calculator or Table 1, we see that $\sqrt{2} \approx 1.414$:

$$\begin{aligned} \frac{-1 + \sqrt{2}}{2} &\approx \frac{-1 + 1.414}{2} \quad \text{or} \quad \frac{-1 - \sqrt{2}}{2} \approx \frac{-1 - 1.414}{2} \\ &\approx \frac{0.414}{2} \quad \text{or} \quad \approx \frac{-2.414}{2} \\ &\approx 0.207 \quad \text{or} \quad \approx -1.207 \end{aligned}$$

The approximate solutions, to the nearest thousandth, are 0.207 and -1.207.

40. $1.207, \ -0.207$

41. *Familiarize.* We will use the formula

$$d = \frac{n^2 - 3n}{2},$$

where d is the number of diagonals and n is the number of sides.

Translate. We substitute 8 for n.

$$d = \frac{8^2 - 3 \cdot 8}{2}$$

Carry out. We do the computation.

$$d = \frac{8^2 - 3 \cdot 8}{2} = \frac{64 - 24}{2} = \frac{40}{2} = 20$$

Check. We can substitute 20 for d in the original formula and see if this yields $n = 8$. This is left to the student.

State. An octagon has 20 diagonals.

42. 35

43. *Familiarize.* We will use the formula

$$d = \frac{n^2 - 3n}{2},$$

where d is the number of diagonals and n is the number of sides.

Translate. We substitute 14 for d.

$$14 = \frac{n^2 - 3n}{2}$$

Carry out. We solve the equation.

$$\begin{aligned} \frac{n^2 - 3n}{2} &= 14 \\ n^2 - 3n &= 28 \qquad \text{Multiplying by 2} \\ n^2 - 3n - 28 &= 0 \\ (n - 7)(n + 4) &= 0 \end{aligned}$$

$$\begin{aligned} n - 7 = 0 \quad &\text{or} \quad n + 4 = 0 \\ n = 7 \quad &\text{or} \qquad n = -4 \end{aligned}$$

Check. Since the number of sides cannot be negative, -4 cannot be a solution. To check 7, we substitute 7 for n in the original formula and determine if this yields $d = 14$. This is left to the student.

State. The polygon has 7 sides.

44. 4

45. *Familiarize.* We will use the formula $s = 16t^2$.

Translate. We substitute 640 for s.

$$640 = 16t^2$$

Carry out. We solve the equation.

$$\begin{aligned} 640 &= 16t^2 \\ 40 &= t^2 \qquad \text{Dividing by 16} \end{aligned}$$

$$t = \sqrt{40} \quad \text{or} \quad t = -\sqrt{40}$$

$$t \approx 6.32 \quad \text{or} \quad t \approx -6.32$$

Check. The number -6.32 cannot be a solution, because time cannot be negative in this situation. We substitute 6.32 in the original equation.

$$s = 16(6.32)^2 = 16(39.9424) = 639.0784$$

This is close. Remember that we approximated a solution. Thus we have a check.

State. It takes about 6.32 sec for an object to fall to the ground from the top of the Gateway Arch.

46. 7.95 sec

47. Familiarize. We will use the formula $s = 16t^2$.

Translate. We substitute 175 for s.

$$175 = 16t^2$$

Carry out. We solve the equation.

$$175 = 16t^2$$

$$\frac{175}{16} = t^2$$

$$\sqrt{\frac{175}{16}} = t \quad \text{or} \quad -\sqrt{\frac{175}{16}} = t$$

$$\frac{\sqrt{175}}{4} = t \quad \text{or} \quad -\frac{\sqrt{175}}{4} = t$$

$$\frac{5\sqrt{7}}{4} = t \quad \text{or} \quad -\frac{5\sqrt{7}}{4} = t \quad \left(\sqrt{175} = \right.$$
$$\left. \sqrt{25 \cdot 7} = 5\sqrt{7}\right)$$

$$3.31 \approx t \quad \text{or} \quad -3.31 \approx t$$

Check. The number -3.31 cannot be a solution because time cannot be negative in this situation. We substitute 3.31 in the original equation.

$$s = 16(3.31)^2 = 16(10.9561) = 175.2976$$

This is close. Remember that we approximated a solution. Thus we have a check.

State. The fall took about 3.31 sec.

48. About 4.41 sec

49. Familiarize. We first make a drawing and label it. We let x represent the length of one leg. Then $x + 17$ represents the length of the other leg.

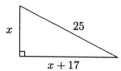

Translate. We use the Pythagorean equation.

$$x^2 + (x + 17)^2 = 25^2$$

Carry out. We solve the equation.

$$x^2 + x^2 + 34x + 289 = 625$$
$$2x^2 + 34x - 336 = 0$$
$$x^2 + 17x - 168 = 0 \quad \text{Multiplying by } \frac{1}{2}$$
$$(x - 7)(x + 24) = 0$$

$$x - 7 = 0 \quad \text{or} \quad x + 24 = 0$$
$$x = 7 \quad \text{or} \quad x = -24$$

Check. Since the length of a leg cannot be negative, -24 does not check. But 7 does check. If the smaller leg is 7, the

other leg is $7 + 17$, or 24. Then, $7^2 + 24^2 = 49 + 576 = 625$, and $\sqrt{625} = 25$, the length of the hypotenuse.

State. The legs measure 7 ft and 24 ft.

50. 10 yd, 24 yd

51. Familiarize. We consider the drawing in the text, where w represents the width of the rectangle and $w + 2$ represents the length.

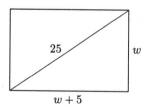

Translate. The area is length $\times$ width. Thus, we have two expressions for the area of the rectangle: $(w + 2)w$ and 80. This gives us a translation.

$$(w + 2)w = 80$$

Carry out. We solve the equation.

$$w^2 + 2w = 80$$
$$w^2 + 2w - 80 = 0$$
$$(w + 10)(w - 8) = 0$$

$$w + 10 = 0 \quad \text{or} \quad w - 8 = 0$$
$$w = -10 \quad \text{or} \quad w = 8$$

Check. Since the length of a side cannot be negative, -10 does not check. But 8 does check. If the width is 8, then the length is $8 + 2$, or 10. The area is 10×8, or 80. This checks.

State. The length is 10 cm, and the width is 8 cm.

52. Length: 9 m, width: 8 m

53. Familiarize. We make a drawing. We let $w =$ the width of the yard. Then $w + 5 =$ the length.

Translate. We use the Pythagorean equation.

$$w^2 + (w + 5)^2 = 25^2$$

Carry out. We solve the equation.

$$w^2 + w^2 + 10w + 25 = 625$$
$$2w^2 + 10w - 600 = 0$$
$$w^2 + 5w - 300 = 0$$
$$(w + 20)(w - 15) = 0$$

$$w + 20 = 0 \quad \text{or} \quad w - 15 = 0$$
$$w = -20 \quad \text{or} \quad w = 15$$

Check. Since the width cannot be negative, -20 does not check. But 15 does check. If the width is 15, then the length is $15 + 5$, or 20, and $15^2 + 20^2 = 225 + 400 = 625 = 25^2$.

State. The yard is 15 m by 20 m.

54. 24 ft

55. *Familiarize*. We make a drawing. Let x = the shorter leg of the right triangle. Then $x + 2.5$ = the longer leg.

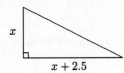

Translate. Using the formula $A = \frac{1}{2}bh$, we substitute 13 for A, $x + 2.5$ for b, and x for h.

$$13 = \frac{1}{2}(x + 2.5)(x)$$

Carry out. We solve the equation.

$$13 = \frac{1}{2}(x + 2.5)(x)$$
$$26 = (x + 2.5)(x) \qquad \text{Multiplying by 2}$$
$$26 = x^2 + 2.5x$$
$$0 = x^2 + 2.5x - 26$$
$$0 = 10x^2 + 25x - 260 \qquad \text{Multiplying by 10 to}$$
$$\qquad\qquad\qquad\qquad\qquad\text{clear the decimal}$$
$$0 = 5(2x^2 + 5x - 52)$$
$$0 = 5(2x + 13)(x - 4)$$

$$2x + 13 = 0 \quad \text{or} \quad x - 4 = 0$$
$$2x = -13 \quad \text{or} \quad x = 4$$
$$x = -6.5 \quad \text{or} \quad x = 4$$

Check. Since the length cannot be negative, -6.5 does not check. But 4 does check. If the shorter leg is 4, then the longer leg is $4 + 2.5$, or 6.5, and $A = \frac{1}{2}(6.5)(4) = 13$.

State. The legs are 4 m and 6.5 m.

56. 5 cm, 6.2 cm

57. *Familiarize*. We make a drawing. We let x represent the the width and $x + 2$ the length.

$$\boxed{20 \text{ in}^2} \;\; x$$
$$x + 2$$

Translate. The area is length × width. We have two expressions for the area of the rectangle: $(x + 2)x$ and 20. This gives us a translation.

$$(x + 2)x = 20$$

Carry out. We solve the equation.

$$x^2 + 2x = 20$$
$$x^2 + 2x - 20 = 0$$
$$a = 1,\ b = 2,\ c = -20$$
$$x = \frac{-2 \pm \sqrt{2^2 - 4 \cdot 1 \cdot (-20)}}{2 \cdot 1}$$
$$x = \frac{-2 \pm \sqrt{4 + 80}}{2} = \frac{-2 \pm \sqrt{84}}{2}$$
$$x = \frac{-2 \pm \sqrt{4 \cdot 21}}{2} = \frac{-2 \pm 2\sqrt{21}}{2}$$
$$x = \frac{2(-1 \pm \sqrt{21})}{2} = -1 \pm \sqrt{21}$$

Using a calculator or Table 1 we find that $\sqrt{21} \approx 4.58$:

$$-1 + \sqrt{21} \approx -1 + 4.58 \quad \text{or} \quad -1 - \sqrt{21} \approx -1 - 4.58$$
$$\approx 3.58 \qquad\qquad \text{or} \qquad\qquad \approx -5.58$$

Check. Since the length of a side cannot be negative, -5.58 does not check. But 3.58 does check. If the width is 3.58, then the length is $3.58 + 2$, or 5.58. The area is $5.58(3.58)$, or $19.9764 \approx 20$.

State. The length is about 5.58 in., and the width is about 3.58 in.

58. Length: 5.65 ft, width: 2.65 ft

59. *Familiarize*. We first make a drawing. We let x represent the width and $2x$ the length.

$$\boxed{10 \text{ m}^2} \;\; x$$
$$2x$$

Translate. The area is length × width. We have two expressions for the area of the rectangle: $2x \cdot x$ and 10. This gives us a translation.

$$2x \cdot x = 10$$

Carry out. We solve the equation.

$$2x^2 = 10$$
$$x^2 = 5$$

$$x = \sqrt{5} \quad \text{or} \quad x = -\sqrt{5}$$
$$x \approx 2.24 \quad \text{or} \quad x \approx -2.24 \quad \text{Using a calculator or}$$
$$\qquad\qquad\qquad\qquad\qquad\qquad\qquad\qquad \text{Table 1}$$

Check. Since the length cannot be negative, -2.24 does not check. But 2.24 does check. If the width is $\sqrt{5}$ m, then the length is $2(\sqrt{5})$, or 4.47 m. The area is $(4.47)(2.24)$, or $10.0128 \approx 10$.

State. The length is about 4.47 m, and the width is about 2.24 m.

60. Length: 6.32 cm, width: 3.16 cm

61. *Familiarize*. We will use the formula $A = P(1 + r)^t$ where P is \$1000, A is \$1210, and t is 2.

Translate. We make the substitutions.

$$A = P(1 + r)^t$$
$$1210 = 1000(1 + r)^2$$

Carry out. We solve the equation.

$$1210 = 1000(1 + r)^2$$
$$\frac{1210}{1000} = (1 + r)^2$$
$$\frac{121}{100} = (1 + r)^2$$

$$\frac{11}{10} = 1 + r \quad \text{or} \quad -\frac{11}{10} = 1 + r$$

Principle of square roots

$$\frac{11}{10} - \frac{10}{10} = r \quad \text{or} \quad -\frac{11}{10} - \frac{10}{10} = r$$
$$\frac{1}{10} = r \quad \text{or} \quad -\frac{21}{10} = r$$

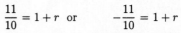

Check. Since the interest rate cannot be negative,

$$\frac{1}{10} = r$$
$$0.1 = r$$
$$10\% = r.$$

We check 10%, or 0.1, in the formula.

$$1000(1 + 0.1)^2 = 1000(1.1)^2 = 1000(1.21) = 1210$$

Our answer checks.

State. The interest rate is 10%.

62. 20%

63. Familiarize. We will use the formula $A = P(1+r)^t$ where P is $2560, A is $2890, and t is 2.

Translate. We make the substitutions.

$$A = P(1+r)^t$$
$$2890 = 2560(1+r)^2$$

Carry out. We solve the equation.

$$2890 = 2560(1+r)^2$$
$$\frac{2890}{2560} = (1+r)^2$$
$$\frac{289}{256} = (1+r)^2$$
$$\frac{17}{16} = 1+r \quad \text{or} \quad -\frac{17}{16} = 1+r$$

$$\text{Principle of square roots}$$

$$\frac{17}{16} - \frac{16}{16} = r \quad \text{or} \quad -\frac{17}{16} - \frac{16}{16} = r$$
$$\frac{1}{16} = r \quad \text{or} \quad -\frac{33}{16} = r$$

Check. Since the interest rate cannot be negative,

$$\frac{1}{16} = r$$
$$0.0625 = r$$
$$6.25\% = r.$$

We check 6.25%, or 0.0625, in the formula.

$$2560(1 + 0.0625)^2 = 2560(1.0625)^2 = 2890$$

Our answer checks.

State. The interest rate is 6.25%.

64. 5%

65. Familiarize. We will use the formula $A = P(1+r)^t$ where P is $6250, A is $7290, and t is 2.

Translate. We make the substitutions.

$$A = P(1+r)^t$$
$$7290 = 6250(1+r)^2$$

Carry out. We solve the equation.

$$7290 = 6250(1+r)^2$$
$$\frac{7290}{6250} = (1+r)^2$$
$$\frac{729}{625} = (1+r)^2$$

$$\frac{27}{25} = 1+r \quad \text{or} \quad -\frac{27}{25} = 1+r$$
$$\frac{27}{25} - \frac{25}{25} = r \quad \text{or} \quad -\frac{27}{25} - \frac{25}{25} = r$$
$$\frac{2}{25} = r \quad \text{or} \quad -\frac{52}{25} = r$$

Check. Since the interest rate cannot be negative,

$$\frac{2}{25} = r$$
$$0.08 = r$$
$$8\% = r.$$

We check 8%, or 0.08, in the formula.

$$6250(1+0.08)^2 = 6250(1.08)^2 = 6250(1.1664) = 7290$$

Our answer checks.

State. The interest rate is 8%.

66. 4%

67. Familiarize. Let $d =$ the diameter (or width) of the flower garden. Then $\frac{d}{2} =$ the radius. We will use the formula for the area of a circle, $A = \pi r^2$.

Translate. We substitute 250 for A, and $\frac{d}{2}$ for r in the formula. Since we will use the π key on a calculator, we do not substitute directly for π.

$$250 = \pi\left(\frac{d}{2}\right)^2$$
$$250 = \pi\left(\frac{d^2}{4}\right)$$

Carry out. We solve the equation.

$$250 = \pi\left(\frac{d^2}{4}\right)$$
$$1000 = \pi d^2 \qquad \text{Multiplying by 4}$$
$$\frac{1000}{\pi} = d^2$$
$$17.84 \approx d \quad \text{or} \quad -17.84 \approx d$$

Check. Since the diameter cannot be negative, -17.84 cannot be a solution. If $d = 17.84$, then $A = \pi\left(\frac{17.84}{2}\right)^2 = 249.9652177 \approx 250$. The answer checks.

State. The width of the largest circular garden Laura can cover with the mulch is about 17.84 ft.

(Note that this is the result when the π key on a calculator is used. If your calculator has no π key and you use 3.14 to approximate π, the result is 17.85 ft.)

68. About 159.58 m

(Note that this is the result when the π key on the calculator is used. If your calculator has no π key and you use 3.14 to approximate π, the result is 159.62 m.)

69.
$$\frac{1}{x} + \frac{2}{x+1} = \frac{3}{x}, \text{ LCD is}$$
$$x(x+1)$$
$$x(x+1)\left(\frac{1}{x} + \frac{2}{x+1}\right) = x(x+1) \cdot \frac{3}{x}$$
$$x(x+1) \cdot \frac{1}{x} + x(x+1) \cdot \frac{2}{x+1} = 3(x+1)$$
$$x + 1 + 2x = 3x + 3$$
$$3x + 1 = 3x + 3$$
$$1 = 3 \quad \text{Subtracting}$$
$$3x$$

We get a false equation. There is no solution.

70. -20

71.
$$\sqrt{40} - 2\sqrt{10} + \sqrt{90} = \sqrt{4 \cdot 10} - 2\sqrt{10} + \sqrt{9 \cdot 10}$$
$$= \sqrt{4}\ \sqrt{10} - 2\sqrt{10} + \sqrt{9}\ \sqrt{10}$$
$$= 2\sqrt{10} - 2\sqrt{10} + 3\sqrt{10}$$
$$= (2 - 2 + 3)\sqrt{10}$$
$$= 3\sqrt{10}$$

72. $30x^5\sqrt{10}$

73.

74.

75.
$$5x + x(x - 7) = 0$$
$$5x + x^2 - 7x = 0$$
$$x^2 - 2x = 0 \quad \text{We can factor.}$$
$$x(x - 2) = 0$$
$$x = 0 \quad \text{or} \quad x - 2 = 0$$
$$x = 0 \quad \text{or} \quad x = 2$$

The solutions are 0 and 2.

76. $0, -\dfrac{4}{3}$

77.
$$(y + 4)(y + 3) = 15$$
$$y^2 + 7y + 12 = 15$$
$$y^2 + 7y - 3 = 0 \quad \text{Standard form}$$

We use the quadratic formula.
$$a = 1,\ b = 7,\ c = -3$$
$$y = \frac{-7 \pm \sqrt{7^2 - 4 \cdot 1 \cdot (-3)}}{2 \cdot 1}$$
$$y = \frac{-7 \pm \sqrt{61}}{2}$$

78. $\dfrac{-2 \pm \sqrt{10}}{2}$

79.
$$(x + 2)^2 + (x + 1)^2 = 0$$
$$x^2 + 4x + 4 + x^2 + 2x + 1 = 0$$
$$2x^2 + 6x + 5 = 0 \quad \text{Standard form}$$

We use the quadratic formula.
$$a = 2,\ b = 6,\ c = 5$$
$$x = \frac{-6 \pm \sqrt{6^2 - 4 \cdot 2 \cdot 5}}{2 \cdot 2}$$

$$x = \frac{-6 \pm \sqrt{-4}}{4} = \frac{-6 \pm 2i}{4}$$
$$x = \frac{2(-3 \pm i)}{2 \cdot 2} = \frac{-3 \pm i}{2}$$

80. $-2 \pm i$

81. $x^2 + x - \sqrt{2} = 0$

We use the quadratic formula.
$$a = 1,\ b = 1,\ c = -\sqrt{2}$$
$$x = \frac{-1 \pm \sqrt{1^2 - 4 \cdot 1 \cdot (-\sqrt{2})}}{2 \cdot 1}$$
$$x = \frac{-1 \pm \sqrt{1 + 4\sqrt{2}}}{2}$$

82. $\dfrac{-\sqrt{5} \pm \sqrt{5 + 4\sqrt{3}}}{2}$

83.
$$\frac{x^2}{x - 4} - \frac{7}{x - 4} = 0, \text{ LCD is } x - 4$$
$$(x - 4)\left(\frac{x^2}{x - 4} - \frac{7}{x - 4}\right) = (x - 4) \cdot 0$$
$$x^2 - 7 = 0$$
$$x^2 = 7$$

$$x = \sqrt{7} \text{ or } x = -\sqrt{7} \quad \text{Principle of square roots}$$

Both numbers check. The solutions are $\sqrt{7}$ and $-\sqrt{7}$.

84. $2 \pm \sqrt{34}$

85.
$$x^3 - 8 = 0$$
$$(x - 2)(x^2 + 2x + 4) = 0$$
$$x - 2 = 0 \text{ or } x^2 + 2x + 4 = 0$$
$$x = 2 \text{ or } \quad x = \frac{-2 \pm \sqrt{2^2 - 4 \cdot 1 \cdot 4}}{2 \cdot 1}$$
$$x = 2 \text{ or } \quad x = \frac{-2 \pm \sqrt{-12}}{2}$$
$$x = 2 \text{ or } \quad x = \frac{-2 \pm 2i\sqrt{3}}{2}$$
$$x = 2 \text{ or } \quad x = -1 \pm i\sqrt{3}$$

86. $-1, \dfrac{1 \pm i\sqrt{3}}{2}$

87. *Familiarize*. From the drawing in the text we see that we have a right triangle whose legs are both r and whose hypotenuse is $r + 1$.

Translate. We use the Pythagorean equation.
$$r^2 + r^2 = (r + 1)^2$$
$$2r^2 = r^2 + 2r + 1$$
$$r^2 - 2r - 1 = 0$$

Carry out. We solve the equation.
$$a = 1,\ b = -2,\ c = -1$$

$$r = \frac{-(-2) \pm \sqrt{(-2)^2 - 4 \cdot 1 \cdot (-1)}}{2 \cdot 1}$$

$$r = \frac{2 \pm \sqrt{4 + 4}}{2} = \frac{2 \pm \sqrt{8}}{2}$$

$$r = \frac{2 \pm \sqrt{4 \cdot 2}}{2} = \frac{2 \pm 2\sqrt{2}}{2}$$

$$r = \frac{2(1 \pm \sqrt{2})}{2} = 1 \pm \sqrt{2}$$

Check. Since $1 - \sqrt{2}$ is negative, it cannot be the length of a leg. Thus it cannot be a solution. If the length of a leg is $1 + \sqrt{2}$, then the length of the hypotenuse is $2 + \sqrt{2}$. We check using the Pythagorean equation.

$$\begin{array}{ll} (1 + \sqrt{2})^2 + (1 + \sqrt{2})^2 & (2 + \sqrt{2})^2 \\ = 2(1 + 2\sqrt{2} + 2) & = 4 + 4\sqrt{2} + 2 \\ = 2(3 + 2\sqrt{2}) & = 6 + 4\sqrt{2} \\ = 6 + 4\sqrt{2} & \end{array}$$

Thus, $(1 + \sqrt{2})^2 + (1 + \sqrt{2})^2 = (2 + \sqrt{2})^2$. The value checks.

State. In the figure, $r = 1 + \sqrt{2} \approx 2.41$ cm.

88. 12 and 13 or -13 and -12

89. From the drawing we see that we have a problem similar to Exercise 87.

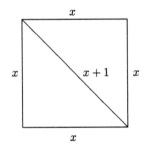

$$x^2 + x^2 = (x + 1)^2$$

From Exercise 87, we know that $x = 1 + \sqrt{2}$.

Then the area is $(1 + \sqrt{2})(1 + \sqrt{2}) = 1 + 2\sqrt{2} + 2 = 3 + 2\sqrt{2} \approx 5.828$ square units.

90. 7.5 ft from the bottom

91. Familiarize. The radius of a 10-in. pizza is $\frac{10}{2}$, or 5 in.

The radius of a d-in. pizza is $\frac{d}{2}$ in. The area of a circle is πr^2.

Translate.

$$\underbrace{\text{Area of } d\text{-in. pizza}}_{\pi\left(\frac{d}{2}\right)^2} = \underbrace{\text{Area of } 10\text{-in. pizza}}_{\pi \cdot 5^2} + \underbrace{\text{Area of } 10\text{-in. pizza}}_{\pi \cdot 5^2}$$

Carry out. We solve the equation.

$$\frac{d^2}{4}\pi = 25\pi + 25\pi$$

$$\frac{d^2}{4}\pi = 50\pi$$

$$\frac{d^2}{4} = 50 \qquad \text{Multiplying by } \frac{1}{\pi}$$

$$d^2 = 200$$

$$d = \sqrt{200} \quad \text{or} \quad d = -\sqrt{200}$$
$$d = 10\sqrt{2} \quad \text{or} \quad d = -10\sqrt{2}$$
$$d \approx 14.14 \quad \text{or} \quad d \approx -14.14 \quad \text{Using a calculator or}$$
$$\text{Table 1}$$

Check. Since the diameter cannot be negative, -14.14 is not a solution. If $d = 10\sqrt{2}$, or 14.14, then $r = 5\sqrt{2}$ and the area is $\pi(5\sqrt{2})^2$, or 50π. The area of the two 10" pizzas is $2 \cdot \pi \cdot 5^2$, or 50π. The value checks.

State. The diameter of the pizza should be $10\sqrt{2}$, or about 14.14 in.

The area of two 10" pizzas is approximately the same as a 14" pizza. Thus, you get more to eat with two 10" pizzas than with a 13-in. pizza.

92. $3 + 3\sqrt{2}$, or about 7.2 cm

93. Familiarize. We will use the formula $A = P(1 + r)^t$ where P is \$4000, A is \$5267.03, and t is 2.

Translate. We make the substitutions.

$$A = P(1 + r)^t$$
$$5267.03 = 4000(1 + r)^2$$

Carry out. We solve the equation.

$$5267.03 = 4000(1 + r)^2$$

$$\frac{5267.03}{4000} = (1 + r)^2$$

$$\sqrt{\frac{5267.03}{4000}} = 1 + r \quad \text{or} \quad -\sqrt{\frac{5267.03}{4000}} = 1 + r$$

$$1.1475 \approx 1 + r \quad \text{or} \quad -1.1475 \approx 1 + r$$
$$\text{Using a calculator}$$
$$0.1475 \approx r \quad \text{or} \quad -2.1475 \approx r$$

Check. Since the interest rate cannot be negative, we check 0.1475 in the formula.

$$4000(1 + 0.1475)^2 = 4000(1.1475)^2 = 5267.03$$

Our answer checks.

State. The interest rate is about 0.1475, or 14.75%.

94. \$2239.13

95. Familiarize. The area of the actual strike zone is $15(40)$, so the area of the enlarged zone is $15(40) + 0.4(15)(40)$, or $1.4(15)(40)$. From the drawing in the text we see that the dimensions of the enlarged strike zone are $15 + 2x$ by $40 + 2x$.

Translate. Using the formula $A = lw$, we write an equation for the area of the enlarged strike zone.

$$1.4(15)(40) = (15 + 2x)(40 + 2x)$$

Carry out. We solve the equation.

$$1.4(15)(40) = (15 + 2x)(40 + 2x)$$

$840 = 600 + 110x + 4x^2$ Multiplying on both sides

$$0 = 4x^2 + 110x - 240$$

$$0 = 2x^2 + 55x - 120$$ Dividing by 2

$a = 2$, $b = 55$, $c = -120$

$$x = \frac{-55 \pm \sqrt{55^2 - 4 \cdot 2 \cdot (-120)}}{2 \cdot 2}$$

$$x = \frac{-55 \pm \sqrt{3985}}{4}$$

$x \approx 2.0$ or $x \approx -29.5$

Check. Since the measurement cannot be negative, -29.5 cannot be a solution. If $x = 2$, then the dimensions of the enlarged strike zone are $15 + 2 \cdot 2$, or 19, by $40 + 2 \cdot 2$, or 44, and the area is $19 \cdot 44 = 836 \approx 840$. The answer checks.

State. The dimensions of the enlarged strike zone are 19 in. by 44 in.

96.

Exercise Set 11.4

1. $\dfrac{1}{x} = \dfrac{x - 2}{24}$, LCD is $24x$

$24x \cdot \dfrac{1}{x} = 24x \cdot \dfrac{x - 2}{24}$ Multiplying by the LCD

$$24 = x(x - 2)$$

$$24 = x^2 - 2x$$

$$0 = x^2 - 2x - 24 \quad \text{Standard form}$$

$$0 = (x - 6)(x + 4) \quad \text{Factoring}$$

$x = 6$ or $x = -4$ Principle of zero products

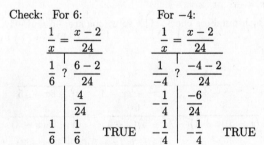

The solutions are 6 and -4.

2. -7, 4

3. $\dfrac{1}{2x - 1} - \dfrac{1}{2x + 1} = \dfrac{1}{4}$,

LCD is $4(2x - 1)(2x + 1)$

$4(2x-1)(2x+1)\left(\dfrac{1}{2x-1} - \dfrac{1}{2x+1} \right) = 4(2x-1)(2x+1) \cdot \dfrac{1}{4}$

$$4(2x + 1) - 4(2x - 1) = (2x - 1)(2x + 1)$$

$$8x + 4 - 8x + 4 = 4x^2 - 1$$

$$8 = 4x^2 - 1$$

$$9 = 4x^2$$

$$\frac{9}{4} = x^2$$

$x = \dfrac{3}{2}$ or $x = -\dfrac{3}{2}$ Principle of square roots

Both numbers check. The solutions are $\dfrac{3}{2}$ and $-\dfrac{3}{2}$, or $\pm\dfrac{3}{2}$.

4. -8, 2

5. $\dfrac{50}{x} - \dfrac{50}{x - 5} = -\dfrac{1}{2}$, LCD is $2x(x - 5)$

$2x(x - 5)\left(\dfrac{50}{x} - \dfrac{50}{x - 5} \right) = 2x(x - 5)\left(-\dfrac{1}{2} \right)$

$$100(x - 5) - 100x = -x(x - 5)$$

$$100x - 500 - 100x = -x^2 + 5x$$

$$x^2 - 5x - 500 = 0$$

$$(x - 25)(x + 20) = 0$$

$x = 25$ or $x = -20$ Principle of zero products

Both numbers check. The solutions are 25 and -20.

6. 2, -1

7. $\dfrac{x + 2}{x} = \dfrac{x - 1}{2}$, LCD is $2x$

$2x \cdot \dfrac{x + 2}{x} = 2x \cdot \dfrac{x - 1}{2}$

$$2(x + 2) = x(x - 1)$$

$$2x + 4 = x^2 - x$$

$$0 = x^2 - 3x - 4$$

$$0 = (x - 4)(x + 1)$$

$x = 4$ or $x = -1$ Principle of zero products

Both numbers check. The solutions are 4 and -1.

8. 6, -3

9. $x - 6 = \dfrac{1}{x + 6}$, LCD is $x + 6$

$(x + 6)(x - 6) = (x + 6) \cdot \dfrac{1}{x + 6}$

$$x^2 - 36 = 1$$

$$x^2 = 37$$

$x = \sqrt{37}$ or $x = -\sqrt{37}$

Both numbers check. The solutions are $\pm\sqrt{37}$.

10. $\pm 5\sqrt{2}$

11.
$$\frac{2}{x} = \frac{x+3}{5}, \text{ LCD is } 5x$$
$$5x \cdot \frac{2}{x} = 5x \cdot \frac{x+3}{5}$$
$$5 \cdot 2 = x(x+3)$$
$$10 = x^2 + 3x$$
$$0 = x^2 + 3x - 10$$
$$0 = (x+5)(x-2)$$
$$x = -5 \text{ or } x = 2$$

Both numbers check. The solutions are -5 and 2.

12. $\dfrac{7 \pm \sqrt{85}}{2}$

13.
$$x + 5 = \frac{3}{x-5}, \text{ LCD is } x - 5$$
$$(x-5)(x+5) = (x-5) \cdot \frac{3}{x-5}$$
$$x^2 - 25 = 3$$
$$x^2 = 28$$
$$x = \pm\sqrt{28}, \text{ or } \pm 2\sqrt{7}$$

Both numbers check. The solutions are $\pm 2\sqrt{7}$.

14. $\pm\sqrt{65}$

15.
$$\frac{40}{x} - \frac{20}{x-3} = \frac{8}{7}, \text{ LCD is } 7x(x-3)$$
$$7x(x-3)\left(\frac{40}{x} - \frac{20}{x-3}\right) = 7x(x-3) \cdot \frac{8}{7}$$
$$280(x-3) - 140x = 8x(x-3)$$
$$280x - 840 - 140x = 8x^2 - 24x$$
$$0 = 8x^2 - 164x + 840$$
$$0 = 2x^2 - 41x + 210$$
$$0 = (2x - 21)(x - 10)$$
$$x = \frac{21}{2} \text{ or } x = 10$$

Both numbers check. The solutions are $\dfrac{21}{2}$ and 10.

16. $-\dfrac{11}{9},\ 2$

17.
$$\frac{5}{x+2} + \frac{3x}{x+6} = 2,$$
$$\text{LCD is } (x+2)(x+6)$$
$$(x+2)(x+6)\left(\frac{5}{x+2} + \frac{3x}{x+6}\right) = (x+2)(x+6) \cdot 2$$
$$5(x+6) + 3x(x+2) = 2(x^2 + 8x + 12)$$
$$5x + 30 + 3x^2 + 6x = 2x^2 + 16x + 24$$
$$x^2 - 5x + 6 = 0$$
$$(x-3)(x-2) = 0$$
$$x = 3 \text{ or } x = 2$$

Both numbers check. The solutions are 3 and 2.

18. $-\dfrac{21}{4},\ -1$

19.
$$\frac{3}{3x+1} + \frac{6x}{11x-1} = 1,$$
$$\text{LCD is } (3x+1)(11x-1)$$
$$(3x+1)(11x-1)\left(\frac{3}{3x+1} + \frac{6x}{11x-1}\right) = (3x+1)(11x-1) \cdot 1$$
$$3(11x-1) + 6x(3x+1) = (3x+1)(11x-1)$$
$$33x - 3 + 18x^2 + 6x = 33x^2 + 8x - 1$$
$$0 = 15x^2 - 31x + 2$$
$$0 = (15x - 1)(x - 2)$$
$$x = \frac{1}{15} \text{ or } x = 2$$

Both numbers check. The solutions are $\dfrac{1}{15}$ and 2.

20. $\dfrac{3}{2},\ 1$

21.
$$\frac{16}{5(x-2)(x+2)} + \frac{9}{5(x+2)(x+3)} = \frac{-x}{(x-2)(x+3)},$$
$$\text{LCD is } 5(x-2)(x+2)(x+3)$$
$$5(x-2)(x+2)(x+3)\left(\frac{16}{5(x-2)(x+2)} + \frac{9}{5(x+2)(x+3)}\right) =$$
$$5(x-2)(x+2)(x+3)\left(\frac{-x}{(x-2)(x+3)}\right)$$
$$16(x+3) + 9(x-2) = -5x(x+2)$$
$$16x + 48 + 9x - 18 = -5x^2 - 10x$$
$$5x^2 + 35x + 30 = 0$$
$$x^2 + 7x + 6 = 0 \quad \text{Multiplying}$$
$$\text{by } \frac{1}{5}$$
$$(x+6)(x+1) = 0$$
$$x = -6 \text{ or } x = -1$$

Both numbers check. The solutions are -6 and -1.

22. 4

23.
$$\frac{6}{x^2 - 4x - 5} - \frac{6}{x^2 - 2x - 3} = \frac{x}{x^2 - 8x + 15}$$
$$\frac{6}{(x-5)(x+1)} - \frac{6}{(x-3)(x+1)} = \frac{x}{(x-5)(x-3)},$$
$$\text{LCD is } (x-5)(x+1)(x-3)$$
$$(x-5)(x+1)(x-3)\left(\frac{6}{(x-5)(x+1)} - \frac{6}{(x-3)(x+1)}\right) =$$
$$(x-5)(x+1)(x-3)\left(\frac{x}{(x-5)(x-3)}\right)$$
$$6(x-3) - 6(x-5) = x(x+1)$$
$$6x - 18 - 6x + 30 = x^2 + x$$
$$0 = x^2 + x - 12$$
$$0 = (x+4)(x-3)$$
$$x = -4 \text{ or } x = 3$$

Only -4 checks. The solution is -4.

24. $5,\ -4$

25. *Familiarize*. We first make a drawing, labeling it with the known and unknown information. We can also organize the information in a table. We let r represent the speed and t the time for the first part of the trip.

r km/h t hr $r - 4$ km/h $8 - t$ hr
          60 km   24 km

Canoe trip	Distance	Speed	Time
1st part	60	r	t
2nd part	24	$r - 4$	$8 - t$

Translate. Using $r = \dfrac{d}{t}$, we get two equations from the table, $r = \dfrac{60}{t}$ and $r - 4 = \dfrac{24}{8 - t}$.

Carry out. We substitute $\dfrac{60}{t}$ for r in the second equation and solve for t.

$$\frac{60}{t} - 4 = \frac{24}{8 - t}, \text{ LCD is } t(8 - t)$$

$$t(8 - t)\left(\frac{60}{t} - 4\right) = t(8 - t) \cdot \frac{24}{8 - t}$$

$$60(8 - t) - 4t(8 - t) = 24t$$

$$480 - 60t - 32t + 4t^2 = 24t$$

$$4t^2 - 116t + 480 = 0 \quad \text{Standard form}$$

$$t^2 - 29t + 120 = 0 \quad \text{Multiplying by } \frac{1}{4}$$

$$(t - 24)(t - 5) = 0$$

$t = 24$ or $t = 5$

Check. Since the time cannot be negative (If $t = 24$, $8 - t = -16$.), we only check 5 hr. If $t = 5$, then $8 - t = 3$. The speed of the first part is $\dfrac{60}{5}$, or 12 km/h. The speed of the second part is $\dfrac{24}{3}$, or 8 km/h. The speed of the second part is 4 km/h slower than the first part. The value checks.

State. The speed of the first part was 12 km/h, and the speed of the second part was 8 km/h.

26. First part: 60 mph, second part: 50 mph

27. *Familiarize*. We first make a drawing. We also organize the information in a table. We let r = the speed and t = the time of the slower trip.

280 mi r mph t hr
280 mi $r + 5$ mph $t - 1$ hr

Trip	Distance	Speed	Time
Slower	280	r	t
Faster	280	$r + 5$	$t - 1$

Translate. Using $t = d/r$, we get two equations from the table:

$$t = \frac{280}{r} \text{ and } t - 1 = \frac{280}{r + 5}$$

Carry out. We substitute $\dfrac{280}{r}$ for t in the second equation and solve for r.

$$\frac{280}{r} - 1 = \frac{280}{r + 5}, \text{ LCD is } r(r + 5)$$

$$r(r + 5)\left(\frac{280}{r} - 1\right) = r(r + 5) \cdot \frac{280}{r + 5}$$

$$280(r + 5) - r(r + 5) = 280r$$

$$280r + 1400 - r^2 - 5r = 280r$$

$$0 = r^2 + 5r - 1400$$

$$0 = (r - 35)(r + 40)$$

$r = 35$ or $r = -40$

Check. Since negative speed has no meaning in this problem, we only check 35. If $r = 35$, then the time for the slower trip is $\dfrac{280}{35}$, or 8 hours. If $r = 35$, then $r + 5 = 40$ and the time for the faster trip is $\dfrac{280}{40}$, or 7 hours. This is 1 hour less time than the slower trip took, so we have an answer to the problem.

State. The speed is 35 mph.

28. 40 mph

29. *Familiarize*. We make a drawing and then organize the information in a table. We let r = the speed and t = the time of the super-prop.

2800 mi r mph t hr
2000 mi $r + 50$ mph $t - 3$ hr

Plane	Distance	Speed	Time
Super-prop	2800	r	t
Turbo-jet	2000	$r + 50$	$t - 3$

Translate. Using $r = d/t$, we get two equations from the table:

$$r = \frac{2800}{t} \text{ and } r + 50 = \frac{2000}{t - 3}$$

Carry out. We substitute $\dfrac{2800}{t}$ for r in the second equation and solve for t.

$$\frac{2800}{t} + 50 = \frac{2000}{t - 3},$$
$$\text{LCD is } t(t - 3)$$

$$t(t - 3)\left(\frac{2800}{t} + 50\right) = t(t - 3) \cdot \frac{2000}{t - 3}$$

$$2800(t - 3) + 50t(t - 3) = 2000t$$

$$2800t - 8400 + 50t^2 - 150t = 2000t$$

$$50t^2 + 650t - 8400 = 0$$

$$t^2 + 13t - 168 = 0$$

$$(t + 21)(t - 8) = 0$$

$t = -21$ or $t = 8$

Check. Since negative time has no meaning in this problem, we only check 8 hours. If $t = 8$, then $t - 3 = 5$. The speed of the super-prop is $\dfrac{2800}{8}$, or 350 mph. The speed of the turbo-jet is $\dfrac{2000}{5}$, or 400 mph. Since the speed of the turbo-jet is 50 mph faster than the speed of the super-prop, the value checks.

State. The speed of the super-prop is 350 mph; the speed of the turbo-jet is 400 mph.

30. Cessna: 200 mph, Beechcraft: 250 mph; or
Cessna: 150 mph, Beechcraft: 200 mph

31. *Familiarize*. We make a drawing and then organize the information in a table. We let r represent the speed and t the time of the trip to Richmond.

Richmond
600 mi r mph t hr

600 mi $r - 10$ mph $22 - t$ hr

Trip	Distance	Speed	Time
To Richmond	600	r	t
Return	600	$r - 10$	$22 - t$

Translate. Using $t = \dfrac{d}{r}$, we get two equations from the table,

$$t = \frac{600}{r} \text{ and } 22 - t = \frac{600}{r - 10}.$$

Carry out. We substitute $\dfrac{600}{r}$ for t in the second equation and solve for r.

$$22 - \frac{600}{r} = \frac{600}{r - 10},$$
$$\text{LCD is } r(r - 10)$$
$$r(r - 10)\left(22 - \frac{600}{r}\right) = r(r - 10) \cdot \frac{600}{r - 10}$$
$$22r(r - 10) - 600(r - 10) = 600r$$
$$22r^2 - 220r - 600r + 6000 = 600r$$
$$22r^2 - 1420r + 6000 = 0$$
$$11r^2 - 710r + 3000 = 0$$
$$(11r - 50)(r - 60) = 0$$

$$r = \frac{50}{11} \text{ or } r = 60$$

Check. Since negative speed has no meaning in this problem (If $r = \dfrac{50}{11}$, then $r - 10 = -\dfrac{60}{11}$.), we only check 60 mph. If $r = 60$, then the time of the trip to Richmond is $\dfrac{600}{60}$, or 10 hr. The speed of the return trip is $60 - 10$, or 50 mph, and the time is $\dfrac{600}{50}$, or 12 hr. The total time for the round trip is $10 + 12$, or 22 hr. The value checks.

State. The speed to Richmond was 60 mph, and the speed of the return trip was 50 mph.

32. To Hillsboro: 10 mph, return trip: 4 mph

33. *Familiarize*. We make a drawing and organize the information in a table. Let r represent the speed of the boat

in still water, and let t represent the time of the trip upstream.

24 mi $r - 2$ mph t hr   → Upstream

Downstream ←   24 mi $r + 2$ mph $5 - t$ hr

Trip	Distance	Speed	Time
Upstream	24	$r - 2$	t
Downstream	24	$r + 2$	$5 - t$

Translate. Using $t = \dfrac{d}{r}$, we get two equations from the table,

$$t = \frac{24}{r - 2} \text{ and } 5 - t = \frac{24}{r + 2}.$$

Carry out. We substitute $\dfrac{24}{r - 2}$ for t in the second equation and solve for r.

$$5 - \frac{24}{r - 2} = \frac{24}{r + 2},$$
$$\text{LCD is } (r - 2)(r + 2)$$
$$(r - 2)(r + 2)\left(5 - \frac{24}{r - 2}\right) = (r - 2)(r + 2) \cdot \frac{24}{r + 2}$$
$$5(r - 2)(r + 2) - 24(r + 2) = 24(r - 2)$$
$$5r^2 - 20 - 24r - 48 = 24r - 48$$
$$5r^2 - 48r - 20 = 0$$
$$(5r + 2)(r - 10) = 0$$

$$r = -\frac{2}{5} \text{ or } r = 10$$

Check. Since negative speed has no meaning in this problem, we only check 10 mph. If $r = 10$, then the speed upstream is $10 - 2$, or 8 mph, and the time is $\dfrac{24}{8}$, or 3 hr. The speed downstream is $10 + 2$, or 12 mph, and the time is $\dfrac{24}{12}$, or 2 hr. The total time of the round trip is $3 + 2$, or 5 hr. The value checks.

State. The speed of the boat in still water is 10 mph.

34. 15 mph

35. *Familiarize*. Let x represent the time it takes the smaller pipe to fill the tank. Then $x - 3$ represents the time it takes the larger pipe to fill the tank. It takes them 2 hr to fill the tank when both pipes are working together, so they can fill $\dfrac{1}{2}$ of the tank in 1 hr. The smaller pipe will fill $\dfrac{1}{x}$ of the tank in 1 hr, and the larger pipe will fill $\dfrac{1}{x - 3}$ of the tank in 1 hr.

Translate. We have an equation.

$$\frac{1}{x} + \frac{1}{x - 3} = \frac{1}{2}$$

Carry out. We solve the equation.

We multiply by the LCD, $2x(x - 3)$.

$$2x(x-3)\left(\frac{1}{x} + \frac{1}{x+3}\right) = 2x(x-3) \cdot \frac{1}{2}$$
$$2(x-3) + 2x = x(x-3)$$
$$2x - 6 + 2x = x^2 - 3x$$
$$0 = x^2 - 7x + 6$$
$$0 = (x-6)(x-1)$$

$x = 6$ or $x = 1$

Check. Since negative time has no meaning in this problem, 1 is not a solution $(1 - 3 = -2)$. We only check 6 hr. This is the time it would take the smaller pipe working alone. Then the larger pipe would take $6 - 3$, or 3 hr working alone. The larger pipe would fill $2\left(\frac{1}{3}\right)$, or $\frac{2}{3}$, of the tank in 2 hr, and the smaller pipe would fill $2\left(\frac{1}{6}\right)$, or $\frac{1}{3}$, of the tank in 2 hr. Thus in 2 hr they would fill $\frac{2}{3} + \frac{1}{3}$ of the tank. This is all of it, so the numbers check.

State. It takes the smaller pipe, working alone, 6 hr to fill the tank.

36. 12 hr

37. Familiarize. We make a drawing and then organize the information in a table. We let r represent Ellen's speed in still water. Then $r - 2$ is the speed upstream and $r + 2$ is the speed downstream. Using $t = \frac{d}{r}$, we let $\frac{1}{r-2}$ represent the time upstream and $\frac{1}{r+2}$ represent the time downstream.

Trip	Distance	Speed	Time
Upstream	1	$r-2$	$\dfrac{1}{r-2}$
Downstream	1	$r+2$	$\dfrac{1}{r+2}$

Translate. The time for the round trip is 1 hour. We now have an equation.

$$\frac{1}{r-2} + \frac{1}{r+2} = 1$$

Carry out. We solve the equation. We multiply by the LCD, $(r-2)(r+2)$.

$$(r-2)(r+2)\left(\frac{1}{r-2} + \frac{1}{r+2}\right) = (r-2)(r+2) \cdot 1$$
$$(r+2) + (r-2) = (r-2)(r+2)$$
$$2r = r^2 - 4$$
$$0 = r^2 - 2r - 4$$
$$a = 1, \ b = -2, \ c = -4$$

$$r = \frac{-(-2) \pm \sqrt{(-2)^2 - 4 \cdot 1(-4)}}{2 \cdot 1}$$
$$r = \frac{2 \pm \sqrt{4+16}}{2} = \frac{2 \pm \sqrt{20}}{2}$$
$$r = \frac{2 \pm 2\sqrt{5}}{2} = 1 \pm \sqrt{5}$$
$$1 + \sqrt{5} \approx 1 + 2.236 \approx 3.24$$
$$1 - \sqrt{5} \approx 1 - 2.236 \approx -1.24$$

Check. Since negative speed has no meaning in this problem, we only check 3.24 mph. If $r \approx 3.24$, then $r - 2 \approx 1.24$ and $r + 2 \approx 5.24$. The time it takes to travel upstream is approximately $\frac{1}{1.24}$, or 0.806 hr, and the time it takes to travel downstream is approximately $\frac{1}{5.24}$, or 0.191 hr. The total time is 0.997 which is approximately 1 hour. The value checks.

State. Ellen's speed in still water is approximately 3.24 mph.

38. 9.34 km/h

39.
$$\sqrt{3x+1} = \sqrt{2x-1} + 1$$
$$3x+1 = 2x - 1 + 2\sqrt{2x+1} + 1$$
Squaring both sides
$$x + 1 = 2\sqrt{2x-1}$$
$$x^2 + 2x + 1 = 4(2x-1) \quad \text{Squaring both sides again}$$
$$x^2 + 2x + 1 = 8x - 4$$
$$x^2 - 6x + 5 = 0$$
$$(x-1)(x-5) = 0$$
$$x = 1 \ \text{ or } \ x = 5$$

Both numbers check. The solutions are 1 and 5.

40. $\dfrac{1}{x-2}$

41. $\sqrt[3]{18y^3} \ \sqrt[3]{4x^2} = \sqrt[3]{72x^2y^3} = \sqrt[3]{8y^3 \cdot 9x^2} = 2y\sqrt[3]{9x^2}$

42. ◈

43. ◈

44. $\dfrac{-3 \pm \sqrt{57}}{2}$

45.
$$\frac{x^2}{x-2} - \frac{x+4}{2} + \frac{2-4x}{x-2} + 1 = 0,$$
$$\text{LCD is } 2(x-2)$$
$$2(x-2)\left(\frac{x^2}{x-2} - \frac{x+4}{2} + \frac{2-4x}{x-2} + 1\right) = 2(x-2) \cdot 0$$
$$2x^2 - (x-2)(x+4) + 2(2-4x) + 2(x-2) = 0$$
$$2x^2 - x^2 - 2x + 8 + 4 - 8x + 2x - 4 = 0$$
$$x^2 - 8x + 8 = 0$$

Use the quadratic formula.

$$x = \frac{-(-8) \pm \sqrt{(-8)^2 - 4 \cdot 1 \cdot 8}}{2 \cdot 1}$$

$$x = \frac{8 \pm \sqrt{32}}{2} = \frac{8 \pm 4\sqrt{2}}{2} = 4 \pm 2\sqrt{2}$$

Both numbers check. The solutions are $4 \pm 2\sqrt{2}$.

46. $\dfrac{-i \pm \sqrt{23}}{4}$

47.
$$\frac{1}{a-1} = a+1, \text{ LCD is } a-1$$

$$(a-1) \cdot \frac{1}{a-1} = (a-1)(a+1)$$

$$1 = a^2 - 1$$

$$2 = a^2$$

$$\pm\sqrt{2} = a$$

Both values check. The solution is $\pm\sqrt{2}$.

48. $2.50

49.

Exercise Set 11.5

1. $x^2 - 6x + 9 = 0$

$a = 1$, $b = -6$, $c = 9$

We substitute and compute the discriminant.
$$b^2 - 4ac = (-6)^2 - 4 \cdot 1 \cdot 9$$
$$= 36 - 36 = 0$$

Since $b^2 - 4ac = 0$, there is just one solution, and it is a real number.

2. One real

3. $x^2 + 7 = 0$

$a = 1$, $b = 0$, $c = 7$

We substitute and compute the discriminant.
$$b^2 - 4ac = 0^2 - 4 \cdot 1 \cdot 7$$
$$= -28$$

Since $b^2 - 4ac < 0$, there are two imaginary-number solutions.

4. Two imaginary

5. $x^2 - 2 = 0$

$a = 1$, $b = 0$, $c = -2$

We substitute and compute the discriminant.
$$b^2 - 4ac = 0^2 - 4 \cdot 1 \cdot (-2)$$
$$= 8$$

Since $b^2 - 4ac > 0$, there are two real solutions.

6. Two real

7. $4x^2 - 12x + 9 = 0$

$a = 4$, $b = -12$, $c = 9$

We substitute and compute the discriminant.
$$b^2 - 4ac = (-12)^2 - 4 \cdot 4 \cdot 9$$
$$= 144 - 144$$
$$= 0$$

Since $b^2 - 4ac = 0$, there is just one solution, and it is a real number.

8. Two real

9. $x^2 - 2x + 4 = 0$

$a = 1$, $b = -2$, $c = 4$

We substitute and compute the discriminant.
$$b^2 - 4ac = (-2)^2 - 4 \cdot 1 \cdot 4$$
$$= 4 - 16$$
$$= -12$$

Since $b^2 - 4ac < 0$, there are two imaginary-number solutions.

10. Two imaginary

11. $a^2 - 10a + 21 = 0$

$a = 1$, $b = -10$, $c = 21$

We substitute and compute the discriminant.
$$b^2 - 4ac = (-10)^2 - 4 \cdot 1 \cdot 21$$
$$= 100 - 84 = 16$$

Since $b^2 - 4ac > 0$, there are two real solutions.

12. One real

13. $6x^2 + 5x - 4 = 0$

$a = 6$, $b = 5$, $c = -4$

We substitute and compute the discriminant.
$$b^2 - 4ac = 5^2 - 4 \cdot 6 \cdot (-4)$$
$$= 25 + 96 = 121$$

Since $b^2 - 4ac > 0$, there are two real solutions.

14. Two real

15. $9t^2 - 3t = 0$

$a = 9$, $b = -3$, $c = 0$

We substitute and compute the discriminant.
$$b^2 - 4ac = (-3)^2 - 4 \cdot 9 \cdot 0$$
$$= 9 - 0$$
$$= 9$$

Since $b^2 - 4ac > 0$, there are two real solutions.

16. Two real

17. $x^2 + 5x = 7$

$x^2 + 5x - 7 = 0$ Standard form

$a = 1$, $b = 5$, $c = -7$

We substitute and compute the discriminant.

$$b^2 - 4ac = 5^2 - 4 \cdot 1 \cdot (-7)$$
$$= 25 + 28 = 53$$

Since $b^2 - 4ac > 0$, there are two real solutions.

18. Two imaginary

19. $y^2 = \frac{1}{2}y - \frac{3}{5}$

$$y^2 - \frac{1}{2}y + \frac{3}{5} = 0 \qquad \text{Standard form}$$

$$a = 1, \ b = -\frac{1}{2}, \ c = \frac{3}{5}$$

We substitute and compute the discriminant.

$$b^2 - 4ac = \left(-\frac{1}{2}\right)^2 - 4 \cdot 1 \cdot \frac{3}{5}$$
$$= \frac{1}{4} - \frac{12}{5}$$
$$= -\frac{43}{20}$$

Since $b^2 - 4ac < 0$, there are two imaginary-number solutions.

20. Two real

21. $4x^2 - 4\sqrt{3}x + 3 = 0$

$$a = 4, \ b = -4\sqrt{3}, \ c = 3$$

We substitute and compute the discriminant.

$$b^2 - 4ac = (-4\sqrt{3})^2 - 4 \cdot 4 \cdot 3$$
$$= 48 - 48$$
$$= 0$$

Since $b^2 - 4ac = 0$, there is just one solution, and it is a real number.

22. Two real

23. The solutions are -11 and 9.

$$x = -11 \ \text{ or } \quad x = 9$$
$$x + 11 = 0 \quad \text{ or } \ x - 9 = 0$$
$$(x + 11)(x - 9) = 0 \quad \text{Principle of zero products}$$
$$x^2 + 2x - 99 = 0 \quad \text{FOIL}$$

24. $x^2 - 16 = 0$

25. The only solution is 7. It must be a "double" solution.

$$x = 7 \ \text{ or } \quad x = 7$$
$$x - 7 = 0 \ \text{ or } \ x - 7 = 0$$
$$(x - 7)(x - 7) = 0 \quad \text{Principle of zero products}$$
$$x^2 - 14x + 49 = 0 \quad \text{FOIL}$$

26. $x^2 + 10x + 25 = 0$

27. The solutions are -3 and -5.

$$x = -3 \ \text{ or } \quad x = -5$$
$$x + 3 = 0 \quad \text{ or } \ x + 5 = 0$$
$$(x + 3)(x + 5) = 0$$
$$x^2 + 8x + 15 = 0$$

28. $x^2 + 9x + 14 = 0$

29. The solutions are 4 and $\frac{2}{3}$.

$$x = 4 \ \text{ or } \qquad x = \frac{2}{3}$$
$$x - 4 = 0 \ \text{ or } \ x - \frac{2}{3} = 0$$
$$(x - 4)\left(x - \frac{2}{3}\right) = 0$$
$$x^2 - \frac{2}{3}x - 4x + \frac{8}{3} = 0$$
$$x^2 - \frac{14}{3}x + \frac{8}{3} = 0$$
$$3x^2 - 14x + 8 = 0 \quad \text{Multiplying by 3}$$

30. $4x^2 - 23x + 15 = 0$

31. The solutions are $\frac{1}{2}$ and $\frac{1}{3}$.

$$x = \frac{1}{2} \ \text{ or } \qquad x = \frac{1}{3}$$
$$x - \frac{1}{2} = 0 \ \text{ or } \ x - \frac{1}{3} = 0$$
$$\left(x - \frac{1}{2}\right)\left(x - \frac{1}{3}\right) = 0$$
$$x^2 - \frac{1}{3}x - \frac{1}{2}x + \frac{1}{6} = 0$$
$$x^2 - \frac{5}{6}x + \frac{1}{6} = 0$$
$$6x^2 - 5x + 1 = 0 \quad \text{Multiplying by 6}$$

32. $8x^2 + 6x + 1 = 0$

33. The solutions are $-\frac{2}{5}$ and $\frac{6}{5}$.

$$x = -\frac{2}{5} \ \text{ or } \qquad x = \frac{6}{5}$$
$$x + \frac{2}{5} = 0 \quad \text{ or } \ x - \frac{6}{5} = 0$$
$$\left(x + \frac{2}{5}\right)\left(x - \frac{6}{5}\right) = 0 \quad \text{Principle of zero products}$$
$$x^2 - \frac{4}{5}x - \frac{12}{25} = 0$$
$$25x^2 - 20x - 12 = 0 \quad \text{Multiplying by 25}$$

34. $49x^2 + 7x - 6 = 0$

35. The solutions are $\sqrt{2}$ and $3\sqrt{2}$.

$$x = \sqrt{2} \ \text{ or } \qquad x = 3\sqrt{2}$$
$$x - \sqrt{2} = 0 \ \text{ or } \ x - 3\sqrt{2} = 0$$
$$(x - \sqrt{2})(x - 3\sqrt{2}) = 0$$
$$x^2 - 4\sqrt{2}x + 6 = 0$$

36. $x^2 - \sqrt{3}x - 6 = 0$

37. The solutions are $-\sqrt{5}$ and $-2\sqrt{5}$.
$$x = -\sqrt{5} \quad \text{or} \qquad x = -2\sqrt{5}$$
$$x + \sqrt{5} = 0 \quad \text{or} \quad x + 2\sqrt{5} = 0$$
$$(x + \sqrt{5})(x + 2\sqrt{5}) = 0$$
$$x^2 + 3\sqrt{5}x + 10 = 0$$

38. $x^2 + 4\sqrt{6}x + 18 = 0$

39. The solutions are $3i$ and $-3i$.
$$x = 3i \quad \text{or} \qquad x = -3i$$
$$x - 3i = 0 \quad \text{or} \quad x + 3i = 0$$
$$(x - 3i)(x + 3i) = 0$$
$$x^2 - (3i)^2 = 0$$
$$x^2 + 9 = 0$$

40. $x^2 + 16 = 0$

41. The solutions are $5 - 2i$ and $5 + 2i$.
$$x = 5 - 2i \quad \text{or} \qquad x = 5 + 2i$$
$$x - 5 + 2i = 0 \quad \text{or} \quad x - 5 - 2i = 0$$
$$[x + (-5 + 2i)][x + (-5 - 2i)] = 0$$
$$x^2 + x(-5-2i) + x(-5+2i) + (-5+2i)(-5-2i) = 0$$
$$x^2 - 5x - 2xi - 5x + 2xi + 25 - 4i^2 = 0$$
$$x^2 - 10x + 29 = 0$$
$$(i^2 = -1)$$

42. $x^2 - 4x + 53 = 0$

43. The solutions are $\dfrac{1 + 3i}{2}$ and $\dfrac{1 - 3i}{2}$.
$$x = \frac{1 + 3i}{2} \quad \text{or} \qquad x = \frac{1 - 3i}{2}$$
$$x - \frac{1 + 3i}{2} = 0 \quad \text{or} \quad x - \frac{1 - 3i}{2} = 0$$
$$\left(x - \frac{1 + 3i}{2}\right)\left(x - \frac{1 - 3i}{2}\right) = 0$$
$$x^2 - \frac{x - 3xi}{2} - \frac{x + 3xi}{2} + \frac{1 - 9i^2}{4} = 0$$
$$x^2 - \frac{x}{2} + \frac{3xi}{2} - \frac{x}{2} - \frac{3xi}{2} + \frac{10}{4} = 0$$
$$x^2 - x + \frac{5}{2} = 0$$
$$2x^2 - 2x + 5 = 0$$

44. $9x^2 - 12x + 5 = 0$

45. *Familiarize*. Let x and y represent the number of 30-sec and 60-sec commercials, respectively. Then the amount of time for the 30-sec commercials was $30x$ sec, or $\dfrac{30x}{60} = \dfrac{x}{2}$ min. The amount of time for the 60-sec commercials was $60x$ sec, or $\dfrac{60x}{60} = x$ min.

Translate. Rewording, we write two equations. We will express time in minutes.

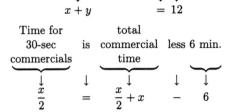

Carry out. Solving the system of equations we get $(6, 6)$.

Check. If there are six 30-sec and six 60-sec commercials, the total number of commercials is 12. The amount of time for six 30-sec commercials is 180 sec, or 3 min, and for six 60-sec commercials is 360 sec, or 6 min. The total commercial time is 9 min, and the amount of time for 30-sec commercials is 6 min less than this. The numbers check.

State. There were six 30-sec and six 60-sec commercials.

46.

47.

48. -1

49. The graph includes the points $(-3, 0)$, $(0, -3)$, and $(1, 0)$. Substituting in $y = ax^2 + bx + c$, we have three equations.
$$0 = 9a - 3b + c,$$
$$-3 = \qquad\qquad c,$$
$$0 = a + b + c$$
The solution of this system of equations is $a = 1$, $b = 2$, $c = -3$.

50. a) 2

     b) $\dfrac{11}{2}$

51. a) $kx^2 - 2x + k = 0$; one solution is -3

We first find k by substituting -3 for x.
$$k(-3)^2 - 2(-3) + k = 0$$
$$9k + 6 + k = 0$$
$$10k = -6$$
$$k = -\frac{6}{10}$$
$$k = -\frac{3}{5}$$

b) Now substitute $-\dfrac{3}{5}$ for k in the original equation.

$$-\dfrac{3}{5}x^2 - 2x + \left(-\dfrac{3}{5}\right) = 0$$

$$3x^2 + 10x + 3 = 0 \quad \text{Multiplying by } -5$$

$$(3x+1)(x+3) = 0$$

$$x = -\dfrac{1}{3} \text{ or } x = -3$$

The other solution is $-\dfrac{1}{3}$.

52. a) 2

b) $1 - i$

53. a) $x^2 - (6+3i)x + k = 0$; one solution is 3.

We first find k by substituting 3 for x.

$$3^2 - (6+3i)3 + k = 0$$

$$9 - 18 - 9i + k = 0$$

$$-9 - 9i + k = 0$$

$$k = 9 + 9i$$

b) Now we substitute $9 + 9i$ for k in the original equation.

$$x^2 - (6+3i)x + (9+9i) = 0$$

$$x^2 - (6+3i)x + 3(3+3i) = 0$$

$$[x - (3+3i)][x-3] = 0$$

$$x = 3 + 3i \text{ or } x = 3$$

The other solution is $3 + 3i$.

54. $h = -36$, $k = 15$

55. From Exercise 47 we know that

$$-\dfrac{b}{a} = \sqrt{3} \text{ and } \dfrac{c}{a} = 8.$$

Multiplying $ax^2 + bx + c = 0$ by $\dfrac{1}{a}$ we have:

$$x^2 + \dfrac{b}{a}x + \dfrac{c}{a} = 0$$

$$x^2 - \left(-\dfrac{b}{a}\right)x + \dfrac{c}{a} = 0$$

$$x^2 - \sqrt{3}x + 8 = 0 \quad \text{Substituting}$$

56. a) 4, -2

b) $1 \pm \sqrt{26}$

57. We substitute $(-3, 0)$, $\left(\dfrac{1}{2}, 0\right)$, and $(0, -12)$ in $f(x) = ax^2 + bx + c$ and get three equations.

$$0 = 9a - 3b + c,$$

$$0 = \dfrac{1}{4}a + \dfrac{1}{2}b + c,$$

$$-12 = c$$

The solution of this system of equations is $a = 8$, $b = 20$, $c = -12$.

58. 2

59. $x^2 + kx + 8 = 0 \qquad x^2 - kx + 8 = 0$

$$x = \dfrac{-k \pm \sqrt{k^2 - 4 \cdot 1 \cdot 8}}{2 \cdot 1} \quad x = \dfrac{-(-k) \pm \sqrt{(-k)^2 - 4 \cdot 1 \cdot 8}}{2 \cdot 1}$$

$$x = \dfrac{-k \pm \sqrt{k^2 - 32}}{2} \quad x = \dfrac{k \pm \sqrt{k^2 - 32}}{2}$$

$$\dfrac{-k + \sqrt{k^2 - 32}}{2} + 6 = \dfrac{k + \sqrt{k^2 - 32}}{2}$$

$$-k + \sqrt{k^2 - 32} + 12 = k + \sqrt{k^2 - 32}$$

$$-k + 12 = k$$

$$12 = 2k$$

$$6 = k$$

Exercise Set 11.6

1. $x^4 - 10x^2 + 25 = 0$

Let $u = x^2$ and think of x^4 as $(x^2)^2$.

$$u^2 - 10u + 25 = 0 \quad \text{Substituting } u \text{ for } x^2$$

$$(u-5)(u-5) = 0$$

$$u - 5 = 0 \text{ or } u - 5 = 0$$

$$u = 5 \text{ or } \quad u = 5$$

Now we substitute x^2 for u and solve the equation.

$$x^2 = 5$$

$$x = \pm\sqrt{5}$$

Both $\sqrt{5}$ and $-\sqrt{5}$ check. They are the solutions.

2. $\pm\sqrt{2}$, ± 1

3. $x^4 - 12x^2 + 27 = 0$

Let $u = x^2$.

$$u^2 - 12u + 27 = 0 \quad \text{Substituting } u \text{ for } x^2$$

$$(u-9)(u-3) = 0$$

$$u = 9 \text{ or } u = 3$$

Now we substitute x^2 for u and solve these equations:

$$x^2 = 9 \quad \text{ or } x^2 = 3$$

$$x = \pm 3 \text{ or } \quad x = \pm\sqrt{3}$$

The numbers 3, -3, $\sqrt{3}$, and $-\sqrt{3}$ check. They are the solutions.

4. ± 2, $\pm\sqrt{5}$

5. $9x^4 - 14x^2 + 5 = 0$

Let $u = x^2$.

$$9u^2 - 14u + 5 = 0 \quad \text{Substituting } u \text{ for } x^2$$

$$(9u-5)(u-1) = 0$$

$$u = \dfrac{5}{9} \quad \text{ or } \quad u = 1$$

$$x^2 = \dfrac{5}{9} \quad \text{ or } x^2 = 1 \quad \text{Substituting } x^2 \text{ for } u$$

$$x = \pm\dfrac{\sqrt{5}}{3} \text{ or } \quad x = \pm 1$$

The numbers $\dfrac{\sqrt{5}}{3}$, $-\dfrac{\sqrt{5}}{3}$, 1, and -1 check. They are the solutions.

6. $\pm\dfrac{\sqrt{3}}{2}$, ± 2

7. $x - 10\sqrt{x} + 9 = 0$

Let $u = \sqrt{x}$ and think of x as $(\sqrt{x})^2$.

$u^2 - 10u + 9 = 0$ Substituting u for $\sqrt{x}$

$(u - 9)(u - 1) = 0$

$u - 9 = 0$ or $u - 1 = 0$

$u = 9$ or $u = 1$

Now we substitute $\sqrt{x}$ for u and solve these equations:

$\sqrt{x} = 9$ or $\sqrt{x} = 1$

$x = 81$ or $x = 1$

The numbers 81 and 1 both check. They are the solutions.

8. $\dfrac{1}{4}$, 16

9. $3x + 10\sqrt{x} - 8 = 0$

Let $u = \sqrt{x}$.

$3u^2 + 10u - 8 = 0$ Substituting u for $\sqrt{x}$

$(3u - 2)(u + 4) = 0$

$u = \dfrac{2}{3}$ or $u = -4$

$\sqrt{x} = \dfrac{2}{3}$ or $\sqrt{x} = -4$ Substituting $\sqrt{x}$ for u

Squaring the first equation, we get $x = \dfrac{4}{9}$. Note that $\sqrt{x} = -4$ has no real solution. The number $\dfrac{4}{9}$ checks and is the solution.

10. $\dfrac{4}{25}$

11. $(x^2 - 9)^2 + 3(x^2 - 9) + 2 = 0$

Let $u = x^2 - 9$.

$u^2 + 3u + 2 = 0$ Substituting u for $x^2 - 9$

$(u + 2)(u + 1) = 0$

$u = -2$ or $u = -1$

$x^2 - 9 = -2$ or $x^2 - 9 = -1$ Substituting $x^2 - 9$ for u

$x^2 = 7$ or $x^2 = 8$

$x = \pm\sqrt{7}$ or $x = \pm\sqrt{8}$

$x = \pm\sqrt{7}$ or $x = \pm 2\sqrt{2}$

The numbers $\sqrt{7}$, $-\sqrt{7}$, $2\sqrt{2}$, and $-2\sqrt{2}$ check. They are the solutions.

12. ± 1, $\pm\sqrt{2}$

13. $(x^2 - 6x)^2 - 2(x^2 - 6x) - 35 = 0$

Let $u = x^2 - 6x$.

$u^2 - 2u - 35 = 0$ Substituting u for $x^2 - 6x$

$(u - 7)(u + 5) = 0$

$u - 7 = 0$ or $u + 5 = 0$

$u = 7$ or $u = -5$

Now we substitute $x^2 - 6x$ for u and solve these equations:

$x^2 - 6x = 7$ or $x^2 - 6x = -5$

$x^2 - 6x - 7 = 0$ or $x^2 - 6x + 5 = 0$

$(x - 7)(x + 1) = 0$ or $(x - 5)(x - 1) = 0$

$x = 7$ or $x = -1$ or $x = 5$ or $x = 1$

The numbers -1, 1, 5, and 7 check. They are the solutions.

14. $\dfrac{3 \pm \sqrt{33}}{2}$, 4, -1

15. $(3 + \sqrt{x})^2 - 3(3 + \sqrt{x}) - 10 = 0$

Let $u = 3 + \sqrt{x}$.

$u^2 - 3u - 10 = 0$ Substituting u for $3 + \sqrt{x}$

$(u - 5)(u + 2) = 0$

$u = 5$ or $u = -2$

$3 + \sqrt{x} = 5$ or $3 + \sqrt{x} = -2$ Substituting $3 + \sqrt{x}$ for u

$\sqrt{x} = 2$ or $\sqrt{x} = -5$

$x = 4$ No real solution

The number 4 checks and is the solution.

16. 1

17. $x^{-2} - x^{-1} - 6 = 0$

Let $u = x^{-1}$ and think of x^{-2} as $(x^{-1})^2$.

$u^2 - u - 6 = 0$ Substituting u for x^{-1}

$(u - 3)(u + 2) = 0$

$u = 3$ or $u = -2$

Now we substitute x^{-1} for u and solve these equations:

$x^{-1} = 3$ or $x^{-1} = -2$

$\dfrac{1}{x} = 3$ or $\dfrac{1}{x} = -2$

$\dfrac{1}{3} = x$ or $-\dfrac{1}{2} = x$

Both $\dfrac{1}{3}$ and $-\dfrac{1}{2}$ check. They are the solutions.

18. $\dfrac{4}{5}$, -1

19. $2x^{-2} + x^{-1} - 1 = 0$

Let $u = x^{-1}$.

$2u^2 + u - 1 = 0$ Substituting u for x^{-1}

$(2u - 1)(u + 1) = 0$

$2u = 1$ or $u = -1$

$u = \dfrac{1}{2}$ or $u = -1$

$x^{-1} = \dfrac{1}{2}$ or $x^{-1} = -1$ Substituting x^{-1} for u

$\dfrac{1}{x} = \dfrac{1}{2}$ or $\dfrac{1}{x} = -1$

$x = 2$ or $x = -1$

Both 2 and -1 check. They are the solutions.

20. $-\dfrac{1}{10}, 1$

21. $t^{2/3} + t^{1/3} - 6 = 0$

Let $u = t^{1/3}$ and think of $t^{2/3}$ as $(t^{1/3})^2$.

$u^2 + u - 6 = 0$ Substituting u for $t^{1/3}$

$(u + 3)(u - 2) = 0$

$u = -3$ or $u = 2$

Now we substitute $t^{1/3}$ for u and solve these equations:

$t^{1/3} = -3$ or $t^{1/3} = 2$

$t = (-3)^3$ or $t = 2^3$ Raising to the third power

$t = -27$ or $t = 8$

Both -27 and 8 check. They are the solutions.

22. $64, -8$

23. $z^{1/2} - z^{1/4} - 2 = 0$

Let $u = z^{1/4}$.

$u^2 - u - 2 = 0$ Substituting u for $z^{1/4}$

$(u - 2)(u + 1) = 0$

$u = 2$ or $u = -1$

$z^{1/4} = 2$ or $z^{1/4} = -1$ Substituting $z^{1/4}$ for u

$\sqrt[4]{z} = 2$ or $\sqrt[4]{z} = -1$

$z = 16$ This equation has no real solution since principal fourth roots are never negative.

The number 16 checks, so it is the solution.

24. 729

25. $x^{2/5} + x^{1/5} - 6 = 0$

Let $u = x^{1/5}$.

$u^2 + u - 6 = 0$ Substituting u for $x^{1/5}$

$(u + 3)(u - 2) = 0$

$u = -3$ or $u = 2$

$x^{1/5} = -3$ or $x^{1/5} = 2$ Substituting $x^{1/5}$ for u

$x = -243$ or $x = 32$ Raising to the fifth power

Both -243 and 32 check. They are the solutions.

26. 81

27. $t^{1/3} + 2t^{1/6} = 3$

$t^{1/3} + 2t^{1/6} - 3 = 0$

Let $u = t^{1/6}$.

$u^2 + 2u - 3 = 0$ Substituting u for $t^{1/6}$

$(u + 3)(u - 1) = 0$

$u = -3$ or $u = 1$

$t^{1/6} = -3$ or $t^{1/6} = 1$ Substituting $t^{1/6}$ for u

No real solution $t = 1$

The number 1 checks and is the solution.

28. $81, 16$

29. $\left(\dfrac{x+3}{x-3}\right)^2 - \left(\dfrac{x+3}{x-3}\right) - 6 = 0$

Let $u = \dfrac{x+3}{x-3}$.

$u^2 - u - 6 = 0$ Substituting u for $\dfrac{x+3}{x-3}$

$(u - 3)(u + 2) = 0$

$u = 3$ or $u = -2$

$\dfrac{x+3}{x-3} = 3$ or $\dfrac{x+3}{x-3} = -2$ Substituting $\dfrac{x+3}{x-3}$ for u

$x + 3 = 3(x - 3)$ or $x + 3 = -2(x - 3)$

Multiplying by $(x - 3)$

$x + 3 = 3x - 9$ or $x + 3 = -2x + 6$

$-2x = -12$ or $3x = 3$

$x = 6$ or $x = 1$

Both 6 and 1 check. They are the solutions.

30. $-\dfrac{11}{6}, -\dfrac{1}{6}$

31. $9\left(\dfrac{x+2}{x+3}\right)^2 - 6\left(\dfrac{x+2}{x+3}\right) + 1 = 0$

Let $u = \dfrac{x+2}{x+3}$.

$9u^2 - 6u + 1 = 0$ Substituting u for $\dfrac{x+2}{x+3}$

$(3u - 1)(3u - 1) = 0$

$3u - 1 = 0$ or $3u - 1 = 0$

$3u = 1$ or $3u = 1$

$u = \dfrac{1}{3}$ or $u = \dfrac{1}{3}$

Now we substitute $\dfrac{x+2}{x+3}$ for u and solve the equation:

$\dfrac{x+2}{x+3} = \dfrac{1}{3}$

$3(x + 2) = x + 3$ Multiplying by $3(x + 3)$

$3x + 6 = x + 3$

$2x = -3$

$x = -\dfrac{3}{2}$

The number $-\dfrac{3}{2}$ checks. It is the solution.

32. $\dfrac{12}{5}$

33. $\left(\dfrac{y^2-1}{y}\right)^2 - 4\left(\dfrac{y^2-1}{y}\right) - 12 = 0$

Let $u = \dfrac{y^2-1}{y}$.

$u^2 - 4u - 12 = 0$ Substituting u for $\dfrac{y^2-1}{y}$

$(u - 6)(u + 2) = 0$

$$u = 6 \quad \text{or} \quad u = -2$$

$$\frac{y^2 - 1}{y} = 6 \quad \text{or} \quad \frac{y^2 - 1}{y} = -2$$

Substituting $\frac{y^2 - 1}{y}$ for u

$$y^2 - 1 = 6y \quad \text{or} \quad y^2 - 1 = -2y$$

Multiplying by y

$$y^2 - 6y - 1 = 0 \quad \text{or} \quad y^2 + 2y - 1 = 0$$

$$y = \frac{-(-6) \pm \sqrt{(-6)^2 - 4 \cdot 1 \cdot (-1)}}{2 \cdot 1} \quad \text{or}$$

$$y = \frac{-2 \pm \sqrt{2^2 - 4 \cdot 1 \cdot (-1)}}{2 \cdot 1}$$

$$y = \frac{6 \pm \sqrt{40}}{2} \quad \text{or} \quad y = \frac{-2 \pm \sqrt{8}}{2}$$

$$y = \frac{6 \pm 2\sqrt{10}}{2} \quad \text{or} \quad y = \frac{-2 \pm 2\sqrt{2}}{2}$$

$$y = 3 \pm \sqrt{10} \quad \text{or} \quad y = -1 \pm \sqrt{2}$$

All four numbers check. They are the solutions.

34. $\frac{9 \pm \sqrt{89}}{2}, \ -1 \pm \sqrt{3}$

35. $\sqrt{3x^2}\sqrt{3x^3} = \sqrt{3x^2 \cdot 3x^3} = \sqrt{9x^5} = \sqrt{9x^4 \cdot x} = 3x^2\sqrt{x}$

36. 4 L of A, 8 L of B

37. $\frac{x+1}{x-1} - \frac{x+1}{x^2+x+1}, \ \text{LCD} = (x-1)(x^2+x+1)$

$$= \frac{x+1}{x-1} \cdot \frac{x^2+x+1}{x^2+x+1} - \frac{x+1}{x^2+x+1} \cdot \frac{x-1}{x-1}$$

$$= \frac{(x^3 + 2x^2 + 2x + 1) - (x^2 - 1)}{(x-1)(x^2+x+1)}$$

$$= \frac{x^3 + x^2 + 2x + 2}{x^3 - 1}$$

38.

39. Solve $f(x) = 0$.

$x^4 - 8x^2 + 7 = 0$

Let $u = x^2$.

$$u^2 - 8u + 7 = 0$$

$$(u - 7)(u - 1) = 0$$

$$u = 7 \quad \text{or} \quad u = 1$$

$$x^2 = 7 \quad \text{or} \quad x^2 = 1$$

$$x = \pm\sqrt{7} \quad \text{or} \quad x = \pm 1$$

All four numbers check. The x-intercepts are $(\sqrt{7}, 0)$, $(-\sqrt{7}, 0)$, $(1, 0)$, and $(-1, 0)$.

40. 28.45

41. $\pi x^4 - \pi^2 x^2 - \sqrt{99.3} = 0$

Let $u = x^2$.

$$\pi u^2 - \pi^2 u - \sqrt{99.3} = 0$$

$$u = \frac{-(-\pi)^2 \pm \sqrt{(-\pi^2)^2 - 4 \cdot \pi \cdot (-\sqrt{99.3})}}{2 \cdot \pi}$$

$$u \approx \frac{9.8696 \pm 14.9209}{6.2832}$$

$$u \approx 3.9455 \quad \text{or} \quad u \approx -0.8039$$

$$x^2 \approx 3.9455 \quad \text{or} \quad x^2 \approx -0.8039$$

$$x \approx \pm 1.99 \quad \text{or} \quad \text{No real solution}$$

The numbers 1.99 and -1.99 check.

42. $\frac{100}{99}$

43. $\left(\sqrt{\frac{x}{x-3}}\right)^2 - 24 = 10\sqrt{\frac{x}{x-3}}$

Let $u = \sqrt{\frac{x}{x-3}}$, substitute, and write in standard form.

$$u^2 - 10u - 24 = 0$$

$$(u - 12)(u + 2) = 0$$

$$u = 12 \quad \text{or} \quad u = -2$$

$$\sqrt{\frac{x}{x-3}} = 12 \quad \text{or} \quad \sqrt{\frac{x}{x-3}} = -2$$

$$\frac{x}{x-3} = 144 \quad \text{or} \quad \text{No real solution}$$

$$x = 144x - 432$$

$$432 = 143x$$

$$\frac{432}{143} = x$$

This number checks.

44. 259

45. $a^3 - 26a^{3/2} - 27 = 0$

Let $u = a^{3/2}$.

$$u^2 - 26u - 27 = 0$$

$$(u - 27)(u + 1) = 0$$

$$u = 27 \quad \text{or} \quad u = -1$$

$$a^{3/2} = 27 \quad \text{or} \quad a^{3/2} = -1$$

$$a = 27^{2/3} \qquad \text{No real solution}$$

$$a = (3^3)^{2/3}$$

$$a = 9$$

The number 9 checks.

46. 3, 1

47. $x^6 + 7x^3 - 8 = 0$

Let $u = x^3$.

$$u^2 + 7u - 8 = 0$$

$$(u + 8)(u - 1) = 0$$

$$u = -8 \quad \text{or} \quad u = 1$$

$$x^3 = -8 \quad \text{or} \quad x^3 = 1$$

$$x = -2 \quad \text{or} \quad x = 1$$

Both numbers check.

48.

49. $-3, -1, 1, 4$

Exercise Set 11.7

1. $f(x) = x^2$

See Example 1 in the text.

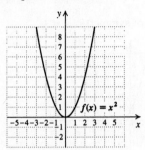

2.

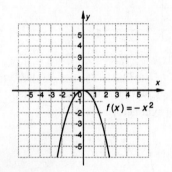

3. $f(x) = -4x^2$

We choose some numbers for x and compute $f(x)$ for each one. Then we plot the ordered pairs $(x, f(x))$ and connect them with a smooth curve.

x	$f(x) = -4x^2$
0	0
1	-4
2	-16
-1	-4
-2	-16

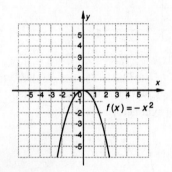

4.

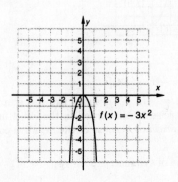

5. $g(x) = \frac{1}{4}x^2$

x	$g(x) = \frac{1}{4}x^2$
0	0
1	$\frac{1}{4}$
2	1
3	$\frac{9}{4}$
-1	$\frac{1}{4}$
-2	1
-3	$\frac{9}{4}$

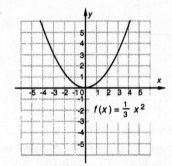

6.

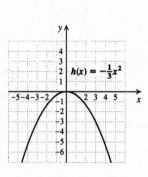

7. $h(x) = -\frac{1}{3}x^2$

x	$h(x) = -\frac{1}{3}x^2$
0	0
1	$-\frac{1}{3}$
2	$-\frac{4}{3}$
3	-3
-1	$-\frac{1}{3}$
-2	$-\frac{4}{3}$
-3	-3

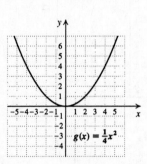

8.

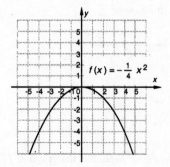

9. $f(x) = \dfrac{3}{2}x^2$

x	$f(x) = \dfrac{3}{2}x^2$
0	0
1	$\dfrac{3}{2}$
2	6
−1	$\dfrac{3}{2}$
−2	6

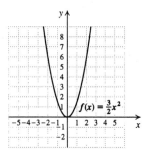

10.

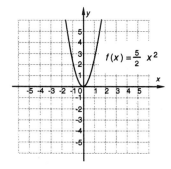

11. $g(x) = (x+1)^2 = [x-(-1)]^2$

We know that the graph of $g(x) = (x+1)^2$ looks like the graph of $f(x) = x^2$ (see Exercise 1) but moved to the left 1 unit.

Vertex: $(-1,0)$, line of symmetry: $x = -1$

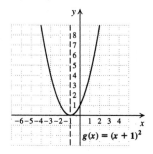

12. Vertex: $(-4,0)$, line of symmetry: $x = -4$

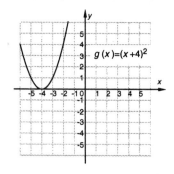

13. $f(x) = (x-4)^2$

The graph of $f(x) = (x-4)^2$ looks like the graph of $f(x) = x^2$ (see Exercise 1) but moved to the right 4 units.

Vertex: $(4,0)$, line of symmetry: $x = 4$

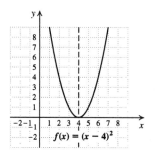

14. Vertex: $(1,0)$, line of symmetry: $x = 1$

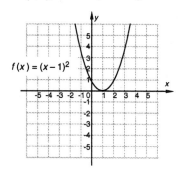

15. $h(x) = (x-3)^2$

The graph of $h(x) = (x-3)^2$ looks like the graph of $f(x) = x^2$ (see Exercise 1) but moved to the right 3 units.

Vertex: $(3,0)$, line of symmetry: $x = 3$

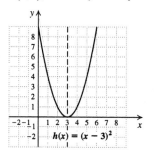

16. Vertex: $(7,0)$, line of symmetry: $x = 7$

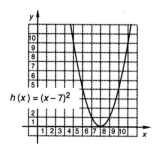

17. $f(x) = -(x+4)^2 = -[x-(-4)]^2$

The graph of $f(x) = -(x+4)^2$ looks like the graph of $f(x) = x^2$ (see Exercise 1) but moved to the left 4 units. It will also open downward because of the negative coefficient, -1.

Vertex: $(-4,0)$, line of symmetry: $x = -4$

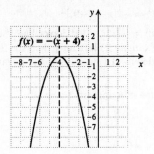

18. Vertex: $(2, 0)$, line of symmetry: $x = 2$

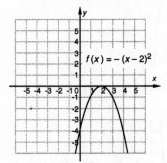

19. $g(x) = -(x - 1)^2$

The graph of $g(x) = -(x - 1)^2$ looks like the graph of $f(x) = x^2$ (see Exercise 1) but moved to the right 1 unit. It will also open downward because of the negative coefficient, -1.

Vertex: $(1, 0)$, line of symmetry: $x = 1$

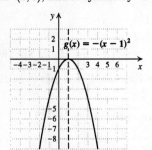

20. Vertex: $(-5, 0)$, line of symmetry: $x = -5$

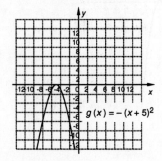

21. $f(x) = 2(x - 1)^2$

The graph of $f(x) = 2(x - 1)^2$ looks like the graph of $h(x) = 2x^2$ (see Example 1) but moved to the right 1 unit.

Vertex: $(1, 0)$, line of symmetry: $x = 1$

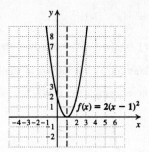

22. Vertex: $(-4, 0)$, line of symmetry: $x = -4$

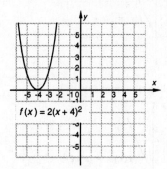

23. $h(x) = -\frac{1}{2}(x - 3)^2$

The graph of $h(x) = -\frac{1}{2}(x - 3)^2$ looks like the graph of $g(x) = \frac{1}{2}x^2$ (see Example 1) but moved to the right 3 units. It will also open downward because of the negative coefficient, $-\frac{1}{2}$.

Vertex: $(3, 0)$, line of symmetry: $x = 3$

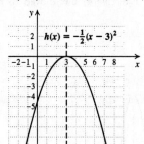

24. Vertex: $(2, 0)$, line of symmetry: $x = 2$

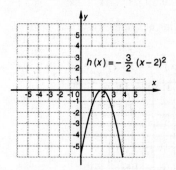

25. $f(x) = \frac{1}{2}(x+1)^2 = \frac{1}{2}[x-(-1)]^2$

The graph of $f(x) = \frac{1}{2}(x+1)^2$ looks like the graph of $g(x) = \frac{1}{2}x^2$ (see Example 1) but moved to the left 1 unit.

Vertex: $(-1,0)$, line of symmetry: $x = -1$

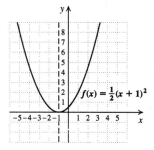

26. Vertex: $(-2,0)$, line of symmetry: $x = -2$

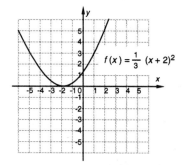

27. $g(x) = -3(x-2)^2$

The graph of $g(x) = -3(x-2)^2$ looks like the graph of $f(x) = -3x^2$ (see Exercise 4) but moved to the right 2 units.

Vertex: $(2,0)$, line of symmetry: $x = 2$

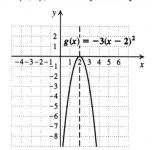

28. Vertex: $(7,0)$, line of symmetry: $x = 7$

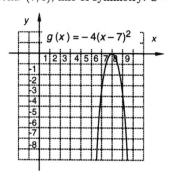

29. $f(x) = -2(x+9)^2 = -2[x-(-9)]^2$

The graph of $f(x) = -2(x+9)^2$ looks like the graph of $h(x) = 2x^2$ (see Example 1) but moved to the left 9 units. It will also open downward because of the negative coefficient, -2.

Vertex: $(-9,0)$, line of symmetry: $x = -9$

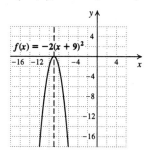

30. Vertex: $(-7,0)$, line of symmetry: $x = -7$

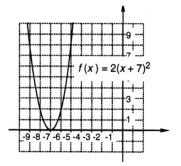

31. $h(x) = -3\left(x-\frac{1}{2}\right)^2$

The graph of $h(x) = -3\left(x-\frac{1}{2}\right)^2$ looks like the graph of $f(x) = -3x^2$ (see Exercise 4) but moved to the right $\frac{1}{2}$ unit.

Vertex: $\left(\frac{1}{2},0\right)$, line of symmetry: $x = \frac{1}{2}$

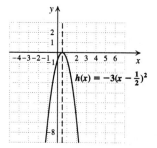

32. Vertex: $\left(-\frac{1}{2}, 0\right)$, line of symmetry: $x = -\frac{1}{2}$

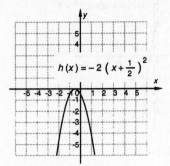

33. $f(x) = (x - 3)^2 + 1$

We know that the graph looks like the graph of $f(x) = x^2$ (see Example 1) but moved to the right 3 units and up 1 unit. The vertex is $(3, 1)$, and the line of symmetry is $x = 3$. Since the coefficient of $(x - 3)^2$ is positive $(1 > 0)$, there is a minimum function value, 1.

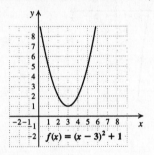

34. Vertex: $(-2, -3)$, line of symmetry: $x = -2$

Minimum: -3

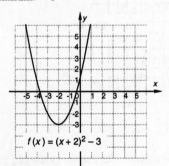

35. $f(x) = (x + 1)^2 - 2$

We know that the graph looks like the graph of $f(x) = x^2$ (see Example 1) but moved to the left 1 unit and down 2 units. The vertex is $(-1, -2)$, and the line of symmetry is $x = -1$. Since the coefficient of $(x + 1)^2$ is positive $(1 > 0)$, there is a minimum function value, -2.

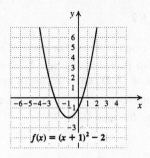

36. Vertex: $(1, 2)$, line of symmetry: $x = 1$

Minimum: 2

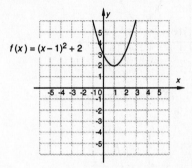

37. $g(x) = (x + 4)^2 + 1$

We know that the graph looks like the graph of $f(x) = x^2$ (see Example 1) but moved to the left 4 units and up 1 unit. The vertex is $(-4, 1)$, and the line of symmetry is $x = -4$. Since the coefficient of $(x + 4)^2$ is positive $(1 > 0)$, there is a minimum function value, 1.

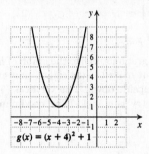

38. Vertex: $(2, -4)$, line of symmetry: $x = 2$

Maximum: -4

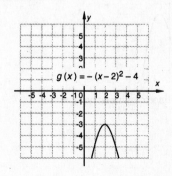

39. $f(x) = \frac{1}{2}(x-5)^2 + 2$

We know that the graph looks like the graph of $g(x) = \frac{1}{2}x^2$ (see Example 1) but moved to the right 5 units and up 2 units. The vertex is $(5, 2)$, and the line of symmetry is $x = 5$. Since the coefficient of $(x-5)^2$ is positive $\left(\frac{1}{2} > 0\right)$, there is a minimum function value, 2.

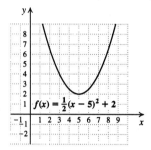

40. Vertex: $(-1, -2)$, line of symmetry: $x = -1$

Minimum: -2

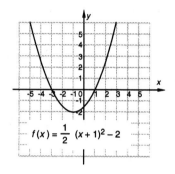

41. $h(x) = -2(x-1)^2 - 3$

We know that the graph looks like the graph of $h(x) = 2x^2$ (see Example 1) but moved to the right 1 unit and down 3 units and turned upside down. The vertex is $(1, -3)$, and the line of symmetry is $x = 1$. The maximum function value is -3.

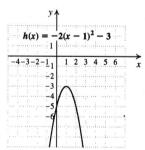

42. Vertex: $(-1, 4)$, line of symmetry: $x = -1$

Maximum: 4

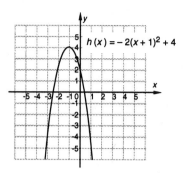

43. $f(x) = -3(x+4)^2 + 1$

We know that the graph looks like the graph of $f(x) = -3x^2$ (see Exercise 4) but moved to the left 4 units and up 1 unit. The vertex is $(-4, 1)$, the line of symmetry is $x = -4$, and the maximum function value is 1.

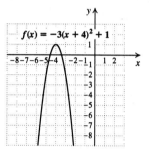

44. Vertex: $(5, -3)$, line of symmetry: $x = 5$

Maximum: -3

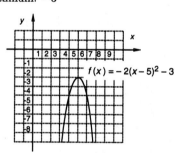

45. $g(x) = -\frac{3}{2}(x-1)^2 + 2$

We know that the graph looks like the graph of $f(x) = \frac{3}{2}x^2$ (see Exercise 9) but moved to the right 1 unit and up 2 units and turned upside down. The vertex is $(1, 2)$, the line of symmetry is $x = 1$, and the maximum function value is 2.

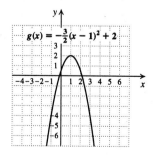

46. Vertex: $(-2, -1)$, line of symmetry: $x = -2$

Minimum: -1

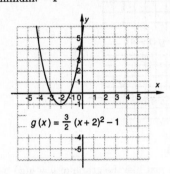

$$g(x) = \frac{3}{2}(x+2)^2 - 1$$

47. $f(x) = 8(x - 9)^2 + 5$

This function is of the form $f(x) = a(x - h)^2 + k$ with $a = 8$, $h = 9$, and $k = 5$. The vertex is (h, k), or $(9, 5)$. The line of symmetry is $x = h$, or $x = 9$. Since $a > 0$, then k, or 5, is the minimum function value.

48. Vertex: $(-5, -8)$

Line of symmetry: $x = -5$

Minimum: -8

49. $h(x) = -\frac{2}{7}(x + 6)^2 + 11$

This function is of the form $f(x) = a(x - h)^2 + k$ with $a = -\frac{2}{7}$, $h = -6$, and $k = 11$. The vertex is (h, k), or $(-6, 11)$. The line of symmetry is $x = h$, or $x = -6$. Since $a < 0$, then k, or 11, is the maximum function value.

50. Vertex: $(7, -9)$

Line of symmetry: $x = 7$

Maximum: -9

51. $f(x) = 5\left(x + \frac{1}{4}\right)^2 - 13$

This function is of the form $f(x) = a(x-h)^2+k$ with $a = 5$, $h = -\frac{1}{4}$, and $k = -13$. The vertex is (h, k), or $\left(-\frac{1}{4}, -13\right)$. The line of symmetry is $x = h$, or $x = -\frac{1}{4}$. Since $a > 0$, then k, or -13, is the minimum function value.

52. Vertex: $\left(\frac{1}{4}, 19\right)$

Line of symmetry: $x = \frac{1}{4}$

Minimum: 19

53. $f(x) = -7(x - 10)^2 - 20$

This function is of the form $f(x) = a(x - h)^2 + k$ with $a = -7$, $h = 10$, and $k = -20$. The vertex is (h, k), or $(10, -20)$. The line of symmetry is $x = h$, or $x = 10$. Since $a < 0$, then k, or -20, is the maximum function value.

54. Vertex: $(-12, 23)$

Line of symmetry: $x = -12$

Maximum: 23

55. $f(x) = \sqrt{2}(x + 4.58)^2 + 65\pi$

This function is of the form $f(x) = a(x - h)^2 + k$ with $a = \sqrt{2}$, $h = -4.58$, and $k = 65\pi$. The vertex is (h, k), or $(-4.58, 65\pi)$. The line of symmetry is $x = h$, or $x = -4.58$. Since $a > 0$, then k, or 65π, is the minimum function value.

56. Vertex: $(38.2, -\sqrt{34})$

Line of symmetry: $x = 38.2$

Minimum: $-\sqrt{34}$

57. $500 = 4a + 2b + c,$ (1)

   $300 = a + b + c,$ (2)

   $0 = c$ (3)

Substitute 0 for c in (1) and (2).

   $500 = 4a + 2b$ (4)

   $300 = a + b$ (5)

Multiply (5) by -2 and add the resulting equation to (4).

$$\begin{array}{rl} 500 = & 4a + 2b \\ -600 = & -2a - 2b \\ \hline -100 = & 2a \\ -50 = & a \end{array} \quad \text{Adding}$$

Substitute -50 for a in (5).

   $300 = -50 + b$

   $350 = b$

The solution is $(-50, 350, 0)$.

58. $\frac{1}{6}$, 2

59. ◈

60. ◈

61. Since there is a maximum at $(0, 4)$, the parabola will have the same shape as $f(x) = -2x^2$. It will be of the form $f(x) = -2(x - h)^2 + k$ with $h = 0$ and $k = 4$: $f(x) = -2x^2 + 4$

62. $f(x) = 2(x - 2)^2$

63. Since there is a minimum at $(6, 0)$, the parabola will have the same shape as $f(x) = 2x^2$. It will be of the form $f(x) = 2(x-h)^2+k$ with $h = 6$ and $k = 0$: $f(x) = 2(x-6)^2$

64. $f(x) = -2x^2 + 3$

65. Since there is a maximum at $(3, 8)$, the parabola will have the same shape as $f(x) = -2x^2$. It will be of the form $f(x) = -2(x - h)^2 + k$ with $h = 3$ and $k = 8$: $f(x) = -2(x - 3)^2 + 8$

66. $f(x) = 2(x + 2)^2 + 3$

67. Since there is a minimum at $(-3, 6)$, the parabola will have the same shape as $f(x) = 2x^2$. It will be of the form $f(x) = 2(x - h)^2 + k$ with $h = -3$ and $k = 6$: $f(x) = 2[x - (-3)]^2 + 6$, or $f(x) = 2(x + 3)^2 + 6$.

68. $f(x) = 6(x-4)^2$

69. Since the parabola has the same shape as

$f(x) = -\dfrac{1}{2}(x-2)^2 + 4$, it is of the form

$g(x) = -\dfrac{1}{2}(x-h)^2 + k$. Since it has a maximum value at
the same point as $f(x) = -2(x-1)^2 - 6$, its vertex is
$(1, -6)$. That is, $h = 1$ and $k = -6$. The equation is
$g(x) = -\dfrac{1}{2}(x-1)^2 - 6$.

70.

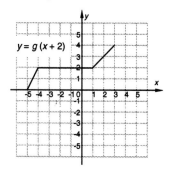

71. $y = g(x-3)$

The graph looks just like the graph of $y = g(x)$ but moved
3 units to the right.

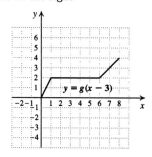

72.

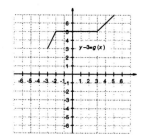

73. $y = g(x) + 4$

The graph looks just like the graph of $y = g(x)$ but moved
up 4 units.

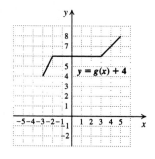

74.

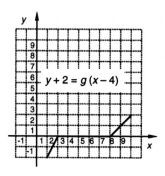

75. $y = g(x-2) + 3$

The graph looks just like the graph of $y = g(x)$ but moved
to the right 2 units and up 3 units.

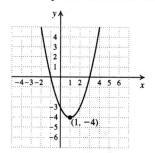

76. ![]

77. ![]

Exercise Set 11.8

1. $f(x) = x^2 - 2x - 3$

$\quad = (x^2 - 2x + 1 - 1) - 3 \quad$ Adding $1 - 1$

$\quad = (x^2 - 2x + 1) - 1 - 3 \quad$ Regrouping

$\quad = (x-1)^2 - 4$

The vertex is $(1, -4)$, the line of symmetry is $x = 1$, and
the graph opens upward since the coefficient 1 is positive.
We plot a few points as a check and draw the curve.

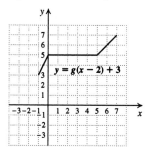

2. Vertex: $(-1, -6)$, line of symmetry: $x = -1$

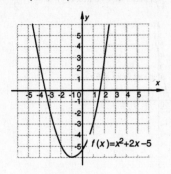

$f(x) = x^2 + 2x - 5$

3. $g(x) = x^2 + 6x + 13$

$\quad = (x^2 + 6x + 9 - 9) + 13$ Adding $9 - 9$

$\quad = (x^2 + 6x + 9) - 9 + 13$ Regrouping

$\quad = (x + 3)^2 + 4$

The vertex is $(-3, 4)$, the line of symmetry is $x = -3$, and the graph opens upward since the coefficient 1 is positive. We plot a few points as a check and draw the curve.

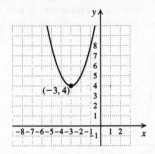

$(-3, 4)$

4. Vertex: $(2, 1)$, line of symmetry: $x = 2$

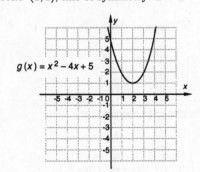

$g(x) = x^2 - 4x + 5$

5. $f(x) = x^2 + 4x - 1$

$\quad = (x^2 + 4x + 4 - 4) - 1$ Adding $4 - 4$

$\quad = (x^2 + 4x + 4) - 4 - 1$ Regrouping

$\quad = (x + 2)^2 - 5$

The vertex is $(-2, -5)$, the line of symmetry is $x = -2$, and the graph opens upward since the coefficient 1 is positive.

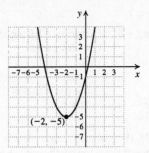

$(-2, -5)$

6. Vertex: $(5, -4)$, line of symmetry: $x = 5$

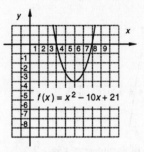

$f(x) = x^2 - 10x + 21$

7. $h(x) = 2x^2 + 16x + 25$

$\quad = 2(x^2 + 8x) + 25$ Factoring 2 from the first two terms

$\quad = 2(x^2 + 8x + 16 - 16) + 25$ Adding $16 - 16$ inside the parentheses

$\quad = 2(x^2 + 8x + 16) + 2(-16) + 25$ Distributing to obtain a trinomial square

$\quad = 2(x + 4)^2 - 7$

The vertex is $(-4, -7)$, the line of symmetry is $x = -4$, and the graph opens upward since the coefficient 2 is positive.

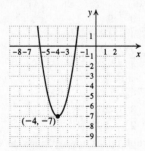

$(-4, -7)$

8. Vertex: $(4, -9)$, line of symmetry: $x = 4$

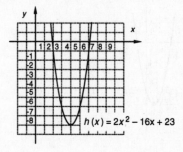

$h(x) = 2x^2 - 16x + 23$

9. $f(x) = -x^2 + 4x + 6$

$\qquad = -(x^2 - 4x) + 6$ Factoring -1 from the
$\qquad\qquad\qquad\qquad\qquad\quad$ first two terms

$\qquad = -(x^2 - 4x + 4 - 4) + 6$
$\qquad\qquad\qquad\qquad\qquad$ Adding $4 - 4$ inside
$\qquad\qquad\qquad\qquad\qquad$ the parentheses

$\qquad = -(x^2 - 4x + 4) - (-4) + 6$

$\qquad = -(x - 2)^2 + 10$

The vertex is $(2, 10)$, the line of symmetry is $x = 2$, and the graph opens downward since the coefficient -1 is negative.

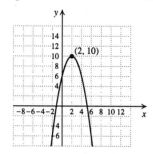

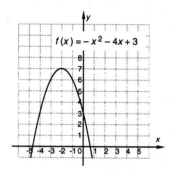

10.   Vertex: $(-2, 7)$, line of symmetry: $x = -2$

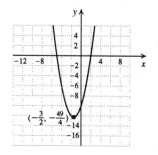

11. $g(x) = x^2 + 3x - 10$

$\qquad = \left(x^2 + 3x + \dfrac{9}{4} - \dfrac{9}{4}\right) - 10$

$\qquad = \left(x^2 + 3x + \dfrac{9}{4}\right) - \dfrac{9}{4} - 10$

$\qquad = \left(x + \dfrac{3}{2}\right)^2 - \dfrac{49}{4}$

The vertex is $\left(-\dfrac{3}{2}, -\dfrac{49}{4}\right)$, the line of symmetry is $x = -\dfrac{3}{2}$, and the graph opens upward since the coefficient 1 is positive.

12.   Vertex: $\left(-\dfrac{5}{2}, -\dfrac{9}{4}\right)$, line of symmetry: $x = -\dfrac{5}{2}$

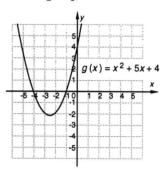

13. $f(x) = 3x^2 - 24x + 50$

$\qquad = 3(x^2 - 8x) + 50$ Factoring

$\qquad = 3(x^2 - 8x + 16 - 16) + 50$
$\qquad\qquad\qquad\qquad\qquad$ Adding $16 - 16$ inside
$\qquad\qquad\qquad\qquad\qquad$ the parentheses

$\qquad = 3(x^2 - 8x + 16) - 3 \cdot 16 + 50$

$\qquad = 3(x - 4)^2 + 2$

The vertex is $(4, 2)$, the line of symmetry is $x = 4$, and the graph opens upward since the coefficient 3 is positive.

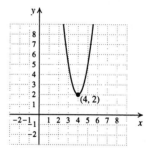

14.   Vertex: $(-1, -7)$, line of symmetry: $x = -1$

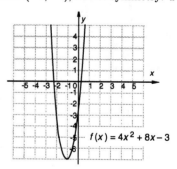

15. $h(x) = x^2 - 9x$

$\qquad = \left(x^2 - 9x + \dfrac{81}{4}\right) - \dfrac{81}{4}$

$\qquad = \left(x - \dfrac{9}{2}\right)^2 - \dfrac{81}{4}$

The vertex is $\left(\dfrac{9}{2}, -\dfrac{81}{4}\right)$, the line of symmetry is $x = \dfrac{9}{2}$, and the graph opens upward since the coefficient 1 is positive.

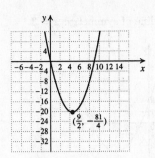

16. Vertex: $\left(-\dfrac{1}{2}, -\dfrac{1}{4}\right)$, line of symmetry: $x = -\dfrac{1}{2}$

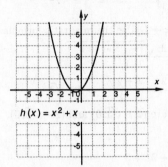

17. $f(x) = -2x^2 - 4x - 6$

$= -2(x^2 + 2x) - 6$ Factoring

$= -2(x^2 + 2x + 1 - 1) - 6$

         Adding $1 - 1$ inside

         the parentheses

$= -2(x^2 + 2x + 1) - 2(-1) - 6$

$= -2(x + 1)^2 - 4$

The vertex is $(-1, -4)$, the line of symmetry is $x = -1$, and the graph opens downward since the coefficient -2 is negative.

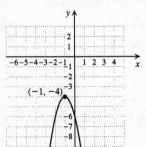

18. Vertex: $(1, 5)$, line of symmetry: $x = 1$

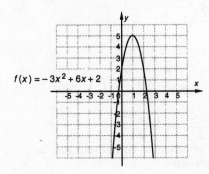

19. $g(x) = 2x^2 - 10x + 14$

$= 2(x^2 - 5x) + 14$ Factoring

$= 2\left(x^2 - 5x + \dfrac{25}{4} - \dfrac{25}{4}\right) + 14$

         Adding $\dfrac{25}{4} - \dfrac{25}{4}$ inside

         the parentheses

$= 2\left(x^2 - 5x + \dfrac{25}{4}\right) + 2\left(-\dfrac{25}{4}\right) + 14$

$= 2\left(x - \dfrac{5}{2}\right)^2 + \dfrac{3}{2}$

The vertex is $\left(\dfrac{5}{2}, \dfrac{3}{2}\right)$, the line of symmetry is $x = \dfrac{5}{2}$, and the graph opens upward since the coefficient 2 is positive.

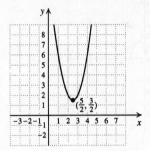

20. Vertex: $\left(-\dfrac{3}{2}, \dfrac{7}{2}\right)$, line of symmetry: $x = -\dfrac{3}{2}$

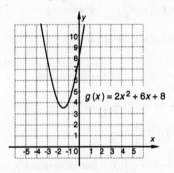

21. $f(x) = -3x^2 - 3x + 1$

$= -3(x^2 + x) + 1$ Factoring

$= -3\left(x^2 + x + \dfrac{1}{4} - \dfrac{1}{4}\right) + 1$

         Adding $\dfrac{1}{4} - \dfrac{1}{4}$ inside

         the parentheses

$= -3\left(x^2 + x + \dfrac{1}{4}\right) - 3\left(-\dfrac{1}{4}\right) + 1$

$= -3\left(x + \dfrac{1}{2}\right)^2 + \dfrac{7}{4}$

The vertex is $\left(-\dfrac{1}{2}, \dfrac{7}{4}\right)$, the line of symmetry is $x = -\dfrac{1}{2}$, and the graph opens downward since the coefficient -3 is negative.

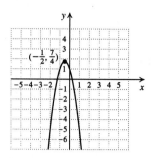

22. Vertex: $\left(\frac{1}{2}, \frac{3}{2}\right)$, line of symmetry: $x = \frac{1}{2}$

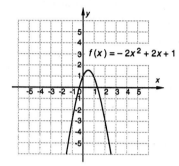

23. $h(x) = \frac{1}{2}x^2 + 4x + \frac{19}{3}$

$= \frac{1}{2}(x^2 + 8x) + \frac{19}{3}$ Factoring

$= \frac{1}{2}(x^2 + 8x + 16 - 16) + \frac{19}{3}$

           Adding $16 - 16$ inside
           the parentheses

$= \frac{1}{2}(x^2 + 8x + 16) + \frac{1}{2}(-16) + \frac{19}{3}$

$= \frac{1}{2}(x + 4)^2 - \frac{5}{3}$

The vertex is $\left(-4, -\frac{5}{3}\right)$, the line of symmetry is $x = -4$, and the graph opens upward since the coefficient $\frac{1}{2}$ is positive.

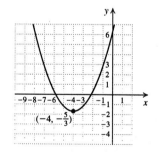

24. Vertex: $\left(3, -\frac{5}{2}\right)$, line of symmetry: $x = 3$

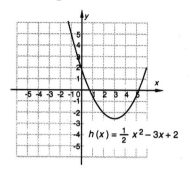

25. Solve $0 = x^2 - 4x + 1$. Use the quadratic formula.

$x = \dfrac{-(-4) \pm \sqrt{(-4)^2 - 4 \cdot 1 \cdot 1}}{2 \cdot 1}$

$x = \dfrac{4 \pm \sqrt{12}}{2} = \dfrac{4 \pm 2\sqrt{3}}{2} = 2 \pm \sqrt{3}$

The x-intercepts are $(2 + \sqrt{3}, 0)$ and $(2 - \sqrt{3}, 0)$.

26. None

27. Solve: $0 = -x^2 + 2x + 3$

$0 = x^2 - 2x - 3$ Multiplying by -1

$0 = (x - 3)(x + 1)$

$x = 3$ or $x = -1$

The x-intercepts are $(3, 0)$ and $(-1, 0)$.

28. $(1 + \sqrt{6}, 0)$, $(1 - \sqrt{6}, 0)$

29. Solve: $0 = x^2 - 3x - 4$

$0 = (x - 4)(x + 1)$

$x = 4$ or $x = -1$

The x-intercepts are $(4, 0)$ and $(-1, 0)$.

30. $(4 + \sqrt{11}, 0)$, $(4 - \sqrt{11}, 0)$

31. Solve: $0 = -x^2 + 3x - 2$

$0 = x^2 - 3x + 2$

$0 = (x - 2)(x - 1)$

$x = 2$ or $x = 1$

The x-intercepts are $(2, 0)$ and $(1, 0)$.

32. None

33. Solve $0 = 2x^2 + 4x - 1$. Using the quadratic formula we get $x = \dfrac{-2 \pm \sqrt{6}}{2}$. The x-intercepts are $\left(\dfrac{-2 + \sqrt{6}}{2}, 0\right)$ and $\left(\dfrac{-2 - \sqrt{6}}{2}, 0\right)$.

34. None

35. Solve: $0 = x^2 - x + 1$

$x = \dfrac{-(-1) \pm \sqrt{(-1)^2 - 4 \cdot 1 \cdot 1}}{2 \cdot 1}$

$x = \dfrac{1 \pm \sqrt{-3}}{2} = \dfrac{1 \pm i\sqrt{3}}{2}$

Since the equation has no real solutions, no x-intercepts exist.

36. $\left(-\dfrac{3}{2}, 0\right)$

37.
$$\sqrt{4x-4} = \sqrt{x+4} + 1$$
$$4x - 4 = x + 4 + 2\sqrt{x+4} + 1$$
$$\text{Squaring both sides}$$
$$3x - 9 = 2\sqrt{x+4}$$
$$9x^2 - 54x + 81 = 4(x+4) \quad \text{Squaring both}$$
$$\text{sides again}$$
$$9x^2 - 54x + 81 = 4x + 16$$
$$9x^2 - 58x + 65 = 0$$
$$(9x - 13)(x - 5) = 0$$
$$x = \frac{13}{9} \quad \text{or} \quad x = 5$$

Check: For $x = \dfrac{13}{9}$:

$$\begin{array}{c|c}
\multicolumn{2}{c}{\sqrt{4x-4} = \sqrt{x+4}+1} \\
\hline
\sqrt{4\left(\frac{13}{9}\right)-4} \ ? & \sqrt{\frac{13}{9}+4}+1 \\
\sqrt{\frac{16}{9}} & \sqrt{\frac{49}{9}}+1 \\
\frac{4}{3} & \frac{7}{3}+1 \\
\frac{4}{3} & \frac{10}{3} \qquad \text{FALSE}
\end{array}$$

For $x = 5$:

$$\begin{array}{c|c}
\multicolumn{2}{c}{\sqrt{4x-4} = \sqrt{x+4}+1} \\
\hline
\sqrt{4 \cdot 5 - 4} \ ? & \sqrt{5+4}+1 \\
\sqrt{16} & \sqrt{9}+1 \\
4 & 3+1 \\
4 & 4 \qquad \text{TRUE}
\end{array}$$

5 checks, but $\dfrac{13}{9}$ does not. The solution is 5.

38. 4

39.

40.

41. a) $f(x) = 2.31x^2 - 3.135x - 5.89$

$$= 2.31(x^2 - 1.357142857x) - 5.89$$
$$= 2.31(x^2 - 1.357142857x +$$
$$0.460459183 - 0.460459183) - 5.89$$
$$= 2.31(x^2 - 1.357142857x + 0.460459183) +$$
$$2.31(-0.460459183) - 5.89$$
$$= 2.31(x - 0.678571428)^2 - 6.953660714$$

Since the coefficient 2.31 is positive, the function has a minimum value. It is
-6.953660714.

b) $2.31x^2 - 3.135x - 5.89 = 0$

$$x = \frac{-(-3.135) \pm \sqrt{(-3.135)^2 - 4(2.31)(-5.89)}}{2(2.31)}$$
$$x \approx \frac{3.135 \pm 8.015723611}{4.62}$$
$$x \approx -1.056433682 \quad \text{or} \quad x \approx 2.413576539$$

The x-intercepts are $(-1.056433682, 0)$ and
$(2.413576539, 0)$.

42. a) Maximum: 7.01412766

b) $(-0.400174191, 0)$, $(0.821450786, 0)$

43. $f(x) = x^2 - x - 6$

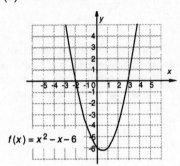

a) $x^2 - x - 6 = 2$

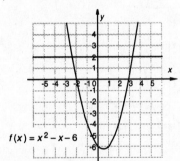

The solutions are approximately -2.4 and 3.4.

b) $x^2 - x - 6 = -3$

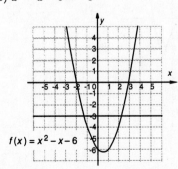

The solutions are approximately -1.3 and 2.3.

44. a) $-3, 1$

b) $-4.5, 2.5$

c) $-5.5, 3.5$

45. $f(x) = mx^2 - nx + p$

$\qquad = m\left(x^2 - \dfrac{n}{m}x\right) + p$

$\qquad = m\left(x^2 - \dfrac{n}{m}x + \dfrac{n^2}{4m^2} - \dfrac{n^2}{4m^2}\right) + p$

$\qquad = m\left(x - \dfrac{n}{2m}\right)^2 - \dfrac{n^2}{4m} + p$

$\qquad = m\left(x - \dfrac{n}{2m}\right)^2 + \dfrac{-n^2 + 4mp}{4m},$ or

$\qquad\quad m\left(x - \dfrac{n}{2m}\right)^2 + \dfrac{4mp - n^2}{4m}$

46. $f(x) = 3\left[x - \left(-\dfrac{m}{6}\right)\right]^2 + \dfrac{11m^2}{12}$

47. $f(x) = |x^2 - 1|$

We plot some points and draw the curve. Note that it will lie entirely on or above the x−axis since absolute value is never negative. For positive values of $x^2 - 1$ it will look like the graph of $f(x) = x^2$ but moved down 1 unit.

x	$f(x) = \lvert x^2 - 1 \rvert$
−2	3
−1	0
0	1
1	0
2	3

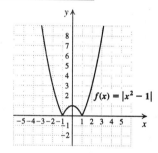

48. $f(x) = |3 - 2x - x^2|$

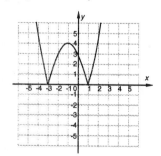

49.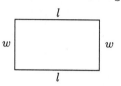

Exercise Set 11.9

1. **Familiarize.** We make a drawing and label it.

Perimeter: $2l + 2w = 76$ ft

Area: $A = l \cdot w$

Translate. We have a system of equations.

$\qquad 2l + 2w = 76$

$\qquad A = lw$

Carry out. Solving the first equation for l, we get $l = 38 - w$. Substituting for l in the second equation we get a quadratic function A:

$\qquad A = (38 - w)w$

$\qquad A = -w^2 + 38w$

Completing the square, we get

$\qquad A = -(w - 19)^2 + 361.$

The maximum function value is 361. It occurs when w is 19. When $w = 19$, $l = 38 - 19$, or 19.

Check. We check a function value for w less than 19 and for w greater than 19.

$\qquad A(18) = -(18)^2 + 38 \cdot 18 = 360$

$\qquad A(20) = -(20)^2 + 38 \cdot 20 = 360$

Since 361 is greater than these numbers, it looks as though we have a maximum.

State. The maximum area of 361 ft^2 occurs when the dimensions are 19 ft by 19 ft.

2. Dimensions: 17 ft by 17 ft; area: 289 ft^2

3. **Familiarize.** We let x and y represent the numbers, and we let P represent their product.

Translate. We have two equations.

$\qquad x + y = 16,$

$\qquad P = xy$

Carry out. Solving the first equation for y, we get $y = 16 - x$. Substituting for y in the second equation we get a quadratic function P:

$\qquad P = x(16 - x)$

$\qquad P = -x^2 + 16x$

Completing the square, we get

$\qquad P = -(x - 8)^2 + 64.$

The maximum function value is 64. It occurs when $x = 8$. When $x = 8$, $y = 16 - 8$, or 8.

Check. We can check a function value for x less than 8 and for x greater than 8.

$\qquad P(7) = -(7)^2 + 16 \cdot 7 = 63$

$\qquad P(9) = -(9)^2 + 16 \cdot 9 = 63$

Since 64 is greater than these numbers, it looks as though we have a maximum.

State. The maximum product of 64 occurs for the numbers 8 and 8.

4. Maximum product: 196; numbers: 14 and 14

5. Familiarize. We let x and y represent the two numbers, and we let P represent their product.

Translate. We have two equations.

$$x + y = 22,$$
$$P = xy$$

Carry out. Solve the first equation for y.

$$y = 22 - x$$

Substitute for y in the second equation.

$$P = x(22 - x)$$
$$P = -x^2 + 22x$$

Completing the square, we get

$$P = -(x - 11)^2 + 121$$

The maximum function value is 121. It occurs when $x = 11$. When $x = 11$, $y = 22 - 11$, or 11.

Check. Check a function value for x less than 11 and for x greater than 11.

$$P(10) = -(10)^2 + 22 \cdot 10 = 120$$
$$P(12) = -(12)^2 + 22 \cdot 12 = 120$$

Since 121 is greater than these numbers, it looks as though we have a maximum.

State. The maximum product of 121 occurs for the numbers 11 and 11.

6. Maximum product: $\dfrac{2025}{4}$; numbers: $\dfrac{45}{2}$ and $\dfrac{45}{2}$

7. Familiarize. We let x and y represent the two numbers, and we let P represent their product.

Translate. We have two equations.

$$x - y = 4,$$
$$P = xy$$

Carry out. Solve the first equation for x.

$$x = 4 + y$$

Substitute for x in the second equation.

$$P = (4 + y)y$$
$$P = y^2 + 4y$$

Completing the square, we get

$$P = (y + 2)^2 - 4.$$

The minimum function value is -4. It occurs when $y = -2$. When $y = -2$, $x = 4 + (-2)$, or 2.

Check. Check a function value for y less than -2 and for y greater than -2.

$$P(-3) = (-3)^2 + 4(-3) = -3$$
$$P(-1) = (-1)^2 + 4(-1) = -3$$

Since -4 is less than these numbers, it looks as though we have a minimum.

State. The minimum product of -4 occurs for the numbers 2 and -2.

8. Minimum product: -25; numbers: 5 and -5

9. Familiarize. We let x and y represent the two numbers, and we let P represent their product.

Translate. We have two equations.

$$x - y = 9,$$
$$P = xy$$

Carry out. Solve the first equation for x.

$$x = y + 9$$

Substitute for x in the second equation.

$$P = (y + 9)y$$
$$P = y^2 + 9y$$

Completing the square, we get

$$P = \left(y + \frac{9}{2}\right)^2 - \frac{81}{4}.$$

The minimum function value is $-\dfrac{81}{4}$. It occurs when $y = -\dfrac{9}{2}$. When $y = -\dfrac{9}{2}$, $x = -\dfrac{9}{2} + 9 = \dfrac{9}{2}$.

Check. Check a function value for y less than $-\dfrac{9}{2}$ and for y greater than $-\dfrac{9}{2}$.

$$P(-5) = (-5)^2 + 9(-5) = -20$$
$$P(-4) = (-4)^2 + 9(-4) = -20$$

Since $-\dfrac{81}{4}$ is less than these numbers, it looks as though we have a minimum.

State. The minimum product of $-\dfrac{81}{4}$ occurs for the numbers $\dfrac{9}{2}$ and $-\dfrac{9}{2}$.

10. Minimum product: $-\dfrac{49}{4}$; numbers: $\dfrac{7}{2}$ and $-\dfrac{7}{2}$

11. Familiarize. We let x and y represent the two numbers, and we let P represent their product.

Translate. We have two equations.

$$x + y = -7,$$
$$P = xy$$

Carry out. Solve the first equation for y.

$$y = -x - 7$$

Substitute for y in the second equation.

$$P = x(-x - 7)$$
$$P = -x^2 - 7x$$

Completing the square, we get

$$P = -\left(x + \frac{7}{2}\right)^2 + \frac{49}{4}.$$

The maximum function value is $\dfrac{49}{4}$. It occurs when $x = -\dfrac{7}{2}$. When $x = -\dfrac{7}{2}$, $y = -\left(-\dfrac{7}{2}\right) - 7 = -\dfrac{7}{2}$.

Check. Check a function value for x less than $-\frac{7}{2}$ and for x greater than $-\frac{7}{2}$.

$$P(-4) = -(-4)^2 - 7(-4) = 12$$
$$P(-3) = -(-3)^2 - 7(-3) = 12$$

Since $\frac{49}{4}$ is greater than these numbers, it looks as though we have a maximum.

State. The maximum product of $\frac{49}{4}$ occurs for the numbers $-\frac{7}{2}$ and $-\frac{7}{2}$.

12. Maximum product: $\frac{81}{4}$; numbers: $-\frac{9}{2}$ and $-\frac{9}{2}$

13. Familiarize. We make a drawing and label it.

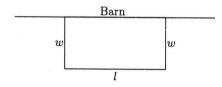

Translate. We have two equations.

$$l + 2w = 40,$$
$$A = lw$$

Carry out. Solve the first equation for l.

$$l = 40 - 2w$$

Substitute for l in the second equation.

$$A = (40 - 2w)w$$
$$A = -2w^2 + 40w$$

Completing the square, we get

$$A = -2(w - 10)^2 + 200.$$

The maximum function value of 200 occurs when $w = 10$. When $w = 10$, $l = 40 - 2 \cdot 10 = 20$.

Check. Check a function value for w less than 10 and for w greater than 10.

$$A(9) = -2 \cdot 9^2 + 40 \cdot 9 = 198$$
$$A(11) = -2 \cdot 11^2 + 40 \cdot 11 = 198$$

Since 200 is greater than these numbers, it looks as though we have a maximum.

State. The maximum area of 200 ft^2 will occur when the dimensions are 10 ft by 20 ft.

14. Maximum area: 450 ft^2; dimensions: 15 ft by 30 ft

15. Familiarize. We make an overhead drawing and label it.

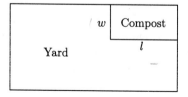

Translate. We have two equations.

$$l + w = 8,$$
$$V = l \cdot w \cdot 3, \text{ or } V = 3lw$$

Carry out. Solve the first equation for l.

$$l = 8 - w$$

Substitute for l in the second equation.

$$V = 3(8 - w)w$$
$$V = -3w^2 + 24w$$

Completing the square, we get

$$V = -3(w - 4)^2 + 48.$$

The maximum function value of 48 occurs when $w = 4$. When $w = 4$, $l = 8 - 4 = 4$.

Check. Check a function value for w less than 4 and for w greater than 4.

$$V(3) = -3 \cdot 3^2 + 24 \cdot 3 = 45$$
$$V(5) = -3 \cdot 5^2 + 24 \cdot 5 = 45$$

Since 48 is greater than these numbers, it looks as though we have a maximum.

State. The dimensions of the base that will maximize the container's volume are 4 ft by 4 ft.

16. 3.5 in.

17. We look for a function of the form $f(x) = ax^2 + bx + c$. Substituting the data points, we get

$$4 = a(1)^2 + b(1) + c,$$
$$-2 = a(-1)^2 + b(-1) + c,$$
$$13 = a(2)^2 + b(2) + c,$$

or

$$4 = a + b + c,$$
$$-2 = a - b + c,$$
$$13 = 4a + 2b + c.$$

Solving this system, we get

$$a = 2, b = 3, \text{ and } c = -1.$$

Therefore the function we are looking for is

$$f(x) = 2x^2 + 3x - 1.$$

18. $f(x) = 3x^2 - x + 2$

19. We look for a function of the form $f(x) = ax^2 + bx + c$. Substituting the data points, we get

$$0 = a(2)^2 + b(2) + c,$$
$$3 = a(4)^2 + b(4) + c,$$
$$-5 = a(12)^2 + b(12) + c,$$

or

$$0 = 4a + 2b + c,$$
$$3 = 16a + 4b + c,$$
$$-5 = 144a + 12b + c.$$

Solving this system, we get

$$a = -\frac{1}{4}, b = 3, c = -5.$$

Therefore the function we are looking for is

$$f(x) = -\frac{1}{4}x^2 + 3x - 5.$$

20. $f(x) = -\frac{1}{3}x^2 + 5x - 12$

21. a) **Familiarize**. We look for a function of the form $f(x) = ax^2 + bx + c$, where $f(x)$ represents the earnings for week x.

Translate. We substitute the given values of x and $f(x)$.

$$38 = a(1)^2 + b(1) + c,$$
$$66 = a(2)^2 + b(2) + c,$$
$$86 = a(3)^2 + b(3) + c,$$
or
$$38 = a + b + c,$$
$$66 = 4a + 2b + c,$$
$$86 = 9a + 3b + c.$$

Carry out. Solving the system of equations, we get

$a = -4$, $b = 40$, $c = 2$.

Check. Recheck the calculations.

State. The function $f(x) = -4x^2 + 40x + 2$ fits the data.

b) Find $f(4)$.

$$f(4) = -4(4)^2 + 40(4) + 2 = 98$$

The predicted earnings for the fourth week are $98.

22. a) $f(x) = 2500x^2 - 6500x + 5000$

b) $19,000

23. a) **Familiarize**. We look for a function of the form $A(s) = as^2 + bs + c$, where $A(s)$ represents the number of daytime accidents (for every 200 million km) and s represents the travel speed (in km/h).

Translate. We substitute the given values of s and $A(s)$.

$$100 = a(60)^2 + b(60) + c,$$
$$130 = a(80)^2 + b(80) + c,$$
$$200 = a(100)^2 + b(100) + c,$$
or
$$100 = 3600a + 60b + c,$$
$$130 = 6400a + 80b + c,$$
$$200 = 10,000a + 100b + c.$$

Carry out. Solving the system of equations, we get

$a = 0.05$, $b = -5.5$, $c = 250$.

Check. Recheck the calculations.

State. The function $A(s) = 0.05s^2 - 5.5s + 250$ fits the data.

b) Find $A(50)$.

$$A(50) = 0.05(50)^2 - 5.5(50) + 250 = 100$$

100 accidents occur at 50 km/h.

24. a) $A(s) = \frac{3}{16}s^2 - \frac{135}{4}s + 1750$

b) 531.25

25. a) **Familiarize**. We look for a function of the form $P(d) = ad^2 + bd + c$, where $P(d)$ represents the price for a pizza with diameter d.

Translate. Substitute the given values of d and $P(d)$.

$$6 = a(8)^2 + b(8) + c,$$
$$8.50 = a(12)^2 + b(12) + c,$$
$$11.50 = a(16)^2 + b(16) + c,$$
or
$$6 = 64a + 8b + c,$$
$$8.50 = 144a + 12b + c,$$
$$11.50 = 256a + 16b + c.$$

Carry out. Solving the system of equations, we get

$$a = \frac{1}{64}, \; b = \frac{5}{16}, \; c = \frac{5}{2}.$$

Check. Recheck the calculations.

State. The function $P(d) = \frac{1}{64}d^2 + \frac{5}{16}d + \frac{5}{2}$ fits the data.

b) Find $P(14)$.

$$P(14) = \frac{1}{64}(14)^2 + \frac{5}{16}(14) + \frac{5}{2}$$
$$= 9.9375$$

The price of a 14-in. pizza is approximately $9.94.

26. $P(x) = -(x - 490)^2 + 237,1000$; $237,100 at $x = 490$

27. Find the total profit:

$$P(x) = R(x) - C(x)$$
$$P(x) = (200x - x^2) - (5000 + 8x)$$
$$P(x) = -x^2 + 192x - 5000$$

To find the maximum value of the total profit and the value of x at which it occurs we complete the square:

$$P(x) = -(x^2 - 192x) - 5000$$
$$= -(x^2 - 192x + 9216 - 9216) - 5000$$
$$= -(x^2 - 192x + 9216) - (-9216) - 5000$$
$$= -(x - 96)^2 + 4216$$

The maximum profit of $4216 occurs at $x = 96$.

28. $P(x) = -(x - 110)^2 + 12,050$; $12,050 at $x = 110$

29.
$$A = 6s^2$$
$$\frac{A}{6} = s^2 \qquad \text{Dividing by 6}$$
$$\sqrt{\frac{A}{6}} = s \qquad \text{Taking the positive square root}$$

30. $r = \frac{1}{2}\sqrt{\frac{A}{\pi}}$

31.
$$F = \frac{Gm_1m_2}{r^2}$$

$$Fr^2 = Gm_1m_2 \qquad \text{Multiplying by } r^2$$

$$r^2 = \frac{Gm_1m_2}{F} \qquad \text{Dividing by } F$$

$$r = \sqrt{\frac{Gm_1m_2}{F}} \qquad \text{Taking the positive square root}$$

32. $s = \sqrt{\dfrac{kQ_1Q_2}{N}}$

33.
$$E = mc^2$$

$$\frac{E}{m} = c^2 \qquad \text{Dividing by } m$$

$$\sqrt{\frac{E}{m}} = c \qquad \text{Taking the square root}$$

34. $r = \sqrt{\dfrac{A}{\pi}}$

35.
$$a^2 + b^2 = c^2$$

$$b^2 = c^2 - a^2 \qquad \text{Subtracting } a^2$$

$$b = \sqrt{c^2 - a^2} \qquad \text{Taking the square root}$$

36. $c = \sqrt{d^2 - a^2 - b^2}$

37.
$$N = \frac{k^2 - 3k}{2}$$

$$2N = k^2 - 3k$$

$$0 = k^2 - 3k - 2N \qquad \text{Standard form}$$

$$a = 1, \ b = -3, \ c = -2N$$

$$k = \frac{-(-3) \pm \sqrt{(-3)^3 - 4 \cdot 1 \cdot (-2N)}}{2 \cdot 1} \qquad \begin{array}{l}\text{Using the}\\ \text{quadratic formula}\end{array}$$

$$k = \frac{3 \pm \sqrt{9 + 8N}}{2}$$

Since taking the negative square root would result in a negative answer, we take the positive one.

$$k = \frac{3 + \sqrt{9 + 8N}}{2}$$

38. $t = \dfrac{-v_0 + \sqrt{v_0^2 + 2gs}}{g}$

39. $A = 2\pi r^2 + 2\pi rh$

$$0 = 2\pi r^2 + 2\pi rh - A \qquad \text{Standard form}$$

$$a = 2\pi, \ b = 2\pi h, \ c = -A$$

$$r = \frac{-2\pi h \pm \sqrt{(2\pi h)^2 - 4 \cdot 2\pi \cdot (-A)}}{2 \cdot 2\pi} \qquad \begin{array}{l}\text{Using the}\\ \text{quadratic formula}\end{array}$$

$$r = \frac{-2\pi h \pm \sqrt{4\pi^2 h^2 + 8\pi A}}{4\pi}$$

$$r = \frac{-2\pi h \pm 2\sqrt{\pi^2 h^2 + 2\pi A}}{4\pi}$$

$$r = \frac{-\pi h \pm \sqrt{\pi^2 h^2 + 2\pi A}}{2\pi}$$

Since taking the negative square root would result in a negative answer, we take the positive one.

$$r = \frac{-\pi h + \sqrt{\pi^2 h^2 + 2\pi A}}{2\pi}$$

40. $r = \dfrac{-\pi s + \sqrt{\pi^2 s^2 + 4\pi A}}{2\pi}$

41.
$$N = \frac{1}{2}(n^2 - n)$$

$$N = \frac{1}{2}n^2 - \frac{1}{2}n$$

$$0 = \frac{1}{2}n^2 - \frac{1}{2}n - N$$

$$a = \frac{1}{2}, \ b = -\frac{1}{2}, \ c = -N$$

$$n = \frac{-\left(-\dfrac{1}{2}\right) \pm \sqrt{\left(-\dfrac{1}{2}\right)^2 - 4 \cdot \dfrac{1}{2} \cdot (-N)}}{2\left(\dfrac{1}{2}\right)}$$

$$n = \frac{1}{2} \pm \sqrt{\frac{1}{4} + 2N}$$

$$n = \frac{1}{2} \pm \sqrt{\frac{1 + 8N}{4}}$$

$$n = \frac{1}{2} \pm \frac{1}{2}\sqrt{1 + 8N}$$

Since taking the negative square root would result in a negative answer, we take the positive one.

$$n = \frac{1}{2} + \frac{1}{2}\sqrt{1 + 8N}, \text{ or } \frac{1 + \sqrt{1 + 8N}}{2}$$

42. $r = 1 - \sqrt{\dfrac{A}{A_0}}$

43.
$$A = 2w^2 + 4lw$$

$$0 = 2w^2 + 4lw - A$$

$$a = 2, \ b = 4l, \ c = -A$$

$$w = \frac{-4l \pm \sqrt{(4l)^2 - 4 \cdot 2 \cdot (-A)}}{2 \cdot 2}$$

$$w = \frac{-4l \pm \sqrt{16l^2 + 8A}}{4}$$

$$w = \frac{-4l \pm 2\sqrt{4l^2 + 2A}}{4}$$

$$w = \frac{-2l \pm \sqrt{4l^2 + 2A}}{2}$$

Since taking the negative square root would result in a negative answer, we take the positive one.

$$w = \frac{-2l + \sqrt{4l^2 + 2A}}{2}$$

44. $r = \dfrac{-\pi h + \sqrt{\pi^2 h^2 + 4\pi A}}{4\pi}$

45.
$$T = 2\pi\sqrt{\frac{l}{g}}$$

$$\frac{T}{2\pi} = \sqrt{\frac{l}{g}} \qquad \text{Multiplying by } \frac{1}{2\pi}$$

$$\frac{T^2}{4\pi^2} = \frac{l}{g} \qquad \text{Squaring}$$

$$gT^2 = 4\pi^2 l \qquad \text{Multiplying by } 4\pi^2 g$$

$$g = \frac{4\pi^2 l}{T^2} \qquad \text{Multiplying by } \frac{1}{T^2}$$

46. $L = \dfrac{1}{W^2 C}$

47. $A = P_1(1+r)^2 + P_2(1+r)$

$$0 = P_1(1+r)^2 + P_2(1+r) - A$$

Let $u = 1 + r$.

$$0 = P_1 u^2 + P_2 u - A \qquad \text{Substituting}$$

$$u = \frac{-P_2 \pm \sqrt{P_2^2 - 4(P_1)(-A)}}{2P_1}$$

$$u = \frac{-P_2 + \sqrt{P_2^2 + 4AP_1}}{2P_1} \qquad \substack{\text{Simplifying}}$$

and taking the positive square root

$$1 + r = \frac{-P_2 + \sqrt{P_2^2 + 4AP_1}}{2P_1} \qquad \text{Substituting}$$

$$1 + r \text{ for } u$$

$$r = -1 + \frac{-P_2 + \sqrt{P_2^2 + 4AP_1}}{2P_1}$$

48. $r = -2 + \dfrac{-P_2 + \sqrt{P_2^2 + 4AP_1}}{P_1}$

49.
$$m = \frac{m_0}{\sqrt{1 - \dfrac{v^2}{c^2}}}$$

$$m^2 = \frac{m_0^2}{1 - \dfrac{v^2}{c^2}} \qquad \text{Principle of powers}$$

$$m^2\left(1 - \frac{v^2}{c^2}\right) = m_0^2$$

$$m^2 - \frac{m^2 v^2}{c^2} = m_0^2$$

$$m^2 - m_0^2 = \frac{m^2 v^2}{c^2}$$

$$c^2(m^2 - m_0^2) = m^2 v^2$$

$$\frac{c^2(m^2 - m_0^2)}{m^2} = v^2$$

$$\sqrt{\frac{c^2(m^2 - m_0^2)}{m^2}} = v$$

$$\frac{c}{m}\sqrt{m^2 - m_0^2} = v$$

50. $c = \dfrac{mv}{\sqrt{m^2 - m_0^2}}$

51. a) *Familiarize and Translate.* From Example 6, we know
$$t = \frac{-v_0 + \sqrt{v_0^2 + 19.6s}}{9.8}.$$

Carry out. Substituting 75 for s and 0 for v_0, we have
$$t = \frac{0 + \sqrt{0^2 + 19.6(75)}}{9.8}$$
$$t \approx 3.9$$

Check. Substitute 3.9 for t and 0 for v_0 in the original formula. (See Example 6.)
$$s = 4.9t^2 + v_0 t = 4.9(3.9)^2 + 0 \cdot (3.9)^2$$
$$\approx 75$$

The answer checks.

State. It takes about 3.9 sec to reach the ground.

b) *Familiarize and Translate.* From Example 6, we know
$$t = \frac{-v_0 + \sqrt{v_0^2 + 19.6s}}{9.8}.$$

Carry out. Substitute 75 for s and 30 for v_0.
$$t = \frac{-30 + \sqrt{30^2 + 19.6(75)}}{9.8}$$
$$t \approx 1.9$$

Check. Substitute 30 for v_0 and 1.9 for t in the original formula. (See Example 6.)
$$s = 4.9t^2 + v_0 t = 4.9(1.9)^2 + (30)(1.9)$$
$$\approx 75$$

The answer checks.

State. It takes about 1.9 sec to reach the ground.

c) *Familiarize and Translate.* We will use the formula $s = 4.9t^2 + v_0 t$.

Carry out. Substitute 2 for t and 30 for v_0.
$$s = 4.9(2)^2 + 30(2) = 79.6$$

Check. We can substitute 30 for v_0 and 79.6 for s in the form of the formula we used in part b).
$$t = \frac{-v_0 + \sqrt{v_0^2 + 19.6s}}{9.8}$$
$$= \frac{-30 + \sqrt{(30)^2 + 19.6(79.6)}}{9.8} = 2$$

The answer checks.

State. The object will fall 79.6 m.

52. a) 10.1 sec

b) 7.49 sec

c) 272.5 m

53. *Familiarize and Translate*. From Example 5, we know

$$T = \frac{\sqrt{3V}}{12}.$$

Carry out. Substituting 38 for V, we have

$$T = \frac{\sqrt{3 \cdot 38}}{12} = \frac{\sqrt{114}}{12}$$

$$T \approx 0.89$$

Check. Substitute 0.89 in the original formula. (See Example 5.)

$$48T^2 = V$$

$$48(0.89)^2 = V$$

$$38 \approx V$$

The answer checks.

State. The hang time is about 0.89 sec.

54. 1 m

55. *Familiarize and Translate*. We will use the formula in Exercise 41. (If we did not know that this is the correct formula, we could derive it ourselves.)

$$N = \frac{1}{2}(n^2 - n)$$

Carry out. Substitute 91 for N and solve for n.

$$91 = \frac{1}{2}(n^2 - n)$$

$$182 = n^2 - n$$

$$0 = n^2 - n - 182$$

$$0 = (n - 14)(n + 13)$$

$$n = 14 \quad \text{or} \quad n = -13$$

Check. Since a negative solution has no meaning in this problem, we check 14. Substitute 14 for n in the original formula.

$$N = \frac{1}{2}(14^2 - 14) = \frac{1}{2}(182) = 91$$

The solution checks.

State. There are 14 teams in the league.

56. 12

57. *Familiarize and Translate*. From Example 6, we know

$$t = \frac{-v_0 + \sqrt{v_0^2 + 19.6s}}{9.8}.$$

Carry out. Substituting 40 for s and 0 for v_0 we have

$$t = \frac{0 + \sqrt{0^2 + 19.6(40)}}{9.8}$$

$$t \approx 2.9$$

Check. Substitute 2.9 for t and 0 for v_0 in the original formula. (See Example 6.)

$$s = 4.9t^2 + v_0 t = 4.9(2.9)^2 + 0(2.9)$$

$$\approx 40$$

The answer checks.

State. He will be falling for about 2.9 sec.

58. 30.625 m

59. *Familiarize and Translate*. We will use the formula in Example 6, $s = 4.9t^2 + v_0 t$.

Carry out. Solve the formula for v_0.

$$s - 4.9t^2 = v_0 t$$

$$\frac{s - 4.9t^2}{t} = v_0$$

Now substitute 51.6 for s and 3 for t.

$$\frac{51.6 - 4.9(3)^2}{3} = v_0$$

$$2.5 = v_0$$

Check. Substitute 3 for t and 2.5 for v_0 in the original formula.

$$s = 4.9(3)^2 + 2.5(3) = 51.6$$

The solution checks.

State. The initial velocity is 2.5 m/sec.

60. 3.2 m/sec

61. *Familiarize and Translate*. From Exercise 47 we know that

$$r = -1 + \frac{-P_2 + \sqrt{P_2^2 + 4P_1 A}}{2P_1},$$

where A is the total amount in the account after two years, P_1 is the amount of the original deposit, P_2 is deposited at the begining of the second year, and r is the annual interest rate.

Carry out. Substitute 3000 for P_1, 1700 for P_2, amd 5253.70 for A.

$$r = -1 + \frac{-1700 + \sqrt{(1700)^2 + 4(3000)(5253.70)}}{2(3000)}$$

Using a calculator, we have $r = 0.07$.

Check. Substitute in the original formula in Exercise 47.

$$P_1(1 + r)^2 + P_2(1 + r) = A$$

$$3000(1.07)^2 + 1700(1.07) = A$$

$$5253.70 = A$$

The answer checks.

State. The annual interest rate is 0.07, or 7%.

62. 8.5%

63.

$$3x + 5 < 5x - 8$$

$$-2x + 5 < -8 \qquad \text{Subtracting } 5x$$

$$-2x < -13 \qquad \text{Subtracting } 5$$

$$x > \frac{13}{2} \qquad \begin{array}{l}\text{Dividing by } -2 \text{ and reversing} \\ \text{the inequality symbol}\end{array}$$

The solution set is $\left\{ x \middle| x > \frac{13}{2} \right\}$, or $\left(\frac{13}{2}, \infty \right)$.

64. $\{x | x \le -7\}$, or $(-\infty, -7]$

65. *Familiarize*. Recall that the formula for the area of a triangle is $A = \frac{1}{2}bh$.

Translate. We have two equations.

$$b + h = 38,$$

$$A = \frac{1}{2}bh$$

Carry out. Solve the first equation for h.

$$h = 38 - b$$

Substitute for h in the second equation.

$$A = \frac{1}{2}b(38 - b)$$

$$A = -\frac{1}{2}b^2 + 19b$$

Completing the square, we get

$$A = -\frac{1}{2}(b - 19)^2 + \frac{361}{2}$$

The maximum function value of $\frac{361}{2}$, or 180.5, occurs when $b = 19$. When $b = 19$, $h = 38 - 19$, or 19.

Check. Check a function value for b less than 19 and for b greater than 19.

$$A(18) = -\frac{1}{2}(18)^2 + 19(18) = 180$$

$$A(20) = -\frac{1}{2}(20)^2 + 19(20) = 180$$

Since 180.5 is greater than these numbers, it looks as though we have a maximum.

State. The maximum area of 180.5 cm^2 occurs when the base and height are both 19 cm.

66. $11\sqrt{2}$ ft

67. $mn^4 - r^2pm^3 - r^2n^2 + p = 0$

Let $u = n^2$. Substitute and rearrange.

$$mu^2 - r^2u - r^2pm^3 + p = 0$$

$$a = m, \ b = -r^2, \ c = -r^2pm^3 + p$$

$$u = \frac{-(-r^2) \pm \sqrt{(-r^2)^2 - 4 \cdot m(-r^2pm^3 + p)}}{2 \cdot m}$$

$$u = \frac{r^2 \pm \sqrt{r^4 + 4m^4r^2p - 4mp}}{2m}$$

$$n^2 = \frac{r^2 \pm \sqrt{r^4 + 4m^4r^2p - 4mp}}{2m}$$

$$n = \pm\sqrt{\frac{r^2 \pm \sqrt{r^4 + 4m^4r^2p - 4mp}}{2m}}$$

68. $t = \dfrac{r + s \pm \sqrt{r^2 + 2rs + s^2 - 4r^2s^2 + 4s^3r}}{2r - 2s}$

69. ***Familiarize***. Let x represent the number of 10¢ increases in the admission price. Then $2.00 + 0.1x$ represents the admission price and $100 - x$ represents the corresponding average attendance.

Translate. Since total revenue is admission price times the attendance, we have the following function for the revenue.

$$R(x) = (2.00 + 0.1x)(100 - x)$$

$$R(x) = -0.1x^2 + 8x + 200$$

Carry out. Completing the square, we get

$$R(x) = -0.1(x - 40)^2 + 360$$

The maximum function value of 360 occurs when $x = 40$. When $x = 40$ the admission price is $2.00 + 0.1(40)$, or \$6.

Check. We check a function value for x less than 40 and for x greater than 40.

$$R(39) = -0.1(39)^2 + 8(39) + 200 = 359.9$$

$$R(41) = -0.1(41)^2 + 8(41) + 200 = 359.9$$

Since 360 is greater than these numbers, it looks as though we have a maximum.

State. In order to maximize revenue the theater owner should charge \$6.00 for admission.

70. 30

71. ***Familiarize***. We want to find the maximum value of a function of the form $h(t) = at^2 + bt + c$ that fits the following data.

Time (sec)	Height (ft)
0	64
3	64
3+2, or 5	0

Translate. Substitute the given values for t and $h(t)$.

$$64 = a(0)^2 + b(0) + c,$$

$$64 = a(3)^2 + b(3) + c,$$

$$0 = a(5)^2 + b(5) + c,$$

or

$$64 = c,$$

$$64 = 9a + 3b + c,$$

$$0 = 25a + 5b + c.$$

Carry out. Solving the system of equations, we get $a = -6.4$, $b = 19.2$, $c = 64$. The function $h(t) = -6.4t^2 + 19.2t + 64$ fits the data.

Completing the square, we get

$$h(t) = -6.4(t - 1.5)^2 + 78.4.$$

The maximum funtion value of 78.4 occurs at $t = 1.5$.

Check. Recheck the calculations. Also check a function value for t less than 1.5 and for t greater than 1.5.

$$h(1) = -6.4(1)^2 + 19.2(1) + 64 = 76.8$$

$$h(2) = -6.4(2)^2 + 19.2(2) + 64 = 76.8$$

Since 78.4 is greater than these numbers, it looks as though we have a maximum.

State. The maximum height is 78.4 ft.

72. 158 ft

73. $A = 2\pi r^2 + 2\pi rh$ (See Exercise 39.)

Substitute $\dfrac{d}{2}$ for r.

$$A = 2\pi\left(\dfrac{d}{2}\right)^2 + 2\pi\left(\dfrac{d}{2}\right)h$$

$$A = \dfrac{\pi}{2}d^2 + \pi dh$$

$$2A = \pi d^2 + 2\pi dh \qquad \text{Clearing the fraction}$$

$$0 = \pi d^2 + 2\pi dh - 2A$$

$$a = \pi, \ b = 2\pi h, \ c = -2A$$

$$d = \dfrac{-2\pi h \pm \sqrt{(2\pi h)^2 - 4\pi(-2A)}}{2\pi}$$

$$d = \dfrac{-2\pi h \pm \sqrt{4\pi^2 h^2 + 8\pi A}}{2\pi}$$

We take the positive square root and simplify.

$$d = \dfrac{-\pi h + \sqrt{\pi^2 h^2 + 2\pi A}}{\pi}$$

(Since $d = 2r$, we could also have multiplied the result of Exercise 39 by 2 to find this formula.)

74. $L = \sqrt{3V^{2/3}}$, or $\sqrt{3}\,\sqrt[3]{V}$

75. Let s represent the length of a side of the cube, let A represent the surface area of the cube, and let L represent the length of the cube's three-dimensional diagonal. From the Pythagorean formula in three dimensions (See Exercise 36.) we know that $L^2 = s^2 + s^2 + s^2$, or $L^2 = 3s^2$. From the formula for the surface area of a cube (See Exercise 29.) we know that $A = 6s^2$, so $\dfrac{A}{2} = 3s^2$ and $L^2 = \dfrac{A}{2}$. Then $L = \sqrt{\dfrac{A}{2}}$.

Exercise Set 11.10

---

1. $(x - 5)(x + 3) > 0$

The solutions of $(x - 5)(x + 3) = 0$ are 5 and -3. They are not solutions of the inequality, but they divide the real-number line in a natural way. The product $(x - 5)(x + 3)$ is positive or negative, for values other than 5 and -3, depending on the signs of the factors $x - 5$ and $x + 3$.

$x - 5 > 0$ when $x > 5$ and $x - 5 < 0$ when $x < 5$.

$x + 3 > 0$ when $x > -3$ and $x + 3 < 0$ when $x < -3$.

We make a diagram.

Sign of $x - 5$ $- - - -|- - - -|+ + + +$
Sign of $x + 3$ $- - - -|+ + + +|+ + + +$
Sign of product $+ + + +|- - - -|+ + + +$

$$\xrightarrow{\hspace{2cm}\underset{-3}{|}\hspace{1cm}\underset{5}{|}\hspace{2cm}}$$

For the product $(x - 5)(x + 3)$ to be positive, both factors must be positive or both factors must be negative. We see from the diagram that numbers satisfying $x < -3$ or $x > 5$ are solutions. The solution set of the inequality is $\{x | x < -3$ or $x > 5\}$, or $(-\infty, -3) \cup (5, \infty)$.

2. $\{x | x < -1$ or $x > 4\}$, or $(-\infty, -1) \cup (4, \infty)$

3. $(x + 1)(x - 2) \le 0$

The solutions of $(x + 1)(x - 2) = 0$ are -1 and 2. They divide the number line into three intervals as shown:

$$\xleftarrow{\hspace{1cm}\overset{A}{\frown}\hspace{0.5cm}|\hspace{0.5cm}\overset{B}{\frown}\hspace{0.5cm}|\hspace{0.5cm}\overset{C}{\frown}\hspace{0.5cm}}\rightarrow$$
$$\hspace{2.5cm}\underset{-1}{}\hspace{1cm}\underset{2}{}$$

We try test numbers in each interval.

A: Test -2, $f(-2) = (-2 + 1)(-2 - 2) = 4$

B: Test 0, $f(0) = (0 + 1)(0 - 2) = -2$

C: Test 3, $f(3) = (3 + 1)(3 - 2) = 4$

Since $f(0)$ is negative, the function value will be negative for all numbers in the interval containing 0. The inequality symbol is $\le$, so we need to include the intercepts.

The solution set is $\{x | -1 \le x \le 2\}$, or $[-1, 2]$.

4. $\{x | 3 \le x \le 5\}$, or $[-3, 5]$

5. $x^2 - x - 2 < 0$

$(x + 1)(x - 2) < 0$ Factoring

See the diagram and test numbers in Exercise 3. The solution set is $\{x | -1 < x < 2\}$, or $(-1, 2)$.

6. $\{x | -2 < x < 1\}$, or $(-2, 1)$

7. $9 - x^2 \le 0$

$(3 - x)(3 + x) \le 0$

The solutions of $(3 - x)(3 + x) = 0$ are 3 and -3. They divide the real-number line in a natural way. The product $(3 - x)(3 + x)$ is positive or negative, for values other than 3 and -3, depending on the signs of the factors $3 - x$ and $3 + x$.

$3 - x > 0$ when $x < 3$ and $3 - x < 0$ when $x > 3$.

$3 + x > 0$ when $x > -3$ and $3 + x < 0$ when $x < -3$.

We make a diagram.

Sign of $3 - x$ $+ + + +|+ + + +|- - - -$
Sign of $3 + x$ $- - - -|+ + + +|+ + + +$
Sign of product $- - - -|+ + + +|- - - -$

$$\xrightarrow{\hspace{2cm}\underset{-3}{|}\hspace{1cm}\underset{3}{|}\hspace{2cm}}$$

For the product $(3 - x)(3 + x)$ to be negative, one factor must be positive and the other negative. We see from the diagram that numbers satisfying $x < -3$ or $x > 3$ are solutions. The intercepts are also solutions, because the inequality symbol is $\le$. The solution set of the inequality is $\{x | x \le -3$ or $x \ge 3\}$, or $(-\infty, -3] \cup [3, \infty)$.

8. $\{x | -2 \le x \le 2\}$, or $[-2, 2]$

9. $x^2 - 2x + 1 \ge 0$

$(x - 1)^2 \ge 0$

The solution of $(x - 1)^2 = 0$ is 1. For all real-number values of x except 1, $(x - 1)^2$ will be positive. Thus the solution set is $\{x | x$ is a real number$\}$, or $(-\infty, \infty)$.

10. $\emptyset$

11.
$$x^2 + 8 < 6x$$
$$x^2 - 6x + 8 < 0$$
$$(x - 4)(x - 2) < 0$$

The solutions of $(x-4)(x-2) = 0$ are 4 and 2. They are not solutions of the inequality, but they divide the real-number line in a natural way. The product $(x-4)(x-2)$ is positive or negative, for values other than 4 and 2, depending on the signs of the factors $x - 4$ and $x - 2$.

$x - 4 > 0$ when $x > 4$ and $x - 4 < 0$ when $x < 4$.

$x - 2 > 0$ when $x > 2$ and $x - 2 < 0$ when $x < 2$.

We make a diagram.

Sign of $x - 4$ $- - - -|- - - - -|+ + + +$
Sign of $x - 2$ $- - - -|+ + + +|+ + + +$
Sign of product $+ + + +|- - - - -|+ + + +$

$$\hspace 2 \qquad\qquad 4$$

For the product $(x - 4)(x - 2)$ to be negative, one factor must be positive and the other negative. The only situation in the diagram for which this happens is when $2 < x < 4$. The solution set of the inequality is $\{x|2 < x < 4\}$, or $(2, 4)$.

12. $\{x|x < -2 \text{ or } x > 6\}$, or $(-\infty, -2) \cup (6, \infty)$

13. $3x(x + 2)(x - 2) < 0$

The solutions of $3x(x + 2)(x - 2) = 0$ are 0, -2, and 2. They divide the real-number line into four intervals as shown:

We try test numbers in each interval.

A: Test -3, $f(-3) = 3(-3)(-3 + 2)(-3 - 2) = -45$

B: Test -1, $f(-1) = 3(-1)(-1 + 2)(-1 - 2) = 9$

C: Test 1, $f(1) = 3(1)(1 + 2)(1 - 2) = -9$

D: Test 3, $f(3) = 3(3)(3 + 2)(3 - 2) = 45$

Since $f(-3)$ amd $f(1)$ are negative, the function value will be negative for all numbers in the intervals containing -3 and 1. The solution set is $\{x|x < -2 \text{ or } 0 < x < 2\}$, or $(-\infty, -2) \cup (0, 2)$.

14. $\{x| -1 < x < 0 \text{ or } x > 1\}$, or $(-1, 0) \cup (1, \infty)$

15. $(x + 3)(x - 2)(x + 1) > 0$

The solutions of $(x + 3)(x - 2)(x + 1) = 0$ are -3, 2, and -1. They are not solutions of the inequality, but they divide the real-number line in a natural way. The product $(x+3)(x-2)(x+1)$ is positive or negative, for values other than -3, 2, and -1, depending on the signs of the factors $x + 3$, $x - 2$, and $x + 1$.

$x + 3 > 0$ when $x > -3$ and $x + 3 < 0$ when $x < -3$.

$x - 2 > 0$ when $x > 2$ and $x - 2 < 0$ when $x < 2$.

$x + 1 > 0$ when $x > -1$ and $x + 1 < 0$ when $x < -1$.

We make a diagram.

Sign of $x + 3$ $- - -| + + +| + + +| + + +$
Sign of $x - 2$ $- - -| - - -| - - -| + + +$
Sign of $x + 1$ $- - -| - - -| + + +| + + +$
Sign of product $- - -| + + +| - - -| + + +$

$$\qquad -3 \qquad -1 \qquad 2$$

The product of three numbers is positive when all three are positive or two are negative and one is positive. We see from the diagram that numbers satisfying $-3 < x < -1$ or $x > 2$ are solutions. The solution set of the inequality is $\{x| -3 < x < -1 \text{ or } x > 2\}$, or $(-3, -1) \cup (2, \infty)$.

16. $\{x|x < -2 \text{ or } 1 < x < 4\}$, or $(-\infty, -2) \cup (1, 4)$

17. $(x + 3)(x + 2)(x - 1) < 0$

The solutions of $(x + 3)(x + 2)(x - 1) = 0$ are -3, -2, and 1. They divide the real-number line into four intervals as shown:

We try test numbers in each interval.

A: Test -4, $f(-4) = (-4 + 3)(-4 + 2)(-4 - 1) = -10$

B: Test $-\frac{5}{2}$, $f\left(-\frac{5}{2}\right) = \left(-\frac{5}{2}+3\right)\left(-\frac{5}{2}+2\right)\left(-\frac{5}{2}-1\right) = \frac{7}{8}$

C: Test 0, $f(0) = (0 + 3)(0 + 2)(0 - 1) = -6$

D: Test 2, $f(2) = (2 + 3)(2 + 2)(2 - 1) = 20$

The function value will be negative for all numbers in intervals A and C. The solution set is $\{x|x < -3 \text{ or } -2 < x < 1\}$, or $(-\infty, -3) \cup (-2, 1)$.

18. $\{x|x < -1 \text{ or } 2 < x < 3\}$, or $(-\infty, -1) \cup (2, 3)$

19. $\frac{1}{x - 4} < 0$

We write the related equation by changing the $<$ symbol to $=$:

$$\frac{1}{x - 4} = 0$$

We solve the related equation.

$$(x - 4) \cdot \frac{1}{x - 4} = (x - 4) \cdot 0$$
$$1 = 0$$

The related equation has no solution.

Next we find the values that make the denominator 0 by setting the denominate equal to 0 and solving:

$$x - 4 = 0$$
$$x = 4$$

We use 4 to divide the number line into two intervals as shown:

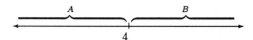

We try test numbers in each interval.

A: Test 0, $\dfrac{1}{0-4} = \dfrac{1}{-4} = -\dfrac{1}{4} < 0$

The number 0 is a solution of the inequality, so the interval A is part of the solution set.

B: Test 5, $\dfrac{1}{5-4} = \dfrac{1}{1} = 1 \not< 0$

The number 5 is not a solution of the inequality, so the interval B is not part of the solution set. The solution set is $\{x | x < 4\}$, or $(-\infty, 4)$.

20. $\{x | x > -5\}$, or $(-5, \infty)$

21. $\dfrac{x+1}{x-3} > 0$

Solve the related equation.

$$\frac{x+1}{x-3} = 0$$

$$x + 1 = 0$$

$$x = -1$$

Find the values that make the denominator 0.

$$x - 3 = 0$$

$$x = 3$$

Use the numbers -1 and 3 to divide the number line into intervals as shown:

```
         A           B           C
  ◄──────┬─────┬─────┬─────┬──────►
        -1           3
```

Try test numbers in each interval.

A: Test -2, $\dfrac{-2+1}{-2-3} = \dfrac{-1}{-5} = \dfrac{1}{5} > 0$

The number -2 is a solution of the inequality, so the interval A is part of the solution set.

B: Test 0, $\dfrac{0+1}{0-3} = \dfrac{1}{-3} = -\dfrac{1}{3} \not> 0$

The number 0 is not a solution of the inequality, so the interval B is not part of the solution set.

C: Test 4, $\dfrac{4+1}{4-3} = \dfrac{5}{1} = 5 > 0$

The number 4 is a solution of the inequality, so the interval C is part of the solution set. The solution set is $\{x | x < -1 \text{ or } x > 3\}$, or $(-\infty, -1) \cup (3, \infty)$.

22. $\{x| -5 < x < 2\}$, or $(-5, 2)$

23. $\dfrac{3x+2}{x-3} \le 0$

Solve the related equation.

$$\frac{3x+2}{x-3} = 0$$

$$3x + 2 = 0$$

$$3x = -2$$

$$x = -\frac{2}{3}$$

Find the values that make the denominator 0.

$$x - 3 = 0$$

$$x = 3$$

Use the numbers $-\dfrac{2}{3}$ and 3 to divide the number line into intervals as shown:

```
         A           B           C
  ◄──────┬─────┬─────┬─────┬──────►
        -2/3         3
```

Try test numbers in each interval.

A: Test -1, $\dfrac{3(-1)+2}{-1-3} = \dfrac{-1}{-4} = \dfrac{1}{4} \not\le 0$

The number -1 is not a solution of the inequality, so the interval A is not part of the solution set.

B: Test 0, $\dfrac{3 \cdot 0 + 2}{0-3} = \dfrac{2}{-3} = -\dfrac{2}{3} \le 0$

The number 0 is a solution of the inequality, so the interval B is part of the solution set.

C: Test 4, $\dfrac{3 \cdot 4 + 2}{4-3} = \dfrac{14}{1} = 14 \not\le 0$

The number 4 is not a solution of the inequality, so the interval C is not part of the solution set. The solution set includes the interval B. The number $-\dfrac{2}{3}$ is also included since the inequality symbol is $\le$ and $-\dfrac{2}{3}$ is the solution of the related equation. The number 3 is not included since $\dfrac{3x+2}{x-3}$ is undefined for $x = 3$. The solution set is $\left\{x \middle| -\dfrac{2}{3} \le x < 3\right\}$, or $\left[-\dfrac{2}{3}, 3\right)$.

24. $\left\{x \middle| x < -\dfrac{3}{4} \text{ or } x \ge \dfrac{5}{2}\right\}$, or $\left(-\infty, -\dfrac{3}{4}\right) \cup \left[\dfrac{5}{2}, \infty\right)$

25. $\dfrac{x-1}{x-2} > 3$

Solve the related equation.

$$\frac{x-1}{x-2} = 3$$

$$x - 1 = 3(x - 2)$$

$$x - 1 = 3x - 6$$

$$5 = 2x$$

$$\frac{5}{2} = x$$

Find the values that make the denominator 0.

$$x - 2 = 0$$

$$x = 2$$

Use the numbers $\dfrac{5}{2}$ and 2 to divide the number line into intervals as shown:

```
         A           B           C
  ◄──────┬─────┬─────┬─────┬──────►
         2           5
                     ─
                     2
```

Try test numbers in each interval.

A: Test 0, $\dfrac{0-1}{0-2} = \dfrac{-1}{-2} = \dfrac{1}{2} \not> 3$

The number 0 is not a solution of the inequality, so the interval A is not part of the solution set.

B: Test $\dfrac{9}{4}$, $\dfrac{\frac{9}{4}-1}{\frac{9}{4}-2} = \dfrac{\frac{5}{4}}{\frac{1}{4}} = 5 > 3$

The number $\dfrac{9}{4}$ is a solution of the inequality, so the interval B is part of the solution set.

C: Test 3, $\dfrac{3-1}{3-2} = \dfrac{2}{1} = 2 \not> 3$

The number 3 is not a solution of the inequality, so the interval C is not part of the solution set. The solution set is $\left\{x \middle| 2 < x < \dfrac{5}{2}\right\}$, or $\left(2, \dfrac{5}{2}\right)$.

26. $\left\{x \middle| x < \dfrac{3}{2} \text{ or } x > 4\right\}$, or $\left(-\infty, \dfrac{3}{2}\right) \cup (4, \infty)$

27. $\dfrac{(x-2)(x+1)}{x-5} < 0$

Solve the related equation.

$$\dfrac{(x-2)(x+1)}{x-5} = 0$$

$$(x-2)(x+1) = 0$$

$$x = 2 \text{ or } x = -1$$

Find the values that make the denominator 0.

$$x - 5 = 0$$

$$x = 5$$

Use the numbers 2, -1, and 5 to divide the number line into intervals as shown:

Try test numbers in each interval.

A: Test -2, $\dfrac{(-2-2)(-2+1)}{-2-5} = \dfrac{-4(-1)}{-7} =$

$-\dfrac{4}{7} < 0$

Interval A is part of the solution set.

B: Test 0, $\dfrac{(0-2)(0+1)}{0-5} = \dfrac{-2 \cdot 1}{-5} = \dfrac{2}{5} \not< 0$

Interval B is not part of the solution set.

C: Test 3, $\dfrac{(3-2)(3+1)}{3-5} = \dfrac{1 \cdot 4}{-2} = -2 < 0$

Interval C is part of the solution set.

D: Test 6, $\dfrac{(6-2)(6+1)}{6-5} = \dfrac{4 \cdot 7}{1} = 28 \not< 0$

Interval D is not part of the solution set.

The solution set is $\{x | x < -1 \text{ or } 2 < x < 5\}$, or $(-\infty, -1) \cup (2, 5)$.

28. $\{x | -4 < x < -3 \text{ or } x > 1\}$, or $(-4, -3) \cup (1, \infty)$

29. $\dfrac{x}{x-2} \geq 0$

Solve the related equation.

$$\dfrac{x}{x-2} = 0$$

$$x = 0$$

Find the values that make the denominator 0.

$$x - 2 = 0$$

$$x = 2$$

Use the numbers 0 and 2 to divide the number line into intervals as shown.

Try test numbers in each interval.

A: Test -1, $\dfrac{-1}{-1-2} = \dfrac{-1}{-3} = \dfrac{1}{3} \geq 0$

Interval A is part of the solution set.

B: Test 1, $\dfrac{1}{1-2} = \dfrac{1}{-1} = -1 \not\geq 0$

Interval B is not part of the solution set.

C: Test 3, $\dfrac{3}{3-2} = \dfrac{3}{1} = 3 \geq 0$

The interval C is part of the solution set.

The solution set includes intervals A and C. The number 0 is also included since the inequality symbol is $\geq$ and 0 is the solution of the related equation. The number 2 is not included since $\dfrac{x}{x-2}$ is undefined for $x = 2$. The solution set is $\{x | x \leq 0 \text{ or } x > 2\}$, or $(-\infty, 0] \cup (2, \infty)$.

30. $\{x | -3 \leq x < 0\}$, or $[-3, 0)$

31. $\dfrac{x-5}{x} < 1$

Solve the related equation.

$$\dfrac{x-5}{x} = 1$$

$$x - 5 = x$$

$$-5 = 0$$

The related equation has no solution.

Find the values that make the denominator 0.

$$x = 0$$

Use the number 0 to divide the number line into two intervals as shown.

Try test numbers in each interval.

A: Test -1, $\dfrac{-1-5}{-1} = \dfrac{-6}{-1} = 6 \not< 1$

Interval A is not part of the solution set.

B: Test 1, $\dfrac{1-5}{1} = \dfrac{-4}{1} = -4 < 1$

Interval B is part of the solution set.

The solution set is $\{x | x > 0\}$, or $(0, \infty)$.

32. $\{x|1 < x < 2\}$, or $(1, 2)$

33. $\dfrac{x-1}{(x-3)(x+4)} < 0$

Solve the related equation.

$$\dfrac{x-1}{(x-3)(x+4)} = 0$$

$$x - 1 = 0$$

$$x = 1$$

Find the values that make the denominator 0.

$$(x - 3)(x + 4) = 0$$

$$x = 3 \text{ or } x = -4$$

Use the numbers 1, 3, and -4 to divide the number line into intervals as shown:

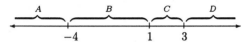

Try test numbers in each interval.

A: Test -5, $\dfrac{-5-1}{(-5-3)(-5+4)} = \dfrac{-6}{-8(-1)} =$

$$-\dfrac{3}{4} < 0$$

Interval A is part of the solution set.

B: Test 0, $\dfrac{0-1}{(0-3)(0+4)} = \dfrac{-1}{-3\cdot 4} = \dfrac{1}{12} \not< 0$

Interval B is not part of the solution set.

C: Test 2, $\dfrac{2-1}{(2-3)(2+4)} = \dfrac{1}{-1\cdot 6} = -\dfrac{1}{6} < 0$

Interval C is part of the solution set.

D: Test 4, $\dfrac{4-1}{(4-3)(4+4)} = \dfrac{3}{1\cdot 8} = \dfrac{3}{8} \not< 0$

Interval D is not part of the solution set.

The solution set is $\{x|x < -4 \text{ or } 1 < x < 3\}$, or $(-\infty, -4)\cup(1, 3)$.

34. $\{x| -7 < x < -2 \text{ or } x > 2\}$, or $(-7, -2) \cup (2, \infty)$

35. $2 < \dfrac{1}{x}$

Solve the related equation.

$$2 = \dfrac{1}{x}$$

$$x = \dfrac{1}{2}$$

Find the values that make the denominator 0.

$$x = 0$$

Use the numbers $\dfrac{1}{2}$ and 0 to divide the number line into intervals as shown.

Try test numbers in each interval.

A: Test -1, $\dfrac{1}{-1} = -1 \not> 2$

Interval A is not part of the solution set.

B: Test $\dfrac{1}{4}$, $\dfrac{1}{\frac{1}{4}} = 4 > 2$

Interval B is part of the solution set.

C: Test 1, $\dfrac{1}{1} = 1 \not> 2$

Interval C is not part of the solution set.

The solution set is $\left\{x\middle|0 < x < \dfrac{1}{2}\right\}$, or $\left(0, \dfrac{1}{2}\right)$.

36. $\left\{x\middle|x < 0 \text{ or } x \geq \dfrac{1}{3}\right\}$, or $(-\infty, 0) \cup \left[\dfrac{1}{3}, \infty\right)$

37. $\sqrt[5]{a^2 b}\sqrt[3]{ab^2} = (a^2 b)^{1/5}(ab^2)^{1/3}$ Converting to
exponential notation

$$= a^{2/5}b^{1/5}a^{1/3}b^{2/3}$$

$$= a^{2/5+1/3}b^{1/5+2/3}$$

$$= a^{11/15}b^{13/15}$$

$$= (a^{11}b^{13})^{1/15}$$

$$= \sqrt[15]{a^{11}b^{13}}$$ Converting back
to radical notation

38. 6

39.

40. $\{x|x < -1 - \sqrt{5} \text{ or } x > -1 + \sqrt{5}\}$, or $(-\infty, -1 - \sqrt{5}) \cup (-1 + \sqrt{5}, \infty)$

41. $x^4 + 2x^2 \geq 0$

$$x^2(x^2 + 2) \geq 0$$

$x^2 > 0$ for all values of x, and $x^2 + 2 > 0$ for all values of x. Then $x^2(x^2 + 2) > 0$ for all values of x. The solution set is the set of all real numbers.

42. $\{0\}$

43. $\left|\dfrac{x+2}{x-1}\right| < 3$

$$-3 < \dfrac{x+2}{x-1} < 3$$

First we solve $-3 < \dfrac{x+2}{x-1}$, or $\dfrac{x+2}{x-1} > -3$

Solve the related equation.

$$\dfrac{x+2}{x-1} = -3$$

$$x + 2 = -3x + 3$$

$$4x = 1$$

$$x = \dfrac{1}{4}$$

Find the values that make the denominator 0.

$$x - 1 = 0$$

$$x = 1$$

Use the numbers $\dfrac{1}{4}$ and 1 to divide the number line into intervals as shown:

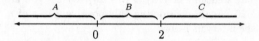

Try test numbers in each interval.

A: Test 0, $\dfrac{0+2}{0-1} = \dfrac{2}{-1} = -2 > -3$

Interval A is in the solution set.

B: Test $\dfrac{1}{2}$, $\dfrac{\frac{1}{2}+2}{\frac{1}{2}-1} = \dfrac{\frac{5}{2}}{-\frac{1}{2}} = -5 \not> -3$

Interval B is not in the solution set.

C: Test 2, $\dfrac{2+2}{2-1} = \dfrac{4}{1} = 4 > -3$

Interval C is in the solution set.

The solution set for this portion of the compound inequality is $\left\{ x \middle| x < \dfrac{1}{4} \ \text{ or } \ x > 1 \right\}$.

Now we solve $\dfrac{x+2}{x-1} < 3$.

Solve the related equation.

$$\dfrac{x+2}{x-1} = 3$$
$$x + 2 = 3x - 3$$
$$5 = 2x$$
$$\dfrac{5}{2} = x$$

As above, the denominator is 0 for $x = 1$.

Use the numbers $\dfrac{5}{2}$ and 1 to divide the number line into intervals as shown:

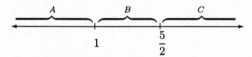

Try test numbers in each interval.

A: Test 0, $\dfrac{0+2}{0-1} = \dfrac{2}{-1} = -2 < 3$

Interval A is in the solution set.

B: Test 2, $\dfrac{2+2}{2-1} = \dfrac{4}{1} = 4 \not< 3$

Interval B is not in the solution set.

C: Test 3, $\dfrac{3+2}{3-1} = \dfrac{5}{2} < 3$

Interval C is in the solution set.

The solution set of for this portion of the compound inequality is $\left\{ x \middle| x < 1 \ \text{ or } \ x > \dfrac{5}{2} \right\}$.

The solution set for the original inequality is

$\left\{ x \middle| x < \dfrac{1}{4} \ \text{or} \ x > 1 \right\} \cap \left\{ x \middle| x < 1 \ \text{or} \ x > \dfrac{5}{2} \right\}$, or

$\left\{ x \middle| x < \dfrac{1}{4} \ \text{or} \ x > \dfrac{5}{2} \right\}$.

44. a) The company makes a profit for values of x such that $10 < x < 200$, or for values of x in the interval $(10, 200)$.

b) The company loses money for values of x such that $0 \le x < 10$ or $x > 200$, or for values of x in the interval $[0, 10) \cup (200, \infty)$.

45. a) $-16t^2 + 32t + 1920 > 1920$
$$-16t^2 + 32t > 0$$
$$t^2 - 2t < 0$$
$$t(t-2) < 0$$

The solutions of $f(t) = t(t-2) = 0$ are 0 and 2. They divide the number line into three intervals as shown:

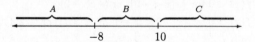

Try test numbers in each interval.

A: Test -1, $f(-1) = -1(-1-2) = 3$

B: Test 1, $f(1) = 1(1-2) = -1$

C: Test 3, $f(3) = 3(3-2) = 3$

The height is greater than 1920 ft for $\{t | 0 < t < 2\}$, or for values of t in the interval $(0, 2)$.

b) $-16t^2 + 32t + 1920 < 640$
$$-16t^2 + 32t + 1280 < 0$$
$$t^2 - 2t - 80 > 0$$
$$(t-10)(t+8) > 0$$

The solutions of $f(t) = (t-10)(t+8) = 0$ are 10 and -8. They divide the number line into three intervals as shown:

Try test numbers in each interval.

A: Test -10, $f(-10) = (-10-10)(-10+8) = 40$

B: Test 0, $f(0) = (0-10)(0+8) = -80$

C: Test 20, $f(20) = (20-10)(20+8) = 80 = 280$

Since negative values of t have no meaning in this problem, the solution set is $\{t | t > 10\}$, or $(10, \infty)$.

46. $\{n | 12 \le n \le 25\}$, or $[12, 25]$

47. Find the values of n for which $D \ge 27$ and $D \le 230$.

For $D \ge 27$:
$$\dfrac{n(n-3)}{2} \ge 27$$
$$n(n-3) \ge 54$$
$$n^2 - 3n - 54 \ge 0$$
$$(n-9)(n+6) \ge 0$$

The solutions of $f(n) = (n-9)(n+6) = 0$ are 9 and -6. Considering only positive values of n, we use the intervals below:

54.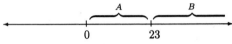

A: Test 1, $f(1) = 1^2 - 3 \cdot 1 - 54 = -56$

B: Test 10, $f(10) = 10^2 - 3 \cdot 10 - 54 = 16$

$D(n) \geq 27$ for $\{n | n \geq 9\}$.

For $D \leq 230$:

$$\frac{n(n-3)}{2} \leq 230$$

$$n(n-3) \leq 460$$

$$n^2 - 3n - 460 \leq 0$$

$$(n-23)(n+20) \leq 0$$

The solutions of $f(n) = (n-23)(n+20) = 0$ are 23 and -20. Again considering only positive values of n, we use the intervals below:

A: Test 1, $f(1) = 1^2 - 3 \cdot 1 - 460 = -462$

B: Test 25, $f(25) = 25^2 - 3 \cdot 25 - 460 = 90$

$D(n) \leq 230$ (and $n > 0$) for $\{n | 0 < n \leq 23\}$.

Then $27 \leq D \leq 230$ for $\{n | 9 \leq n \leq 23\}$, or for all values of n in the interval $[9, 23]$.

48. The solutions of $f(x) = 0$ are -2, 1, and 3.

The solution of $f(x) < 0$ is $\{x | x < -2 \ or \ 1 < x < 3\}$, or $(-\infty, -2) \cup (1, 3)$.

The solution of $f(x) > 0$ is $\{x | -2 < x < 1 \ or \ x > 3\}$, or $(-2, 1) \cup (3, \infty)$.

49. $f(x) = 0$ for $x = -2$ or $x = 1$;

$f(x) < 0$ for $\{x | x < -2\}$, or $(-\infty, -2)$;

$f(x) > 0$ for $\{x | -2 < x < 1 \ or \ x > 1\}$, or $(-2, 1) \cup (1, \infty)$

50. $f(x)$ has no zeros.

The solutions $f(x) < 0$ are $\{x | x < 0\}$, or $(-\infty, 0)$.

The solutions of $f(x) > 0$ are $\{x | x > 0\}$, or $(0, \infty)$.

51. $f(x) = 0$ for $x = 0$ or $x = 1$;

$f(x) < 0$ for $\{x | 0 < x < 1\}$, or $(0, 1)$;

$f(x) > 0$ for $\{x | x > 1\}$, or $(1, \infty)$

52. The solutions of $f(x) = 0$ are -2, 1, 2, and 3.

The solutions of $f(x) < 0$ are $\{x | -2 < x < 1 \ or \ 2 < x < 3\}$, or $(-2, 1) \cup (2, 3)$.

The solutions of $f(x) > 0$ are $\{x | x < -2 \ or \ 1 < x < 2 \ or \ x > 3\}$, or $(-\infty, -2) \cup (1, 2) \cup (3, \infty)$.

53. $f(x) = 0$ for $x = -2$, $x = 0$, or $x = 1$;

$f(x) < 0$ for $\{x | x < -3 \ or \ -2 < x < 0 \ or \ 1 < x < 2\}$, or $(-\infty, -3) \cup (-2, 0) \cup (1, 2)$;

$f(x) > 0$ for $\{x | -3 < x < -2 \ or \ 0 < x < 1 \ or \ x > 2\}$, or $(-3, -2) \cup (0, 1) \cup (2, \infty)$

Chapter 12

Exponential and Logarithmic Functions

Exercise Set 12.1

1. Graph: $y = 2^x$

We compute some function values, thinking of y as $f(x)$, and keep the results in a table.

$f(0) = 2^0 = 1$

$f(1) = 2^1 = 2$

$f(2) = 2^2 = 4$

$f(-1) = 2^{-1} = \dfrac{1}{2^1} = \dfrac{1}{2}$

$f(-2) = 2^{-2} = \dfrac{1}{2^2} = \dfrac{1}{4}$

x	y, or $f(x)$
0	1
1	2
2	4
-1	$\dfrac{1}{2}$
-2	$\dfrac{1}{4}$

Next we plot these points and connect them with a smooth curve.

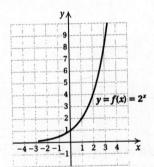

2.

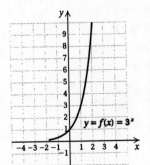

3. Graph: $y = 5^x$

We compute some function values, thinking of y as $f(x)$, and keep the results in a table.

$f(0) = 5^0 = 1$

$f(1) = 5^1 = 5$

$f(2) = 5^2 = 25$

$f(-1) = 5^{-1} = \dfrac{1}{5^1} = \dfrac{1}{5}$

$f(-2) = 5^{-2} = \dfrac{1}{5^2} = \dfrac{1}{25}$

x	y, or $f(x)$
0	1
1	5
2	25
-1	$\dfrac{1}{5}$
-2	$\dfrac{1}{25}$

Next we plot these points and connect them with a smooth curve.

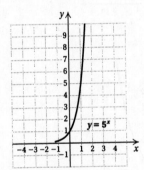

4.

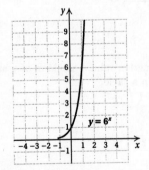

5. Graph: $y = 2^{x+1}$

We compute some function values, thinking of y as $f(x)$, and keep the results in a table.

$f(0) = 2^{0+1} = 2^1 = 2$

$f(-1) = 2^{-1+1} = 2^0 = 1$

$f(-2) = 2^{-2+1} = 2^{-1} = \dfrac{1}{2^1} = \dfrac{1}{2}$

$f(-3) = 2^{-3+1} = 2^{-2} = \dfrac{1}{2^2} = \dfrac{1}{4}$

$f(1) = 2^{1+1} = 2^2 = 4$

$f(2) = 2^{2+1} = 2^3 = 8$

x	y, or $f(x)$
0	2
-1	1
-2	$\dfrac{1}{2}$
-3	$\dfrac{1}{4}$
1	4
2	8

x	y, or $f(x)$
0	$\dfrac{1}{9}$
1	$\dfrac{1}{3}$
2	1
3	3
4	9
-1	$\dfrac{1}{27}$
-2	$\dfrac{1}{81}$

Next we plot these points and connect them with a smooth curve.

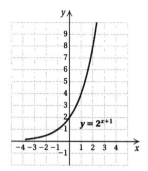

Next we plot these points and connect them with a smooth curve.

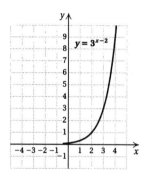

6.

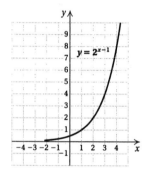

8.

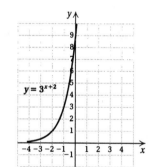

7. Graph: $y = 3^{x-2}$

We compute some function values, thinking of y as $f(x)$, and keep the results in a table.

$f(0) = 3^{0-2} = 3^{-2} = \dfrac{1}{3^2} = \dfrac{1}{9}$

$f(1) = 3^{1-2} = 3^{-1} = \dfrac{1}{3^1} = \dfrac{1}{3}$

$f(2) = 3^{2-2} = 3^0 = 1$

$f(3) = 3^{3-2} = 3^1 = 3$

$f(4) = 3^{4-2} = 3^2 = 9$

$f(-1) = 3^{-1-2} = 3^{-3} = \dfrac{1}{3^3} = \dfrac{1}{27}$

$f(-2) = 3^{-2-2} = 3^{-4} = \dfrac{1}{3^4} = \dfrac{1}{81}$

9. Graph: $y = 2^x - 3$

We construct a table of values, thinking of y as $f(x)$. Then we plot the points and connect them with a smooth curve.

$f(0) = 2^0 - 3 = 1 - 3 = -2$

$f(1) = 2^1 - 3 = 2 - 3 = -1$

$f(2) = 2^2 - 3 = 4 - 3 = 1$

$f(3) = 2^3 - 3 = 8 - 3 = 5$

$f(-1) = 2^{-1} - 3 = \dfrac{1}{2} - 3 = -\dfrac{5}{2}$

$f(-2) = 2^{-2} - 3 = \dfrac{1}{4} - 3 = -\dfrac{11}{4}$

x	y, or $f(x)$
0	-2
1	-1
2	1
3	5
-1	$-\dfrac{5}{2}$
-2	$-\dfrac{11}{4}$

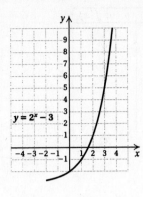

$y = 2^x - 3$

12.

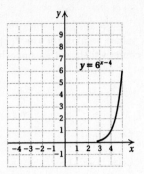

$y = 6^{x-4}$

10.

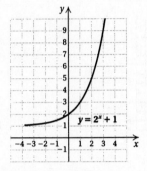

$y = 2^x + 1$

13. Graph: $y = \left(\dfrac{1}{2}\right)^x$

We construct a table of values, thinking of y as $f(x)$. Then we plot the points and connect them with a smooth curve.

$f(0) = \left(\dfrac{1}{2}\right)^0 = 1$

$f(1) = \left(\dfrac{1}{2}\right)^1 = \dfrac{1}{2}$

$f(2) = \left(\dfrac{1}{2}\right)^2 = \dfrac{1}{4}$

$f(3) = \left(\dfrac{1}{2}\right)^3 = \dfrac{1}{8}$

$f(-1) = \left(\dfrac{1}{2}\right)^{-1} = \dfrac{1}{\left(\dfrac{1}{2}\right)^1} = \dfrac{1}{\dfrac{1}{2}} = 2$

$f(-2) = \left(\dfrac{1}{2}\right)^{-2} = \dfrac{1}{\left(\dfrac{1}{2}\right)^2} = \dfrac{1}{\dfrac{1}{4}} = 4$

$f(-3) = \left(\dfrac{1}{2}\right)^{-3} = \dfrac{1}{\left(\dfrac{1}{2}\right)^3} = \dfrac{1}{\dfrac{1}{8}} = 8$

11. Graph: $y = 5^{x+3}$

We construct a table of values, thinking of y as $f(x)$. Then we plot the points and connect them with a smooth curve.

$f(0) = 5^{0+3} = 5^3 = 125$

$f(-1) = 5^{-1+3} = 5^2 = 25$

$f(-2) = 5^{-2+3} = 5^1 = 5$

$f(-3) = 5^{-3+3} = 5^0 = 1$

$f(-4) = 5^{-4+3} = 5^{-1} = \dfrac{1}{5}$

$f(-5) = 5^{-5+3} = 5^{-2} = \dfrac{1}{25}$

x	y, or $f(x)$
0	125
-1	25
-2	5
-3	1
-4	$\dfrac{1}{5}$
-5	$\dfrac{1}{25}$

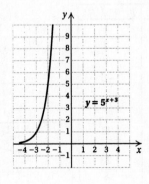

$y = 5^{x+3}$

x	y, or $f(x)$
0	1
1	$\dfrac{1}{2}$
2	$\dfrac{1}{4}$
3	$\dfrac{1}{8}$
-1	2
-2	4
-3	8

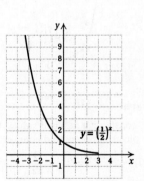

$y = \left(\dfrac{1}{2}\right)^x$

14.

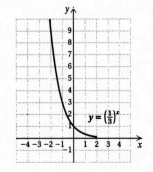

$y = \left(\dfrac{1}{3}\right)^x$

15. Graph: $y = \left(\dfrac{1}{5}\right)^x$

We construct a table of values, thinking of y as $f(x)$. Then we plot the points and connect them with a smooth curve.

$f(0) = \left(\dfrac{1}{5}\right)^0 = 1$

$f(1) = \left(\dfrac{1}{5}\right)^1 = \dfrac{1}{5}$

$f(2) = \left(\dfrac{1}{5}\right)^2 = \dfrac{1}{25}$

$f(-1) = \left(\dfrac{1}{5}\right)^{-1} = \dfrac{1}{\frac{1}{5}} = 5$

$f(-2) = \left(\dfrac{1}{5}\right)^{-2} = \dfrac{1}{\frac{1}{25}} = 25$

x	y, or $f(x)$
0	1
1	$\dfrac{1}{5}$
2	$\dfrac{1}{25}$
-1	5
-2	25

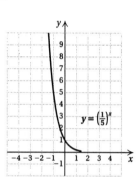

16.

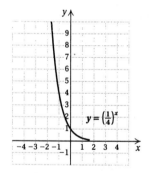

17. Graph: $y = 2^{2x-1}$

We construct a table of values, thinking of y as $f(x)$. Then we plot the points and connect them with a smooth curve.

$f(0) = 2^{2 \cdot 0 - 1} = 2^{-1} = \dfrac{1}{2}$

$f(1) = 2^{2 \cdot 1 - 1} = 2^1 = 2$

$f(2) = 2^{2 \cdot 2 - 1} = 2^3 = 8$

$f(-1) = 2^{2(-1)-1} = 2^{-3} = \dfrac{1}{8}$

$f(-2) = 2^{2(-2)-1} = 2^{-5} = \dfrac{1}{32}$

x	y, or $f(x)$
0	$\dfrac{1}{2}$
1	2
2	8
-1	$\dfrac{1}{8}$
-2	$\dfrac{1}{32}$

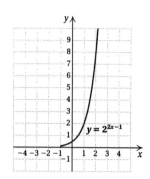

18.

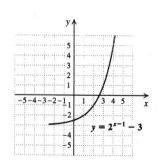

19. Graph: $y = 2^{x-1} - 3$

We construct a table of values, thinking of y as $f(x)$. Then we plot the points and connect them with a smooth curve.

$f(0) = 2^{0-1} - 3 = 2^{-1} - 3 = \dfrac{1}{2} - 3 = -\dfrac{5}{2}$

$f(1) = 2^{1-1} - 3 = 2^0 - 3 = 1 - 3 = -2$

$f(2) = 2^{2-1} - 3 = 2^1 - 3 = 2 - 3 = -1$

$f(3) = 2^{3-1} - 3 = 2^2 - 3 = 4 - 3 = 1$

$f(4) = 2^{4-1} - 3 = 2^3 - 3 = 8 - 3 = 5$

$f(-1) = 2^{-1-1} - 3 = 2^{-2} - 3 = \dfrac{1}{4} - 3 = -\dfrac{11}{4}$

$f(-2) = 2^{-2-1} - 3 = 2^{-3} - 3 = \dfrac{1}{8} - 3 = -\dfrac{23}{8}$

x	y, or $f(x)$
0	$-\dfrac{5}{2}$
1	-2
2	-1
3	1
4	5
-1	$-\dfrac{11}{4}$
-2	$-\dfrac{23}{8}$

20.

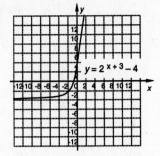

$y = 2^{x+3} - 4$

22.

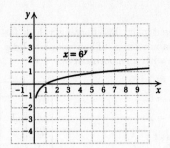

$x = 6^y$

21. Graph: $x = 3^y$

We can find ordered pairs by choosing values for y and then computing values for x.

For $y = 0$, $x = 3^0 = 1$.

For $y = 1$, $x = 3^1 = 3$.

For $y = 2$, $x = 3^2 = 9$.

For $y = 3$, $x = 3^3 = 27$.

For $y = -1$, $x = 3^{-1} = \dfrac{1}{3^1} = \dfrac{1}{3}$.

For $y = -2$, $x = 3^{-2} = \dfrac{1}{3^2} = \dfrac{1}{9}$.

For $y = -3$, $x = 3^{-3} = \dfrac{1}{3^3} = \dfrac{1}{27}$.

x	y
1	0
3	1
9	2
27	3
$\dfrac{1}{3}$	-1
$\dfrac{1}{9}$	-2
$\dfrac{1}{27}$	-3

↑ ↳ (1) Choose values for y.

— (2) Compute values for x.

We plot the points and connect them with a smooth curve.

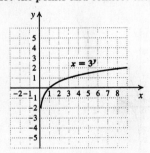

$x = 3^y$

23. Graph: $x = \left(\dfrac{1}{2}\right)^y$

We can find ordered pairs by choosing values for y and then computing values for x. Then we plot these points and connect them with a smooth curve.

For $y = 0$, $x = \left(\dfrac{1}{2}\right)^0 = 1$.

For $y = 1$, $x = \left(\dfrac{1}{2}\right)^1 = \dfrac{1}{2}$.

For $y = 2$, $x = \left(\dfrac{1}{2}\right)^2 = \dfrac{1}{4}$.

For $y = 3$, $x = \left(\dfrac{1}{2}\right)^3 = \dfrac{1}{8}$.

For $y = -1$, $x = \left(\dfrac{1}{2}\right)^{-1} = \dfrac{1}{\frac{1}{2}} = 2$.

For $y = -2$, $x = \left(\dfrac{1}{2}\right)^{-2} = \dfrac{1}{\frac{1}{4}} = 4$.

For $y = -3$, $x = \left(\dfrac{1}{2}\right)^{-3} = \dfrac{1}{\frac{1}{8}} = 8$.

x	y
1	0
$\dfrac{1}{2}$	1
$\dfrac{1}{4}$	2
$\dfrac{1}{8}$	3
2	-1
4	-2
8	-3

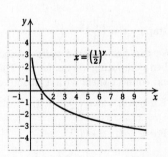

$x = \left(\dfrac{1}{2}\right)^y$

24.

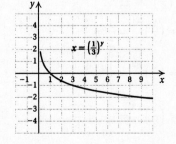

$x = \left(\dfrac{1}{3}\right)^y$

25. Graph: $x = 5^y$

We can find ordered pairs by choosing values for y and then computing values for x. Then we plot these points and connect them with a smooth curve.

For $y = 0$, $x = 5^0 = 1$.

For $y = 1$, $x = 5^1 = 5$.

For $y = 2$, $x = 5^2 = 25$.

For $y = -1$, $x = 5^{-1} = \dfrac{1}{5}$.

For $y = -2$, $x = 5^{-2} = \dfrac{1}{25}$.

x	y
1	0
5	1
25	2
$\dfrac{1}{5}$	-1
$\dfrac{1}{25}$	-2

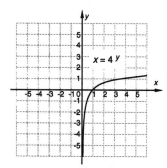

26.

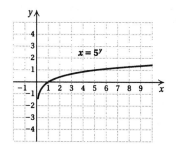

27. Graph: $x = \left(\dfrac{2}{3}\right)^y$

We can find ordered pairs by choosing values for y and then computing values for x. Then we plot these points and connect them with a smooth curve.

For $y = 0$, $x = \left(\dfrac{2}{3}\right)^0 = 1$.

For $y = 1$, $x = \left(\dfrac{2}{3}\right)^1 = \dfrac{2}{3}$.

For $y = 2$, $x = \left(\dfrac{2}{3}\right)^2 = \dfrac{4}{9}$.

For $y = 3$, $x = \left(\dfrac{2}{3}\right)^3 = \dfrac{8}{27}$.

For $y = -1$, $x = \left(\dfrac{2}{3}\right)^{-1} = \dfrac{1}{\frac{2}{3}} = \dfrac{3}{2}$.

For $y = -2$, $x = \left(\dfrac{2}{3}\right)^{-2} = \dfrac{1}{\frac{4}{9}} = \dfrac{9}{4}$.

For $y = -3$, $x = \left(\dfrac{2}{3}\right)^{-3} = \dfrac{1}{\frac{8}{27}} = \dfrac{27}{8}$.

x	y
1	0
$\dfrac{2}{3}$	1
$\dfrac{4}{9}$	2
$\dfrac{8}{27}$	3
$\dfrac{3}{2}$	-1
$\dfrac{9}{4}$	-2
$\dfrac{27}{8}$	-3

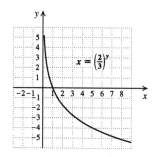

28.

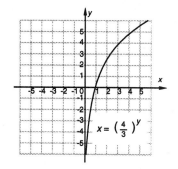

29. Graph $y = 2^x$ (see Exercise 1) and $x = 2^y$ (see Example 4) using the same set of axes.

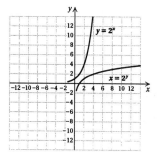

30.

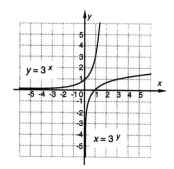

31. Graph $y = \left(\frac{1}{2}\right)^x$ (see Exercise 13) and $x = \left(\frac{1}{2}\right)^y$ (see Exercise 23) using the same set of axes.

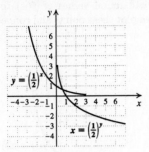

32.

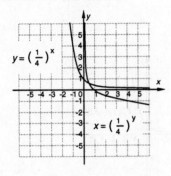

33. Keep in mind that t represents the number of years after 1989.

a) For 1993, $t = 1993 - 1989$, or 4:

$$N(4) = 100,000(1.4)^4$$
$$= 384,160$$

b) For 1998, $t = 1998 - 1989$, or 9:

$$N(9) = 100,000(1.4)^9$$
$$= 2,066,105$$

c) We use the function values computed in parts (a) and (b), and others if we wish, to draw the graph. Note that the axes are scaled differently because of the large numbers.

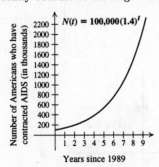

34. a) 4243; 6000; 8485; 12,000; 24,000

b)

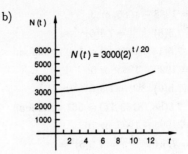

35. a) Substitute for t.

$$N(0) = 250,000\left(\frac{2}{3}\right)^0 = 250,000 \cdot 1 = 250,000;$$

$$N(1) = 250,000\left(\frac{2}{3}\right)^1 = 250,000 \cdot \frac{2}{3} = 166,667;$$

$$N(4) = 250,000\left(\frac{2}{3}\right)^4 = 250,000 \cdot \frac{16}{81} \approx 49,383;$$

$$N(10) = 250,000\left(\frac{2}{3}\right)^{10} = 250,000 \cdot \frac{1024}{59,049} \approx 4335$$

b) We use the function values computed in part (a) to draw the graph of the function. Note that the axes are scaled differently because of the large function values.

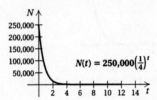

36. a) \$5200; \$4160; \$3328; \$1703.94; \$558.35

b)

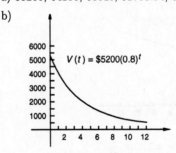

37. a) Keep in mind that t represents the number of years after 1985 and that N is given in millions.

For 1985, $t = 0$:

$$N(0) = 7.5(6)^{0.5(0)} = 7.5(6)^0 = 7.5(1) =$$
$$7.5 \text{ million};$$

For 1986, $t = 1$:

$$N(1) = 7.5(6)^{0.5(1)} = 7.5(6)^{0.5} \approx$$
$$7.5(2.449489743) \approx 18.4 \text{ million};$$

For 1988, $t = 1988 - 1985$, or 3:
$$N(3) = 7.5(6)^{0.5(3)} = 7.5(6)^{1.5} \approx$$
$$7.5(14.69693846) \approx 110.2 \text{ million};$$

For 1990, $t = 1990 - 1985$, or 5:
$$N(5) = 7.5(6)^{0.5(5)} = 7.5(6)^{2.5} \approx$$
$$7.5(88.18163074) \approx 661.4 \text{ million};$$

For 1995, $t = 1995 - 1985$, or 10:
$$N(10) = 7.5(6)^{0.5(10)} = 7.5(6)^{5} =$$
$$7.5(7776) = 58,320 \text{ million};$$

For 2000, $t = 2000 - 1985$, or 15:
$$N(15) = 7.5(6)^{0.5(15)} = 7.5(6)^{7.5} \approx$$
$$7.5(685,700.3606) \approx$$
$$5,142,752.7 \text{ million};$$

b) Use the function values compared in part (a) to draw the graph of the function.

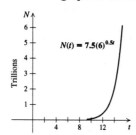

38. a) 2.6 lb; 3.7 lb; 10.9 lb; 14.6 lb; 22.6 lb

    b) 22.6 lb

    c)

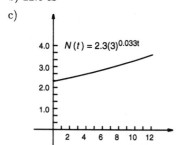

39. $x^{-5} \cdot x^{3} = x^{-5+3} = x^{-2}$, or $\dfrac{1}{x^{2}}$

40. x^{-12}, or $\dfrac{1}{x^{12}}$

41. $\dfrac{x^{-3}}{x^{4}} = x^{-3-4} = x^{-7}$, or $\dfrac{1}{x^{7}}$

42. 1

43.

44.

45. Since the bases are the same, the one with the larger exponent is the larger number. Thus $\pi^{2.4}$ is larger.

46. $8^{\sqrt{3}}$

47. Graph: $f(x) = (2.3)^{x}$

Use a calculator with a power key to construct a table of values. (We will round values of $f(x)$ to the nearest hundredth.) Then plot these points and connect them with a smooth curve.

x	y
0	1
1	2.3
2	5.29
3	12.17
-1	0.43
-2	0.19

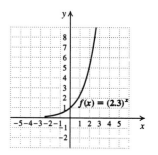

48.

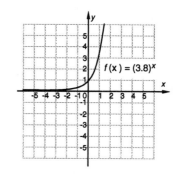

49. Graph: $g(x) = (0.125)^{x}$

Use the procedure described in Exercise 47, rounding values of $g(x)$ to the nearest thousandth.

x	y
0	1
1	0.125
2	0.016
3	0.002
-1	8
-2	64

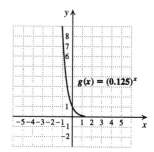

50.

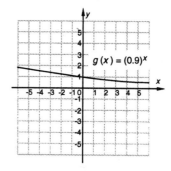

51. Graph: $y = 2^{x} + 2^{-x}$

Construct a table of values, thinking of y as $f(x)$. Then plot these points and connect them with a curve.

$f(0) = 2^0 + 2^{-0} = 1 + 1 = 2$

$f(1) = 2^1 + 2^{-1} = 2 + \dfrac{1}{2} = 2\dfrac{1}{2}$

$f(2) = 2^2 + 2^{-2} = 4 + \dfrac{1}{4} = 4\dfrac{1}{4}$

$f(3) = 2^3 + 2^{-3} = 8 + \dfrac{1}{8} = 8\dfrac{1}{8}$

$f(-1) = 2^{-1} + 2^{-(-1)} = \dfrac{1}{2} + 2 = 2\dfrac{1}{2}$

$f(-2) = 2^{-2} + 2^{-(-2)} = \dfrac{1}{4} + 4 = 4\dfrac{1}{4}$

$f(-3) = 2^{-3} + 2^{-(-3)} = \dfrac{1}{8} + 8 = 8\dfrac{1}{8}$

x	y, or $f(x)$
0	2
1	$2\dfrac{1}{2}$
2	$4\dfrac{1}{4}$
3	$8\dfrac{1}{8}$
-1	$2\dfrac{1}{2}$
-2	$4\dfrac{1}{4}$
-3	$8\dfrac{1}{8}$

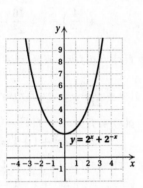

52.

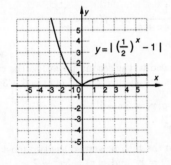

53. Graph: $y = 3^x + 3^{-x}$

We construct a table of values, thinking of y as $f(x)$. Then plot these points and connect them with a curve.

$f(0) = 3^0 + 3^{-0} = 1 + 1 = 2$

$f(1) = 3^1 + 3^{-1} = 3 + \dfrac{1}{3} = 3\dfrac{1}{3}$

$f(2) = 3^2 + 3^{-2} = 9 + \dfrac{1}{9} = 9\dfrac{1}{9}$

$f(-1) = 3^{-1} + 3^{-(-1)} = \dfrac{1}{3} + 3 = 3\dfrac{1}{3}$

$f(-2) = 3^{-2} + 3^{-(-2)} = \dfrac{1}{9} + 9 = 9\dfrac{1}{9}$

x	y, or $f(x)$
0	2
1	$3\dfrac{1}{3}$
2	$9\dfrac{1}{9}$
-1	$3\dfrac{1}{3}$
-2	$9\dfrac{1}{9}$

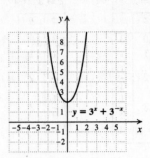

54.

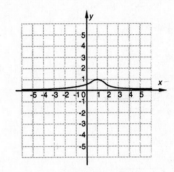

55. Graph: $y = |2x^2 - 1|$

We construct a table of values, thinking of y as $f(x)$. Then we plot these points and connect them with a curve.

$f(0) = |2^{0^2} - 1| = |1 - 1| = 0$

$f(1) = |2^{1^2} - 1| = |2 - 1| = 1$

$f(2) = |2^{2^2} - 1| = |16 - 1| = 15$

$f(-1) = |2^{(-1)^2} - 1| = |2 - 1| = 1$

$f(-2) = |2^{(-2)^2} - 1| = |16 - 1| = 15$

x	y, or $f(x)$
0	0
1	1
2	15
-1	1
-2	15

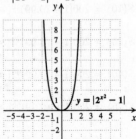

56.

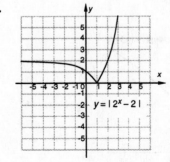

57. $y = 3^{-(x-1)}$ $x = 3^{-(y-1)}$

x	y
0	3
1	1
2	$\dfrac{1}{3}$
3	$\dfrac{1}{9}$
−1	9

x	y
3	0
1	1
$\dfrac{1}{3}$	2
$\dfrac{1}{9}$	3
9	−1

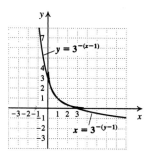

58.

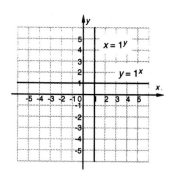

59. a) $S(10) = 200[1 - (0.99)^{10}] \approx 19$ words per minute;

  $S(20) = 200[1 - (0.99)^{20}] \approx 36$ words per minute;

  $S(40) = 200[1 - (0.99)^{40}] \approx 66$ words per minute;

  $S(85) = 200[1 - (0.99)^{85}] \approx 115$ words per minute

b) Use the function values calculated in part (a) to draw the graph.

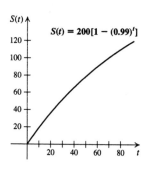

Exercise Set 12.2

1. $f \circ g(x) = f(g(x)) = f(2x - 1) =$
$$3(2x - 1)^2 + 2 =$$
$$3(4x^2 - 4x + 1) + 2 =$$
$$12x^2 - 12x + 3 + 2 = 12x^2 - 12x + 5$$

  $g \circ f(x) = g(f(x)) = g(3x^2 + 2) =$
$$2(3x^2 + 2) - 1 = 6x^2 + 4 - 1 =$$
$$6x^2 + 3$$

2. $f \circ g(x) = 8x^2 - 17$; $g \circ f(x) = 32x^2 + 48x + 13$

3.
$$f \circ g(x) = f(g(x)) = f\left(\frac{2}{x}\right) = 4\left(\frac{2}{x}\right)^2 - 1 =$$
$$4\left(\frac{4}{x^2}\right) - 1 = \frac{16}{x^2} - 1$$

  $g \circ f(x) = g(f(x)) = g(4x^2 - 1) = \dfrac{2}{4x^2 - 1}$

4. $f \circ g(x) = \dfrac{3}{2x^2 + 3}$; $g \circ f(x) = \dfrac{18}{x^2} + 3$

5. $f \circ g(x) = f(g(x)) = f(x^2 - 1) = (x^2 - 1)^2 + 1 =$
$$x^4 - 2x^2 + 1 + 1 = x^4 - 2x^2 + 2$$

  $g \circ f(x) = g(f(x)) = g(x^2 + 1) = (x^2 + 1)^2 - 1 =$
$$x^4 + 2x^2 + 1 - 1 = x^4 + 2x^2$$

6. $f \circ g(x) = \dfrac{1}{x^2 + 4x + 4}$; $g \circ f(x) = \dfrac{1}{x^2} + 2$

7. $h(x) = (5 - 3x)^2$

This is $5 - 3x$ raised to the second power, so the two most obvious functions are $f(x) = x^2$ and $g(x) = 5 - 3x$.

8. $f(x) = 4x^2 + 9$, $g(x) = 3x - 1$

9. $h(x) = (3x^2 - 7)^5$

This is $3x^2 - 7$ to the fifth power, so the two most obvious functions are $f(x) = x^5$ and $g(x) = 3x^2 - 7$.

10. $f(x) = \sqrt{x}$, $g(x) = 5x + 2$

11. $h(x) = \dfrac{1}{x - 1}$

This is the reciprocal of $x - 1$, so the two most obvious functions are $f(x) = \dfrac{1}{x}$ and $g(x) = x - 1$.

12. $f(x) = x + 4$, $g(x) = \dfrac{3}{x}$

13. $h(x) = \dfrac{1}{\sqrt{7x + 2}}$

This is the reciprocal of the square root of $7x + 2$. Two functions that can be used are $f(x) = \dfrac{1}{\sqrt{x}}$ and $g(x) = 7x + 2$.

14. $f(x) = \sqrt{x} - 3$, $g(x) = x - 7$

15. $h(x) = \dfrac{x^3 + 1}{x^3 - 1}$

Two functions that can be used are $f(x) = \dfrac{x + 1}{x - 1}$ and $g(x) = x^3$.

16. $f(x) = x^4$, $g(x) = \sqrt{x} + 5$

17. The graph of $f(x) = 3x - 4$ is shown below.

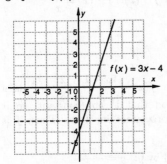

$f(x) = 3x - 4$

Since there is no horizontal line that crosses the graph more than once, the function is one-to-one.

18. Yes

19. The graph of $f(x) = x^2 - 3$ is shown below.

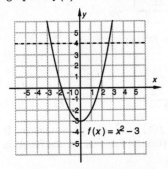

$f(x) = x^2 - 3$

There are many horizontal lines that cross the graph more than once. In particular, the line $y = 4$ crosses the graph more than once. The function is not one-to-one.

20. No

21. The graph of $g(x) = 3^x$ is shown below.

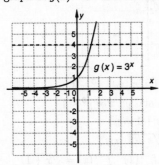

$g(x) = 3^x$

Since no horizontal line crosses the graph more than once, the function is one-to-one.

22. Yes

23. The graph of $g(x) = |x|$ is shown below.

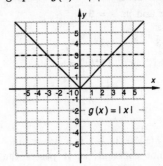

$g(x) = |x|$

There are many horizontal lines that cross the graph more than once. In particular, the line $y = 3$ crosses the graph more than once. The function is not one-to-one.

24. No

25. a) The function $f(x) = x + 4$ is a linear function that is not constant, so it passes the horizontal-line test. Thus, f is one-to-one.

  b) Replace $f(x)$ by y: $y = x + 4$

  Interchange x and y: $x = y + 4$

  Solve for y: $x - 4 = y$

  Replace y by $f^{-1}(x)$: $f^{-1}(x) = x - 4$

26. a) Yes

  b) $f^{-1}(x) = x - 7$

27. a) The function $f(x) = 5 - x$ is a linear function that is not constant, so it passes the horizontal-line test. Thus, f is one-to-one.

  b) Replace $f(x)$ by y: $y = 5 - x$

  Interchange x and y: $x = 5 - y$

  Solve for y: $y = 5 - x$

  Replace y by $f^{-1}(x)$: $f^{-1}(x) = 5 - x$

28. a) Yes

  b) $f^{-1}(x) = 9 - x$

29. a) The function $g(x) = x - 5$ is a linear function that is not constant, so it passes the horizontal-line test. Thus, g is one-to-one.

  b) Replace $g(x)$ by y: $y = x - 5$

  Interchange x and y: $x = y - 5$

  Solve for y: $x + 5 = y$

  Replace y by $g^{-1}(x)$: $g^{-1}(x) = x + 5$

30. a) Yes

  b) $g^{-1}(x) = x + 8$

31. a) The function $f(x) = 3x$ is a linear function that is not constant, so it passes the horizontal-line test. Thus, f is one-to-one.

b) Replace $f(x)$ by y: $y = 3x$

Interchange x and y: $x = 3y$

Solve for y: $\dfrac{x}{3} = y$

Replace y by $f^{-1}(x)$: $f^{-1}(x) = \dfrac{x}{3}$

32. a) Yes

b) $f^{-1}(x) = \dfrac{x}{4}$

33. a) The function $g(x) = 3x + 2$ is a linear function that is not constant, so it passes the horizontal-line test. Thus, g is one-to-one.

b) Replace $g(x)$ by y: $y = 3x + 2$

Interchange variables: $x = 3y + 2$

Solve for y: $x - 2 = 3y$

$\dfrac{x - 2}{3} = y$

Replace y by $g^{-1}(x)$: $g^{-1}(x) = \dfrac{x - 2}{3}$

34. a) Yes

b) $g^{-1}(x) = \dfrac{x - 7}{4}$

35. a) The graph of $h(x) = 7$ is shown below. The horizontal line $y = 7$ crosses the graph more than once, so the function is not one-to-one.

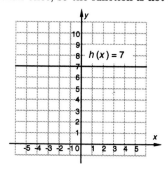

36. a) No

37. a) The graph of $f(x) = \dfrac{1}{x}$ is shown below. It passes the horizontal-line test, so the function is one-to-one.

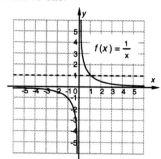

b) Replace $f(x)$ by y: $y = \dfrac{1}{x}$

Interchange x and y: $x = \dfrac{1}{y}$

Solve for y: $xy = 1$

$y = \dfrac{1}{x}$

Replace y by $f^{-1}(x)$: $f^{-1}(x) = \dfrac{1}{x}$

38. a) Yes

b) $f^{-1}(x) = \dfrac{3}{x}$

39. a) The function $f(x) = \dfrac{2x + 1}{3} = \dfrac{2}{3}x + \dfrac{1}{3}$ is a linear function that is not constant, so it passes the horizontal- line test. Thus, f is one-to-one.

b) Replace $f(x)$ by y: $y = \dfrac{2x + 1}{3}$

Interchange x and y: $x = \dfrac{2y + 1}{3}$

Solve for y: $3x = 2y + 1$

$3x - 1 = 2y$

$\dfrac{3x - 1}{2} = y$

Replace y by $f^{-1}(x)$: $f^{-1}(x) = \dfrac{3x - 1}{2}$

40. a) Yes

b) $f^{-1}(x) = \dfrac{5x - 2}{3}$

41. a) The graph of $f(x) = x^3 - 1$ is shown below. It passes the horizontal-line test, so the function is one-to-one.

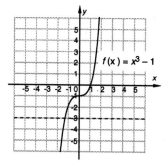

b) Replace $f(x)$ by y: $y = x^3 - 1$

Interchange x and y: $x = y^3 - 1$

Solve for y: $x + 1 = y^3$

$\sqrt[3]{x + 1} = y$

Replace y by $f^{-1}(x)$: $f^{-1}(x) = \sqrt[3]{x + 1}$

42. a) Yes

b) $f^{-1}(x) = \sqrt[3]{x - 5}$

43. a) The graph of $g(x) = (x - 2)^3$ is shown below. It passes the horizontal-line test, so the function is one-to-one.

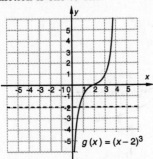

b) Replace $g(x)$ by y: $y = (x - 2)^3$

Interchange x and y: $x = (y - 2)^3$

Solve for y: $\sqrt[3]{x} = y - 2$

$\sqrt[3]{x} + 2 = y$

Replace y by $g^{-1}(x)$: $g^{-1}(x) = \sqrt[3]{x} + 2$

44. a) Yes

b) $g^{-1}(x) = \sqrt[3]{x} - 7$

45. a) The graph of $f(x) = \sqrt{x}$ is shown below. It passes the horizontal-line test, so the function is one-to-one.

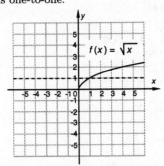

b) Replace $f(x)$ by y: $y = \sqrt{x}$ (Note that $f(x) \geq 0$.)

Interchange x and y: $x = \sqrt{y}$

Solve for y: $x^2 = y$

Replace y by $f^{-1}(x)$: $f^{-1}(x) = x^2$, $x \geq 0$

46. a) Yes

b) $f^{-1}(x) = x^2 + 1$, $x \geq 0$

47. a) The graph of $f(x) = 2x^2 + 3$, $x \geq 0$, is shown below. It passes the horizontal-line test, so the function is one-to-one.

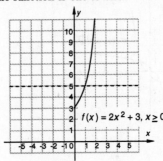

b) Replace $f(x)$ by y: $y = 2x^2 + 3$

Interchange x and y: $x = 2y^2 + 3$

Solve for y: $x - 3 = 2y^2$

$$\frac{x - 3}{2} = y^2$$

$$\sqrt{\frac{x - 3}{2}} = y$$

(We take the principal square root since $y \geq 0$.)

Replace y by $f^{-1}(x)$: $f^{-1}(x) = \sqrt{\dfrac{x - 3}{2}}$

48. a) Yes

b) $f^{-1}(x) = \sqrt{\dfrac{x + 2}{3}}$

49. First graph $f(x) = \dfrac{1}{2}x - 3$. Then graph the inverse function by reflecting the graph of $f(x) = \dfrac{1}{2}x - 3$ across the line $y = x$. The graph of the inverse function can also be found by first finding a formula for the inverse, substituting to find function values, and then plotting points.

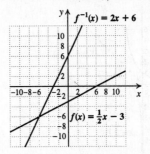

50.

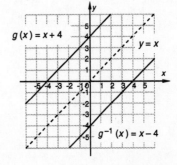

51. Follow the procedure described in Exercise 49 to graph the function and its inverse.

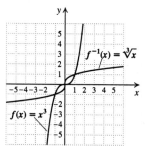

52.

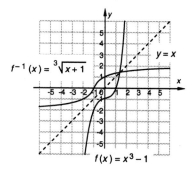

53. Use the procedure described in Exercise 49 to graph the function and its inverse.

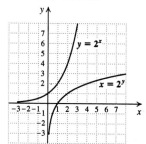

54.

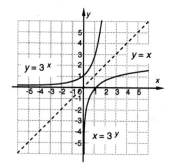

55. Use the procedure described in Exercise 49 to graph the function and its inverse.

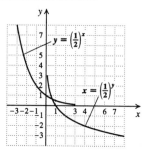

56.

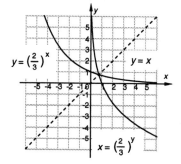

57. Use the procedure described in Exercise 49 to graph the function and its inverse.

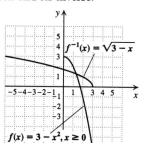

58.

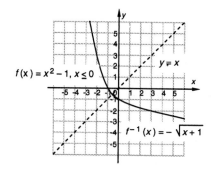

59. We check to see that $f^{-1} \circ f(x) = x$ and $f \circ f^{-1}(x) = x$.

a) $f^{-1} \circ f(x) = f^{-1}(f(x)) = f^{-1}\left(\frac{4}{5}x\right) =$

$\frac{5}{4} \cdot \frac{4}{5}x = x$

b) $f \circ f^{-1}(x) = f(f^{-1}(x)) = f\left(\frac{5}{4}x\right) =$

$\frac{4}{5} \cdot \frac{5}{4}x = x$

60. a) $f^{-1} \circ f(x) = 3\left(\dfrac{x+7}{3}\right) - 7 = x + 7 - 7 = x$

b) $f \circ f^{-1}(x) = \dfrac{(3x-7)+7}{3} = \dfrac{3x}{3} = x$

61. We check to see that $f^{-1} \circ f(x) = x$ and $f \circ f^{-1}(x) = x$.

a) $f^{-1} \circ f(x) = f^{-1}(f(x)) = f^{-1}\left(\dfrac{1-x}{x}\right) =$

$\dfrac{1}{\dfrac{1-x}{x}+1} = \dfrac{1}{\dfrac{1-x}{x}+1} \cdot \dfrac{x}{x} = \dfrac{x}{1-x+x} =$

$\dfrac{x}{1} = x$

b) $f \circ f^{-1}(x) = f(f^{-1}(x)) = f\left(\dfrac{1}{x+1}\right) =$

$\dfrac{1 - \dfrac{1}{x+1}}{\dfrac{1}{x+1}} = \dfrac{1 - \dfrac{1}{x+1}}{\dfrac{1}{x+1}} \cdot \dfrac{x+1}{x+1} =$

$\dfrac{x+1-1}{1} = \dfrac{x}{1} = x$

62. a) $f^{-1} \circ f(x) = \sqrt[3]{x^3 - 5 + 5} = \sqrt[3]{x^3} = x$

b) $f \circ f^{-1}(x) = (\sqrt[3]{x+5})^3 - 5 = x + 5 - 5 = x$

63. a) $f(8) = 8 + 32 = 40$

Size 40 in France corresponds to size 8 in the U.S.

$f(10) = 10 + 32 = 42$

Size 42 in France corresponds to size 10 in the U.S.

$f(14) = 14 + 32 = 46$

Size 46 in France corresponds to size 14 in the U.S.

$f(18) = 18 + 32 = 50$

Size 50 in France corresponds to size 18 in the U.S.

b) The function $f(x) = x + 32$ is a linear function that is not constant, so it passes the horizontal-line test. Thus, f is one-to-one and, hence, has an inverse that is a function. We now find a formula for the inverse.

Replace $f(x)$ by y: $y = x + 32$

Interchange x and y: $x = y + 32$

Solve for y: $x - 32 = y$

Replace y by $f^{-1}(x)$: $f^{-1}(x) = x - 32$

c) $f^{-1}(40) = 40 - 32 = 8$

Size 8 in the U.S. corresponds to size 40 in France.

$f^{-1}(42) = 42 - 32 = 10$

Size 10 in the U.S. corresponds to size 42 in France.

$f^{-1}(46) = 46 - 32 = 14$

Size 14 in the U.S. corresponds to size 46 in France.

$f^{-1}(50) = 50 - 32 = 18$

Size 18 in the U.S. corresponds to size 50 in France.

64. a) 40, 44, 52, 60

b) $f^{-1}(x) = \dfrac{x - 24}{2}$, or $\dfrac{x}{2} - 12$

c) 8, 10, 14, 18

65. $y = kx$

$7.2 = k(0.8)$ Substituting

$9 = k$ Variation constant

$y = 9x$ Equation of variation

66. $y = \dfrac{21.35}{x}$

67.

68.

69. From Exercise 64(b), we know that a function that converts dress sizes in Italy to those in the United States is $g(x) = \dfrac{x-24}{2}$. From Exercise 63, we know that a function that converts dress sizes in the United States to those in France is $f(x) = x + 32$. Then a function that converts dress sizes in Italy to those in France is

$h(x) = f \circ g(x)$

$h(x) = f\left(\dfrac{x-24}{2}\right)$

$h(x) = \dfrac{x-24}{2} + 32$

$h(x) = \dfrac{x}{2} - 12 + 32$

$h(x) = \dfrac{x}{2} + 20.$

70. No

71. Yes

72. Yes

73. No

Exercise Set 12.3

1. Graph: $y = \log_2 x$

The equation $y = \log_2 x$ is equivalent to $2^y = x$. We can find ordered pairs by choosing values for y and computing the corresponding x-values.

For $y = 0$, $x = 2^0 = 1$.

For $y = 1$, $x = 2^1 = 2$.

For $y = 2$, $x = 2^2 = 4$.

For $y = 3$, $x = 2^3 = 8$.

For $y = -1$, $x = 2^{-1} = \dfrac{1}{2}$.

For $y = -2$, $x = 2^{-2} = \dfrac{1}{4}$.

x, or 2^y	y
1	0
2	1
4	2
8	3
$\dfrac{1}{2}$	-1
$\dfrac{1}{4}$	-2

$\uparrow$
$\qquad$ (1) Select y.

$\qquad$ (2) Compute x.

We plot the set of ordered pairs and connect the points with a smooth curve.

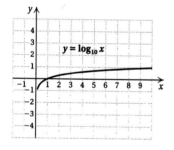

2.

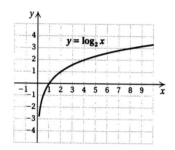

3. Graph: $y = \log_6 x$

The equation $y = \log_6 x$ is equivalent to $6^y = x$. We can find ordered pairs by choosing values for y and computing the corresponding x-values.

For $y = 0$, $x = 6^0 = 1$.

For $y = 1$, $x = 6^1 = 6$.

For $y = 2$, $x = 6^2 = 36$.

For $y = -1$, $x = 6^{-1} = \dfrac{1}{6}$.

For $y = -2$, $x = 6^{-2} = \dfrac{1}{36}$.

x, or 6^y	y
1	0
6	1
36	2
$\dfrac{1}{6}$	-1
$\dfrac{1}{36}$	-2

We plot the set of ordered pairs and connect the points with a smooth curve.

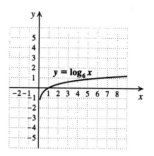

4.

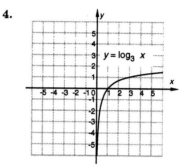

5. Graph: $f(x) = \log_4 x$

Think of $f(x)$ as y. Then $y = \log_4 x$ is equivalent to $4^y = x$. We find ordered pairs by choosing values for y and computing the corresponding x-values. Then we plot the points and connect them with a smooth curve.

For $y = 0$, $x = 4^0 = 1$.

For $y = 1$, $x = 4^1 = 4$.

For $y = 2$, $x = 4^2 = 16$.

For $y = -1$, $x = 4^{-1} = \dfrac{1}{4}$.

For $y = -2$, $x = 4^{-2} = \dfrac{1}{16}$.

x, or 4^y	y
1	0
4	1
16	2
$\dfrac{1}{4}$	-1
$\dfrac{1}{16}$	-2

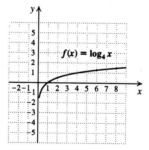

6.

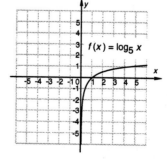

7. Graph: $f(x) = \log_{1/2} x$

Think of $f(x)$ as y. Then $y = \log_{1/2} x$ is equivalent to $\left(\dfrac{1}{2}\right)^y = x$. We construct a table of values, plot these points and connect them with a smooth curve.

For $y = 0$, $x = \left(\dfrac{1}{2}\right)^0 = 1$.

For $y = 1$, $x = \left(\dfrac{1}{2}\right)^1 = \dfrac{1}{2}$.

For $y = 2$, $x = \left(\dfrac{1}{2}\right)^2 = \dfrac{1}{4}$.

For $y = -1$, $x = \left(\dfrac{1}{2}\right)^{-1} = 2$.

For $y = -2$, $x = \left(\dfrac{1}{2}\right)^{-2} = 4$.

For $y = -3$, $x = \left(\dfrac{1}{2}\right)^{-3} = 8$.

x, or $\left(\dfrac{1}{2}\right)^y$	y
1	0
$\dfrac{1}{2}$	1
$\dfrac{1}{4}$	2
2	-1
4	-2
8	-3

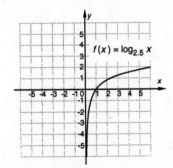

8.

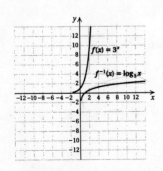

9. Graph $f(x) = 3^x$ (see Exercise Set 12.1, Exercise 2) and $f^{-1}(x) = \log_3 x$ (see Exercise 4 above) on the same set of axes.

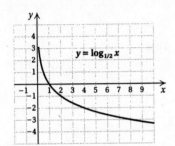

10.

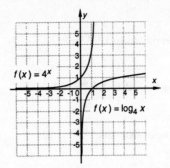

11.

The exponent is the logarithm.

$10^3 = 1000 \Rightarrow 3 = \log_{10} 1000$

The base remains the same.

12. $2 = \log_{10} 100$

13.

The exponent is the logarithm.

$5^{-3} = \dfrac{1}{125} \Rightarrow -3 = \log_5 \dfrac{1}{125}$

The base remains the same.

14. $-5 = \log_4 \dfrac{1}{1024}$

15. $8^{1/3} = 2$ is equivalent to $\dfrac{1}{3} = \log_8 2$.

16. $\dfrac{1}{4} = \log_{16} 2$

17. $10^{0.3010} = 2$ is equivalent to $0.3010 = \log_{10} 2$.

18. $0.4771 = \log_{10} 3$

19. $e^2 = t$ is equivalent to $2 = \log_e t$.

20. $k = \log_p 3$

21. $Q^t = x$ is equivalent to $t = \log_Q x$.

22. $m = \log_p V$

23. $e^2 = 7.3891$ is equivalent to $2 = \log_e 7.3891$.

24. $3 = \log_e 20.0855$

25. $e^{-2} = 0.1353$ is equivalent to $-2 = \log_e 0.1353$.

26. $-4 = \log_e 0.0183$

27.

The base remains the same.

$t = \log_3 8 \Rightarrow 3^t = 8$

The logarithm is the exponent.

28. $7^h = 10$

29.

The logarithm is the exponent.

$\log_5 25 = 2 \Rightarrow 5^2 = 25$

The base remains the same.

30. $6^1 = 6$

31. $\log_{10} 0.1 = -1$ is equivalent to $10^{-1} = 0.1$.

32. $10^{-2} = 0.01$

33. $\log_{10} 7 = 0.845$ is equivalent to $10^{0.845} = 7$.

34. $10^{0.4771} = 3$

35. $\log_e 20 = 2.9957$ is equivalent to $e^{2.9957} = 20$.

36. $e^{2.3026} = 10$

37. $\log_t Q = k$ is equivalent to $t^k = Q$.

38. $m^a = P$

39. $\log_e 0.25 = -1.3863$ is equivalent to $e^{-1.3863} = 0.25$.

40. $e^{-0.0111} = 0.989$

41. $\log_r T = -x$ is equivalent to $r^{-x} = T$.

42. $c^{-w} = M$

43. $\log_3 x = 2$

$\quad\quad 3^2 = x \quad$ Converting to an exponential equation

$\quad\quad\; 9 = x \quad$ Computing 3^2

44. 64

45. $\log_x 125 = 3$

$\quad\quad x^3 = 125 \quad$ Converting to an exponential equation

$\quad\quad\;\; x = 5 \quad$ Taking cube roots

46. 4

47. $\log_2 16 = x$

$\quad\quad 2^x = 16 \quad$ Converting to an exponential equation

$\quad\quad 2^x = 2^4$

$\quad\quad\; x = 4 \quad$ The exponents must be the same.

48. 2

49. $\log_3 27 = x$

$\quad\quad 3^x = 27 \quad$ Converting to an exponential equation

$\quad\quad 3^x = 3^3$

$\quad\quad\; x = 3 \quad$ The exponents must be the same.

50. 2

51. $\log_x 13 = 1$

$\quad\quad x^1 = 13 \quad$ Converting to an exponential equation

$\quad\quad\; x = 13 \quad$ Simplifying x^1

52. 23

53. $\log_6 x = 0$

$\quad\quad 6^0 = x \quad$ Converting to an exponential equation

$\quad\quad\; 1 = x \quad$ Computing 6^0

54. 9

55. $\log_2 x = -1$

$\quad\quad 2^{-1} = x \quad$ Converting to an exponential equation

$\quad\quad \frac{1}{2} = x \quad$ Simplifying

56. $\frac{1}{9}$

57. $\log_8 x = \frac{1}{3}$

$\quad\quad 8^{1/3} = x$

$\quad\quad\;\; 2 = x$

58. 2

59. Let $\log_{10} 100 = x$. Then

$\quad 10^x = 100$

$\quad 10^x = 10^2$

$\quad\;\; x = 2.$

Thus, $\log_{10} 100 = 2$.

60. 5

61. Let $\log_{10} 0.1 = x$. Then

$\quad 10^x = 0.1 = \frac{1}{10}$

$\quad 10^x = 10^{-1}$

$\quad\;\; x = -1.$

Thus, $\log_{10} 0.1 = -1$.

62. -3

63. Let $\log_{10} 1 = x$. Then

$\quad 10^x = 1$

$\quad 10^x = 10^0 \quad (10^0 = 1)$

$\quad\;\; x = 0.$

Thus, $\log_{10} 1 = 0$.

64. 1

65. Let $\log_5 625 = x$. Then

$\quad 5^x = 625$

$\quad 5^x = 5^4$

$\quad\; x = 4.$

Thus, $\log_5 625 = 4$.

66. 6

67. Let $\log_5 \frac{1}{25} = x$. Then

$$5^x = \frac{1}{25}$$
$$5^x = 5^{-2}$$
$$x = -2.$$

Thus, $\log_5 \frac{1}{25} = -2$.

68. -4

69. Let $\log_3 1 = x$. Then

$$3^x = 1$$
$$3^x = 3^0 \qquad (3^0 = 1)$$
$$x = 0.$$

Thus, $\log_3 1 = 0$.

70. 1

71. Let $\log_e e = x$. Then

$$e^x = e$$
$$e^x = e^1$$
$$x = 1.$$

Thus, $\log_e e = 1$.

72. 0

73. Let $\log_{27} 9 = x$. Then

$$27^x = 9$$
$$(3^3)^x = 3^2$$
$$3^{3x} = 3^2$$
$$3x = 2$$
$$x = \frac{2}{3}.$$

Thus, $\log_{27} 9 = \frac{2}{3}$.

74. $\frac{1}{3}$

75. Let $\log_e e^3 = x$. Then

$$e^x = e^3$$
$$x = 3.$$

Thus, $\log_e e^3 = 3$.

76. -4

77. Let $\log_{10} 10^t = x$. Then

$$10^x = 10^t$$
$$x = t.$$

Thus, $\log_{10} 10^t = t$.

78. p

79. $\dfrac{\dfrac{3}{x} - \dfrac{2}{xy}}{\dfrac{2}{x^2} + \dfrac{1}{xy}}$

The LCD of all the denominators is $x^2 y$. We multiply numerator and denominator by the LCD.

$$\frac{\dfrac{3}{x} - \dfrac{2}{xy}}{\dfrac{2}{x^2} + \dfrac{1}{xy}} \cdot \frac{x^2 y}{x^2 y} = \frac{\left(\dfrac{3}{x} - \dfrac{2}{xy}\right) x^2 y}{\left(\dfrac{2}{x^2} + \dfrac{1}{xy}\right) x^2 y}$$

$$= \frac{\dfrac{3}{x} \cdot x^2 y - \dfrac{2}{xy} \cdot x^2 y}{\dfrac{2}{x^2} \cdot x^2 y + \dfrac{1}{xy} \cdot x^2 y}$$

$$= \frac{3xy - 2x}{2y + x}, \text{ or}$$

$$\frac{x(3y - 2)}{2y + x}$$

80. $\dfrac{x+2}{x+1}$

81. $8^{-4} = \dfrac{1}{8^4}$, or $\dfrac{1}{4096}$

82. $\sqrt[5]{x^4}$

83. $t^{-2/3} = \dfrac{1}{t^{2/3}} = \dfrac{1}{\sqrt[3]{t^2}}$

84. 5

85. ◈

86. ◈

87. Graph: $y = \left(\dfrac{3}{2}\right)^x$ \qquad Graph: $y = \log_{3/2} x$, or

$$x = \left(\dfrac{3}{2}\right)^y$$

x	y, or $\left(\dfrac{3}{2}\right)^x$
0	1
1	$\dfrac{3}{2}$
2	$\dfrac{9}{4}$
3	$\dfrac{27}{8}$
-1	$\dfrac{2}{3}$
-2	$\dfrac{4}{9}$

x, or $\left(\dfrac{3}{2}\right)^y$	y
1	0
$\dfrac{3}{2}$	1
$\dfrac{9}{4}$	2
$\dfrac{27}{8}$	3
$\dfrac{2}{3}$	-1
$\dfrac{4}{9}$	-2

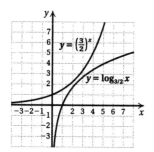

88.

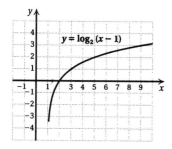

89. Graph: $y = \log_3 |x + 1|$

x	y
0	0
2	1
8	2
-2	0
-4	1
-9	2

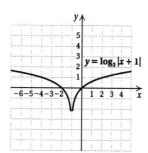

90. $27, \ \dfrac{1}{27}$

91. $\log_{125} x = \dfrac{2}{3}$

$$125^{2/3} = x$$
$$(5^3)^{2/3} = x$$
$$5^2 = x$$
$$25 = x$$

92. 6

93. $\log_8 (2x + 1) = -1$

$$8^{-1} = 2x + 1$$
$$\frac{1}{8} = 2x + 1$$
$$1 = 16x + 8 \quad \text{Multiplying by 8}$$
$$-7 = 16x$$
$$-\frac{7}{16} = x$$

94. $-25, \ 4$

95. Let $\log_{1/4} \dfrac{1}{64} = x$. Then

$$\left(\frac{1}{4}\right)^x = \frac{1}{64}$$
$$\left(\frac{1}{4}\right)^x = \left(\frac{1}{4}\right)^3$$
$$x = 3.$$

Thus, $\log_{1/4} \dfrac{1}{64} = 3$.

96. 1

97. $\log_{10} (\log_4 (\log_3 81))$

     $= \log_{10} (\log_4 4) \qquad (\log_3 81 = 4)$

     $= \log_{10} 1 \qquad\qquad (\log_4 4 = 1)$

     $= 0$

98. 1

99. Let $\log_{1/5} 25 = x$. Then

$$\left(\frac{1}{5}\right)^x = 25$$
$$(5^{-1})^x = 25$$
$$5^{-x} = 5^2$$
$$-x = 2$$
$$x = -2.$$

Thus, $\log_{1/5} 25 = -2$.

Exercise Set 12.4

---

1. $\log_2 (32 \cdot 8) = \log_2 32 + \log_2 8$ Using the product rule

2. $\log_3 27 + \log_3 81$

3. $\log_4 (64 \cdot 16) = \log_4 64 + \log_4 16$ Using the product rule

4. $\log_5 25 + \log_5 125$

5. $\log_c Bx = \log_c B + \log_c x$ Using the product rule

6. $\log_t 5 + \log_t Y$

7. $\log_a 6 + \log_a 70 = \log_a (6 \cdot 70)$ Using the product rule

8. $\log_b (65 \cdot 2)$

9. $\log_c K + \log_c y = \log_c (K \cdot y)$ Using the product rule

10. $\log_t HM$

11. $\log_a x^3 = 3 \log_a x$ Using the power rule

12. $5 \log_b t$

13. $\log_c y^6 = 6 \log_c y$ Using the power rule

14. $7 \log_{10} y$

15. $\log_b C^{-3} = -3 \log_b C$ Using the power rule

16. $-5 \log_c M$

17. $\log_a \dfrac{67}{5} = \log_a 67 - \log_a 5$ Using the quotient rule

18. $\log_t T - \log_t 7$

19. $\log_b \dfrac{3}{4} = \log_b 3 - \log_b 4$ Using the quotient rule

20. $\log_a y - \log_a x$

21. $\log_a 15 - \log_a 7 = \log_a \dfrac{15}{7}$ Using the quotient rule

22. $\log_b \dfrac{42}{7}$, or $\log_b 6$

23.
$$\log_a x^2 y^3 z$$
$$= \log_a x^2 + \log_a y^3 + \log_a z$$ Using the product rule
$$= 2 \log_a x + 3 \log_a y + \log_a z$$ Using the power rule

24. $\log_a x + 4 \log_a y + 3 \log_a z$

25.
$$\log_b \dfrac{xy^2}{z^3}$$
$$= \log_b xy^2 - \log_b z^3$$ Using the quotient rule
$$= \log_b x + \log_b y^2 - \log_b z^3$$ Using the product rule
$$= \log_b x + 2 \log_b y - 3 \log_b z$$ Using the power rule

26. $2 \log_b x + 5 \log_b y - 4 \log_b w - 7 \log_b z$

27.
$$\log_c \sqrt[3]{\dfrac{x^4}{y^3 z^2}}$$
$$= \log_c \left(\dfrac{x^4}{y^3 z^2} \right)^{1/3}$$
$$= \dfrac{1}{3} \log_c \dfrac{x^4}{y^3 z^2}$$ Using the power rule
$$= \dfrac{1}{3} \left(\log_c x^4 - \log_c y^3 z^2 \right)$$ Using the quotient rule
$$= \dfrac{1}{3} [\log_c x^4 - (\log_c y^3 + \log_c z^2)]$$ Using the product rule
$$= \dfrac{1}{3} \left(\log_c x^4 - \log_c y^3 - \log_c z^2 \right)$$ Removing parentheses
$$= \dfrac{1}{3} (4 \log_c x - 3 \log_c y - 2 \log_c z)$$ Using the power rule

28. $\dfrac{1}{2} (6 \log_a x - 5 \log_a y - 8 \log_a z)$

29.
$$\log_a \sqrt[4]{\dfrac{x^8 y^{12}}{a^3 z^5}}$$
$$= \dfrac{1}{4} \left(\log_a \dfrac{x^8 y^{12}}{a^3 z^5} \right)$$ Using the power rule
$$= \dfrac{1}{4} (\log_a x^8 y^{12} - \log_a a^3 z^5)$$ Using the quotient rule
$$= \dfrac{1}{4} [\log_a x^8 + \log_a y^{12} - (\log_a a^3 + \log_a z^5)]$$ Using the product rule
$$= \dfrac{1}{4} (\log_a x^8 + \log_a y^{12} - \log_a a^3 - \log_a z^5)$$ Removing parentheses
$$= \dfrac{1}{4} (\log_a x^8 + \log_a y^{12} - 3 - \log_a z^5)$$ 3 is the power to which you raise a to get a^3.
$$= \dfrac{1}{4} (8 \log_a x + 12 \log_a y - 3 - 5 \log_a z)$$ Using the power rule

30. $\dfrac{1}{3} (6 \log_a x + 3 \log_a y - 2 - 7 \log_a z)$

31.
$$\dfrac{2}{3} \log_a x - \dfrac{1}{2} \log_a y$$
$$= \log_a x^{2/3} - \log_a y^{1/2}$$ Using the power rule
$$= \log_a \sqrt[3]{x^2} - \log_a \sqrt{y}$$
$$= \log_a \dfrac{\sqrt[3]{x^2}}{\sqrt{y}}, \text{ or}$$ Using the quotient rule
$$\log_a \dfrac{\sqrt[3]{x^2}\sqrt{y}}{y}$$ Multiplying by $\dfrac{\sqrt{y}}{\sqrt{y}}$

32. $\log_a \dfrac{\sqrt{x} y^3}{x^2}$

33.
$$\log_a 2x + 3(\log_a x - \log_a y)$$
$$= \log_a 2x + 3 \log_a x - 3 \log_a y$$
$$= \log_a 2x + \log_a x^3 - \log_a y^3$$ Using the power rule
$$= \log_a 2x^4 - \log_a y^3$$ Using the product rule
$$= \log_a \dfrac{2x^4}{y^3}$$ Using the quotient rule

34. $\log_a x$

35.
$$\log_a \dfrac{a}{\sqrt{x}} - \log_a \sqrt{ax}$$
$$= \log_a ax^{-1/2} - \log_a a^{1/2} x^{1/2}$$
$$= \log_a \dfrac{ax^{-1/2}}{a^{1/2} x^{1/2}}$$ Using the quotient rule
$$= \log_a \dfrac{a^{1/2}}{x}$$
$$= \log_a \dfrac{\sqrt{a}}{x}$$

36. $\log_a (x + 2)$

37. $\log_b 15 = \log_b (3 \cdot 5)$

$\qquad = \log_b 3 + \log_b 5 \qquad$ Using the product rule

$\qquad = 1.099 + 1.609$

$\qquad = 2.708$

38. -0.51

39. $\log_b \dfrac{5}{3} = \log_b 5 - \log_b 3 \qquad$ Using the quotient rule

$\qquad = 1.609 - 1.099$

$\qquad = 0.51$

40. -1.099

41. $\log_b \dfrac{1}{5} = \log_b 1 - \log_b 5 \qquad$ Using the quotient rule

$\qquad = 0 - 1.609 \qquad (\log_b 1 = 0)$

$\qquad = -1.609$

42. $\dfrac{1}{2}$

43. $\log_b \sqrt{b^3} = \log_b b^{3/2} = \dfrac{3}{2} \qquad$ 3/2 is the power to which you raise b to get $b^{3/2}$.

44. 2.099

45. $\log_b 5b = \log_b 5 + \log_b b \qquad$ Using the product rule

$\qquad = 1.609 + 1 \qquad (\log_b b = 1)$

$\qquad = 2.609$

46. 2.198

47. $\log_b 25 = \log_b 5^2$

$\qquad = 2 \log_b 5 \qquad$ Using the power rule

$\qquad = 2(1.609)$

$\qquad = 3.218$

48. 4.317

49. $\log_t t^9 = 9 \qquad$ 9 is the power to which you raise t to get t^9.

50. 4

51. $\log_e e^m = m$

52. -2

53. $\log_3 3^4 = x$

$\qquad 4 = x$

54. 7

55. $\log_e e^x = -7$

$\qquad x = -7$

56. 2.7

57. $i^{29} = i^{28} \cdot i = (i^4)^7 \cdot i = 1^7 \cdot i = 1 \cdot i = i$

58. 5

59. $\dfrac{2+i}{2-i} = \dfrac{2+i}{2-i} \cdot \dfrac{2+i}{2+i} = \dfrac{4+4i+i^2}{4-i^2} = \dfrac{4+4i-1}{4-(-1)} =$

$\dfrac{3+4i}{5} = \dfrac{3}{5} + \dfrac{4}{5}i$

60. $23 - 18i$

61.

62.

63. $\log_a (x^8 - y^8) - \log_a (x^2 + y^2)$

$\qquad = \log_a \dfrac{x^8 - y^8}{x^2 + y^2} \qquad$ Using the quotient rule

$\qquad = \log_a \dfrac{(x^4 + y^4)(x^2 + y^2)(x+y)(x-y)}{x^2 + y^2}$

$\qquad\qquad\qquad\qquad\qquad$ Factoring

$\qquad = \log_a [(x^4 + y^4)(x^2 - y^2)] \qquad$ Simplifying

$\qquad = \log_a (x^6 - x^4 y^2 + x^2 y^4 - y^6) \qquad$ Multiplying

64. $\log_a (x^3 + y^3)$

65. $\log_a \sqrt{1 - s^2}$

$\qquad = \log_a (1 - s^2)^{1/2}$

$\qquad = \dfrac{1}{2} \log_a (1 - s^2)$

$\qquad = \dfrac{1}{2} \log_a [(1 - s)(1 + s)]$

$\qquad = \dfrac{1}{2} \log_a (1 - s) + \dfrac{1}{2} \log_a (1 + s)$

66. $\dfrac{1}{2} \log_a (c - d) - \dfrac{1}{2} \log_a (c + d)$

67. $\log_a \dfrac{\sqrt[3]{x^2 z}}{\sqrt[3]{y^2 z^{-2}}}$

$\qquad = \log_a \left(\dfrac{x^2 z^3}{y^2} \right)^{1/3}$

$\qquad = \dfrac{1}{3} (\log_a x^2 z^3 - \log_a y^2)$

$\qquad = \dfrac{1}{3} (2 \log_a x + 3 \log_a z - 2 \log_a y)$

$\qquad = \dfrac{1}{3} [2 \cdot 2 + 3 \cdot 4 - 2 \cdot 3] \qquad$ Substituting

$\qquad = \dfrac{1}{3} (10)$

$\qquad = \dfrac{10}{3}$

68. -2

69. $\log_a x = 2$ Given

    $a^2 = x$ Definition

Let $\log_{1/a} x = n$ and solve for n.

    $\log_{1/a} a^2 = n$ Substituting a^2 for x

      $\left(\dfrac{1}{a}\right)^n = a^2$

      $(a^{-1})^n = a^2$

        $a^{-n} = a^2$

         $-n = 2$

           $n = -2$

Thus, $\log_{1/a} x = -2$ when $\log_a x = 2$.

70. False. For a counterexample, let $a = 10$, $P = 100$, and $Q = 10$.

71. The statement is false. For example, let $a = 10$, $P = 100$, and $Q = 10$. Then

$\dfrac{\log_a P}{\log_a Q} = \dfrac{\log_{10} 100}{\log_{10} 10} = \dfrac{2}{1} = 2$, but

$\log_a P - \log_a Q = \log_{10} 100 - \log_{10} 10 = 2 - 1 = 1$.

72. True

73. The statement is false. For example, let $a = 3$ and $x = 9$. Then

$\log_a 3x = \log_3 3 \cdot 9 = \log_3 27 = 3$, but

$3\log_a x = 3\log_3 9 = 3 \cdot 2 = 6$.

74. False. For a counterexample, let $a = 2$, $P = 1$, and $Q = 1$.

75. The statement is true, by the power rule.

76. $\log_a \dfrac{x + \sqrt{x^2 - 3}}{3} =$

$\log_a \left(\dfrac{x + \sqrt{x^2 - 3}}{3} \cdot \dfrac{x - \sqrt{x^2 - 3}}{x - \sqrt{x^2 - 3}}\right) =$

$\log_a \left(\dfrac{3}{3(x - \sqrt{x^2 - 3})}\right) = \log_a \left(\dfrac{1}{x - \sqrt{x^2 - 3}}\right) =$

$\log_a 1 - \log_a (x - \sqrt{x^2 - 3}) =$

$0 - \log_a (x - \sqrt{x^2 - 3}) = -\log_a (x - \sqrt{x^2 - 3})$

Exercise Set 12.5

1. 0.3010

2. 0.6990

3. 0.8021

4. 0.7007

5. 1.6532

6. 1.8692

7. 2.6405

8. 2.4698

9. 4.1271

10. 4.9689

11. -1.2840

12. -0.4123

13. 1000

14. 100,000

15. 501.1872

16. 6.3096×10^{14}

17. 3.0001

18. 1.1623

19. 0.2841

20. 0.4567

21. 0.0011

22. 79,104.2833

23. 0.6931

24. 1.0986

25. 4.1271

26. 3.4012

27. 8.3814

28. 6.8037

29. -5.0832

30. -7.2225

31. 36.7890

32. 138.5457

33. 0.0023

34. 0.1002

35. 1.0057

36. 1.0112

37. 5.8346×10^{14}

38. 2.0917×10^{24}

39. 7.6331

40. 0.2520

41. We will use common logarithms for the conversion. Let $a = 10$, $b = 6$, and $M = 100$ and substitute in the change-of-base formula.

$\log_b M = \dfrac{\log_a M}{\log_a b}$

$\log_6 100 = \dfrac{\log_{10} 100}{\log_{10} 6}$

$\approx \dfrac{2}{0.7782}$

≈ 2.5702

42. 2.6309

43. We will use common logarithms for the conversion. Let $a = 10$, $b = 2$, and $M = 10$ and substitute in the change-of-base formula.

$$\log_2 10 = \frac{\log_{10} 10}{\log_{10} 2}$$

$$\approx \frac{1}{0.3010}$$

$$\approx 3.3219$$

44. 2.0104

45. We will use natural logarithms for the conversion. Let $a = e$, $b = 200$, and $M = 30$ and substitute in the change-of-base formula.

$$\log_{200} 30 = \frac{\ln 30}{\ln 200}$$

$$\approx \frac{3.4012}{5.2983}$$

$$\approx 0.6419$$

46. 0.7386

47. We will use natural logarithms for the conversion. Let $a = e$, $b = 0.5$, and $M = 5$ and substitute in the change-of-base formula.

$$\log_{0.5} 5 = \frac{\ln 5}{\ln 0.5}$$

$$\approx \frac{1.6094}{-0.6931}$$

$$\approx -2.3219$$

48. -0.4771

49. We will use common logarithms for the conversion. Let $a = 10$, $b = 2$, and $M = 0.2$ and substitute in the change-of-base formula.

$$\log_2 0.2 = \frac{\log_{10} 0.2}{\log_{10} 2}$$

$$\approx \frac{-0.6990}{0.3010}$$

$$\approx -2.3219$$

50. -3.6439

51. We will use natural logarithms for the conversion. Let $a = e$, $b = \pi$, and $M = 58$ and substitute in the change-of-base formula.

$$\log_\pi 58 = \frac{\ln 58}{\ln \pi}$$

$$\approx \frac{4.0604}{1.1447}$$

$$\approx 3.5471$$

52. 4.6284

53. Graph: $f(x) = e^x$

We find some function values with a calculator. We use these values to plot points and draw the graph.

x	e^x
0	1
1	2.7
2	7.4
3	20.1
-1	0.4
-2	0.1

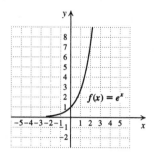

54.

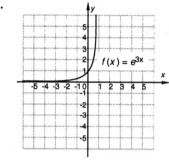

55. Graph: $f(x) = e^{-3x}$

We find some function values, plot points, and draw the graph.

x	e^{-3x}
0	1
1	0.05
2	0.002
-1	20.1
-2	403.4

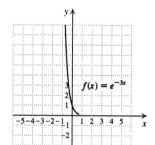

56.

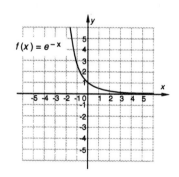

57. Graph: $f(x) = e^{x-1}$

We find some function values, plot points, and draw the graph.

x	e^{x-1}
0	0.4
1	1
2	2.7
3	7.4
4	20.1
−1	0.1
−2	0.05

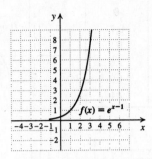

58.

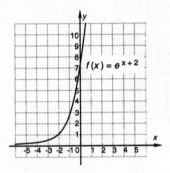

59. Graph: $f(x) = e^{-x} - 3$

We find some function values, plot points, and draw the graph.

x	$e^{-x} - 3$
0	−2
1	−2.6
2	−2.9
3	−3.0
−1	−0.3
−2	4.4
−3	17.1

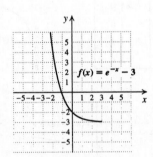

60.

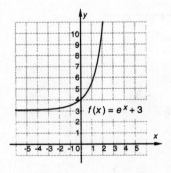

61. Graph: $f(x) = 5e^{0.2x}$

We find some function values, plot points, and draw the graph.

x	$5e^{0.2x}$
0	5
1	6.1
2	7.5
3	9.1
4	11.1
−1	4.1
−2	3.3
−3	2.7
−4	2.2

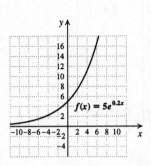

62.

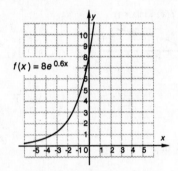

63. Graph: $f(x) = 20e^{-0.5x}$

We find some function values, plot points, and draw the graph.

x	$20e^{-0.5x}$
0	20
1	12.1
2	7.4
3	4.5
4	2.7
−1	33.0
−2	54.4
−3	89.6
−4	147.8

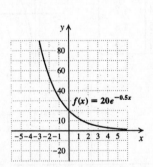

64.

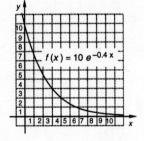

65. Graph: $f(x) = \ln(x+4)$

x	$\ln(x+4)$
0	1.4
1	1.6
2	1.8
3	1.9
4	2.1
−1	1.1
−2	0.7
−3	0
−4	Undefined

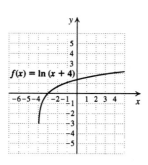

66.

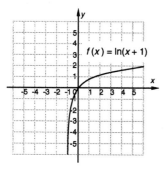

67. Graph: $f(x) = 3\ln x$

x	$3\ln x$
0.5	−2.1
1	0
2	2.1
3	3.3
4	4.2
5	4.8
6	5.4

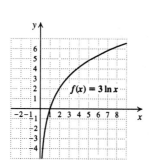

68.

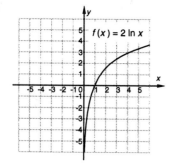

69. Graph: $f(x) = \ln(x-2)$

x	$\ln(x-2)$
2.5	−0.7
3	0
4	0.7
5	1.1
7	1.6
9	1.9
10	2.1

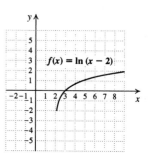

70.

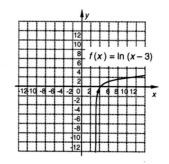

71. Graph: $f(x) = \ln x - 3$

x	$\ln x - 3$
0.5	−3.7
1	−3
2	−2.3
3	−1.9
4	−1.6
5	−1.4
6	−1.2

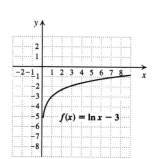

72.

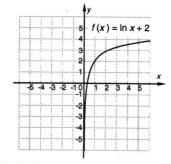

73.
$$4x^2 - 25 = 0$$
$$(2x+5)(2x-5) = 0$$
$$2x+5 = 0 \quad \text{or} \quad 2x-5 = 0$$
$$2x = -5 \quad \text{or} \quad 2x = 5$$
$$x = -\frac{5}{2} \quad \text{or} \quad x = \frac{5}{2}$$

The solutions are $-\dfrac{5}{2}$ and $\dfrac{5}{2}$.

74. $0, \dfrac{7}{5}$

75. $x^{1/2} - 6x^{1/4} + 8 = 0$

Let $u = x^{1/4}$.

$u^2 - 6u + 8 = 0$ Substituting

$(u - 4)(u - 2) = 0$

$u = 4$ or $u = 2$

$x^{1/4} = 4$ or $x^{1/4} = 2$

$x = 256$ or $x = 16$ Raising both sides to the fourth power

Both numbers check. The solutions are 256 and 16.

76. $\dfrac{1}{4}, 9$

77.

78.

79. Use the change-of-base formula with $a = 10$ and $b = e$. We obtain

$$\ln M = \frac{\log M}{\log e}.$$

80. $\log M = \dfrac{\ln M}{\ln 10}$

81. $\log 374x = 4.2931$

$374x = 10^{4.2931}$

$374x \approx 19{,}638.12409$

$x \approx 52.5084$

82. ± 3.3112

83. $\log 692 + \log x = \log 3450$

$\log x = \log 3450 - \log 692$

$\log x = \log \dfrac{3450}{692}$ Using the quotient rule

$x = \dfrac{3450}{692}$

$x \approx 4.9855$

84. 1.5893

Exercise Set 12.6

1. $2^x = 8$

$2^x = 2^3$

$x = 3$ The exponents must be the same.

2. 4

3. $4^x = 256$

$4^x = 4^4$

$x = 4$ The exponents must be the same.

4. 3

5. $2^{2x} = 32$

$2^{2x} = 2^5$

$2x = 5$

$x = \dfrac{5}{2}$

6. 1

7. $3^{5x} = 27$

$3^{5x} = 3^3$

$5x = 3$

$x = \dfrac{3}{5}$

8. $\dfrac{4}{7}$

9. $2^x = 9$

$\log 2^x = \log 9$ Taking the common logarithm on both sides

$x \log 2 = \log 9$ Using the power rule

$x = \dfrac{\log 9}{\log 2}$ Solving for x

$x \approx 3.170$ Using a calculator

10. 4.907

11. $2^x = 10$

$\log 2^x = \log 10$ Taking the common logarithm on both sides

$x \log 2 = \log 10$ Using the power rule

$x = \dfrac{\log 10}{\log 2}$ Solving for x

$x \approx 3.322$ Using a calculator

12. 5.044

13. $5^{4x-7} = 125$

$5^{4x-7} = 5^3$

$4x - 7 = 3$ The exponents must be the same.

$4x = 10$

$x = \dfrac{10}{4}$, or $\dfrac{5}{2}$

14. -1

15. $3^{x^2} \cdot 3^{4x} = \dfrac{1}{27}$

$3^{x^2+4x} = 3^{-3}$

$x^2 + 4x = -3$

$x^2 + 4x + 3 = 0$

$(x + 3)(x + 1) = 0$

$x = -3$ or $x = -1$

16. $\dfrac{1}{2}, -3$

17.
$$4^x = 7$$
$$\log 4^x = \log 7$$
$$x \log 4 = \log 7$$
$$x = \frac{\log 7}{\log 4}$$
$$x \approx 1.404$$

18. 1.107

19.
$$e^t = 100$$
$\ln e^t = \ln 100$　Taking ln on both sides
$t = \ln 100$　Finding the logarithm of the base to a power
$t \approx 4.605$　Using a calculator

20. 6.908

21.
$$e^{-t} = 0.1$$
$\ln e^{-t} = \ln 0.1$　Taking ln on both sides
$-t = \ln 0.1$　Finding the logarithm of the base to a power
$-t \approx -2.303$
$t \approx 2.303$

22. 4.605

23.
$$e^{-0.02t} = 0.06$$
$\ln e^{-0.02t} = \ln 0.06$　Taking ln on both sides
$-0.02t = \ln 0.06$　Finding the logarithm of the base to a power
$$t = \frac{\ln 0.06}{-0.02}$$
$$t \approx \frac{-2.8134}{-0.02}$$
$$t \approx 140.671$$

24. 9.902

25.
$$2^x = 3^{x-1}$$
$$\log 2^x = \log 3^{x-1}$$
$$x \log 2 = (x - 1) \log 3$$
$$x \log 2 = x \log 3 - \log 3$$
$$\log 3 = x \log 3 - x \log 2$$
$$\log 3 = x(\log 3 - \log 2)$$
$$\frac{\log 3}{\log 3 - \log 2} = x$$
$$\frac{0.4771}{0.4771 - 0.3010} \approx x$$
$$-2.710 \approx x$$

26. 7.452

27.
$$(2.8)x = 41$$
$$\log (2.8)^x = \log 41$$
$$x \log 2.8 = \log 41$$
$$x = \frac{\log 41}{\log 2.8}$$
$$x \approx \frac{1.6128}{0.4472}$$
$$x \approx 3.607$$

28. 3.581

29.
$$20 - (1.7)^x = 0$$
$$20 = (1.7)^x$$
$$\log 20 = \log (1.7)^x$$
$$\log 20 = x \log 1.7$$
$$\frac{\log 20}{\log 1.7} = x$$
$$\frac{1.3010}{0.2304} \approx x$$
$$5.646 \approx x$$

30. 3.210

31.
$$\log_3 x = 3$$
$x = 3^3$　Writing an equivalent exponential equation
$$x = 27$$

32. 625

33.
$$\log_2 x = -3$$
$x = 2^{-3}$　Writing an equivalent exponential equation
$$x = \frac{1}{8}$$

34. 2

35.
$\log x = 1$　The base is 10.
$$x = 10^1$$
$$x = 10$$

36. 1000

37.
$\log x = -2$　The base is 10.
$$x = 10^{-2}$$
$$x = \frac{1}{100}$$

38. $\dfrac{1}{1000}$

39.
$$\ln x = 2$$
$$x = e^2 \approx 7.389$$

40. 2.718

41.
$$\ln x = -1$$
$$x = e^{-1}$$
$$x = \frac{1}{e} \approx 0.368$$

42. 0.050

43. $\log_5 (2x - 7) = 3$

$2x - 7 = 5^3$

$2x - 7 = 125$

$2x = 132$

$x = 66$

The answer checks. The solution is 66.

44. $-\dfrac{25}{6}$

45. $\log x + \log (x - 9) = 1$ The base is 10.

$\log_{10} [x(x - 9)] = 1$ Using the product rule

$x(x - 9) = 10^1$

$x^2 - 9x = 10$

$x^2 - 9x - 10 = 0$

$(x - 10)(x + 1) = 0$

$x = 10 \text{ or } x = -1$

Check: For 10:

$$\frac{\log x + \log (x - 9) = 1}{\log 10 + \log (10 - 9) \; ? \; 1}$$
$$\begin{array}{c|c} \log 10 + \log 1 & \\ 1 + 0 & \\ 1 & 1 \quad \text{TRUE} \end{array}$$

For -1:

$$\frac{\log x + \log (x - 9) = 1}{\log(-1) + \log (-1 - 9) \; ? \; 1} \quad \text{FALSE}$$

The number -1 does not check, because negative numbers do not have logarithms. The solution is 10.

46. 1

47. $\log x - \log (x + 3) = -1$ The base is 10.

$\log_{10} \dfrac{x}{x + 3} = -1$ Using the quotient rule

$\dfrac{x}{x + 3} = 10^{-1}$

$\dfrac{x}{x + 3} = \dfrac{1}{10}$

$10x = x + 3$

$9x = 3$

$x = \dfrac{1}{3}$

The answer checks. The solution is $\dfrac{1}{3}$.

48. 1

49. $\log_2 (x + 1) + \log_2 (x - 1) = 3$

$\log_2[(x + 1)(x - 1)] = 3$ Using the product rule

$(x + 1)(x - 1) = 2^3$

$x^2 - 1 = 8$

$x^2 = 9$

$x = \pm 3$

The number 3 checks, but -3 does not. The solution is 3.

50. $\dfrac{83}{15}$

51. $\log_4 (x + 6) - \log_4 x = 2$

$\log_4 \dfrac{x + 6}{x} = 2$ Using the quotient rule

$\dfrac{x + 6}{x} = 4^2$

$\dfrac{x + 6}{x} = 16$

$x + 6 = 16x$

$6 = 15x$

$\dfrac{2}{5} = x$

The answer checks. The solution is $\dfrac{2}{5}$.

52. 4

53. $\log_4 (x + 3) + \log_4 (x - 3) = 2$

$\log_4[(x + 3)(x - 3)] = 2$ Using the product rule

$(x + 3)(x - 3) = 4^2$

$x^2 - 9 = 16$

$x^2 = 25$

$x = \pm 5$

The number 5 checks, but -5 does not. The solution is 5.

54. $\sqrt{41}$

55. $(125x^7 y^{-2} z^6)^{-2/3} =$

$(5^3)^{-2/3}(x^7)^{-2/3}(y^{-2})^{-2/3}(z^6)^{-2/3} =$

$5^{-2} x^{-14/3} y^{4/3} z^{-4} = \dfrac{1}{25} x^{-14/3} y^{4/3} z^{-4}$, or

$\dfrac{y^{4/3}}{25 x^{14/3} z^4}$

56. $-i$

57. $E = mc^2$

$\dfrac{E}{m} = c^2$

$\sqrt{\dfrac{E}{m}} = c$ Taking the principal square root

58. $\pm 10, \pm 2$

59. ◈

60.

61.　　$8^x = 16^{3x+9}$

　　　　$(2^3)^x = (2^4)^{3x+9}$

　　　　$2^{3x} = 2^{12x+36}$

　　　　　$3x = 12x + 36$

　　　　　$-36 = 9x$

　　　　　$-4 = x$

62. $\dfrac{12}{5}$

63.　$\log_6 (\log_2 x) = 0$

　　　　　　$\log_2 x = 6^0$

　　　　　　$\log_2 x = 1$

　　　　　　　　$x = 2^1$

　　　　　　　　$x = 2$

64. $\sqrt[3]{3}$

65.　$\log_5 \sqrt{x^2 - 9} = 1$

　　　　　$\sqrt{x^2 - 9} = 5^1$

　　　　　　$x^2 - 9 = 25$　　　Squaring both sides

　　　　　　　　$x^2 = 34$

　　　　　　　　　$x = \pm\sqrt{34}$

Both numbers check. The solutions are $\pm\sqrt{34}$.

66. -1

67.　$\log (\log x) = 5$　　　　The base is 10.

　　　　　$\log x = 10^5$

　　　　　$\log x = 100{,}000$

　　　　　　　$x = 10^{100{,}000}$

The number checks. The solution is $10^{100{,}000}$.

68. $-3, -1$

69.　$\log x^2 = (\log x)^2$

　　　$2 \log x = (\log x)^2$

　　　　　　$0 = (\log x)^2 - 2 \log x$

let $u = \log x$.

　　　$0 = u^2 - 2u$

　　　$0 = u(u - 2)$

　　　$u = 0$　　or　　$u = 2$

　$\log x = 0$　or　$\log x = 2$

　　$x = 10^0$　or　　$x = 10^2$

　　$x = 1$　　or　　$x = 100$

Both numbers check. The solutions are 1 and 100.

70. $625, -625$

71.　　$\log x^{\log x} = 25$

　　$\log x (\log x) = 25$　　Using the power rule

　　　　$(\log x)^2 = 25$

　　　　　$\log x = \pm5$

　$x = 10^5$　　　or　$x = 10^{-5}$

　$x = 100{,}000$　or　$x = \dfrac{1}{100{,}000}$

Both numbers check.　The solutions are 100,000 and $\dfrac{1}{100{,}000}$.

72. $\dfrac{1}{2}$, 5000

Both numbers check.

73.　$(81^{x-2}(27^{x+1}) = 9^{2x-3}$

　$[(3^4)^{x-2}][(3^3)^{x+1}] = (3^2)^{2x-3}$

　　$(3^{4x-8}(3^{3x+3}) = 3^{4x-6}$

　　　　　$3^{7x-5} = 3^{4x-6}$

　　　　　$7x - 5 = 4x - 6$

　　　　　　$3x = -1$

　　　　　　　$x = -\dfrac{1}{3}$

74. 1.465, 1

75.　$3^{2x} - 3^{2x-1} = 18$

　　$3^{2x}(1 - 3^{-1}) = 18$　　Factoring

　　$3^{2x}\left(1 - \dfrac{1}{3}\right) = 18$

　　　$3^{2x}\left(\dfrac{2}{3}\right) = 18$

　　　　　$3^{2x} = 27$　　Multiplying by $\dfrac{3}{2}$

　　　　　$3^{2x} = 3^3$

　　　　　$2x = 3$

　　　　　　$x = \dfrac{3}{2}$

76. 38

77. $\log_5 125 = 3$ and $\log_{125} 5 = \dfrac{1}{3}$, so $x = (log_{125}5)^{log_5 125}$ is equivalent to $x = \left(\dfrac{1}{3}\right)^3 = \dfrac{1}{27}$. Then $\log_3 x = \log_3 \dfrac{1}{27} = -3$.

78.

Exercise Set 12.7

1. a) One billion is 1000 million, so we set
$N(t) = 1000$ and solve for t:

$$1000 = 7.5(6)^{0.5t}$$

$$\frac{1000}{7.5} = (6)^{0.5t}$$

$$\log \frac{1000}{7.5} = \log(6)^{0.5t}$$

$$\log 1000 - \log 7.5 = 0.5t \, \log 6$$

$$t = \frac{\log 1000 - \log 7.5}{0.5 \, \log 6}$$

$$t \approx \frac{3 - 0.87506}{0.5(0.77815)} \approx 5.5$$

After about 5.5 years, one billion compact discs will be sold in a year.

b) When $t = 0$, $N(t) = 7.5(6)^{0.5(0)} = 7.5(6)^0 = 7.5(1) = 7.5$. Twice this initial number is 15, so we set $N(t) = 15$ and solve for t:

$$15 = 7.5(6)^{0.5t}$$

$$2 = (6)^{0.5t}$$

$$\log 2 = \log(6)^{0.5t}$$

$$\log 2 = 0.5t \, \log 6$$

$$t = \frac{\log 2}{0.5 \, \log 6} \approx \frac{0.30103}{0.5(0.77815)} \approx 0.8$$

The doubling time is about 0.8 year.

2. a) 3.6 days

b) 0.6 days

3. a) We set $A(t) = \$450,000$ and solve for t:

$$450,000 = 50,000(1.06)^t$$

$$\frac{450,000}{50,000} = (1.06)^t$$

$$9 = (1.06)^t$$

$$\log 9 = \log (1.06)^t \quad \text{Taking the common}$$
$$\text{logarithm on both sides}$$

$$\log 9 = t \, \log 1.06 \quad \text{Using the power rule}$$

$$t = \frac{\log 9}{\log 1.06} \approx \frac{0.95424}{0.02531} \approx 37.7$$

It will take about 37.7 years for the \$50,000 to grow to \$450,000.

b) We set $A(t) = \$100,000$ and solve for t:

$$100,000 = 50,000(1.06)^t$$

$$\frac{100,000}{50,000} = (1.06)^t$$

$$2 = (1.06)^t$$

$$\log 2 = \log (1.06)^t \quad \text{Taking the common}$$
$$\text{logarithm on both sides}$$

$$\log 2 = t \, \log 1.06 \quad \text{Using the power rule}$$

$$t = \frac{\log 2}{\log 1.06} \approx \frac{0.30103}{0.02531} \approx 11.9$$

The doubling time is about 11.9 years.

4. a) 59.7 yr

b) 19.1 yr

5. a) We set $N(t) = 60,000$ and solve for t:

$$60,000 = 250,000\left(\frac{2}{3}\right)^t$$

$$\frac{60,000}{250,000} = \left(\frac{2}{3}\right)^t$$

$$0.24 = \log \left(\frac{2}{3}\right)^t$$

$$\log 0.24 = \log \left(\frac{2}{3}\right)^t$$

$$\log 0.24 = t \, \log \frac{2}{3}$$

$$t = \frac{\log 0.24}{\log \frac{2}{3}} \approx \frac{-0.61979}{-0.17609} \approx 3.5$$

After about 3.5 years 60,000 cans will still be in use.

b) We set $N(t) = 1000$ and solve for t.

$$1000 = 250,000\left(\frac{2}{3}\right)^t$$

$$\frac{1000}{250,000} = \left(\frac{2}{3}\right)^t$$

$$0.004 = \log \left(\frac{2}{3}\right)^t$$

$$\log 0.004 = \log \left(\frac{2}{3}\right)^t$$

$$\log 0.004 = t \, \log \frac{2}{3}$$

$$t = \frac{\log 0.004}{\log \frac{2}{3}} \approx \frac{-2.39794}{-0.17609} \approx 13.6$$

After about 13.6 years 1000 cans will still be in use.

6. a) 6.6 yr

b) 3.1 yr

7. a) Substitute 0.09 for k:

$$P(t) = P_0 \, e^{0.09t}$$

b) To find the balance after one year, set $P_0 = 1000$ and $t = 1$. We find $P(1)$:

$$P(1) = 1000 \, e^{0.09(1)} = 1000 \, e^{0.09} \approx$$
$$1000(1.094174282) \approx \$1094.17$$

To find the balance after 2 years, set $P_0 = 1000$ and $t = 2$. We find $P(2)$:

$$P(2) = 1000 \, e^{0.09(2)} = 1000 \, e^{0.18} \approx$$
$$1000(1.197217363) \approx \$1197.22$$

c) To find the doubling time, we set $P_0 = 1000$ and $P(t) = 2000$ and solve for t.

$$2000 = 1000\,e^{0.09t}$$
$$2 = e^{0.09t}$$

$\ln\,2 = \ln\,e^{0.09t}$ Taking the natural logarithm on both sides

$\ln\,2 = 0.09t$ Finding the logarithm of the base to a power

$$\frac{\ln\,2}{0.09} = t$$
$$7.7 \approx t$$

The investment will double in about 7.7 years.

8. a) $P(t) = P_0\,e^{0.1t}$

b) \$22,103.42, \$24,428.06

c) 6.9 years

9. We start with the exponential growth equation
$$P(t) = P_0\,e^{kt}.$$

Substitute $2\,P_0$ for $P(t)$ and 0.01 for k and solve for t.

$$2\,P_0 = P_0\,e^{0.01t}$$

$2 = e^{0.01t}$ Dividing by P_0 on both sides

$\ln\,2 = \ln\,e^{0.01t}$ Taking the natural logarithm on both sides

$\ln\,2 = 0.01t$ Finding the logarithm of the base to a power

$$\frac{\ln\,2}{0.01} = t$$
$$69.3 \approx t$$

The doubling time is about 69.3 years.

10. About 19.8 years

11. a) The equation $P(t) = P_0\,e^{kt}$ can be used to model population growth. At $t = 0$ (1990), the population was 5.2 billion. We substitute 5.2 for P_0 and 1.6%, or 0.016, for k to obtain the exponential growth function:

$P(t) = 5.2\,e^{0.016t}$, where t is the number of years after 1990.

b) In 2000, $t = 2000 - 1990$, or 10. To estimate the population in 2000, we find $P(10)$.

$$P(10) = 5.2\,e^{0.016(10)}$$
$$= 5.2\,e^{0.16}$$

$\approx 5.2(1.17351)$ Finding $e^{0.16}$ using a calculator

$$\approx 6.1$$

The population of the world in 2000 will be about 6.1 billion.

Percent to decimal

$$\frac{1.6}{100}$$

c) We set $P(t) = 8$ and solve for t.

$$8 = 5.2\,e^{0.016t}$$
$$\frac{8}{5.2} = e^{0.016t}$$
$$\ln\left(\frac{8}{5.2}\right) = \ln\,e^{0.016t}$$
$$\ln\left(\frac{8}{5.2}\right) = 0.016t$$
$$\frac{\ln\left(\frac{8}{5.2}\right)}{0.016} = t$$
$$26.9 \approx t$$

The population will be 8.0 billion about 26.9 years after 1990, or in 2017.

12. a) 86.4 minutes

b) 300.5 minutes

c) 20 minutes

13. a) $S(0) = 68 - 20\log\,(0 + 1) = 68 - 20\,\log\,1 = 68 - 20(0) = 68\%$

b) $S(4) = 68 - 20\log\,(4 + 1) = 68 - 20\,\log\,5 \approx 68 - 20(0.69897) \approx 54\%$

$S(24) = 68 - 20\log\,(24 + 1) = 68 - 20\,\log\,25 \approx 68 - 20\,(1.39794) \approx 40\%$

c) Using the values we computed in parts (a) and (b) and any others we wish to calculate, we sketch the graph:

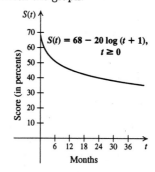

d) We set $S(t) = 50$ and solve for t:

$$50 = 68 - 20\,\log\,(t + 1)$$
$$-18 = -20\,\log\,(t + 1)$$
$$0.9 = \log\,(t + 1)$$

$10^{0.9} = t + 1$ Using the definition of logarithms or taking the antilogarithm

$$7.9 \approx t + 1$$
$$6.9 \approx t$$

After about 6.9 months, the average score was 50.

14. a) 2000

b) 2452

c)

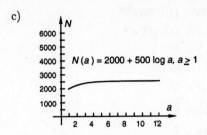

$N(a) = 2000 + 500 \log a, \; a \geq 1$

d) \$1,000,000 thousand, or \$1,000,000,000

15. We substitute 6.3×10^{-7} for H^+ in the formula for pH.

$\text{pH} = -\log [H^+] = -\log [6.3 \times 10^{-7}] =$

$-[\log 6.3 + \log 10^{-7}] \approx -[0.7993 - 7] =$

$-[-6.2007] = 6.2007 \approx 6.2$

The pH of a common brand of mouthwash is about 6.2.

16. 7.4

17. We substitute 1.6×10^{-8} for H^+ in the formula for pH.

$\text{pH} = -\log [H^+] = -\log [1.6 \times 10^{-8}] =$

$-[\log 1.6 + \log 10^{-8}] \approx -[0.2041 - 8] =$

$-[-7.7959] = 7.7959 \approx 7.8$

The pH of eggs is about 7.8.

18. 4.2

19. We substitute 7 for pH in the formula and solve for H^+.

$7 = -\log [H^+]$

$-7 = \log [H^+]$

$10^{-7} = H^+$ Using the definition of logarithm or taking the antilogarithm

The hydrogen ion concentration of tap water is 10^{-7} moles per liter.

20. 4.0×10^{-6} moles per liter

21. We substitute 3.2 for pH in the formula and solve for H^+.

$3.2 = -\log [H^+]$

$-3.2 = \log [H^+]$

$10^{-3.2} = H^+$ Using the definition of logarithm or taking the antilogarithm

$0.00063 \approx H^+$

$6.3 \times 10^{-4} \approx H^+$ Writing in scientific notation

The hydrogen ion concentration of orange juice is about 6.3×10^{-4} moles per liter.

22. 1.6×10^{-5} moles per liter

23. We substitute into the formula.

$R = \log \dfrac{10^{8.25} I_0}{I_0} = \log 10^{8.25} = 8.25$

24. 5

25. We start with the exponential growth equation

$D(t) = D_0 e^{kt}$, where t is the number of years after 1993.

Substitute $2 D_0$ for $D(t)$ and 0.1 for k and solve for t.

$2 D_0 = D_0 e^{0.1t}$

$2 = e^{0.1t}$

$\ln 2 = 0.1t$

$\ln 2 = \ln e^{0.1t}$

$\dfrac{\ln 2}{0.1} = t$

$6.9 \approx t$

The demand will be double that of 1993 about 6.9 years after 1993, or in 2000.

26. 2011

27. a) We start with the exponential growth equation

$N(t) = N_0 e^{kt}$, where t is the number of years after 1967.

Substituting 1 for N_0, we have

$N(t) = e^{kt}$.

To find the exponential growth rate k, observe that there were 1418 heart transplants in 1987, or 20 years after 1967. We substitute and solve for k.

$N(20) = e^{k \cdot 20}$

$1418 = e^{20k}$

$\ln 1418 = \ln e^{20k}$

$\ln 1418 = 20k$

$\dfrac{\ln 1418}{20} = k$

$0.363 \approx k$

Thus the exponential growth function is

$N(t) = e^{0.363t}$, where t is the number of years after 1967.

b) In 1988, $t = 1998 - 1967$, or 31. We find $N(31)$.

$N(31) = e^{0.363(31)} \approx 77,111$

The number of heart transplants in 1998 will be about 77,111.

28. a) 28.2¢

b) 40¢

c) 2025

29. We will use the function derived in Example 7:

$$P(t) = P_0 e^{-0.00012t}$$

If the tusk has lost 20% of its carbon-14 from an initial amount P_0, then $80\%(P_0)$ is the amount present. To find the age of the tusk t, we substitute $80\%(P_0)$, or $0.8P_0$, for $P(t)$ in the function above and solve for t.

$$0.8P_0 = P_0 e^{-0.00012t}$$
$$0.8 = e^{-0.00012t}$$
$$\ln 0.8 = \ln e^{-0.00012t}$$
$$-0.2231 \approx -0.00012t$$
$$t \approx \frac{-0.2231}{-0.00012} \approx 1860$$

The tusk is about 1860 years old.

30. 878 yr

31. The function $P(t) = P_0 e^{-kt}$, $k > 0$, can be used to model decay. For iodine-131, $k = 9.6\%$, or 0.096. To find the half-life we substitute 0.096 for k and $\frac{1}{2} P_0$ for $P(t)$, and solve for t.

$$\frac{1}{2} P_0 = P_0 e^{-0.096t}, \text{ or } \frac{1}{2} = e^{-0.096t}$$
$$\ln \frac{1}{2} = \ln e^{-0.096t} = -0.096t$$
$$t = \frac{\ln 0.5}{-0.096} \approx \frac{-0.6931}{-0.096} \approx 7.2 \text{ days}$$

32. 11 years

33. We use the function $P(t) = P_0 e^{-kt}$, $k > 0$. When $t = 3$, $P(t) = \frac{1}{2} P_0$. We substitute and solve for k.

$$\frac{1}{2} P_0 = P_0 e^{-k(3)}, \text{ or } \frac{1}{2} = e^{-3k}$$
$$\ln \frac{1}{2} = \ln e^{-3k} = -3k$$
$$k = \frac{\ln 0.5}{-3} \approx \frac{-0.6931}{-3} \approx 0.23$$

The decay rate is 0.23, or 23%, per minute.

34. 3.2% per year

35. a) We start with the exponential growth equation

$$V(t) = V_0 e^{kt}, \text{ where } t \text{ is the number of years after 1986.}$$

Substituting 110,000 for V_0, we have

$$V(t) = 110,000 e^{kt}.$$

To find the exponential growth rate k, observe that the card sold for $451,000 in 1991, or 5 years after 1986. We substitute and solve for k.

$$V(5) = 110,000 e^{k \cdot 5}$$
$$451,000 = 110,000 e^{5k}$$
$$4.1 = e^{5k}$$
$$\ln 4.1 = \ln e^{5k}$$
$$\ln 4.1 = 5k$$
$$\frac{\ln 4.1}{5} = k$$
$$0.2822 \approx k$$

Thus the exponential growth function is $V(t) = 110,000 e^{0.2822t}$, where t is the number of years after 1986.

b) In 1998, $t = 1998 - 1986$, or 12.
$$V(12) = 110,000 e^{0.2822(12)} \approx 3,251,528$$
The card's value in 1998 will be about $3,251,528.

c) Substitute $220,000 for $V(t)$ and solve for t.
$$220,000 = 110,000 e^{0.2822t}$$
$$2 = e^{0.2822t}$$
$$\ln 2 = \ln e^{0.2822t}$$
$$\ln 2 = 0.2822t$$
$$\frac{\ln 2}{0.2822} = t$$
$$2.5 \approx t$$

The doubling time is about 2.5 years.

d) Substitute $9,000,000 for $V(t)$ and solve for t.
$$9,000,000 = 110,000 e^{0.2822t}$$
$$\frac{900}{11} = e^{0.2822t}$$
$$\ln \frac{900}{11} = \ln e^{0.2822t}$$
$$\ln \frac{900}{11} = 0.2822t$$
$$\frac{\ln \frac{900}{11}}{0.2822} = t$$
$$15.6 \approx t$$

The value of the card will be $9,000,000 about 15.6 years after 1986, or in 2002.

36. a) $k \approx 0.16$; $V(t) = 84,000e^{0.16t}$, where V is in thousands of dollars and t is the number of years after 1947.

b) $250,400 thousand, or $250,400,000

c) 4.3 yr

d) About 58.7 years after 1947

37. $\dfrac{\dfrac{x-5}{x+3}}{\dfrac{x}{x-3}+\dfrac{2}{x+3}}$

$$= \dfrac{\dfrac{x-5}{x+3}}{\dfrac{x}{x-3}\cdot\dfrac{x+3}{x+3}+\dfrac{2}{x+3}\cdot\dfrac{x-3}{x-3}} \qquad \text{Finding the LCD and}$$

adding in the denominator

$$= \dfrac{\dfrac{x-5}{x+3}}{\dfrac{x^2+3x+2x-6}{(x-3)(x+3)}}$$

$$= \dfrac{\dfrac{x-5}{x+3}}{\dfrac{x^2+5x-6}{(x-3)(x+3)}}$$

$$= \dfrac{x-5}{x+3}\cdot\dfrac{(x-3)(x+3)}{x^2+5x-6} \qquad \begin{array}{l}\text{Multiplying by the}\\ \text{reciprocal of the de-}\\ \text{nominator}\end{array}$$

$$= \dfrac{(x-5)(x-3)(x+3)}{(x+3)\,(x+6)(x-1)} \qquad \begin{array}{l}\text{Factoring}\\ x^2+5x-6\text{ and mul-}\\ \text{tiplying}\end{array}$$

$$= \dfrac{(x-5)(x-3)(\cancel{x+3})}{(\cancel{x+3})\,(x+6)(x-1)}$$

$$= \dfrac{(x-5)(x-3)}{(x+6)(x-1)}, \text{ or } \dfrac{(x-5)(x-3)}{x^2+5x-6}$$

38. $\dfrac{3ab^2+5a^2b}{2b^2-4a^2}$

39. ◈

40. ◈

41. ◈

42. $(6, \$403)$

Chapter 13

Conic Sections

1. $d = \sqrt{(x_2 - x_1)^2 + (y_2 + y_1)^2}$ Distance formula

$\qquad = \sqrt{(6-9)^2 + (1-5)^2}$ Substituting

$\qquad = \sqrt{(-3)^2 + (-4)^2}$

$\qquad = \sqrt{25} = 5$

2. 10

3. $d = \sqrt{(x_2 - x_1)^2 + (y_2 - y_1)^2}$ Distance formula

$\qquad = \sqrt{(3-0)^2 + [-4-(-7)]^2}$ Substituting

$\qquad = \sqrt{3^2 + 3^2}$

$\qquad = \sqrt{18} \approx 4.243$ Simplifying and approximating

4. 10

(Since these points are on a vertical line, we could have found the distance between them by computing $|-8-2|$.)

5. $d = \sqrt{(x_2 - x_1)^2 + (y_2 - y_1)^2}$

$\qquad = \sqrt{(-2-2)^2 + (-2-2)^2}$

$\qquad = \sqrt{32} \approx 5.657$

6. $\sqrt{464} \approx 21.541$

7. $d = \sqrt{(x_2 - x_1)^2 + (y_2 - y_1)^2}$

$\qquad = \sqrt{(-9.2 - 8.6)^2 + [-3.4 - (-3.4)]^2}$

$\qquad = \sqrt{(-17.8)^2 + 0^2}$

$\qquad = \sqrt{316.84} = 17.8$

(Since these points are on a horizontal line, we could have found the distance between them by finding $|x_2 - x_1| = |-9.2 - 8.6| = |-17.8| = 17.8$.)

8. $\sqrt{98.93} \approx 9.946$

9. $d = \sqrt{(x_2 - x_1)^2 + (y_2 - y_1)^2}$

$\qquad d = \sqrt{\left(\frac{5}{7} - \frac{1}{7}\right)^2 + \left(\frac{1}{14} - \frac{11}{14}\right)^2}$

$\qquad = \sqrt{\left(\frac{4}{7}\right)^2 + \left(-\frac{5}{7}\right)^2}$

$\qquad = \sqrt{\frac{16}{49} + \frac{25}{49}}$

$\qquad = \sqrt{\frac{41}{49}}$

$\qquad = \frac{\sqrt{41}}{7} \approx 0.915$

10. $\sqrt{13} \approx 3.606$

11. $d = \sqrt{(x_2 - x_1)^2 + (y_2 - y_1)^2}$

$\qquad = \sqrt{[56 - (-23)]^2 + (-17 - 10)^2}$

$\qquad = \sqrt{79^2 + (-27)^2} = \sqrt{6970} \approx 83.487$

12. $\sqrt{6800} \approx 82.462$

13. $d = \sqrt{(x_2 - x_1)^2 + (y_2 - y_1)^2}$

$\qquad = \sqrt{(a-0)^2 + (b-0)^2}$

$\qquad = \sqrt{a^2 + b^2}$

14. $\sqrt{p^2 + q^2}$

15. $d = \sqrt{(x_2 - x_1)^2 + (y_2 - y_1)^2}$

$\qquad = \sqrt{(-1-6)^2 + (3k - 2k)^2}$

$\qquad = \sqrt{49 + k^2}$

16. $\sqrt{a^2 + 64}$

17. $d = \sqrt{(x_2 - x_1)^2 + (y_2 - y_1)^2}$

$\qquad = \sqrt{(-\sqrt{7} - \sqrt{2})^2 + [\sqrt{5} - (-\sqrt{3})]^2}$

$\qquad = \sqrt{7 + 2\sqrt{14} + 2 + 5 + 2\sqrt{15} + 3}$

$\qquad = \sqrt{17 + 2\sqrt{14} + 2\sqrt{15}} \approx 5.677$

18. $\sqrt{22 + 2\sqrt{40} + 2\sqrt{18}} \approx 6.568$

19. $d = \sqrt{(x_2 - x_1)^2 + (y_2 - y_1)^2}$

$\qquad = \sqrt{[1000 - (-2000)]^2 + (-240 - 580)^2}$

$\qquad = \sqrt{9,672,400} \approx 3110.048$

20. $\sqrt{8,044,900} \approx 2836.353$

21. First we find the squares of the distances between the points.

$d_1^2 = [9 - (-1)]^2 + (6-2)^2$

$\qquad = 10^2 + 4^2 = 116$

(The distance between $(9,6)$ and $(-1,2)$ is $\sqrt{116}$.)

$d_2^2 = (9-1)^2 + [6 - (-3)]^2$

$\qquad = 8^2 + 9^2 = 145$

(The distance between $(9,6)$ and $(1,-3)$ is $\sqrt{145}$.)

$d_3^2 = (-1-1)^2 + [2 - (-3)]^2$

$\qquad = (-2)^2 + 5^2 = 29$

(The distance between $(-1,2)$ and $(1,-3)$ is $\sqrt{29}$.) Since $d_1^2 + d_3^2 = d_2^2$, the points are vertices of a right triangle.

22. No

23. First we find the squares of the distances between the points.
$$d_1^2 = [-5 - (-1)^2] + [1 - (-2)]^2$$
$$= (-4)^2 + 3^2 = 25$$
(The distance between $(-5, 1)$ and $(-1, -2)$ is $\sqrt{25}$, or 5.)
$$d_2^2 = (-5 - 4)^2 + (1 - 10)^2$$
$$= (-9)^2 + (-9)^2 = 162$$
(The distance between $(-5, 1)$ and $(4, 10)$ is $\sqrt{162}$.)
$$d_3^2 = (-1 - 4)^2 + (-2 - 10)^2$$
$$= (-5)^2 + (-12)^2 = 169$$
(The distance between $(-1, -2)$ and $(4, 10)$ is $\sqrt{169}$, or 13.)

Now the largest of these numbers is d_3^2, or 169, but $d_1^2 + d_2^2 \neq d_3^2$, or $25 + 162 \neq 169$. Thus the points are not vertices of a right triangle.

24. Yes

25. We use the midpoint formula:
$$\left(\frac{x_1 + x_2}{2}, \frac{y_1 + y_2}{2}\right) = \left(\frac{-3 + 2}{2}, \frac{6 + (-8)}{2}\right), \text{ or}$$
$$\left(\frac{-1}{2}, \frac{-2}{2}\right), \text{ or } \left(-\frac{1}{2}, -1\right)$$

26. $\left(\dfrac{13}{2}, -1\right)$

27. We use the midpoint formula:
$$\left(\frac{x_1 + x_2}{2}, \frac{y_1 + y_2}{2}\right) = \left(\frac{8 + (-1)}{2}, \frac{5 + 2}{2}\right), \text{ or}$$
$$\left(\frac{7}{2}, \frac{7}{2}\right)$$

28. $\left(0, -\dfrac{1}{2}\right)$

29. We use the midpoint formula:
$$\left(\frac{x_1 + x_2}{2}, \frac{y_1 + y_2}{2}\right) = \left(\frac{-8 + 6}{2}, \frac{-5 + (-1)}{2}\right), \text{ or}$$
$$\left(\frac{-2}{2}, \frac{-6}{2}\right), \text{ or } (-1, -3)$$

30. $\left(\dfrac{5}{2}, 1\right)$

31. We use the midpoint formula:
$$\left(\frac{x_1 + x_2}{2}, \frac{y_1 + y_2}{2}\right) = \left(\frac{-3.4 + 2.9}{2}, \frac{8.1 + (-8.7)}{2}\right),$$
$$\text{or } \left(\frac{-0.5}{2}, \frac{-0.6}{2}\right), \text{ or } (-0.25, -0.3)$$

32. $(4.65, 0)$

33. We use the midpoint formula:
$$\left(\frac{x_1 + x_2}{2}, \frac{y_1 + y_2}{2}\right) = \left(\frac{-a + a}{2}, \frac{b + b}{2}\right), \text{ or}$$
$$\left(\frac{0}{2}, \frac{2b}{2}\right), \text{ or } (0, b)$$

34. $(0, 0)$

35. We use the midpoint formula:
$$\left(\frac{x_1 + x_2}{2}, \frac{y_1 + y_2}{2}\right) = \left(\frac{\frac{1}{6} + \left(-\frac{1}{3}\right)}{2}, \frac{-\frac{3}{4} + \frac{5}{6}}{2}\right),$$
$$\text{or } \left(\frac{-\frac{1}{6}}{2}, \frac{\frac{1}{12}}{2}\right), \text{ or } \left(-\frac{1}{12}, \frac{1}{24}\right)$$

36. $\left(-\dfrac{27}{80}, \dfrac{1}{24}\right)$

37. We use the midpoint formula:
$$\left(\frac{x_1 + x_2}{2}, \frac{y_1 + y_2}{2}\right) = \left(\frac{\sqrt{2} + \sqrt{3}}{2}, \frac{-1 + 4}{2}\right), \text{ or}$$
$$\left(\frac{\sqrt{2} + \sqrt{3}}{2}, \frac{3}{2}\right)$$

38. $\left(\dfrac{5}{2}, \dfrac{7\sqrt{3}}{2}\right)$

39. $x^2 + 2x - 1 = 0$

We use the quadratic formula.
$$a = 1, \, b = 2, \, c = -1$$
$$x = \frac{-2 \pm \sqrt{2^2 - 4 \cdot 1 \cdot (-1)}}{2 \cdot 1}$$
$$x = \frac{-2 \pm \sqrt{8}}{2} = \frac{-2 \pm \sqrt{4 \cdot 2}}{2}$$
$$x = \frac{-2 \pm 2\sqrt{2}}{2} = \frac{2(-1 \pm \sqrt{2})}{2}$$
$$x = -1 \pm \sqrt{2}$$
The solutions are $-1 + \sqrt{2}$ and $-1 - \sqrt{2}$.

40. $\dfrac{-1 \pm i\sqrt{19}}{2}$

41. $\quad 2a^2 - a - 3 = 0$
$$(2a - 3)(a + 1) = 0$$
$$2a - 3 = 0 \quad \text{or} \quad a + 1 = 0$$
$$2a = 3 \quad \text{or} \qquad a = -1$$
$$a = \frac{3}{2} \quad \text{or} \qquad a = -1$$
The solutions are $\dfrac{3}{2}$ and -1.

42. $\dfrac{-1 \pm \sqrt{41}}{2}$

43.

44.

45. Let $(0, y)$ be the point on the y-axis that is equidistant from $(2, 10)$ and $(6, 2)$. Then the distance between $(2, 10)$ and $(0, y)$ is the same as the distance between $(6, 2)$ and $(0, y)$.

$$\sqrt{(0-2)^2 + (y-10)^2} = \sqrt{(0-6)^2 + (y-2)^2}$$
$$(-2)^2 + (y-10)^2 = (-6)^2 + (y-2)^2$$

Squaring both sides

$$4 + y^2 - 20y + 100 = 36 + y^2 - 4y + 4$$
$$64 = 16y$$
$$4 = y$$

This number checks. The point is $(0, 4)$.

46. $(-5, 0)$

47. $\left(\dfrac{x_1 + x_2}{2}, \dfrac{y_1 + y_2}{2}\right) =$

$\left(\dfrac{(2 - \sqrt{3}) + (2 + \sqrt{3})}{2}, \dfrac{5\sqrt{2} + 3\sqrt{2}}{2}\right)$, or $\left(\dfrac{4}{2}, \dfrac{8\sqrt{2}}{2}\right)$,

or $(2, 4\sqrt{2})$

48. $8\sqrt{m^2 + n^2}$

49. $d = \sqrt{(x_2 - x_1)^2 + (y_2 - y_1)^2}$

$= \sqrt{(\sqrt{d} - \sqrt{d})^2 + [\sqrt{3c} - (-\sqrt{3c})]^2}$

$= \sqrt{0^2 + (2\sqrt{3c})^2} = \sqrt{4 \cdot 3c}$

$= 2\sqrt{3c}$

(Since these points are on a vertical line, we could have found the distance between them by finding $|y_2 - y_1| = |-\sqrt{3c} - \sqrt{3c}| = |-2\sqrt{3c}| = 2\sqrt{3c}$.)

50. $\sqrt{72} \approx 8.485$

51. $d = \sqrt{(x_2 - x_1)^2 + (y_2 - y_1)^2}$

$= \sqrt{(3.712 - 5.989)^2 + (-7.784 - 2.001)^2}$

$= \sqrt{100.930954} \approx 10.046$

52. Let $P_1 = (x_1, y_1)$, $P_2 = (x_2, y_2)$, and

$M = \left(\dfrac{x_1 + x_2}{2}, \dfrac{y_1 + y_2}{2}\right)$. Let $d(AB)$ denote the

distance from point A to point B.

i) $\quad d(P_1 M)$

$= \sqrt{\left(\dfrac{x_1 + x_2}{2} - x_1\right)^2 + \left(\dfrac{y_1 + y_2}{2} - y_1\right)^2}$

$= \dfrac{1}{2}\sqrt{(x_2 - x_1)^2 + (y_2 - y_1)^2};$

$\quad d(P_2 M)$

$= \sqrt{\left(\dfrac{x_1 + x_2}{2} - x_2\right)^2 + \left(\dfrac{y_1 + y_2}{2} - y_2\right)^2}$

$= \dfrac{1}{2}\sqrt{(x_1 - x_2)^2 + (y_1 - y_2)^2}$

$= \dfrac{1}{2}\sqrt{(x_2 - x_1)^2 + (y_2 - y_1)^2} = d(P_1 M).$

ii) $\quad d(P_1 M) + d(P_2 M)$

$= \dfrac{1}{2}\sqrt{(x_2 - x_1)^2 + (y_2 - y_1)^2} +$

$\quad \dfrac{1}{2}\sqrt{(x_2 - x_1)^2 + (y_2 - y_1)^2}$

$= \sqrt{(x_2 - x_1)^2 + (y_2 - y_1)^2}$

$= d(P_1 P_2).$

Exercise Set 13.2

1. $y = x^2$

a) This is equivalent to $y = (x - 0)^2 + 0$. The vertex is $(0, 0)$.

b) We choose some x-values on both sides of the vertex and compute the corresponding values of y. The graph opens upward, because the coefficient of x^2, 1, is positive.

x	y
0	0
1	1
2	4
-1	1
-2	4

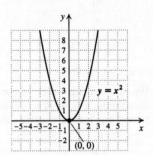

2. $x = y^2$

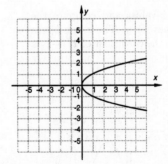

3. $x = y^2 + 4y + 1$

a) We find the vertex by completing the square.

$x = (y^2 + 4y + 4) + 1 - 4$

$x = (y + 2)^2 - 3$

The vertex is $(-3, -2)$.

b) To find ordered pairs, we choose values for y and compute the corresponding values of x. The graph opens to the right, because the coefficient of y^2, 1, is positive.

x	y
-3	-2
-2	-3
-2	-1
1	-4
1	0

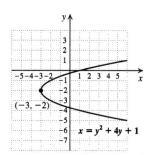

$x = y^2 + 4y + 1$

4. $y = x^2 - 2x + 3$

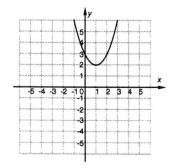

5. $y = -x^2 + 4x - 5$

a) We can find the vertex by computing the first coordinate, $x = -b/2a$, and then substituting to find the second coordinate:
$$x = -\frac{b}{2a} = -\frac{4}{2(-1)} = 2$$
$$y = -x^2 + 4x - 5 = -(2)^2 + 4(2) - 5 = -1$$
The vertex is $(2, -1)$.

b) We choose some x-values and compute the corresponding values for y. The graph opens downward because the coefficient of x^2, -1, is negative.

x	y
2	-1
3	-2
4	-5
1	-2
0	-5

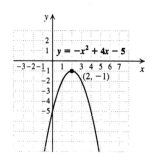

$y = -x^2 + 4x - 5$

$(2, -1)$

6. $x = 4 - 3y - y^2$

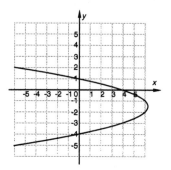

7. $x = y^2 + 1$

a) $x = (y - 0)^2 + 1$

The vertex is $(1, 0)$.

b) To find the ordered pairs, we choose y-values and compute the corresponding values for x. The graph opens to the right, because the coefficient of y^2, 1, is positive.

x	y
1	0
2	1
5	2
2	-1
5	-2

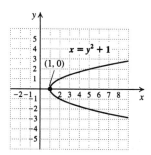

$x = y^2 + 1$

$(1, 0)$

8. $x = 2y^2$

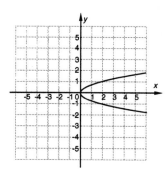

9. $x = -1 \cdot y^2$

a) $x = -1 \cdot (y - 0)^2 + 0$

The vertex is $(0, 0)$.

b) We choose y-values and compute the corresponding values for x. The graph opens to the left, because the coefficient of y^2, -1, is negative.

x	y
0	0
-1	1
-4	2
-1	-1
-4	-2

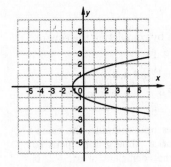

10. $x = y^2 - 1$

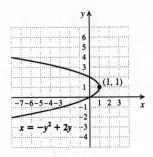

11. $x = -y^2 + 2y$

   a) We find the vertex by computing the second coordinate, $y = -b/2a$, and then substituting to find the first coordinate:

$$y = -\frac{b}{2a} = -\frac{2}{2(-1)} = 1$$
$$x = -y^2 + 2y = -(1)^2 + 2(1) = 1$$

   The vertex is $(1, 1)$.

   b) We choose y-values and compute the corresponding values for x. The graph opens to the left, because the coefficient of y^2, -1, is negative.

x	y
1	1
0	0
-3	-1
0	2
-3	3

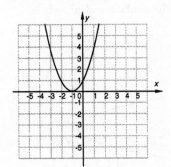

12. $x = y^2 + y - 6$

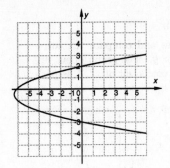

13. $x = 8 - y - y^2$

   a) We find the vertex by completing the square.

$$x = -(y^2 + y) + 8$$
$$x = -\left(y^2 + y + \frac{1}{4}\right) + 8 + \frac{1}{4}$$
$$x = -\left(y + \frac{1}{2}\right)^2 + \frac{33}{4}$$

   The vertex is $\left(\dfrac{33}{4}, -\dfrac{1}{2}\right)$.

   b) We choose y-values and compute the corresponding values for x. The graph opens to the left, because the coefficient of y^2, -1, is negative.

x	y
$\dfrac{33}{4}$	$-\dfrac{1}{2}$
8	0
6	1
2	2
8	-1
6	-2
2	-3

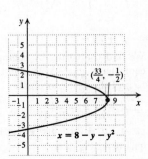

14. $y = x^2 + 2x + 1$

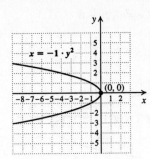

15. $y = x^2 - 2x + 1$

    a) $y = (x-1)^2 + 0$

    The vertex is $(1, 0)$.

    b) We choose x-values and compute the corresponding values for y. The graph opens upward, because the coefficient of x^2, 1, is positive.

x	y
1	0
0	1
-1	4
2	1
3	4

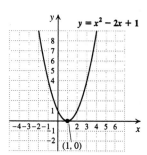

16. $y = -\dfrac{1}{2}x^2$

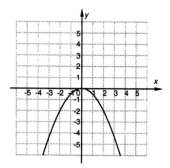

17. $x = -y^2 + 2y + 3$

    a) We find the vertex by computing the second coordinate, $y = -b/2a$, and then substituting to find the first coordinate.

$$y = -\frac{b}{2a} = -\frac{2}{2(-1)} = 1$$
$$x = -y^2 + 2y + 3 = -(1)^2 + 2(1) + 3 = 4$$

    The vertex is $(4, 1)$.

    b) We choose y-values and compute the corresponding values for x. The graph opens to the left, because the coefficient of y^2, -1, is negative.

x	y
4	1
3	0
0	-1
3	2
0	3

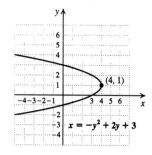

18. $x = -y^2 - 2y + 3$

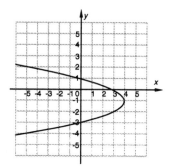

19. $x = -2y^2 - 4y + 1$

    a) We find the vertex by completing the square.

$$x = -2(y^2 + 2y) + 1$$
$$x = -2(y^2 + 2y + 1) + 1 + 2$$
$$x = -2(y+1)^2 + 3$$

    The vertex is $(-3, -1)$.

    b) We choose y-values and compute the corresponding values for x. The graph opens to the left, because the coefficient of y^2, -2, is negative.

x	y
3	-1
1	-2
-5	-3
1	0
-5	1

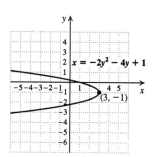

20. $x = 2y^2 + 4y - 1$

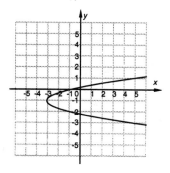

21. $\quad (x-h)^2 + (y-k)^2 = r^2 \quad$ Standard form

$\quad (x-0)^2 + (y-0)^2 = 7^2 \quad$ Substituting

$\qquad\qquad\quad x^2 + y^2 = 49 \quad$ Simplifying

22. $x^2 + y^2 = 16$

23. $\quad (x-h)^2 + (y-k)^2 = r^2 \quad$ Standard form

$\quad [x-(-2)]^2 + (y-7)^2 = 3^2 \quad$ Substituting

$\qquad\quad (x+2)^2 + (y-7)^2 = 9$

24. $(x - 5)^2 + (y - 6)^2 = 1$

25. $\qquad (x - h)^2 + (y - k)^2 = r^2 \qquad$ Standard form
$\qquad [x - (-4)]^2 + (y - 3)^2 = (\sqrt{5})^2 \quad$ Substituting
$\qquad (x + 4)^2 + (y - 3)^2 = 5$

26. $(x + 2)^2 + (y - 7)^2 = 3$

27. $\qquad (x - h)^2 + (y - k)^2 = r^2$
$\qquad [x - (-7)]^2 + [y - (-2)]^2 = (5\sqrt{2})^2$
$\qquad (x + 7)^2 + (y + 2)^2 = 50$
$\qquad\qquad\qquad [(5\sqrt{2})^2 = 25 \cdot 2 = 50]$

28. $(x + 5)^2 + (y + 8)^2 = 20$

29. Since the center is $(0, 0)$, we have
$$(x - 0)^2 + (y - 0)^2 = r^2 \text{ or } x^2 + y^2 = r^2$$
The circle passes through $(-3, 4)$. We find r^2 by substituting -3 for x and 4 for y.
$$(-3)^2 + 4^2 = r^2$$
$$9 + 16 = r^2$$
$$25 = r^2$$
Then $x^2 + y^2 = 25$ is an equation of the circle.

30. $(x - 3)^2 + (y + 2)^2 = 64$

31. Since the center is $(-4, 1)$, we have
$$[x - (-4)]^2 + (y - 1)^2 = r^2, \text{ or}$$
$$(x + 4)^2 + (y - 1)^2 = r^2.$$
The circle passes through $(-2, 5)$. We find r^2 by substituting -2 for x and 5 for y.
$$(-2 + 4)^2 + (5 - 1)^2 = r^2$$
$$4 + 16 = r^2$$
$$20 = r^2$$
Then $(x + 4)^2 + (y - 1)^2 = 20$ is an equation of the circle.

32. $x^2 + y^2 = 10$

33. We write standard form.
$$(x - 0)^2 + (y - 0)^2 = 6^2$$
The center is $(0, 0)$, and the radius is 6.

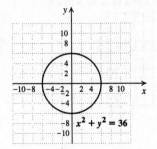

34. Center: $(0, 0)$; radius: 5

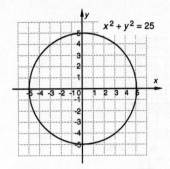

35. $\qquad (x + 1)^2 + (y + 3)^2 = 4$
$\qquad [x - (-1)]^2 + [y - (-3)]^2 = 2^2 \quad$ Standard form

The center is $(-1, -3)$, and the radius is 2.

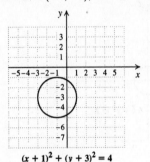

36. Center: $(2, -3)$
Radius: 1

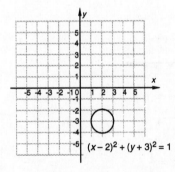

37. $\qquad (x - 8)^2 + (y + 3)^2 = 40$
$\qquad (x - 8)^2 + [y - (-3)]^2 = (2\sqrt{10})^2 \quad (\sqrt{40} = 2\sqrt{10})$

The center is $(8, -3)$, and the radius is $2\sqrt{10}$.

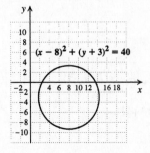

38. Center: $(-5, 1)$

Radius: $5\sqrt{3}$

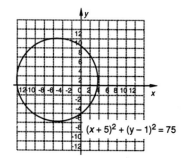

39. $$x^2 + y^2 = 2$$
$$(x-0)^2 + (y-0)^2 = (\sqrt{2})^2 \quad \text{Standard form}$$

The center is $(0, 0)$, and the radius is $\sqrt{2}$.

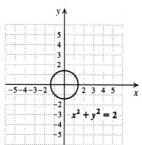

40. Center: $(0, 0)$

Radius: $\sqrt{3}$

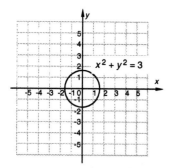

41. $$(x-5)^2 + y^2 = \frac{1}{4}$$
$$(x-5)^2 + (y-0)^2 = \left(\frac{1}{2}\right)^2 \quad \text{Standard form}$$

The center is $(5, 0)$, and the radius is $\frac{1}{2}$.

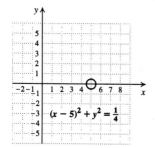

42. Center: $(0, 1)$

Radius: $\dfrac{1}{5}$

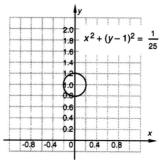

43. $$x^2 + y^2 + 8x - 6y - 15 = 0$$
$$x^2 + 8x + y^2 - 6y = 15$$
$$(x^2 + 8x + 16) + (y^2 - 6y + 9) = 15 + 16 + 9$$
$$\text{Completing the square twice}$$
$$(x + 4)^2 + (y - 3)^2 = 40$$
$$[x - (-4)]^2 + (y - 3)^2 = (2\sqrt{10})^2$$
$$\text{Standard form}$$

The center is $(-4, 3)$, and the radius is $2\sqrt{10}$.

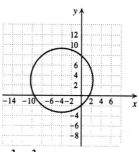

$$x^2 + y^2 + 8x - 6y - 15 = 0$$

44. Center: $(-3, 2)$

Radius: $2\sqrt{7}$

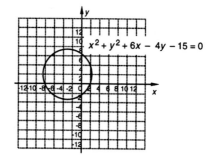

45.
$$x^2 + y^2 - 8x + 2y + 13 = 0$$
$$x^2 - 8x + y^2 + 2y = -13$$
$$(x^2 - 8x + 16) + (y^2 + 2y + 1) = -13 + 16 + 1$$

Completing the square twice

$$(x - 4)^2 + (y + 1)^2 = 4$$
$$(x - 4)^2 + [y - (-1)]^2 = 2^2$$

Standard form

The center is $(4, -1)$, and the radius is 2.

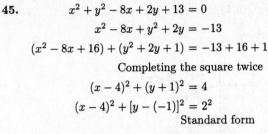

$x^2 + y^2 - 8x + 2y + 13 = 0$

46. Center: $(-3, -2)$

Radius: 1

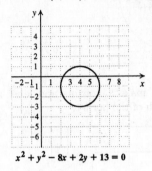

$x^2 + y^2 + 6x + 4y + 12 = 0$

47.
$$x^2 + y^2 - 4x = 0$$
$$x^2 - 4x + y^2 = 0$$
$$(x^2 - 4x + 4) + y^2 = 4 \quad \text{Completing the square}$$
$$(x - 2)^2 + (y - 0)^2 = 2^2 \quad \text{Standard form}$$

The center is $(2, 0)$, and the radius is 2.

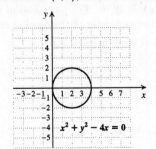

$x^2 + y^2 - 4x = 0$

48. Center: $(-3, 0)$

Radius: 3

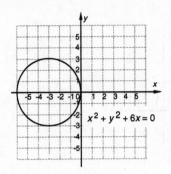

$x^2 + y^2 + 6x = 0$

49.
$$x^2 + y^2 + 10y - 75 = 0$$
$$x^2 + y^2 + 10y = 75$$
$$x^2 + (y^2 + 10y + 25) = 75 + 25$$
$$(x - 0)^2 + (y + 5)^2 = 100$$
$$(x - 0)^2 + [y - (-5)]^2 = 10^2$$

The center is $(0, -5)$, and the radius is 10.

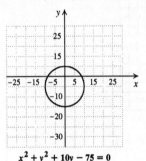

$x^2 + y^2 + 10y - 75 = 0$

50. Center: $(4, 0)$

Radius: 10

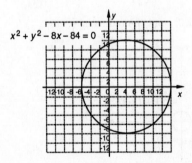

$x^2 + y^2 - 8x - 84 = 0$

51.
$$x^2 + y^2 + 7x - 3y - 10 = 0$$
$$x^2 + 7x + y^2 - 3y = 10$$
$$\left(x^2 + 7x + \frac{49}{4}\right) + \left(y^2 - 3y + \frac{9}{4}\right) = 10 + \frac{49}{4} + \frac{9}{4}$$
$$\left(x + \frac{7}{2}\right)^2 + \left(y - \frac{3}{2}\right)^2 = \frac{98}{4}$$
$$\left[x - \left(-\frac{7}{2}\right)\right]^2 + \left(y - \frac{3}{2}\right)^2 = \left(\sqrt{\frac{98}{4}}\right)^2$$

The center is $\left(-\dfrac{7}{2}, \dfrac{3}{2}\right)$, and the radius is $\sqrt{\dfrac{98}{4}}$, or $\dfrac{\sqrt{98}}{2}$, or $\dfrac{7\sqrt{2}}{2}$.

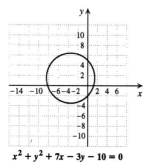

$x^2 + y^2 + 7x - 3y - 10 = 0$

52. Center: $\left(\dfrac{21}{2}, \dfrac{33}{2}\right)$

Radius: $\dfrac{\sqrt{1462}}{2}$

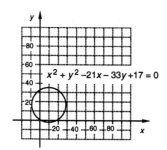

$x^2 + y^2 - 21x - 33y + 17 = 0$

53.
$$4x^2 + 4y^2 = 1$$
$$x^2 + y^2 = \dfrac{1}{4} \qquad \text{Multiplying by } \dfrac{1}{4} \text{ on both sides}$$
$$(x - 0)^2 + (y - 0)^2 = \left(\dfrac{1}{2}\right)^2$$

The center is $(0, 0)$, and the radius is $\dfrac{1}{2}$.

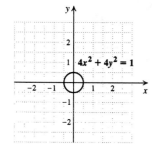

$4x^2 + 4y^2 = 1$

54. Center: $(0, 0)$

Radius: $\dfrac{1}{5}$

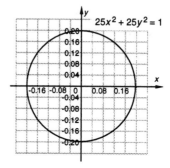

$25x^2 + 25y^2 = 1$

55. *Familiarize*. We make a drawing and label it. Let x represent the width of the border.

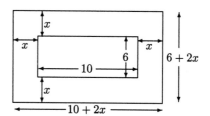

The perimeter of the larger rectangle is

$2(10 + 2x) + 2(6 + 2x)$, or $8x + 32$.

The perimeter of the smaller rectangle is

$2(10) + 2(6)$, or 32.

Translate. The perimeter of the larger rectangle is twice the perimeter of the smaller rectangle.

$$8x + 32 = 2 \cdot 32$$

Carry out. We solve the equation.

$$8x + 32 = 64$$
$$8x = 32$$
$$x = 4$$

Check. If the width of the border is 4 in., then the length and width of the larger rectangle are 18 in. and 14 in. Thus its perimeter is $2(18) + 2(14)$, or 64 in. The perimeter of the smaller rectangle is 32 in. The perimeter of the larger rectangle is twice the perimeter of the smaller rectangle.

State. The width of the border is 4 in.

56. 2640 mi

57. ◈

58. ◈

59. $x = y^2 - y - 6$

$x = \left(y^2 - y + \dfrac{1}{4}\right) - 6 - \dfrac{1}{4}$

$x = \left(y - \dfrac{1}{2}\right)^2 - \dfrac{25}{4}$

The vertex is $\left(-\dfrac{25}{4}, \dfrac{1}{2}\right)$.

x	y
$-\dfrac{25}{4}$	$\dfrac{1}{2}$
-6	1
-4	2
0	3
-6	0
-4	-1
0	-2

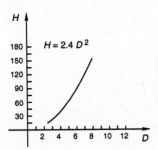

a) Graph $x = 2$ on the same set of axes as $x = y^2 - y - 6$ and approximate the y-coordinates of the points of intersection. (See the graph above.) The solutions are approximately 3.4 and -2.4.

b) Graph $x = -3$ on the same set of axes as $x = y^2 - y - 6$ and approximate the y-coordinates of the points of intersection. (See the graph above.) The solutions are approximately 2.3 and -1.3.

60.

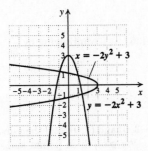

61.

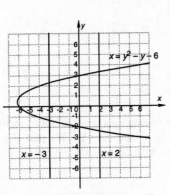

Reflect one graph across the line $y = x$ to obtain the other.

62.

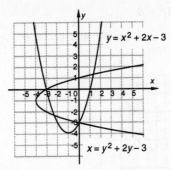

Reflect one graph across the line $y = x$ to obtain the other.

63. $\qquad 4x^2 + 4y^2 + 4x - 8y + 1 = 0$

$4x^2 + 4x + 4y^2 - 8y = -1$

$4(x^2 + x) + 4(y^2 - 2y) = -1$

$x^2 + x + y^2 - 2y = -\dfrac{1}{4}$ $\quad$ Dividing by 4

$\left(x^2 + x + \dfrac{1}{4}\right) + (y^2 - 2y + 1) = -\dfrac{1}{4} + \dfrac{1}{4} + 1$

Completing the square twice

$\left(x + \dfrac{1}{2}\right)^2 + (y - 1)^2 = 1$

$\left[x - \left(-\dfrac{1}{2}\right)\right]^2 + (y - 1)^2 = 1^2$

Standard form

The center is $\left(-\dfrac{1}{2}, 1\right)$, and the radius is 1.

64. Center: $\left(-1, -\dfrac{1}{3}\right)$

Radius: 2

65. a) When the circle is positioned on a coordinate system as shown in the text, the center lies on the y-axis. To find the center, we will find the point on the y-axis that is equidistant from $(-4, 0)$ and $(0, 2)$. Let $(0, y)$ be this point.

$\sqrt{[0 - (-4)]^2 + (y - 0)^2} = \sqrt{(0 - 0)^2 + (y - 2)^2}$

$4^2 + y^2 = 0^2 + (y - 2)^2$

Squaring both sides

$16 + y^2 = y^2 - 4y + 4$

$12 = -4y$

$-3 = y$

The center of the circle is $(0, -3)$.

b) We find the radius of the circle.

$(x - 0)^2 + [y - (-3)]^2 = r^2$ $\quad$ Standard form

$x^2 + (y + 3)^2 = r^2$

$(-4)^2 + (0 + 3)^2 = r^2$ $\quad$ Substituting $(-4, 0)$ for (x, y)

$16 + 9 = r^2$

$25 = r^2$

$5 = r$

The radius is 5 ft.

66. $(x-3)^2 + (y+5)^2 = 9$

67. We make a drawing of a circle with center $(-7,-4)$ and tangent to the x-axis.

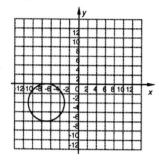

We see that the circle touches the x-axis at $(-7,0)$. Hence, the radius is the distance between $(-7,0)$ and $(-7,-4)$, or $|-4-0|$, or 4. Now we write the equation of the circle.
$$(x-h)^2 + (y-k)^2 = r^2$$
$$[x-(-7)]^2 + [y-(-4)]^2 = 4^2$$
$$(x+7)^2 + (y+4)^2 = 16$$

68. $(x-3)^2 + y^2 = 25$

69. We use the formula $C = 2\pi r$ to find the radius.
$$2\pi r = 8\pi$$
$$r = 4$$

Write the equation of a circle with a center $(-3,5)$ and radius 4.
$$[x-(-3)]^2 + (y-5)^2 = 4^2$$
$$(x+3)^2 + (y-5)^2 = 16$$

70. $x^2 + (y-30.6)^2 = 590.49$

Exercise Set 13.3

1. $\dfrac{x^2}{4} + \dfrac{y^2}{1} = 1$

$\dfrac{x^2}{2^2} + \dfrac{y^2}{1^2} = 1$

The x-intercepts are $(2,0)$ and $(-2,0)$, and the y-intercepts are $(0,1)$ and $(0,-1)$. We plot these points and connect them with an oval-shaped curve.

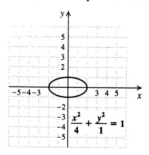

2. $\dfrac{x^2}{1} + \dfrac{y^2}{4} = 1$

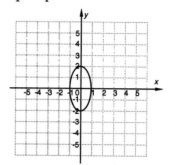

3. $\dfrac{x^2}{16} + \dfrac{y^2}{25} = 1$

$\dfrac{x^2}{4^2} + \dfrac{y^2}{5^2} = 1$

The x-intercepts are $(4,0)$ and $(-4,0)$, and the y-intercepts are $(0,5)$ and $(0,-5)$. We plot these points and connect them with an oval-shaped curve.

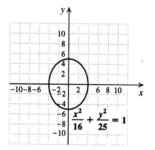

4. $\dfrac{x^2}{9} + \dfrac{y^2}{25} = 1$

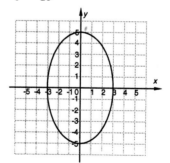

5. $4x^2 + 9y^2 = 36$

$\dfrac{1}{36}(4x^2 + 9y^2) = \dfrac{1}{36}(36)$ Multiplying by $\dfrac{1}{36}$

$\dfrac{x^2}{9} + \dfrac{y^2}{4} = 1$

$\dfrac{x^2}{3^2} + \dfrac{y^2}{2^2} = 1$

The x-intercepts are $(-3,0)$ and $(3,0)$, and the y-intercepts are $(0,-2)$ and $(0,2)$. We plot these points and connect them with an oval-shaped curve.

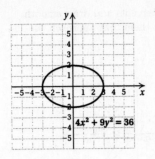

6. $9x^2 + 4y^2 = 36$

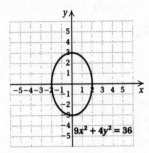

7. $16x^2 + 9y^2 = 144$

$$\frac{x^2}{9} + \frac{y^2}{16} = 1 \quad \text{Multiplying by } \frac{1}{144}$$

$$\frac{x^2}{3^2} + \frac{y^2}{4^2} = 1$$

The x-intercepts are $(3, 0)$ and $(-3, 0)$, and the y-intercepts are $(0, 4)$ and $(0, -4)$. We plot these points and connect them with an oval-shaped curve.

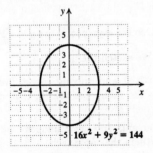

8. $9x^2 + 16y^2 = 144$

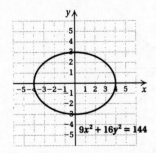

9. $$2x^2 + 3y^2 = 6$$

$$\frac{x^2}{3} + \frac{y^2}{2} = 1 \quad \text{Multiplying by } \frac{1}{6}$$

$$\frac{x^2}{(\sqrt{3})^2} + \frac{y^2}{(\sqrt{2})^2} = 1$$

The x-intercepts are $(\sqrt{3}, 0)$ and $(-\sqrt{3}, 0)$, and the y-intercepts are $(0, \sqrt{2})$ and $(0, -\sqrt{2})$. We plot these points and connect them with an oval-shaped curve.

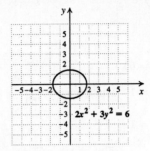

10. $5x^2 + 7y^2 = 35$

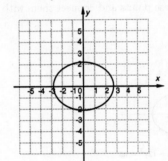

11. $$4x^2 + 9y^2 = 1$$

$$\frac{x^2}{\frac{1}{4}} + \frac{y^2}{\frac{1}{9}} = 1 \qquad 4x^2 = \frac{x^2}{\frac{1}{4}},\ 9y^2 = \frac{y^2}{\frac{1}{9}}$$

$$\frac{x^2}{\left(\frac{1}{2}\right)^2} + \frac{y^2}{\left(\frac{1}{3}\right)^2} = 1$$

The x-intercepts are $\left(\frac{1}{2}, 0\right)$, and $\left(-\frac{1}{2}, 0\right)$, and the y-intercepts are $\left(0, \frac{1}{3}\right)$ and $\left(0, -\frac{1}{3}\right)$. We plot these points and connect them with an oval-shaped curve.

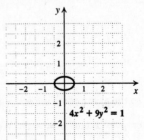

12. $25x^2 + 16y^2 = 1$

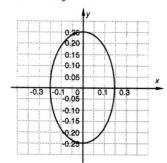

13. $\qquad 5x^2 + 12y^2 = 60$

$$\frac{x^2}{12} + \frac{y^2}{5} = 1 \quad \text{Multiplying by } \frac{1}{60}$$

$$\frac{x^2}{(\sqrt{12})^2} + \frac{y^2}{(\sqrt{5})^2} = 1$$

The x-intercepts are $(\sqrt{12}, 0)$ and $(-\sqrt{12}, 0)$, or $(2\sqrt{3}, 0)$ and $(-2\sqrt{3}, 0)$ and the y-intercepts are $(0, \sqrt{5})$ and $(0, -\sqrt{5})$. We plot these points and connect them with an oval-shaped curve.

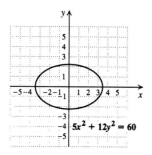

14. $8x^2 + 3y^2 = 24$

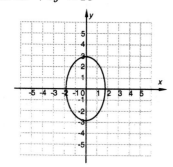

15. $\dfrac{x^2}{16} - \dfrac{y^2}{16} = 1$

$$\frac{x^2}{4^2} - \frac{y^2}{4^2} = 1$$

a) $a = 4$ and $b = 4$, so the asymptotes are $y = \dfrac{4}{4}x$ and $y = -\dfrac{4}{4}x$, or $y = x$ and $y = -x$. We sketch them.

b) Replacing y with 0 and solving for x, we get $x = \pm 4$, so the intercepts are $(4, 0)$ and $(-4, 0)$.

c) We plot the intercepts and draw smooth curves through them that approach the asymptotes.

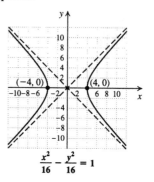

16. $\dfrac{y^2}{9} - \dfrac{x^2}{9} = 1$

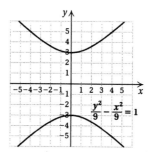

17. $\dfrac{y^2}{16} - \dfrac{x^2}{9} = 1$

$$\frac{y^2}{4^2} - \frac{x^2}{3^2} = 1$$

a) $a = 3$ and $b = 4$, so the asymptotes are $y = \dfrac{4}{3}x$ and $y = -\dfrac{4}{3}x$. We sketch them.

b) Replacing x with 0 and solving for y, we get $y = \pm 4$, so the intercepts are $(0, 4)$ and $(0, -4)$.

c) We plot the intercepts and draw smooth curves through them that approach the asymptotes.

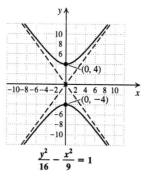

18. $\dfrac{x^2}{9} - \dfrac{y^2}{4} = 1$

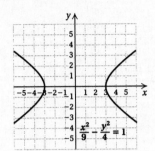

19. $\dfrac{x^2}{25} - \dfrac{y^2}{36} = 1$

$\dfrac{x^2}{5^2} - \dfrac{y^2}{6^2} = 1$

a) $a = 5$ and $b = 6$, so the asymptotes are $y = \dfrac{6}{5}x$

and $y = -\dfrac{6}{5}x$. We sketch them.

b) Replacing y with 0 and solving for x, we get $x = \pm 5$, so the intercepts are $(5, 0)$ and $(-5, 0)$.

c) We plot the intercepts and draw smooth curves through them that approach the asymptotes.

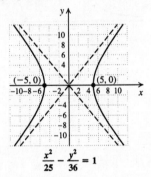

20. $\dfrac{y^2}{9} - \dfrac{x^2}{25} = 1$

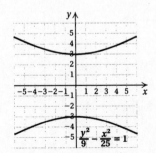

21. $x^2 - y^2 = 4$

$\dfrac{x^2}{4} - \dfrac{y^2}{4} = 1$ Multiplying by $\dfrac{1}{4}$

$\dfrac{x^2}{2^2} - \dfrac{y^2}{2^2} = 1$

a) $a = 2$ and $b = 2$, so the asymptotes are $y = \dfrac{2}{2}x$ and $y = -\dfrac{2}{2}x$, or $y = x$ and $y = -x$. We sketch them.

b) Replacing y with 0 and solving for x, we get $x = \pm 2$, so the intercepts are $(2, 0)$ and $(-2, 0)$.

c) We plot the intercepts and draw smooth curves through them that approach the asymptotes.

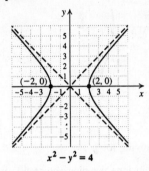

22. $y^2 - x^2 = 25$

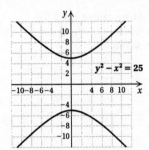

23. $4y^2 - 9x^2 = 36$

$\dfrac{y^2}{9} - \dfrac{x^2}{4} = 1$ Multiplying by $\dfrac{1}{36}$

$\dfrac{y^2}{3^2} - \dfrac{x^2}{2^2} = 1$

a) $a = 2$ and $b = 3$, so the asymptotes are $y = \dfrac{3}{2}x$ and $y = -\dfrac{3}{2}x$. We sketch them.

b) Replacing x with 0 and solving for y, we get $y = \pm 3$, so the intercepts are $(0, 3)$ and $(0, -3)$.

c) We plot the intercepts and draw smooth curves through them that approach the asymptotes.

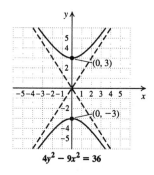

$$4y^2 - 9x^2 = 36$$

24. $25x^2 - 16y^2 = 400$

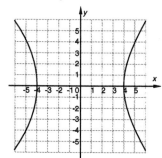

25. $xy = 6$

  $y = \dfrac{6}{x}$ Solving for y

We find some solutions, keeping the results in a table.

x	y
1	6
2	3
3	2
6	1
$\frac{1}{2}$	12
$\frac{1}{3}$	18
-1	-6
-2	-3
-3	-2
-6	-1
$-\frac{1}{2}$	-12
$-\frac{1}{3}$	-18

Note that we cannot use 0 for x. The x-axis and the y-axis are the asymptotes.

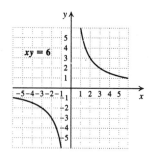

26. $xy = -4$

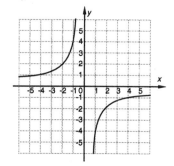

27. $xy = -9$

  $y = -\dfrac{9}{x}$ Solving for y

x	y
1	-9
3	-3
9	-1
$\frac{1}{2}$	-18
-1	9
-3	3
-9	1
$-\frac{1}{2}$	18

Note that we cannot use 0 for x. The x-axis and the y-axis are the asymptotes.

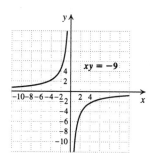

28. $xy = 3$

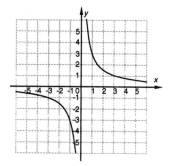

29. $xy = -1$

  $y = -\dfrac{1}{x}$ Solving for y

x	y
1	-1
2	$-\frac{1}{2}$
4	$-\frac{1}{4}$
$\frac{1}{2}$	-2
$\frac{1}{4}$	-4
-1	1
-2	$\frac{1}{2}$
-4	$\frac{1}{4}$
$-\frac{1}{2}$	2
$-\frac{1}{4}$	4

Note that we cannot use 0 for x. The x-axis and the y-axis are the asymptotes.

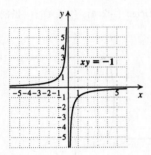

30. $xy = -2$

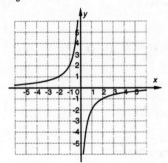

31. $xy = 2$

$$y = \frac{2}{x} \qquad \text{Solving for } x$$

x	y
1	2
2	1
4	$\frac{1}{2}$
$\frac{1}{2}$	4
-1	-2
-2	-1
-4	$-\frac{1}{2}$
$-\frac{1}{2}$	-4

Note that we cannot use 0 for x. The x-axis and the y-axis are the asymptotes.

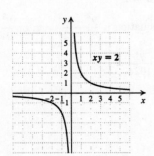

32. $xy = 1$

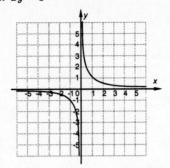

33. $x^2 + y^2 - 10x + 8y - 40 = 0$

Completing the square twice, we obtain an equivalent equation:

$$(x^2 - 10x) + (y^2 + 8y) = 40$$
$$(x^2 - 10x + 25) + (y^2 + 8y + 16) = 40 + 25 + 16$$
$$(x - 5)^2 + (y + 4)^2 = 81$$

The graph is a circle.

34. Parabola

35. $9x^2 - 4y^2 - 36 = 0$

$$9x^2 - 4y^2 = 36$$
$$\frac{x^2}{4} - \frac{y^2}{9} = 1$$

The graph is a hyperbola.

36. Parabola

37. $4x^2 + 25y^2 - 100 = 0$

$$4x^2 + 25y^2 = 100$$
$$\frac{x^2}{25} + \frac{y^2}{4} = 1$$

The graph is an ellipse.

38. Circle

39.
$$x^2 + y^2 = 2x + 4y + 4$$
$$x^2 - 2x + y^2 - 4y = 4$$
$$(x^2 - 2x + 1) + (y^2 - 4y + 4) = 4 + 1 + 4$$
$$(x - 1)^2 + (y - 2)^2 = 9$$

The graph is a circle.

40. Circle

41.
$$4x^2 = 64 - y^2$$
$$4x^2 + y^2 = 64$$
$$\frac{x^2}{16} + \frac{y^2}{64} = 1$$

The graph is an ellipse.

42. Hyperbola

43. $x - \dfrac{3}{y} = 0$

$$x = \frac{3}{y}$$
$$xy = 3$$

The graph is a hyperbola.

44. Parabola

45. $y + 6x = x^2 + 6$

$$y = x^2 - 6x + 6$$

The graph is a parabola.

46. Hyperbola

47. $9y^2 = 36 + 4x^2$

$$9y^2 - 4x^2 = 36$$

$$\frac{y^2}{4} - \frac{x^2}{9} = 1$$

The graph is a hyperbola.

48. Circle

49. $\sqrt[3]{125t^{15}} = \sqrt[3]{5^3 \cdot (t^5)^3} = 5t^5$

50. $\pm i\sqrt{5}$

51. $\dfrac{4\sqrt{2} - 5\sqrt{3}}{6\sqrt{3} - 8\sqrt{2}} = \dfrac{4\sqrt{2} - 5\sqrt{3}}{6\sqrt{3} - 8\sqrt{2}} \cdot \dfrac{6\sqrt{3} + 8\sqrt{2}}{6\sqrt{3} + 8\sqrt{2}}$

$$= \frac{24\sqrt{6} + 32 \cdot 2 - 30 \cdot 3 - 40\sqrt{6}}{36 \cdot 3 - 64 \cdot 2}$$

$$= \frac{-26 - 16\sqrt{6}}{-20}$$

$$= \frac{-2(13 + 8\sqrt{6})}{-2 \cdot 10}$$

$$= \frac{13 + 8\sqrt{6}}{10}$$

52. Smaller plane: 400 mph, larger plane: 720 mph

53.

54. 2.134×10^8 mi

55. a), b) See the answer section in the text.

56.

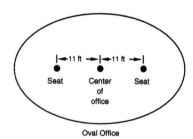

Oval Office

57. For the given ellipse, $a = 6/2$, or 3, and $b = 2/2$, or 1.
The patient's mouth should be at a distance 2c from the
light source, where the coordinates of the foci of the ellipse
are $(-c, 0)$ and $(c, 0)$. From Exercise 55(b), we know $b^2 = a^2 - c^2$. We use this to find c.

$$b^2 = a^2 - c^2$$

$$1^2 = 3^2 - c^2 \quad \text{Substituting}$$

$$c^2 = 8$$

$$c = \sqrt{8}$$

Then $2c = 2\sqrt{8} \approx 5.66$. The patient's mouth should be
about 5.66 ft from the light source.

58. $\dfrac{x^2}{81} + \dfrac{y^2}{121} = 1$

59. $16x^2 + y^2 + 96x - 8y + 144 = 0$

$$16x^2 + 96x + y^2 - 8y = -144$$

$$16(x^2 + 6x) + (y^2 - 8y) = -144$$

$$16(x^2 + 6x + 9) + (y^2 - 8y + 16) = -144 + 144 + 16$$

$$16(x + 3)^2 + (y - 4)^2 = 16$$

$$\frac{(x + 3)^2}{1} + \frac{(y - 4)^2}{16} = 1 \quad \begin{array}{l}\text{Standard} \\ \text{form}\end{array}$$

$h = -3$, $k = 4$, $a = 1$, $b = 4$

Center: $(-3, 4)$

Vertices: $(-2, 4)$, $(-4, 4)$, $(-3, 8)$, $(-3, 0)$

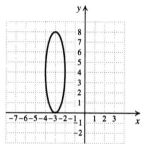

$16x^2 + y^2 + 96x - 8y + 144 = 0$

60. Center: $(1, -1)$

Vertices: $(6, -1)$, $(-4, -1)$, $(1, 1)$, $(1, -3)$

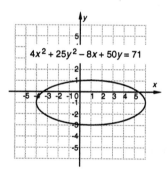

$4x^2 + 25y^2 - 8x + 50y = 71$

61. Since the intercepts are $(0, 8)$ and $(0, -8)$, we know that
the hyperbola is of the form $\dfrac{y^2}{b^2} - \dfrac{x^2}{a^2} = 1$ and that $b = 8$.
The equations of the asymptotes tell us that $b/a = 4$, so

$$\frac{8}{a} = 4$$

$$a = 2.$$

The equation is $\dfrac{y^2}{8^2} - \dfrac{x^2}{2^2} = 1$, or $\dfrac{y^2}{64} - \dfrac{x^2}{4} = 1$.

62. $\dfrac{x^2}{64} - \dfrac{y^2}{1024} = 1$

63. Center: $(2, -1)$

Vertices: $(-1, -1)$, $(5, -1)$

Asymptotes: $y + 1 = \dfrac{4}{3}(x - 2)$, $y + 1 = -\dfrac{4}{3}(x - 2)$

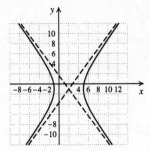

64. $\dfrac{(x + 3)^2}{1} - \dfrac{(y - 2)^2}{4} = 1$

Center: $(-3, 2)$

Vertices: $(-4, 2)$, $(-2, 2)$

Asymptotes: $y - 2 = 2(x + 3)$, $y - 2 = -2(x + 3)$

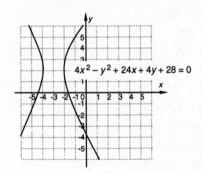

65.
$$4y^2 - 25x^2 - 8y - 100x - 196 = 0$$
$$4(y^2 - 2y) - 25(x^2 + 4x) = 196$$
$$4(y^2 - 2y + 1) - 25(x^2 + 4x + 4) = 196 + 4 - 100$$
$$4(y - 1)^2 - 25(x + 2)^2 = 100$$
$$\dfrac{(y - 1)^2}{25} - \dfrac{(x + 2)^2}{4} = 1 \quad \text{Standard form}$$

Center: $(-2, 1)$

Vertices: $(-2, 6)$, $(-2, -4)$

Asymptotes: $y - 1 = \dfrac{5}{2}(x + 2)$, $y - 1 = -\dfrac{5}{2}(x + 2)$

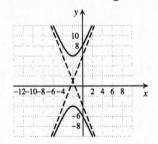

66. $\dfrac{(x - 2)^2}{9} - \dfrac{(y + 1)^2}{9} = 1$

Center: $(2, -1)$

Vertices: $(5, -1)$, $(-1, -1)$

Asymptotes: $y + 1 = x - 2$, $y + 1 = -(x - 2)$

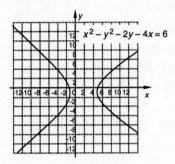

Exercise Set 13.4

1. $x^2 + y^2 = 25$,   (1)

$y - x = 1$ (2)

First solve Eq. (2) for y.

$y = x + 1$ (3)

Then substitute $x + 1$ for y in Eq. (1) and solve for x.

$$x^2 + y^2 = 25$$
$$x^2 + (x + 1)^2 = 25$$
$$x^2 + x^2 + 2x + 1 = 25$$
$$2x^2 + 2x - 24 = 0$$
$$x^2 + x - 12 = 0 \quad \text{Multiplying by } \dfrac{1}{2}$$
$$(x + 4)(x - 3) = 0 \quad \text{Factoring}$$
$$x + 4 = 0 \quad \text{or} \quad x - 3 = 0 \quad \text{Principle of zero}$$
$$\text{products}$$
$$x = -4 \quad \text{or} \qquad x = 3$$

Now substitute these numbers into Eq. (3) and solve for y.

$y = -4 + 1 = -3$

$y = 3 + 1 = 4$

The pairs $(-4, -3)$ and $(3, 4)$ check, so they are the solutions.

2. $(-8, -6)$, $(6, 8)$

3. $4x^2 + 9y^2 = 36$,   (1)

$3y + 2x = 6$ (2)

First solve Eq. (2) for y.

$$3y = -2x + 6$$
$$y = -\dfrac{2}{3}x + 2 \qquad (3)$$

Then substitute $-\frac{2}{3}x + 2$ for y in Eq. (1) and solve for x.

$$4x^2 + 9y^2 = 36$$

$$4x^2 + 9\left(-\frac{2}{3}x + 2\right)^2 = 36$$

$$4x^2 + 9\left(\frac{4}{9}x^2 - \frac{8}{3}x + 4\right) = 36$$

$$4x^2 + 4x^2 - 24x + 36 = 36$$

$$8x^2 - 24x = 0$$

$$x^2 - 3x = 0$$

$$x(x - 3) = 0$$

$x = 0$ or $x = 3$

Now substitute these numbers in Eq. (3) and solve for y.

$$y = -\frac{2}{3} \cdot 0 + 2 = 2$$

$$y = -\frac{2}{3} \cdot 3 + 2 = 0$$

The pairs $(0, 2)$ and $(3, 0)$ check, so they are the solutions.

4. $(2, 0)$, $(0, 3)$

5. $y^2 = x + 3$,   (1)

$2y = x + 4$ (2)

First solve Eq. (2) for x.

$2y - 4 = x$ (3)

Then substitute $2y - 4$ for x in Eq. (1) and solve for y.

$$y^2 = x + 3$$

$$y^2 = (2y - 4) + 3$$

$$y^2 = 2y - 1$$

$$y^2 - 2y + 1 = 0$$

$$(y - 1)(y - 1) = 0$$

$y - 1 = 0$ or $y - 1 = 0$

$y = 1$ or $y = 1$

Now substitute 1 for y in Eq. (3) and solve for x.

$2 \cdot 1 - 4 = x$

$-2 = x$

The pair $(-2, 1)$ checks. It is the solution.

6. $(2, 4)$, $(1, 1)$

7. $x^2 - xy + 3y^2 = 27$,   (1)

$x - y = 2$ (2)

First solve Eq. (2) for y.

$x - 2 = y$ (3)

Then substitute $x - 2$ for y in Eq. (1) and solve for x.

$$x^2 - xy + 3y^2 = 27$$

$$x^2 - x(x - 2) + 3(x - 2)^2 = 27$$

$$x^2 - x^2 + 2x + 3x^2 - 12x + 12 = 27$$

$$3x^2 - 10x - 15 = 0$$

$$x = \frac{-(-10) \pm \sqrt{(-10)^2 - 4(3)(-15)}}{2 \cdot 3}$$

$$x = \frac{10 \pm \sqrt{100 + 180}}{6} = \frac{10 \pm \sqrt{280}}{6}$$

$$x = \frac{10 \pm 2\sqrt{70}}{6} = \frac{5 \pm \sqrt{70}}{3}$$

Now substitute these numbers in Eq. (3) and solve for y.

$$y = \frac{5 + \sqrt{70}}{3} - 2 = \frac{-1 + \sqrt{70}}{3}$$

$$y = \frac{5 - \sqrt{70}}{3} - 2 = \frac{-1 - \sqrt{70}}{3}$$

The pairs $\left(\frac{5 + \sqrt{70}}{3}, \frac{-1 + \sqrt{70}}{3}\right)$ and

$\left(\frac{5 - \sqrt{70}}{3}, \frac{-1 - \sqrt{70}}{3}\right)$ check, so they are the solutions.

8. $\left(\frac{11}{4}, -\frac{9}{8}\right)$, $(1, -2)$

9. $x^2 + 4y^2 = 25$, (1)

$x + 2y = 7$ (2)

First solve Eq. (2) for x.

$x = -2y + 7$ (3)

Then substitute $-2y + 7$ for x in Eq. (1) and solve for y.

$$x^2 + 4y^2 = 25$$

$$(-2y + 7)^2 + 4y^2 = 25$$

$$4y^2 - 28y + 49 + 4y^2 = 25$$

$$8y^2 - 28y + 24 = 0$$

$$2y^2 - 7y + 6 = 0$$

$$(2y - 3)(y - 2) = 0$$

$y = \frac{3}{2}$ or $y = 2$

Now substitute these numbers in Eq. (3) and solve for x.

$$x = -2 \cdot \frac{3}{2} + 7 = 4$$

$$x = -2 \cdot 2 + 7 = 3$$

The pairs $\left(4, \frac{3}{2}\right)$ and $(3, 2)$ check, so they are the solutions.

10. $\left(-\frac{5}{3}, -\frac{13}{3}\right)$, $(3, 5)$

11. $x^2 - xy + 3y^2 = 5$,   (1)

$x - y = 2$ (2)

First solve Eq. (2) for y.

$x - 2 = y$ (3)

Then substitute $x - 2$ for y in Eq. (1) and solve for x.

$$x^2 - xy + 3y^2 = 5$$

$$x^2 - x(x - 2) + 3(x - 2)^2 = 5$$

$$x^2 - x^2 + 2x + 3x^2 - 12x + 12 = 5$$

$$3x^2 - 10x + 7 = 0$$

$$(3x - 7)(x - 1) = 0$$

$x = \frac{7}{3}$ or $x = 1$

Now substitute these numbers in Eq. (3) and solve for y.

$y = \frac{7}{3} - 2 = \frac{1}{3}$

$y = 1 - 2 = -1$

The pairs $\left(\frac{7}{3}, \frac{1}{3}\right)$ and $(1, -1)$ check, so they are the solutions.

12. $\left(\frac{3 + \sqrt{7}}{2}, \frac{-1 + \sqrt{7}}{2}\right)$, $\left(\frac{3 - \sqrt{7}}{2}, \frac{-1 - \sqrt{7}}{2}\right)$

13. $3x + y = 7,$ (1)

  $4x^2 + 5y = 24$ (2)

First solve Eq. (1) for y.

  $y = 7 - 3x$ (3)

Then substitute $7 - 3x$ for y in Eq. (2) and solve for x.

$$4x^2 + 5y = 24$$
$$4x^2 + 5(7 - 3x) = 24$$
$$4x^2 + 35 - 15x = 24$$
$$4x^2 - 15x + 11 = 0$$
$$(4x - 11)(x - 1) = 0$$

$x = \frac{11}{4}$ or $x = 1$

Now substitute these numbers into Eq. (3) and solve for y.

$y = 7 - 3 \cdot \frac{11}{4} = -\frac{5}{4}$

$y = 7 - 3 \cdot 1 = 4$

The pairs $\left(\frac{11}{4}, -\frac{5}{4}\right)$ and $(1, 4)$ check, so they are the solutions.

14. $\left(-3, \frac{5}{2}\right)$, $(3, 1)$

15. $a + b = 7,$ (1)

  $ab = 4$ (2)

First solve Eq. (1) for a.

$a = -b + 7$ (3)

Then substitute $-b + 7$ for a in Eq. (2) and solve for b.

$(-b + 7)b = 4$

  $-b^2 + 7b = 4$

    $0 = b^2 - 7b + 4$

$b = \frac{-(-7) \pm \sqrt{(-7)^2 - 4 \cdot 1 \cdot 4}}{2 \cdot 1}$

$b = \frac{7 \pm \sqrt{33}}{2}$

Now substitute these numbers in Eq. (3) and solve for a.

$a = -\left(\frac{7 + \sqrt{33}}{2}\right) + 7 = \frac{7 - \sqrt{33}}{2}$

$a = -\left(\frac{7 - \sqrt{33}}{2}\right) + 7 = \frac{7 + \sqrt{33}}{2}$

The pairs $\left(\frac{7 - \sqrt{33}}{2}, \frac{7 + \sqrt{33}}{2}\right)$ and

$\left(\frac{7 + \sqrt{33}}{2}, \frac{7 - \sqrt{33}}{2}\right)$ check, so they are the

solutions.

16. $(1, -7)$, $(-7, 1)$

17. $2a + b = 1,$ (1)

   $b = 4 - a^2$ (2)

Eq. (2) is already solved for b. Substitute $4 - a^2$ for b in Eq. (1) and solve for a.

$2a + 4 - a^2 = 1$

    $0 = a^2 - 2a - 3$

    $0 = (a - 3)(a + 1)$

$a = 3$ or $a = -1$

Substitute these numbers in Eq. (2) and solve for b.

$b = 4 - 3^2 = -5$

$b = 4 - (-1)^2 = 3$

The pairs $(3, -5)$ and $(-1, 3)$ check.

18. $(3, 0)$, $\left(-\frac{9}{5}, \frac{8}{5}\right)$

19. $a^2 + b^2 = 89,$ (1)

   $a - b = 3$ (2)

First solve Eq. (2) for a.

  $a = b + 3$ (3)

Then substitute $b + 3$ for a in Eq. (1) and solve for b.

  $(b + 3)^2 + b^2 = 89$

  $b^2 + 6b + 9 + b^2 = 89$

   $2b^2 + 6b - 80 = 0$

    $b^2 + 3b - 40 = 0$

   $(b + 8)(b - 5) = 0$

$b = -8$ or $b = 5$

Substitute these numbers in Eq. (3) and solve for a.

  $a = -8 + 3 = -5$

  $a = 5 + 3 = 8$

The pairs $(-5, -8)$ and $(8, 5)$ check.

20. $(1, 4)$, $(4, 1)$

21. $x^2 + y^2 = 25,$ (1)

   $y^2 = x + 5$ (2)

We substitute $x + 5$ for y^2 in Eq. (1) and solve for x.

   $x^2 + y^2 = 25$

  $x^2 + (x + 5) = 25$

   $x^2 + x - 20 = 0$

  $(x + 5)(x - 4) = 0$

$x + 5 = 0$ or $x - 4 = 0$

  $x = -5$ or $x = 4$

We substitute these numbers for x in either Eq. (1) or Eq. (2) and solve for y. Here we use Eq. (2).

$y^2 = -5 + 5 = 0$ and $y = 0$.

$y^2 = 4 + 5 = 9$ and $y = \pm 3$.

The pairs $(-5, 0)$, $(4, 3)$ and $(4, -3)$ check. They are the solutions.

22. $(0,0)$, $(1,1)$, $\left(-\dfrac{1}{2}+\dfrac{\sqrt{3}}{2}i, -\dfrac{1}{2}-\dfrac{\sqrt{3}}{2}i\right)$,

$\left(-\dfrac{1}{2}-\dfrac{\sqrt{3}}{2}i, -\dfrac{1}{2}+\dfrac{\sqrt{3}}{2}i\right)$

23.
$$x^2 + y^2 = 9, \qquad (1)$$
$$x^2 - y^2 = 9 \qquad (2)$$

Here we use the elimination method.

$$
\begin{aligned}
x^2 + y^2 &= 9 \qquad (1)\\
\underline{x^2 - y^2 = 9} &\qquad \quad (2)\\
2x^2 = 18 &\qquad \text{Adding}\\
x^2 = 9 &\\
x = \pm 3 &
\end{aligned}
$$

If $x = 3$, $x^2 = 9$, and if $x = -3$, $x^2 = 9$, so substituting 3 or -3 in Eq. (1) gives us

$$x^2 + y^2 = 9$$
$$9 + y^2 = 9$$
$$y^2 = 0$$
$$y = 0.$$

The pairs $(3,0)$ and $(-3,0)$ check. They are the solutions.

24. $(0,2)$, $(0,-2)$

25.
$$x^2 + y^2 = 25, \quad (1)$$
$$xy = 12 \qquad (2)$$

First we solve Eq. (2) for y.

$$xy = 12$$
$$y = \frac{12}{x}$$

Then we substitute $\dfrac{12}{x}$ for y in Eq. (1) and solve for x.

$$x^2 + y^2 = 25$$
$$x^2 + \left(\frac{12}{x}\right)^2 = 25$$
$$x^2 + \frac{144}{x^2} = 25$$
$$x^4 + 144 = 25x^2 \quad \text{Multiplying by } x^2$$
$$x^4 - 25x^2 + 144 = 0$$
$$u^2 - 25u + 144 = 0 \qquad \text{Letting } u = x^2$$
$$(u-9)(u-16) = 0$$
$$u = 9 \ \text{ or } \ u = 16$$

We now substitute x^2 for u and solve for x.

$$x^2 = 9 \quad \text{or} \quad x^2 = 16$$
$$x = \pm 3 \quad \text{or} \quad x = \pm 4$$

Since $y = 12/x$, if $x = 3$, $y = 4$; if $x = -3$, $y = -4$; if $x = 4$, $y = 3$; and if $x = -4$, $y = -3$. The pairs $(3,4)$, $(-3,-4)$, $(4,3)$, and $(-4,-3)$ check. They are the solutions.

26. $(-5,3)$, $(-5,-3)$, $(4,0)$

27.
$$x^2 + y^2 = 4, \qquad (1)$$
$$16x^2 + 9y^2 = 144 \quad (2)$$

$$
\begin{aligned}
-9x^2 - 9y^2 &= -36 \quad \text{Multiplying (1) by } -9\\
\underline{16x^2 + 9y^2 = 144} &\\
7x^2 = 108 &\quad \text{Adding}\\
x^2 = \frac{108}{7} &
\end{aligned}
$$

$$x = \pm\sqrt{\frac{108}{7}} = \pm 6\sqrt{\frac{3}{7}}$$
$$x = \pm\frac{6\sqrt{21}}{7} \quad \text{Rationalizing the denominator}$$

Substituting $\dfrac{6\sqrt{21}}{7}$ or $-\dfrac{6\sqrt{21}}{7}$ for x in Eq. (1) gives us

$$\frac{36 \cdot 21}{49} + y^2 = 4$$
$$y^2 = 4 - \frac{108}{7}$$
$$y^2 = -\frac{80}{7}$$
$$y = \pm\sqrt{-\frac{80}{7}} = \pm 4i\sqrt{\frac{5}{7}}$$
$$y = \pm\frac{4i\sqrt{35}}{7}. \quad \text{Rationalizing the denominator}$$

The pairs $\left(\dfrac{6\sqrt{21}}{7}, \dfrac{4i\sqrt{35}}{7}\right)$,

$\left(\dfrac{6\sqrt{21}}{7}, -\dfrac{4i\sqrt{35}}{7}\right)$, $\left(-\dfrac{6\sqrt{21}}{7}, \dfrac{4i\sqrt{35}}{7}\right)$, and

$\left(-\dfrac{6\sqrt{21}}{7}, -\dfrac{4i\sqrt{35}}{7}\right)$ check. They are the solutions.

28. $(0,5)$, $(0,-5)$

29.
$$x^2 + y^2 = 16, \qquad x^2 + y^2 = 16, \quad (1)$$
$$\text{or}$$
$$y^2 - 2x^2 = 10 \qquad -2x^2 + y^2 = 10 \quad (2)$$

Here we use the elimination method.

$$
\begin{aligned}
2x^2 + 2y^2 &= 32 \quad \text{Multiplying (1) by 2}\\
\underline{-2x^2 + y^2 = 10} &\\
3y^2 &= 42 \quad \text{Adding}\\
y^2 &= 14\\
y &= \pm\sqrt{14}
\end{aligned}
$$

Substituting $\sqrt{14}$ or $-\sqrt{14}$ for y in Eq. (1) gives us

$$x^2 + 14 = 16$$
$$x^2 = 2$$
$$x = \pm\sqrt{2}$$

The pairs $(-\sqrt{2}, -\sqrt{14})$, $(-\sqrt{2}, \sqrt{14})$, $(\sqrt{2}, -\sqrt{14})$, and $(\sqrt{2}, \sqrt{14})$ check. They are the solutions.

30. $(-3, -\sqrt{5})$, $(-3, \sqrt{5})$, $(3, -\sqrt{5})$, $(3, \sqrt{5})$

31. $x^2 + y^2 = 5$, (1)
$xy = 2$ (2)

First we solve Eq. (2) for y.

$xy = 2$

$y = \dfrac{2}{x}$

Then we substitute $\dfrac{2}{x}$ for y in Eq. (1) and solve for x.

$x^2 + y^2 = 5$

$x^2 + \left(\dfrac{2}{x}\right)^2 = 5$

$x^2 + \dfrac{4}{x^2} = 5$

$x^4 + 4 = 5x^2$ Multiplying by x^2

$x^4 - 5x^2 + 4 = 0$

$u^2 - 5u + 4 = 0$ Letting $u = x^2$

$(u - 4)(u - 1) = 0$

$u = 4$ or $u = 1$

We now substitute x^2 for u and solve for x.

$x^2 = 4$ or $x^2 = 1$

$x = \pm 2$ $x = \pm 1$

Since $y = 2/x$, if $x = 2$, $y = 1$; if $x = -2$, $y = -1$; if $x = 1$, $y = 2$; and if $x = -1$, $y = -2$. The pairs $(2,1)$, $(-2,-1)$, $(1,2)$, and $(-1,-2)$ check. They are the solutions.

32. $(4,2)$, $(-4,-2)$, $(2,4)$, $(-2,-4)$

33. $x^2 + y^2 = 13$, (1)
$xy = 6$ (2)

First we solve Equation (2) for y.

$xy = 6$

$y = \dfrac{6}{x}$

Then we substitute $\dfrac{6}{x}$ for y in Eq. (1) and solve for x.

$x^2 + y^2 = 13$

$x^2 + \left(\dfrac{6}{x}\right)^2 = 13$

$x^2 + \dfrac{36}{x^2} = 13$

$x^4 + 36 = 13x^2$ Multiplying by x^2

$x^4 - 13x^2 + 36 = 0$

$u^2 - 13u + 36 = 0$ Letting $u = x^2$

$(u - 9)(u - 4) = 0$

$u = 9$ or $u = 4$

We now substitute x^2 for u and solve for x.

$x^2 = 9$ or $x^2 = 4$

$x = \pm 3$ or $x = \pm 2$

Since $y = 6/x$, if $x = 3$, $y = 2$; if $x = -3$, $y = -2$; if $x = 2$, $y = 3$; and if $x = -2$, $y = -3$. The pairs $(3,2)$, $(-3,-2)$, $(2,3)$, and $(-2,-3)$ check. They are the solutions.

34. $(4,1)$, $(-4,-1)$, $(2,2)$, $(-2,-2)$

35. $3xy + x^2 = 34$, (1)
$2xy - 3x^2 = 8$ (2)

$\begin{aligned} 6xy + 2x^2 &= 68 \quad \text{Multiplying (1) by 2} \\ -6xy + 9x^2 &= -24 \quad \text{Multiplying (2) by } -3 \\ \hline 11x^2 &= 44 \quad \text{Adding} \\ x^2 &= 4 \\ x &= \pm 2 \end{aligned}$

Substitute for x in Eq. (1) and solve for y.

When $x = 2$: $3 \cdot 2 \cdot y + 2^2 = 34$

$6y + 4 = 34$

$6y = 30$

$y = 5$

When $x = -2$: $3(-2)(y) + (-2)^2 = 34$

$-6y + 4 = 34$

$-6y = 30$

$y = -5$

The pairs $(2,5)$ and $(-2,-5)$ check. They are the solutions.

36. $(2,1)$, $(-2,-1)$

37. $xy - y^2 = 2$, (1)
$2xy - 3y^2 = 0$ (2)

$\begin{aligned} -2xy + 2y^2 &= -4 \quad \text{Multiplying (1) by } -2 \\ 2xy - 3y^2 &= 0 \\ \hline -y^2 &= -4 \quad \text{Adding} \\ y^2 &= 4 \\ y &= \pm 2 \end{aligned}$

We substitute for y in Eq. (1) and solve for x.

When $y = 2$: $x \cdot 2 - 2^2 = 2$

$2x - 4 = 2$

$2x = 6$

$x = 3$

When $y = -2$: $x(-2) - (-2)^2 = 2$

$-2x - 4 = 2$

$-2x = 6$

$x = -3$

The pairs $(3,2)$ and $(-3,-2)$ check. They are the solutions.

38. $\left(2, -\dfrac{4}{5}\right)$, $\left(-2, -\dfrac{4}{5}\right)$, $(5,2)$, $(-5,2)$

39. $x^2 - y = 5$, (1)
$x^2 + y^2 = 25$ (2)

We solve Eq. (1) for y.

$x^2 - 5 = y$ (3)

Substitute $x^2 - 5$ for y in Eq. (2) and solve for x.

$$x^2 + (x^2 - 5)^2 = 25$$
$$x^2 + x^4 - 10x^2 + 25 = 25$$
$$x^4 - 9x^2 = 0$$
$$u^2 - 9u = 0 \quad \text{Letting } u = x^2$$
$$u(u - 9) = 0$$

$u = 0$ or $u = 9$

$x^2 = 0$ or $x^2 = 9$

$x = 0$ or $x = \pm 3$

Substitute in Eq. (3) and solve for y.

When $x = 0$: $y = 0^2 - 5 = -5$

When $x = 3$ or -3 : $y = 9 - 5 = 4$

The pairs $(0, -5)$, $(3, 4)$, and $(-3, 4)$ check. They are the solutions.

(This exercise could also be solved using the elimination method.)

40. $(-\sqrt{2}, \sqrt{2})$, $(\sqrt{2}, -\sqrt{2})$

41. ***Familiarize***. We first make a drawing. We let l and w represent the length and width, respectively.

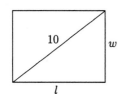

Translate. The perimeter is 28 cm.

$2l + 2w = 28$, or $l + w = 14$

Using the Pythagorean theorem we have another equation.

$l^2 + w^2 = 10^2$, or $l^2 + w^2 = 100$

Carry out. We solve the system:

$l + w = 14$,   (1)

$l^2 + w^2 = 100$ (2)

First solve Eq. (1) for w.

$w = 14 - l$ (3)

Then substitute $14 - l$ for w in Eq. (2) and solve for l.

$$l^2 + w^2 = 100$$
$$l^2 + (14 - l)^2 = 100$$
$$l^2 + 196 - 28l + l^2 = 100$$
$$2l^2 - 28l + 96 = 0$$
$$l^2 - 14l + 48 = 0$$
$$(l - 8)(l - 6) = 0$$

$l = 8$ or $l = 6$

If $l = 8$, then $w = 14 - 8$, or 6. If $l = 6$, then $w = 14 - 6$, or 8. Since the length is usually considered to be longer than the width, we have the solution $l = 8$ and $w = 6$, or $(8, 6)$.

Check. If $l = 8$ and $w = 6$, then the perimeter is $2 \cdot 8 + 2 \cdot 6$, or 28. The length of a diagonal is $\sqrt{8^2 + 6^2}$, or $\sqrt{100}$, or 10. The numbers check.

State. The length is 8 cm, and the width is 6 cm.

42. 2 in. by 1 in.

43. ***Familiarize***. We first make a drawing. Let l = the length and w = the width of the rectangle.

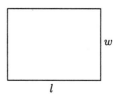

Translate.

Area: $lw = 20$

Perimeter: $2l + 2w = 18$, or $l + w = 9$

Carry out. We solve the system:

Solve the second equation for l: $l = 9 - w$

Substitute $9 - w$ for l in the first equation and solve for w.

$$(9 - w)w = 20$$
$$9w - w^2 = 20$$
$$0 = w^2 - 9w + 20$$
$$0 = (w - 5)(w - 4)$$

$w = 5$ or $w = 4$

If $w = 5$, then $l = 9 - w$, or 4. If $w = 4$, then $l = 9 - 4$, or 5. Since length is usually considered to be longer than width, we have the solution $l = 5$ and $w = 4$, or $(5, 4)$.

Check. If $l = 5$ and $w = 4$, the area is $5 \cdot 4$, or 20. The perimeter is $2 \cdot 5 + 2 \cdot 4$, or 18. The numbers check.

State. The length is 5 in. and the width is 4 in.

44. 2 yd by 1 yd

45. ***Familiarize***. We make a drawing of the field. Let l = the length and w = the width.

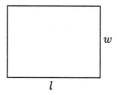

Since it takes 210 yd of fencing to enclose the field, we know that the perimeter is 210 yd.

Translate.

Perimeter: $2l + 2w = 210$, or $l + w = 105$

Area: $lw = 2250$

Carry out. We solve the system:

Solve the first equation for l: $l = 105 - w$

Substitute $105 - w$ for l in the second equation and solve for w.

$$(105 - w)w = 2250$$
$$105w - w^2 = 2250$$
$$0 = w^2 - 105w + 2250$$
$$0 = (w - 30)(w - 75)$$

$w = 30$ or $w = 75$

If $w = 30$, then $l = 105 - 30$, or 75. If $w = 75$, then $l = 105 - 75$, or 30. Since length is usually considered to be longer than width, we have the solution $l = 75$ and $w = 30$, or $(75, 30)$.

Check. If $l = 75$ and $w = 30$, the perimeter is $2 \cdot 75 + 2 \cdot 30$, or 210. The area is $75(30)$, or 2250. The numbers check.

State. The length is 75 yd and the width is 30 yd.

46. 12 ft by 5 ft

47. Familiarize. We make a drawing and label it. Let x and y represent the lengths of the legs of the triangle.

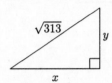

Translate. The product of the lengths of the legs is 156, so we have:

$$xy = 156$$

We use the Pythagorean theorem to get a second equation:

$$x^2 + y^2 = (\sqrt{313})^2, \text{ or } x^2 + y^2 = 313$$

Carry out. We solve the system of equations.

$$xy = 156, \qquad (1)$$
$$x^2 + y^2 = 313 \qquad (2)$$

First solve Equation (1) for y.

$$xy = 156$$

$$y = \frac{156}{x}$$

Then we substitute $\dfrac{156}{x}$ for y in Eq. (2) and solve for x.

$$x^2 + y^2 = 313 \qquad (2)$$

$$x^2 + \left(\frac{156}{x}\right)^2 = 313$$

$$x^2 + \frac{24,336}{x^2} = 313$$

$$x^4 + 24,336 = 313x^2$$

$$x^4 - 313x^2 + 24,336 = 0$$

$$u^2 - 313u + 24,336 = 0 \qquad \text{Letting } u = x^2$$

$$(u - 169)(u - 144) = 0$$

$$u - 169 = 0 \quad \text{or} \quad u - 144 = 0$$

$$u = 169 \quad \text{or} \quad u = 144$$

We now substitute x^2 for u and solve for x.

$$x = \pm 13 \quad \text{or} \quad x = \pm 12$$

Since $y = 156/x$, if $x = 13$, $y = 12$; if $x = -13$, $y = -12$; if $x = 12$, $y = 13$; and if $x = -12$, $y = -13$. The possible solutions are $(13, 12)$, $(-13, -12)$, $(12, 13)$, and $(-12, -13)$.

Check. Since measurements cannot be negative, we consider only $(13, 12)$ and $(12, 13)$. Since both possible solutions give the same pair of legs, we only need to check $(13, 12)$. If $x = 13$ and $y = 12$, their product is 156. Also,

$\sqrt{13^2 + 12^2} = \sqrt{313}$. The numbers check.

State. The lengths of the legs are 13 and 12.

48. 6 and 10 or -6 and -10

49. Familiarize. We let $x = $ the length of a side of one peanut bed and $y = $ the length of a side of the other peanut bed. Make a drawing.

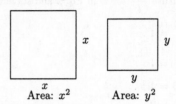

Area: x^2 Area: y^2

Translate.

The sum of the areas is 832 ft^2.

$$x^2 + y^2 = 832$$

The difference of the areas is 320 ft^2.

$$x^2 - y^2 = 320$$

Carry out. We solve the system of equations.

$$
\begin{array}{rl}
x^2 + y^2 = & 832 \\
x^2 - y^2 = & 320 \\
\hline
2x^2 \quad\quad = & 1152 \quad \text{Adding} \\
x^2 = & 576 \\
x = & \pm 24
\end{array}
$$

Since measurements cannot be negative, we consider only $x = 24$. Substitute 24 for x in the first equation and solve for y.

$$24^2 + y^2 = 832$$

$$576 + y^2 = 832$$

$$y^2 = 256$$

$$y = \pm 16$$

Again, we consider only the positive value, 16. The possible solution is $(24, 16)$.

Check. The areas of the peanut beds are 24^2, or 576, and 16^2, or 256. The sum of the areas is $576 + 256$, or 832. The difference of the areas is $576 - 256$, or 320. The values check.

State. The lengths of the beds are 24 ft and 16 ft.

50. $125, 6\%$

51. Familiarize. We first make a drawing. Let $l = $ the length and $w = $ the width.

Translate.

Area: $lw = \sqrt{3}$ (1)

From the Pythagorean theorem: $l^2 + w^2 = 2^2$ (2)

Carry out. We solve the system of equations.

We first solve Eq. (1) for w.

$$lw = \sqrt{3}$$

$$w = \frac{\sqrt{3}}{l}$$

Then we substitute $\frac{\sqrt{3}}{l}$ for w in Eq. 2 and solve for l.

$$l^2 + \left(\frac{\sqrt{3}}{l}\right)^2 = 4$$

$$l^2 + \frac{3}{l^2} = 4$$

$$l^4 + 3 = 4l^2$$

$$l^4 - 4l^2 + 3 = 0$$

$$u^2 - 4u + 3 = 0 \quad \text{Letting } u = l^2$$

$$(u - 3)(u - 1) = 0$$

$$u = 3 \quad \text{or} \quad u = 1$$

We now substitute l^2 for u and solve for l.

$$l^2 = 3 \quad \text{or} \quad l^2 = 1$$

$$l = \pm\sqrt{3} \quad \text{or} \quad l = \pm 1$$

Measurements cannot be negative, so we only need to consider $l = \sqrt{3}$ and $l = 1$. Since $w = \sqrt{3}/l$, if $l = \sqrt{3}$, $w = 1$ and if $l = 1$, $w = \sqrt{3}$. Length is usually considered to be longer than width, so we have the solution $l = \sqrt{3}$ and $w = 1$, or $(\sqrt{3}, 1)$.

Check. If $l = \sqrt{3}$ and $w = 1$, the area is $\sqrt{3} \cdot 1 = \sqrt{3}$. Also $(\sqrt{3})^2 + 1^2 = 3 + 1 = 4 = 2^2$. The numbers check.

State. The length is $\sqrt{3}$ m, and the width is 1 m.

52. $\sqrt{2}$ m by 1 m

53. $\sqrt{48} = \sqrt{16 \cdot 3} = \sqrt{16}\sqrt{3} = 4\sqrt{3}$

54. $2a^6d^2 \sqrt[4]{2d}$

55. *Familiarize*. Let $r =$ the speed of the boat in still water and $t =$ the time of the trip upstream. Organize the information in a table.

	Speed	Time	Distance
Upstream	$r - 2$	t	4
Downstream	$r + 2$	$3 - t$	4

Recall that $rt = d$, or $t = d/r$.

Translate. From the first line of the table we obtain $t = \frac{4}{r - 2}$. From the second line we obtain $3 - t = \frac{4}{r + 2}$.

Carry out. Substitute $\frac{4}{r - 2}$ for t in the second equation and solve for r.

$$3 - \frac{4}{r - 2} = \frac{4}{r + 2},$$

LCD is $(r - 2)(r + 2)$

$$(r-2)(r+2)\left(3 - \frac{4}{r-2}\right) = (r-2)(r+2) \cdot \frac{4}{r+2}$$

$$3(r - 2)(r + 2) - 4(r + 2) = 4(r - 2)$$

$$3(r^2 - 4) - 4r - 8 = 4r - 8$$

$$3r^2 - 12 - 4r - 8 = 4r - 8$$

$$3r^2 - 8r - 12 = 0$$

$$r = \frac{-(-8) \pm \sqrt{(-8)^2 - 4(3)(-12)}}{2 \cdot 3}$$

$$r = \frac{8 \pm \sqrt{208}}{6} = \frac{8 \pm 4\sqrt{13}}{6}$$

$$r = \frac{4 \pm 2\sqrt{13}}{3}$$

Since negative speed has no meaning in this problem, we consider only the positive square root.

$$r = \frac{4 + 2\sqrt{13}}{3} \approx 3.7$$

Check. The value checks. The check is left to the student.

State. The speed of the boat in still water is approximately 3.7 mph.

56. $\dfrac{x - 2\sqrt{xh} + h}{x - h}$

57.

58. One piece should be 61.52 cm long, and then the other will be 38.48 cm long.

59. Let (h, k) represent the point on the line $5x + 8y = -2$ which is the center of a circle that passes through the points $(-2, 3)$ and $(-4, 1)$. The distance between (h, k) and $(-2, 3)$ is the same as the distance between (h, k) and $(-4, 1)$. This gives us one equation:

$$\sqrt{[h - (-2)]^2 + (k - 3)^2} = \sqrt{[h - (-4)]^2 + (k - 1)^2}$$

$$(h + 2)^2 + (k - 3)^2 = (h + 4)^2 + (k - 1)^2$$

$$h^2 + 4h + 4 + k^2 - 6k + 9 = h^2 + 8h + 16 + k^2 - 2k + 1$$

$$4h - 6k + 13 = 8h - 2k + 17$$

$$-4h - 4k = 4$$

$$h + k = -1$$

We get a second equation by substituting (h, k) in $5x + 8y = -2$.

$$5h + 8k = -2$$

We now solve the following system:

$$h + k = -1,$$

$$5h + 8k = -2$$

The solution, which is the center of the circle, is $(-2, 1)$.

Next we find the length of the radius. We can find the distance between either $(-2, 3)$ or $(-4, 1)$ and the center $(-2, 1)$. We use $(-2, 3)$.

$r = \sqrt{[-2 - (-2)]^2 + (1 - 3)^2}$

$r = \sqrt{0^2 + (-2)^2}$

$r = \sqrt{4} = 2$

We can write the equation of the circle with center $(-2, 1)$ and radius 2.

$(x - h)^2 + (y - k)^2 = r^2$

$[x - (-2)]^2 + (y - 1)^2 = 2^2$

$(x + 2)^2 + (y - 1)^2 = 4$

60. $4x^2 + 3y^2 = 43$

61. *Familiarize*. We let x and y represent the length and width of the base of the box, respectively. Then the dimensions of the metal sheet are $x + 10$ and $y + 10$. Make a drawing.

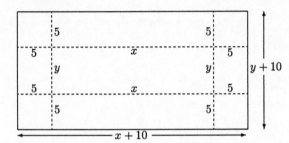

Translate. The area of the metal sheet is 340 in^2, so we have:

$(x + 10)(y + 10) = 340$

The volume of the box is 350 in^3, so we have another equation:

$x \cdot y \cdot 5 = 350$

Carry out. Solve the system

$(x + 10)(y + 10) = 340,$

$x \cdot y \cdot 5 = 350.$

The solutions are $(10, 7)$ and $(7, 10)$.

Check. Since length is usually considered to be longer than width, we check $l = 10$ and $w = 7$, or $(10, 7)$. The numbers check.

State. The dimensions of the box are 10 in. by 7 in. by 5 in.

62. 30 and 50

63. $p^2 + q^2 = 13,$ (1)

$\dfrac{1}{pq} = -\dfrac{1}{6}$ (2)

Solve Eq. (2) for p.

$\dfrac{1}{q} = -\dfrac{p}{6}$

$-\dfrac{6}{q} = p$

Substitute $-6/q$ for p in Eq. (1) and solve for q.

$\left(-\dfrac{6}{q} \right)^2 + q^2 = 13$

$\dfrac{36}{q^2} + q^2 = 13$

$36 + q^4 = 13q^2$

$q^4 - 13q^2 + 36 = 0$

$u^2 - 13u + 36 = 0$ Letting $u = q^2$

$(u - 9)(u - 4) = 0$

$u = 9$ or $u = 4$

$x^2 = 9$ or $x^2 = 4$

$x = \pm 3$ or $x = \pm 2$

Since $p = -6/q$, if $q = 3$, $p = -2$; if $q = -3$, $p = 2$; if $q = 2$, $p = -3$; and if $q = -2$, $p = 3$. The pairs $(-2, 3)$, $(2, -3)$, $(-3, 2)$, and $(3, -2)$ check. They are the solutions.

64. $\left(\dfrac{1}{3}, \dfrac{1}{2} \right), \left(\dfrac{1}{2}, \dfrac{1}{3} \right)$

Chapter 14

Sequences, Series, and Combinatorics

1. $a_n = 3n + 1$

$a_1 = 3 \cdot 1 + 1 = 4,$

$a_2 = 3 \cdot 2 + 1 = 7,$

$a_3 = 3 \cdot 3 + 1 = 10,$

$a_4 = 3 \cdot 4 + 1 = 13;$

$a_{10} = 3 \cdot 10 + 1 = 31;$

$a_{15} = 3 \cdot 15 + 1 = 46$

2. 2, 5, 8, 11; 29; 44

3. $a_n = \dfrac{n}{n+1}$

$a_1 = \dfrac{1}{1+1} = \dfrac{1}{2},$

$a_2 = \dfrac{2}{2+1} = \dfrac{2}{3},$

$a_3 = \dfrac{3}{3+1} = \dfrac{3}{4},$

$a_4 = \dfrac{4}{4+1} = \dfrac{4}{5};$

$a_{10} = \dfrac{10}{10+1} = \dfrac{10}{11};$

$a_{15} = \dfrac{15}{15+1} = \dfrac{15}{16}$

4. 2, 5, 10, 17; 101; 226

5. $a_n = n^2 - 2n$

$a_1 = 1^2 - 2 \cdot 1 = -1,$

$a_2 = 2^2 - 2 \cdot 2 = 0,$

$a_3 = 3^2 - 2 \cdot 3 = 3,$

$a_4 = 4^2 - 2 \cdot 4 = 8;$

$a_{10} = 10^2 - 2 \cdot 10 = 80;$

$a_{15} = 15^2 - 2 \cdot 15 = 195$

6. $0, \dfrac{3}{5}, \dfrac{4}{5}, \dfrac{15}{17}; \dfrac{99}{101}; 15 = \dfrac{112}{113}$

7. $a_n = n + \dfrac{1}{n}$

$a_1 = 1 + \dfrac{1}{1} = 2,$

$a_2 = 2 + \dfrac{1}{2} = 2\dfrac{1}{2},$

$a_3 = 3 + \dfrac{1}{3} = 3\dfrac{1}{3},$

$a_4 = 4 + \dfrac{1}{4} = 4\dfrac{1}{4};$

$a_{10} = 10 + \dfrac{1}{10} = 10\dfrac{1}{10};$

$a_{15} = 15 + \dfrac{1}{15} = 15\dfrac{1}{15}$

8. $1, -\dfrac{1}{2}, \dfrac{1}{4}, -\dfrac{1}{8}; -\dfrac{1}{512}; \dfrac{1}{16,384}$

9. $a_n = (-1)^n n^2$

$a_1 = (-1)^1 1^2 = -1,$

$a_2 = (-1)^2 2^2 = 4,$

$a_3 = (-1)^3 3^2 = -9,$

$a_4 = (-1)^4 4^2 = 16;$

$a_{10} = (-1)^{10} 10^2 = 100;$

$a_{15} = (-1)^{15} 15^2 = -225$

10. $-4, 5, -6, 7; 13; -18$

11. $a_n = (-1)^{n+1}(3n - 5)$

$a_1 = (-1)^{1+1}(3 \cdot 1 - 5) = -2,$

$a_2 = (-1)^{2+1}(3 \cdot 2 - 5) = -1,$

$a_3 = (-1)^{3+1}(3 \cdot 3 - 5) = 4,$

$a_4 = (-1)^{4+1}(3 \cdot 4 - 5) = -7;$

$a_{10} = (-1)^{10+1}(3 \cdot 10 - 5) = -25;$

$a_{15} = (-1)^{15+1}(3 \cdot 15 - 5) = 40$

12. 0, 7, −26, 63; 999; −3374

13. $a_n = 4n - 7$

$a_8 = 4 \cdot 8 - 7 = 32 - 7 = 25$

14. 56

15. $a_n = (3n + 4)(2n - 5)$

$a_7 = (3 \cdot 7 + 4)(2 \cdot 7 - 5) = 25 \cdot 9 = 225$

16. 400

17. $a_n = (-1)^{n-1}(3.4n - 17.3)$

$a_{12} = (-1)^{12-1}[3.4(12) - 17.3] = -23.5$

18. $-37,916,508.16$

19. $a_n = 5n^2(4n - 100)$

$a_{11} = 5(11)^2(4 \cdot 11 - 100) = 5(121)(-56) = -33,880$

20. 528, 528

21. $a_n = \left(1 + \dfrac{1}{n}\right)^2$

$a_{20} = \left(1 + \dfrac{1}{20}\right)^2 = \left(\dfrac{21}{20}\right)^2 = \dfrac{441}{400}$

22. $\dfrac{2744}{3375}$

23. $a_n = \log 10^n$

$a_{43} = \log 10^{43} = 43$

24. 67

25. 1, 3, 5, 7, 9, . . .

These are odd integers, so the general term may be $2n - 1$.

26. 3^n

27. $-2, 6, -18, 54, \ldots$

We can see a pattern if we write the sequence as

$-1 \cdot 2 \cdot 1, \; 1 \cdot 2 \cdot 3, \; -1 \cdot 2 \cdot 9, \; 1 \cdot 2 \cdot 27, \ldots$

The general term may be $(-1)^n 2(3)^{n-1}$.

28. $5n - 7$

29. $\dfrac{2}{3}, \dfrac{3}{4}, \dfrac{4}{5}, \dfrac{5}{6}, \dfrac{6}{7}, \ldots$

These are fractions in which the denominator is 1 greater than the numerator. Also, each numerator is 1 greater than the preceding numerator. The general term may be $\dfrac{n+1}{n+2}$.

30. $\sqrt{2n}$

31. $\sqrt{3}, 3, 3\sqrt{3}, 9, 9\sqrt{3}, \ldots$

These are powers of $\sqrt{3}$. The general term may be $(\sqrt{3})^n$, or $3^{n/2}$.

32. $n(n+1)$

33. $-1, -4, -7, -10, -13, \ldots$

Each term is 3 less than the preceding term. The general term may be $-1 - 3(n-1)$. After removing parentheses and simplifying, we can express the general term as $-3n + 2$, or $-(3n - 2)$.

34. $\log 10^{n-1}$, or $n - 1$

35. 1, 2, 3, 4, 5, 6, 7, . . .

$S_7 = 1 + 2 + 3 + 4 + 5 + 6 + 7 = 28$

36. -8

37. 2, 4, 6, 8, . . .

$S_5 = 2 + 4 + 6 + 8 + 10 = 30$

38. $\dfrac{5269}{3600}$

39. $\displaystyle\sum_{k=1}^{5} \dfrac{1}{2k} = \dfrac{1}{2 \cdot 1} + \dfrac{1}{2 \cdot 2} + \dfrac{1}{2 \cdot 3} + \dfrac{1}{2 \cdot 4} + \dfrac{1}{2 \cdot 5}$

$= \dfrac{1}{2} + \dfrac{1}{4} + \dfrac{1}{6} + \dfrac{1}{8} + \dfrac{1}{10}$

$= \dfrac{60}{120} + \dfrac{30}{120} + \dfrac{20}{120} + \dfrac{15}{120} + \dfrac{12}{120}$

$= \dfrac{137}{120}$

40. $\dfrac{1}{3} + \dfrac{1}{5} + \dfrac{1}{7} + \dfrac{1}{9} + \dfrac{1}{11} + \dfrac{1}{13} = \dfrac{43,024}{45,045}$

41. $\displaystyle\sum_{k=0}^{5} 2^k = 2^0 + 2^1 + 2^2 + 2^3 + 2^4 + 2^5$

$= 1 + 2 + 4 + 8 + 16 + 32$

$= 63$

42. $\sqrt{7} + \sqrt{9} + \sqrt{11} + \sqrt{13} \approx 12.5679$

43. $\displaystyle\sum_{k=7}^{10} \log k = \log 7 + \log 8 + \log 9 + \log 10 \approx$

3.7024

44. $0 + \pi + 2\pi + 3\pi + 4\pi \approx 31.4159$

45. $\displaystyle\sum_{k=1}^{8} \dfrac{k}{k+1} = \dfrac{1}{1+1} + \dfrac{2}{2+1} + \dfrac{3}{3+1} + \dfrac{4}{4+1} +$

$\dfrac{5}{5+1} + \dfrac{6}{6+1} + \dfrac{7}{7+1} + \dfrac{8}{8+1}$

$= \dfrac{1}{2} + \dfrac{2}{3} + \dfrac{3}{4} + \dfrac{4}{5} + \dfrac{5}{6} + \dfrac{6}{7} + \dfrac{7}{8} + \dfrac{8}{9}$

$= \dfrac{15,551}{2520}$

46. $-\dfrac{1}{4} + 0 + \dfrac{1}{6} + \dfrac{2}{7} = \dfrac{17}{84}$

47. $\displaystyle\sum_{k=1}^{5} (-1)^k$

$= (-1)^1 + (-1)^2 + (-1)^3 + (-1)^4 + (-1)^5$

$= -1 + 1 - 1 + 1 - 1$

$= -1$

48. $1 - 1 + 1 - 1 + 1 = 1$

49. $\displaystyle\sum_{k=1}^{8} (-1)^{k+1} 3^k = (-1)^2 3^1 + (-1)^3 3^2 + (-1)^4 3^3 +$

$(-1)^5 3^4 + (-1)^6 3^5 + (-1)^7 3^6 +$

$(-1)^8 3^7 + (-1)^9 3^8$

$= 3 - 9 + 27 - 81 + 243 - 729 +$

$2187 - 6561$

$= -4920$

50. $-4^2 + 4^3 - 4^4 + 4^5 - 4^6 + 4^7 - 4^8 = -52,432$

51. $\displaystyle\sum_{k=0}^{5} (k^2 - 2k + 3)$

$= (0^2 - 2 \cdot 0 + 3) + (1^2 - 2 \cdot 1 + 3) +$

$(2^2 - 2 \cdot 2 + 3) + (3^2 - 2 \cdot 3 + 3) +$

$(4^2 - 2 \cdot 4 + 3) + (5^2 - 2 \cdot 5 + 3)$

$= 3 + 2 + 3 + 6 + 11 + 18$

$= 43$

52. $4 + 2 + 2 + 4 + 8 + 14 = 34$

53. $\displaystyle\sum_{k=1}^{10} \frac{1}{k(k+1)} = \frac{1}{1(1+1)} + \frac{1}{2(2+1)} + \frac{1}{3(3+1)} +$

$\displaystyle\qquad \frac{1}{4(4+1)} + \frac{1}{5(5+1)} + \frac{1}{6(6+1)} +$

$\displaystyle\qquad \frac{1}{7(7+1)} + \frac{1}{8(8+1)} + \frac{1}{9(9+1)} +$

$\displaystyle\qquad \frac{1}{10(10+1)}$

$\displaystyle\quad = \frac{1}{1 \cdot 2} + \frac{1}{2 \cdot 3} + \frac{1}{3 \cdot 4} + \frac{1}{4 \cdot 5} + \frac{1}{5 \cdot 6} +$

$\displaystyle\qquad \frac{1}{6 \cdot 7} + \frac{1}{7 \cdot 8} + \frac{1}{8 \cdot 9} + \frac{1}{9 \cdot 10} + \frac{1}{10 \cdot 11}$

$\displaystyle\quad = \frac{1}{2} + \frac{1}{6} + \frac{1}{12} + \frac{1}{20} + \frac{1}{30} + \frac{1}{42} + \frac{1}{56} +$

$\displaystyle\qquad \frac{1}{72} + \frac{1}{90} + \frac{1}{110}$

$\displaystyle\quad = \frac{10}{11}$

54. $\displaystyle\frac{2}{3} + \frac{4}{5} + \frac{8}{9} + \frac{16}{17} + \frac{32}{33} + \frac{64}{65} + \frac{128}{129} + \frac{256}{257} +$

$\displaystyle\qquad \frac{512}{513} + \frac{1024}{1025}$

55. $\displaystyle\frac{1}{2} + \frac{2}{3} + \frac{3}{4} + \frac{4}{5} + \frac{5}{6} + \frac{6}{7}$

This is a sum of fractions in which the denominator is one greater than the numerator. Also, each numerator is 1 greater than the preceding numerator. Sigma notation is

$$\sum_{k=1}^{6} \frac{k}{k+1}.$$

56. $\displaystyle\sum_{k=1}^{5} 3k$

57. $-2 + 4 - 8 + 16 - 32 + 64$

This is a sum of powers of 2 with alternating signs. Sigma notation is

$$\sum_{k=1}^{6} (-1)^k 2^k, \quad \text{or} \quad \sum_{k=1}^{6} (-2)^k.$$

58. $\displaystyle\sum_{k=1}^{5} \frac{1}{k^2}$

59. $4 - 9 + 16 - 25 + \ldots + (-1)^n n^2$

This is a sum of terms of the form $(-1)^k k^2$, beginning with $k = 2$ and continuing through $k = n$. Sigma notation is

$$\sum_{k=2}^{n} (-1)^k k^2.$$

60. $\displaystyle\sum_{k=3}^{n} (-1)^{k+1} k^2$

61. $5 + 10 + 15 + 20 + 25 + \ldots$

This is a sum of multiples of 5, and it is an infinite series. Sigma notation is

$$\sum_{k=1}^{\infty} 5k.$$

62. $\displaystyle\sum_{k=1}^{\infty} 7k$

63. $\displaystyle\frac{1}{1 \cdot 2} + \frac{1}{2 \cdot 3} + \frac{1}{3 \cdot 4} + \frac{1}{4 \cdot 5} + \ldots$

This is a sum of fractions in which the numerator is 1 and the denominator is a product of two consecutive integers. The larger integer in each product is smaller integer in the succeeding product. It is an infinite series. Sigma notation is

$$\sum_{k=1}^{\infty} \frac{1}{k(k+1)}.$$

64. $\displaystyle\sum_{k=1}^{\infty} \frac{1}{k(k+1)^2}$

65. $\log_3 3 = 1$

1 is the power to which you raise 3 to get 3.

66. 0

67. $\log_3 3^7 = 7$

7 is the power to which you raise 3 to get 3^7.

68. 1

69.

70.

71. $a_n = \displaystyle\frac{1}{2^n} \log 1000^n$

$a_1 = \displaystyle\frac{1}{2^1} \log 1000^1 = \frac{1}{2} \log 10^3 = \frac{1}{2} \cdot 3 = \frac{3}{2}$

$a_2 = \displaystyle\frac{1}{2^2} \log 1000^2 = \frac{1}{4} \log (10^3)^2 = \frac{1}{4} \log 10^6 =$

$\displaystyle\qquad \frac{1}{4} \cdot 6 = \frac{3}{2}$

$a_3 = \displaystyle\frac{1}{2^3} \log 1000^3 = \frac{1}{8} \log (10^3)^3 = \frac{1}{8} \log 10^9 =$

$\displaystyle\qquad \frac{1}{8} \cdot 9 = \frac{9}{8}$

$a_4 = \displaystyle\frac{1}{2^4} \log 1000^4 = \frac{1}{16} \log (10^3)^4 =$

$\displaystyle\qquad \frac{1}{16} \log 10^{12} = \frac{1}{16} \cdot 12 = \frac{3}{4}$

$a_5 = \displaystyle\frac{1}{2^5} \log 1000^5 = \frac{1}{32} \log (10^3)^5 =$

$\displaystyle\qquad \frac{1}{32} \log 10^{15} = \frac{1}{32} \cdot 15 = \frac{15}{32}$

$S_5 = \displaystyle\frac{3}{2} + \frac{3}{2} + \frac{9}{8} + \frac{3}{4} + \frac{15}{32} = \frac{171}{32}$

72. $i, -1, -i, 1, i; i$

73. $a_n = \ln(1 \cdot 2 \cdot 3 \cdots n)$

$a_1 = \ln 1 = 0$

$a_2 = \ln(1 \cdot 2 = \ln 2$

$a_3 = \ln(1 \cdot 2 \cdot 3) = \ln 6$

$a_4 = \ln(1 \cdot 2 \cdot 3 \cdot 4) = \ln 24$

$a_5 = \ln(1 \cdot 2 \cdot 3 \cdot 4 \cdot 5) = \ln 120$

$S_5 = 0 + \ln 2 + \ln 6 + \ln 24 + \ln 120$

$\quad = \ln(2 \cdot 6 \cdot 24 \cdot 120) = \ln 34,560$

74. 0.414214, 0.317837, 0.267949, 0.236068, 0.213422, 0.196262

75. $a_n = \left(1 + \dfrac{1}{n}\right)^n$

$a_1 = \left(1 + \dfrac{1}{1}\right)^1 = 2$

$a_2 = \left(1 + \dfrac{1}{2}\right)^2 = (1.5)^2 = 2.25$

$a_3 = \left(1 + \dfrac{1}{3}\right)^3 = 2.370370$

$a_4 = \left(1 + \dfrac{1}{4}\right)^4 = 2.441406$

$a_5 = \left(1 + \dfrac{1}{5}\right)^5 = 2.488320$

$a_6 = \left(1 + \dfrac{1}{6}\right)^6 = 2.521626$

76. 1, 1, 1, 1, 1, 1

77. $a_1 = 0$, $a_{n+1} = a_n^2 + 4$

$a_1 = 0$

$a_2 = a_1^2 + 4 = 0^2 + 4 = 4$

$a_3 = a_2^2 + 4 = 4^2 + 4 = 20$

$a_4 = a_3^2 + 4 = 20^2 + 4 = 404$

$a_5 = a_4^2 + 4 = 404^2 + 4 = 163,220$

$a_6 = a_5^2 + 4 = 163,220^2 + 4 = 26,640,768,404$

78. 1, 2, 4, 8, 16, 32, 64, 128, 256, 512, 1024, 2048, 4096, 8192, 16,384, 32,768, 65,536

79. Find each term by multiplying the preceding term by 0.75:

$5200, $3900, $2925, $2193.75, $1645.31,

$1233.98, $925.49, $694.12, $520.59, $390.44

80. $6.20, $6.60, $7.00, $7.40, $7.80, $8.20, $8.60, $9.00, $9.40, $9.80

Exercise Set 14.2

1. 2, 7, 12, 17, . . .

$a_1 = 2$

$d = 5$ $\quad (7 - 2 = 5, \ 12 - 7 = 5, \ 17 - 12 = 5)$

2. $a_1 = 1.06$, $d = 0.06$

3. 7, 3, −1, −5, . . .

$a_1 = 7$

$d = -4$ $\quad (3 - 7 = -4, -1 - 3 = -4,$
$\qquad\qquad\quad -5 - (-1) = -4)$

4. $a_1 = -9$, $d = 3$

5. $\dfrac{3}{2}, \dfrac{9}{4}, 3, \dfrac{15}{4}, \ldots$

$a_1 = \dfrac{3}{2}$

$d = \dfrac{3}{4}$ $\quad \left(\dfrac{9}{4} - \dfrac{3}{2} = \dfrac{3}{4}, \ 3 - \dfrac{9}{4} = \dfrac{3}{4}\right)$

6. $a_1 = \dfrac{3}{5}$, $d = -\dfrac{1}{2}$

7. $2.12, $2.24, $2.36, $2.48, . . .

$a_1 = 2.12

$d = 0.12 $\quad ($2.24 - $2.12 = $0.12, \ $2.36-$
$\qquad\qquad\quad $2.24 = $0.12, \ $2.48 - $2.36 =$
$\qquad\qquad\quad $0.12)$

8. $a_1 = 214, $d = -$3$

9. 2, 6, 10, . . .

$a_1 = 2$, $d = 4$, and $n = 12$

$a_n = a_1 + (n - 1)d$

$a_{12} = 2 + (12 - 1)4 = 2 + 11 \cdot 4 = 2 + 44 = 46$

10. 0.57

11. 7, 4, 1, . . .

$a_1 = 7$, $d = -3$, and $n = 17$

$a_n = a_1 + (n - 1)d$

$a_{17} = 7 + (17 - 1)(-3) = 7 + 16(-3) =$
$\qquad 7 - 48 = -41$

12. $-\dfrac{17}{3}$

13. $1200, $964.32, $728.64, . . .

$a_1 = 1200, $d = $964.32 - $1200 = -$235.68$,

$\quad$ and $n = 13$

$a_n = a_1 + (n - 1)d$

$a_{13} = $1200 + (13 - 1)(-$235.68) =$

$\qquad $1200 + 12(-$235.68) = $1200 - $2828.16 =$

$\qquad -$1628.16$

14. $7941.62

15. $a_1 = 2$, $d = 4$

$a_n = a_1 + (n - 1)d$

Let $a_n = 106$, and solve for n.

$106 = 2 + (n - 1)(4)$

$106 = 2 + 4n - 4$

$108 = 4n$

$27 = n$

The 27th term is 106.

16. 33rd

17.
$$a_1 = 7, \; d = -3$$
$$a_n = a_1 + (n-1)d$$
$$-296 = 7 + (n-1)(-3)$$
$$-296 = 7 - 3n + 3$$
$$-306 = -3n$$
$$102 = n$$
The 102nd term is -296.

18. 46th

19.
$$a_n = a_1 + (n-1)d$$
$$a_{17} = 5 + (17-1)6 \quad \text{Substituting 17 for } n,$$
$$\qquad\qquad\qquad\quad 5 \text{ for } a_1, \text{ and 6 for } d$$
$$= 5 + 16 \cdot 6$$
$$= 5 + 96$$
$$= 101$$

20. -43

21.
$$a_n = a_1 + (n-1)d$$
$$33 = a_1 + (8-1)4 \quad \text{Substituting 33 for } a_8,$$
$$\qquad\qquad\qquad\quad 8 \text{ for } n, \text{ and 4 for } d$$
$$33 = a_1 + 28$$
$$5 = a_1$$
(Note that this procedure is equivalent to subtracting d from a_8 seven times to get a_1: $33 - 7(4) = 33 - 28 = 5$)

22. -54

23.
$$a_n = a_1 + (n-1)d$$
$$-76 = 5 + (n-1)(-3) \quad \text{Substituting } -76 \text{ for}$$
$$\qquad\qquad\qquad\qquad a_n, \, 5 \text{ for } a_1, \text{ and } -3$$
$$\qquad\qquad\qquad\qquad \text{for } d$$
$$-76 = 5 - 3n + 3$$
$$-76 = 8 - 3n$$
$$-84 = -3n$$
$$28 = n$$

24. 39

25. We know that $a_{17} = -40$ and $a_{28} = -73$. We would have to add d eleven times to get from a_{17} to a_{28}. That is,
$$-40 + 11d = -73$$
$$11d = -33$$
$$d = -3.$$
Since $a_{17} = -40$, we subtract d sixteen times to get to a_1.
$$a_1 = -40 - 16(-3) = -40 + 48 = 8$$
We write the first five terms of the sequence:
$$8, \, 5, \, 2, \, -1, \, -4$$

26. $\dfrac{1}{3}, \, \dfrac{5}{6}, \, \dfrac{4}{3}, \, \dfrac{11}{6}, \, \dfrac{7}{3}$

27. $5 + 8 + 11 + 14 + \ldots$

Note that $a_1 = 5$, $d = 3$, and $n = 20$. Before using Formula 2, we find a_{20}:
$$a_{20} = 5 + (20-1)3 \quad \text{Substituting into}$$
$$\qquad\qquad\qquad\qquad \text{Formula 1}$$
$$= 5 + 19 \cdot 3 = 62$$
Then
$$S_{20} = \frac{20}{2}(5 + 62) \quad \text{Using Formula 2}$$
$$= 10(67) = 670.$$

28. -210

29. The sum is $1 + 2 + 3 + \ldots + 299 + 300$. This is the sum of the arithmetic sequence for which $a_1 = 1$, $a_n = 300$, and $n = 300$. We use Formula 2.
$$S_n = \frac{n}{2}(a_1 + a_n)$$
$$S_{300} = \frac{300}{2}(1 + 300) = 150(301) = 45,150$$

30. 80,200

31. The sum is $2 + 4 + 6 + \ldots + 98 + 100$. This is the sum of the arithmetic sequence for which $a_1 = 2$, $a_n = 100$, and $n = 50$. We use Formula 2.
$$S_n = \frac{n}{2}(a_1 + a_n)$$
$$S_{50} = \frac{50}{2}(2 + 100) = 25(102) = 2550$$

32. 2500

33. The sum is $7 + 14 + 21 + \ldots + 91 + 98$. This is the sum of the arithmetic sequence for which $a_1 = 7$, $a_n = 98$, and $n = 14$. We use Formula 2.
$$S_n = \frac{n}{2}(a_1 + a_n)$$
$$S_{14} = \frac{14}{2}(7 + 98) = 7(105) = 735$$

34. 34,036

35. Before using Formula 2, we find a_{20}:
$$a_{20} = 2 + (20-1)5 \quad \text{Substituting into}$$
$$\qquad\qquad\qquad\qquad \text{Formula 1}$$
$$= 2 + 19 \cdot 5 = 97$$
Then
$$S_{20} = \frac{20}{2}(2 + 97) \quad \text{Using Formula 2}$$
$$= 10(99) = 990.$$

36. -1264

37. We first find how many plants will be in the last row.

Familiarize. The sequence is 35, 31, 27, ... It is an arithmetic sequence with $a_1 = 35$ and $d = -4$. Since each row must contain a positive number of plants, we must determine how many times we can add -4 to 35 and still have a positive result.

Translate. We find the largest integer x for which $35 + x(-4) > 0$. Then we evaluate the expression $35 - 4x$ for that value of x.

Carry out. We solve the inequality.

$$35 - 4x > 0$$
$$35 > 4x$$
$$\frac{35}{4} > x$$
$$8\frac{3}{4} > x$$

The integer we are looking for is 8. Thus $35 - 4x = 35 - 4(8) = 3$.

Check. If we add -4 to 35 eight times we get 3, a positive number, but if we add -4 to 35 more than eight times we get a negative number.

State. There will be 3 plants in the last row.

Next we find how many plants there are altogether.

Familiarize. We want to find the sum $35 + 31 + 27 + \ldots + 3$. We know $a_1 = 35$ $a_n = 3$, and, since we add -4 to 35 eight times, $n = 9$. (There are 8 terms after a_1, for a total of 9 terms.) We will use the formula $S_n = \frac{n}{2}(a_1 + a_n)$.

Translate. We want to find the sum of the first 9 terms of an arithmetic sequence in which $a_1 = 35$ and $a_9 = 3$.

Carry out. Substituting into Formula 2 we have,

$$S_9 = \frac{9}{2}(35 + 3)$$
$$= \frac{9}{2} \cdot 38 = 171$$

Check. We can check the calculations by doing them again. We could also do the entire addition:

$$35 + 31 + 27 + \ldots + 3.$$

State. There are 171 plants altogether.

38. 62; 950

39. *Familiarize*. We go from 50 poles in a row, down to six poles in the top row, so there must be 45 rows. We want the sum $50 + 49 + 48 + \ldots + 6$. Thus we want the sum of an arithmetic sequence. We will use the formula $S_n = \frac{n}{2}(a_1 + a_n)$.

Translate. We want to find the sum of the first 45 terms of an arithmetic sequence with $a_1 = 50$ and $a_{45} = 6$.

Carry out. Substituting into Formula 2, we have

$$S_{45} = \frac{45}{2}(50 + 6)$$
$$= \frac{45}{2} \cdot 56 = 1260$$

Check. We can do the calculation again, or we can do the entire addition:

$$50 + 49 + 48 + \ldots + 6.$$

State. There will be 1260 poles in the pile.

40. $49.60

41. *Familiarize*. We want to find the sum of an arithmetic sequence with $a_1 = \$600$, $d = \$100$, and $n = 20$. We will use Formula 1 to find a_{20}, and then we will use Formula 2 to find S_{20}.

Translate. Substituting into Formula 1, we have

$$a_{20} = 600 + (20 - 1)(100).$$

Carry out. We first find a_{20}.

$$a_{20} = 600 + 19 \cdot 100 = 600 + 1900 = 2500$$

Then we use Formula 2 to find S_{20}.

$$S_{20} = \frac{20}{2}(600 + 2500) = 10(3100) = 31,000$$

Check. We can do the calculation again.

State. They save $31,000 (disregarding interest).

42. $10,230

43. *Familiarize*. We want to find the sum of an arithmetic sequence with $a_1 = 28$, $d = 4$, and $n = 50$. We will use Formula 1 to find a_{50}, and then we will use Formula 2 to find S_{50}.

Translate. Substituting into Formula 1, we have

$$a_{50} = 28 + (50 - 1)(4).$$

Carry out. We find a_{50}.

$$a_{50} = 28 + 49 \cdot 4 = 28 + 196 = 224$$

Then we use Formula 2 to find S_{50}.

$$S_{50} = \frac{50}{2}(28 + 224) = 25 \cdot 252 = 6300$$

Check. We can do the calculation again.

State. There are 6300 seats.

44. $462,500

45. The logarithm is the exponent.

$$\log_a P = k \qquad a^k = P$$

The base does not change.

46. $e^a = t$

47. Standard form for the equation of a circle with center (h, k) and radius r is

$$(x - h)^2 + (y - k)^2 = r^2.$$

We substitute 0 for h, 0 for k, and 9 for r:

$$(x - 0)^2 + (y - 0)^2 = 9^2$$
$$x^2 + y^2 = 81$$

48. $(x + 2)^2 + (y - 5)^2 = 18$

49.

50.

51. *Familiarize*. Let x represent the first number in the sequence, and let d represent the common difference. Then the three numbers in the sequence are x, $x + d$, and $x + 2d$.

Translate.

The sum of the first and third numbers is 10.

$$x + x + 2d \qquad\qquad = 10$$

The product of the first
and second numbers   is   15.
$$x(x + d) \qquad\qquad = 15$$

Carry out. Solving the system of equations we get $x = 3$ and $d = 2$. Thus the numbers are 3, 5, and 7.

Check. The numbers are in an arithmetic sequence. Also $3 + 7 = 10$ and $3 \cdot 5 = 15$. The numbers check.

State. The numbers are 3, 5, and 7.

52. $S_n = n^2$

53.

$a_1 = \$8760$

$a_2 = \$8760 + (-\$798.23) = \$7961.77$

$a_3 = \$8760 + 2(-\$798.23) = \$7163.54$

$a_4 = \$8760 + 3(-\$798.23) = \$6365.31$

$a_5 = \$8760 + 4(-\$798.23) = \$5567.08$

$a_6 = \$8760 + 5(-\$798.23) = \$4768.85$

$a_7 = \$8760 + 6(-\$798.23) = \$3970.62$

$a_8 = \$8760 + 7(-\$798.23) = \$3172.39$

$a_9 = \$8760 + 8(-\$798.23) = \$2374.16$

$a_{10} = \$8760 + 9(-\$798.23) = \$1575.93$

54. $51,679.65$

55. See the answer section in the text.

56. a) $a_t = \$5200 - \$512.50t$

b) $5200, \$4687.50, \$4175, \$3662.50, \$3150, \$1612.50, \$1100$

Exercise Set 14.3

1. $2, 4, 8, 16, \ldots$

$\dfrac{4}{2} = 2, \quad \dfrac{8}{4} = 2, \quad \dfrac{16}{8} = 2$

$r = 2$

2. $-\dfrac{1}{3}$

3. $1, -1, 1, -1, \ldots$

$\dfrac{-1}{1} = -1, \quad \dfrac{1}{-1} = -1, \quad \dfrac{-1}{1} = -1$

$r = -1$

4. 0.1

5. $\dfrac{1}{2}, -\dfrac{1}{4}, \dfrac{1}{8}, -\dfrac{1}{16}, \ldots$

$\dfrac{-\frac{1}{4}}{\frac{1}{2}} = -\dfrac{1}{4} \cdot \dfrac{2}{1} = -\dfrac{2}{4} = -\dfrac{1}{2}$

$\dfrac{\frac{1}{8}}{-\frac{1}{4}} = \dfrac{1}{8} \cdot \left(-\dfrac{4}{1}\right) = -\dfrac{4}{8} = -\dfrac{1}{2}$

$r = -\dfrac{1}{2}$

6. -2

7. $75, 15, 3, \dfrac{3}{5}, \ldots$

$\dfrac{15}{75} = \dfrac{1}{5}, \quad \dfrac{3}{15} = \dfrac{1}{5}, \quad \dfrac{\frac{3}{5}}{3} = \dfrac{3}{5} \cdot \dfrac{1}{3} = \dfrac{1}{5}$

$r = \dfrac{1}{5}$

8. 0.1

9. $\dfrac{1}{x}, \dfrac{1}{x^2}, \dfrac{1}{x^3}, \ldots$

$\dfrac{\frac{1}{x^2}}{\frac{1}{x}} = \dfrac{1}{x^2} \cdot \dfrac{x}{1} = \dfrac{x}{x^2} = \dfrac{1}{x}$

$\dfrac{\frac{1}{x^3}}{\frac{1}{x^2}} = \dfrac{1}{x^3} \cdot \dfrac{x^2}{1} = \dfrac{x^2}{x^3} = \dfrac{1}{x}$

$r = \dfrac{1}{x}$

10. $\dfrac{m}{2}$

11. $\$780, \$858, \$943.80, \$1038.18, \ldots$

$\dfrac{\$858}{\$780} = 1.1, \quad \dfrac{\$943.80}{\$858} = 1.1,$

$\dfrac{\$1038.18}{\$943.80} = 1.1$

$r = 1.1$

12. 0.95

13. $2, 4, 8, 16, \ldots$

$a_1 = 2, \ n = 6,$ and $r = \dfrac{4}{2},$ or 2.

We use the formula $a_n = a_1 r^{n-1}$.

$a_6 = 2(2)^{6-1} = 2 \cdot 2^5 = 2 \cdot 32 = 64$

14. $781, 250$

15. $2, 2\sqrt{3}, 6, \ldots$

$a_1 = 2, \ n = 9,$ and $r = \dfrac{2\sqrt{3}}{2},$ or $\sqrt{3}$

$a_n = a_1 r^{n-1}$

$a_9 = 2(\sqrt{3})^{9-1} = 2(\sqrt{3})^8 = 2 \cdot 81 = 162$

16. 1

17. $\dfrac{8}{243}, \dfrac{8}{81}, \dfrac{8}{27}, \ldots$

$a_1 = \dfrac{8}{243}, \ n = 10,$ and $r = \dfrac{\frac{8}{81}}{\frac{8}{243}} = \dfrac{8}{81} \cdot \dfrac{243}{8} = 3$

$a_n = a_1 r^{n-1}$

$a_{10} = \dfrac{8}{243}(3)^{10-1} = \dfrac{8}{243}(3)^9 = \dfrac{8}{243} \cdot 19,683 = 648$

18. $2,734,375$

19. $1000, $1080, $1166.40, . . .

$a_1 = \$1000$, $n = 12$, and $r = \dfrac{\$1080}{\$1000} = 1.08$

$a_n = a_1 r^{n-1}$

$a_{12} = \$1000(1.08)^{12-1} \approx \$1000(2.331638997) \approx$
     $\$2331.64$

20. $1967.15

21. $1, 3, 9, . . .$

$a_1 = 1$ and $r = \dfrac{3}{1}$, or 3

$a_n = a_1 r^{n-1}$

$a_n = 1(3)^{n-1} = 3^{n-1}$

22. 5^{3-n}

23. $1, -1, 1, -1, . . .$

$a_1 = 1$ and $r = \dfrac{-1}{1} = -1$

$a_n = a_1 r^{n-1}$

$a_n = 1(-1)^{n-1} = (-1)^{n-1}$

24. 2^n

25. $\dfrac{1}{x}, \dfrac{1}{x^2}, \dfrac{1}{x^2}, . . .$

$a_1 = \dfrac{1}{x}$ and $r = \dfrac{1}{x}$ (See Exercise 9.)

$a_n = a_1 r^{n-1}$

$a_n = \dfrac{1}{x}\left(\dfrac{1}{x}\right)^{n-1} = \dfrac{1}{x} \cdot \dfrac{1}{x^{n-1}} = \dfrac{1}{x^{1+n-1}} = \dfrac{1}{x^n}$

26. $5\left(\dfrac{m}{2}\right)^{n-1}$

27. $6 + 12 + 24 + . . .$

$a_1 = 6$, $n = 7$, and $r = \dfrac{12}{6}$, or 2

$S_n = \dfrac{a_1(1 - r^n)}{1 - r}$

$S_7 = \dfrac{6(1 - 2^7)}{1 - 2} = \dfrac{6(1 - 128)}{-1} = \dfrac{6(-127)}{-1} = 762$

28. $\dfrac{21}{2}$, or 10.5

29. $\dfrac{1}{18} - \dfrac{1}{6} + \dfrac{1}{2} - . . .$

$a_1 = \dfrac{1}{18}$, $n = 7$, and $r = \dfrac{-\dfrac{1}{6}}{\dfrac{1}{18}} = -\dfrac{1}{6} \cdot \dfrac{18}{1} = -3$

$S_n = \dfrac{a_1(1 - r^n)}{1 - r}$

$S_7 = \dfrac{\dfrac{1}{18}\left[1 - (-3)^7\right]}{1 - (-3)} = \dfrac{\dfrac{1}{18}(2 + 2187)}{4} = \dfrac{\dfrac{1}{18}(2188)}{4} =$
     $\dfrac{1}{18}(2188)\left(\dfrac{1}{4}\right) = \dfrac{547}{18}$

30. 6.6666

31. $1 + x + x^2 + x^3 + . . .$

$a_1 = 1$, $n = 8$, and $r = \dfrac{x}{1}$, or x

$S_n = \dfrac{a_1(1 - r^n)}{1 - r}$

$S_8 = \dfrac{1(x - x^8)}{1 - x} = \dfrac{(1 + x^4)(1 - x^4)}{1 - x} =$

$\dfrac{(1 + x^4)(1 + x^2)(1 - x^2)}{1 - x} =$

$\dfrac{(1 + x^4)(1 + x^2)(1 + x)(1 - x)}{1 - x} =$

$(1 + x^4)(1 + x^2)(1 + x)$

32. $\dfrac{1 - x^{20}}{1 - x^2}$

33. $200, $200(1.06), $200(1.06)^2, . . .$

$a_1 = \$200$, $n = 16$, and $r = \dfrac{\$200(1.06)}{\$200} = 1.06$

$S_n = \dfrac{a_1(1 - r^n)}{1 - r}$

$S_{16} = \dfrac{\$200[1 - (1.06)^{16}]}{1 - 1.06} \approx$
     $\dfrac{\$200(1 - 2.540351685)}{-0.06} \approx \$5134.51$

34. $60,893.30

35. $4 + 2 + 1 + . . .$

$|r| = \left|\dfrac{2}{4}\right| = \left|\dfrac{1}{2}\right| = \dfrac{1}{2}$, and since $|r| < 1$, the series
does have a sum.

$S_\infty = \dfrac{a_1}{1 - r} = \dfrac{4}{1 - \dfrac{1}{2}} = \dfrac{4}{\dfrac{1}{2}} = 4 \cdot \dfrac{2}{1} = 8$

36. $\dfrac{49}{4}$

37. $25 + 20 + 16 + . . .$

$|r| = \left|\dfrac{20}{25}\right| = \left|\dfrac{4}{5}\right| = \dfrac{4}{5}$, and since $|r| < 1$, the
series does have a sum.

$S_\infty = \dfrac{a_1}{1 - r} = \dfrac{25}{1 - \dfrac{4}{5}} = \dfrac{25}{\dfrac{1}{5}} = 25 \cdot \dfrac{5}{1} = 125$

38. 48

39. $100 - 10 + 1 - \dfrac{1}{10} + . . .$

$|r| = \left|\dfrac{-10}{100}\right| = \left|-\dfrac{1}{10}\right| = \dfrac{1}{10}$, and since $|r| < 1$,
the series does have a sum.

$S_\infty = \dfrac{a_1}{1 - r} = \dfrac{100}{1 - \left(-\dfrac{1}{10}\right)} = \dfrac{100}{\dfrac{11}{10}} =$
     $100 \cdot \dfrac{10}{11} = \dfrac{1000}{11}$

40. No

41. $8 + 40 + 200 + \cdots$

$|r| = \left|\dfrac{40}{8}\right| = |5| = 5$, and since $|r| \not< 1$ the series does not have a sum.

42. -4

43. $0.3 + 0.03 + 0.003 + \cdots$

$|r| = \left|\dfrac{0.03}{0.3}\right| = |0.1| = 0.1$, and since $|r| < 1$ the series does have a sum.

$S_\infty = \dfrac{a_1}{1-r} = \dfrac{0.3}{1-0.1} = \dfrac{0.3}{0.9} = \dfrac{3}{9} = \dfrac{1}{3}$

44. $\dfrac{37}{99}$

45. $\$500(1.02)^{-1} + \$500(1.02)^{-2} + \$500(1.02)^{-3} + \cdots$

$|r| = \left|\dfrac{\$500(1.02)^{-2}}{\$500(1.02)^{-1}}\right| = |(1.02)^{-1}| = (1.02)^{-1}$, or $\dfrac{1}{1.02}$, and since $|r| < 1$, the series does have a sum.

$S_\infty = \dfrac{a_1}{1-r} = \dfrac{\$500(1.02)^{-1}}{1-\left(\dfrac{1}{1.02}\right)} = \dfrac{\dfrac{\$500}{1.02}}{\dfrac{0.02}{1.02}} =$

$\dfrac{\$500}{1.02} \cdot \dfrac{1.02}{0.02} = \$25,000$

46. $\$12,500$

47. $0.4444\ldots = 0.4 + 0.04 + 0.004 + 0.0004 + \cdots$

This is an infinite geometric series with $a_1 = 0.4$.

$|r| = \left|\dfrac{0.04}{0.4}\right| = |0.1| = 0.1 < 1$, so the series has a sum.

$S_\infty = \dfrac{a_1}{1-r} = \dfrac{0.4}{1-0.1} = \dfrac{0.4}{0.9} = \dfrac{4}{9}$

Fractional notation for $0.4444\ldots$ is $\dfrac{4}{9}$.

48. 10

49. $0.55555 = 0.5 + 0.05 + 0.005 + 0.0005 + \cdots$

This is an infinite geometric series with $a_1 = 0.5$.

$|r| = \left|\dfrac{0.05}{0.5}\right| = |0.1| = 0.1 < 1$, so the series has a sum.

$S_\infty = \dfrac{a_1}{1-r} = \dfrac{0.5}{1-0.1} = \dfrac{0.5}{0.9} = \dfrac{5}{9}$

50. $\dfrac{2}{3}$

51. $0.15151515\ldots = 0.15 + 0.0015 + 0.000015 + \cdots$

This is an infinite geometric series with $a_1 = 0.15$.

$|r| = \left|\dfrac{0.0015}{0.15}\right| = |0.01| = 0.01 < 1$, so the series has a sum.

$S_\infty = \dfrac{a_1}{1-r} = \dfrac{0.15}{1-0.01} = \dfrac{0.15}{0.99} = \dfrac{15}{99} = \dfrac{5}{33}$

52. $\dfrac{4}{33}$

53. *Familiarize.* The rebound distances form a geometric sequence:

$\dfrac{1}{4} \times 16, \quad \left(\dfrac{1}{4}\right)^2 \times 16, \quad \left(\dfrac{1}{4}\right)^3 \times 16, \ldots,$

or $4, \quad \dfrac{1}{4} \times 4, \quad \left(\dfrac{1}{4}\right)^2 \times 4, \ldots$

The height of the 6th rebound is the 6th term of the sequence.

Translate. We will use the formula $a_n = a_1 r^{n-1}$, with $a_1 = 4$, $r = \dfrac{1}{4}$, and $n = 6$:

$a_6 = 4\left(\dfrac{1}{4}\right)^{6-1}$

Carry out. We calculate to obtain $a_6 = \dfrac{1}{256}$.

Check. We can do the calculation again.

State. It rebounds $\dfrac{1}{256}$ ft the 6th time.

54. $5\dfrac{1}{3}$ ft

55. *Familiarize.* In one year, the population will be $100,000 + 0.03(100,000)$, or $(1.03)100,000$. In two years, the population will be $(1.03)100,000 + 0.03(1.03)100,000$, or $(1.03)^2 100,000$. Thus the populations form a geometric sequence:

$100,000, \quad (1.03)100,000, \quad (1.03)^2 100,000, \ldots$

The population in 15 years will be the 16th term of the sequence.

Translate. We will use the formula $a_n = a_1 r^{n-1}$ with $a_1 = 100,000$, $r = 1.03$, and $n = 16$:

$a_{16} = 100,000(1.03)^{16-1}$

Carry out. We calculate to obtain $a_{16} \approx 155,797$.

Check. We can do the calculation again.

State. In 15 years the population will be about 155,797.

56. About 24 years

57. *Familiarize.* The amounts owed at the beginning of successive years form a geometric sequence:

$\$1200, \quad (1.12)\$1200, \quad (1.12)^2\$1200,$

$(1.12)^3\$1200, \ldots$

The amount to be repaid at the end of 13 years is the amount owed at the beginning of the 14th year.

Translate. We use the formula $a_n = a_1 r^{n-1}$ with $a_1 = 1200$, $r = 1.12$, and $n = 14$:

$a_{14} = 1200(1.12)^{14-1}$

Carry out. We calculate to obtain $a_{14} \approx 5236.19$.

Check. We can do the calculation again.

State. At the end of 13 years, $\$5236.19$ will be repaid.

58. $10,485.76$ in.

59. *Familiarize.* The lengths of the falls form a geometric sequence:

$$556, \quad \left(\frac{3}{4}\right)556, \quad \left(\frac{3}{4}\right)^2 556, \quad \left(\frac{3}{4}\right)^3 556, \ldots$$

The total length of the first 6 falls is the sum of the first six terms of this sequence. The heights of the rebounds also form a geometric sequence:

$$\left(\frac{3}{4}\right)556, \quad \left(\frac{3}{4}\right)^2 556, \quad \left(\frac{3}{4}\right)^3 556, \ldots, \quad \text{or}$$

$$417, \quad \left(\frac{3}{4}\right)417, \quad \left(\frac{3}{4}\right)^2 417, \ldots$$

When the ball hits the ground for the 6th time, it will have rebounded 5 times. Thus the total length of the rebounds is the sum of the first five terms of this sequence.

Translate. We use the formula $S_n = \dfrac{a_1(1 - r^n)}{1 - r}$ twice, once with $a_1 = 556$, $r = \dfrac{3}{4}$, and $n = 6$ and a second time with $a_1 = 417$, $r = \dfrac{3}{4}$, and $n = 5$.

$D = $ Length of falls + length of rebounds

$$= \frac{556\left[1 - \left(\frac{3}{4}\right)^6\right]}{1 - \frac{3}{4}} + \frac{417\left[1 - \left(\frac{3}{4}\right)^5\right]}{1 - \frac{3}{4}}.$$

Carry out. We use a calculator to obtain $D \approx 3100.35$.

Check. We can do the calculations again.

State. The ball will have traveled about 3100.35 ft.

60. 3892 ft

61. *Familiarize.* The amounts form a geometric series:

$$\$0.01, \ \$0.01(2), \ \$0.01(2)^2, \ \$0.01(2^3) + \ldots$$

$$(\$0.01)(2)^{27}$$

Translate. We use the formula $S_n = \dfrac{a_1(1 - r^n)}{1 - r}$ to find the sum of the geometric series with $a_1 = 0.01$, $r = 2$, and $n = 28$:

$$S_{28} = \frac{0.01(1 - 2^{28})}{1 - 2}$$

Carry out. We use a calculator to obtain $S_{28} = \$2,684,354.55$.

Check. We can do the calculation again.

State. You would earn $2,684,354.55.

62. $645,826.93

63.
$$5x - 2y = -3, \quad (1)$$
$$2x + 5y = -24 \quad (2)$$

Multiply Eq. (1) by 5 and Eq. (2) by 2 and add.

$$\begin{array}{r} 25x - 10y = -15 \\ 4x + 10y = -48 \\ \hline 29x \quad\quad\;\; = -63 \end{array}$$

$$x = -\frac{63}{29}$$

Substitute $-\dfrac{63}{29}$ for x in the second equation and solve for y.

$$2\left(-\frac{63}{29}\right) + 5y = -24$$

$$-\frac{126}{29} + 5y = -24$$

$$5y = -\frac{570}{29}$$

$$y = -\frac{114}{29}$$

The solution is $\left(-\dfrac{63}{29}, -\dfrac{114}{29}\right)$.

64. $(-1, 2, 3)$

65.

66.

67. $1 + x + x^2 + \ldots$

This is a geometric series with $a_1 = 1$ and $r = x$.

$$S_n = \frac{a_1(1 - r^n)}{1 - r} = \frac{1(1 - x^n)}{1 - x} = \frac{1 - x^n}{1 - x}$$

68. $\dfrac{x^2[1 - (-x)^n]}{1 + x}$

69. *Familiarize.* The length of a side of the first square is 16 cm. The length of a side of the next square is the length of the hypotenuse of a right triangle with legs 8 cm and 8 cm, or $8\sqrt{2}$ cm. The length of a side of the next square is the length of the hypotenuse of a right triangle with legs $4\sqrt{2}$ cm and $4\sqrt{2}$ cm, or 8 cm. The areas of the squares form a sequence:

$$(16)^2, \quad (8\sqrt{2})^2, \quad (8)^2, \ldots, \quad \text{or}$$

$$256, \quad 128, \quad 64, \ldots.$$

This is a geometric sequence with $a_1 = 256$ and $r = \dfrac{1}{2}$.

Translate. We find the sum of the infinite geometric series $256 + 128 + 64 + \ldots$

$$S_\infty = \frac{a_1}{1 - r}$$

$$S_\infty = \frac{256}{1 - \dfrac{1}{2}}$$

Carry out. We calculate to obtain $S_\infty = 512$.

Check. We can do the calculation again.

State. The sum of the areas is 512 cm².

Exercise Set 14.4

1. $9! = 9 \cdot 8 \cdot 7 \cdot 6 \cdot 5 \cdot 4 \cdot 3 \cdot 2 \cdot 1 = 362,880$

2. $3,628,800$

3. $11! = 11 \cdot 10 \cdot 9 \cdot 8 \cdot 7 \cdot 6 \cdot 5 \cdot 4 \cdot 3 \cdot 2 \cdot 1 = 39,916,800$

4. $479,001,600$

5. 0! is defined to be 1.

6. 1

7. $\dfrac{7!}{4!} = \dfrac{7 \cdot 6 \cdot 5 \cdot 4!}{4!} = 7 \cdot 6 \cdot 5 = 210$

8. 56

9. $\dfrac{9!}{5!} = \dfrac{9 \cdot 8 \cdot 7 \cdot 6 \cdot 5!}{5!} = 9 \cdot 8 \cdot 7 \cdot 6 = 3024$

10. 720

11. $(8-3)! = 5! = 5 \cdot 4 \cdot 3 \cdot 2 \cdot 1 = 120$

12. 24

13. $8! - 3! = (8 \cdot 7 \cdot 6 \cdot 5 \cdot 4 \cdot 3 \cdot 2 \cdot 1) - (3 \cdot 2 \cdot 1) =$
$40,320 - 6 = 40,314$

14. $362,760$

15. $_6P_6 = 6! = 6 \cdot 5 \cdot 4 \cdot 3 \cdot 2 \cdot 1 = 720$

16. 120

17. Using formula (1), we have $_4P_3 = 4 \cdot 3 \cdot 2 = 24$.
Using formula (2), we have
$$_4P_3 = \frac{4!}{(4-3)!} = \frac{4!}{1!} = \frac{4 \cdot 3 \cdot 2 \cdot 1}{1} = 24.$$

18. 2520

19. Using formula (1), we have
$_{10}P_7 = 10 \cdot 9 \cdot 8 \cdot 7 \cdot 6 \cdot 5 \cdot 4 = 604,800.$
Using formula (2), we have
$$_{10}P_7 = \frac{10!}{(10-7)!} = \frac{10!}{3!} = \frac{10 \cdot 9 \cdot 8 \cdot 7 \cdot 6 \cdot 5 \cdot 4 \cdot 3!}{3!} =$$
$604,800.$

20. 720

21. Using formula (1), we have $_6P_1 = 6$.
Using formula (2), we have
$$_6P_1 = \frac{6!}{(6-1)!} = \frac{6!}{5!} = \frac{6 \cdot 5!}{5!} = 6.$$

22. 12

23. Using formula (1), we have $_6P_5 = 6 \cdot 5 \cdot 4 \cdot 3 \cdot 2 = 720$.
Using formula (2), we have
$$_6P_5 = \frac{6!}{(6-5)!} = \frac{6!}{1!} = \frac{6 \cdot 5 \cdot 4 \cdot 3 \cdot 2 \cdot 1}{1} = 720.$$

24. $479,001,600$

25. The route from New York to Indianapolis can be chosen in 3 ways, the route from Indianapolis to Denver can be chosen in 2 ways, and the route from Denver to San Francisco can be chosen in 2 ways. By the fundamental counting principle, the total number of different routes is $3 \cdot 2 \cdot 2$, or 12.

26. $5 \cdot 10 \cdot 4 \cdot 4$, or 800

27. There are 3 choices for the first letter, 2 for the second, and 1 for the third. The number of permutations is $3 \cdot 2 \cdot 1$, or $_3P_3$, or 3!, or 6.

28. $_2P_2$, or 2!, or 2

29. There are 7 choices for the first letter, 6 for the second, 5 for the third, 4 for the fourth, 3 for the fifth, 2 for the sixth, and 1 for the seventh. The number of permutations is $7 \cdot 6 \cdot 5 \cdot 4 \cdot 3 \cdot 2 \cdot 1$, or $_7P_7$, or 7!, or 5040.

30. $_5P_5$, or 5!, or 120

31. $_7P_4 = \dfrac{7!}{4!} = \dfrac{7!}{(7-4)!} = \dfrac{7!}{3!} = \dfrac{7 \cdot 6 \cdot 5 \cdot 4 \cdot 3!}{3!} = 840$

32. $_5P_3$, or 60

33. Without repetition, the total is the number of permutations of 5 objects taken 5 at a time:
$$_5P_5 = 5! = 120$$
With repetition each of the 5 digits can be chosen in 5 ways:
$$5 \cdot 5 \cdot 5 \cdot 5 \cdot 5 = 3125$$

34. $_4P_4$, or 24;
$4 \cdot 4 \cdot 4 \cdot 4$, or 256

35. The number of arrangements is the number of permutations of 5 objects taken 5 at a time:
$$_5P_5 = 5! = 120$$

36. 7!, or 5040

37. There are only 9 choices for the first digit since 0 is excluded. There are also 9 choices for the second digit since 0 can be included and the first digit cannot be repeated. Because no digit is used more than once there are only 8 choices for the third digit, 7 for the fourth, 6 for the fifth, 5 for the sixth, and 4 for the seventh. By the fundamental counting principle the total number of permutations is
$9 \cdot 9 \cdot 8 \cdot 7 \cdot 6 \cdot 5 \cdot 4$, or 544,320.
Thus 544,320 7-digit phone numbers can be formed.

38. $_{12}P_4$, or 11,880

39. a) The number of ways in which the coins can be lined up is $_5P_5 = 5! = 120$.

b) There are 5 choices for the first coin and 2 possibilities (head or tail) for each choice. This results in a total of 10 choices for the first selection. There are 4 choices (no coin can be used more than once) for the second coin and 2 possibilities (head or tail) for each choice. This results in a total of 8 choices for the second selection. There are 3 choices for the third coin and 2 possibilities (head or tail) for each choice. This results in a total of 6 choices for the third selection. Likewise there are 4 choices for the fourth selection and 2 choices for the fifth selection. Using the fundamental counting principle we know there are
$$10 \cdot 8 \cdot 6 \cdot 4 \cdot 2, \text{ or } 3840$$
ways the coins can be lined up.

40. a) $_4P_4$, or 24

b) $8 \cdot 6 \cdot 4 \cdot 2$, or 384

41. $_{52}P_4 = \dfrac{52!}{(52-4)!} = \dfrac{52!}{48!} = \dfrac{52 \cdot 51 \cdot 50 \cdot 49 \cdot 48!}{48!} =$
6,497,400

42. 254,251,200

43. There are 80 choices for the number of the county, 26 choices for the letter of the alphabet, and 9999 choices for the number that follows the letter. By the fundamental counting principle we know there are $80 \cdot 26 \cdot 9999$, or 20,797,920 possible license plates.

44. a) $26 \cdot 10 \cdot 26 \cdot 10 \cdot 26 \cdot 10$, or $17,576,600$

b) No

45. a) Since repetition is allowed, each of the 5 digits can be chosen in 10 ways. The number of zip-codes possible is $10 \cdot 10 \cdot 10 \cdot 10 \cdot 10$, or 100,000.

b) Since there are 100,000 possible zip-codes, there could be 100,000 post offices.

46. a) $10 \cdot 10 \cdot 10 \cdot 10 \cdot 10 \cdot 10 \cdot 10 \cdot 10 \cdot 10$, or 1,000,000,000

b) Yes

47. a) Since repetition is allowed, each digit can be chosen in 10 ways. There can be
$10 \cdot 10 \cdot 10 \cdot 10 \cdot 10 \cdot 10 \cdot 10 \cdot 10 \cdot 10$, or 1,000,000,000 social security numbers.

b) Since more than 261 million social security numbers are possible, each person can have a social security number.

48. Approximately 41,466 yr

49. We find the number of distinct circular permutations of 12 objects:
$$(12-1)! = 11! = 39,916,800$$

50. $(13-1)!$, or $479,001,600$

51. The number of ways 6 students can be arranged in a straight line is $_6P_6$, or $6!$, or 720.

The number of ways 6 students can be arranged in a circle is $(6-1)!$, or $5!$, or 120.

52. $_7P_7$, or 5040;

$6!$, or 720

53. $\log_b 35 = \log_b (5 \cdot 7)$
$$= \log_b 5 + \log_b 7$$
$$= 1.609 + 1.946$$
$$= 3.555$$

54. -0.337

55. $\log_b 49 = \log_b 7^2$
$$= 2 \log_b 7$$
$$= 2(1.946)$$
$$= 3.892$$

56. 10

57. $\log_b b^{-5} = -5$

(The power to which you raise b in order to get b^{-5} is -5).

58. m

59.

60.

61.
$$_nP_5 = 7 \cdot_n P_4$$

$\dfrac{n!}{(n-5)!} = 7 \cdot \dfrac{n!}{(n-4)!}$ Formula (2)

$\dfrac{n!}{7(n-5)!} = \dfrac{n!}{(n-4)!}$ Dividing by 7

$7(n-5)! = (n-4)!$ The denominators must be the same.

$7(n-5)! = (n-4)(n-5)!$

$7 = n-4$ Dividing by $(n-5)!$

$11 = n$

62. 8

63.
$$_nP_5 = 9 \cdot_{n-1} P_4$$

$\dfrac{n!}{(n-5)!} = 9 \cdot \dfrac{(n-1)!}{(n-1-4)!}$ Formula (2)

$\dfrac{n!}{(n-5)!} = 9 \cdot \dfrac{(n-1)!}{(n-5)!}$

$n! = 9(n-1)!$ Multiplying by $(n-5)!$

$n(n-1)! = 9(n-1)!$

$n = 9$ Dividing by $(n-1)!$

64. 11

65. a) $_6P_6 = 6!720$

b) If a man is placed in the first seat, we have $_3P_3 \cdot_3 P_3 = 3! \cdot 3!$.

Similarly, if a woman is placed in the first seat, we have $_3P_3 \cdot_3 P_3 = 3! \cdot 3!$.

The total number of arrangements is $2 \cdot 3! \cdot 3!$, or 72.

c) The man and woman who must sit together can be seated in 5 different pairs of chairs (the first and second chairs, or the second and third chairs, and so on to the fifth and sixth chairs). In addition, they can be seated in $_2P_2$, or $2!$, or 2 ways and the 4 remaining people can be seated in $_4P_4$, or $4!$, or 24 ways.

Then the total number of arrangements is $5 \cdot 2! \cdot 4!$, or 240.

d) The number of arrangements if a particular man and woman must not sit together is the difference between the number of arrangements with no seating restrictions and the number of arrangements when a particular man and woman must sit together. Using the results of parts (a) and (c), we have $720 - 240$, or 480 arrangements.

66. a)$_6P_6 = 720$

b) $4 \cdot 2! \cdot 4!$, or 192

c) $8 \cdot 2! \cdot 2! \cdot 2!$, or 64

67. We will only consider factorizations in which the factors of a are both positive. Since b and c are both positive, we will also consider only positive factors of c. Now 6 has 2 such factorizations ($1 \cdot 6$ and $2 \cdot 3$), and 12 has 3 such factorizations ($1 \cdot 12$, $2 \cdot 6$, and $3 \cdot 4$). For each trial factorization $(px+\ )(rx+\ )$ there are 6 arrangements for the second terms of the factors. These are the 3 pairs of factors of 12 in the order shown above and the 3 pairs formed by taking these factors in the opposite order. Since there are 2 possible pairs of choices for p and r, the number of possible trial factorizations is 2·6, or 12. (Note that if we also reverse the order of p and r to form 2 additional sets of possibilities we only repeat the 12 original factorizations.)

68. $2 \cdot 1$, or 2

Exercise Set 14.5

1. $_{13}C_2 = \dfrac{13!}{(13-2)!2!}$

$= \dfrac{13!}{11!2!} = \dfrac{13 \cdot 12 \cdot 11!}{11! \cdot 2 \cdot 1}$

$= \dfrac{13 \cdot 12}{2 \cdot 1} = \dfrac{13 \cdot 6 \cdot 2}{2 \cdot 1}$

$= 78$

2. 84

3. $\begin{pmatrix} 13 \\ 11 \end{pmatrix} = \dfrac{13!}{(13-11)!11!}$

$= \dfrac{13!}{2!11!}$

$= 78$ (See Exercise 1.)

4. 84

5. $\begin{pmatrix} 7 \\ 1 \end{pmatrix} = \dfrac{7!}{(7-1)!1!}$

$= \dfrac{7!}{6!1!} = \dfrac{7 \cdot 6!}{6! \cdot 1}$

$= 7$

6. 1

7. $\dfrac{_5P_3}{3!} = \dfrac{5 \cdot 4 \cdot 3}{3!}$

$= \dfrac{5 \cdot 4 \cdot 3}{3 \cdot 2 \cdot 1} = \dfrac{5 \cdot 2 \cdot 2 \cdot 3}{3 \cdot 2 \cdot 1}$

$= 5 \cdot 2 = 10$

8. 252

9. $\begin{pmatrix} 6 \\ 0 \end{pmatrix} = \dfrac{6!}{(6-0)!0!}$

$= \dfrac{6!}{6!0!} = \dfrac{6!}{6! \cdot 1}$

$= 1$

10. 20

11. $_{12}C_{11} = \dfrac{12!}{(12-11)!11!}$

$= \dfrac{12!}{1!11!} = \dfrac{12 \cdot 11!}{1 \cdot 11!}$

$= 12$

12. 66

13. $_{20}C_{18} = \dfrac{20!}{(20-18)!18!}$

$= \dfrac{20!}{2!18!} = \dfrac{20 \cdot 19 \cdot 18!}{2 \cdot 1 \cdot 18!}$

$= \dfrac{20 \cdot 19}{2 \cdot 1} = \dfrac{2 \cdot 10 \cdot 19}{2 \cdot 1}$

$= 190$

14. 4060

15. $\begin{pmatrix} 35 \\ 2 \end{pmatrix} = \dfrac{35!}{(35-2)!2!}$

$= \dfrac{35!}{33!2!} = \dfrac{35 \cdot 34 \cdot 33!}{33! \cdot 2 \cdot 1}$

$= \dfrac{35 \cdot 34}{2 \cdot 1} = \dfrac{35 \cdot 2 \cdot 17}{2 \cdot 1}$

$= 595$

16. 780

17. $_{10}C_5 = \dfrac{10!}{(10-5)!5!}$

$= \dfrac{10!}{5!5!} = \dfrac{10 \cdot 9 \cdot 8 \cdot 7 \cdot 6 \cdot 5!}{5 \cdot 4 \cdot 3 \cdot 2 \cdot 1 \cdot 5!}$

$= \dfrac{10 \cdot 9 \cdot 8 \cdot 7 \cdot 6}{5 \cdot 4 \cdot 3 \cdot 2 \cdot 1} = \dfrac{5 \cdot 2 \cdot 3 \cdot 3 \cdot 4 \cdot 2 \cdot 7 \cdot 6}{5 \cdot 4 \cdot 3 \cdot 2 \cdot 1}$

$= 252$

18. 1365

19. $_{23}C_4 = \dfrac{23!}{(23-4)!4!}$

$= \dfrac{23!}{19!4!} = \dfrac{23 \cdot 22 \cdot 21 \cdot 20 \cdot 19!}{19! \cdot 4 \cdot 3 \cdot 2 \cdot 1}$

$= \dfrac{23 \cdot 22 \cdot 21 \cdot 20}{4 \cdot 3 \cdot 2 \cdot 1} = \dfrac{23 \cdot 2 \cdot 11 \cdot 3 \cdot 7 \cdot 4 \cdot 5}{4 \cdot 3 \cdot 2 \cdot 1}$

$= 8855$

20. $_9C_2$, or 36;

$2 \cdot _9C_2$, or 72

21. $_{10}C_6 = \dfrac{10!}{(10-6)!6!}$

$= \dfrac{10!}{4!6!} = \dfrac{10 \cdot 9 \cdot 8 \cdot 7 \cdot 6!}{4 \cdot 3 \cdot 2 \cdot 1 \cdot 6!}$

$= \dfrac{10 \cdot 9 \cdot 8 \cdot 7}{4 \cdot 3 \cdot 2 \cdot 1} = \dfrac{10 \cdot 3 \cdot 3 \cdot 4 \cdot 2 \cdot 7}{4 \cdot 3 \cdot 2 \cdot 1}$

$= 210$

22. $_{11}C_7$, or 330

23. Since two points determine a line and no three of these 8 points are collinear, we need to find the number of combinations of 8 points taken 2 at a time, $_8C_2$.

$$_8C_2 = \binom{8}{2} = \frac{8!}{2!(8-2)!}$$
$$= \frac{8 \cdot 7 \cdot 6!}{2 \cdot 1 \cdot 6!} = \frac{4 \cdot 2 \cdot 7}{2 \cdot 1}$$
$$= 28$$

Thus 28 lines are determined.

Since three noncolinear points determine a triangle, we need to find the number of combinations of 8 points taken 3 at a time, $_8C_3$.

$$_8C_3 = \binom{8}{3} = \frac{8!}{3!(8-3)!}$$
$$= \frac{8 \cdot 7 \cdot 6 \cdot 5!}{3 \cdot 2 \cdot 1 \cdot 5!} = \frac{8 \cdot 7 \cdot 3 \cdot 2}{3 \cdot 2 \cdot 1}$$
$$= 56$$

Thus 56 triangles are determined.

24. a) $_7C_2$, or 21

    b) $_7C_3$, or 35

25. $_{10}C_7 \cdot _5C_3 = \binom{10}{7} \cdot \binom{5}{3}$ Using the fundamental counting principle

$$= \frac{10!}{7!(10-7)!} \cdot \frac{5!}{3!(5-3)!}$$
$$= \frac{10 \cdot 9 \cdot 8 \cdot 7!}{7! \cdot 3!} \cdot \frac{5 \cdot 4 \cdot 3!}{3! \cdot 2!}$$
$$= \frac{10 \cdot 9 \cdot 8}{3 \cdot 2 \cdot 1} \cdot \frac{5 \cdot 4}{2 \cdot 1} = 120 \cdot 10 = 1200$$

26. $_8C_6 \cdot _4C_3$, or 112

27. Using the fundamental counting principle, we have $_{58}C_6 \cdot _{42}C_4$.

28. $_{63}C_8 \cdot _{37}C_{12}$

29. We use the fundamental counting principle. There are 6 choices of seafood, 9 choices of vegetables and 7 choices of sauces.

$$6 \cdot 9 \cdot 7 = 378$$

30. $6 \cdot 6$, or 36

31. $_{52}C_5$

32. $_{52}C_{13}$

33. We use the fundamental counting principle.

$$_5C_2 \cdot _6C_3 \cdot _3C_1$$
$$= \frac{5!}{(5-2)!2!} \cdot \frac{6!}{(6-3)!3!} \cdot \frac{3!}{(3-1)!1!}$$
$$= \frac{5!}{3!2!} \cdot \frac{6!}{3!3!} \cdot \frac{3!}{2!1!}$$
$$= \frac{5 \cdot 4 \cdot 3!}{3! \cdot 2!} \cdot \frac{6 \cdot 5 \cdot 4 \cdot 3!}{3! \cdot 3 \cdot 2 \cdot 1} \cdot \frac{3 \cdot 2!}{2! \cdot 1}$$
$$= \frac{5 \cdot 4}{2} \cdot \frac{6 \cdot 5 \cdot 4}{3 \cdot 2} \cdot \frac{3}{1} = 10 \cdot 20 \cdot 3$$
$$= 600$$

34. $_7C_4 \cdot _5C_3$, or 350

35. The oldest child gets 4 of the 9 books. The middle child gets 3 of the remaining 5 books, and the youngest gets the last 2 books. We use the fundamental counting principle.

$$_9C_4 \cdot _5C_3 \cdot _2C_2$$
$$= \frac{9!}{(9-4)!4!} \cdot \frac{5!}{(5-3)!3!} \cdot \frac{2!}{(2-2)!2!}$$
$$= \frac{9!}{5!4!} \cdot \frac{5!}{2!3!} \cdot \frac{2!}{0!2!}$$
$$= \frac{9 \cdot 8 \cdot 7 \cdot 6 \cdot 5!}{5! \cdot 4 \cdot 3 \cdot 2 \cdot 1} \cdot \frac{5 \cdot 4 \cdot 3!}{2 \cdot 1 \cdot 3!} \cdot \frac{2!}{1 \cdot 2!}$$
$$= \frac{9 \cdot 8 \cdot 7 \cdot 6}{4 \cdot 3 \cdot 2 \cdot 1} \cdot \frac{5 \cdot 4}{2 \cdot 1} \cdot 1 = 126 \cdot 10 \cdot 1$$
$$= 1260$$

36. $\binom{8}{2}\binom{6}{5}\binom{1}{1}$, or 168

37. The pizza can have no toppings or 1 topping or 2 or 3 or 4 or 5 or 6 or 7 or 8 or 9 or 10 toppings. We add these combinations to find the total number possible.

$$\binom{10}{0} + \binom{10}{1} + \binom{10}{2} + \binom{10}{3} + \binom{10}{4} + \binom{10}{5} +$$
$$\binom{10}{6} + \binom{10}{7} + \binom{10}{8} + \binom{10}{9} + \binom{10}{10} = 1024$$

38. $\binom{2}{1} \cdot \binom{3}{1} \cdot \left[\binom{10}{0} + \binom{10}{1} + \binom{10}{2} + \binom{10}{3} + \binom{10}{4} + \right.$
$$\left. \binom{10}{5} + \binom{10}{6} + \binom{10}{7} + \binom{10}{8} + \binom{10}{9} + \binom{10}{10} \right],$$ or
6144

39. In a 52-card deck there are 4 aces and 48 cards that are not aces. We use the fundamental counting principle.

$$\binom{4}{3} \cdot \binom{48}{2} = \frac{4!}{(4-3)!3!} \cdot \frac{48!}{(48-2)!2!}$$
$$= \frac{4!}{1!3!} \cdot \frac{48!}{46!2!}$$
$$= \frac{4 \cdot 3!}{1 \cdot 3!} \cdot \frac{48 \cdot 47 \cdot 46!}{46! \cdot 2 \cdot 1}$$
$$= \frac{4}{1} \cdot \frac{48 \cdot 47}{2 \cdot 1}$$
$$= 4512$$

40. $\binom{4}{2} \cdot \binom{48}{3}$, or 103,776

41. a) If order is considered and repetition is not allowed, we have

$$_{33}P_3 = 33 \cdot 32 \cdot 31 = 32,736.$$

    b) If order is considered and repetition is allowed, each scoop of ice cream can be chosen in 33 ways, so we have

$$33 \cdot 33 \cdot 33 = 35,937.$$

c) If order is not considered and repetition is not allowed, we have

$$_{33}C_3 = \frac{33!}{(33-3)!3!}$$
$$= \frac{33!}{30!3!} = \frac{33 \cdot 32 \cdot 31 \cdot 30!}{30!3 \cdot 2 \cdot 1}$$
$$= 5456.$$

42. a) $_{31}P_2$, or 930

b) $31 \cdot 31$, or 961

c) $_{31}C_2$, or 465

43. $2^x = \frac{1}{4}$

$$2^x = \frac{1}{2^2}$$

$$2^x = 2^{-2}$$

$x = -2$ The exponents must be the same.

The solution is -2.

44. $\frac{3}{2}$

45. $\log_5 (x+1) = 2$

$\qquad 5^2 = x+1$ Writing an equivalent exponential equation

$\qquad 25 = x+1$

$\qquad 24 = x$

The number 24 checks and is the solution.

46. 5

47.

48.

49. $\binom{m}{1} = \dfrac{m!}{(m-1)!1!}$

$$= \frac{m(m-1)!}{(m-1)! \cdot 1}$$

$$= m$$

50. m

51. $\binom{m}{0} = \dfrac{m!}{(m-0)!0!}$

$$= \frac{m!}{m! \cdot 1}$$

$$= 1$$

52. $\dfrac{m(m-1)}{2}$

53.

$$\binom{n+1}{3} = 2 \cdot \binom{n}{2}$$

$$\frac{(n+1)!}{(n+1-3)!3!} = 2 \cdot \frac{n!}{(n-2)!2!}$$

$$\frac{(n+1)!}{(n-2)!3!} = 2 \cdot \frac{n!}{(n-2)!2!}$$

$$\frac{(n+1)(n)(n-1)(n-2)!}{(n-2)!3 \cdot 2 \cdot 1} = 2 \cdot \frac{n(n-1)(n-2)!}{(n-2)! \cdot 2 \cdot 1}$$

$$\frac{(n+1)(n)(n-1)}{6} = n(n-1)$$

$$\frac{n^3 - n}{6} = n^2 - n$$

$$n^3 - n = 6n^2 - 6n$$

$$n^3 - 6n^2 + 5n = 0$$

$$n(n^2 - 6n + 5) = 0$$

$$n(n-5)(n-1) = 0$$

$$n = 0 \ \text{ or } \ n-5 = 0 \ \text{ or } \ n-1 = 0$$

$$n = 0 \ \text{ or } \qquad n = 5 \ \text{ or } \qquad n = 1$$

Only 5 checks. The solution is 5.

54. 4

55.

$$\binom{n+2}{4} = 6 \cdot \binom{n}{2}$$

$$\frac{(n+2)!}{(n+2-4)!4!} = 6 \cdot \frac{n!}{(n-2)!2!}$$

$$\frac{(n+2)!}{(n-2)!4!} = 6 \cdot \frac{n!}{(n-2)!2!}$$

$$\frac{(n+2)!}{4!} = 6 \cdot \frac{n!}{2!}$$ Multiplying by $(n{-}2)!$

$$4! \cdot \frac{(n+2)!}{4!} = 4! \cdot 6 \cdot \frac{n!}{2!}$$

$$(n+2)! = 72 \cdot n!$$

$$(n+2)(n+1)n! = 72 \cdot n!$$

$$(n+2)(n+1) = 72$$ Dividing by $n!$

$$n^2 + 3n + 2 = 72$$

$$n^2 + 3n - 70 = 0$$

$$(n+10)(n-7) = 0$$

$$n+10 = 0 \quad \text{ or } \ n-7 = 0$$

$$n = -10 \ \text{ or } \qquad n = 7$$

Only 7 checks. The solution is 7.

56. 6

57. There is one losing team per game. In order to leave one tournament winner there must be $n-1$ losers produced in $n-1$ games.

58. $2n-1$

59. $\binom{m}{3} = \dfrac{m!}{(m-3)!3!}$

$\phantom{\binom{m}{3}} = \dfrac{m(m-1)(m-2)(m-3)!}{(m-3)! \cdot 3 \cdot 2 \cdot 1}$

$\phantom{\binom{m}{3}} = \dfrac{m(m-1)(m-2)}{6}$

60. $\binom{m}{2}\binom{n}{2}$

61. See the answer section in the text.

Exercise Set 14.6

1. Expand $(m+n)^5$.

Form 1: The expansion of $(m+n)^5$ has $5+1$, or 6 terms. The sum of the exponents in each term is 5. The exponents of m start with 5 and decrease to 0. The last term has no factor of m. The first term has no factor of n. The exponents of n start in the second term with 1 and increase to 5. We get the coefficients from the 6th row of Pascal's triangle.

$$
\begin{array}{ccccccccccc}
 & & & & & 1 & & & & & \\
 & & & & 1 & & 1 & & & & \\
 & & & 1 & & 2 & & 1 & & & \\
 & & 1 & & 3 & & 3 & & 1 & & \\
 & 1 & & 4 & & 6 & & 4 & & 1 & \\
1 & & 5 & & 10 & & 10 & & 5 & & 1
\end{array}
$$

$(m+n)^5 = 1 \cdot m^5 + 5 \cdot m^4 n^1 + 10 \cdot m^3 \cdot n^2 +$
$\qquad 10 \cdot m^2 \cdot n^3 + 5 \cdot m \cdot n^4 + 1 \cdot n^5$
$\qquad = m^5 + 5m^4 n + 10 m^3 n^2 + 10 m^2 n^3 +$
$\qquad 5mn^4 + n^5$

Form 2: We have $a = m$, $b = n$, and $n = 5$.

$(m+n)^5 = \binom{5}{0} m^5 + \binom{5}{1} m^4 n + \binom{5}{2} m^3 n^2 +$

$\qquad \binom{5}{3} m^2 n^3 + \binom{5}{4} mn^4 + \binom{5}{5} n^5$

$\qquad = \dfrac{5!}{5!0!} m^5 + \dfrac{5!}{4!1!} m^4 n + \dfrac{5!}{3!2!} m^3 n^2 +$

$\qquad \dfrac{5!}{2!3!} m^2 n^3 + \dfrac{5!}{1!4!} mn^4 + \dfrac{5!}{0!5!} m^5$

$\qquad = m^5 + 5m^4 n + 10 m^3 n^2 + 10 m^2 n^3 +$
$\qquad 5mn^4 + n^5$

2. $a^3 - 3a^2 b + 3ab^2 - b^3$

3. Expand $(x-y)^6$.

Form 1: The expansion of $(x-y)^6$ has $6+1$, or 7 terms. The sum of the exponents in each term is 6. The exponents of x start with 6 and decrease to 0. The last term has no factor of x. The first term has no factor of $-y$. The exponents of $-y$ start in the second term with 1 and increase to 6. We get the coefficients from the 7th row of Pascal's triangle.

$$
\begin{array}{ccccccccccccc}
 & & & & & & 1 & & & & & & \\
 & & & & & 1 & & 1 & & & & & \\
 & & & & 1 & & 2 & & 1 & & & & \\
 & & & 1 & & 3 & & 3 & & 1 & & & \\
 & & 1 & & 4 & & 6 & & 4 & & 1 & & \\
 & 1 & & 5 & & 10 & & 10 & & 5 & & 1 & \\
1 & & 6 & & 15 & & 20 & & 15 & & 6 & & 1
\end{array}
$$

$(x-y)^6 = 1 \cdot x^6 + 6 \cdot x^5 \cdot (-y) + 15 \cdot x^4 \cdot (-y)^2 +$
$\qquad 20 \cdot x^3 \cdot (-y)^3 + 15 \cdot x^2 \cdot (-y)^4 +$
$\qquad 6 \cdot x \cdot (-y)^5 + 1 \cdot (-y)^6$
$\qquad = x^6 - 6x^5 y + 15 x^4 y^2 - 20 x^3 y^3 +$
$\qquad 15 x^2 y^4 - 6xy^5 + y^6$

Form 2: We have $a = x$, $b = -y$, and $n = 6$.

$(x-y)^6 = \binom{6}{0} x^6 + \binom{6}{1} x^5 (-y) + \binom{6}{2} x^4 (-y)^2 +$

$\qquad \binom{6}{3} x^3 (-y)^3 + \binom{6}{4} x^2 (-y)^4 +$

$\qquad \binom{6}{5} x(-y)^5 + \binom{6}{6} (-y)^6$

$\qquad = \dfrac{6!}{6!0!} x^6 + \dfrac{6!}{5!1!} x^5 (-y) + \dfrac{6!}{4!2!} x^4 y^2 +$

$\qquad \dfrac{6!}{3!3!} x^3 (-y^3) + \dfrac{6!}{2!4!} x^2 y^4 + \dfrac{6!}{1!5!} x(-y^5) +$

$\qquad \dfrac{6!}{0!6!} y^6$

$\qquad = x^6 - 6x^5 y + 15 x^4 y^2 - 20 x^3 y^3 +$
$\qquad 15 x^2 y^4 - 6xy^5 + y^6$

4. $p^4 + 4p^3 q + 6p^2 q^2 + 4pq^3 + q^4$

5. Expand $(x^2 - 3y)^5$.

We have $a = x^2$, $b = -3y$, and $n = 5$.

Form 1: We get the coefficients from the 6th row of Pascal's triangle. From Exercise 1 we know that the coefficients are

$\qquad 1 \qquad 5 \qquad 10 \qquad 10 \qquad 5 \qquad 1.$

$(x^2 - 3y)^5 = 1 \cdot (x^2)^5 + 5 \cdot (x^2)^4 \cdot (-3y) +$
$\qquad 10 \cdot (x^2)^3 \cdot (-3y)^2 + 10 \cdot (x^2)^2 \cdot (-3y)^3 +$
$\qquad 5 \cdot (x^2) \cdot (-3y)^4 + 1 \cdot (-3y)^5$
$\qquad = x^{10} - 15 x^8 y + 90 x^6 y^2 - 270 x^4 y^3 +$
$\qquad 405 x^2 y^4 - 243 y^5$

Form 2:

$(x^2 + 3y)^5 = \binom{5}{0} (x^2)^5 + \binom{5}{1} (x^2)^4 (-3y) +$

$\qquad \binom{5}{2} (x^2)^3 (-3y)^2 + \binom{5}{3} (x^2)^2 (-3y)^3 +$

$\qquad \binom{5}{4} x^2 (-3y)^4 + \binom{5}{5} (-3y)^5$

$\qquad = \dfrac{5!}{5!0!} x^{10} + \dfrac{5!}{4!1!} x^8 (-3y) + \dfrac{5!}{3!2!} x^6 (9y^2) +$

$\qquad \dfrac{5!}{2!3!} x^4 (-27y^3) + \dfrac{5!}{1!4!} x^2 (81y^4) +$

$\qquad \dfrac{5!}{0!5!} (-243y^5)$

$$= x^{10} - 15x^8 y + 90x^6 y^2 - 270x^4 y^3 +$$
$$405x^2 y^4 - 243y^5$$

6. $2187c^7 - 5103c^6 d + 5103c^5 d^2 - 2835c^4 d^3 + 945c^3 d^4 - 189c^2 d^5 + 21cd^6 - d^7$

7. Expand $(3c - d)^6$.

We have $a = 3c$, $b = -d$, and $n = 6$.

Form 1: We get the coefficients from the 7th row of Pascal's triangle. From Exercise 3 we know that the coefficients are

$$\begin{array}{ccccccc} 1 & 6 & 15 & 20 & 15 & 6 & 1. \end{array}$$

$$(3c - d)^6 = 1 \cdot (3c)^6 + 6 \cdot (3c)^5 \cdot (-d) +$$
$$15 \cdot (3c)^4 \cdot (-d)^2 + 20 \cdot (3c)^3 \cdot (-d)^3 +$$
$$15 \cdot (3c)^2 \cdot (-d)^4 + 6 \cdot (3c) \cdot (-d)^5 +$$
$$1 \cdot (-d)^6$$
$$= 3^6 c^6 - 6 \cdot 3^5 c^5 d + 15 \cdot 3^4 c^4 d^2 -$$
$$20 \cdot 3^3 c^3 d^3 + 15 \cdot 3^2 c^2 d^4 - 6 \cdot 3cd^5 + d^6$$
$$= 729c^6 - 6 \cdot 243c^5 d + 15 \cdot 81c^4 d^2 -$$
$$20 \cdot 27c^3 d^3 + 15 \cdot 9c^2 d^4 - 6 \cdot 3cd^5 + d^6$$
$$= 729c^6 - 1458c^5 d + 1215c^4 d^2 - 540c^3 d^3 +$$
$$135c^2 d^4 - 18cd^5 + d^6$$

Form 2:

$$(3c - d)^6 = \binom{6}{0}(3c)^6 + \binom{6}{1}(3c)^5(-d) +$$
$$\binom{6}{2}(3c)^4(-d)^2 + \binom{6}{3}(3c)^3(-d)^3 +$$
$$\binom{6}{4}(3c)^2(-d)^4 + \binom{6}{5}(3c)(-d)^5 +$$
$$\binom{6}{6}(-d)^6$$
$$= \frac{6!}{6!0!}(729c^6) + \frac{6!}{5!1!}(243c^5)(-d) +$$
$$\frac{6!}{4!2!}(81c^4)(d^2) + \frac{6!}{3!3!}(27c^3)(-d^3) +$$
$$\frac{6!}{2!4!}(9c^2)(d^4) + \frac{6!}{1!5!}(3c)(-d^5) +$$
$$\frac{6!}{0!6!}d^6$$
$$= 729c^6 - 1458c^5 d + 1215c^4 d^2 - 540c^3 d^3 +$$
$$135c^2 d^4 - 18cd^5 + d^6$$

8. $t^{-12} + 12t^{-10} + 60t^{-8} + 160t^{-6} + 240t^{-4} + 192t^{-2} + 64$

9. Expand $(x - y)^7$.

We have $a = x$, $b = -y$, and $n = 7$.

Form 1: We get the coefficients from the 8th row of Pascal's triangle.

$$\begin{array}{ccccccccccccc}
 & & & & & & 1 & & & & & & \\
 & & & & & 1 & & 1 & & & & & \\
 & & & & 1 & & 2 & & 1 & & & & \\
 & & & 1 & & 3 & & 3 & & 1 & & & \\
 & & 1 & & 4 & & 6 & & 4 & & 1 & & \\
 & 1 & & 5 & & 10 & & 10 & & 5 & & 1 & \\
1 & & 6 & & 15 & & 20 & & 15 & & 6 & & 1
\end{array}$$

$$\begin{array}{ccccccccccccc}
1 & & 7 & & 21 & & 35 & & 35 & & 21 & & 7 & & 1
\end{array}$$

$$(x - y)^7$$
$$= 1 \cdot x^7 + 7x^6(-y) + 21x^5(-y)^2 + 35x^4(-y)^3 +$$
$$35x^3(-y)^4 + 21x^2(-y)^5 + 7x(-y)^6 + 1 \cdot (-y)^7$$
$$= x^7 - 7x^6 y + 21x^5 y^2 - 35x^4 y^3 + 35x^3 y^4 -$$
$$21x^2 y^5 + 7xy^6 - y^7$$

Form 2:

$$(x - y)^7$$
$$= \binom{7}{0}x^7 + \binom{7}{1}x^6(-y) + \binom{7}{2}x^5(-y)^2 +$$
$$\binom{7}{3}x^4(-y)^3 + \binom{7}{4}x^3(-y)^4 +$$
$$\binom{7}{5}x^2(-y)^5 + \binom{7}{6}x(-y)^6 + \binom{7}{7}(-y)^7$$
$$= \frac{7!}{7!0!}x^7 + \frac{7!}{6!1!}x^6(-y) + \frac{7!}{5!2!}x^5 y^2 +$$
$$\frac{7!}{4!3!}x^4(-y^3) + \frac{7!}{3!4!}x^3 y^4 + \frac{7!}{2!5!}x^2(-y^5) +$$
$$\frac{7!}{1!6!}xy^6 + \frac{7!}{0!7!}(-y^7)$$
$$= x^7 - 7x^6 y + 21x^5 y^2 - 35x^4 y^3 + 35x^3 y^4 -$$
$$21x^2 y^5 + 7xy^6 - y^7$$

10. $x^5 - 5x^4 y + 10x^3 y^2 - 10x^2 y^3 + 5xy^4 - y^5$

11. Expand $\left(\dfrac{1}{x} + y\right)^7$.

We have $a = \dfrac{1}{x}$, $b = y$, and $n = 7$.

Form 1: We get the coefficients from the 8th row of Pascal's triangle. From Exercise 9 we know that the coefficients are

$$\begin{array}{ccccccc} 1 & 7 & 21 & 35 & 35 & 21 & 7 & 1. \end{array}$$

$$\left(\frac{1}{x} + y\right)^7 = 1 \cdot \left(\frac{1}{x}\right)^7 + 7\left(\frac{1}{x}\right)^6 \cdot y + 21\left(\frac{1}{x}\right)^5 \cdot y^2 +$$
$$35\left(\frac{1}{x}\right)^4 \cdot y^3 + 35\left(\frac{1}{x}\right)^3 \cdot y^4 +$$
$$21\left(\frac{1}{x}\right)^2 \cdot y^5 + 7\left(\frac{1}{x}\right) \cdot y^6 + 1 \cdot y^7$$
$$= x^{-7} + 7x^{-6} y + 21x^{-5} y^2 + 35x^{-4} y^3 +$$
$$35x^{-3} y^4 + 21x^{-2} y^5 + 7x^{-1} y^6 + y^7$$

Form 2:

$$\left(\frac{1}{x} + y\right)^7$$

$$= \binom{7}{0}\left(\frac{1}{x}\right)^7 + \binom{7}{1}\left(\frac{1}{x}\right)^6 y + \binom{7}{2}\left(\frac{1}{x}\right)^5 y^2 +$$

$$\binom{7}{3}\left(\frac{1}{x}\right)^4 y^3 + \binom{7}{4}\left(\frac{1}{x}\right)^3 y^4 + \binom{7}{5}\left(\frac{1}{x}\right)^2 y^5 +$$

$$\binom{7}{6}\left(\frac{1}{x}\right) y^6 + \binom{7}{7} y^7$$

$$= \frac{7!}{7!0!}\left(\frac{1}{x}\right)^7 + \frac{7!}{6!1!}\left(\frac{1}{x}\right)^6 y + \frac{7!}{5!2!}\left(\frac{1}{x}\right)^5 y^2 +$$

$$\frac{7!}{4!3!}\left(\frac{1}{x}\right)^4 y^3 + \frac{7!}{3!4!}\left(\frac{1}{x}\right)^3 y^4 + \frac{7!}{2!5!}\left(\frac{1}{x}\right)^2 y^5 +$$

$$\frac{7!}{1!6!}\left(\frac{1}{x}\right) y^6 + \frac{7!}{0!7!} y^7$$

$$= x^{-7} + 7x^{-6}y + 21x^{-5}y^2 + 35x^{-4}y^3 + 35x^{-3}y^4 +$$

$$21x^{-2}y^5 + 7x^{-1}y^6 + y^7$$

12. $8s^3 - 36s^2t^2 + 54st^4 - 27t^6$

13. Expand $\left(a - \frac{2}{a}\right)^9$.

We have $a = a$, $b = -\frac{2}{a}$, and $n = 9$.

Form 1: We get the coefficients from the 10th row of Pascal's triangle.

```
                    1
                 1     1
              1     2     1
           1     3     3     1
        1     4     6     4     1
     1     5    10    10     5     1
  1     6    15    20    15     6     1
1    7    21    35    35    21     7     1
1  8   28   56    70    56    28    8    1
1 9  36   84   126   126   84   36   9   1
```

$$\left(a - \frac{2}{a}\right)^9 = 1 \cdot a^9 + 9a^8\left(-\frac{2}{a}\right) + 36a^7\left(-\frac{2}{a}\right)^2 +$$

$$84a^6\left(-\frac{2}{a}\right)^3 + 126a^5\left(-\frac{2}{a}\right)^4 +$$

$$126a^4\left(-\frac{2}{a}\right)^5 + 84a^3\left(-\frac{2}{a}\right)^6 +$$

$$36a^2\left(-\frac{2}{a}\right)^7 + 9a\left(-\frac{2}{a}\right)^8 + 1 \cdot \left(-\frac{2}{a}\right)^9$$

$$= a^9 - 18a^7 + 144a^5 - 672a^3 + 2016a -$$

$$4032a^{-1} + 5376a^{-3} - 4608a^{-5} +$$

$$2304a^{-7} - 512a^{-9}$$

Form 2:

$$\left(a - \frac{2}{a}\right)^9$$

$$= \binom{9}{0}a^9 + \binom{9}{1}a^8\left(-\frac{2}{a}\right) + \binom{9}{2}a^7\left(-\frac{2}{a}\right)^2 +$$

$$\binom{9}{3}a^6\left(-\frac{2}{a}\right)^3 + \binom{9}{4}a^5\left(-\frac{2}{a}\right)^4 +$$

$$\binom{9}{5}a^4\left(-\frac{2}{a}\right)^5 + \binom{9}{6}a^3\left(-\frac{2}{a}\right)^6 +$$

$$\binom{9}{7}a^2\left(-\frac{2}{a}\right)^7 + \binom{9}{8}a\left(-\frac{2}{a}\right)^8 +$$

$$\binom{9}{9}\left(-\frac{2}{a}\right)^9$$

$$= \frac{9!}{9!0!}a^9 + \frac{9!}{8!1!}a^8\left(-\frac{2}{a}\right) + \frac{9!}{7!2!}a^7\left(\frac{4}{a^2}\right) +$$

$$\frac{9!}{6!3!}a^6\left(-\frac{8}{a^3}\right) + \frac{9!}{5!4!}a^5\left(\frac{16}{a^4}\right) +$$

$$\frac{9!}{4!5!}a^4\left(-\frac{32}{a^5}\right) + \frac{9!}{3!6!}a^3\left(\frac{64}{a^6}\right) +$$

$$\frac{9!}{2!7!}a^2\left(-\frac{128}{a^7}\right) + \frac{9!}{1!8!}a\left(\frac{256}{a^8}\right) +$$

$$\frac{9!}{0!9!}\left(-\frac{512}{a^9}\right)$$

$$= a^9 - 9(2a^7) + 36(4a^5) - 84(8a^3) + 126(16a) -$$

$$126(32a^{-1}) + 84(64a^{-3}) - 36(128a^{-5}) +$$

$$9(256a^{-7}) - 512a^{-9}$$

$$= a^9 - 18a^7 + 144a^5 - 672a^3 + 2016a - 4032a^{-1} +$$

$$5376a^{-3} - 4608a^{-5} + 2304a^{-7} - 512a^{-9}$$

14. $512x^9 + 2304x^7 + 4608x^5 + 5376x^3 + 4032x + 2016x^{-1} + 672x^{-3} + 144x^{-5} + 18x^{-7} + x^{-9}$

15. Expand $(a^2 + 2b^3)^4$.

We have $a = a^2$, $b = 2b^3$, and $n = 4$.

Form 1: We get the coefficients from the fifth row of Pascal's triangle.

```
            1
          1   1
        1   2   1
      1   3   3   1
    1   4   6   4   1
```

$$(a^2 + 2b^3)^4$$

$$= 1 \cdot (a^2)^4 + 4(a^2)^3(2b^3) + 6(a^2)^2(2b^3)^2 +$$

$$4(a^2)(2b^3)^3 + 1 \cdot (2b^3)^4$$

$$= a^8 + 8a^6b^3 + 24a^4b^6 + 32a^2b^9 + 16b^{12}$$

Form 2:

$$(a^2 + 2b^3)^4 = \binom{4}{0}(a^2)^4 + \binom{4}{1}(a^2)^3(2b^3) +$$

$$\binom{4}{2}(a^2)^2(2b^3)^2 + \binom{4}{3}(a^2)(2b^3)^3 +$$

$$\binom{4}{4}(2b^3)^4$$

$$= \frac{4!}{4!0!}(a^8) + \frac{4!}{3!1!}(a^6)(2b^3) +$$

$$\frac{4!}{2!2!}(a^4)(4b^6) + \frac{4!}{1!3!}(a^2)(8b^9) +$$

$$\frac{4!}{0!4!}(16b^{12})$$

$$= a^8 + 8a^6b^3 + 24a^4b^6 + 32a^2b^9 + 16b^{12}$$

16. $x^{18} + 12x^{15} + 60x^{12} + 160x^9 + 240x^6 + 192x^3 + 64$

17. Expand $(\sqrt{3} - t)^5$.

We have $a = \sqrt{3}$, $b = -t$, and $n = 5$.

Form 1: We get the coefficients from the sixth row of Pascal's triangle. From Exercise 1 we know that the coefficients are

$$1 \quad 5 \quad 10 \quad 10 \quad 5 \quad 1.$$

$$(\sqrt{3} - t)^5 = 1 \cdot (\sqrt{3})^5 + 5(\sqrt{3})^4(-t) +$$

$$10(\sqrt{3})^3(-t)^2 + 10(\sqrt{3})^2(-t)^3 +$$

$$5(\sqrt{3})(-t)^4 + 1 \cdot (-t)^5$$

$$= 9\sqrt{3} - 45t + 30\sqrt{3}t^2 - 30t^3 +$$

$$5\sqrt{3}t^4 - t^5$$

Form 2:

$$(\sqrt{3} - t)^5 = \binom{5}{0}(\sqrt{3})^5 + \binom{5}{1}(\sqrt{3})^4(-t) +$$

$$\binom{5}{2}(\sqrt{3})^3(-t)^2 + \binom{5}{3}(\sqrt{3})^2(-t)^3 +$$

$$\binom{5}{4}(\sqrt{3})(-t)^4 + \binom{5}{5}(-t)^5$$

$$= \frac{5!}{5!0!}(9\sqrt{3}) + \frac{5!}{4!1!}(9)(-t) +$$

$$\frac{5!}{3!2!}(3\sqrt{3})(t^2) + \frac{5!}{2!3!}(3)(-t^3) +$$

$$\frac{5!}{1!4!}(\sqrt{3})(t^4) + \frac{5!}{0!5!}(-t^5)$$

$$= 9\sqrt{3} - 45t + 30\sqrt{3}t^2 - 30t^3 + 5\sqrt{3}t^4 - t^5$$

18. $125 + 150\sqrt{5}\,t + 375t^2 + 100\sqrt{5}\,t^3 + 75t^4 + 6\sqrt{5}\,t^5 + t^6$

19. Expand $(x^{-2} + x^2)^4$.

We have $a = x^{-2}$, $b = x^2$, and $n = 4$.

Form 1: We get the coefficients from the fifth row of Pascal's triangle. From Exercise 15 we know that the coefficients are

$$1 \quad 4 \quad 6 \quad 4 \quad 1.$$

$$(x^{-2} + x^2)^4$$

$$= 1 \cdot (x^{-2})^4 + 4(x^{-2})^3(x^2) + 6(x^{-2})^2(x^2)^2 +$$

$$4(x^{-2})(x^2)^3 + 1 \cdot (x^2)^4$$

$$= x^{-8} + 4x^{-4} + 6 + 4x^4 + x^8$$

Form 2:

$$(x^{-2} + x^2)^4$$

$$= \binom{4}{0}(x^{-2})^4 + \binom{4}{1}(x^{-2})^3(x^2) +$$

$$\binom{4}{2}(x^{-2})^2(x^2)^2 + \binom{4}{3}(x^{-2})(x^2)^3 +$$

$$\binom{4}{4}(x^2)^4$$

$$= \frac{4!}{4!0!}(x^{-8}) + \frac{4!}{3!1!}(x^{-6})(x^2) + \frac{4!}{2!2!}(x^{-4})(x^4) +$$

$$\frac{4!}{1!3!}(x^{-2})(x^6) + \frac{4!}{0!4!}(x^8)$$

$$= x^{-8} + 4x^{-4} + 6 + 4x^4 + x^8$$

20. $x^{-3} - 6x^{-2} + 15x^{-1} - 20 + 15x - 6x^2 + x^3$

21. Find the 3rd term of $(a + b)^6$.

First, we note that $3 = 2 + 1$, $a = a$, $b = b$, and $n = 6$. Then the 3rd term of the expansion of $(a+b)^6$ is

$$\binom{6}{2}a^{6-2}b^2, \text{ or } \frac{6!}{4!2!}a^4b^2, \text{ or } 15a^4b^2.$$

22. $21x^2y^5$

23. Find the 12th term of $(a - 2)^{14}$.

First, we note that $12 = 11 + 1$, $a = a$, $b = -2$, and $n = 14$. Then the 12th term of the expansion of $(a - 2)^{14}$ is

$$\binom{14}{11}a^{14-11} \cdot (-2)^{11} = \frac{14!}{3!11!}a^3(-2048)$$

$$= 364a^3(-2048)$$

$$= -745,472a^3$$

24. $3,897,234x^2$

25. Find the 5th term of $(2x^3 - \sqrt{y})^8$.

First, we note that $5 = 4 + 1$, $a = 2x^3$, $b = -\sqrt{y}$, and $n = 8$. Then the 5th term of the expansion of $(2x^3 - \sqrt{y})^8$ is

$$\binom{8}{4}(2x^3)^{8-4}(-\sqrt{y})^4$$

$$= \frac{8!}{4!4!}(2x^3)^4(-\sqrt{y})^4$$

$$= 70(16x^{12})(y^2)$$

$$= 1120x^{12}y^2$$

26. $\dfrac{35}{27}b^{-5}$

27. The expansion of $(2u - 3v^2)^{10}$ has 11 terms so the 6th term is the middle term. Note that $6 = 5 + 1$, $a = 2u$, $b = -3v^2$, and $n = 10$. Then the 6th term of the expansion of $(2u - 3v^2)^{10}$ is

$$\binom{10}{5}(2u)^{10-5}(-3v^2)^5$$

$$= \frac{10!}{5!5!}(2u)^5(-3v^2)^5$$

$$= 252(32u^5)(-243v^{10})$$

$$= -1,959,552u^5v^{10}$$

28. $30x\sqrt{x}$, $30x\sqrt{3}$

29. $x^2 + (y-1)^2 = 49$

      $(x-0)^2 + (y-1)^2 = 7^2$ Standard form

The center is $(0,1)$, and the radius is 7.

30. Center: $(-3,0)$

Radius: $2\sqrt{3}$

31. $2\log_a x - 3\log_a y - \frac{1}{2}\log_a x$

$$= \frac{3}{2}\log_a x - 3\log_a y \quad \text{Collecting like terms}$$

$$= \log_a x^{3/2} - \log_a y^3 \quad \text{Power rule}$$

$$= \log_a \frac{x^{3/2}}{y^3} \qquad \text{Quotient rule}$$

This result could also be expressed as $\log_a \frac{x\sqrt{x}}{y^3}$.

32. $\log_a \dfrac{x^2}{2}$

33.

34.

35. Find the third term of $(0.313 + 0.687)^5$:

$$\binom{5}{2}(0.313)^{5-2}(0.687)^2 = \frac{5!}{3!2!}(0.313)^3(0.687)^2 \approx$$
$$0.145$$

36. $\binom{8}{5}(0.15)^3(0.85)^5 \approx 0.084$

37. Find and add the 3rd through 6th terms of
$(0.313 + 0.687)^5$:

$$\binom{5}{2}(0.313)^3(0.687)^2 + \binom{5}{3}(0.313)^2(0.687)^3 +$$

$$\binom{5}{4}(0.313)(0.687)^4 + \binom{5}{5}(0.687)^5 \approx 0.964$$

38. $\binom{8}{6}(0.15)^2(0.85)^6 + \binom{8}{7}(0.15)(0.85)^7 +$

$$\binom{8}{8}(0.85)^8 \approx 0.89$$

39. Expand $(1+i)^4$.

We have $a = 1$, $b = i$, and $n = 4$.

Form 1: We get the coefficients from the fifth row of Pascal's triangle. From Exercise 15 we know that the coefficients are

$$1 \quad 4 \quad 6 \quad 4 \quad 1.$$
$$(1+i)^4 = 1 \cdot 1^4 + 4 \cdot 1^3 \cdot i + 6 \cdot 1^2 \cdot i^2 +$$
$$4 \cdot 1 \cdot i^3 + 1 \cdot i^4$$
$$= 1 + 4i - 6 - 4i + 1$$
$$= -4$$

Form 2:

$$(1+i)^4 = \binom{4}{0}\cdot 1^4 + \binom{4}{1}\cdot 1^3 \cdot i + \binom{4}{2}\cdot 1^2 \cdot i^2 +$$

$$\binom{4}{3}\cdot 1 \cdot i^3 + \binom{4}{4}\cdot i^4$$

$$= \frac{4!}{4!0!}\cdot 1 + \frac{4!}{3!1!}\cdot i + \frac{4!}{2!2!}(-1) +$$

$$\frac{4!}{1!3!}(-i) + \frac{4!}{0!4!}\cdot 1$$

$$= 1 + 4i - 6 - 4i + 1$$

$$= -4$$

40. $-4 + 4i$

41. The $(r+1)$st term of $\left(\dfrac{3x^2}{2} - \dfrac{1}{3x}\right)^{12}$ is

$\binom{12}{r}\left(\dfrac{3x^2}{2}\right)^{12-r}\left(-\dfrac{1}{3x}\right)^r$. In the term which does not contain x, the exponent of x in the numerator is equal to the exponent of x in the denominator.

$$2(12-r) = r$$
$$24 - 2r = r$$
$$24 = 3r$$
$$8 = r$$

Find the $(8+1)$st, or 9th term:

$$\binom{12}{8}\left(\frac{3x^2}{2}\right)^4\left(-\frac{1}{3x}\right)^8 = \frac{12!}{4!8!}\left(\frac{3^4x^8}{2^4}\right)\left(\frac{1}{3^8x^8}\right) = \frac{55}{144}$$

42. $-4320x^6y^{9/2}$

43.
$$\dfrac{\binom{5}{3}(p^2)^2\left(-\frac{1}{2}p\sqrt[3]{q}\right)^3}{\binom{5}{2}(p^2)^3\left(-\frac{1}{2}p\sqrt[3]{q}\right)^2} = \dfrac{-\frac{1}{8}p^7q}{\frac{1}{4}p^8\sqrt[3]{q^2}} = -\dfrac{\sqrt[3]{q}}{2p}$$

44. $-35x^{-1/6}$, or $-\dfrac{35}{x^{1/6}}$

45. The degree of $(x^2+3)^4$ is the degree of $(x^2)^4 = x^8$, or 8.

Exercise Set 14.7

1. 100 people were surveyed. 57 wore either glasses or contacts. $100 - 57$, or 43 wore neither glasses nor contacts.

We use Principle P.

The probability that a person wears either glasses or contacts is P where

$$P = \frac{57}{100}, \text{ or } 0.57.$$

The probability that a person wears neither glasses nor contacts is P where

$$P = \frac{43}{100}, \text{ or } 0.43.$$

2. $\frac{18}{100}, \frac{24}{100}, \frac{23}{100}, \frac{23}{100}, \frac{12}{100}$

People tend not to choose the first or last numbers.

3. There was a total of 1044 letters.

A occurred 78 times.

E occurred 140 times.

I occurred 60 times.

O occurred 74 times.

U occurred 31 times.

We use Principle P.

The probability of the occurrence of an A is

$$\frac{78}{1044} \approx 0.075.$$

The probability of the occurrence of an E is

$$\frac{140}{1044} \approx 0.134.$$

The probability of the occurrence of an I is

$$\frac{60}{1044} \approx 0.057.$$

The probability of the occurrence of an O is

$$\frac{74}{1044} \approx 0.071.$$

The probability of the occurrence of a U is

$$\frac{31}{1044} \approx 0.030.$$

4. 0.367

5. There was a total of 1044 letters.

The total number of vowels was $78 + 140 + 60 + 74 + 31$, or 383.

The total number of consonants was $1044 - 383$, or 661.

The probability of a consonant occurring is

$$\frac{661}{1044} \approx 0.633.$$

6. Z; 0.999

7. There are 52 equally likely outcomes.

8. $\frac{1}{13}$

9. Since there are 52 equally likely outcomes and there are 13 ways to obtain a heart, by Principle P we have

$$P(\text{drawing a heart}) = \frac{13}{52}, \text{ or } \frac{1}{4}.$$

10. $\frac{1}{4}$

11. Since there are 52 equally likely outcomes and there are 4 ways to obtain a 4, by Principle P we have

$$P(\text{drawing a 4}) = \frac{4}{52}, \text{ or } \frac{1}{13}.$$

12. $\frac{1}{2}$

13. Since there are 52 equally likely outcomes and there are 26 ways to obtain a black card, by Principle P we have

$$P(\text{drawing a black card}) = \frac{26}{52}, \text{ or } \frac{1}{2}.$$

14. $\frac{2}{13}$

15. Since there are 52 equally likely outcomes and there are 8 ways to obtain a 9 or a king (four 9's and four kings), we have, by Principle P,

$$P(\text{drawing a 9 or a king}) = \frac{8}{52} = \frac{2}{13}.$$

16. $\frac{2}{7}$

17. Since there are 14 equally likely ways of selecting a marble from a bag containing 4 red marbles and 10 green marbles, we have, by Principle P,

$$P(\text{selecting a green marble}) = \frac{10}{14} = \frac{5}{7}.$$

18. 0

19. There are 14 equally likely ways of selecting any marble from a bag containing 4 red marbles and 10 green marbles. Since the bag does not contain any white marbles, there are 0 ways of selecting a white marble. By Principle P, we have

$$P(\text{selecting a white marble}) = \frac{0}{14} = 0.$$

20. $\frac{11}{4165}$

21. The number of ways of drawing 4 cards from a deck of 52 cards is $_{52}C_4$. Now 13 of the 52 cards are hearts, so the number of ways of drawing 4 hearts is $_{13}C_4$. Thus,

$$P(\text{getting 4 hearts}) = \frac{_{13}C_4}{_{52}C_4}, \text{ or } \frac{11}{4165}.$$

22. $\frac{60}{143}$

23. The number of ways of selecting 4 people from a group of 15 is $_{15}C_4$. Two men can be selected in $_8C_2$ ways, and 2 women can be selected in $_7C_2$ ways. By the fundamental counting principle, the number of ways of selecting 2 men and 2 women is $_8C_2 \cdot _7C_2$. Thus,

$$P(\text{2 men and 2 women are chosen}) = \frac{_8C_2 \cdot _7C_2}{_{15}C_4},$$

or $\frac{28}{65}$.

24. $\frac{5}{36}$

25. On each die there are 6 possible outcomes. The outcomes are paired so there are $6 \cdot 6$, or 36 possible ways in which the two can fall. The pairs that total 3 are $(1, 2)$ and $(2, 1)$. Thus there are 2 possible ways of getting a total of 3, so the probability is $\frac{2}{36}$, or $\frac{1}{18}$.

26. $\dfrac{1}{36}$

27. On each die there are 6 possible outcomes. The outcomes are paired so there are $6 \cdot 6$, or 36 possible ways in which the two can fall. There is only 1 way of getting a total of 12, the pair $(6,6)$, so the probability is $\dfrac{1}{36}$.

28. $\dfrac{245}{1938}$

29. The number of ways of selecting 6 coins from a bag containing 20 coins is $_{20}C_6$. Three nickels can be selected in $_6C_3$ ways since the bag contains 6 nickels. Two dimes can be selected in $_{10}C_2$ ways since the bag contains 10 dimes. One quarter can be selected in $_4C_1$ ways since the bag contains 4 quarters. By the fundamental counting principle the number of ways of selecting 3 nickels, 2 dimes, and 1 quarter is $_6C_3 \cdot_{10}C_2 \cdot_4C_1$. Thus,

$P(\text{getting 3 nickels, 2 dimes, and 1 quarter})$

$$= \frac{_6C_3 \cdot_{10}C_2 \cdot_4C_1}{_{20}C_6}, \text{ or } \frac{30}{323}.$$

30. $\dfrac{343}{2925}$

31. The number of ways of selecting 4 cans from a box of 25 cans is $_{25}C_4$. Using the fundamental counting principle, we find that one can of meat, one can of vegetables, one can of fruit, and one can of soup can be selected in $_4C_1 \cdot_{10}C_1 \cdot_6C_1 \cdot_5C_1$ ways since there are 4 cans of meat, 10 cans of vegetables, 6 cans of fruit, and 5 cans of soup. Then

$$P = \frac{_4C_1 \cdot_{10}C_1 \cdot_6C_1 \cdot_5C_1}{_{25}C_4}, \text{ or } \frac{24}{253}$$

32. $\dfrac{9}{19}$

33. The roulette wheel contains 38 equally likely slots. Eighteen of the 38 slots are colored red. Thus, by Principle P,

$P(\text{the ball falls in a red slot}) = \dfrac{18}{38} = \dfrac{9}{19}$.

34. $\dfrac{18}{19}$

35. The roulette wheel contains 38 equally likely slots. Only 1 slot is numbered 00. Then, by Principle P,

$P(\text{the ball falls in the 00 slot}) = \dfrac{1}{38}$.

36. $\dfrac{1}{38}$

37. The roulette wheel contains 38 equally likely slots, 2 of which are numbered 00 and 0. Thus, using Principle P,

$$P\left(\begin{array}{c}\text{the ball falling in either}\\ \text{the 00 or the 0 slot}\end{array}\right) = \frac{2}{38} = \frac{1}{19}.$$

38. $\dfrac{9}{19}$

39. $2x + 5y = 7$,   (1)
    $3x + 2y = 16$ (2)

Multiply Equation (1) by 2, multiply Equation (2) by -5, and add.

$$\begin{array}{r} 4x + 10y = 14 \\ -15x - 10y = -80 \\ \hline -11x = -66 \\ x = 6 \end{array}$$

Substitute 6 for x in Equation (2) and solve for y.

$$3 \cdot 6 + 2y = 16$$
$$18 + 2y = 16$$
$$2y = -2$$
$$y = -1$$

The solution is $(6, -1)$.

40. -1

41. $\log_a \dfrac{x^2 y}{z^3}$

$= \log_a x^2 y - \log_a z^3$ Quotient rule

$= \log_a x^2 + \log_a y - \log_a z^3$ Product rule

$= 2\log_a x + \log_a y - 3\log_a z$ Power rule

42. $x^2 + (y + 3)^2 = 12$

43.

44.

45. $_{52}C_5 = \dfrac{52!}{47!5!} = \dfrac{52 \cdot 51 \cdot 50 \cdot 49 \cdot 48 \cdot 47!}{47! \cdot 5 \cdot 4 \cdot 3 \cdot 2 \cdot 1}$

$\phantom{_{52}C_5} = 26 \cdot 17 \cdot 10 \cdot 49 \cdot 12$

$\phantom{_{52}C_5} = 2,598,960$

46. a) 4

b) $\dfrac{4}{2,598,960} \approx 0.0000015$

47. Consider a suit

A K Q J 10 9 8 7 6 5 4 3 2

A straight flush can be any of the following combinations in the same suit.

K	Q	J	10	9
Q	J	10	9	8
J	10	9	8	7
10	9	8	7	6
9	8	7	6	5
8	7	6	5	4
7	6	5	4	3
6	5	4	3	2
5	4	3	2	A

Remember a straight flush does not include A K Q J 10 which is a royal flush.

a) Since there are 9 straight flushes per suit, there are 36 straight flushes in all 4 suits.

b) Since 2,598,960, or $_{52}C_5$, poker hands can be dealt from a standard 52-card deck and 36 of those hands are straight flushes, the probability of getting a straight flush is $\dfrac{36}{2,598,960}$, or 0.0000139.

b) $\dfrac{123,552}{_{52}C_5} = \dfrac{123,552}{2,598,960} \approx 0.0475$

54. a) $10 \cdot 4 \cdot 4 \cdot 4 \cdot 4 \cdot 4 - 4 - 36$, or 10,200

b) $\dfrac{10,200}{2,598,960} \approx 0.00392$

48. a) $13 \cdot 48$, or 624

b) $\dfrac{624}{2,598,960} \approx 0.00024$

49. a) There are 13 ways to select a denomination. Then from that denomination there are $_4C_3$ ways to pick 3 of the 4 cards in that denomination. Now there are 12 ways to select any one of the remaining 12 denominations and $_4C_2$ ways to pick 2 cards from the 4 cards in that denomination. Thus the number of full houses is $(13 \cdot _4C_3) \cdot (12 \cdot _4C_2)$ or 3744.

b) $\dfrac{3744}{_{52}C_5} = \dfrac{3744}{2,598,960} \approx 0.00144$

50. a) $13 \begin{pmatrix} 4 \\ 2 \end{pmatrix} \begin{pmatrix} 12 \\ 3 \end{pmatrix} \begin{pmatrix} 4 \\ 1 \end{pmatrix} \begin{pmatrix} 4 \\ 1 \end{pmatrix} \begin{pmatrix} 4 \\ 1 \end{pmatrix}$, or 1,098,240

b) $\dfrac{1,098,240}{2,598,960} \approx 0.423$

51. a) There are 13 ways to select a denomination and then $\begin{pmatrix} 4 \\ 3 \end{pmatrix}$ ways to choose 3 of the 4 cards in that denomination. Now there are $\begin{pmatrix} 48 \\ 2 \end{pmatrix}$ ways to choose 2 cards from the 12 remaining denominations ($4 \cdot 12$, or 48 cards). But these combinations include the 3744 hands in a full house like Q-Q-Q-4-4 (Exercise 49), so these must be subtracted. Thus the number of three of a kind hands is $13 \cdot \begin{pmatrix} 4 \\ 3 \end{pmatrix} \cdot \begin{pmatrix} 48 \\ 2 \end{pmatrix} - 3744$, or 54,912.

b) $\dfrac{54,912}{_{52}C_5} = \dfrac{54,912}{2,598,960} \approx 0.0211$

52. a) $4 \cdot \begin{pmatrix} 13 \\ 5 \end{pmatrix} - 4 - 36$, or 5108

b) $\dfrac{5108}{2,598,960} \approx 0.00197$

53. a) There are $\begin{pmatrix} 13 \\ 2 \end{pmatrix}$ ways to select 2 denominations from the 13 denominations. Then in each denomination there are $\begin{pmatrix} 4 \\ 2 \end{pmatrix}$ ways to choose 2 of the 4 cards. Finally there are $\begin{pmatrix} 44 \\ 1 \end{pmatrix}$ ways to choose the fifth card from the 11 remaining denominations ($4 \cdot 11$, or 44 cards). Thus the number of two pairs hands is $\begin{pmatrix} 13 \\ 2 \end{pmatrix} \cdot \begin{pmatrix} 4 \\ 2 \end{pmatrix} \cdot \begin{pmatrix} 4 \\ 2 \end{pmatrix} \cdot \begin{pmatrix} 44 \\ 1 \end{pmatrix}$, or 123,552.